GELS

IN THE CONSERVATION OF ART

GELS

IN THE CONSERVATION OF ART

Edited by Lora V. Angelova, Bronwyn Ormsby,
Joyce H. Townsend, and Richard Wolbers

Archetype
Publications
www.archetype.co.uk

First published 2017 by Archetype Publications Ltd

Archetype Publications Ltd
c/o International Academic Projects
1 Birdcage Walk
London SW1H 9JJ
www.archetype.co.uk

ISBN: 978-1-909492-50-9

British Library Cataloguing in Publication Data
A catalogue record for this book is available from the British Library.

The organizers of the conference, IAP and Tate, are extremely grateful to Tru-Vue Inc. for its very generous support of this publication.

TRU VUE®

Printed on acid-free paper

Designed by Marcus Nichols at PDQ Digital Media Solutions Ltd.
Typeset by PDQ Digital Media Solutions Ltd, Bungay
Printed in the UK by Hobbs the Printers Ltd, Totton, Hampshire

FSC www.fsc.org MIX Paper from responsible sources FSC® C020438

Contents

Foreword

It gives me great pleasure to introduce this preprints volume, arising from the October 2017 *Gels in Conservation* conference organized jointly by International Academic Projects (IAP) and Tate. It is extraordinary to see so many submissions from such a wide range of conservators, conservation scientists and academics spanning Europe and the Americas. Numerous conservation disciplines are represented here and many practising conservators, as well as students from conservation programmes worldwide. The breadth and diversity of the contributions alone deserves a commendation.

Gels are a relatively new class of material to conservation. This pioneering conference presents a global snapshot of current theory, research and practice on the use of gel materials for treatment processes. The range of gels featured is impressive, including examples drawn from the biological, cosmetic and food industries, as well as emerging products created specifically for conservation purposes.

The myriad of applications described in this volume points towards the growing importance of these materials, and their positive impact on conservation and environmental practice is clearly evidenced. Employing gels and therefore fewer solvent-based materials represents one of the conservation profession's greatest contributions to the green movement. Additionally, some of the gel treatments discussed make for significantly safer working conditions, and all of them manage and reduce risks to artworks and objects undergoing treatment.

This publication is a showcase and testament to the professionalism of conservation as a discipline – combining an open-minded, holistic approach, while remaining necessarily cautious and meticulous in the application of these materials to works of art and heritage objects. What strikes me most about the content is the collaborative, borderless search to ensure that our cultural heritage is conserved in as low risk a way as possible, and the shared intent to preserve both the material makeup and the integrity of a given work of art. Such collective endeavour is vital and deeply impressive, and will undoubtedly inspire others.

A conference of this magnitude requires inspiration, planning, passion and consistent application, and it is clear that both the International Academic Projects and Tate teams have these qualities in abundance. On that note I would like to thank the conference committee: Jim Black, Rachel Gallen of IAP, Richard Wolbers of the University of Delaware Conservation Program/Winterthur Museum; Bronwyn Ormsby and Joyce Townsend from the Tate Conservation Science team, Lora Angelova (formerly Tate, now The National Archives), with support from the Tate Collection Care division, including Rosemary Lynch, Deborah Potter, Yunsun Choi, Rachel Barker and Natasha Trenwith. IAP has ensured the smooth running of the organization and registration for this conference, as well as maintaining communication with some 75 authors of the subject papers, shorter papers and poster summaries, which form a highly impressive permanent record of this event. I have no doubt that this publication will serve as a key reference for students, practitioners and researchers for many years to come.

Maria Balshaw, Director, Tate.
June 2017

Keynote paper

Gels, green chemistry, gurus and guides

Richard Wolbers

In conservation, we often tend to see ourselves as slightly separated from other technologies in the 'real world'. In fact, we do work under a very different, and specific, set of practical and ethical constraints. We are a field driven to protect, preserve, and alter as little as possible, the artefacts that fall under our care. The real world is not as constrained as we are in this regard. Much of the real world wants to sell ever more of its products, and to create new markets. Cleaning has always been a specific, potentially high risk operation for us, and taking no action at all is always one of the options to be considered – unless 'inherent vice' means that damage will occur in future as a result of our inaction now. We are incredibly careful about the types of materials we use, and the manner in which we bring cleaning agents into contact with artefacts.

But, despite our idiosyncratic nature, we do, in many ways, share choices of cleaning materials with 'real world' applications. We are such a small field that we have no other choice but to draw in materials sometimes designed for very different applications, to meet our specific needs. Few materials are produced solely for conservation use, or at least not with continuous investment in product development. And despite our differences from the rest of the technological world around us, we do share some common challenges. Ultimately, we can and should dip into the multitude of materials *they* produce to meet their own challenges. In terms of cleaning materials, specifically, whether we know it or not, our development in this area has been driven by economics (again, like the real world), but also by the tenets of green chemistry, and, I suspect, it will continue to be so driven for the foreseeable future.

The trajectory which the conservation field has taken very much follows from the 12 principles of 'green chemistry', as expressed originally by Anastas and Warner (1998: 30). Their core principles that are pertinent here (6 of the 12) are, in their own words:

- waste minimization at source
- use of catalysts in place of reagents
- improved atom efficiency
- use of non-toxic or less toxic reagents
- use of renewable materials
- use of solvent-free or recyclable and environmentally benign solvent systems, e.g. water.

At first blush the connection between these principles and a conference on 'gels in conservation' seems strained and distant – but bear with me. If we 'translate' the list to a degree, the connection to our field becomes clearer. In the long run the list provides us with some sense of both where we have been, and the path we are on, with hints at a direction for the future as well.

Waste minimization at source

We should use only the solvents we need, when we need them, and in just the amount we need to remove a particular material and any consequential treatment residues. This guiding principle can be advantageous in so many ways. It is more efficient to use a solvent in a gel, for instance, than in an 'open' or 'free' form that might allow it to evaporate away quickly. Open solvent use is inefficient, and moreover, wasteful in the sense that some of the solvent that could have done work for us is lost to the environment, and we add to the environmental burden of the cleaning process (our carbon footprint), by requiring fume extractors, personal safety equipment, etc. to work a little harder than they might otherwise have had to work. By gelling solvents to slow their evaporation, we can also have a reaction benefit as well: as we slow the evaporation of volatile components in a solvent mixture we automatically maintain more control over complex or multi-component mixtures. We make it possible to have longer 'dwell times', sometimes giving us an advantage for solubilizing higher molecular weight materials that are slow to dissolve. Throughout this conference examples of gels being used in just this way are seen (e.g. cellulose ethers such as methyl cellulose and Klucel G; agarose, gellan, xanthan gum). Solvent gels made with Carbopol/ Ethomeen can also increase dwell time, slow down solvent evaporation, and carry out other functions.

Use of catalysts in place of reagents

We already have very good examples in conservation of the literal use of catalysts. We have used catalysts whenever we have chosen to use an enzyme to break down biopolymers to a more manageable and soluble size. Paper and textile conservators alike have at times used amylase to break down starch paste residues, and proteases to help dissolve stubborn glue residues.

One of the very first enzyme treatments I did as a student in 1983 was to assemble a protease gel to help remove the glue residues from the surface of a Picasso oil painting (*Man with Guitar* 1915) while I was an intern at the Philadelphia Museum of Art, Philadelphia, USA. Soon after graduation I also used an enzyme - lipase, which is an oil-degrading enzyme - to help remove a linseed oil coating from an American portrait of *Anne Ross Hopkins* from about 1800 (Wolbers 2000: 139). This latter case is a good example of the powerful specificity of enzymes. The portrait was painted by an itinerant American artist (Jacob Eichholtz) who worked in the mid-Atlantic states of the USA. There are several surviving examples of his work to be found in institutions and private collections in the USA. Most have been cleaned and restored long ago, so we have very little information now as to how the artist would have finished them, whether they were varnished right away, whether the artist himself applied the varnish, or whether only certain colors would have been varnished. The portrait of *Anne Ross Hopkins* appeared to be unrestored when it came to the studio. There was some evidence of a natural resin varnish on the paint surface, along with an almost pure linseed oil coating applied overall at some later point, no doubt in an attempt to re-saturate an aging paint/varnish surface in the late 19th or early 20th century. Now if I had chosen to apply solvents to the drying oil coating in an attempt to remove it, I would surely have lost the natural resin coating beneath, as it would have been more soluble than the oil coating. This would have been risky in any event: the same solvents that would have swelled and (perhaps) dissolved the oil coating might also have swelled and damaged the oil paints. Therefore, I used an aqueous preparation, gelled (made with neutral methyl cellulose), with an active lipase in it. The gelling medium was chosen so it would not compromise the action of the lipase. The lipase degraded the oil coating, made it water soluble within a minute or two, and I could then remove it completely without removing the natural resin coating beneath.

By replacing a solvent method with an aqueous one, I had gained an advantage – I could glimpse the original artist-applied varnish. It was extraordinary, a revelation in fact, because the original varnish which was evident in normal or white light, and especially under ultraviolet light, appeared to have been applied to specific colours, and the brush-marking was evident. The application followed the form of the drapes behind the sitter, around the head, etc. It was immediately obvious that the varnish had been applied very early in the life of the painting, when it could still mix with the oil paints: it had not been applied evenly and did not cover the entire painting. Today, when we apply a coating to a painting we tend to apply it broadly and evenly, as an overall protective and saturating material. But in this case, the varnish application was clearly localized to specific design elements and intended to even out the gloss and saturation of the surface as an aesthetic effect alone. The recovery of the original varnished surface would not have been possible without the selective removal of the uppermost oil coating, leaving the natural resin intact beneath. This treatment required a technology that was new (at least at that date). If solvents had been used it would have led to entirely the opposite result – the oil would have been swollen, the resin coating dissolved, and the information as to the artist's original intent lost forever. Enzymes represent green chemistry at its best. They are renewable, degradable, and specific for the task at hand, and inherently water-based on account of their evolution in living creatures. For conservators, they can replace a solvent method with an aqueous one.

To date we have really only scratched the surface of 'potential' enzyme use in conservation. Most of the enzymes that have been used thus far would fall into the narrow class of 'hydrolases', that is, enzymes that break down – hydrolyse – biopolymers such as proteins, glues or starches, or polymer-forming materials like drying oils, as part of a process of cleaning and removing non-original materials. But enzymes can, and will, play an increasing role in transforming materials in a variety of ways. Oxidizing, reducing, crosslinking or modifying materials to change their solubility will increasingly play a role in many of our conservation processes in the future.

Here I am going to stretch the green chemistry notion of a 'catalyst' a little. Polymers that form gels in solvents or water can do so much more than simply thicken those materials, or modify the physical properties of the solutions or materials they contain: they can also become part of the cleaning system or process as well. One of the other aspects of green chemistry that we have already started to use is adopting polyfunctional materials that are capable of doing more than one thing. By employing this strategy, it means we can do more, with fewer materials. This also reduces the possibilities for long-term chemical interactions between the artefact and any residues from a multitude of materials. The Carbopol/ Ethomeen solvent gels are a good example. As I suggested above, a number of polymers can be used to increase the viscosity of solvents, or even gel them. Polyacrylic acid can gel water and a range of polar solvents. But as a polyanionic material we can couple cationic moieties to it. It can at the same time be both a polymer, and the scaffolding for other materials and functions, acting not as a catalyst per se, but more like a 'smart' scaffolding from which to work. Ionically combining cationic detergents (Ethomeen C-12 and C-25) with a polyacrylic acid such as Carbopol, as complex salts, allowed me in the mid-1980s not only to gel a variety of solvent and solvent mixtures, but also to carry surfactants in those solvent solutions in a rational and more predictable way. By tying up the surfactant on a polymer like a polyacrylic acid, it is slightly easier to keep track of its whereabouts. Wherever the polymer goes (or does not go) the surfactant stays with it. If we remove the polymer, we remove the surfactant as well (Stulik and Miller 2004: 35–9). The polymer has become polyfunctional by doing this: it is a gel, formed in a variety of solvents, depending on the structure of the surfactant used; it can be rinsed in a wider range of solvents (including weaker or less polar ones than those in the gel); and as surfactant laden-materials the gels can embrace, solubilize, and disperse materials that are not, or are only sparingly soluble in a given solvent, extending the amount of treatment work we can do with any given solvent.

Atom efficiency

Since 2006 we have seen natural polymers (e.g. xanthan gums) and synthetic polymers (such as Pemulen), increasingly used in conservation for gelling aqueous preparations, as well as polymeric emulsifiers. Not exactly 'atom efficiency' as in the strictest green chemistry sense, but our equivalent might read something like 'if I can make a material do more than one thing in *any* cleaning system, I am better off in a number of ways'.

Xanthan, as an example, is a good choice: the gum is renewable; cheap, non-toxic, virtually unaffected by changes to pH or ion concentration; and it has the ability not only to form stable gels in water-based systems and control the application of aqueous materials on surfaces physically, but also to create stable emulsions of solvent in water. It is a gelling material that can suspend soiling particles, and form emulsions without the use of added surfactants (making it a 'self-emulsifier'). The shorter the list of ingredients, the less work we need to do to clear materials from surfaces, and fewer 'residues' will potentially need to be sought and tracked.

Non-toxic or less toxic reagents and aqueous methods

The use of less toxic or non-toxic materials should be a fairly obvious practice to advocate. We are generally a small field, with workers spending their days in space-limited workspaces and laboratories, and in small businesses. For many practitioners, personal contact with cleaning reagents is constant. Yes, we can mask and glove ourselves for protection, but protection can come at a price: less dexterity and control, and more tiring working conditions. Whether we can substitute more toxic solvents for less toxic ones should always be part of our conservation treatment planning. I have often taken cosmetic and pharmacological technologies as a model: in most of the gelling and cleaning systems I have developed over the years I have 'borrowed' heavily from both fields, for materials that might be useful in our day-to-day practice.

Where we can replace solvent use with aqueous or water-borne materials, we should. It was some years ago, in a workshop in Australia, that I made the connection. One of the paintings we were using for testing cleaning materials was a 19th-century landscape. It was an oil painting which had a degraded natural resin coating on its surface, most likely a soft natural resin such as dammar or mastic. It was a typically aged, oxidized, and quite yellowed coating. It was, to all eyes, a coating that we could remove from the painting surface with a simple solvent mixture, for instance a 1:1 (v/v) acetone/white spirits mixture. I had hoped to demonstrate the use of a solvent gel (one of the Ethomeen/ Carbopol gels) as an alternative to free solvent for removing the coating. The first plan was to show that much less polar solvents could be used if a surfactant was incorporated into the solvent mix. My idea was to couple a surfactant to a gelling polymer, to make surfactant chemistry do some of the work of cleaning, while keeping well out of the range of solvent or solvent mixes that could swell oil paint and leach out soluble components. We also tried a 'resin soap' mixture on the degraded coating as well; resin soaps – such as the triethanolamine abietate soaps for which I published case studies in the 1990s, for removal of degraded soft resin films or varnishes (Wolbers 2000: 39–52) – would also have worked. But I also tried a xanthan gel (pH adjusted to 6.5) with sodium citrate, and picked up a considerable amount of the coating material with that simple aqueous gel alone. By adding a small proportion of benzyl alcohol to the gel (remember that xanthan gels are emulsion-forming materials, and benzyl alcohol can be emulsified with simple agitation), something on the order of 2–3% by weight of the solvent, I could remove the coating completely, with little apparent effect on the oil beneath. Moreover, the recovered surface looked much more intact than might be seen after a cleaning test with a free solvent. It was, in a word, a 'shock', because it struck me that for generations we have conceived of only one way of dissolving natural resin materials like dammar and mastic coatings on paintings: with simple free solvents or solvent mixtures.

We have long compensated for the aging and oxidation of natural resins by increasing the polarity of the solvents we apply to them. But with solvent-only methods, at some point we cross a line: aged natural resin coatings begin to need solvents that are so polar, in order to dissolve these low molecular weight degradation products of the original coating, that there is an increased risk of swelling the oil paints beneath, and extracting low molecular weight material from them, thereby altering their chemical and mechanical properties. When we do use solvent-only methods, we are usually not solvating everything in degraded natural resin varnishes: we can dissolve most of the resin mass, but often we are left with a slight hazy, whitish residue. It is the 'baggage' left over from using solvent-only methods, because the most oxidized (and acidic) fractions of resin materials as well as some of the higher molecular weight 'waxy' or aliphatic fractions of natural resin materials may not be soluble in simple solvent mixtures. We will have removed a large proportion of the natural resin coating, but the most oxidized fractions are the very materials we want to remove, before they oxidize further (van der Doelen and Boon 2000). These acidic materials are often simply varnished over, that is, re-saturated, and forgotten. But these acidic materials are the very materials we really should be concerned about in terms of the overall health and preservation of the paint we recover after cleaning, and the subsequent coatings we might apply to these surfaces. These acidic materials can force the further deterioration and discoloration of oils and resins alike in time through increased unsaturation (Feiser and Feiser 1959: 15–25, 98–400). Cleaning or removing a varnish is an aesthetic act, but it is also part of a long-term preservation plan for the paint itself. The 'shock' that I felt was this: I could readily remove a degraded natural resin varnish from an oil paint with a water-based gel (2% xanthan gum in water, with a small volume of solvent in it). The gel was 97% water, at a pH of 6.5, and with about a 3% w/w amount of solvent emulsified in the gel. This material benefits both the painting and the conservator working on it.

But it was a 'shock' as well to realize that so much of a deteriorated (therefore polar) natural resin film could be so water-soluble. For years I had been so 'immersed' in solvent methods for removing varnishes from oil paints that I thought that this was the paradigm for cleaning – use solvents, mixtures of solvents, and look for additional methods only when the solvent methods failed.

Today, my first thoughts are: how degraded is the coating or material? How much of it is potentially water-soluble? Can I actually start with an aqueous method rather than 'resort' to it when solvent methods fail? There is another advantage: with solvent-only methods, when we dissolve resin coatings there is really no 'brake' or stopping point in some cases. As we dissolve the coating and approach the oil paint surface, the concentration of oil intermixed with the paint surface increases. There is an 'interactive zone' or overlap of resin into oil and oil into resin. With solvent methods we can easily move into, and disrupt, that boundary. Solvation in general slows as the oil concentration rises, but by then we have removed some material from the paint surface. With aqueous or predominantly aqueous methods the effect may be just the opposite. Water and water at a pH of around 6.0 is an especially poor solvent for dried oil films, and natural resin materials, since it is beginning to approach the pK_a of common resin and oil fatty acids in each (Holmbom *et al.* 2010). As an aqueous preparation gets closer and closer to an historic oil surface the solvation effect may slow appreciably, and as a result I think we have a better chance of preserving structure, chemistry, and aesthetics at that surface when we work with an aqueous system.

Regarding resin soaps, I introduced them in workshops as early as *c.*1992 in Paris, France. They represent one stop along the way towards an ever greener cleaning methodology by avoiding traditional surfactants (low critical micelle concentration or 'CMC' structures) that might be harder to clear from oil paint surfaces (Wolbers 1992). I mention them here to remind us that they can indeed be used as an aqueous method to remove degraded natural resin coatings, but the core principle that allows them to work can be further refined. (Even gurus evolve!) Today, rather than a 'resin soap', I might simply use a xanthan gel with a small solvent-phase addition to do the same thing that the resin soaps did. This is a good example of the 'engineering' pathway I have followed. I am endlessly looking for and trying new things and new ways to clean fine art surfaces. But as I formulate a clearer notion of what I am aiming for, I become more demanding of the materials I work with, continually refining and looking for easier, better, and safer (for the artefact) ways to make use of aqueous materials and gels. I am not worried that 10 years ago I did one thing, but today I might do another. Engineering is like that: as new materials come along we naturally want to see what they can do. As we understand more about how older materials work in terms of chemistry, we sometimes choose to stop using them and adopt better materials. A healthy sign for our field is a continued evaluation of materials and methods, and a steady acquisition of new materials and methods into our core practical technologies. The same ethical principles that have guided us thus far will guide us in the future, even as we adopt new materials and methods.

Skill will always be important to what we do. We are as much a 'craft' profession as we are an innovative field. These are not *competing* ideas, merely complementary ones that reinforce each other. I want to use even the newest of materials as skilfully as possible and to learn as much as possible about them as I can, before I adopt them. But I am not afraid to change when something better comes along. Innovation challenges us, but in a good way I think, to be continually critical of what we are doing. I act as a guide when I teach with these ideas in mind. This conference is a forum for discussing innovative practice, as well as the use of new materials.

Environmentally benign materials

The notion of 'down-the-drain' disposal, and disposal of materials through evaporation, is also important in engineering safer and greener cleaning materials. Like every other field, we should be putting only 'biodegradable' materials down the drain, with the 'grey water' that includes the materials that can legally flow into the drainage system from kitchen and bathroom sinks in the home, as well as small businesses, and museum and academic premises. Grey water ends up in the nation's sewage system, and ultimately in ground water or sea water. The materials that go down our own drain eventually make contact with many other materials. Biodegradation is a complicated science, but we can make some sense of the materials we are putting down the drain by using simple software packages available to us as consumers. The Environmental Protection Agency (EPA) in the USA has made available software to do just that. Biowin 1, 2 are EPA software packages that ultimately assign a number (between 0 and 1) to the overall biodegradability of a material, and are quite easy to use. We can enter into its database virtually any chemical we are interested in, and quickly assess its likelihood of degrading fully in grey water. Generally those materials that have a Biowin 1 number greater than about 0.5 can be considered appropriately biodegradable. I think we should consider that number every time we put something down the drain, either as a conservator or as a consumer.

Let me provide a tangible example with surfactant chemistry: virtually every building and monument in Washington DC has had its exterior or surface cleaned at some point with water and detergent cleaners. Commercial materials are 'designed' or specified for these large-scale exterior applications. Many monuments have been cleaned specifically with 'ivory soap' in water – a simple (natural) soap or sodium palmitate solution in water – quite benign in the sense that sodium palmitate has a Biowin number of about 0.80. But with stone and building materials in particular, we need to look to a second important number: the octanol/water coefficient, or 'log P' value, for any material. This is a number that designates how polar/non-polar a material or solvent is. It is a logarithmic value, so every time we increase the solubility tenfold in octanol, log P increases by 1, and every time the solubility decreases tenfold in water log P goes down by 1. As an example, a material with a low number (say 1 or below) is water-soluble, available, as it were, to aqueous degradation chemistry in ground water at the end of the 'down the drain' disposal option. Materials with log P numbers greater than about 5–6 are not so water-soluble, being strongly absorbed for instance on silicate and other non-polar surfaces, and are thus not readily available to be biodegraded. Materials that have a log P value in the middle, in the range of about 1–5, are weakly absorbed, but hard to rinse effectively with water. Therein lies the problem: a material with an octanol/water number of between 1 and 6 would be difficult to rinse with water from a silicate surface, and tends to create residue problems in the long run. Sodium palmitate has a log P value of about 3.2: it is in the worst possible position for creating a residue that is hard to rinse with water. But more importantly, recall that its biodegradation number (Biowin 1 number) is 0.8. So we have in the same material simultaneously: a high adsorption value, difficulty in rinsing with water, *and* a high biodegradation property. The result is that retained sodium palmitate in buildings in Washington DC has fostered bio-colonization in a massive way. Almost every monument cleaned in that way foams the minute it is first re-wetted with (rain)water; moreover, virtually every building exterior and monument has some biological material growing on its surface (for example the blackish material on the exterior of the Jefferson Memorial). The problem is really threefold: how do we now relieve the burden of old surfactant residues in these buildings and monuments? How do we now effectively remove a sometimes very intractable surface regrowth of material (which, by the way, is self-compounding: as one organism dies, it becomes 'food' for the next and so on until there are endless and hugely complicated organic materials on these surfaces, far beyond what is found in normal environmental soiling aerosols)? And, lastly, do we have the cleaning technologies to do both of these operations without adding to the burden of materials on these surfaces?

Whether or not we consciously acknowledge the thought, we grew up and were trained in an ocean of solvent use that essentially came from varnish and paint-making technologies. The solvents we have traditionally had in our studio arsenals are aliphatic/aromatic petroleum distillate materials, such as xylene, toluene, acetone, alcohols, etc. But in any other 'real world' technology, for example cosmetics, pharmaceuticals, etc., we would be looking at a very different list of materials with which to work. Non-polar solvents in cosmetics might include isopropyl palmitate (IPP) or myristate (IPM). IPP and IPM are good examples of greener solvents that could perform in a similar solvent way to petroleum-derived aliphatics (they have the same Hanson parameter), but have the advantage of being biosourced, and more importantly, biodegradable, materials. They are simply hydrolysed to isopropanol and a fatty acid in grey water waste systems, both of which can easily be degraded further biologically.

I sometimes think the same thing about surfactants in general: in cleaning 'soils' from most surfaces we tend to reach for surfactants as our go-to 'cleaners', largely because of their history of use in the field. Wet cleaning in textiles is a good example: textile conservators will often use a surfactant in one form or another to aid in wet cleaning. But over the years I have tended to de-emphasize the use of surfactants in conservation cleaning endeavours, especially in the wet cleaning of textiles. Surfactant chemistry is difficult, to say the least (we have to manage the hydrophile-lipophile balance (HLB) number, critical micelle concentration (CMC), cloud point temperatures, and their changes on real surfaces). But surfactants are also difficult to remove completely from some surfaces and structures, particularly those that are porous, have a high surface area or are hydrophobic (that is, water-repelling). The use of surfactants engenders environmental concerns when they are flushed down the drain. Moreover, they may not address the cleaning issues at hand. For example, when we see greyness in textiles on aging, it is usually from a build-up of calcium salts in the textile structure from washing during its service lifetime (Chi and Obendorf 1998). When a fabric is washed again and again in traditional soaps in hard water, we are likely to see residual insoluble calcium soaps accrue. In time higher and higher solubility product calcium salts tend to form from these residual soaps. Removal of the greyness is a chelation problem ultimately, and is not amenable to treatment with surfactants. Trying to perpetually wet clean textiles with surfactants is simply missing the mark.

Green chemistry should also imply adopting the best tool for the job: anything less and we are needlessly using a material that adds to an already burgeoning environmental problem.

Sustainability and efficiency

Several novel, surfactant-free methods or forms for creating emulsions and solvent gels are currently available to us. These new materials have come straight from the huge and profitable industrial sector of cosmetics, and the current need to reduce or eliminate surfactant dependence in these systems for environmental and health reasons. One benefit of trailing after other technologies to find materials we can use is that the time, expense, and research has already been done by others in large part, and we can draw from their range of materials and experience to aid us in our cleaning endeavours. Micro- and nano-particle technologies can, in a number of ways, be 'greener' strategies (and there are several papers on this in this book) because they can increase the efficiency of cleaning materials, and thereby cut down on waste.

As one example: particle emulsifiers ('Pickering' particles) for water-sensitive materials have been used in conservation since about 2007. That was the date of a workshop in Trento, Italy, where I introduced the Velvesil Plus material as a water-in-oil (w/o) type of particle emulsifier. And more recently they have been used by conservators of modern acrylic paints, which are extremely water-sensitive materials, to build simple w/o emulsions that are surfactant-free and easier to clear from these fragile surfaces, and can control exquisitely the amount of water brought to these surfaces (Reeves *et al.* 2015; Lagalante and Wolbers 2015).

There is no end to the challenges we face in cleaning fine art surfaces. Here, in the early 21st century, we are now inheriting ageing artefacts to care for and conserve from the latter part of the 20th century, replete with an ever-expanding catalogue of synthetic materials. Our ethics will not change – we still want to preserve artefacts and their cultural and artistic aspects for as long as possible. But our tools will become greener. We need to consider green chemistry whenever we are tackling new materials and their problems. Like other sectors, we must replace or minimize solvent use where we can. Gels are one way, but we need to prepare ourselves for continual revision and updating of our treatments, as new materials are made available to us. The emulsion-forming polymers have been a revelation because they get us out of the trap of using surfactants for forming emulsions. But we cannot always 'borrow' from other technologies for our critical tools. In the ideal world we would be self-sustaining (that is, generate through our own research and practical channels the materials and methods we need). In this book you will read papers on novel cleaning materials that have been largely generated *within* the conservation field. And it is a good start. But I am afraid we are far too small a field to support, entirely from within, the necessary level of research and engineering that we will ultimately need, given the range of materials in our care. Looking solely for 'outside' industrial materials to solve our problems is also a two-edged sword however: they *should* ultimately produce and provide materials that are less toxic and safer to handle for humans. But industries such as cosmetics or pharmaceuticals have their own agendas: they are for-profit, and will never have concerns as to the long-term effects on us or the artefacts in our care. Nor will they ever be 'transparent' in the way we demand now within our profession. I want to know exactly what is in a material I am using to clean the surface of an artefact surface: its properties, its stability as a material, and the likely chemistry associated with its aging. Our research efforts have to meet engineering needs that are never going to be of interest to any major industry, such as the effects their products might have after decades and centuries, on a surface whose significance is aesthetic as much as it is tangible. And we are *always* going to remain interested in products that have been completely superseded and forgotten by their makers, if we know they have been in contact with an artefact and might have affected the way it will age or change in the future.

The tenets of green chemistry can be a *guide* for us for sustainability: of course we can find better and less toxic materials to replace the more toxic materials used to date. But we can also reduce the reliance on the toxic materials that we still have to use by limiting those to just the levels we need in treatments. Change the form or way we use things (yes, gels). But we can be better at using materials that are appropriate to the task at hand (emulsions when we have both water-soluble and solvent-soluble mixtures to pick up; chelators and surfactant-free preparations when we need them), and finally, we can ask any material to do more than one thing, and be more efficient in that way.

We can do all this and still meet the core ethical demands of our rather idiosyncratic field. We will always place the safety of the artefact over the speed or cheapness of the treatment, yet as responsible professionals, we will also seek out products that are safe for our own use. This conference is essentially about cleaning materials, of course, but there may be a hint here of future directions in other areas. Reducing the burden of solvents and their ultimate disposal, and replacing them with aqueous systems, has another consequence which is perhaps even broader. We have grown up as a profession, 'spawned' in an ocean where solvent chemistry was at the core of our ateliers. Perhaps it will not be so in the future? As we use more and more less toxic 'water options' to clean surfaces, we should think outside the box about other conservation processes. What about newer, better, biosourced consolidants? Or, dare I suggest it, biosourced and water-applied and removable coatings? One can – and should – always dream.

References

Anastas, P. and Warner, J. 1998. *Chemistry: Theory and Practice.* Oxford: Oxford University Press.

Chi, Y.-S. and Obendorf, S.K. 1998. 'Aging of oily soils on textile materials: a literature review', *Journal of Surfactants and Detergents* 1(3): 407–18.

Feiser, L. and Feiser, M. 1959. *Steroids.* New York: Reinhold.

Holmbom, B., Sundberg, A. and Strand, A. 2010. 'Surface active compounds as forest industry byproducts', in M. Kjellin and I. Johansson (eds), *Surfactants from Renewable Sources.* Chichester: John Wiley & Sons Ltd, 48–52.

Lagalante, A. and Wolbers, R. 2015. 'The cleaning of acrylic paintings: new particle-based water-in-oil emulsifiers', *CESMAR 7 VII Congresso Internazionale Colore e Conservazione 13–14 Novembre 2015, Politecnico Di Milano, Aula Rogers.*

Reeves, M., Brown, A.T., Schofield, A.B., Cates, M.E. and Thijssen, J.H.J. 2015. 'Particle-size effects in the formation of bi-continuous Pickering emulsions', *Physics Review E* 92: 32308.

Stulik, D. and Miller, D. 2004. 'Research into solvent gel residues', in V. Dorge (ed.), *Solvent Gels for the Cleaning of Works of Art*, Los Angeles: Getty Conservation Institute, 18–65.

van der Doelen, G.A. and Boon, J.J. 2000. 'Artificial ageing of varnish triterpenoids in solution', *Journal of Photochemistry and Photobiology A: Chemistry* 134: 45–57.

Wolbers, R. 1992. 'A radio-isotope assay for the direct measurement of residual cleaning materials on a paint film', in J.S. Mills and P. Smith (eds), *Cleaning, Retouching and Coatings: Technology and Practice for Easel Paintings and Polychrome Sculptures.* London: IIC, 119–25.

Wolbers, R. 2000. *Cleaning Painted Surfaces: Aqueous Methods.* London: Archetype Publications.

Author's address

Richard Wolbers, University of Delaware, Department of Art Conservation, Newark, DE, USA (doczepp@gmail.com)

Polysaccharides: agar, gellan, xanthan and methyl celluloses

A study of commercial agar gels as cleaning materials

Moira Bertasa, Oscar Chiantore, Tommaso Poli, Chiara Riedo, Valeria di Tullio, Carmen Canevali, Antonio Sansonetti and Dominique Scalarone

Introduction

Gels are soft materials consisting of interconnected long polymer chains dispersed in a fluid (water or/and organic solvents), forming a three-dimensional network (Fratini and Carretti 2013). Gels are present in our everyday life and recently they have gained an important role as a cleaning and consolidant tool for items of cultural heritage. They are 'structured' cleaning systems swelled by water or organic solvents or mixtures. The advantages of using gels are mainly the restricted release of liquid phases, reducing the diffusion of solvents into the material that needs to be cleaned and minimizing the aggressiveness of solvents. The use of gels as a cleaning tool has increased enormously in recent decades, thanks to their versatility and effectiveness. A variety of polymeric gels can be used. The literature highlights the application and superb efficacy of a polysaccharide gel, so-called agar gel. Agar was the first phytocolloid extracted from red seaweed of the class *Rhodophyceae*, in particular in *Gracilariales* and *Gelidiales* species. It is a natural polymer commonly employed as a biological culture medium and as a food thickener. According to the literature, the agarose and agaropectin contents vary depending on the seaweed source, which affect the physicochemical, mechanical and rheological properties of the gel (Nishinara and Watase 1983). In 1956 Araki identified the two main components of agar: agarose and agaropectin. The first is a linear polymer consisting of alternating β-D-galactose and 3,6-anhydro-L-galactose units linked by glycosidic bonds; the second is a heterogeneous agarose heavily modified with acidic side-groups, such as sulphate and pyruvate, whose main effect is a reduction of gelling properties. Like other seaweed-derived products, when the *Rhodophyta* undergo different alkaline treatments, the sulphate groups at position C6 on the carbon backbone are converted to 3,6-anhydro-L-galactose, thereby increasing gel strength (Rees *et al.* 1969; Duckworth *et al.* 1971). Agar forms a hydrophilic and thermoreversible semi-rigid gel after a heating and cooling process. Agar is insoluble at room temperature, but above 85 °C thermal agitation overcomes its tendency to form double helices, and agar exhibits random coils in solution. Only below 43 °C and with further cooling will agar form a 3D network in which double helices work as junction points among the different polymer chains (Mohammed *et al.* 1998).

Gels – and in particular agar gels – satisfy the fundamental requirements for a cleaning method in terms of effectiveness, selectivity and safety for the surface of the artwork. Nevertheless, empirical experience from conservators, and the cleaning results on real objects, has raised many scientific questions about the cleaning mechanism. Agar gels are easily removable, therefore minimize mechanical stress at the surface of the artwork, but they show different cleaning performances at different concentrations.

This work presents the first results of a wider comparative study of four different raw agar products. A multi-analytical approach has been used in order to identify possible compositional differences and to understand if these are responsible, at least in part, for the varied performance of different agar gels. In particular, morphological characterization and compositional analysis are reported, together with a discussion on the gel's mode of binding water and rheological properties . Powdery raw materials and lyophilized gels were observed at high magnification by scanning electron microscopy (SEM); elemental and molecular analyses were carried out with various analytical techniques: an energy dispersive X-ray analysis (EDX) probe; vibrational spectroscopies including Fourier transform infrared (FTIR) and Raman; and pyrolysis–gas chromatography–mass spectrometry (Py–GC–MS). For the water binding study, agar gels at different concentrations were studied with unilateral NMR (NMR-MOUSE), and results were compared with those obtained by capillarity absorption measurements carried out according to the Italian standard

UNI-Normal 10859-2000 (water absorption by capillarity), slightly modified. Rheological properties were measured with an oscillatory rheometer.

The experimental results obtained by studying agar powders and laboratory-prepared agar gels were applied to a series of conservation treatments for the removal of staining due to metallic copper, and conservators observed an interesting extraction phenomenon. Model marble samples were prepared to simulate the typical greenish colour of copper staining. To assess the effectiveness of copper stain removal by different formulations of agar gels, the model samples and the marble base of Antonio Canova's bronze figure of Napoleon (Brera Gallery Courtyard, Milan, Italy) were cleaned using pure agar gels and agar gels with added chelating solutions, under the same experimental conditions. Colour measurements and quantitative analyses were carried out and compared in order to propose a preliminary protocol for copper stain removal with agar gels.

Experimental

Materials

Four agar powders from different suppliers were selected and characterized: AgarArt (CTS, Italy) and Agar Purissimo (Bresciani, Italy) are among the most frequently used gel materials by Italian conservators. Sigma Aldrich agar (A7002_CAS:9002-18-0) is mostly used for biological applications. The last agar powder, referred to here as Agar Food, is used in the food industry and was imported from the UK. For pyrolysis analyses, different reference materials were purchased from Sigma Aldrich, such as agarose (CAS:9012-36-6), D-(+)-galactose (CAS:59-23-4), 1,6-anhydro-D-glucose (CAS:498-07-7) and D-(+)-glucose (CAS:50-99-7).

Model samples (5 × 5 × 2 cm) of Sivec dolomitic marble, quarried in Prilep (Macedonia) were prepared and stained by keeping them in contact with a 0.1 M solution of $CuSO_4{\cdot}5H_2O$ (Carlo Erba) in Mill-Q water. The Canova case study has a marble base with a copper stain that was resistant to removal. Using Raman spectroscopy, the stain was identified as brochantite.

Methods

Preliminary morphological characterization was carried out using SEM. The morphological studies of agar in powder form were performed directly on the samples, which were first coated with carbon, using a JEOL 5910LV SEM and backscattered electron (BSE) and secondary electron imaging (SEI), with EDX for elemental analysis in a high vacuum (HV). A Zeiss EVO60 SEM equipped with an EDX microprobe (Oxford PentaFET) was used for the morphological study of lyophilized agar gels. All sample sets were analysed in low vacuum mode without a coating. In both cases, an accelerating voltage of 20 kV and magnifications of ×69 and ×500 were used to observe the internal morphology of the gels.

Spectroscopic characterization was performed with an FTIR spectrometer (Thermo Nicolet Nexus) equipped with an Olympus Continuμm IR microscope with a mercury cadmium telluride detector (MCT/A) cooled with liquid nitrogen (spectral range 4000–650 cm^{-1}). The spectral resolution was 2 cm^{-1}, with 256 scans. Samples were mounted in a diamond anvil cell.

For Py–GC–MS, agar powders were derivatized with thermally assisted hydrolysis and methylation (THM), using tetramethylammonium hydroxide (TMAH) in aqueous solution at 25% w/w (Sigma Aldrich, Italy). Samples were prepared by dissolving 0.1 mg powder in 50 μL TMAH solution in an oil bath at 90 °C for about 30 minutes. Then 8 μL of the solution was spread onto quartz wool inserted in a quartz tube for pyrolysis. A CDS Pyroprobe 1500 (Analytical Inc., USA) filament pyrolyzer directly connected to a GC–MS system was used. The GC was a 6890N Network GC System (Agilent Technologies, USA) gas chromatograph with a methyl-phenyl-polysiloxane crosslinked 5% phenyl methyl silicone (30 m, 0.25 mm i.d., 0.25 μm film thickness) capillary column. The temperature of pyrolysis was 600 °C, the temperature of the pyrolysis interface was 300 °C, and the temperature of the injector was kept at 280 °C. The carrier gas was helium (1.0 mL/min) and the split ratio was 1/20 of the total flow. The MS coupled to the GC apparatus was a 5973 Network Mass Selective Detector (Agilent Technologies, USA). Mass spectra were recorded under electron impact at 70 eV, scan range 40–600 m/z. The interface was kept at 280 °C, the ion source at 230 °C and the quadrupole mass analyser at 150 °C. All instruments were controlled by Enhanced Chem Station (ver. 9.00.00.38) software. The mass assignment was made using the Wiley 138 and NIST1992 libraries, and by comparison with data in the literature.

Water binding was investigated with a portable NMR instrument from Bruker Biospin interfaced with a purpose-built single-sided sensor (RWTH Aachen University, Aachen, Germany). Longitudinal and transverse relaxation times (T_1 and T_2) were measured, and H^1 depth profiles were made at 13.62 MHz. In particular, longitudinal relaxation times T_1 were measured with a Saturation Recovery pulse sequence followed by a CPMG-train in the detection period to increase the sensitivity. Due to the presence in the portable NMR of a strong inhomogeneous magnetic field (G=20 T/m) the transverse relaxation time values are shortened with respect to those measured in a homogeneous field. The effective transverse relaxation times T_{2eff} were measured with a CPMG pulse sequence and 4,096 echoes were recorded at a depth of 1 mm inside the sample with an echo time 2τ of 71.2 μs. The experimental data obtained applying the CPMG pulse sequence were normalized and fitted using the following equation:

$$Y(t) = \sum_{i=1}^{n} W_i e^{-\left(\frac{t}{T_{2i}}\right)}$$

where n is the number of components of the decay, W_i is the weight of the ith component, and T_{2i} is the transverse relaxation time of the ith component.

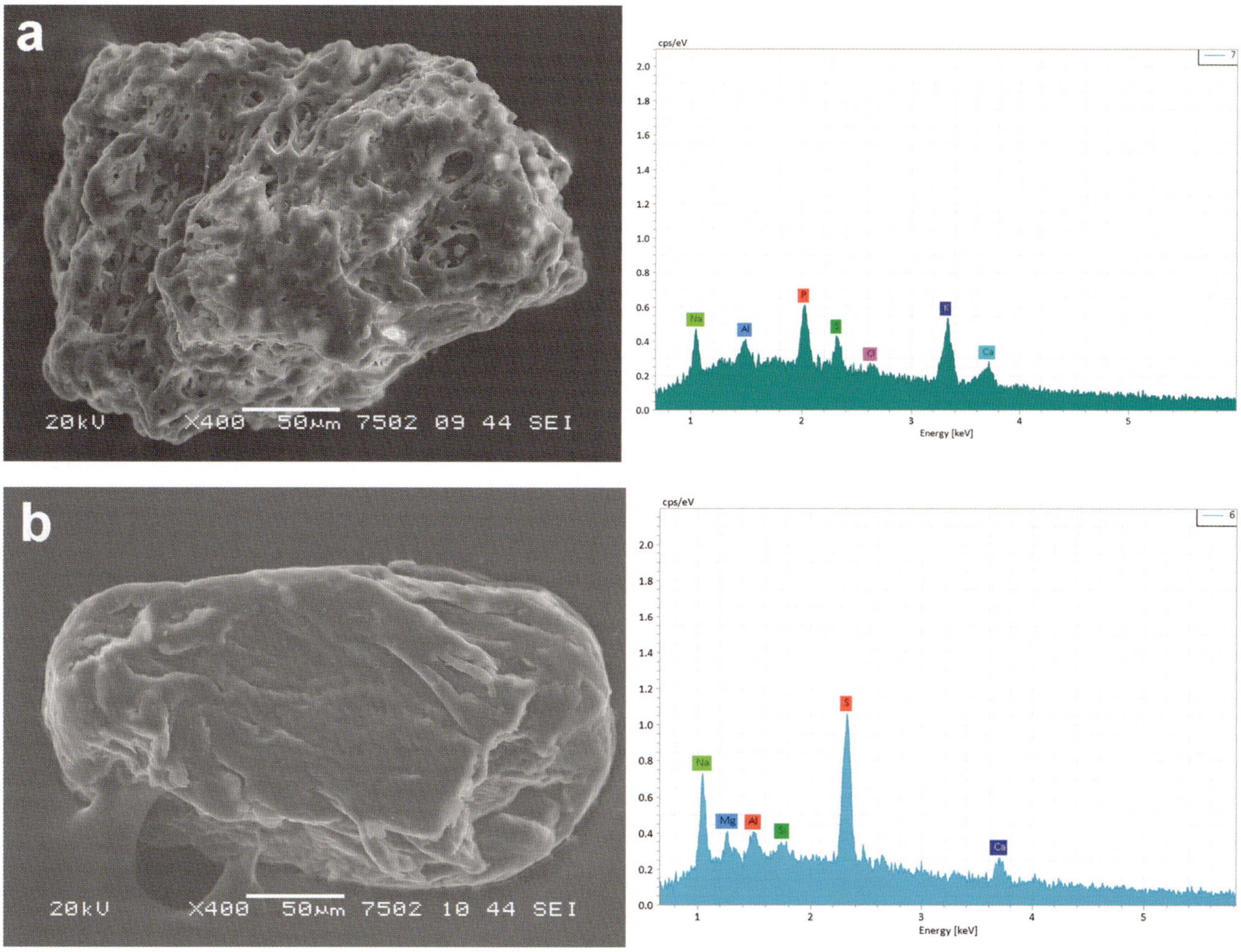

FIG. 1 SEM images and EDX analysis of agar powder particles: (a) Bresciani and (b) Sigma.

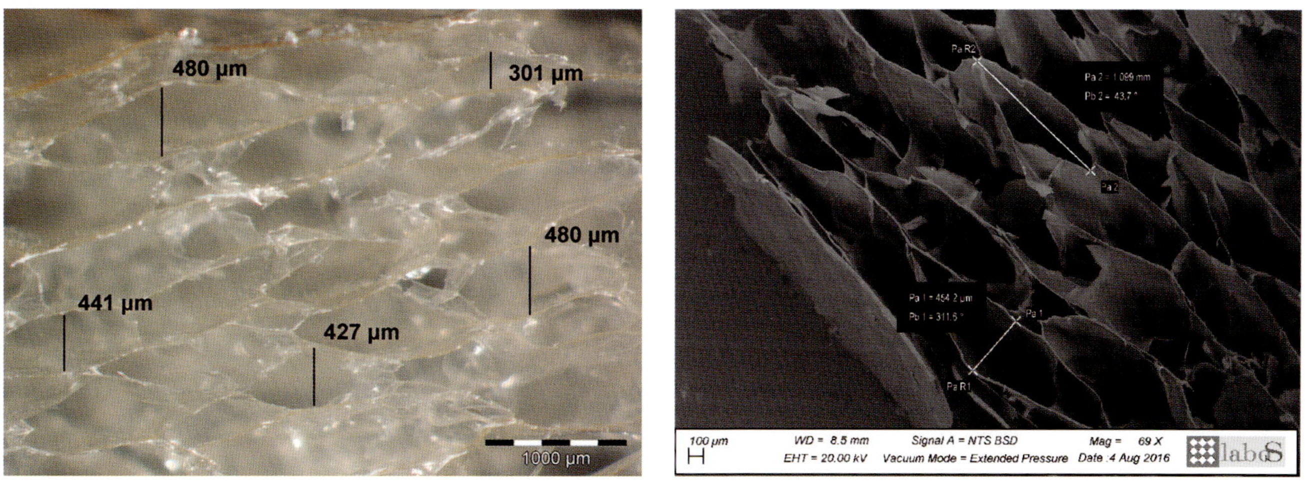

FIG. 2 Stereomicroscope and SEM images of freeze-dried CTS agar at 3% (w/v), with corresponding cell measurements.

Rheological properties of gels at different concentrations and after a progressive heating process were investigated with an oscillatory rheometer (TA Instruments Discovery HR) equipped with 20 mm plate geometry and Peltier plate temperature control. Frequency sweeps were performed over the frequency range of 0.05–16 Hz and stress amplitude 1% (Labropoulos *et al.* 2002a,b). Samples were loaded at 90 °C in order to favour the correct arrangement of the sample, then equilibrated for 3 min at 25 °C prior to the run. The viscoelastic properties of agar gels at different concentrations and after a double heating process were observed and monitored. Three measurements were performed for each sample and the mean values were used together with their standard deviations and presented in the histograms.

On model samples and in the case study, colour measurements were made using a Konica Minolta Chromameter

TABLE 1 Cell size and morphology of freeze-dried agar gels.

Agar type	Cell size (μm)	Cell shape
	range	
CTS	300–1000	Regular and elongated
BS	450–820	Round-regular
Sigma	450–750	Round-regular and elongated
Food	–	undefined

CM-700d with D65 source, d/8° geometry, CIE L*a*b* 1976 colour space, measuring a circular area of 6 mm diameter. The parameter L* represents the lightness of colours from 0 (black) to 100 (white); a* denotes the red/green values and b* the yellow/blue values, both ranging from +60 to −60. For each specimen, 25 measurements were taken. Given ΔL*, Δa*, Δb*, the changes in L* etc, before and after cleaning, the total colour difference can be stated by a single number $\Delta E^* = [(\Delta L^*)^2 + (\Delta a^*)^2 + (\Delta b^*)^2]^{1/2}$, representing the distance between two points in CIE L*a*b* space.

To determine the amount of copper extracted from substrates, agar gels were dissolved in HNO_3 (70%, Carlo Erba), diluted in Milli-Q water and analysed by an inductively coupled plasma atomic emission spectrometer (ICP-AES, 38 Sequential, Horiba Jobin Yvon), using the copper analyte line at 327.369 nm, and argon (99.999%, Sapio) as the carrier gas. Calibration was performed using copper standard solutions (Sigma Aldrich) diluted in Milli-Q water.

Results

Morphological observation and elemental analysis

At first glance, the different agar particles seem to have no particular differences in terms of morphology (Fig. 1). They have heterogeneous morphology with rounded and sharp shapes when observed in the SEM. EDX allowed for the classification of the raw materials in two classes: one group, the Bresciani and CTS AgarArt powders, has alkaline metals such as K, Na and Ca present, and also a significant amount of P, with a low S content that was absent in the Sigma and Food powders. These last two form a second group, showing a strong signal for Ca, Na and Fe, a lower amount of K, and the unexpected presence of Br. Moreover, all powders contain Cl, except CTS powder, probably because of a different purification process. The morphology of the freeze-dried agar gel at 3% w/v is shown in Fig. 2 and the values of cell size are reported in Table 1.

In general, all agar samples have a round and regular cell, except Food. The largest cell dimension, around 1 mm, is found in CTS AgarArt. Sigma agar forms the most regular gel in terms of cell size and shape. Moreover, both gels from CTS and Sigma agar have an elliptical shape, not observed in the other samples.

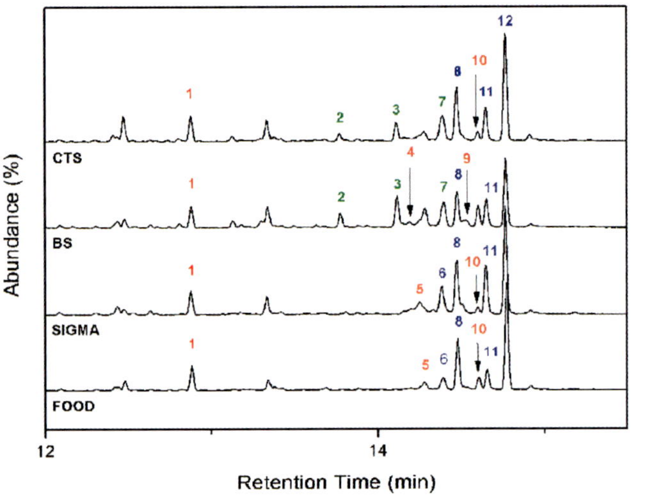

FIG. 3 Pyrograms of agar powders with peaks assigned: trimethoxybenzene (red), D-galactose (blue), 4-O-substituted glucose (green), and glucose not chemically bonded (orange).

FTIR

All spectra are very similar and, as expected from data in the literature, they reveal the presence of characteristic absorption bands of polysaccharides from red seaweed. Beside the stretching vibration of OH groups (3400–3300 cm^{-1}) and weak signals of CH stretching modes (2900–2970 cm^{-1}), FTIR spectra show a band for sulphates in the region between 1430 and 1200 cm^{-1}. Several other peaks are assigned to sulphate groups: the weak absorption at 870 cm^{-1} is assigned to the C-O-SO_3^- vibration of the sulphate group in position C6 of D-galactose (Matsuhiro 1996), and the shoulder present at 820 cm^{-1} is related to the presence of the sulphate group linked to the C6 of D-galactose units, which are converted to 3,6-anhydrogalactose by the alkaline treatment used during the extraction process (Souza *et al.* 2013). The most remarkable band in the IR spectra of agar powders, at 1070–1080 cm^{-1}, is assigned to the C-O stretching of 3,6-anhydro-L-galactose, with the contribution of the C-C-O out-of-phase stretching of primary and secondary alcohols of saccharide rings. The medium strong peak at 930 and 890 cm^{-1} can be assigned to 3,6-anhydro-L-galactose and galactose units respectively (Chopin *et al.* 1999; Pereira *et al.* 2009). The region between 770 and 740 cm^{-1} presents other characteristic absorption bands assigned to the skeletal bending of pyranose rings.

Py–GC–MS

Some interesting results were collected by Py–GC–MS (Bertasa *et al.* 2017). Using the THM method it was possible to identify characteristic markers of mono- and di-saccharides (Aspinall *et al.* 1975; Riedo *et al.* 2010). Agarose, anhydro-glucose and glucose reference standards were analysed in order to compare their pyrograms. In addition to trimethoxybenzene, which is a typical THM product from mono- and polysaccharides, the pyrograms of all agar powders confirm the presence of galactose, according to peaks 6-8-11-12, which correspond to the

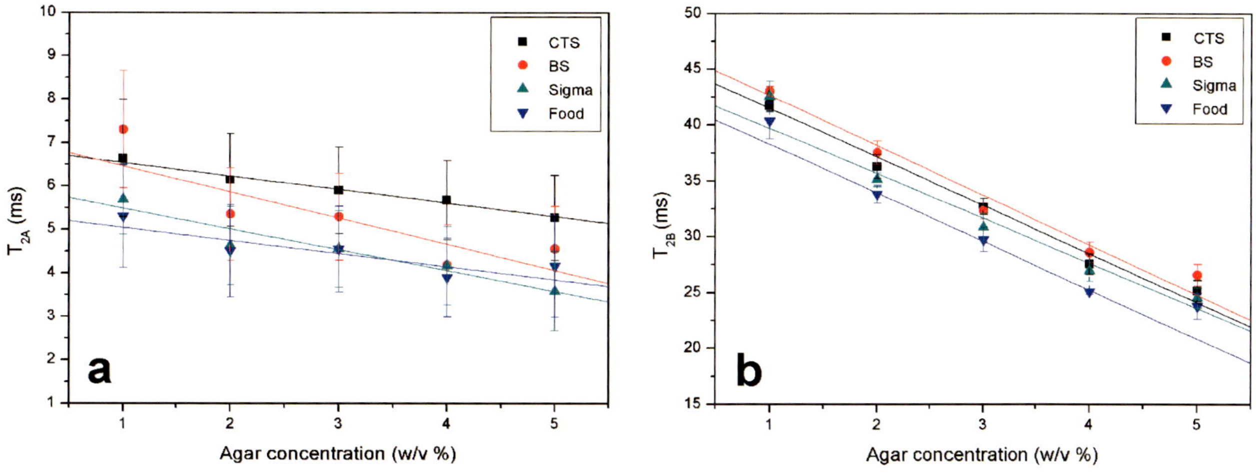

FIG. 4 Transverse relaxation time of water in agar gels as a function of agar concentration. The solid line was obtained by fitting data to a straight line. (a) T2A (R2 = 0.97). (b) T2B (R2 = 0.98).

permethylated 3-deoxy-hexonic acid methyl ester (8–12) and to the partially methylated one (6–11). These peaks are markers of D-galactose linked on position C3, and are present in both agarose and agaropectin (Fig. 3).

The pyrograms also underline the presence of glucose derivatives not chemically bonded to the polysaccharide, such as permethylated glucose derivatives (peaks 5 and 10) and partially methylated ones (peaks 4 and 9). The most interesting results were those for the CTS and Bresciani agars, where three additional peaks were detected and identified as partially methylated (2–3) and permethylated (7) isosaccharinic acid derived from 4-O-substituted glucose. These compounds are typical of cellulose, and their presence is indicative of incomplete purification of these products. These cellulose markers are absent in the agarose standard and in the Sigma sample.

Binding of water

An important step in this research is the study of the binding of water in agar gels at different concentrations. It is generally accepted that these states in hydrogels should be divided into three categories: non-freezable bound, freezable bound and free water (Watase *et al.* 1988; Ping *et al.* 2001). By modifying a UNI standard for the study of water transport by capillarity (Italian Standard UNI 10859, 2000), the penetration of water into a substrate has been studied on Noto stone specimens kept in contact with agar gels prepared at different concentrations. The amount of water absorbed by capillarity decreases when the agar gel concentration increases. The data from a 1% w/v agar gel are similar to those obtained from a multilayer of filter paper, demonstrating its similarity to a reservoir of 'free' water.

These results have been confirmed by NMR, which can provide detailed information on molecular mobility by measuring the relaxation motions of water (rotational and translational molecular motions). The two major NMR relaxation phenomena are longitudinal relaxation (characterized by a relaxation time T_1) and transverse relaxation (characterized by T_2). A multi-exponential magnetization decay for water has been observed in many porous systems (Brownstein and Tarr 1977, 1979; Bottomley *et al.* 1987). As reported in literature (Homans *et al.* 1992), the transverse relaxation of a nucleus is sensitive to both slow and fast motions, while the longitudinal relaxation is affected by fast motions only. As well known in bulk water, T_1 and T_2 are of order of seconds and show a mono-exponential trend. On the contrary, in agar gel the presence of agarose macromolecules shortens the longitudinal and transverse relaxation times of water with respect to those of bulk water. Therefore, the observation that the transverse relaxation time T_2 (milliseconds) was much smaller than T_1 (seconds) indicates that water also has a slow motion in agar gel.

To explain the transverse relaxation times in agar gel, two proton species must be taken into account: one with a short transverse relaxation time, T_{2A}, related to 'bound' water, and another with a long transverse relaxation time, T_{2B}, related to 'free' water. Protons related to 'bound' water are possibly affected by hydrogen bonding interaction and/or chemical exchange with agar macromolecules. Moreover, the literature (Ablett *et al.* 1978) demonstrates that an additional species characterized by a T_2 of the order of microseconds is present in agarose gel and is attributed to the non-exchangeable protons from the polysaccharide macromolecule. Unfortunately, this contribution could not be collected due to a long dead time in NMR-MOUSE. However, in all samples the amount of 'bound' water was found to be very small (4–7%) with respect to free water (96–93%). Many studies on agarose (Arnott *et al.* 1974; Hickson *et al.* 1968) indicated that macromolecules have a double helical conformation maintained by hydrogen bonding and, as the agar concentration increases, the interaction between water molecules and agar macromolecules also increases. As shown in Fig. 4, both relaxation times decreased with increased agar concentration. However the slope of the line of best fit for the plot of T_{2B} (free water)

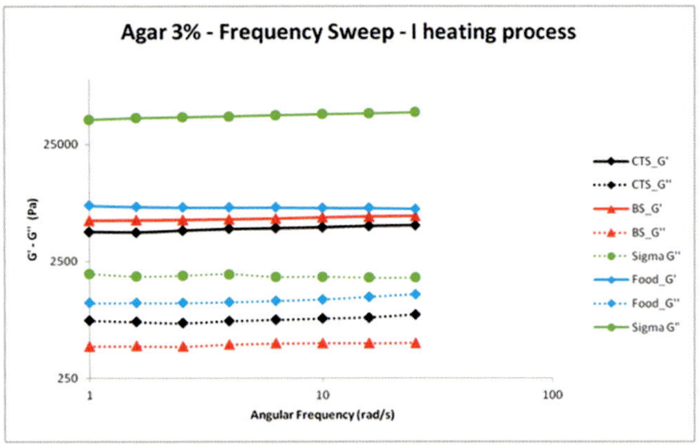

FIG. 5 Frequency dependence of dynamic moduli of agar gels at 3% w/v after one heating cycle: G' > G" for the whole frequency range analysed.

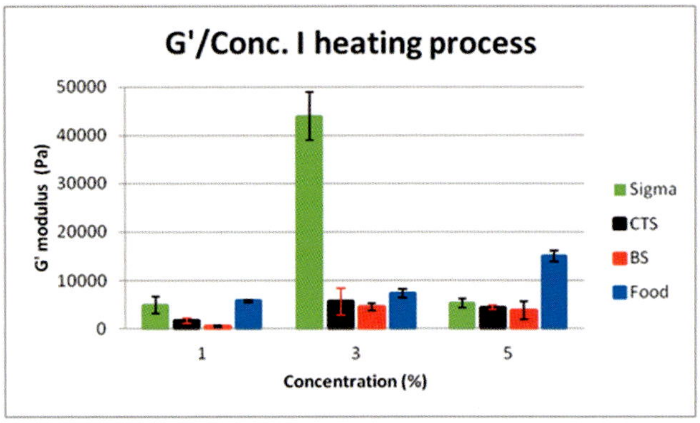

FIG. 6 Histograms of the mean storage moduli (G') of all sets of gels after one heating cycle, at different concentrations

TABLE 2 Formulations of agar gels.

Agar (CTS)	Concentration (%w/v)	Additive (%w/v)
A	1	–
B	3	–
C	5	–
D	**3**	**EDTA (1%)**
E	3	EDTA (1%) CA (0.2%)
F	**3**	**TAC (1%)**

against agar concentration was found to be steeper than that of T_{2A} (bound). Therefore, the agar concentration affects the mobility of 'free' water more than that of 'bound' water, indicating that by increasing the population of double helices and their interconnection, the diffusivity of water may be increasingly hindered as the average probability of collision between water molecules and the gel network increases.

Rheology

All agar gels exhibit frequency-independence of dynamic modulus for the entire frequency range (0.05–16 Hz) probed and G'>G". Consequently from the rheological point of view they are considered to be real gels (Fig. 5) (Burchard and Ross-Murphy 1990; Almdal *et al.* 1993). From conservators' experience, agar gels assume a more rigid consistency with increasing concentration and increased cycles of heating and cooling. The storage modulus (G') trend was investigated at different concentrations and with repeated heating and cooling cycles.

The histograms in Fig. 6 describe G' as a function of agar concentration after one heating cycle. G' increases for all agar gels from 1% to 3% w/v and decreases for 5% w/v gels, except for Food. It is possible that, up to 3% concentration, the physical entanglements between double helices permit a structural stiffening, which is partially lost at a concentration of 5%. For example, the Sigma gel exhibits a drastic decrease in the storage modulus at 5% w/v, which could be caused by a structural collapse due to a significant increase in pore size. On the contrary, the Food gel shows a different trend compared to the other samples: G' increases gradually with the agar concentration, so at 5% w/v it seems to be more rigid than the others, which exhibit a progressive decrease. According to the literature (Nishinara and Watase 1983), it seems that the alkaline treatment used during the extraction and purification of the agar decreases the storage modulus as a function of the alkaline concentration. Results of the second heating process show different trends in G' for each sample and G' does not increase with increased concentration. From these preliminary investigations, only Bresciani gel seems to be close to the empirical experience of conservators. It is still difficult to understand the mechanical dynamics of these gels compared to their performance during the cleaning process.

Copper stain removal on model samples and the case study

Several studies have demonstrated that the effectiveness of salt removal from stone is governed by stone porosity and salt-related factors, such as solubility, concentration and distribution (Heritage *et al.* 2013). Agar concentration and gel–stone contact conditions play an important role in salt extraction as well. Conservators often prepare agar gels in the range 0.5–5% w/w, and add agents designed to chelate on copper stains, such as ethylenediaminetetraacetic acid (EDTA), ammonium carbonate (AC) or triammonium citrate (TAC) (Macchia *et al.* 2011; Sansonetti *et al.* 2014).

To assess the appropriate cleaning conditions, model samples were stained and cleaned in well-controlled and reproducible conditions. The same cleaning procedure was applied to the Canova case study. Table 2 shows agar formulations and the best results obtained for model samples (highlighted in green) and for the case study (highlighted in orange). The effectiveness of copper stain removal was assessed by colour measurements and measurements on the copper extracted after a 1 hour agar gel application.

The results obtained on the model samples are both interesting and promising. For pure agar gels, small ΔE* values were recorded (≈1) whereas for chelate-containing gels, ΔE* was higher (≈3) and Δa* showed a significant decrease of the green component. Chelate-containing gels extracted twice the quantity of copper compared to pure gels (Canevali *et al.* 2016). On the statue base, the measured ΔE* values were approximately 1, except for the gel containing TAC, where higher ΔE* values and a weaker decrease of the greenish component were recorded. Moreover, pure gels did not show a noticeable extracting capacity, while the extraction trend of additive-containing gels was comparable to that recorded on the model samples.

Conclusion

The multi-analytical approach proposed provides a chemical and physical characterization of four different agar products available on the market for different applications – the conservation of cultural heritage, the food industry and biological applications. Analyses carried out so far demonstrate interesting and significant differences in their overall composition, water content and rheological properties. Elemental analyses permit the classification of raw agar powders into two groups, whereas FTIR spectral patterns are too similar to discern differences in composition. Py–GC–MS proved to be a powerful tool to distinguish agar products intended for conservation treatments from the other agar samples. In addition, Py–GC–MS data show that agar powders used for cleaning treatments contain impurities probably derived from cellulosic residues.

With regard to agar gels prepared from the four powders, the morphological characterization shows some differences in their porosity. NMR data indicated that the mobility of water inside the polysaccharidic network of agar gels and the water content in the pores strictly depend on the polysaccharide concentration. Increasing the population of double helices and their interconnection decreases the free water values. This is also confirmed by the results obtained by capillarity, and it is probably correlated with the porosity of the different gels. Rheological analyses demonstrated that agar gels can be considered real gels because of frequency-independency of dynamic moduli (G'>G"). G' modulus increases for all agar gels from 1% to 3% w/v, but decreases for samples 5% w/v in concentration. As already mentioned, the physical entanglement between double helices permits a structural stiffening up to a certain concentration (in this case 3%), which is completely lost when the agar concentration is increased further. Unfortunately, this result does not confirm the empirical experience, where, by increasing the concentration, the gels became more rigid. Further analyses will be carried out using different methods to better understand these contradictory results.

When considering the application of agar gels in the case studies, this research underlines that the diffusion of free water at the gel–stone interface plays an important role in the transport phenomena of copper salts, as does copper coordination by the agar network and additives (Canevali *et al.* 2016). On model samples, copper stains were reduced significantly by both pure and chelator-containing gels – the best results in terms of colour change and quantitative removal of copper were obtained using gels prepared with EDTA. In contrast, for the Canova case study, the colour measurements indicated useful results only for the gel containing TAC, while agar gels containing EDTA showed ΔE* ≈1, comparable to pure gels. Additive-containing gels showed extractions of higher amounts of copper with respect to the pure gels, with no significant differences among the additives used in this study.

Future work envisages an in-depth investigation into the effect of concentration and porosity on the cleaning capability of agar gel for the removal of copper stains caused by different compounds, and on other types of substrate.

Acknowledgements

Compagnia di San Paolo and the University of Torino are gratefully acknowledged for funding Project Torino_call2014_L2_b181 through *Bando per il finanziamento di progetti di ricerca di Ateneo – anno 2014* (project title *Polymer Gels for Cultural Heritage*).

References

Ablett, S., Lillford, J., Baghdadi, M.A. and Derbyshire, W. 1978. 'Nuclear magnetic resonance investigations of polysaccharide films, sols, and gels: I. Agarose', *Journal of Colloid and Interface Science* 67(2): 355–77.

Almdal, K., Dyre, J., Hvidt, S. and Kramer, O. 1993. 'Towards a phenomenological definition of the term "gel" polymer', *Polymer Gels and Networks* 1: 5–17.

Araki, C. 1956. 'Structure of agarose constituent of agar-agar', *Bulletin of the Chemical Society of Japan* 29: 43–4.

Arnott, S., Fulmer, A., Scott, W.E., Dea, I.C.M., Moorhouse, R. and Rees, D.A. 1974. 'The agarose double helix and its function in agarose gel structure', *Journal of Molecular Biology* 90(2): 269–84.

Aspinall, G.O., Krishnamurthy, T.N., Furda, I. and Khan, R. 1975. 'Base catalyzed degradations of carbohydrates. VII. Alkaline degradation of 3,6-di-O-subsituted hexoses', *Canadian Journal of Chemistry* 53(14): 2171–7.

Bertasa, M., Botteon, A., Brambilla, L., Riedo C., Chiantore, O., Poli, T., Sansonetti, A. and Scalarone, D. 2017 'Cleaning materials: a multi-analytical characterization of commercial agar powders', *Journal of Analytical and Applied Pyrolyis* [submitted].

Bottomley, P.A., Hardy, C.J., Argersinger, R.E. and Allen-Moore, G. 1987. 'A review of ^{1}H nmr relaxation in pathology: are T_1 and T2 diagnostic?', *Medical Physics Journal* 14(1):1–37.

Brownstein, K.R. and Tarr, C.E. 1977. 'Spin-lattice relaxation in a system governed by diffusion', *Journal of Magnetic Resonance* 26(1): 17–24.

Brownstein, K.R. and Tarr, C.E. 1979. 'Importance of classical diffusion in NMR studies of water in biological cells', *Physical Review A* 19(6): 2446–53.

Burchard, W. and Ross-Murphy, S.B. 1990. 'Introduction: physical gels from synthetic and biological macromolecules', in W. Burchard and S.B. Ross-Murphy (eds), *Physical Networks: Polymer and Gels*. London: Elsevier Applied Science, 1–14.

Canevali, C., Fasoli, M., Bertasa, M., Botteon, A., Colombo, A., Di Tullio, V., Capitani, D., Proietti, N., Scalarone, D. and Sansonetti, A. 2016. 'A multi-analytical approach for the study of copper stain removal by agar gels', *Microchemical Journal* 129: 249–58.

Chopin, T., Kerin, B.F. and Mazerolle, R. 1999. 'Phycocolloid chemistry as taxonomic indicator of phylogeny in the *Gigartinales, Rhodophyceae*: a review and current developments using Fourier transform infrared diffuse reflectance spectroscopy', *Phycological Research* 47(3): 167–88.

Duckworth, M., Hong, K.C. and Yaphe, W. 1971. 'The agar polysaccharides of *Gracilaria* species', *Carbohydrate Research* 18: 1–9.

Fratini, E. and Carreti, E. 2013. 'Cleaning IV: gels and polymeric dispersions', in P. Baglioni (ed.), *Nanoscience for the Conservation of Works of Art*, Nanoscience and Nanotechnology series. London: Royal Society of Chemistry, 252–79.

Heritage, A., Heritage, A. and Zezza, F. 2013. *Desalination of Historic Buildings, Stone and Wall Paintings*. London: Archetype Publications.

Hickson, T.G.L. and Polson, A. 1968. 'Some physical characteristic of the agarose molecule', *Biochimica et Biophysica Acta* 165: 43–58.

Homans, S.W. 1992. *A Dictionary of Concepts in NMR*. Oxford and New York: Clarendon Press.

Italian Standard UNI10859. 2000. *Natural and Artificial Stones, Determination of Water Absorption by Capillarity*.

Labropoulos, K.C., Niesz, D.E., Danforth, S.C. and Kevrekidis, P.G. 2002a. 'Dynamic rheology of agar gels: theory and experiments. Part I. Development of rheological model', *Carbohydrate Polymers* 50: 393–406.

Labropoulos, K.C., Niesz, D.E., Danforth, S.C. and Kevrekidis, P.G. 2002b. 'Dynamic rheology of agar gels: theory and experiments. Part II. Gelation behaviour of agar sol and fitting of a theoretical rheological model', *Carbohydrate Polymers* 50: 407–15.

Macchia, A., Sammartino, M.P. and Tabasso, M.L. 2011. 'A new method to remove copper corrosion stains from stone surfaces', *Journal of Archaeological Science* 38(6): 1300–307.

Matsuhiro, B. 1996. 'Vibrational spectroscopy of seaweed galactans', *Hydrobiologia* 326/327: 481–9.

Mohammed, Z.H., Hember, M.W.N., Richardson, R.K. and Morris, E.R. 1998. 'Kinetic and equilibrium processes in the formation and melting of agarose gels', *Carbohydrate Polymers* 36: 15–26.

Nishinara, K. and Watase, M. 1983. 'Effect of alkali pretreatment on the rheological properties of concentrated agar-agar gels', *Carbohydrate Polymers* 3(1): 39–52.

Pereira, L., Amado, A.M., Critchley, A.T., Van de Velde, F. and Ribeiro-Claro, P.J.A. 2009. 'Identification of selected seaweed polysaccharides (phycocolloids) by vibrational spectroscopy (FTIR-ATR and FT-Raman)', *Food Hydrocolloids* 23(7): 1903–9.

Ping, Z.H., Nguyen, Q.T., Chen, S.M., Zhou, J.Q. and Ding, T.D. 2001. 'States of water in different hydrophilic polymers: DSC and FTIR studies', *Polymer* 40(20): 8461–7.

Rees, D.A., Steele, I.W. and Williamson, F.B. 1969 'Conformational analysis of polysaccharides III. The relation between stereochemistry and properties of some natural polysaccharide sulfates', *Journal of Polymer Science. Part C: Polymer Symposia* 28: 261–76.

Riedo, C., Scalarone, D. and Chiantore, O. 2010. 'Advances in identification of plant gums in cultural heritage by thermally assisted hydrolysis and methylation', *Analytical and Bioanalytical Chemistry* 396: 1559–69.

Sansonetti, A., Casati, M., Striova, J., Canevali, C., Anzani, M. and Rabbolini, A. 2014. 'A cleaning method based on the use of agar gels: new tests and perspectives', in G. Wheeler (ed.), *12th International Congress on the Deterioration and Conservation of Stone*, New York: ICOM-CC.

Souza, B.W.S., Cerqueria, M.A., Bourbon, A.I., Pinheiro, A.C., Martins, J.T., Teixeria, J.A., Coibra, M.A. and Vincente, A.A. 2013. 'Chemical characterization and antioxidant activity of sulphated polysaccharide from the red seaweed *Gracilaria birdiae*', *Food Hydrocolloids* 27: 287–92.

Watase, M., Nishinari, K. and Hatakeyama, T. 1988. 'DSC study of properties of water in concentrated agarose gels', *Food Hydrocolloids* 2: 427–38.

Authors' addresses

- Moira Bertasa, Institute for the Conservation and Valorization of Cultural Heritage (ICVBC), National Research Council (CNR), Milan, Italy, and Department of Chemistry and INSTM, University of Turin, Turin, Italy (bertasa@icvbc.cnr.it)
- Oscar Chiantore, Department of Chemistry and INSTM, University of Turin, Turin, Italy
- Tommaso Poli, Department of Chemistry and INSTM, University of Turin, Turin, Italy
- Chiara Riedo, Department of Chemistry and INSTM, University of Turin, Turin, Italy
- Valeria di Tullio, Institute of Chemical Methodologies (IMC), National Research Council (CNR), Research Area RM 1, Montelibretti, Romania
- Carmen Canevali, Department of Materials Science, University of Milan Bicocca, Milan, Italy
- Antonio Sansonetti, Institute for the Conservation and Valorization of Cultural Heritage (ICVBC), National Research Council (CNR), Milan, Italy
- Dominique Scalarone, Department of Chemistry and INSTM, University of Turin, Turin, Italy

Thermo-reversible rigid agar hydrogels: their properties and action in cleaning

Paolo Cremonesi and Antonella Casoli

Introduction

Due to its high surface tension, water is a liquid characterized by limited wetting ability of surfaces and a high rate of vertical and capillary diffusion. Both features may represent a risk factor when using water or aqueous solutions for surface cleaning of artworks, or for more complex procedures such as the removal of film-forming materials (varnishes and coatings, and retouches and overpaints with a variety of binding media). The general risk can be described as the unwanted contact of water with water-sensitive materials or layers. Some materials may be intrinsically sensitive to water in terms of partial solubility, swelling and biodeterioration – cellulose, gelatine and plaster are probably the most representative examples. In other instances, this sensitivity may either be the result of chemical changes during ageing (mainly oxidative and hydrolytic processes occurring, for example in drying oils and natural terpenoid resins) or caused by the presence of various additives used to modify the properties of binding media such as surfactants, gelling agents and fillers – a critical issue for contemporary water-based paint media (Cooper *et al.* 2014).

One of the strategies generally adopted to overcome these limitations relies on the use of gelling agents: polymers that dissolve in an aqueous solution (e.g. cellulose ethers and xanthan gum) or that both dissolve and swell in water (e.g. polyacrylic acid derivatives) and increase its viscosity due to tight crosslinking of their macromolecules (Wolbers 2000; Cremonesi 2012). Although effective in restricting the diffusion of water, this procedure suffers from a major drawback: free water must be used for a proper removal of residual gelling material, and in some instances this could nullify the advantage originally achieved by the choice of using a gel.

Thermo-reversible hydrogels based on agarose do not suffer from this limitation: they combine very limited adhesion to a surface and high retention of water, and one of their most interesting features is that even when tested on one of the most porous and adsorbing surfaces (a plaster cast), no residual gelling material was detectable by analysis on the surface (Anzani *et al.* 2008).

The main limitation of agar gels is their surface morphology: being rigid, the pre-formed gels can only make contact, and consequently act upon, a surface almost completely flat and planar. However, in many instances this limit can be circumvented by applying the gel not as a pre-formed solid pad, but in another form: a semi-solid, still fluid state, ready to turn solid as soon as it is applied onto the surface. In the last decade agar hydrogels have been used extensively for the treatment of artworks of varying nature: paper, textiles, wooden supports, mobile and mural paintings, plaster and stone sculptures as well as metals (Campani *et al.* 2007; Cremonesi 2008, 2013).

The structure, properties and the gelling mechanism of agar have been extensively described elsewhere (Armisén and Galatas 2000). Here it suffices to recall the fact that agar is composed of two polymers: agarose, the non-ionic gelling fraction homopolymer of galactose, and agaropectin, the non-gelling, acidic, structurally more complex fraction, containing sulphate and pyruvate ionizable functional groups.

Preparation and application of agar gels

Agar (2–4 g) is dispersed in 100 ml demineralized water, and the mixture is heated to boiling in a microwave oven, stirring occasionally, and taking care to prevent water from boiling out of the container. Boiling is continued, generally from 30 seconds to 1 minute, until solid particles are no longer detected on the walls of the container. The hot solution is then allowed to cool, and at this stage, depending on the chosen application, two different procedures can be followed:

- pour the solution into a flat container to the desired thickness, and allow to cool to room temperature;

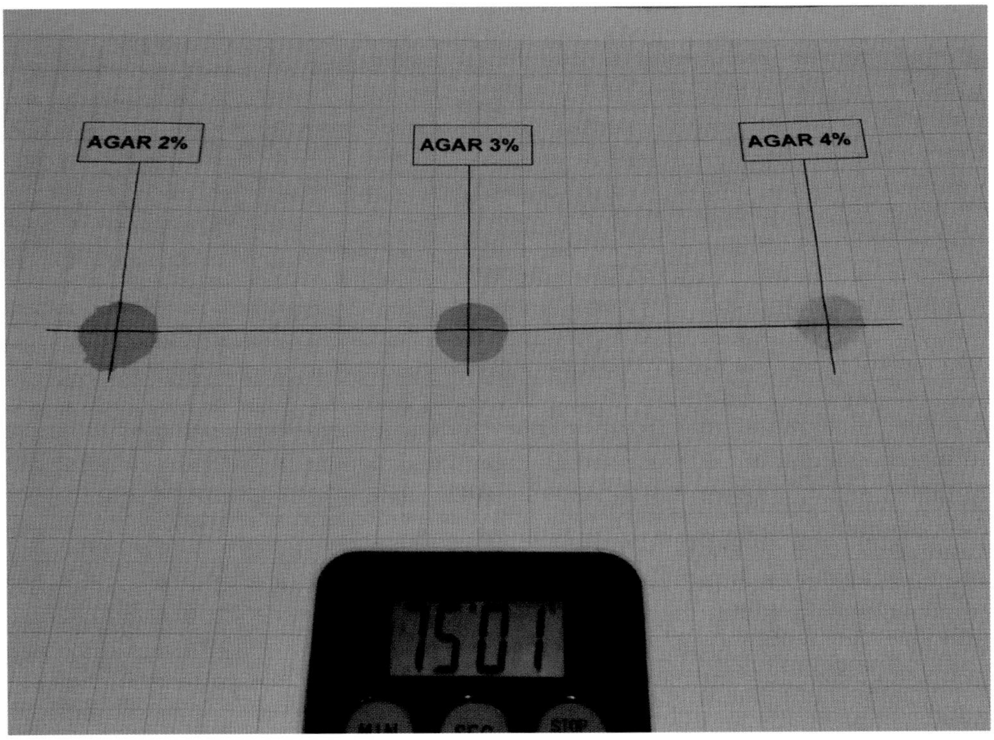

FIG. 1 Comparison of agar gels of different concentration for their release of water.

- cool the solution until a semi-solid paste is yielded and then quickly brush onto the surface of the object to be treated.

Most common types of agar have a gelling temperature around 39–40 °C. When treating sensitive surfaces according to the second procedure listed above, a type of agar with a lower gelling temperature of 29–30 °C (agar type E, Sigma-Aldrich) should be used: the gel in its semi-solid state can then be applied at moderate temperatures of 35–40 °C, which is safe for many traditional paints. When treating highly porous materials (such as plaster) or capillary materials (e.g. paper and textiles) it is best to prepare the gel by two heating–cooling cycles to produce the optimum retention of water in the gel.

If an aqueous solution, rather than pure water, is gelled with agar, all components of the solution must be stable to temperatures of 100 °C, although for only a short time otherwise enzymes and other protein-based materials, for example, would be irreversibly denatured. A different protocol must then be adopted: 75% of the required amount of water is used to prepare the agar gel, while the remaining 25% of water is used to dissolve the sensitive material at an appropriate temperature, around 45–50 °C. Once the agar solution has cooled to 40–50 °C the two solutions can safely be mixed together. A further application has been developed which is particularly effective for treating fragile, sensitive surfaces that require mild aqueous cleaning: a pre-formed rigid gel can be grated into tiny particles, which are then worked onto the surface with a soft brush (Cremonesi 2016).

Besides the actual components of the aqueous solution gelled with agar, different factors will affect the application time: the nature of the material to be removed (i.e. soil or a film-forming material); its thickness; the porosity/capillarity of the surface and the agar concentration. A 2% w/v gel will allow more water to diffuse into the surface, whereas in a 4% w/v gel, the pores are much smaller (i.e. the right size to promote gel capillarity) so water will tend to move into the gel. Fig. 1 compares the amount of water released from agar gels of different concentrations.

Once the desired action is obtained, the gel is carefully lifted from the surface to avoid leaving behind residual material. The clearance process then depends on the type of action that agar has produced, which will in turn depend on the treatment:

- when removing soiling material, the particles become attached to the gel membrane and generally no other action is required: the surface simply has to dry, untouched;
- when the gel has transferred the right amount of water to swell a film-forming material, such as gelatine, requiring removal, the swollen material is collected with a dry cotton swab, thus completing the process;
- when material has migrated into the gel (tannins, coloured cellulose oxidation products from wood or paper supports), likewise a gentle superficial action with a dry cotton swab may complete the process satisfactorily.

In general, no rinsing is required. However, if there is the possibility of leaving residual components (such as buffers or chelators used in aqueous solutions in the gel), then another rigid gel may be applied, at 4 w/v% agar, made with pure water, to act as a 'rinsing gel'. Such a choice should also be carefully evaluated on fragile surfaces in terms of ion concentration, as will be explained later.

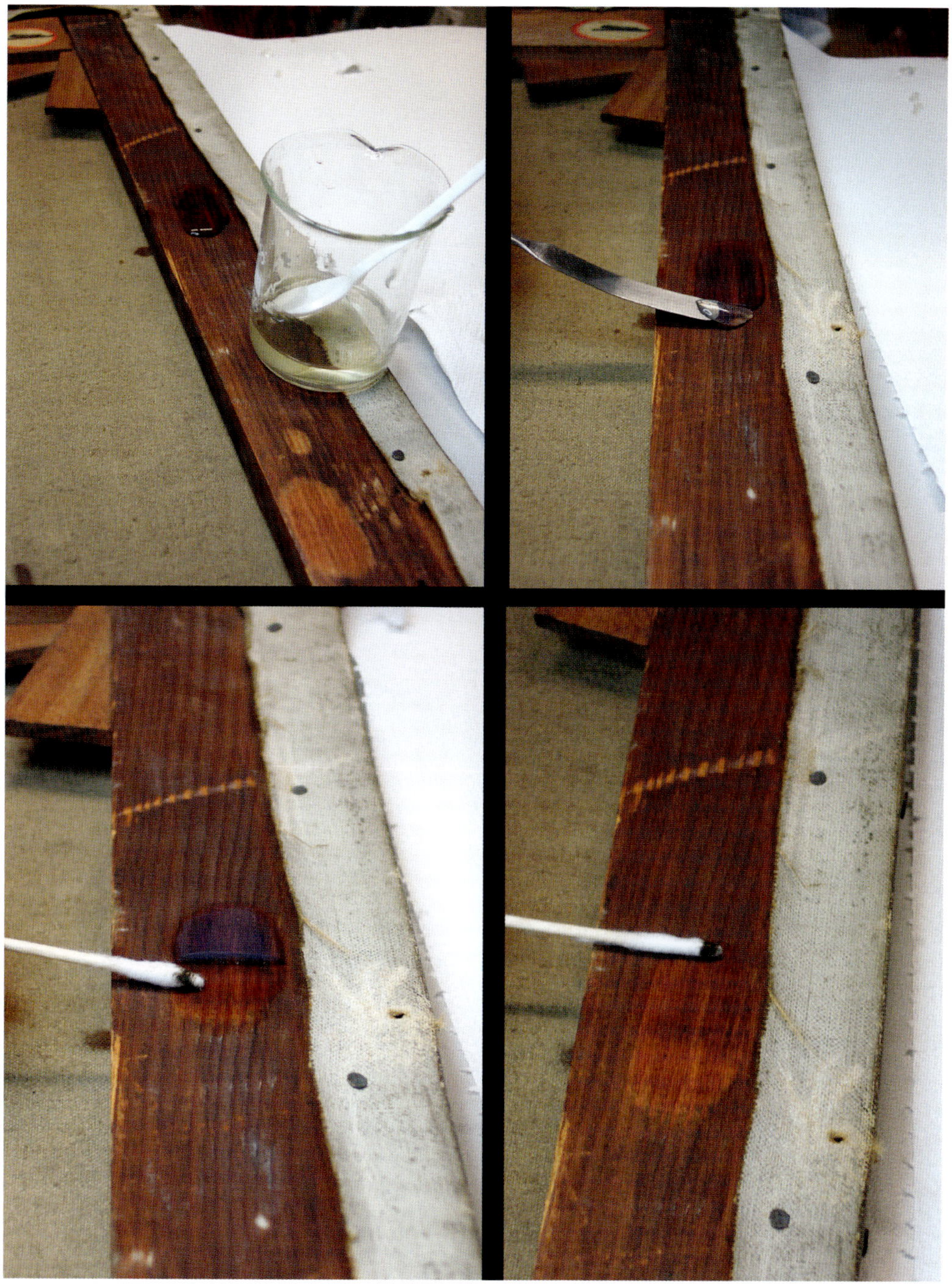

FIG. 2 A semi-solid agar gel used for cleaning a wooden stretcher.

Case study treatment

Agar should be regarded as more than just a gelling material – it can be considered one of the 'safest' available ways for delivering water:

- to surfaces that are extremely sensitive to water, for example those with partially soluble binding media, paper and wood supports, water gilding, etc.;
- whenever it is considered appropriate, for whatever reason, to limit the amount of water in a treatment.

Mural paintings provide one of the best examples when the amount of water needs to be limited: agar gels have been tested and proved to be effective in extracting salts from mock-ups of fresco paintings, purposely saturated with salts (chlorides, nitrates and sulphates). Traditionally, this treatment has been achieved by applying poulticing materials loaded with water. Such treatments are effective, but release a large amount of water into the plaster, which maintains the mobility of the salts for a long time. As a consequence, after the initial successful extraction, salts often tend to migrate slowly back into the surface. In comparison, simple visual examination of wall paintings treated with an agar gel has shown that after several months no salts had migrated back to the surfaces.

A particularly useful application of agar gels is for the cleaning treatment of polychrome painted and sculptural elements on wooden ceilings which were traditionally painted with

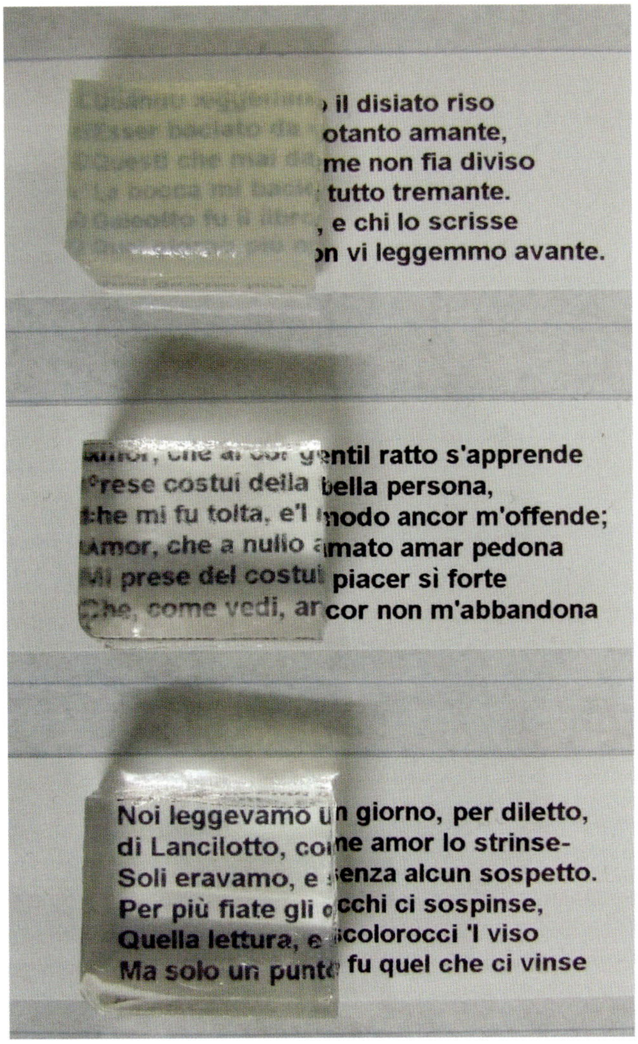

FIG. 3 The optical properties of agar (*top*) and two types of gellan pre-formed gels: (*middle*) Phytagel and (*bottom*) Kelcogel CG-LA.

water-based proteinaceous media such as milk, egg and caseinates which can withstand only limited contact with water – and often no mechanical action at all. Through the centuries, damage has resulted mainly from water seeping through damaged roofs and mobilizing soiling material, tannins and glues. Frequently, whole rooms, including the decorated ceiling, have been painted over with various paints or lime (as a preventive health measure). Recovering the original polychromy is always a complex task – in many instances, safe intervention can only be carried out by mechanically removing the overpaints with a scalpel. Agar gels can provide effective alternatives to this time-consuming and sometimes risky approach.

An example of a useful application is shown in Fig. 2: an agar gel applied in the semi-solid state to a wooden stretcher. After a few minutes' application the gel is lifted, and the soiling material is easily picked up from the surface with dry cotton swabs. No additional amount of water is required for completing the cleaning procedure.

With regard to the treatment of paper, the authors in collaboration with ICRCPAL (Istituto Centrale per il Restauro e la Conservazione del Patrimonio Archivistico e Librario) in Rome, Italy, focused first on agar and then considered other similar gelling materials in comparison. One type of gellan gum, Kelcogel CG-LA, was finally selected as the best-performing material for paper due to its transparency, flexibility and chemical compatibilities (Berzioli *et al.* 2011). Fig. 3 compares the optical properties of agar with two types of gellan: Phytagel and Kelcogel CG-LA.

Analytical studies

The authors first became involved in the study of agar in 2004, when the Italian conservation community was introduced to the use of rigid agar gels by Richard Wolbers during a workshop in Verona. Analytical studies were then started, mainly aimed at addressing the residues issue. The decision was made to test artworks made of three of the most porous, water-sensitive materials: paper; wall paintings and plaster sculptures. Compared to these, most other artworks exhibit much less sensitivity in terms of water diffusion, water acting as a solvent and persistence of residues.

Plaster fragments from sculptures were treated with a 2.5% w/v agar gel for 3 minutes, 20 minutes and until gel dryness. Once dry, a sample (approx. 1 cm^2 and 0.5 g) was then taken from each treated area. Each sample was powdered and extracted in boiling distilled water. The resulting solutions were gradually dried out and each residue was analysed with Fourier transform infrared (FTIR) spectroscopy and gas chromatography–mass spectrometry (GC–MS). Galactose, the main monosaccharide constituting the gels, was detected only in trace amounts (Fig. 4), based on the area ratio of the chromatographic peak of galactose to that of the internal standard sorbitol, thus proving that a very limited amount of the gel permeated into the plaster (Anzani *et al.* 2008).

When applying agar gels to paper, the use of pre-formed gels is the only advisable procedure – in the semi-fluid state the gel would become too strongly adhered to the surface. In this instance, the main concern, besides the permanence of residual gel material, is an issue that is critical in the washing of cellulose-based textiles: water may wash out coloured, soluble oxidation products of cellulose and polysaccharide fragments that still contribute to the material's cohesion. The same process on paper would result in decreased mechanical strength in the treated paper. Seventeenth-century paper samples were tested for mechanical strength, both untreated and treated with a 2% w/v agar gel. The results indicated that treated paper samples do not appear weaker: on the contrary, there was an average increase in strength and in the strain to failure when compared to untreated paper. Scanning electron microscopy (SEM) observation of the treated paper did not reveal residual material on the surface. This was in contrast to an aqueous gel made from methylhydroxyethylcellulose when used as a comparison (Berzioli *et al.* 2011).

Characterization of residual gelling material was also carried out on mock-ups of fresco paintings, comparing the two main application techniques: 2 and 4% w/v pre-formed rigid and semi-solid gels. To achieve the same cleaning results, pre-formed gels had to remain in place for up to 40 minutes, whereas a 10-minute application of semi-solid gels was sufficient. Trace amounts of residual galactose were detected only in the samples treated with pre-formed rigid gels; this can be explained as a consequence of the much longer application time (Selva Bonino 2008).

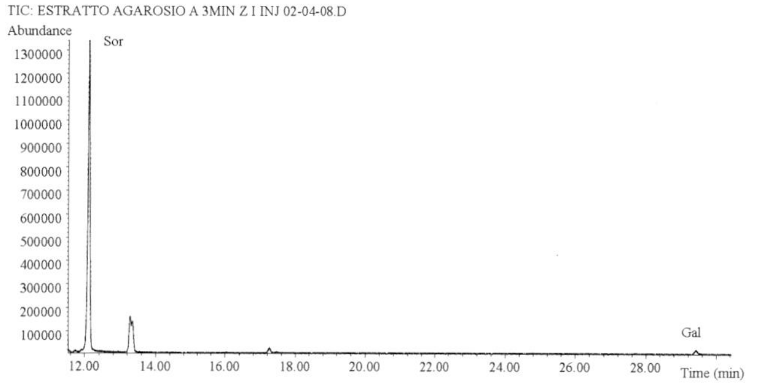

FIG. 4 Chromatogram of the extract from a plaster fragment treated with agar gel. (Sor = sorbitol, used as an internal standard; Gal = galactose).

Considerations on the use of agar gels

Working with an agar rigid gel made with pure water will often give the impression that the action is greater than that expected from pure, non-gelled water. At least two different advantages to using the pure water agar gel may be evident: increased action for insoluble iron corrosion products and the ability to produce similar effects to those generally expected from aqueous solutions with an appreciable ion content. We will now examine the first of these experiences. Treating stone surfaces with agar gels at times led to an unexpected result: spots of iron corrosion products could be appreciably thinned down and in a few instances even completely removed. The surprising fact is that these gels were made with pure water alone; such an action would normally be expected only when using aqueous solutions containing chelators such as citrates, polyphosphates, ethylenediaminetetraacetic acid (EDTA) or diethylenetriaminepentaacetic acid (DTPA).

No specific reference has been found on the ability of agar to act as a chelator for bi- and tri-valent metal ions. However, on a theoretical level the polyhydroxylated structure of agarose would be compatible with chelation: it can act as a polydentate chelator, mostly a mild chelator considering that the molecules bear no charges, due to its non-ionic structure. 'Agar influences the bio-availability of metals by binding to them' is a statement often encountered when searching through biological sources related to the preparation of culture media based on agar. With phenol groups, which are ionizable in alkaline medium, a stronger chelating action might reasonably be expected. However with non-phenolic hydroxyl groups this milder complexing action, independent of pH, is probably more likely.

Another similarity could be hypothesized with adsorptive resins, gel types and microporous resins made of functionalized polymers and capable of exerting chelating activity on metal ions (Sud 2012; Wheaton and Lefevre 2000). These resins generally have an ionizable structure of the aminocarboxylic acid type such as EDTA, and this is the feature missing in agarose. However, another important parameter in the action of these resins is the porous matrix and its ability to retain ions, a feature that agarose possesses. Lastly, agaropectin (the ionic part of agar), has a structure which could be compatible with a chelate-like action towards Fe^{3+} ions: a cation-exchange reaction, performed by the sulphate and pyruvate acidic functional groups. One of these mechanisms – or more likely a contributing factor from each of them – could explain the experimental findings. Specific studies should be devised to clarify this interesting aspect.

Regarding the second type of action that agar displays when it is loaded with pure water – that of more closely resembling the activity of an aqueous solution with some ionic content – a more straightforward explanation can probably be extrapolated from the experimental findings. Unlike pure agarose, agar is an ionic gelling material. Even when loaded with pure distilled or deionized water, an agar rigid gel will have some ionic content, as recent analysis of current commercial products has revealed.[1] The conductivities measured for agar gels of different concentration are: 440 μS/cm for a 2% gel, 540 μS/cm for a 3% and 630 μS/cm for a 4% w/v. When a rigid gel is in contact with a porous surface, it will allow diffusion of a certain amount of water, depending on its concentration and the time of contact. At this point, the system of the rigid gel and wet porous layers will aim to reach equilibrium between the concentrations of ionic species.

Two different processes can theoretically be considered: ion diffusion and osmosis. Ion diffusion occurs when there is an ion concentration gradient. The intrinsic characteristics of the surface itself (porosity and pore size) determine which ions preferentially migrate across the interface between the gel and the surface. Through the painted surface of a painting on canvas, based on organic binding media, simple, small 'mineral' ions rather than ionized macromolecules (e.g. proteins) would be more likely to migrate. Conversely, in an osmotic process, water molecules would migrate against the concentration gradient, trying to dilute the more concentrated side of the system.

However, there is a strict requirement for an osmotic process to occur: in the general description of osmosis, the presence of a semipermeable membrane separating the solutions, i.e. a membrane allowing only water molecules (and at most, some smaller ions) to move across. In our specific context, this would be an organic film-forming material, separating the 'true solution' – the gelled water – from the layers once wet. Without this specific feature, no osmotic processes can be expected, but instead just water and ion diffusion. The lipid bilayer surrounding the cells in living organisms is an example of such a membrane of natural origin.

Is it reasonable to hypothesize the existence of anything similar on the surface of a work of art? The authors have devised a simple but effective test in an attempt to answer this crucial question. It originates from a simple operation: sprinkling salt on cut vegetables before grilling. This draws the water out from the vegetables, by osmotic movement, making them crispy once baked. This is how the 'zucchini test' was adapted to our area of interest (Fig. 5a). A small jar is filled

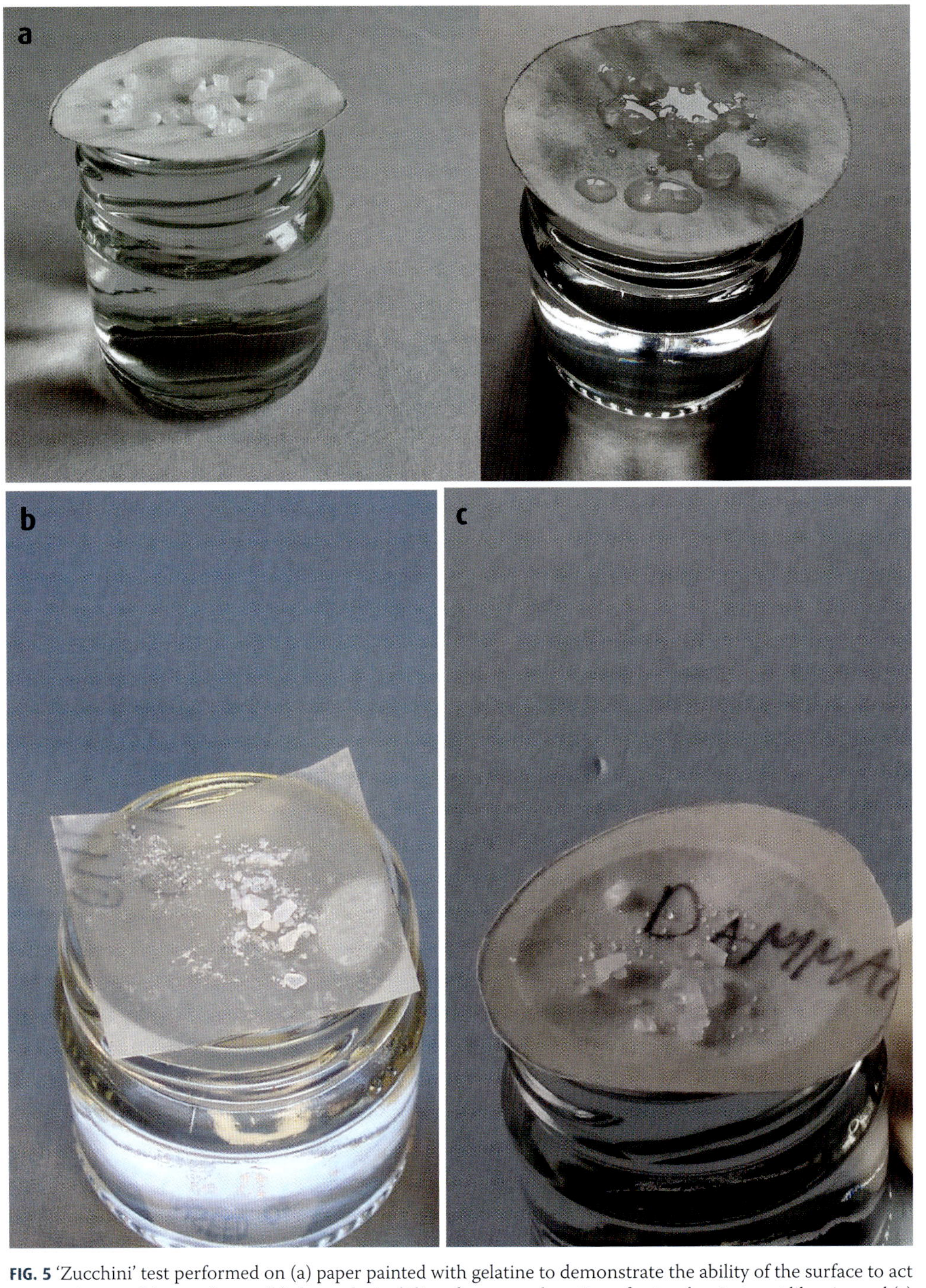

FIG. 5 'Zucchini' test performed on (a) paper painted with gelatine to demonstrate the ability of the surface to act as a semipermeable membrane; (b) shows the inability of an unaged coating of natural resin to act likewise and (c) demonstrates the moderate ability of the coating of natural resin after three years of natural ageing to act likewise.

with demineralized water to the very top. Next, a disc of filter paper, on which a film-forming material has previously been painted, is placed on top of the jar, making contact with the water underneath. Rock salt is then sprinkled on the surface of the coated paper disc. If after some time droplets of water seep through and surround the salt grains, this movement can be taken as an indication of the 'painted paper' acting as a semipermeable membrane. A control paper disc, similarly painted but with no salt applied, did not allow water to penetrate under the same experimental conditions. The polymer that was painted on (in this case, gelatine) therefore has some ability to promote osmosis.

The tests show that all water-soluble natural polymers tested (gelatine, albumen, caseinate, polysaccharide gums and cellulose itself) behave like semipermeable membranes. In contrast, negative results were obtained from linseed oil and natural terpenoid resins. Fig. 5 compares the result obtained from gelatine (Fig. 5a), after 10 minutes with that obtained from dammar in turpentine after as long as 40 minutes: one month after application to the paper (Fig. 5b) and after three years (Fig. 5c) of aging in a domestic environment. A small amount of water only permeated the layer in the last sample. This result can be explained by considering the chemistry of terpenoid natural resins

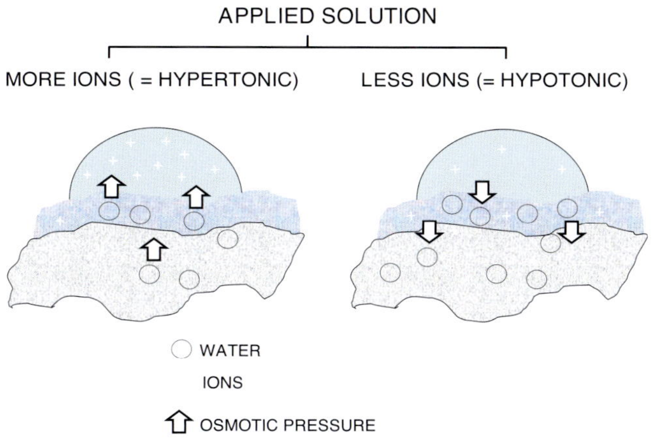

FIG. 6 Schematic representation of the osmotic processes on the surface of a painting treated with an aqueous solution.

(van der Doelen 1999). When applied as a varnish onto a painted surface they are prevalently low molecular weight molecules. Their rather low polarity imparts a definite hydrophobic character to the coating, generally not likely to promote osmosis. As the resinous material ages, a higher molecular weight fraction is formed through oxidation and crosslinking, which is more polar and more hydrophilic in character. In this way, an aged, oxidized natural resin coating may become somewhat more similar to a polymeric, hydrophilic coating, making it more susceptible to osmotic processes. Likewise a young oil film possesses a definite hydrophobic (water-repelling) character. In addition, being unpigmented, this coating can hardly be representative of a true oil-based paint layer.

From a practical point of view, this latter finding may have important consequences. The presence of a natural resin varnish on a painting on canvas/panel – aged but still fairly even, unbroken and compact – would probably represent a 'barrier' to osmotic water movements rather than a semi-permeable membrane. Could this be taken as an indication that there should be less concern with the ion concentration in the cleaning solution when treating a painting coated with a natural resin varnish with aqueous media? Another critical factor should also be considered when evaluating the likelihood of osmosis: the length of the contact time of water with the surface which is required to produce osmotic movements. Does the actual 'cleaning time' match this length? Furthermore, the same could be said about the more recently introduced 'low-molecular weight resins' of synthetic origin, namely aliphatic hydrocarbon and urea-aldehyde resins, which are capable of mimicking in many ways the behaviour of the natural terpenoid resins. These remain open questions until further investigation is carried out.

If the surface of the artwork to be cleaned with an aqueous medium is capable of promoting osmosis, then the ion concentration of the aqueous solution itself becomes an important parameter. More specifically, a concentration imbalance between the ion concentration of the cleaning solution and the ion concentration that water will encounter once applied to the artwork may promote the movement of water. If the osmotic pressure associated with these movements is too high, it could be damaging to the surface. Both sides of the imbalance are potentially damaging (Fig. 6), for example:

- if the ion concentration of the cleaning solution is lower than that of the artwork (hypotonic conditions) – as occurs in general when using demineralized water – water may diffuse into the layers;
- conversely, if the solution has higher concentration than the layers (hypertonic conditions), water that initially wet the layers by diffusion or capillarity will tend to migrate out. In itself, this is a positive effect as it means shorter permanence of water inside the layers. However, if this outbound movement is too intense, the associated pressure could open cracks on the surface.

Ideally, isotonic conditions, namely a similar ion concentration both in the cleaning solution and in the wet layers, would be the most appropriate, allowing free and stress-free water movements in and out of the layers. An even wiser choice would be a 'slightly hypertonic' condition to ensure that water moves out and the layers dry faster.

Based on these considerations we can now return to the discussion on how rigid agar gels behave on different surfaces. First, by severely restricting water diffusion into the surface, these gels may 'mitigate' the potential effects of water, including osmosis: the higher the agar concentration, the more restricted the diffusion. As was shown in Fig. 1, the contact time of the gel with the surface will also affect this.

The second most important factor is the ion concentration of the aqueous phase of the gel, bearing in mind that agar itself contributes to the overall ion content. When a rigid agar gel is applied to an inorganic/mineral surface, such as a plaster sculpture, it may be reasonable to rule out the likelihood of osmotic diffusion of water: the porosity is much too high and, more importantly, there is no film-forming material capable of acting as a semipermeable membrane. There is also a balance between the capillary forces of the plaster and those of the agar pores. The practical response appears to confirm this mechanism: an outbound flow or force seems to detach the soiling material from the surface and move it towards the gel. As a practical example, the conductivity measured on a plaster fragment (wetting and inserting into it the metal electrodes of a conductivity meter) was *c.*70 μS/cm, and that of a 2% w/v rigid agar gel about 450 μS/cm; the gel would then be hypertonic versus the plaster fragment (Stavroudis and Doherty 2013), and the movement of water driven by osmosis out of the fragment could promote the surface cleaning action. Naturally, this takes for granted the presence of a proper semipermeable membrane.

The simple test described above could provide an explanation of this controversial issue: when a rigid agar gel was tested under the same conditions, water was found to seep out of the gel and wet the salt grains within less than 10 minutes (Fig. 7). Agar itself (i.e. the gel matrix) is acting as a semipermeable membrane.

In view of these findings the overall picture changes quite significantly: any time water or an aqueous solution

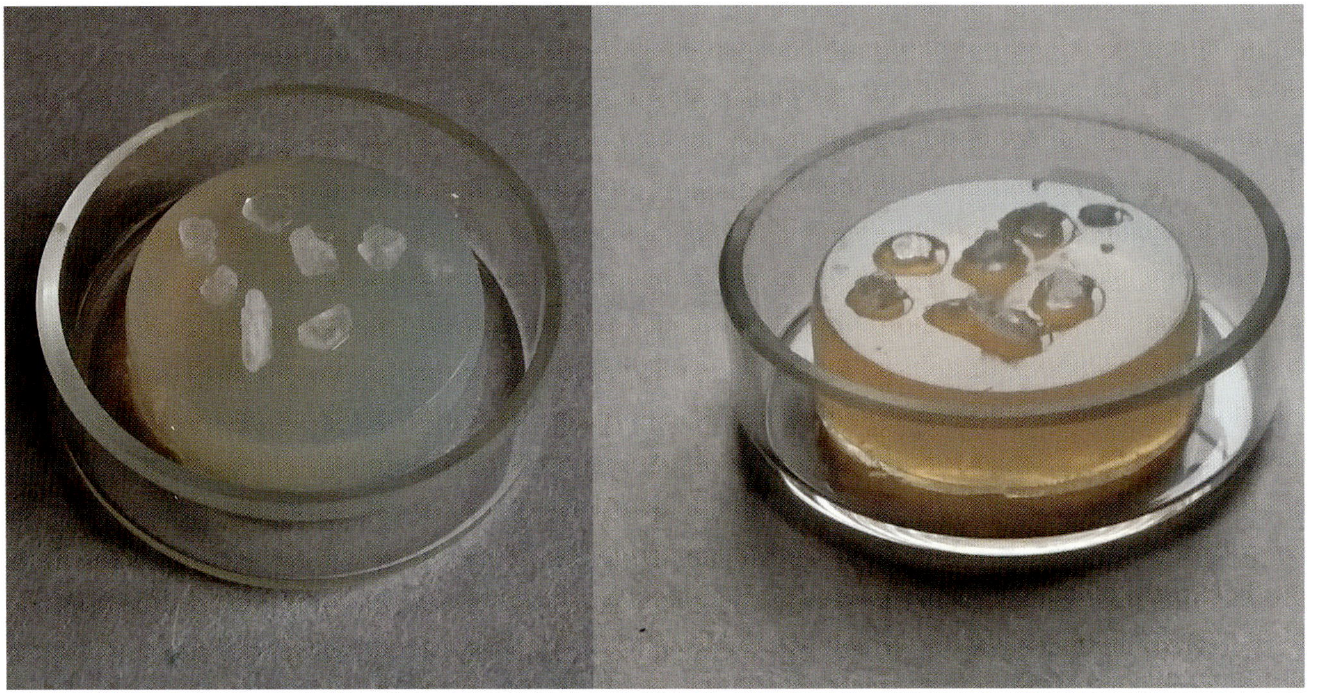

FIG. 7 A rigid agar gel tested for its ability to act as a semipermeable membrane.

TABLE 1 pH and conductivity of water and various aqueous buffered solutions in free form and as 2.5% agar rigid gels. All buffers were obtained by neutralizing the starting acid components with 1M sodium hydroxide solutions.

Solution	Free solution		2.5% agar gel	
	pH	Conductivity (μS/cm)	pH	Conductivity (μS/cm)
Water	6.9	15	6.8–7.1	570
25 mM acetate buffer	5.5	2,000	5.5-5.6	2,200
25 mM phosphate buffer	7.0	3,300	6.9-7.0	3,000
25 mM diphosphate buffer	8.5	5,500	8.2-8.4	4,300
25 mM citrate buffer	6.0	4,500	5.9-6.0	3,800
25 mM citrate, 25 mM diphosphate buffer	8.0	8,200	7.7	6,700
25 mM citrate, 25 mM diphosphate buffer	9.0	8,400	8.6	7,100
25 mM DTPA buffer	8.0	4,800	6.8	4,600
25 mM DTPA buffer	9.0	5,500	8.8	4,900
25 mM DTPA buffer	10.0	6,300	9.7	5,300

is delivered in the form of a rigid agar gel to a surface for cleaning or different purposes, osmotic processes should be expected, independently of the material type of the artwork whether varnished or unvarnished paintings, paper, textiles, wooden, gypsum or stone objects. A critical factor in formulating the cleaning solution will then be its ion concentration. After measuring the conductivity of the surface to be treated in the most accurate way, the conductivity of the cleaning solution should be adjusted according to the measured surface value. In general, hypotonic conditions are considered the most dangerous and should always be avoided: they drive water into the layers and may cause swelling and potential disruption. Isotonic, or better still, moderately hypertonic conditions, are safer under most circumstances. The conductivity of the solution can be lowered as necessary by diluting with water or raised by adding neutral salts such as sodium or potassium chloride.[2]

Conclusion

It is interesting that some of the most innovative cleaning treatments have been made possible by agar, one of the oldest known gelling materials, used for centuries at least by eastern civilizations (Armisén and Galatas 1987). From the start agar looked to be an interesting material for producing aqueous gels suitable for cleaning treatments of various works of art and its effectiveness became immediately obvious. Furthermore, 'structural' interventions, and not just true cleaning procedures, could be carried out with safety margins unattainable with other techniques: for example, removal of residual animal glues from water-sensitive surfaces. It only remains to characterize the residues issue which a number of analytical studies have started to address.

Thanks to the positive results obtained, the use of agar gels on a variety of works of art can be regarded as 'safe', and it has become a standard strategy for applying an aqueous medium to highly sensitive surfaces. The authors remain firmly convinced that, at least on works of art based on organic materials, agar should be used as a gelling material for aqueous media only, in preference to recent recommendations that propose the combination of water and polar organic solvents miscible with water, such as acetone and ethanol. Polar solvents by themselves possess a broad spectrum of actions on fresh and aged film-forming materials; if the ionization/dissociating ability typical of water is also added, effectiveness definitely increases but at the expense of selectivity.

Acknowledgements

The authors wish to thank all the colleagues from different institutions who have been involved with various tasks in these projects, and throughout these years have made a valuable contribution: the Department of Chemistry, University of Parma; *CESMAR7* (Centre for the Study of Materials for Restoration) in Verona, Italy; the Galleria d'Arte Moderna in Milan, Italy; and the Istituto Centrale per il Restauro e la Conservazione del Patrimonio Archivistico e Librario in Rome, Italy.

Materials and supplier

- 05040 agar, gelling temperature 39 °C; A4675 agar type E, gelling temperature 26–28 °C: Sigma-Aldrich Co, St. Louis MO, http://www.sigmaaldrich.com

Notes

1. Moira Bertasa, Oscar Chiantore, Tommaso Poli, Chiara Riedo, Valeria di Tullio, Carmen Canevali, Antonio Sansonetti and Dominique Scalarone, 'A study of commercial agar gels as cleaning materials', in this volume.
2. A final remark on agar gels: the ionic character of agar may modify the parameters of the aqueous solutions that are being turned into rigid gels. Table 1 compares the conductivities of agar gels prepared from pure water and various aqueous solutions. For the measurements of pH and conductivity, small discs (respectively 10 and 3 mm in diameter) cut with plastic cylinders from the cast rigid gels were placed in contact with the flat electrode of a Hanna Instruments Skincheck pH meter and the measuring cell of a Horiba Laqua Twin B771 conductivity meter. In general, buffers seem able to maintain the original pH of the solutions, even in the moderate concentration used (25 mM). DTPA buffered to pH 8 is the most striking exception: the pH of the gel is considerably lower than that of the parent solution. No satisfactory explanation can yet be offered. For conductivity it is also difficult to explain the unexpected finding that the conductivity of the agar gels is lower than that of the parent-free solutions. It may be linked with the restricted mobility of some ions inside the gel matrix or it may simply reflect the experimental difference between the two measurements: one in solution, the other on the rigid gel surface. It is rather difficult to make predictions: it is safer always to re-measure the pH once the rigid gel has been prepared.

References

Anzani, M., Berzioli, M, Cagna, M., Campani, E., Casoli, A., Cremonesi, P., Fratelli, M., Rabbolini, D. and Riggiardi A. 2008. *Gel rigidi di Agar per il trattamento di pulitura di manufatti in gesso/Use of Rigid Agar Gels for Cleaning Plaster Objects*, *Quaderni del Cesmar7*, no. 6. Padua: Il Prato.

Armisén, R. and Galatas, F. 1987. 'Chapter 1. Production, properties and uses of agar', in D.J. McHugh (ed.), *Production and Utilization of Products from Commercial Seaweeds*, Rome: Food and Agriculture Organization of the United Nations. Available at: http://www.fao.org/docrep/x5822e/x5822e03.htm (accessed 20 March 2017).

Armisén, R. and Galatas, F. 2000. 'Agar', in G.O. Phillips and P.A. Williams (eds), *Handbook of Hydrocolloids*. Boca Raton, FL: CRC Press, 22–40.

Berzioli, M., Botti, L., Casoli, A., Corazza, A., Cremonesi, P., Iannuccelli, S., Isca, C., Placido, M., Residori, L., Ruggiero, D., Sotgiu, S. and Tireni, L. 2011. 'Evaluation of cleaning and chemical stabilization of paper treated with a rigid hydrogel of gellan gum by means of chemical and physical analyses', in J. Bridgland (ed.), *Preprints of the ICOM-CC 16th Triennial Congress*. Paris: ICOM, paper 0602. [CD only]

Campani, E., Casoli, A., Cremonesi, P., Saccani, I. and Signorini, E. 2007. *L'uso di agarosio e agar per la preparazione di 'gel rigidi'/Use of Agarose and Agar for Preparing 'Rigid Gels',* D. Kunzelman (tr.), *Quaderni del Cesmar7*, no. 4. Padua: Il Prato.

Cooper, A., Burnstock, A., van den Berg, K. H. and Ormsby, B. 2014. 'Water sensitive oil paints in the 20th century: a study of the distribution of water-soluble degradation products in Winsor & Newton Artists' oil colour paint swatches, with case studies from Tate's collection', in A. Burnstock, M. de Keijzer, J. Krueger, T. Learner, A. de Tagle and G. Heydenreich (eds), *Issues in Contemporary Oil Paint*. Cham: Springer, 295–310.

Cremonesi, P. 2008. 'Applicazione di metodologie di intervento più recenti per la pulitura del materiale cartaceo', in M. Fratelli and E. Signorini (eds), *Problemi conservativi dei manufatti dell'Ottocento. I dipinti, la carta, i gessi*. Padua: Il Prato, 41–8.

Cremonesi, P. 2012. *L'ambiente acquoso per il trattamento di opere policrome*, I Talenti – Metodologie, tecniche e formazione nel mondo del restauro 20. Padua: Il Prato.

Cremonesi, P. 2013. 'Rigid gels and enzyme cleaning', in M.F. Mecklenburg, A.E. Charola and R.J. Koestler (eds), *Cleaning 2010: New Insights into the Cleaning of Paintings*. Washington, DC: Smithsonian Institution Scholarly Press, 179–84.

Cremonesi, P. 2016. 'Surface cleaning? Yes, freshly grated agar gel, please', *Studies in Conservation* 61(6): 362–7.

Sansonetti, A., Casati, M., Striova, J., Carnevali, C., Anzani, M. and Rabbolini, A. 2012. 'A cleaning method based on the use of agar gels: new tests and perspectives', paper present at the *12th International Congress on the Deterioration and Conservation of Stone*. New York: Columbia University.

Selva Bonino, V.E. 2008. *Studio preliminare di gel rigidi a base di agarosio per il trattamento conservativo di pitture murali*, unpublished Master's dissertation, Department of Chemistry, Università degli Studi di Parma.

Stavroudis, C. and Doherty, T. 2013. 'The modular cleaning program in practice: application to acrylic paintings', in M.F. Mecklenburg, A.E. Charola and R.J. Koestler (eds), *Cleaning 2010: New Insights into the Cleaning of Paintings*. Washington, DC: Smithsonian Institution Scholarly Press, 139–45.

Sud, D. 2012. 'Chelating ion exchangers: theory and applications', in A.A.K. Inamuddin and M. Luqman (eds), *Ion Exchange Technology I: Theory and Materials*. New York: Springer Science and Business Media, 372–402.

van der Doelen, G.A. 1999. *Molecular Studies of Fresh and Aged Triterpenoid Varnishes*. Amsterdam: FOM-MOLART.

Wheaton, R.M. and Lefevre, L.J. 2000. *Dowex Ion Exchange Resins: Fundamentals of Ion Exchange*, Midland, MI: The Dow Chemical Company. Available at: http://msdssearch.dow.com/PublishedLiteratureDOWCOM/dh_0032/0901b803800326ca.pdf (accessed 21 March 2017).

Wolbers, R. 2000. *Cleaning Painted Surfaces: Aqueous Methods*. London: Archetype Publications.

Authors' addresses

- Paolo Cremonesi, Lodi, Italy (paolocremonesi57@gmail.com)
- Antonella Casoli, Department of Chemistry, Life Sciences and Environmental Sustainability, University of Parma, Parma, Italy (antonella.casoli@unipr.it)

The treatment of Barnett Newman's *Shining Forth* in the MNAM Pompidou collection

Véronique Sorano-Stedman, Richard Wolbers and Sophie Germond

Introduction

Shining Forth (to George) is an exceptional painting, considered as Barnett Newman's major work (Fig. 1). The artist dedicated this painting to his brother George, who died in February 1961. The Hebrew version of the name George is Zerach, meaning 'to shine' in English, translated by Newman as 'Shining Forth'. The work was purchased in 1979 thanks to an exceptional gift from the Scaler Foundation.

The background of the painting was originally unbleached white, the colour of the canvas support. Three vertical black stripes, called 'zips' by the artist, stand out against this background. The largest is 7.5 cm wide and seems to cut the canvas in two. The other two stripes, approximately 2.5 cm in width, are on either side of the central zip. These three stripes are composed of black paint, and the right was painted in negative using masking tape. They were not drawn with great accuracy and the edges are more or less rough. The negative band is treated with vigorous brushstrokes or perhaps with a palette knife (Fig. 2). Although the location of these stripes may seem to be symmetrical, this is not the case: the middle stripe is not in the centre of the canvas but slightly to the left. Similarly, the distance of the left stripe from the edge is not the same as for the right 'negative' stripe. Furthermore,

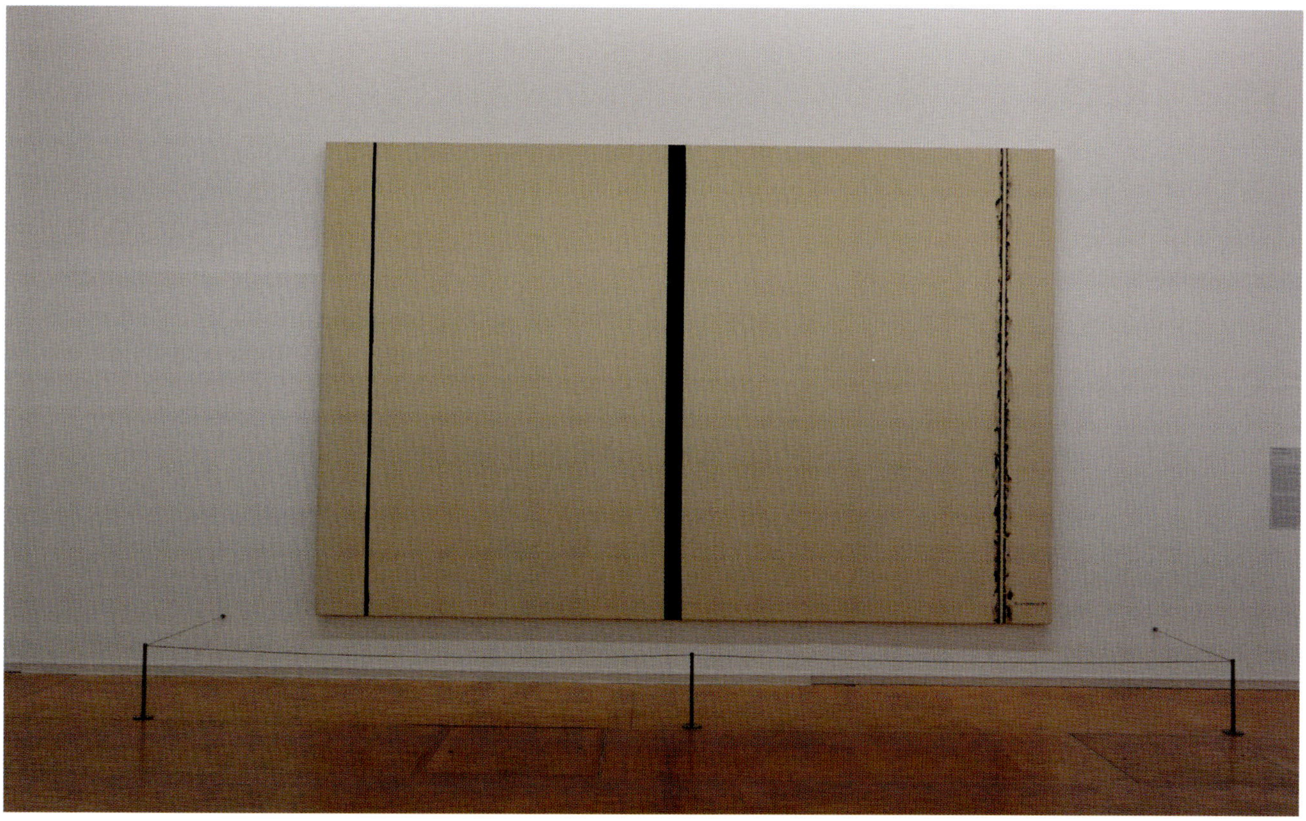

FIG. 1 Barnett Newman, *Shining Forth (to George)*, 1961, oil on canvas, 290 × 442 cm, MNAM Centre National d'Art et de Culture G. Pompidou, inv. no. AM 1978-371: after treatment.

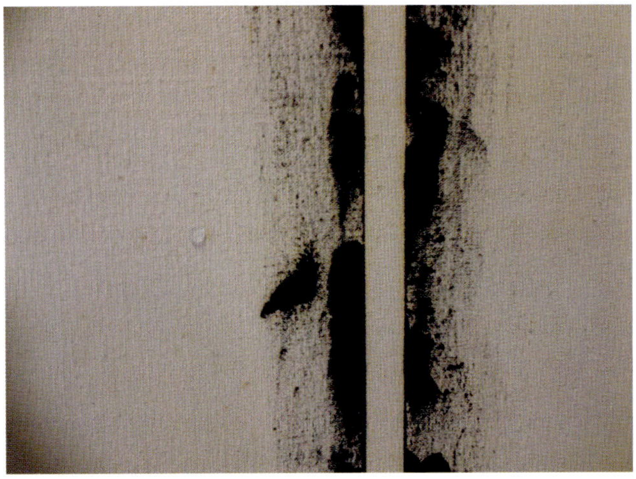

FIG. 2 Barnett Newman, *Shining Forth (to George)*: detail of the zip on the right side.

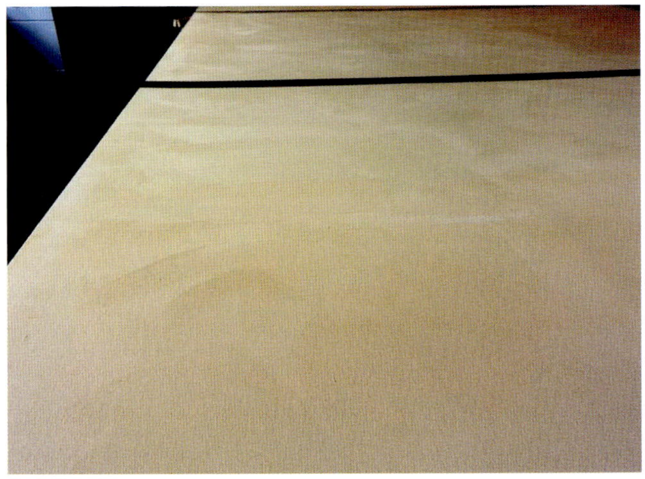

FIG. 3 Barnett Newman, *Shining Forth (to George)*: raking light image, showing the brushstrokes in the Rivit size applied to the canvas.

Shining Forth is one of the few paintings by Newman whose dimensions approximate to the golden section:[1] the long side is roughly equal to one and half times the short side. Newman has accentuated the concept of bareness by leaving large areas of unpainted canvas. This reduction in the pictorial work, coupled with a simplification of the formal elements, illustrates that painting is also a 'mental thing'. All the components of *Shining Forth* participate in this synthesis between the pictorial and conceptual, the sensory and the cerebral.

Identification of materials

Support

Shining Forth was painted on an unbleached cotton duck canvas with a blue border, secured by tacks to a wooden stretcher. The cotton duck is a coarse, mechanically woven fabric with a close weave and a thread count of 20 × 13 threads/cm^2. This type of canvas is not intended for fine art and is mainly used in the maritime sector, but its broadloom manufacture made it popular with artists wanting to work on large formats (Ellsworth Kelly, Jackson Pollock, Morris Louis to name but a few). The tacks on the edges were hidden by the original paper edging tape, which has become very yellowed.

Sizing

The artist applied a transparent size to the canvas, once he had fixed it to the stretcher. Under close examination with raking light the surface reveals very broad brushstrokes, most of which are at an oblique angle to the surface of the painting and extend over the whole surface (Fig. 3). The canvas was very lightly impregnated: sizing was not carried out methodically and the artist missed a few small areas. The slight yellowing of the unpainted areas is not visible on these unsized parts. This suggests that the aging effect mainly concerns the impregnating substance rather than the canvas.

This size is not a material routinely used by painters, but rather by cabinetmakers. It is a product based on unplasticized polyvinyl acetate, in which an arsenic-based fungicide was used at the time. The brand used by Barnett Newman was Rivit, made by the Behlens Company, New York (Germond 2004). The Centre Pompidou was given a sample of this product, which is no longer made, for analysis during the first investigation phase.[2] In addition, after the accident that befell the painting in 1990, the French Museums Research Laboratory (LMRF) took surface samples by scraping and analysed the size, after dissolving it in acetone and filtering it, using Fourier transform infrared (FTIR) spectroscopy and pyrolysis–gas chromatography–mass spectrometry (Py–GC–MS). The results showed that the artist covered the canvas with a polyvinyl acetate solution which, after evaporation and absorption by the cloth, left a thin but probably porous surface film. The quantity deposited by the brush was probably insufficient and the solvent was probably too volatile for the solution to penetrate further than the surface of the canvas. No indications for aromatics or dibutyl phthalate (a commonly used plasticizer) were found by either analytical method. This led us to believe that the polyvinyl acetate deposited was probably not plasticized.

Black paint

The binder of the paint used for the zips, like the size, was analysed by Py–GC–MS after saponification. The result indicates the presence of linseed oil. The absorption of the oil by the canvas is visible on the edges of painted areas, which present more pronounced yellowing. A very small sample was taken from the lower edge directly below the vertical black stripe on the right. Py–GC–MS of the binding agent detected the presence of linseed oil. This confirmed and completed the technical description. The black layer was directly painted on the canvas, the fibres of which were not protected by a size. A coat of polyvinyl acetate impregnation was no doubt applied before the black paint but this could not be detected during the analysis. There was not enough PVA to prevent the oil from spreading slightly to the immediate vicinity of the painted areas. The absorption of the oil by the canvas is apparent at the edges of the painted areas which show the more pronounced yellowing mentioned previously.

Identification of alterations

Oil stains

On 12 June 1990 the unpainted area was accidentally splashed with soiled lubricating oil from a hydraulic hoist in the museum. Seven darkish vertical drip stains appeared, with the oil penetrating into the canvas fibres. Their positions and extent were documented as follows:

1/30 × 9 mm	10 cm from the top and 148 cm from the left
2/31 × 8 mm	48 cm from top and 145 cm from left
3/80 × 8 mm	180 cm from bottom and 155 from left
4/40 × 8 mm	54 cm from bottom and 155 from left
5/200 × 8 mm	54 cm from bottom and 155 from left
6/320 × 10 cm	20 cm from bottom and 49 from left
7/10 × 10 mm	14 cm from bottom and 32 from left

Following the incident, the head conservator Jacques Hourrière immediately took down the painting and laid it horizontally. The stains were covered with highly absorbent, stain-removing powdered clay (*terre de Sommières*).

This oil was later identified as Higol[3] through analysis by LMRF. It is composed of a hydrocarbon mineral oil and two additives: 1.5% polymethacrylate as a viscosity index improver and 0.5% zinc dithiophosphate as an anti-wear agent. A colouring agent consisting of Carburox Red and Organol Red (ortho-toluene/azo-ortho-toluene/azo-meta-naphthol) was also found.

The nature and extent of the remedial treatment required, and the risks involved, were difficult to assess because of the particular circumstances:

- the major importance of the work;
- the risk of affecting its meaning by making even very minor changes to its material composition;
- the difficulty in obtaining an efficient restricted cleaning process without reactivating the spread of the oil further into the canvas;
- the risk of making the treated areas appear mismatched with the rest of the work.

Moreover, as well as a few light stains located on the edges of the painting, the work as a whole has taken on a slight ivory hue with age.

Other stains

Apart from this accidental episode, *Shining Forth* also had a number of small stains with a slightly reddish hue. These small spots, scattered among the cotton fibres, were also visible on the back of the canvas. The particles can also be seen in the reserved (i.e. unpainted) areas of other works by Newman, such as *Station of the Cross No. 10* in the National Gallery of Art (NGA), Washington, DC, USA. This spotting is therefore considered to be related to the natural evolution of the work. However, in several well-localized places on *Shining Forth*, the same type of 'coffee-coloured' spot was found to have a splash shape, suggesting a second accident (Fig. 4), which was subsequently verified and confirmed to be due to a broken pipe in the storage room. Finally, at the bottom of the painting, to the

FIG. 4 Barnett Newman, *Shining Forth (to George)*: brown stains in the central zip area.

FIG. 5 Barnett Newman, *Shining Forth (to George)*: detail of the main stain and one of the oil drops on the right.

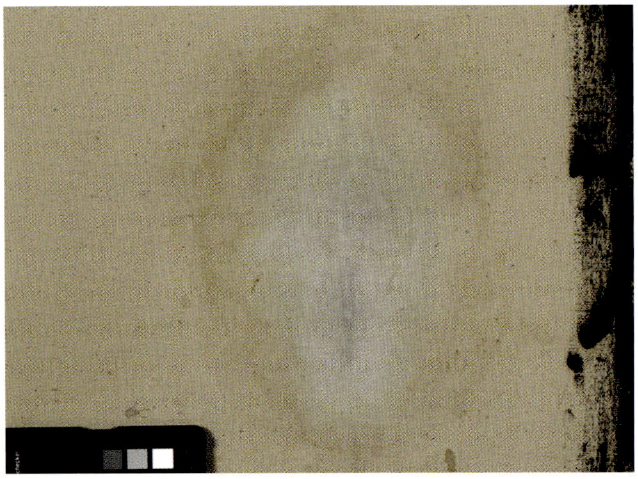

FIG. 6 Barnett Newman, *Shining Forth (to George)*: first cleaning test in 1993.

left of the central zip, there was a large ochre-coloured stain of unknown origin (about 71 × 55 cm) (Fig. 5), associated with a small surface abrasion that is rendered visible in raking light, suggesting an attempt at wiping an accidental spill.

The scattered spotting on the canvas was analysed by the LRMF. The authors of the report noted in 1991 that there were few such spots when samples were taken from the front and back. They were soluble in water and the residue was soluble in acetone. They included the size, since the FTIR spectrum corresponded to that of polyvinyl acetate. Staining tests with amido black 10B showed that the spots were not proteinaceous. However, the Molisch test gave a positive result which indicated the presence of polysaccharides, and this was confirmed by FTIR. Analyses carried out by the NGA on *Station of the Cross No. 10*, which exhibited similar spotting, indicated that the spots on that painting contained both acrylic and phthalate moities, and suggested that they are original components of the cotton duck, introduced during the fabrication of the canvas.

Phase I 1990–2003: cleaning tests and discussion

Cleaning tests

On 14 May 1991, an international group of conservation professionals specializing in 20th-century art met at the request of Germain Viatte, then Director of MNAM. Textile conservators recommended the use of trichlorethylene, white spirit or ethers to treat the oil stains. However trichlorethylene is a solvent for polyvinyl acetate, while the use of ethers could give rise to problems. Nonetheless, it was decided to perform a cleaning test using a related material to the former solvent. This intervention took place on 17 May 1993 at the Musée d'Art Moderne de la Ville de Paris. A stain in the lower right corner near the signature was chosen. The selected tetrachlorethylene solvent was applied with two spray guns, one for the solvent and the other for drying. This treatment was applied using a small perforated suction table lined with paper towels. The stain was first treated from the back and the treated area was then enlarged (18 mm to 22 mm). The treatment produced a decrease in the oil stain, but impurities remained embedded in the canvas fibres.

To understand the results of this first test, the National Centre for Evaluation and Photoprotection (CNEP) in Clermond-Ferrand, France, directed by Professor Jacques Lemaire, was asked to carry out FTIR analysis in 1994. Analysis of the size showed that it had become unusually sensitive to water, even at a neutral pH. The area treated in 1993 was analysed. The size had been removed from the treated area causing bleaching around the spot (Fig. 6). The black particles were found to be tar resulting from the condensation of oxidation products of the mineral oil (organic tars are poorly soluble in organic solvents). In parallel, colour measurements were carried out with a spectrophotometer, after dividing the painting into a grid of 12 zones. This was performed to investigate a possible link between the illumination of the painting and the lightening of the canvas colour, and to try to determine the cause of yellowing (Hourrière and Novion 1993: appendix p. 1).

The CNEP was then asked to perform further tests with cleaning agents in order to design a treatment that would improve the state of the cleaned area. These treatments varied according to the proportion of solvent and detergent used. For each result, an analysis of the size was performed together with an assessment of the success of stain removal. Most of the stain removal attempts were incomplete, and there was complete or partial hydrolysis of the size (polyvinyl acetate was hydrolysed into polyvinyl alcohol). Each aqueous cleaning solution tested contained an oxidizing agent, a surfactant and an optical brightening agent:[4] this last was intended to counteract the yellowing of the canvas over time.

The second treatment was performed on the same stain with an organic solvent, 1,1,1-trifluoro-2,2,2-trichloroethane. A cleaning solution was used during the second phase with which the black particles present in the painting were selectively removed from the surface.

The optical brightener gave a bluish tint to the area where the stain had been removed, which could be destroyed by photo-oxidation with excitation at 365 mm using quasi-monochromatic light. The dissolved sizing agent was replaced by the same type of size. The white area caused by the brightening agents was still highly visible and formed a sort of halo. As the results of this phase were not considered conclusive, the project was temporarily abandoned and the work was put back in the reserve collection.

Use of a mock-up

Alongside the problem of the stains, a study of the general yellowing was performed because the work had been exhibited again in 1995 and had considerably yellowed. A mock-up was made in 1994 with the constituent materials of *Shining Forth*. An accelerated aging study of samples provided by the Conservation Department of the MNAM was performed in an accelerating aging chamber (SEPAP 12/24). Aging at 80 °C for a maximum duration of 1263 hours showed that yellowing occurred both on canvas sized with polyvinyl acetate and on unsized canvas. For the same maximum duration, aging at 60 °C did not cause any significant yellowing. The mock-up was exposed to daylight with a cardboard cover masking parts of the canvas; only the areas that were protected from light turned yellow. A second mask was made in 2002 and used to expose the yellowed part to light while concealing the non-yellowed half. These parts were modified in turn, now creating four distinct areas on the mock-up.

The tendency of *Shining Forth* to yellow in the dark was therefore established. This was demonstrated by the CNEP study in May 2000, which was continued in order to attempt to remove the yellow compounds. Paul Whitmore also carried out bleaching tests in 2001 on *Shining Forth* with a device he had developed: the micro-fade tester (Whitmore 2001; Whitmore *et al.* 2000). The results were detectable with colour measurements, but not with the eye. The large differences in hues present on the whole canvas do not permit a comprehensive treatment, given the large size of the canvas. Published research (Whitmore and Colaluca 1995) has demonstrated the tendency of synthetic paint media to turn yellow in the dark and to lose yellowness in the light.

Intervention phase II: 2011–2015

Hypothesis

In 2010, the possibility of restoring *Shining Forth* was again raised because of its long absence from public view.[5] A new inspection to check its condition took place in the reserve collection after the arrival of a Newman specialist. There had been a marked deterioration in the general appearance of the painting. A large coffee-coloured stain, of the type described earlier, had developed in the central part at the bottom of the painting, together with many similar small stains scattered over the whole surface: these were more disturbing to the eye than the brown oil colours on the right. The painting was not suitable for exhibition in this state and was returned to storage. In the meantime, the problem of the overall yellowing had become secondary since it did not affect the composition. The possibility of an intervention was therefore debated by raising the following questions: Could the presentation of *Shining Forth* be improved by removing the most troubling stains? Do the latest techniques reduce risks sufficiently to enable safe treatment given the sensitivity of the size to solvents and water that are nevertheless necessary to remove the polyvinyl acetate residues responsible for the most recent stains?

Resumption of the *Shining Forth* restoration project was considered in 2011 as a result of new cleaning processes presented at workshops by Richard Wolbers at the National Heritage Institute (INP), Paris. Tests performed on samples kept from the first phase of study confirmed that a more controlled use of solvents was possible by using gels. The most recent research had developed products combining a silicone solvent-based isolating barrier with the use of emulsion-forming gels incorporating a small amount of benzyl alcohol as the dispersed solvent and an aqueous phase containing a buffer (citrate at pH 6.0) and chelating agent diethylenetriaminepentaacetic acid (DTPA). This type of protocol seemed to overcome the major drawbacks that had led to the previous abandonment of the cleaning of *Shining Forth*: the risk of ring marks and the possible reactivation of yellow stains. It was first necessary to check the hypothesis linking the appearance of brown spots to the rising to the surface of surfactants present in the vinyl size. The stains that appeared during the years after the treatment of the canvas by the artist were not produced by the interaction between the canvas and the size. A substance appeared to be the catalyst for these new reactions: an aqueous substance. The size, whether Rivit or another synthetic medium, is a dispersion containing 50–70% water. When this evaporates, the film contains the resin and its additives. Addition of water a few years after the application of the paint may have had the effect of dissolving the adhesive and allowing the additives present in the film to migrate to the surface, creating a new coloured area, this time restricted to the area recently exposed to water. Two major steps preceded the implementation of a cleaning protocol: verification of the nature of the compounds present in the stained areas, and micro-tests on the work itself.

Preliminary tests

In 2012, Nathalie Balcar[6] took samples of the brown yellow material composing these stains. Samples were also taken from the sample bottle of Rivit received from Barnett Newman's widow. Both results confirmed the presence of similar compounds derived from the size. A cleaning test was performed in a zone located near the lower edge of the painting. A xanthan gum gel system was specifically tested as the emulsion-forming gel medium. Xanthan gels can tolerate a wide range of aqueous conditions and agents and maintain gel stability. They form stable emulsions with water-immiscible solvents up to 20% w/w (in a 2% gel w/v in water) without the need for a surfactant. Xanthan gum can additionally suspend non-polar materials in the gel structure through the formation of hydrophobic 'pockets' within the gel structure. A small amount of benzyl alcohol (5%) was incorporated in the gel as a hydrophobic or emulsified solvent phase. Both water and benzyl alcohol are immiscible in silicone solvents such as the cyclomethicones D4 and D5. Therefore applying these solvents first as a masking material to the fabric support could allow us to work safely with a xanthan/benzyl alcohol emulsion gel on the painting surface without fear of penetrating into the fabric support during the application and removal of the emulsion.

Results of preliminary tests and discussion

The yellow colour entirely disappeared in the tested area and no halo developed (Fig. 7). Examination under a binocular microscope before and after treatment confirmed the

FIG. 7 Barnett Newman, *Shining Forth (to George)*: cleaning test in 2014 on the main stain.

disappearance of the coloured residues (Fig. 8). The result however was a notable lightening of the colour, barely perceptible in the areas less affected by yellowing such as the accidentally unsized areas, but clearly standing out in areas adjacent to the stains. Two facts were apparent concerning the risk of creating irregularities in tone:

- The unsized areas that had become visible over time due to a general yellowing of the Rivit already created a heterogeneous appearance.
- The migration of molecules derived from the size to the surface was also irregular in density, depth and colour intensity, and their dissolution did not give a uniform result.

There were two possible responses to the latter unavoidable drawback:

- Perform a general cleaning calibrated on the maximum degree of surface lightening, implying an almost complete elimination of the size, except at the zips.
- Perform localized cleaning and obtain a uniform final appearance of the surface by a subsequent reintegration step.

The second option was chosen according to the precautionary principle and as it was ethically preferable. There was also a technical constraint as the major drawback of the protocol is the lack of visual control during cleaning. The area impregnated by the silicone solvent darkens and this prevents any tonal comparison. The result only appears after complete evaporation of the barrier liquid therefore it was better to progress slowly and selectively.

Implementation of treatment

Cleaning

The work was laid flat for treatment, which started from the right, and the resistance of the zips checked beforehand. Products were applied over areas of 3 × 3 mm. The process was very slow, as after each cleaning step it was necessary to wait for the evaporation of the silicone solvent to determine the results achieved. The surface of the central part where the main stain was located had been modified by the accident responsible for its appearance (local fibre abrasions caused by friction). Special care was taken in this area to avoid increasing this surface modification, which was slightly discernible in raking light. The small stains or brown spots were only removed when their density was visible to the eye at a normal viewing distance.

For the deep-set oil stains (those ostensibly derived from machine lubricating oils), in addition to the benzyl alcohol/xanthan gel described above, we tested a variant of the gel that included an iron chelating material, specifically (4,5-dihydroxy-1,3-benzenedisulfonic acid disodium salt monohydrate (Tiron, supplied by Sigma Aldrich) at 0.5% w/v in the aqueous phase of the gel. The iron chelator did not prove very effective on the deep-set stains although we obtained a superficial result satisfactory enough to allow a retouching treatment to be carried out.

Retouching

After the cleaning process, some irregularities were observed as expected, mainly in the large central stain: lighter zones due to the cleaning process and lighter zones linked to the original unsized areas. Only the former were treated by retouching; it was not considered necessary to treat the latter – this allows the hand of the artist to still be seen. Adjustments to the local colour were made in the areas that appeared too light after cleaning. Different small tests were carried out using pure pigments, pastels, dry pencils and cellulose powder.[7] Cellulose nano-particles tinted by light calcination were deemed visually to be most appropriate retouching method (usually applied to drawings) with respect to the appearance of the surface of the unpainted areas. The selected cellulose Alpha-Cel is more than 99% cellulose, the main structural component. Purified, and then subject to the mechanical disintegration, Alpha-Cel retains its totally organic nature, is pure, safe and biodegradable. The manufacturing process allows for a variety of fibre lengths ranging from 15 to 700+ microns. A fibre length of 5 μm was selected. A range of four different shades of ivory white was obtained after calcination and applied either with a brush or by hand (using a white cotton glove). Dry pencil was also locally used on small greyish areas.

Conclusion

The use of emulsion-forming polymers such as xanthan gum represents major progress in the cleaning treatment for this type of surface and has made it possible to improve the aesthetic status of *Shining Forth*. The painting can now be displayed. A compromise was necessary between an ideal and definitive solution for stain removal without thinning

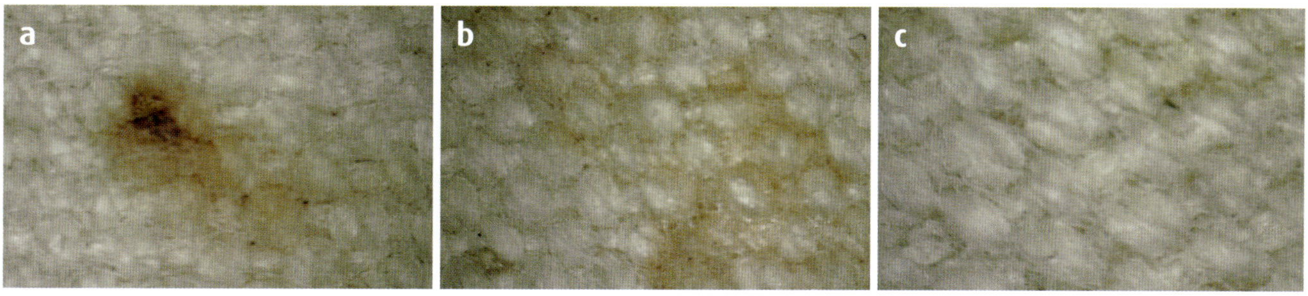

FIG. 8 Barnett Newman, *Shining Forth (to George)*: canvas imaged at ×100. (a) Before cleaning, (b) during cleaning and (c) after cleaning.

or removing the Rivit film, and a protocol was designed for local lightening and rebalancing of the final appearance by light retouching. It would not have been technically feasible to mask the dark spots without covering and creating an extra thickness with a visually unacceptable change in surface appearance. The question of the general yellowing of the work remains unsolved. Studies on this subject show that regular exposure of the work is preferable to its storage in the dark. This treatment was carried out in 2015 and the painting is currently stable.

Notes

1. The golden section is the height to width ratio of a rectangle with sides a and b where a + b is to a as a is to b, that is, a:b = (a+b):a.
2. The sample was provided via Carole Ungaro-Mangusi, Head of Conservation, Whitney Museum, New York, who was a member of the steering committee set up during the first discussion and testing phase of the research. This material was analysed by the LMRF, now the Centre for Research and Restoration of the Museums of France (C2RMF); see Rioux and Colinart 1991: 3.
3. Commercial name: Higol reference-MATIC- 32, Société Higol France.
4. Optical brightening agents are active substances that transform absorbed UV light into visible light, and give the observer the impression that the overall intensity of reflected light is higher than normal; see Petit *et al.* 1999: 307.
5. When Véronique Sorano-Stedman took over from Jacques Hourriere as head of the MNAM conservation department.
6. Nathalie Balcar was then an engineering student at the C2RMF, with expertise in environmental chemistry, biochemistry, and chemical biology. She is now a chemist and engineer in the Conservation Department of the C2RMF.
7. From International Fiber Europe N.V.Eurolaan 5 Industriezone T.T.S Zone BB-9140 Temse, Belgium.

References

Germond, S. 2004. *Vieillissement et altérations des toiles de Cotton duck: Shining Forth de Barnett Newman*, unpublished Master's thesis, Science and Techniques of Conservation-Restoration of Cultural Property, Paris I Panthéon-Sorbonne University, UFR03, France.

Hourrière, J. and Novion, I, 1993. *Report on the Preliminary Studies for the Restoration of* Shining Forth *by Barnett Newman*. Paris: Centre Georges Pompidou.

Morris, H.R., Colaluca, V.G. and Whitmore, P.M. 2003. 'Preventing discoloration in films of acrylic artists' media by exposure to ambient light', *Studies in Conservation* 48: 9–102.

Petit, J., Roire, J. and Valot, H. 1999, *Encyclopédie de la peinture*, vol. 1. Puteaux: EREC.

Rioux, J.P. and Colinart, S. 1991. *Report on the Identification of the Materials used in* Shining Forth. Paris: LMRF.

Whitmore, P.W. 2001. *MNAM Report on the Lightfastness Testing of* Shining Forth. Paris: LMRF.

Whitmore, P.M. and Colaluca, V.G. 1995. 'The natural and accelerated aging of an acrylic artist medium', *Studies in Conservation* 40: 51–64.

Whitmore, P.M., Pan, X. and Bailie, C. 1999. 'Predicting the fading of objects: identification of fugitive colorants through direct non-destructive lightfastness measurements', *Journal of the American Institute for Conservation* 38: 395–409.

Whitmore, P.M., Baillie, C. and Connors, S.A. 2000. 'Micro-fading tests to predict the result of exhibition: progress and prospects', in A. Roy and P. Smith (eds), *Tradition and Innovation: Advances in Conservation*. London: IIC, 200–205.

Whitmore, P.W., Colaluca, V.G. and Morris, H.R. 2002. 'The light bleaching of discolored films of an acrylic artists' medium', *Studies in Conservation* 47: 228–36.

Authors' addresses

- Véronique Sorano-Stedman, MNAM Centre National d'Art et de Culture G. Pompidou, Paris, France (Veronique Sorano-Stedman@centrepompidou.fr)
- Richard Wolbers, University of Delaware Program in Art Conservation, Department of Art Conservation, Newark, DE, USA (doczepp@gmail.com)
- Sophie Germond, Paris, France (sophiegermond1@gmail.com)

Coupling gellan gels and electrochemical biosensors: real-time monitoring of cleaning and enzymatic treatments of Diana Scultori's *The Wedding of Cupid and Psyche*

Silvia Sotgiu, Simonetta Iannuccelli, Serena Dominijanni, Silvia Puteo, Laura Micheli and Claudia Mazzuca

Introduction

Since 2003 the Conservation Department of Istituto Centrale per il Restauro e la Conservazione del Patrimonio Archivistico e Librario (ICRCPAL) in Rome, Italy, has been developing a new method to perform wet treatments on works of art on paper using rigid gellan gels made up in an aqueous solution of calcium acetate and gellan gum. This gel is able to gradually release water and also to absorb water-soluble degradation products present on the paper. The gel technique proved to be safe and effective not only as a paper washing method, but also for chemical stabilization and enzymatic treatments on both handmade and machine-made paper supports. From 2012, in collaboration with the Department of Chemical Sciences and Technologies of 'Tor Vergata' University, Rome (hereafter DCST-UTOV), we have been developing an efficient and simple method to load gellan gels with enzymes, obtaining 'enzymatic gellan gels' to remove linings and adhesive residues from paper supports.

This paper presents the results obtained and shows its application in a case study of an engraving produced in 1613. The paper support showed structural weakness and darkening from a strong oxidative degradation mainly due, in turn, to a previous lining treatment and biodeterioration. The first wet cleaning treatment was monitored by means of a cleaning and diagnostic tool based on a gellan gum hydrogel combined with an electrochemical sensor. This system is suitable to verify *in situ* and in a simple way the efficiency of the cleaning treatment and detect its conclusion.

The print

The innovative diagnostic approach was successfully tested on Diana Scultori's *The Wedding of Cupid and Psyche*, etched on copper plate using a burin in 1575, and printed

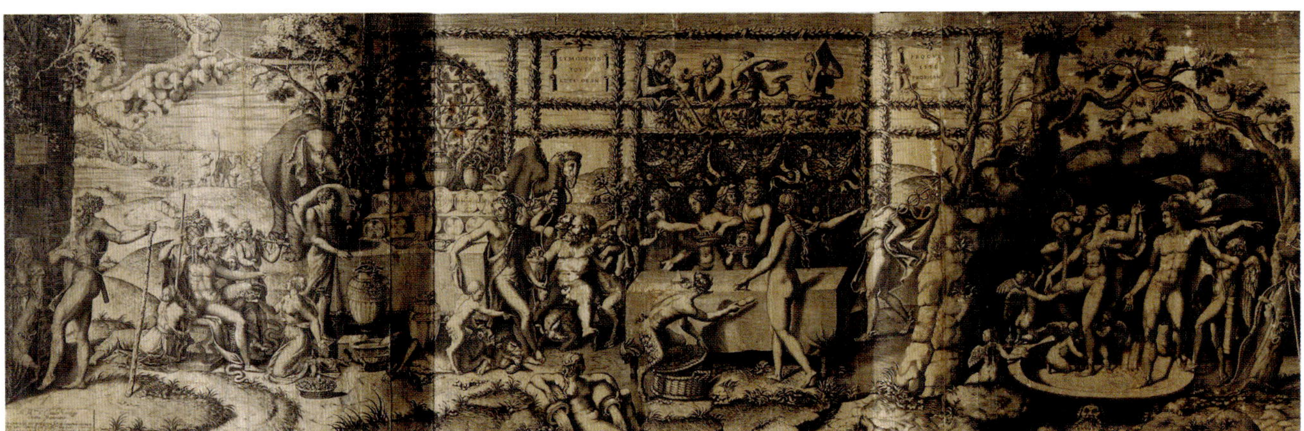

FIG. 1 Diana Scultori, *The Wedding of Cupid and Psyche*, 1575, engraving on paper, 1125 × 373 mm, Lafrery Collection, Archeology and Art History Library (BiASA), Rome XI.147.I24: before treatment. With kind permission of BiASA (protocol no. 143 cl. 28.04.1613).

in 1613 by Antonio Carenzano (Bartsch 1813: 449; Massari 1980: 93; Boorsch and Spike 1986; Bellini 1991: 201).

Diana Scultori (1547–1612) is one of the most important women engravers. Her first professional works were based on the collection of drawings of Giulio Romano, Raffaello's disciple. In 1524, he had been employed by Federico II Gonzaga for the realization of the Palazzo Te frescoes in Mantua. He then gave his drawings to Giovan Battista Scultori, Diana's father, to make engraved copies of them (D'Arco 1840: 10–28; Bellini 1991: 30–31; Calidonna 2010: 1–12). The artist and art critic Giorgio Vasari visited Scultori's workshop in 1566 and, surprised by the ability of the young woman, praised her in the edition of the description of the lives of artists, published in 1568 (Vasari 1966 [1568]). Diana Scultori was also the first woman to receive a Papal privilege by Gregorio XIII in 1575 with the authorization to sign her own compositions and to sell them. It was during this period that she created *The Wedding of Cupid and Psyche.*[1]

The print (Fig. 1) consists of three sheets, corresponding to three copper plates (total width 1125 mm and height 373 mm). The work had dark stains scattered across the verso (Dominijanni *et al.* 2016). Between the 16th and the 18th century, the work had already undergone repair, with the application of a paper lining on the verso intended to hide the dark stains, fill in lacunae and reinforce the paper with a single intervention. The use of an adhesive mixture was probably one of the main causes of cellulose oxidation, leading to an increase of carbonyl and carboxyl groups content, which provoked surface yellowing and an overall loss of structural strength (Mosca Conte *et al.* 2012; Missori *et al.* 2014).

Determination of the adhesive composition

Fourier transform infrared (FTIR) spectra were acquired on a Thermo-Scientific (mod. Is50) instrument (Thermo Scientific Inc., Madison WI), equipped with an attenuated total reflectance (ATR) diamond cell for measurement in the 4000–525 cm^{-1} region, at a resolution of 4 cm^{-1}. In total, 64 scans were collected for each sample. Spectra were collected by placing the paper samples directly on the ATR cell. As shown in Fig. 2, peaks attributable to starch paste, localized at 998 cm^{-1} (Mazzuca *et al.* 2014), and those attributable to animal glue (amide I and II for protein), localized at 1660 and 1570 cm^{-1}, are clearly evident. However, the last two bands can also be attributed to water bound in cellulose and carboxylates (Calvini and Gorassini 2002).

UV-visible (UV-Vis) fluorescence spectra were recorded on a Fluorolog III spectrofluorimeter (Horiba, Japan) operating in a single photon counting mode using an excitation wavelength of 390 nm, by dissolving increasing amounts of glue fragments in fluorescamine (1 μM) solution. Fluorescamine is a dye that is not fluorescent intrinsically but becomes fluorescent when linked to protein (Udenfriend *et al.* 1972). As can be seen from Fig. 3, the emission intensity of fluorescamine is almost zero in the absence of glue fragments, and increases abruptly on adding the glue. Thus measurements on small fragments of the paper support indicated that the lining adhesive was made from both starch paste and animal glue.

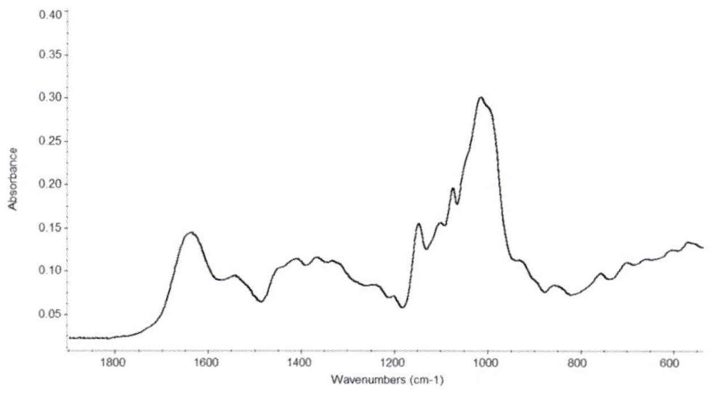

FIG. 2 FTIR spectrum of paper fragments with the adhesive.

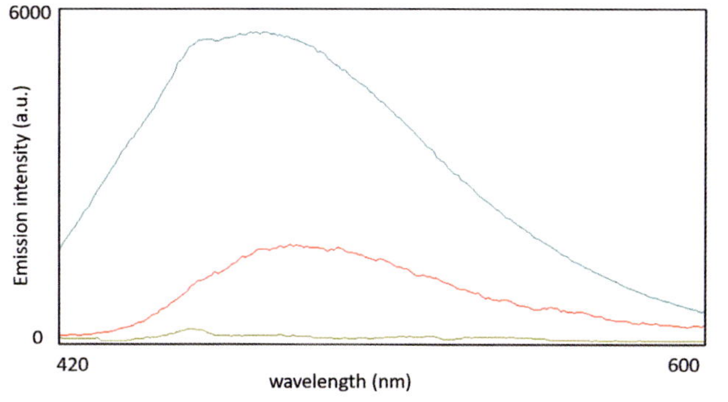

FIG. 3 Emission spectra of: the fluorescamine solution (light brown line); the fluorescamine solution and an adhesive sample (red line); the fluorescamine solution and two adhesive samples (light blue line).

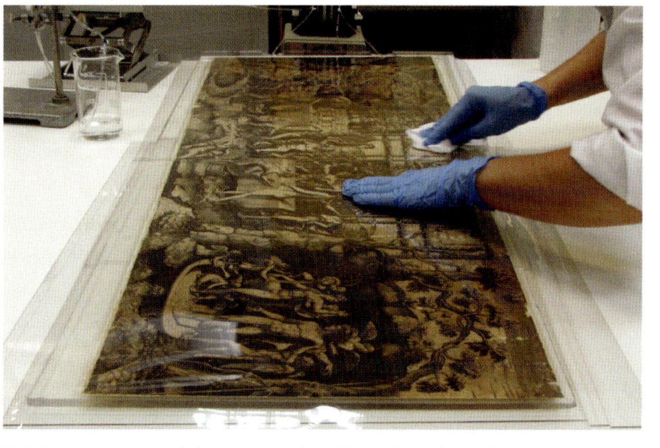

FIG. 4 Application of the enzymatic gellan gel on the print.

Treatment

We planned to detach the lining, to improve the appearance of the work, allowing the support to regain its original opacity and texture. Works on paper are characterized by a single layer – a sheet of paper – providing both the support and the background to the image. Therefore its original features have

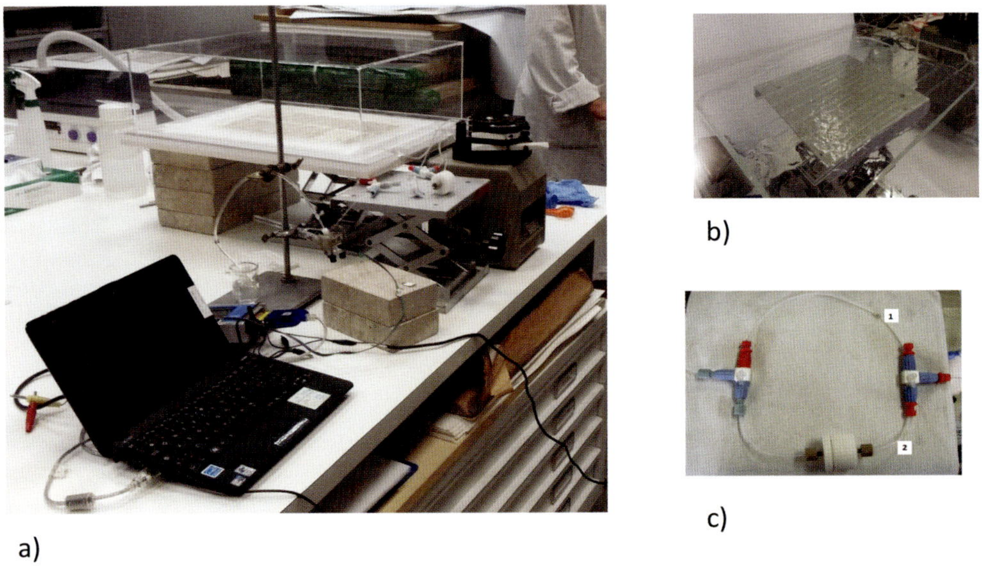

FIG. 5 Real-time monitoring tool: (a) complete system with flow sampling plate/hydrogel/electrochemical biosensor; (b) ad hoc flow sampling plate with 12 channels; (c) two-way channels, 1 for monitoring the endogenous glucose removed by gellan gel during the cleaning process, 2 for glucose obtained through enzymatic hydrolysis of starch paste during adhesive removal by means of the enzymatic gellan gel.

to be preserved for both structural and aesthetic reasons. Artists carefully selected papers for their work, according to specific properties such as opacity, colour, porosity, thickness, brightness and whiteness: any alteration in these compromises the aesthetic appreciation of the artwork as well as its physical and chemical stability. Other aspects to be preserved are the three-dimensional effect due to the mould pressure on the paper during printing, and, thus, the printing size, important for delineating the relationship with the original copper plate (Iannuccelli and Sotgiu 2012).

Wet treatments, necessary in this case to clean the paper and remove the lining, constituted the most delicate part of the treatment process. As we wished to preserve all the morphological and optical characteristics of the paper, we decided to use the gel method as conceived by the ICRCPAL Conservation Department, using rigid gels made up with a water solution of calcium acetate and gellan gum, a microbial exopolysaccharide, that allows a constant and controlled release of water onto the paper. The adoption of this method minimizes the effects of cellulose swelling, ensuring maintenance of morphological characteristics. The exchange mechanism, established between the paper and the gel during treatment, allows for effective removal of surface deposits and acidic degradation by-products from within the paper structure (Iannuccelli and Sotgiu 2012). Gellan gels were used for the cleaning of the paper support and, in combination with enzymes, to remove lining and adhesives residues from the verso (Iannuccelli and Sotgiu 2012; Mazzuca *et al.* 2016).

After carefully dry cleaning the surface to remove powdery deposits that could penetrate into the paper during the subsequent wet treatments, and having tested the solubility of the inks, we initially applied a 2% gellan gel in water.[2] For lining removal, a second gellan gel loaded with enzymes was applied (Fig. 4). The enzymatic gellan gels use the specificity of selected hydrolytic enzymes in relation to different adhesives, reducing the application time and assisting in the removal of the lining and of adhesive residues (Iannuccelli and Sotgiu 2012; Mazzuca *et al.* 2014, 2016).

During the wet treatment, an innovative monitoring system developed at DCST-UTOV was used. The system employs electrochemical biosensors to follow the progress of the treatment with enzymatic gellan gels in real time. Electrochemical biosensors can identify the by-products of enzyme metabolism and, finally, mark the conclusion of the treatment (Micheli *et al.* 2016).

Preparation of gels and treatment monitoring

Hydrogel preparation

We proceeded with the preparation and application of two enzymatic hydrogels on the artwork; the first gel, containing α-amylase enzyme was intended to promote the depolymerization of the starch-based components of the adhesive, while the second, loaded with proteinase K enzyme, a metabolite of proteinaceous materials, would facilitate the hydrolysis of animal glue.

The hydrogels were prepared following a general protocol already described in the literature with slight modifications (Micheli *et al.* 2016; Mazzuca *et al.* 2016). To prepare the hydrogel containing calcium ions, an aqueous solution containing deacylated gellan gum (20 g/L) and calcium acetate (0.40 g/L) was mixed for almost a minute in a microwave at 600 W (Mars Microwave, CEM Corporation, Matthews, NC, USA) until it boiled and became transparent; it was then left to cool on a Petri dish. To prepare gels loaded with proteinase K,

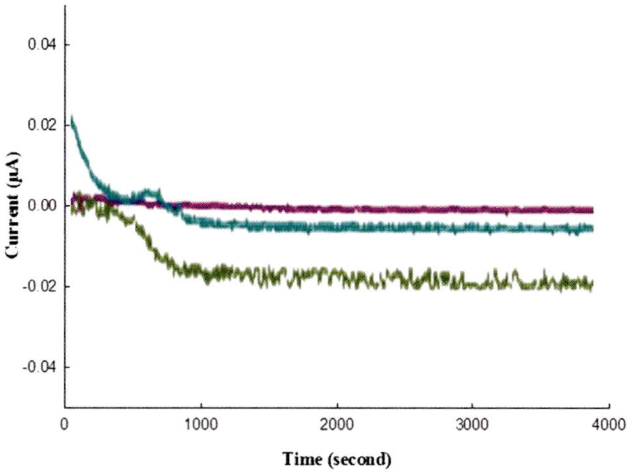

FIG. 6 Real-time monitoring of the cleaning process and depolymerization of the adhesive using amperometric detection of glucose: buffer (pink line), cleaning process using gellan gel (blue), and depolymerization of the adhesive using enzymatic gellan gel (green).

an aliquot of a concentrated enzyme solution (35 μM, pH 8, plus 8 mM $CaCl_2$) was added to gellan gel at 60 °C, while it was cooling in a Petri dish. The final enzyme concentration was 1.2 μM. Similarly, to prepare gels loaded with the α-amylase, an aliquot of a concentrated enzyme solution (200 μM in 8 mM $CaCl_2$, pH 6.9) was added to the gellan gel at 60 °C, while it was cooling in a Petri dish. The final enzyme concentration was 5 μM.

Sampling and monitoring system coupled with the electrochemical biosensor

The combined tool, comprising a flow sampling plate/hydrogel/electrochemical biosensor, is represented in Fig. 5. This tool enables monitoring of the cleaning procedure, and simultaneously, screening of the degradation of the work, combined with the identification – and quantification in some cases – of the degradation products removed by the gel. In particular, the monitoring system is composed of a flow sampling system, specifically designed to be applied directly on gellan gel during the cleaning step. The flow sampling system consists of a plate of Perspex on which a serpentine[3] with 12 channels is present (Fig. 5b). In the curved channels, the buffer flows continuously (using a peristaltic pump MINIPLUS3 – Gilson, USA), up through the gel to an electrochemical thin layer cell, where a screen-printed biosensor is integrated. The thin layer cell is a LCEC (liquid chromatography–electrochemistry cell) from Bio-Analytical Systems (West Lafayette IN, USA), adapted for screen-printed electrodes (SPEs).[4] The diameter of the working electrode was 0.3 cm resulting in a geometric area of 0.07 cm^2.

Between the sampling system and the biosensor there is a two-way system: the first channel allows us to follow the removal of only the cellulose degradation products, while the second channel detects the removal of adhesive residues due to enzymatic activity in the gel. The two-way system can be switched between the paper samples and the paper

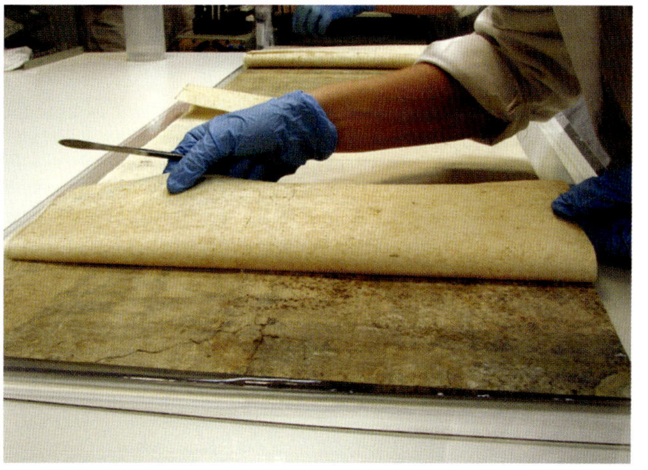

FIG. 7 Lining removal from the print after enzymatic gellan gel treatments.

samples with adhesive by means of a simple system of taps. The working and carrier buffer was 0.05 M phosphate buffer + 0.1 M KCl, pH 6.8. The buffer flow (0.1 mL/min) was regulated by a peristaltic pump, connected directly to the plate. Amperometric measurements were carried out using a PalmSens potentiostat (Palm Instruments BV, Netherlands) connected to a laptop.

Biosensor for glucose determination

The selected cellulose degradation product to be monitored was glucose (Micheli *et al.* 2014) because it is the final product of the hydrolysis of cellulose, an endogenous paper degradation process (Strlic *et al.* 2005). The characteristics and the sensitivity of the glucose biosensor, based on the immobilization of the glucose oxidase (GOx) enzyme on an SPE, are well known (Mahosenaho *et. al.* 2010; Karyakin and Karyakina 2001). The reaction is:

$$\text{Glucose} + O_2 \xrightarrow{\text{Glucose-oxidase}} \text{Gluconic acid} + H_2O_2$$

The product H_2O_2 was detected amperometrically with an SPE modified with Prussian blue (PB-SPE), and used at an applied potential of −50 mV *vs* pseudoreference Ag. By immobilizing the GOx on the working electrode area, a glucose biosensor was assembled and utilized for the detection of glucose.

The cleaning and removal of the lining was monitored for 1 hour (the application time chosen by the conservators). The results in Fig. 6 show that the cleaning process begins immediately when the gel containing α-amylase is applied and the solution is converted through channel 2, with removal of glucose as both a paper degradation product and a product of adhesive degradation. The results confirm the possibility of quantifying the glucose present, indicating the condition of the paper. The monitoring systems designed for this artwork allow monitoring of both processes in real time, providing important data on the processes of degradation of the cellulose, and for estimating the time needed to hydrolyse the lining adhesive.

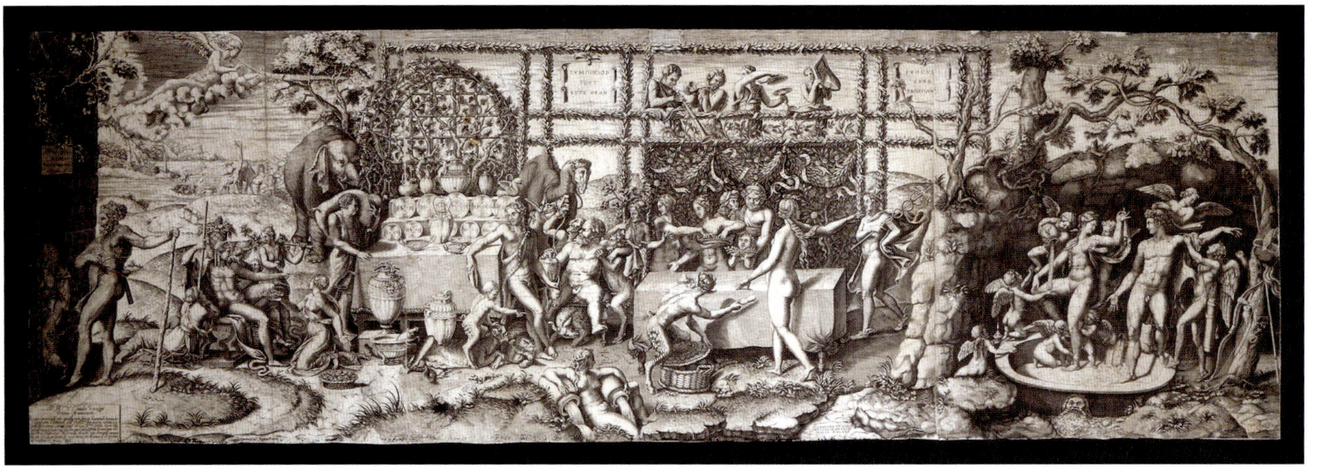

FIG. 8 Diana Scultori, *The Wedding of Cupid and Psyche*, 1575 (**FIG. 1**): the print after treatment shows a decrease in surface darkening. With kind permission of BiASA (protocol no. 143 cl. 28.04.1613).

Lining removal and final phase of treatment

The testing of the method was carried out after a first attempt at removal of the lining from the verso. The lining was removed very easily and without causing physical stress to the original support (Fig. 7), thereby confirming the effectiveness of the enzymatic gel and the appropriately pre-selected duration of the treatment. The treatment continued with temporary repairs of the tears and lacerations of the original paper support, previously held in place by the lining. The second phase of the wet treatment was then started: a structural cleaning of the artwork. The cleaning was carried out with capillary washing by means of a draining fabric (Igortex OL 60), a method that, like gellan gels, minimizes the contact of the paper with liquid water and does not cause excessive 'swelling' of the cellulose (Schalkx *et al.* 2011).

The wet treatments performed on the artwork have removed most of the non-original surface deposits and those organic components produced by the depolymerization of the cellulose chain that, if left in the fibrous structure, could activate further oxidation reactions. Although the treatment has significantly attenuated the browning of the paper support, further investigation[5] (Mosca Conte *et al.* 2012; Missori *et al.* 2014) confirmed the presence of chemical species produced by oxidative degradation, thus necessitating a chemical stabilization treatment with a reducing agent (Sanna *et al.* 2009). The increasing of the pH into the neutrality range precluded the need for deacidification. Following complete drying, the work was consolidated and retouched (Fig. 8).

Conclusion

The Wedding of Cupid and Psyche, traditionally considered to be Diana Scultori's masterpiece, has regained its original aesthetic qualities. The conservation process, supported by an innovative and highly specialized diagnostic tool, was able to minimize the impact of wet treatments, preserving most of the typical appearance of works of art on paper (Iannuccelli and Sotgiu 2010). The gellan gel technique for wet treatments of works of art on paper has been validated. The collaboration with DCST-UTOV has enabled the development of an increasingly powerful methodology that can be used safely on such works: artefacts that, compared to other categories of cultural objects, are characterized by a number of challenging issues due to the inherent fragility of their very structure – paper.

Materials and suppliers

- Proteinase K from *Tritirachium album* [EC-3.4.21.64, ≥30 U/mg], α-amylase [EC-232–560–9; 30500U/ml], calcium acetate, calcium chloride, fluorescamine and phosphate buffer pH 6.8: Sigma-Aldrich, Mo, St. Louis, USA
- Gellan gum Kelcogel CG-LA: CP Kelco, Atlanta Georgia, USA
- Solvents such as methanol, spectroscopic and chromatographic grade, Nafion 5% v\v solution: Carlo Erba Reagenti srl, Milan, Italy
- Glucose oxidase (GOx, EC 1.1.3.4, type II, 15,500 U\g, from *Aspergillus Niger*), D (+)-glucose monohydrate, glutaraldehyde (GAD, 25% v\v aqueous solution): Sigma-Aldrich, St. Louis, MO, USA
- All reagents were of analytical grade and used without further purification: Sigma-Aldich, USA. In all cases, in the preparation of buffer solutions doubly distilled water was used: Millipore, Billerica, MA, USA

Notes

1. Archivio Segreto Vaticano 1575, Arm. XLVII, 28, f. 213.
2. Established by TAPPI Test Method T 458 surface wettability of paper (angle of contact method).
3. The word 'serpentine' is used here as a technical term indicating a long channel drawn on a flat plate with a laser technique. See Liu *et al.* 2000; Wang and Lee 2006; Suzuki and Ho 2002.

4. These electrodes were produced in-house at DCST-UTOV.
5. Investigations carried out with reflectance spectroscopy by the Istituto dei Sistemi Complessi-Consiglio Nazionale delle Ricerche (ISC-CNR) in Rome.

References

Bartsch, A. 1813. *Le peintre graveur* (21 vols), vol. 15. Vienna: Chez Pierre Mechetti.

Bellini, P. 1991. *L'opera incisa di Adamo e Diana Scultori*. Vicenza: Neri Pozza.

Boorsch, S. and Spike, J. 1986. 'Italian artists of the sixteenth century', *The Illustrated Bartsch* (166 vols), vol. 31. New York: Abaris Books, 280–81.

Calidonna, I. 2010. *Diana Scultori Mantovana (1547–1612). La vita, i documenti, le opere*, unpublished cultural heritage specialization thesis, University of Siena, Italy.

Calvini, P. and Gorassini, A. 2002. 'FTIR-deconvolution spectra of paper documents', *Restaurator* 23: 205–21.

D'Arco, C. 1840. *Di cinque valenti incisori mantovani*. Mantua: Elmucci.

Dominijanni, S., Iannuccelli, S., Missori, M., Micheli, L., Mazzuca, C. and Teodonio, L. 2016. 'Diana Scultori e le donne incisori tra XVI e XVII secolo. Il restauro delle opere grafiche *Le Nozze di Psiche* e *Gesù Cristo e l'adultera* monitorato con tecniche diagnostiche sperimentali', paper presented at the conference *I giovani e il restauro. L'arte nel tempo. Significato, trasformazione e conservazione, Rome, 24–26 February 2016*.

Iannuccelli, S. and Sotgiu, S. 2010. 'Wet treatments of works of art on paper with rigid gellan gels', *The Book and Paper Group Annual* 29: 25–39.

Iannuccelli, S. and Sotgiu S. (eds) 2012. *La pulitura a umido di opere d'arte su carta con gel rigidi di Gellano: presupposti teorici, metodologia applicativa e verifica analitica, Quaderni del Cesmar7*, no. 11. Padua: Il Prato.

Karyakin, A.A. and Karyakina, E.E. 2001. 'Electroanalytical application of Prussian blue and its analogs', *Russian Chemical Bulletin* 50(10): 1811–17.

Liu, R.H., Stremler, M.A., Sharp, K.V., Olsen, M.G., Santiago, J.G., Adrian, R.J., Aref, H. and Beebe, D.J. 2000. 'Passive mixing in a three-dimensional serpentine microchannel', *Journal of Microelectromechanical Systems* 9(2): 190–97.

Mahosenaho M., Micheli L., Caprio F., Sesay A.M., Palleschi G. and Virtanen V. 2010. 'A disposable biosensor for the determination of alpha-amylase in human saliva', *Microchimica Acta* 170(3–4): 243–9.

Massari, S. 1980. *Incisori mantovani del '500*. Rome: De Luca.

Mazzuca, C., Micheli, L., Cervelli, E., Basoli, F., Cencetti, C., Coviello, T., Iannuccelli, S., Sotgiu, S. and Palleschi, A. 2014. 'Cleaning of paper artworks: development of an efficient gel-based material able to remove starch paste', *ACS Applied Materials and Interfaces* 6(19): 16519–28.

Mazzuca, C., Micheli, L., Lettieri, R., Cervelli, E., Coviello, T., Cencetti, C., Sotgiu, S., Iannuccelli, S., Palleschi, G. and Palleschi A. 2016. 'How to tune a paper cleaning process by means of modified gellan hydrogels', *Microchemical Journal* 126: 359–67.

Micheli, L., Mazzuca, C., Moscone, D., Palleschi, A. and Palleschi, G. 2014. 'Electrochemical technique for cultural heritage: an example of application of monitoring cleaning process of paper art work', in E. Papa, M.G. Perrone and A. Piazzalunga (eds), *Facciamo germogliare le nostre Idee - Quaderno della III Scuola di Chimica Ambientale e dei Beni Culturali*., Barza-Varese: Società Chimica Italiana, 77–81.

Micheli, L., Mazzuca, C., Palleschi, A. and Palleschi, G. 2016. 'Development of a diagnostic and cleaning tool for paper artworks: a case study', *Microchemical Journal* 126: 32–41.

Missori, M., Teodonio, L., Pulci, O., Violante, C., Kupchak, I., Bagniuk, J., Lojewska, J. and Mosca Conte, A. 2014. 'Optical response of strongly absorbing inhomogeneous materials: application to paper degradation', *Physical Review B* 89: 054201.

Mosca Conte, A., Pulci, O., Knapik, A., Bagniuk, J., Del Sole, R., Lojewska, J. and Missori, M. 2012. 'Role of cellulose oxidation in the yellowing of ancient paper', *Physical Review Letters* PRL 108: 158301.

Sanna, C., Sodo, A., Laguzzi, G., Mancini, G. and Bicchieri, M. 2009. 'Tert-butyl amine borane complex: an unusual application of a reducing agent on model molecule of cellulose based materials', *Journal of Cultural Heritage* 10: 356–61.

Schalkx, H., Iedema P., Reissland B. and van Velzen, B. 2011. 'Aqueous treatment of water-sensitive paper objects: capillary unit, blotter wash or paraprint wash?, *Journal of Paper Conservation* 12(1): 11–20.

Strlič, M., Kolar, J. and Scholten, S. 2005. 'Paper and durability', in M. Strlic and J. Kolar (eds), *Ageing and Stabilization of Paper*. Ljubljana: Ljubljana National and University Library, 1–23.

Suzuki, H. and Ho, C.M. 2002. 'A magnetic force driven chaotic micro-mixer', *Micro Electro Mechanical Systems, Technical Digest. MEMS 2002 IEEE International Conference. Fifteenth IEEE International Conference on Micro Electro Mechanical Systems* (Cat. No.02CH37266), 341–8.

Udenfriend, S., Udenfriend, S., Stein, S., Böhlen, P., Dairman, W., Leimgruber, W., Weigele, M. 1972. 'Fluorescamine; a reagent for assay of amino acids, peptides, proteins, and primary amines in picomole range', *Science* 178: 871–2.

Vasari, G. 1966 [1568]. 'Le vite de' più eccellenti pittori, scultori, e architettori', in R. Bettarini (ed.), *Le vite de' più eccellenti pittori, scultori, e architettori nelle redazioni del 1550 e 1568* (6 vols), vol. 5. Florence: Sansoni, 424.

Wang, C.H. and Lee, G.B. 2006. 'Pneumatically driven peristaltic micropumps utilizing serpentine-shape channels', *Journal of Micromechanics and Microengineering* 16: 341–8.

Authors' addresses

- Silvia Sotgiu, ICRCPAL, Rome, Italy (silvia.sotgiu@beniculturali.it)
- Simonetta Iannuccelli, ICRCPAL, Rome, Italy (simonetta.iannuccelli@beniculturali.it)
- Serena Dominijanni, SAF ICRCPAL, Rome, Italy (serena.dominijanni@gmail.com)
- Silvia Puteo, SAF ICRCPAL, Rome, Italy (silvia.puteo@virgilio.it)
- Laura Micheli, Dipartimento di Scienze e Tecnologie Chimiche, Università di Roma 'Tor Vergata', Rome, Italy (micheli@uniroma2.it)
- Claudia Mazzuca, Dipartimento di Scienze e Tecnologie Chimiche, Università di Roma 'Tor Vergata', Rome, Italy (claudia.mazzuca@uniroma2.it)

Rigid polysaccharide gels for paper conservation: a residue study

Michelle R. Sullivan, Teresa T. Duncan, Barbara H. Berrie and Richard G. Weiss

Introduction

Over the course of the past 10 years, interest in the use of rigid polysaccharide gels – particularly agarose and gellan gum – has steadily increased within the field of paper conservation. Growing interest among paper conservators in rigid gels is easily understood as they afford a high degree of control in a variety of local and overall treatments routinely performed on paper objects: aqueous and solvent-based poulticing for adhesive reduction and attachment removal; stain and tideline reduction; humidification; and bathing. By altering their chemical and physical properties, rigid gels can be tailored to suit the needs of a specific treatment and offer numerous advantages over traditional methods: precise application; restricted lateral solvent flow to confine cleaning; increased dwell time of reagents; and minimized mechanical action when compared to working with a brush, swab, or spatula. They are colorless and clear or translucent, allowing the conservator to monitor the course of treatment and, unlike viscous gels, they can be removed immediately should a sudden, undesirable change occur.

As with any new approach to treatment, however, scientific studies are required to understand fully the immediate and long-term effects of gel treatments on paper. Given the porous nature of paper and its susceptibility to absorption and adsorption of materials, residues are a primary concern of the conservator. During aqueous treatment on any scale, paper fibers swell and become vulnerable to damage and deformation. Generally, paper cannot withstand the physical action of clearance steps that are required to remove residues completely, a goal that may not be achievable. These concerns provided the impetus for the comparative residue study described here.

To date, paper conservation literature related to the use of rigid polysaccharide gels has focused primarily on treatment case studies and application rather than analyses (Botti *et al.* 2011; Hawkes 2013; Hughes and Sullivan 2016; Iannucelli and Sotgiu 2010; Keynan and Hughes 2013; Lennig 2010; Maheux 2015; Sullivan *et al.* 2014; van Dyke 2004; Yip 2015). Articles describing analytical studies have centered on efficiency of gel cleaning, understanding the removal of sizing and degradation products from paper during treatment, and the effect of gel cleaning on the physical and optical properties of paper (Casoli *et al.* 2014; Mazzuca *et al.* 2014a,b,c). In 2007, a former student and staff of the conservation training program at the State University College at Buffalo, New York published a comparative study that included agarose and preliminarily concluded that the rigid polysaccharide gel did not leave significant amounts of residue on paper samples (Warda *et al.* 2007).[1] Other rigid polysaccharide gel residue studies have not included paper samples. A study carried out by Campani *et al.* (2007: 33–51) utilized gas chromatography-mass spectrometry (GC-MS) and Fourier transform infrared (FTIR) spectroscopy to determine that neither agarose nor agar left detectable residues when applied to porous ceramic plates. A study executed by Kavda *et al.* (2016) employed scanning electron microscopy (SEM) and sum frequency generation (SFG) spectroscopy to assess residues left by gellan gum and partially hydrolysed poly(vinyl) acetate-borax gels on poly(methyl methacrylate) samples. In that study, gellan gum residues were clearly visible in secondary electron (SE) images.

With the previous studies in mind, the authors designed the comparative study presented below to determine whether residues remained on papers treated using the rigid polysaccharide gels of agarose and gellan gum and, if so, the relative amounts deposited. The study also aimed to identify treatment variables that can be controlled to minimize residue deposition during conservation treatments to minimize residue deposition. To explore these issues, experiments were carried out using fluorescein-tagged gels that allow much greater sensitivity to the presence of residues than visual inspection alone.[2]

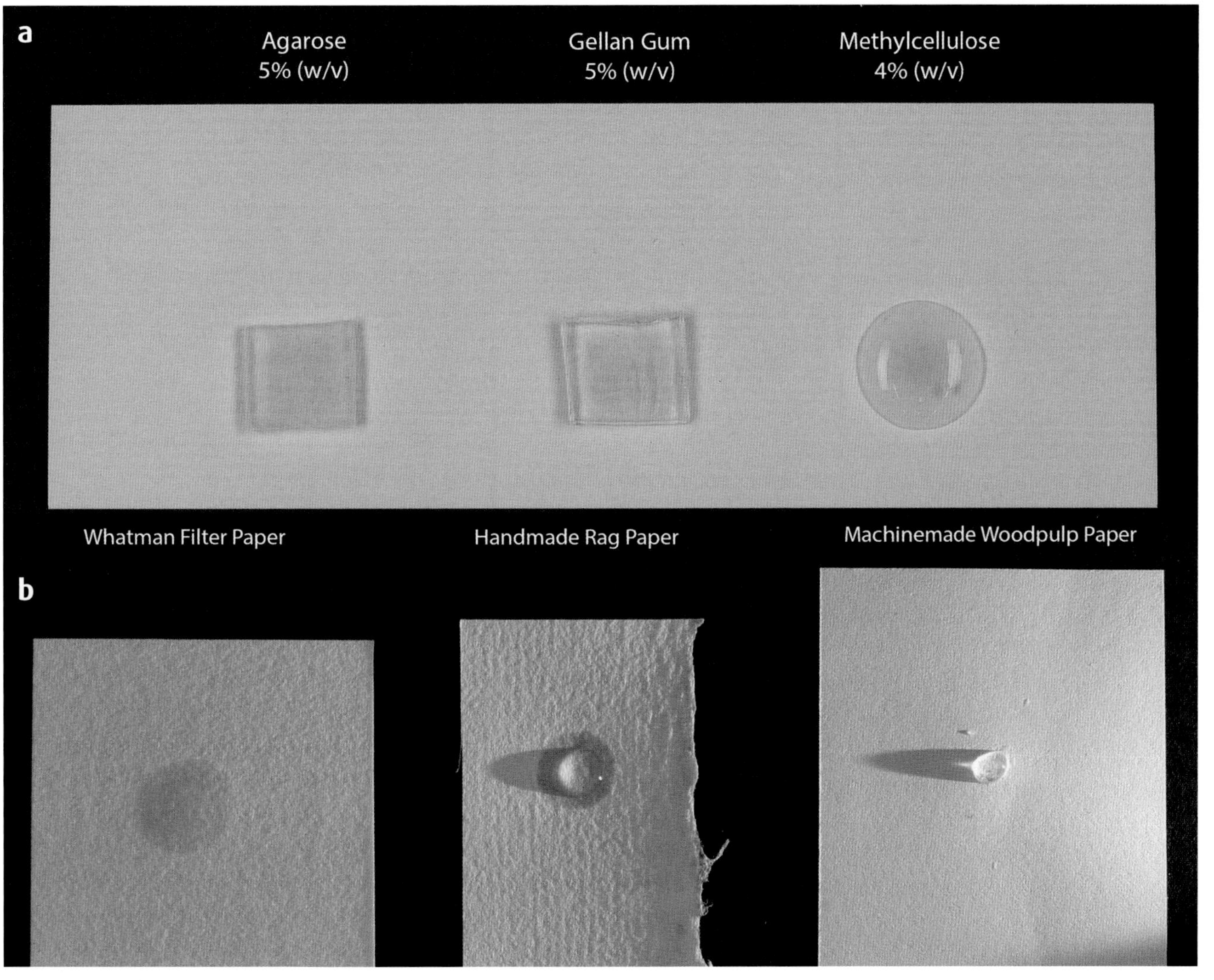

FIG. 1 (a) (*Left to right*) 5% w/v agarose, 5% w/v gellan gum, and 4% w/v methylcellulose applied to handmade rag paper and (b) contact angle test used to determine relative wettability of paper samples. A drop of water was applied to (*left to right*) filter paper, handmade rag paper, and machine-made, chemically purified wood-pulp paper.

Experimental

Experimental design

Gels prepared from agarose, gellan gum, and methylcellulose (Fig. 1a) were placed on three different types of paper for three different time intervals. The papers chosen for this study were Whatman filter paper No. 3 (unsized, soft, cotton), Barcham Green Hayle (gelatin-sized, lightly textured, handmade cotton-linen rag), and Mohawk Superfine White (calendered, machine-made, chemically purified wood pulp).[3] These papers, each with different physical properties, were selected to represent the range typically encountered in a fine art collection. A simple contact angle test, in which a drop of water was placed on the surface of each paper, quickly confirmed the varying wettability of the papers selected (Fig. 1b). The time intervals for gel contact – 10, 20, and 40 minutes – were chosen to approximate actual treatment durations. In some of the experiments, *usu mino* Japanese tissue was placed between the gels and paper samples to evaluate the efficacy of a barrier layer to prevent or minimize gel residues.

Although methylcellulose is not a rigid gel, it was included in the experiments for comparison since it has been widely used in the past. Methycellulose is a viscous, gel-like cellulose ether, long favored by paper conservators for a variety of treatments for which agarose, gellan gum, and other rigid gels are increasingly used. Conservators acknowledge that methylcellulose cannot be fully cleared from porous materials like paper. However, there is a general consensus that these residues are acceptable based upon the results of accelerated aging studies (Feller and Wilt 1990; Warda *et al.* 2007) as well as practical experience.

Agarose, gellan gum, and methylcellulose, being colorless and translucent in gel form, were tagged with fluorescein to give polymer-fl[4] prior to gel preparation to facilitate detection of residue. By covalently attaching fluorescein to the polysaccharide polymers (Fig. 2), it was possible to detect very small amounts of residues on paper samples. The tagged polymers produce bright emission under ultraviolet (UV) and blue light. Visualization and interpretation are straightforward and do not require special equipment. The percent conversion of -OH groups to a fluorescein chromophore

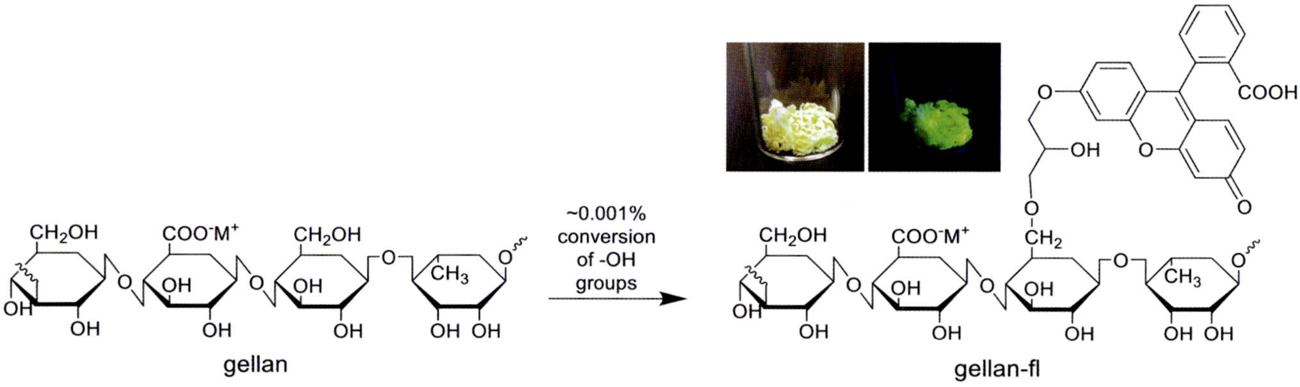

FIG. 2 Chemical structure of gellan before and after covalent attachment of fluorescein with inset photographs of fluorescein-fl under visible light (*left*) and a 365 nm UV lamp (*right*).

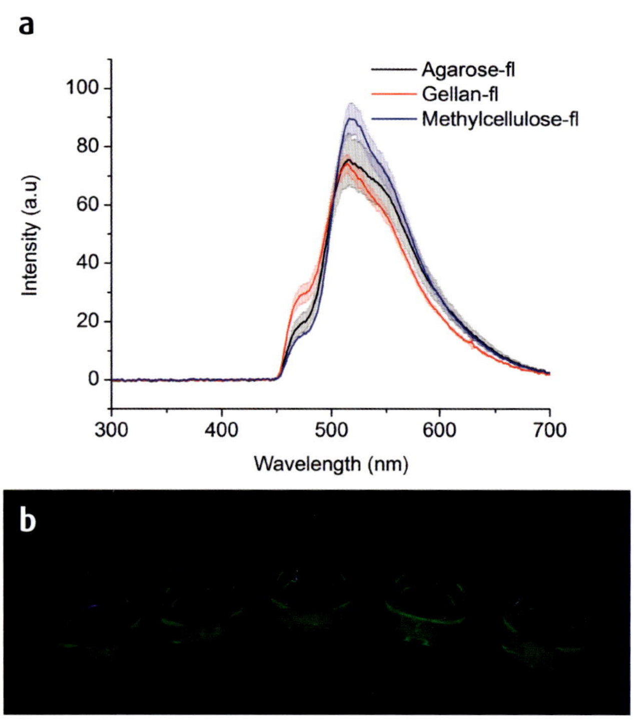

FIG. 3 Luminescence of the gels. (a) Emission spectra of 5% w/v agarose-fl, 5% w/v gellan-fl, and 4% w/v methylcellulose-fl gels after casting; the shaded region represents the standard deviation of three measurements. (b) 0.18 wt% gellan-epf solutions at pH values of 5, 6, 7, 8, and 9 (*left to right*).

was purposely kept low in order minimize any changes in the gelation abilities of the polymers. UV-visible (UV-vis) spectroscopy of polymer-fl solutions was used to calculate the approximate concentration of attached fluorescein and the percent substitution of -OH groups by fluorescein moieties of 0.0002%, 0.002%, and 0.0003% for agarose-fl, gellan-fl, and methylcellulose-fl, respectively.

From UV-vis absorption spectroscopy measurements, gellan-fl had an attached fluorescein concentration 10 times higher than that of agarose-fl and methylcellulose-fl. To ensure that the three gels used for residue studies had similar final bound fluorescein concentrations, gellan-fl and gellan (unmodified) in a 1:9 ratio was dissolved in hot water (adjusted to pH 8 with saturated $Ca(OH)_2$ solution) for a total polymer concentration of 5% w/v. In accordance with the gellan gum preparation described in conservation literature, calcium acetate was also added to the pH-adjusted water (0.4 g/L) used to prepare the gel (Iannuccelli and Sotgiu 2010; Maheux 2015). The hot solution was poured into a Mylar tray, in which it was allowed to rest overnight, resulting in a rigid, clear gel. This procedure was repeated to prepare a 5% w/v agarose-fl, but with no unmodified agarose added, resulting in a rigid, translucent gel. A 4% w/v methylcellulose-fl dispersion was also prepared by stirring the methylcellulose-fl in hot water and letting it rest overnight at room temperature, which resulted in a translucent, viscous dispersion. The weight-percent concentrations for each polymer in the gels were chosen to match those commonly used in actual treatments of works on paper.

Emission spectra (Fig. 3a) of each cast polymer-fl gel show that the fluorescence intensities are similar, indicating that levels of residue can be compared amongst the gels. Subtle differences in the λ_{max} and spectral shapes of each spectrum can be attributed to the presence of different relative concentrations of various pH-dependent fluorescein species (Kubista *et al.* 2004). In order to ensure that any small differences in pH would not lead to differences in the color or intensity (which might in turn lead to incorrect comparisons of the relative amounts of polymer residues on paper supports), solutions were prepared with equal concentrations of gellan-fl at pH values of 5, 6, 7, 8, and 9 (Fig. 3b): because all five solutions appear to be of similar colors and intensities, small deviations in pH between agarose-fl, gellan-fl, and methylcellulose-fl should not lead to any visible differences when analysing the color and intensity of the fluorescence of the residue. The pH of all cast gels was 6 as measured with a Horiba Laqua Twin pH meter.

Synthesis of polymer-fl

3-Epoxypropoxy fluorescein (EPF) was synthesized as reported in the literature (Guan *et al.* 2006), using potassium carbonate as a base rather than sodium hydroxide (Angelova *et al.* 2015).

TABLE 1 Summary of residue experiments

Experiment name	Gels tested	Papers tested	Variables
Residue study	5% w/v agarose 5% w/v gellan gum* 4% w/v methylcellulose	Filter paper Handmade rag paper Machine-made wood-pulp paper	Time: 10, 20, and 40 minute intervals Japanese tissue barrier
Clearance study	5% w/v gellan gum* 4% w/v methylcellulose	Filter paper Handmade rag paper Machine-made wood-pulp paper	Time: 10, 20, and 40 minute intervals Additional clearance with damp cotton swab after gel removal
Calcium acetate study	5% w/v gellan gum	Filter paper Handmade rag paper Machine-made wood-pulp paper	Time: 10, 20, and 40 minute intervals Calcium acetate concentration in water used to prepare gel: 0.0g/L, 0.2g/L, and 1.0g/L

*Unless otherwise noted, gellan gum was prepared using a calcium acetate solution with 0.4g/L concentration

EPF (0.05 g) was reacted with 3 g agarose, gellan, or methylcellulose according to the method reported by Guan *et al.* (2006) by using the selected polymer in place of poly(vinyl alcohol). The product was precipitated from the reaction medium (dimethyl sulfoxide) by adding ethyl acetate. The precipitate was dissolved in water, the solution was acidified with a 10% w/v aqueous HCl in water solution to pH ~6, and the product was precipitated by the addition of acetone. Then, the product was filtered, washed with acetone, and dried under vacuum (~125 mmHg) at room temperature for one week.

UV-vis spectroscopy of polymer-fl

A Cary Bio 300 UV-vis spectrophotometer was used to measure the absorbance of polymer-fl solutions in a plastic cuvette with a 1 cm path length. Solutions of agarose-fl and gellan-fl were prepared in a Tris buffer (pH 8), but methylcellulose-fl was not soluble in the buffer so water was used instead. The approximate concentration of fluorescein chromophores, estimated from the λ_{max} absorbance and the extinction coefficient of 3-*O*-methylfluorescein (26000 $M^{-1}cm^{-1}$ at 470 nm and pH 7, according to Gottlin *et al.* 1996), was used to calculate the approximate conversion of -OH groups to fluorescein groups.

Fluorescence of cast gels

Emission spectra were collected on a CRAIC 508 PV microspectrophotometer on a Leica DMRX base with an I3 filter cube.

Gellan-fl pH studies

Solutions of gellan-fl in water (0.18% w/v) were prepared at a range of pH values by adding water pH-adjusted to 9 with ammonium hydroxide.[5] pH was measured with Merck MColorpHast pH indicator strips.

Residue experiments

Three series of experiments were undertaken to compare the relative amount of residue of each gel on different papers, to measure qualitatively the efficiency of a clearance protocol, and to investigate the effect of added calcium acetate on the amount of residue of gellan gels on paper samples (Table 1). In the *Residue study*, all gels were applied to the paper samples, with and without a Japanese tissue barrier, for all three time intervals. In the *Clearance study*, gellan gum or methylcellulose was applied to the paper samples for all three time intervals, in duplicate. Once the gels were removed, a cotton swab lightly moistened with deionized water was gently rolled over the treated area to lift possible residues. In the *Calcium acetate study*, 5% w/v gellan gum was prepared three times using calcium acetate in different concentrations. The aim of this experiment was to determine the role of divalent cations – which are typically added to the aqueous solution used to prepare gellan gum to strengthen the gel network and increase rigidity by crosslinking – in affecting residue deposition, since earlier studies suggest a link (CP Kelco 2007).

In all experiments, agarose and gellan gum gels were cast to approximately 3 mm thickness (a thickness useful for treatment) and cut into 1.25 × 1.25 cm squares. The squares were transferred onto the paper samples with forceps and/or a microspatula. Evaporation was slowed and contact with the paper samples was encouraged by placing Mylar on top of the gels. To ensure that approximately the same amount of methylcellulose was applied to each of the paper samples, it was transferred with a microspatula to Mylar templates on an analytical balance. Each portion weighed 0.08–0.10 g and was applied to the paper samples by inverting the gel-loaded Mylar template and placing it on the paper. At the end of each test interval, bulk methylcellulose was removed mechanically using a microspatula followed by clearance using a damp cotton swab as described above, consistent with usual practice when using the cellulose ether in a poulticing treatment.

Examination and documentation

Each paper sample was photographed before and after treatment using a Canon EOS 5D Mark III D SLR camera fitted with 50 mm lens and Canon EOS Utility for tethered capture. Samples were imaged under visible light using Horizon lights with 5600 K daylight-balanced fluorescent lamps, and under UV radiation to document UV-induced fluorescence due to polymer residues using SuperBright II 3000 series ultraviolet

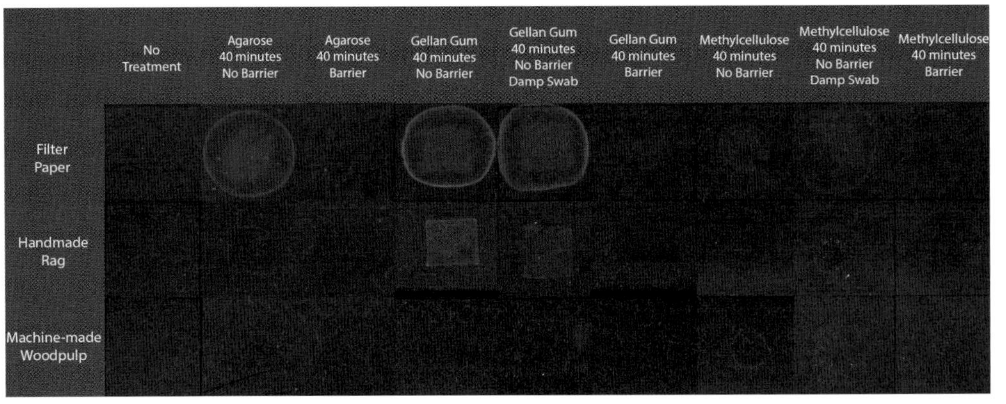

FIG. 4 Images of papers under UV illumination following a 40-minute application of agarose, gellan gum, or methylcellulose, each with and without a Japanese tissue barrier and an additional clearance step using a damp cotton swab.

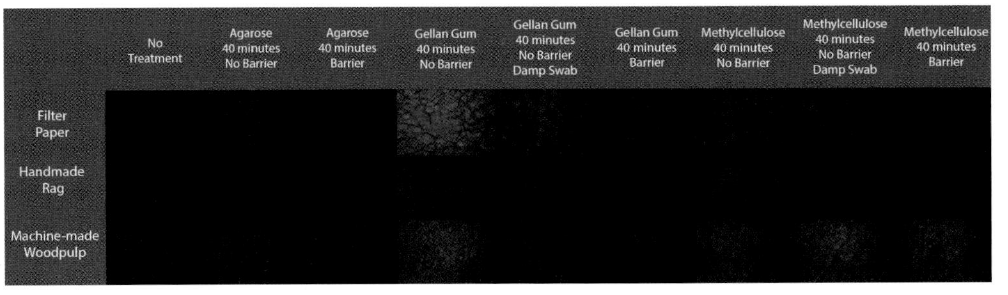

FIG. 5 Fluorescence photomicrographs of papers tested after a 40-minute application of fluorescein-tagged agarose, gellan gum, or methylcellulose, with and without a Japanese tissue barrier and an additional clearance step using a damp cotton swab. See text for conditions.

lamps (368 nm peak emission). Fluorescence microscopy was performed using a Leica DMRX microscope fitted with an nPL 20× objective and a high-pressure Hg lamp (100 W) and an I3 filter cube (450–490 nm excitation, 510 nm dichroic mirror, and 515 nm suppression filter). Photomicrographs were captured using a Canon EOS 5D Mark III SLR camera and Canon EOS Utility. Images were captured in CR and JPEG formats. To correlate fluorescence intensity to relative amounts of residue among the samples, the capture settings were kept constant throughout imaging: 0.5 second exposure and ISO 200.

Results and discussion

Experiments in this study (Table 1) included 10-, 20-, and 40-minute intervals for gel application; however, only images from the 40-minute interval are illustrated (Figs 4–7). The results obtained at 10 and 20 minutes are consistent with those obtained at 40 minutes.

Residue study

To compare the amount of residue deposited by each gel on each paper tested, and the impact of treatment variables (i.e. the use of a barrier layer and additional clearance step with water-dampened cotton swab), we studied fluorescence images captured at both the macro (Fig. 4) and micro (Fig. 5) level. In some cases, trends revealed by macroscopic UV photography were seemingly contradicted by observations made by fluorescence microscopy, raising additional research questions and highlighting the potential for misinterpretation of experimental results.

In the *Residue study*, macro- and microscopic fluorescence imaging suggest that while some gels are more prone to leaving residues, some papers are more susceptible to adhering polymer residues. In the UV macrophotographs, the most intense fluorescence is observed in the papers treated with gellan gum, suggesting greatest deposition of residue. Given the physical properties of the three gels, this was somewhat surprising. The authors anticipated that the viscous methylcellulose would leave the largest deposit of residue, rather than one of the rigid gels.

From the photomacrographs, it appears that the Whatman filter paper is most susceptible to residue retention as these samples exhibit the highest fluorescence intensity for treatments with all three gels tested. The highly fluorescent (white) tideline observed in the Whatman filter paper samples is the result of solubilized material in the paper that has moved with diffusion of water, not small-fraction tagged gel polymer or excess fluorescein that has migrated. This was confirmed by placing a drop of deionized water on a sample of the filter paper, allowing it to dry, and examining it under UV radiation. A similar white-fluorescing tideline was observed. The

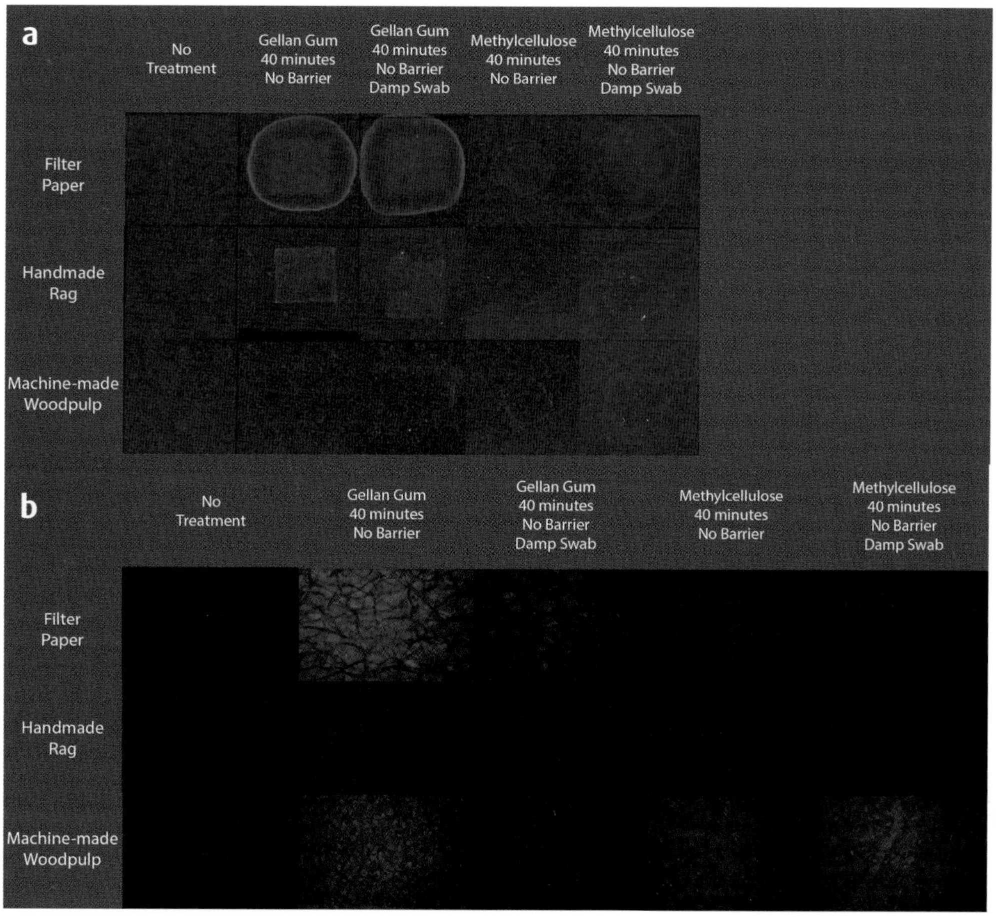

FIG. 6 Images of three papers tested after a 40-minute application of gellan gum and methylcellulose, with and without an additional clearance step using a damp cotton swab. (a) UV images and (b) visible-induced fluorescence photomicrographs. See text for conditions.

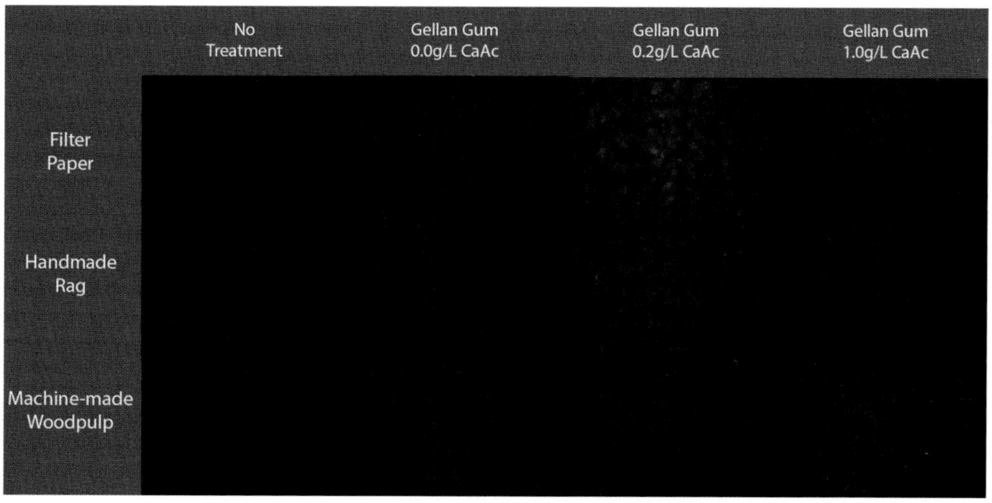

FIG. 7 Visible-induced fluorescence photomicrographs of three papers tested after a 40-minute application of gellan gum prepared with calcium acetate solutions of varied concentration: 0.0g/L, 0.2g/L, and 1.0g/L.

Mohawk Superfine machine-made wood pulp paper exhibits the least fluorescence following treatment with all three gels, suggesting it is the least susceptible to residue deposition. Intuitively, these results make sense to the paper conservator: the filter paper is unsized and soft with fibers readily available to adhere to polysaccharide moieties. The machine-made paper, with its low degree of wettability and slick, calendered, chalk-filled surface would seem to have less affinity for the gel residues.

The images under UV illumination show the efficacy of the Japanese tissue barrier in preventing residue deposition of rigid gels and reducing residue deposition of methylcellulose for all three papers treated. It is interesting to note the discrete shape of the rigid gel residues and, in the case of the handmade rag paper, the removal of gelatin sizing observed as a decrease in autofluorescence, even with a barrier tissue in place. An 'edge effect' was observed on nearly all of the handmade rag and machine-made paper samples in which brighter fluorescence at the border of the applied gel suggests greater residue is deposited along its edges, where they are losing water most rapidly. However, spectra have not been collected to determine if there is a contribution from deposition of fluorescent species in the paper depositing at the evaporation edge. This will be studied further in the future.

Reviewing the fluorescence photomicrographs from the *Residue study*, gellan gum appears to leave the greatest amount of material overall, but surprisingly high fluorescence was observed in the machine-made paper samples treated with methylcellulose. These images suggest that comparatively more residue is present in the paper than is perceived by looking at the UV macrophotographs. The microphotography experiment is better suited to detecting fluorescein (fluorescein absorbs most strongly at ~490 nm), but the discrepancies in observed fluorescence in the macro and micro imaging of treated samples raises some interesting questions:

- Where is the residue located within the paper?
- Is evidence for residue more readily captured by UV macrophotography when it is located at the surface of the paper?
- Can residue that has migrated into the paper be imaged more effectively through fluorescence microscopy?

In general, the fluorescence photomicrographs suggest that comparatively most residue is left on the machine-made paper samples for all gels, including agarose. The handmade rag paper appears the most resistant to residue deposition. With the exception of methylcellulose on machine-made paper, use of the Japanese tissue barrier appears to be very effective in preventing residue deposition. This conclusion is consistent with the inference made from macroscopic UV images, determined from the low intensity of emission from fluorescein-derivatized polymer.

Clearance study

In the UV macrophotographs taken for this experiment, the effect of wet swab clearance in reducing gellan gum and methylcellulose residues is perceptible, though subtle (Fig. 6a). The additional clearance step appears to be more effective on methylcellulose residues than gellan gum residues. In the corresponding fluorescence photomicrographs – with the exception of the machine-made paper – diminished fluorescence resulting from swab clearance suggests removal of some polymer residue (Fig. 6b). This discrepancy between observations made in overall UV images and fluorescence microscopy again raises the question: where in the paper is the residue located? In that regard, does additional clearance with a damp swab remove polysaccharide residues or does it drive them deeper into a porous paper support?

Calcium acetate study

The inclusion of a divalent cation significantly impacts the ability of a lower concentration of gellan gum solutions (i.e. <2 % w/v) to set and form a rigid gel. A rigid gel can be prepared at higher concentrations (4–5%) without including cations. To remain consistent with the experiments described above, all gellan gum prepared for the *Calcium acetate study* was 5% w/v polymer.

From inspection of the fluorescence photomicrographs from this experiment, varying the concentration of calcium acetate added to water does not appear to have an appreciable effect on residue deposition of low-acyl gellan gum prepared at 5% w/v. Fluorescence is observable in all the paper samples treated and comparable when treated with gels containing 0.0 g/L, 0.2 g/L, and 1.0 g/L calcium acetate (Fig. 7). The filter paper and rag paper samples treated with 1.0 g/L-calcium acetate gellan gum appear to exhibit less fluorescence than the other gel-treated samples. These assertions are supported by UV photography, which demonstrates similar fluorescence for all samples regardless of the concentration of calcium acetate used to prepare the gels (not illustrated).

Drawing from practical experience using gellan gum in conservation treatments, the authors find these results somewhat surprising because they expected increasing the calcium acetate concentration would alter mechanical properties of the gels, specifically stiffening the gel, and thus influence residue deposition.

Conclusion

Although the results of the residue studies reported here are not straightforward and raise more research questions, one conclusion can be firmly drawn: agarose and gellan gum are not truly residue-less gels. While the amount of residues they deposit on paper may be small, the highly sensitive fluorescein-tagging technique has unequivocally demonstrated that residues are present. Conservators should bear in mind, however, that the presence of small amounts of residue is not necessarily problematic. Based on aging studies of cellulose ethers, conservators accept that leaving behind methylcellulose residues – or using methylcellulose as an adhesive, consolidant, or sizing agent – is acceptable because of its known, stable aging behavior. Both agarose and gellan gum are also polysaccharides, somewhat similar to cellulose at a chemical level. If it can be demonstrated that these residues age in a manner compatible with the papers on which they remain, conservators may feel confident that their gel treatments will not induce long-term adverse effects. The authors plan to pursue complementary accelerated aging studies in the future, using the gels and paper samples from this set of experiments, to address this point.

The results of this study underscore the differential response of different kinds of papers to a treatment protocol. The properties of the papers employed in the experiments described above played an important role in the amount of residue deposited. Generally, the unsized, open-surfaced filter paper appears to retain the greatest amount of residue. This is not a surprising result: fibers in the filter paper are readily available to interact with the polysaccharide gel polymers. The presence of sizing in paper seems to have an impact on the amount of residue deposited. In the sized, handmade rag paper, gelatin has coated individual fibers and imparted a degree of hydrophobicity, resulting in lower residue deposition.[6]

Finally, this study partially answered the question: can measures be taken to lessen or prevent residue deposition during treatment? Across all of the experiments, a Japanese tissue barrier was highly successful in attenuating the amount of residues of rigid gels. Barrier tissues were also effective in minimizing methylcellulose residues, with the exception for the machine-made paper sample. Other methods for minimizing residue explored in this study – additional clearance with a damp cotton swab and increasing the concentration of divalent cations when preparing gellan gum – require further research to make firm conclusions. It would be of interest to rerun the *Calcium acetate study* with 2% and 3% w/v gellan gum gels in which the presence of a divalent cation is more critical to establish rigidity. In the future, analyses of cross-sectional samples from this study could clarify where gel residues are located within the paper, particularly in samples subjected to an additional 'clearance' step using a damp cotton swab after the gels have been removed.

Without question, rigid polysaccharide gels have tremendous potential for expanding the paper conservator's toolkit. However, there is much to learn about optimizing and tailoring gels to meet the specific needs of paper objects and the short- and long-term effects of these polymers on cellulosic supports. The authors hope that the results of the residue studies presented here will inform future use of agarose and gellan gum within the field of paper conservation and inspire further collaborative studies between conservators and scientists to that end.

Acknowledgements

The authors would like to extend their gratitude to: colleagues in the Departments of Paper Conservation, Photograph Conservation, and Scientific Research at the National Gallery of Art, Washington, DC (NGA); Winterthur/University of Delaware Program in Art Conservation; the Department of Paper Conservation at the J. Paul Getty Museum; and the Department of Chemistry at Georgetown University. With special thanks to Kimberly Schenck, Head of Paper Conservation at NGA, and Michelle Facini, Paper Conservator at NGA, for their support and interest in this research, without which this study would not have been possible. Many thanks to Michael Palmer, Conservation Scientist at NGA, for lending his expertise and characterizing paper fibers from the samples used in the study. Deep thanks to Richard Wolbers, Professor at the University of Delaware, for his knowledge, continued support, and encouragement to pursue this research and develop gel cleaning systems for works on paper. The authors express their gratitude to the Andrew W. Mellon Foundation (for support of Michelle R. Sullivan), the US National Science Foundation (Grant CHE-1502856) and to Achievement Rewards for College Scientists (ARCS) Foundation (for support of Teresa T. Duncan).

Materials and equipment

- Agarose LE (#A1700): Benchmark Scientific, Inc., PO Box 709, Edison, NJ 08818, USA, +001 (908) 769-5555.
- Kelcogel LA gellan gum: CP Kelco US, Inc., 3100 Cumberland Boulevard, Suite 600, Atlanta, GA 30339, USA, +001 (800) 535-2687.
- Calcium acetate hydrate (#5743-26-0): Fisher Scientific, 300 Industry Drive, Pittsburgh, PA 15275, USA, +001 (877) 885-2081.
- Methylcellulose A4M (M0512): Sigma-Aldrich, PO Box 14508, St. Louis, MO 63178, USA, +001 (800) 325-3010.
- Qualitative Filter Paper, Grade 3: Whatman, GE Healthcare Bio-Sciences, P.O. Box 643065 Pittsburgh, PA 15264, USA, +001 (800) 526-3593.
- Hayle handmade, rag paper: J. Barcham Green & Co., Ltd. (no longer trading).
- Mohawk Superfine White (machine-made, wood-pulp paper): Mohawk Fine Papers, Inc., 465 Saratoga Street, Cohoes, NY 12047, USA, +001 (518) 237-1740.
- *Usu mino* HM-54 (Japanese tissue barrier): Hiromi Paper International, 9469 Jefferson Blvd #117, Culver City, CA, USA, +001 (310) 998-0098.
- Horiba Laqua Twin pH meter, Cole-Parmer, 625 East Bunker, Court Vernon Hills, IL 60061, USA, +001 (800) 323-4340, sales@coleparmer.com.
- CRAIC microspectrophotometer: CRAIC Technologies, Inc., 948 N. Amelia Ave.,San Dimas, CA 91773, USA, +001 310 573 8180.
- Leica DMRX fluorescence microscope with nPL objective (20×) and filter cube I3: Leica Microsystems, Inc., Buffalo Grove, IL 60089, USA, +001 (800) 248-0123.
- EOS 5D Mark III DSLR camera; Canon, 1 Canon Plaza, Lake Success, NY 11042, USA, +001 (516) 328-5000.
- Horizon light with 5600 K fluorescent tubes (Model CF2): Photogenic Professional Lighting, 1268 Humbracht Circle, Bartlett, IL 60103, USA, +001 (630) 830-2500.
- SuperBright II 3000 series ultraviolet lamps (368 nm peak emission): UV Systems, Inc., 16605 127th Ave. SE, Renton, WA 98058, USA, +001 (877) 689-5142.

Notes

1. In this study, paper samples were treated with the rigid polysaccharide gel agarose and three viscous gels – Carbopol 940 (polyacrylic acid-based), Laponite RD (silicate-based), and methylcellulose A4M. These paper samples were then subjected to accelerated aging alongside additional samples moistened with free water for comparison. Because dry agarose polymer discolored upon accelerated aging, it was presumed that any residue would also discolor and serve as an indicator of polymer residue. The paper samples treated with agarose and methylcellulose did not exhibit discoloration or changes in UV-induced fluorescence after accelerated aging.

2. Fluorescent tags have been applied in previous studies to track the penetration of gelatin and methylcellulose into paper supports when used as consolidants and resizing materials (Hummert *et al.* 2013; Ream 1995).
3. The fiber content of paper samples was confirmed through polarized light microscopy performed by Michael Palmer, National Gallery of Art, Washington DC, June 2015. The filler in the machine-made paper was determined to be chalk using SEM-EDX.
4. Polymer-fl is a generic term for those polymers that have been tagged with fluorescein. Gellan-fl, agarose-fl, and methylcellulose-fl similarly refer to those specific polymers that have been tagged with fluorescein.
5. To prepare water pH-adjusted to 9, 2 drops of 28-30% ammonium hydroxide were added to 200 mL water.
6. In SE images (not illustrated), fibers in the rag paper samples are visibly coated with gelatin sizing.

References

Angelova, L.V., Berrie, B.H., de Ghetaldi, K., Kerr, A. and Weiss, R.G. 2015. 'Partially hydrolyzed poly(vinyl acetate)-borax-based gel-like materials for conservation of art: characterization and applications', *Studies in Conservation* 60: 227–44.

Botti, L., Corazza, A., Iannuccelli, S., Placido, M., Residori, L., Ruggiero, D., Sotgiu, S., Tireni, L., Berzioli, M., Casoli, A., Isca, C. and Cremonesi, P. 2011. 'Evaluation of cleaning and chemical stabilization of paper treated with a rigid hydrogel of gellan gum by means of chemical and physical analyses', in J. Bridgland (ed.), *Preprints of the ICOM-CC 16th Triennial Meeting*. Almada: Critério, 1–11.

Campani, E., Casoli, A., Cremonesi, P., Saccani, I. and Signorini, E. 2007. *L'uso di agarosio e agar per la preparazione di 'gel rigidi'/Use of Agarose and Agar for Preparing 'Rigid Gels'*, D. Kunzelman (tr.), *Quaderni del Cesmar7*, no. 4. Padua: Il Prato.

Casoli, A., Isca, C., De Iasio, S., Botti, L., Iannuccelli, S., Residori, L., Ruggiero, D. and Sotgiu, S. 2014. 'Analytical evaluation, by GC/MS, of gelatine removal from ancient papers induced by wet cleaning: a comparison between immersion treatment and application of rigid gellan gum gel', *Microchemical Journal* 117: 61–7.

CP Kelco 2007. Kelcogel® gellan gum book, 5th edn. Available from: www.cpkelco.com (accessed 7 July 2015).

Feller, R. and Wilt, M. 1990. *Evaluation of Cellulose Ethers for Conservation*. Marina del Rey: Getty Conservation Institute.

Gottlin, E., Xu, X., Epstein, D., Burke, S., Eckstein, J., Ballou, D. and Dixon, J. 1996. 'Kinetic analysis of the catalytic domain of human cdc25B', *Journal of Biological Chemistry* 271(44): 27445–9.

Guan, X.L., Liu, X.Y., Su, Z.X. and Liu, P. 2006. 'The preparation and photophysical behaviors of temperature/pH-sensitive polymer materials bearing fluorescein', *Reactive and Functional Polymers* 66(11): 1227–39.

Hawkes, R. 2013. 'The use of gels in aqueous conservation of paper', *ICON News* 48: 27–8.

Hughes, A. and Sullivan, M. 2016. 'Targeted cleaning of works on paper: rigid polysaccharide gels and modified aqueous solutions', *The Book and Paper Group Annual* 35: 30–41.

Hummert, E., Henniges, U. and Potthast, A. 2013. 'Fluorescence labeling of gelatin and methylcellulose: monitoring their penetration behavior into paper', *Cellulose* 20: 919–31.

Iannuccelli, S. and Sotgiu, S. 2010. 'Wet treatments of works of art on paper with rigid gellan gels', *The Book and Paper Group Annual* 29: 25–39.

Kavda, S., Dhopatkar, N., Angelova, A., Richardson, E., Golfomitsou, S. and Dhinojwala, A. 2016. 'Surface behavior of PMMA: is gel cleaning the way to go?', *Studies in Conservation* 61: 297–9.

Keynan, D. and Hughes, A. 2013. 'Testing the waters: new technical applications for the cleaning of acrylic paint films and paper supports', *The Book and Paper Group Annual* 32: 43–51.

Kubista, M., Ismail, I. H., Forootan, A. and Sjogreen, B. 2004. 'Determination of protolytic constants by trilinear fluorescence spectroscopy', *Journal of Fluorescence* 14(2): 139–44.

Lennig, H. 2010. 'Solvent gels for removing aged pressure-sensitive tape from paper', *Restaurator* 31: 92–105.

Maheux, A. 2015. 'Cross-disciplinary uses for gellan gum in conservation', *The Book and Paper Group Annual* 34: 69–79.

Mazzuca, C., Micheli, L., Cervelli, E., Basoli, F., Cencetti, C., Coviello, T., Iannuccelli, S., Sotgiu, S. and Palleschi, A. 2014a. 'Cleaning of paper artworks: development of an efficient gel-based material able to remove starch paste', *ACS Applied Materials and Interfaces* 6(19): 16519–28.

Mazzuca, C., Micheli, L., Marini, F., Bevilacqua, M., Bocchinfuso, G., Palleschi, G. and Palleschi, A. 2014b. 'Rheoreversible hydrogels in paper restoration processes: a versatile tool', *Chemistry Central Journal* 8(10). Available at: http://journal.chemistrycentral.com/content/8/1/10 (accessed 7 July 2015).

Mazzuca, C., Micheli, L., Carbone, M., Basoli, F., Cervelli, E., Iannuccelli, S., Sotgiu, S. and Palleschi, A. 2014c. 'Gellan hydrogel as a powerful tool in paper cleaning process: a detailed study', *Journal of Colloid and Interface Science* 416: 205–11.

Ream, J. 1995. 'Observations on the penetration of two consolidants applied to insecure gouache on paper', *The Book and Paper Group Annual* 14: 27–31.

Sullivan, M., Brogdon-Grantham, S. and Taira, K. 2014. 'New approaches to cleaning works on paper and photographs', *Proceedings of the 2014 Conference of the Association of North American Graduate Programs in Conservation of Cultural Heritage*. Available at: http://cool.conservation-us.org/anagpic/2014pdf/anagpic2014_sullivan_etal_paper.pdf (accessed 7 July 2015).

van Dyke, Y. 2004. 'Practical applications of protease enzymes in paper conservation', *The Book and Paper Group Annual* 23: 93–107.

Warda, J., Brückle, I., Bezur, A. and Kushel, D. 2007. 'Analysis of agarose, Carbopol, and Laponite gel poultices in paper conservation', *Journal of the American Institute for Conservation* 46(3): 263–79.

Yip, V. 2015. 'The conservation of two contemporary Chinese woodblock prints using gellan gum', *News in Conservation* 48: 10–13.

Authors' addresses

- Michelle R. Sullivan, Department of Paper Conservation, J. Paul Getty Museum, Los Angeles, CA, USA (msullivan@getty.edu)
- Teresa T. Duncan, Department of Chemistry, Georgetown University, Washington, DC, USA (ttd5@georgetown.edu)
- Barbara H. Berrie, Head of Scientific Research Department, National Gallery of Art, Washington, DC, USA (b-berrie@nga.gov)
- Richard G. Weiss, Department of Chemistry, Georgetown University, Washington, DC, USA (weissr@georgetown.edu)

The use of agar gel for treating water stains on an acrylic canvas

Maggie Barkovic, Olympia Diamond and Maureen Cross

Introduction

Composition, 1963 is a large early acrylic emulsion painting by the British abstract artist Justin Knowles (1935–2004). The brightly coloured, hard line geometric forms are set centrally against an unpigmented, acrylic sized, linen canvas. Both the top and bottom edges had been severely damaged by water, resulting in dark tide stains that rendered the painting unfit for exhibition (Fig. 1).

Treatment was complicated both by a lack of established methods for treating water-stained paintings, and the presence of a water-sensitive p(EA/MMA) acrylic copolymer size layer, which was equally affected by polar solvents. A collaborative research project was undertaken: phase I focused on the technical and art historical impact of water staining on Knowles' work and included trials of surfactants, oxidizing agents and retouching media for the reduction of the stain (Diamond 2014). Phase II further explored tideline formation and studies conducted in the fields of textile and paper conservation, and reviewed recent developments in treatment methods for cleaning acrylic painted surfaces. Phase II also expanded the cleaning trials and explored agar gel as a delivery system. Potential changes to the unpigmented acrylic sized surface and residues from the agar gel were imaged using 3D digital microscopy, atomic force microscopy (AFM), and Fourier transform infrared attentuated total reflection (FTIR-ATR) spectroscopy (Barkovic 2016). Findings led to the creation of a tailored cleaning system and the successful treatment of *Composition*, 1963.

To evaluate methods of stain reduction, replica paintings were created using a similar technique and contemporary materials comparable to those employed by Knowles. Different methods for stain reduction were chosen based on prior conservation treatments involving the reduction of water stains on cellulosic materials (Diamond 2014; Barkovic 2016). The first methods evaluated were dry methods and aqueous solutions. One key consideration was the means used to apply the cleaning solutions. Early tests using dilute Ecosurf EH-9 in deionized water, initially applied from the reverse, proved particularly difficult to control due to a combination of the natural wicking capacity of the linen and increased wetting power, resulting in an undesirable expansion of the cleaning solution and associated transport of materials beyond the stain borders. This prompted an investigation into the use of a gelled system to allow more controlled application.

Justin Knowles and *Composition*, 1963

Justin Knowles created *Composition* in 1963 at the age of 30. This work preceded his decision in 1965 to work full-time as an artist by which time he was already developing ideas and

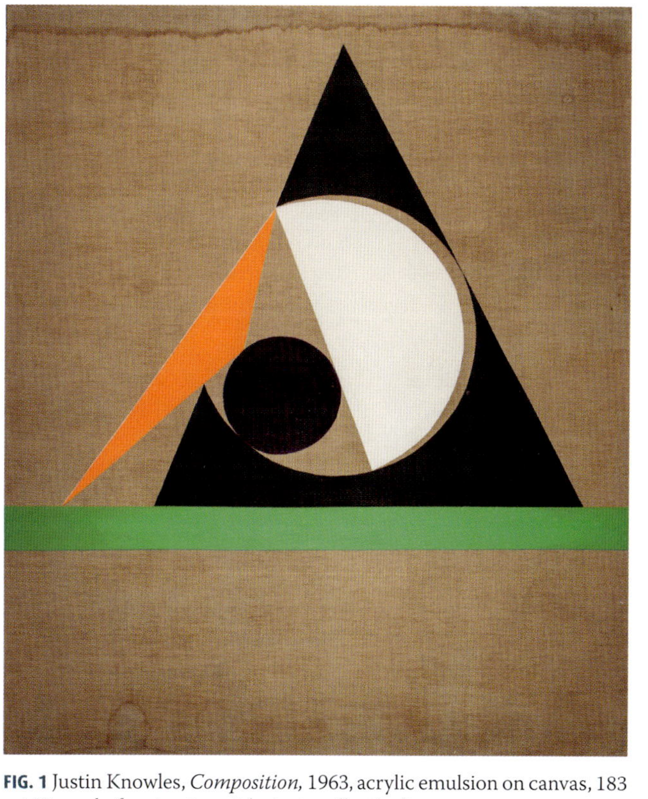

FIG. 1 Justin Knowles, *Composition*, 1963, acrylic emulsion on canvas, 183 × 152 cm: before treatment (private collection).

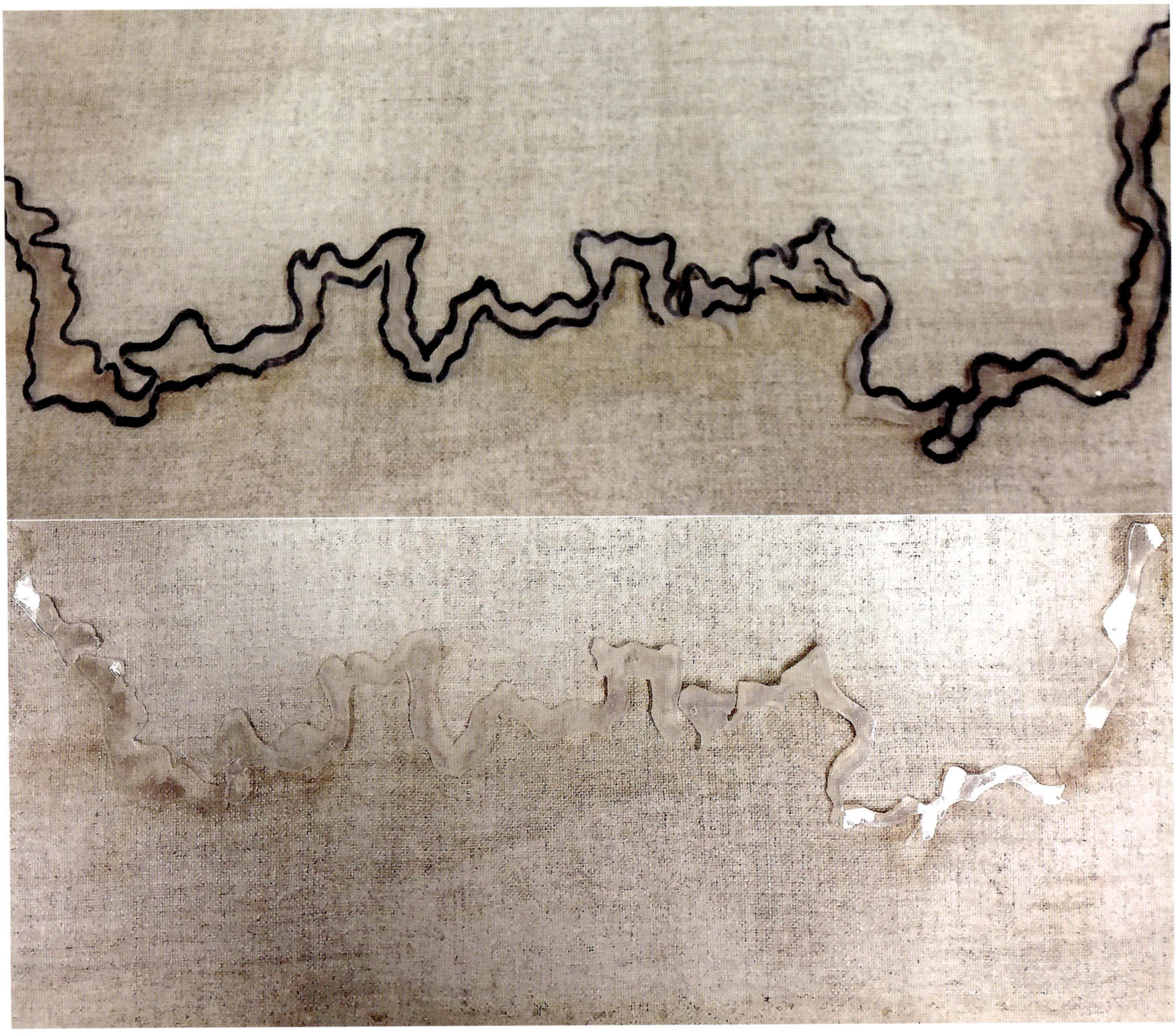

FIG. 2 (*Top*) Tracing agar between two sheets of Melinex and (*bottom*) after tracing, the gel is cut and applied.

a painting technique that aligned with the hard-edge painting style that favoured a 'neatness of surface; and; economy of form'.[1] Once Knowles began exhibiting works, he rapidly developed concepts such as spatial tension and movement, using non-illusionistic geometric compositions that often employed negative space in the form of unpainted canvas. Knowles went on to transform his canvas shapes and explore dimensionality through colour and form, eventually crossing the line between painting and sculpture. *Composition* clearly encompasses these initial ideas, which Knowles continued to develop throughout his career.

Knowles' technique and his choice of materials for *Composition* reflects the importance of the canvas and its relation with the space around it. As he stated, 'natural materials have their own inherent colour characteristics which relate to other colour(s)' (Knowles and Butler 2002), hence the colour of the linen canvas and paint are equally important. The unpainted (acrylic emulsion-sized) canvas naturally acts as a block of colour, where each material works to define the other, and their relationship to space (Diamond 2014). The presence of a water stain obviously disrupts this relationship and lessens impact. Few works survive from this early period of his career due to a fire in 1973 that destroyed Knowles' studio and all of his works (Gooding 2002).

Methodology

Phase I

Knowles used unbleached, unprimed linen, then applied a size layer of an unpigmented waterborne (pEA/MMA) acrylic emulsion, presumably procured from his supplier Rowney & Sons Ltd (Diamond 2014). The replica paintings were created using untreated linen (Rowney) that was stretched and sized with Primal AC-34, an unpigmented (pEA/MMA) acrylic copolymer emulsion. To mimic the artist's technique, the thickness of the size layer was varied slightly between the replica test paintings.

TABLE 1 List of aqueous cleaning solutions tested in Phase II.

Solutions used	pH	Conductivity
Group A Based on Diamond 2014		
Sodium borohydride, 0.5% Ecosurf EH-9	9.0	13.21mS
Group B Evaluating pH in isolation with surfactant		
0.5% boric acid/ with Ecosurf EH-9	6.5	42uS
0.5% boric acid, TEA, Ecosurf EH-9	7.0	177uS
0.5% boric acid, TEA, Ecosurf EH-9	7.5	138uS
0.5% boric acid, TEA, Ecosurf EH-9	8.0	402uS
Group C Evaluating affectivity of chelators		
0.5% boric acid, 0.5% TAC, TEA, Ecosurf EH9	7.0	1077uS
0.5% boric acid, 0.5% TAC, TEA, Ecosurf EH9	7.5	1080uS
0.5% boric acid, 0.5% TAC, TEA, Ecosurf EH9	8.0	1086uS
0.5% boric acid, 0.5% DTPA, TEA, Ecosurf EH9	6.5	1845uS
0.5% boric acid, 0.5% DTPA, TEA, Ecosurf EH9	7.0	1070uS
0.5% boric acid, 0.5% DTPA, TEA, Ecosurf EH9	7.5	1074uS
0.5% boric acid, 0.5% DTPA, TEA, Ecosurf EH9	8.0	1080uS
Group D Evaluating increase in conductivity		
0.5% boric acid, 0.5% DTPA, TEA, Ecosurf EH9	7.0	2.58mS
0.5% boric acid, 0.5% DTPA, TEA, Ecosurf EH9	7.5	2.51mS
0.5% boric acid, 0.5% DTPA, TEA, Ecosurf EH9	8.0	2.57mS
0.5% boric acid, 0.5% TAC, TEA, Ecosurf EH9	8.0	2.77mS
Group E Evaluating Agar as a delivery system		
0.5% boric acid, 0.5% DTPA, TEA, 2.5%Agar	6.5	1940 uS
0.5% boric acid, 0.5% DTPA, TEA, 2.5%Agar	8.0	2.42mS
0.5% boric acid, 0.5% DTPA, TEA, 2.5%Agar	8.0	2.42mS
0.5% boric acid, 0.5% DTPA, TEA, 4% Agar	8.0	2.42 mS
Sodium borohydride, 0.5% Ecosurf EH-9, 2.5% Agar	10	2.6mS
Group F Agar with a more acidic solution		
0.5% boric acid, 0.5% DTPA, TEA, 2.5% agar	5.8	2.5 mS

The age and complex constitution of the water stain was unknown, which limited the replication of the damage; however, based on X-ray fluorescence (XRF) analysis,[2] and empirical testing of the stain at the tideline, a visually similar and tenacious water stain was recreated by submerging the test paintings into a tailored aqueous dirt mixture.[3] After the stains were successfully formed, the canvases were aged in a light box for two years under museum conditions[4] to allow the stain to photo-oxidize and bond with the fabric and size layer.

Phase I cleaning trials with aqueous solutions were largely hindered by limited penetration into the acrylic priming layer. Water remained on the surface, causing the film to swell (Ormsby and Learner 2009; Ormsby *et al.* 2007) and making the process of stain removal from the front of the work difficult.

Phase II

At the start of Phase II, the canvases had also been in dark storage for one year. Both the aqueous solutions and delivery methods were further evaluated and modified. Agar gel was used because of its rigid physical structure and capillary action, which facilitates the concurrent penetration of water into the stain and absorption of solubilized material into the gel (Cremonesi 2010; Warda *et al.* 2007; Keynan and Hughes 2013; Skelton *et al.* 2016). Adjusting concentration controlled the capillary pore size and water penetration rate; the gel thickness was manipulated by pouring molten agar between sheets of Melinex, with staining traced and pieces cut to fit the stains[5] (Fig. 2).

Sixteen different aqueous solutions with varying pH and added chelators were evaluated on the replica paintings by applying liquid solutions using a 000 sable brush over a Mitka suction plate. A higher pH (neutral to slightly alkaline), and higher conductivity options were preferable in terms of reducing the stain and controlling the swelling of the acrylic priming (Table 1, groups A–D). The most successful six solutions were then prepared in various concentrations of agar gel (Table 1, groups E–F). After further assessments on the replicas, the system resulting in the best stain reduction was applied to a small area of *Composition*.

TABLE 2 FTIR–ATR evaluation of possible treatment residues.

Sample	1110 Height (C-O) surfactant	1730 Height (C=O) acrylic carbonyl	Ratio of peaks (1110/1730)
Aged AC			
Sample control	0.083	0.800	1.039
Knowles, control	0.006	0.058	0.107
Knowles, treated for 4 hours	0.000	0.036	0.006
2.5% w/v, 2 minute treatment	0.001	0.103	0.012
2.5% w/v, 5 minute treatment	0.001	0.113	0.006
2.5% w/v, 30 minute treatment	0.002	0.103	0.015
2.5% w/v, 4 hour treatment	0.001	0.564	0.002
2.5% w/v, 8 hour treatment	0.002	0.090	0.018
5.0% w/v, 2 minute treatment	0.002	0.094	0.026
5.0% w/v, 5 minute treatment	0.002	0.112	0.015
5.0% w/v, 30 minute treatment	0.001	0.093	0.011
5.0% w/v, 4 hour treatment	0.001	0.126	0.011
5.0% w/v, 8 hour treatment	0.002	0.116	0.013

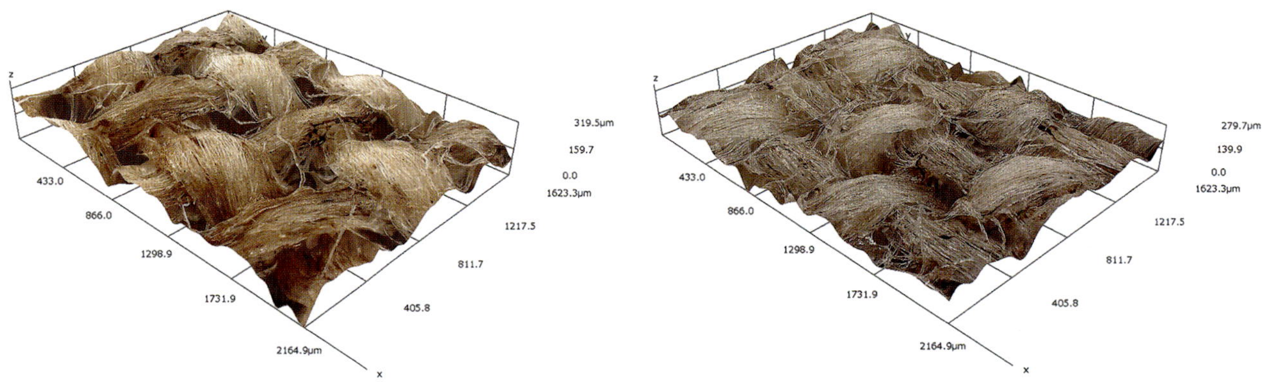

FIG. 3 (*Left*) Stained canvas fibres from test canvas 05 before treatment (×140 magnification). (*Right*) Fibres from test canvas 05 after treatment: cleaning, retouching, and removal of retouching (×140 magnification).

Evaluation methodology

The results on the replica paintings were evaluated using colour measurement[6] and through visual observations by four conservators. The test paintings were viewed on an easel and under the microscope to assess perceptible changes in colour, surface gloss and texture. The samples were also examined using dynamic AFM[7] and high-resolution digital microscopy.[8] In addition, a preliminary investigation into possible residues was carried out using FTIR-ATR. Fibre samples from both the replicas and small samples removed from discrete locations on *Composition* were also evaluated.[9]

Results and discussion

There was a general increase in colour removal from the stain with increased pH (from acidic to slightly alkaline). The combination of the chelator used, higher conductivity of the solutions, and a more alkaline pH helped to solubilize the stain, which was more acidic and had higher conductivity than the unstained areas. There was no perceptible difference in surface gloss after the tests, and any swelling observed was minimal and generally decreased as the conductivity increased, as expected for acrylic emulsion layers. On *Composition*, rusted nails along the tacking margins had caused a deep red staining, possibly ferric hydroxide, but since the water stains in *Composition* are in unpigmented areas, iron-specific chelators could be used. For this reason, triammonium citrate (TAC) and diethylenetriamine-pentaacetic acid (DTPA) were evaluated. DTPA is a much stronger chelator with an octadentate ligand, making it more suitable for chelating Fe^{2+}, Fe^{3+}, and Ca^{2+} which were present in the soiling. DTPA solutions generally resulted in greater stain reduction than TAC.

From the trials carried out on the replicas, the most successful aqueous cleaning solution (boric acid 0.5% w/v, DTPA 0.5% w/v, TEA, pH 8) was then combined with agar gel at 2.5–5.0% w/v. The gels were placed on the stain, blotting paper

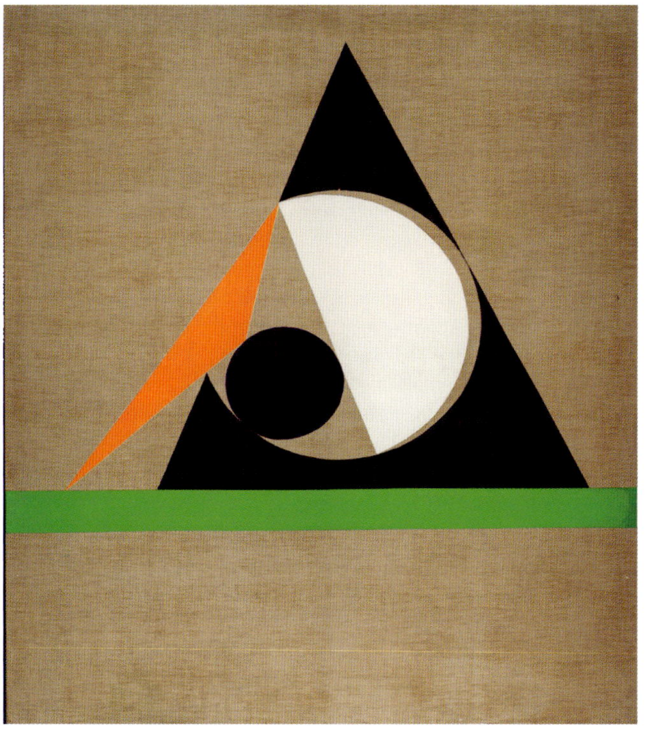

FIG. 4 Justin Knowles, *Composition*, after treatment.

was positioned above the agar and below the stained canvas (on the reverse), and a light sandbag was placed on top. The blotting paper absorbed excess moisture while the weight helped to maintain contact between the gel and the painting surface as the agar gel dried, absorbed the stain, and slowly shrank.

The gel was initially applied in increments of two minutes, however it was quickly clear that longer exposure times were required, up to four hours, occasionally involving multiple applications before the stain was adequately reduced. Although the exposure time had increased, the water absorption remained controlled and the blotting paper was changed every hour. The concentration of agar was also adjusted depending on the absorbency and thickness of the acrylic priming layer. A more rigid concentration of 5% w/v prevented excess liquid from spreading from the gel and bevelling the edges, and/or using thin layers prevented the water from spreading. The gel-cleaned areas were not cleared due to the inherent risk of additional swelling to the acrylic size layer.

The effects on size layer morphology were investigated using AFM and digital microscopy. There were differences in the surface topography within the first two minutes of treatment: the surface was flatter and smoother, indicative of swelling and to be expected as part of this treatment process. The imaged area[10] for AFM was too small to draw absolute conclusions from what is a complex heterogeneous medium film; images captured were consistent with results that showed phase changes in the paint film after two minutes of treatment, as found in previous studies (Kampasakali *et al.* 2011). In addition, high-resolution imaging showed little difference between water-damaged fibres from the replicas and *Composition* after gel treatment (Fig. 3). FTIR-ATR of the replica surfaces after treatment showed either a decrease or complete removal of any migrated surfactant from the acrylic size layer, despite differences in the agar gel concentrations, by calculating the ratio of the peaks from the surfactant acrylic carbonyl functional groups. Any residues, if present, were beyond the detection limits of FTIR-ATR (Table 2).

Conclusion_

This research demonstrates that in practice, a painting's material and physical history may combine to compromise reversibility of a treatment. The success of the treatment for *Composition* resulted from a combination of evaluating Knowles' materials and aesthetic aims, assessing recent research into the effects of conservation treatments on acrylic emulsion materials, preparing replica canvases and trialling a series of treatment options to help refine and reach a tailored treatment strategy which was then taken to the work of art. The use of agar gel as a delivery system permitted the safe reduction of water stains on an unpigmented acrylic emulsion-sized linen canvas. The gel system facilitated the delivery of a tailored aqueous solution including a useful chelating agent set at a pH which promoted the gradual swelling of the size layer and the controlled reduction of the unwanted stain, without adversely affecting surrounding areas. Although the treatment of *Composition* (Fig. 4) ultimately required a combination of cleaning and retouching, the preparation and evaluation of replica test paintings led to a successful stain reduction process which minimized the retouching campaign required to return the work to a displayable condition.

Acknowledgements

Hirox 3D digital microscopy was carried out at Tate Britain, London, the purchase of which was supported in part by the Horizon2020 funded NanoRestArt project (agreement 646063). In addition, FTIR-ATR was done at Tate, with training and assistance from Dr Bronwyn Ormsby. AFM was carried out at Kings College, London.

Materials and suppliers

- Linen canvas: Russell & Chapple, 30–31 Store St, Bloomsbury, London WC1E 7QE, UK, https://www.russellandchapple.co.uk/
- Ecosurf EH 9: The Dow Chemical Company, Diamond House Lotus Park, The Causeway, Staines TW18 3AG, UK, http://campusemeai.dow.com/en-us/locations/united-kingdom
- Agar gel, boric acid, sodium borohydride, DTPA, triethanolamine, TAC: Sigma-Aldrich Ltd., Second Ave, Heatherhouse Industrial Estate, Irvine KA12 8NB, UK, http://www.sigmaaldrich.com/united-kingdom.html

Notes

1. *Hard-Edge Painting*, The Solomon R. Guggenheim Foundation, 2013, http://www.guggenheim.org/new-york/collections/collection-online/movements/195217 (accessed 10 November 2013).
2. A Bruker Tracer 3-SD XRF was used at 2 mm distance to detect iron and calcium.
3. Tap water, 14 rusty nails, 4 pieces of deteriorated linen, 2 pieces of non-deteriorated linen, 4.2 g sawdust, 2 drops of Ecosurf EH 6 non-ionic surfactant, and dirt and dust. The mixture was based on discussions with Stephen Hackney (formerly Tate), Mark van Gelder, conservator, and Jay Krueger, National Gallery Washington (Diamond 2014).
4. At ~ 28 °C with 22% RH and an ultraviolet filter. Assuming reciprocity, 200 lux for 10 hours a day for 730 days (2 years) with an average illuminance of 8000 lux resulted in approx. 1.5M lux-hours of exposure.
5. Samantha Skelton also used blotting paper above the agarose gel and below the canvas, with sandbags (pers. comm. 25 November 2015).
6. A Melinex template was created for each area. A Minolta CM-2600d spectrophotometer and CIELAB176 colorspace was used before and after treatment, and compared to an unstained area of the Knowles, with three measurements averaged, and with the conservators' subjective observations.
7. A Veeco Dimension Icon AFM with ScanAsyst in Dynamic Tapping mode was used at Kings College, London, scanning a 10 × 10 μm^2 area at 1.99–5 Hz using 256 scan lines.
8. A Hirox KH-7800 digital microscope with a ring-light diffuser attachment and ×140 magnification was used.
9. The samples were treated with variable concentrations of agar gel containing the successful cleaning solution (2.5% w/v and 5.0% w/v) at time intervals of 2, 5, 10, 30 minutes and 1, 2, 4, 6 and 8 hours.
10. 18 different locations were measured three times.

References

Barkovic, M. 2016. 'Integrative Solutions for Treating Water Stains on Acrylic Canvases: A Case Study of *Composition*, 1963 by Justin Knowles', unpublished PGDip 3rd year project, Courtauld Institute of Art, London.

Cremonesi, P. 2010. 'Rigid gels and enzyme cleaning', in M. Mecklenburg, A.E Charola and R. Koestler (eds), *Cleaning 2010: New Insights into the Cleaning of Paintings*. Washington, DC: Smithsonian Institution Scholarly Press.

Diamond, O. 2014. 'Treating a Water Stained Canvas: A Case Study of *Composition* 1963 by Justin Knowles', unpublished PGDip 3rd year project, Courtauld Institute of Art, London.

Gooding, M. 2002. 'The work of Justin Knowles', in J. Knowles and J. Butler (eds), *Justin Knowles*. Prague: Bovery Art Books.

Kampasakali, E., Ormsby, B., Cosentino, A., Miliani, C. and Learner, T. 2011. 'A preliminary evaluation of the surfaces of acrylic emulsion paint films and the effects of wet-cleaning treatment by atomic force microscopy (AFM)', *Studies in Conservation* 56: 216–30.

Keynan, D. and Hughes, A. 2013. 'Testing the waters: new technical applications for the cleaning of acrylic paint films and paper supports', *The Book and Paper Group Annual* 32: 43–51.

Knowles, J. and Butler, J. 2002. *Justin Knowles*. Prague: Bovery Art Books.

Ormsby, B. and Learner, T. 2009. 'The effects of wet surface cleaning treatments on acrylic emulsion artists' paints: a review of recent scientific research', *Reviews in Conservation* 10: 29–41.

Ormsby, B., Learner, T., Foster, G., Druzik, J. and Schilling, M. 2007. 'Wet-cleaning acrylic emulsion paint films: an evaluation of physical, chemical and optical changes', in T.J.S. Learner, P. Smithen, J.W. Krueger and M.R. Schilling (eds), *Modern Paints Uncovered*. Los Angeles: Getty Conservation Institute, 187–98.

Skelton, S., Rogge, C. and Veliz Bomford, Z. 2016. 'Testing the limits: the theoretical development and practical reality of a large-scale agarose gel treatment for a discolored Morris Louis', *Studies in Conservation* 61 (suppl. 2): 214–18.

Warda, J., Brückle, I., Bezur, A. and Kushel, D. 2007. 'Analysis of agarose, Carbopol, and Laponite gel poultices in paper conservation', *Journal of the American Institute for Conservation* 46(3): 263–79.

Authors' addresses

- Maggie Barkovic, Williamstown Art Conservation Center, Williamstown, MA USA
- Olympia Diamond, National Museums Liverpool, Liverpool, UK (email:ao.diamond@gmail.com)
- Maureen Cross, Lecturer, Department of Conservation and Technology, Courtauld Institute of Art, London, UK (Maureen.Cross@courtauld.ac.uk)

Gellan gum and agar compared to aqueous immersion for cleaning paper

Céline Delattre, Stéphane Bouvet and Emilie Le Bourg

Introduction: the album and the conservation project

This album (Fig. 1a) was compiled in 1840 by the French painter Félix Philippoteaux. The drawings were made in sketchbooks during the artist's four month period in Algeria where he had followed the French army during a military campaign. When he returned to France, Philippoteaux trimmed his drawings and created an album arranged by topic and date. The drawings depict scenes from life in military camps, the local population, animals and portraits of French soldiers (Fig. 1b).

The album contains 122 drawings made with graphite, watercolour and black ink, spread over 59 pages. They are pasted by four dots of animal glue in the corners. The drawings were mostly in good condition, but the adhesive dots in the corners have created tension, and some brown stains. The paper of the pages was yellowed, especially on the edges, and had many foxing stains with irregular sizes and shapes. The edges of the pages were torn and mechanically weak. Originally, the album was bound with cardboard covered with brown paper, and the spine was covered with brown leather. When the album arrived, the two boards were detached, a few pages were still attached to the spine, but the majority had become detached. It was decided, in agreement with the museum, that the original binding would be restored therefore we removed what was still attached in order to recompose the whole.

The album includes inscriptions in the artist's hand and some sketches drawn directly on the leaves. We had to improve the condition of the drawings while keeping the pages intact. It was first necessary to eliminate both tension in the sheets of paper and the animal glue on the drawings. This required the removal of each sheet of paper individually and its reattachment with an appropriate adhesive, and allowed us to proceed to the cleaning of the background pages while preserving the drawings from exposure to moisture. We were able to clean the album pages individually since they were almost all detached from the binding; we detached the rest. The drawings were in good condition but because water-soluble ink had been used, we decided against aqueous immersion for cleaning. The aim was to consolidate the pages to provide mechanical strength and to water clean them. The drawings would then be in a better condition with respect to future aging and the foxing stains would be mitigated. The aesthetic appearance of the whole object would be enhanced.

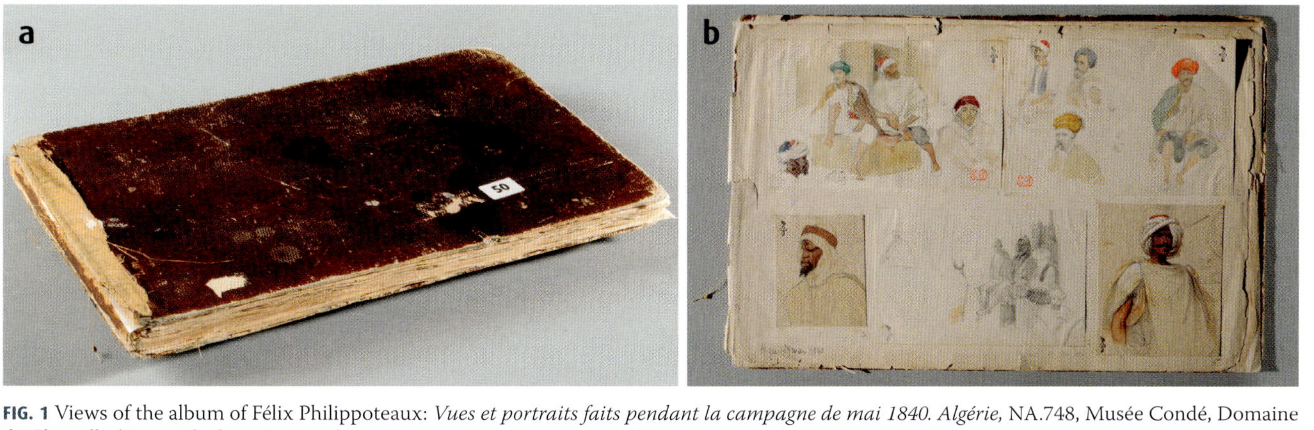

FIG. 1 Views of the album of Félix Philippoteaux: *Vues et portraits faits pendant la campagne de mai 1840. Algérie,* NA.748, Musée Condé, Domaine de Chantilly (© Inp Ghislain Vanneste).

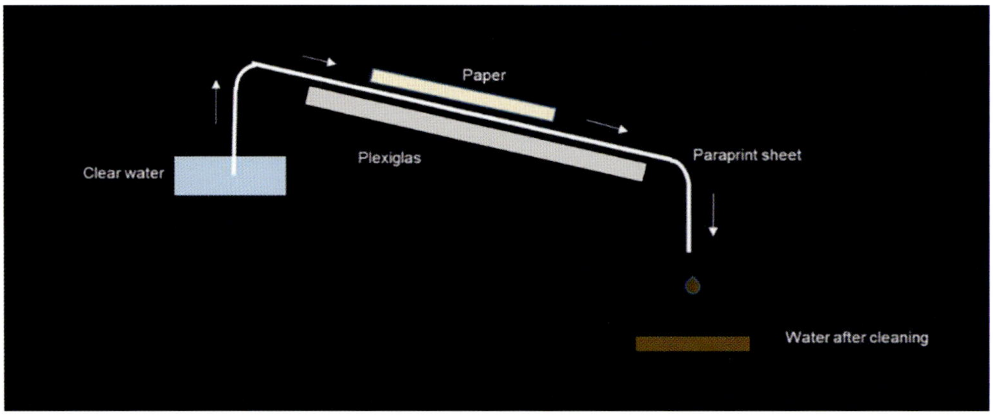

FIG. 2 Schematic of the capillary unit.

TABLE 1 Treatment times selected.

Method	Treatment times
Immersion	30 minutes, 45 minutes, 1 hour
Damp Paraprint	1 hour
Paraprint capillary unit	2 hours
Gellan gum and agar at 2% 3% and 4%	5 minutes, 30 minutes, 1 hour, 2 hours

Experiments on cleaning with water

Cleaning with water aims to extract soluble degradation products formed by hydrolysis or oxidation, and to stabilize the paper chemically. Common washing methods (immersion in water, float washing in a water bath, damp blotter washing and suction table washing) present some potential drawbacks such as sudden water intake into the paper, and the risk of uneven cleaning, as well as numerous manipulations of the wet and therefore weakened paper, which we wished to avoid.

Trials were carried out in order to determine the best method of washing the pages. We conducted a comparative study of washing techniques using Paraprint OL60 and several rigid gels that had been reported in private and public studios, but which had not been extensively studied and whose repercussions for the paper were relatively unknown. They were compared to immersion washing which remains the most widely practised technique in conservation studios in France.

The album pages are a woven machine-made paper, 0.14 mm thick, unstained and slightly grainy. Fibre analyses indicated flax and hemp pulp. The pH, taken on the surface and by aqueous extraction, was approximately 6. The paper had a very low porosity: it was bulk sized with rosin and surface sized with starch. To perform cleaning tests, we located a paper with very similar characteristics in terms of composition, thickness, pH and sizing, and which also included foxing stains. Three cleaning methods were compared:

- Simple immersion wash in a purified water bath, stirring and changing the water regularly to allow diffusion into the fibres;
- The use of Paraprint OL60, a viscose non-woven material;
- The use of rigid gels (gellan gum, agar and agarose) which are often employed as a means of local humidification to remove adhesives.

Paraprint OL60 can be used in a number of different ways. Firstly, to achieve a capillary unit, a method developed at the State Library of South Australia in 1990 and brought to the attention of conservators in 2001 (Kirchner 2001). Two more recent publications (Banik and Brückle 2011; Schalkx *et al.* 2011) have confirmed its effectiveness as a cleaning method. A strip of Paraprint is placed on an inclined Plexiglas plane. One of its ends is placed in a container with purified water, which serves as a reservoir, while the other end hangs above an empty container placed below the filled container that collects water with the degradation products after cleaning. The paper to be treated is placed directly on the strip of Paraprint. By capillarity, a flow is created and the water passes through the Paraprint sheet from the reservoir to the other end, carrying with it the soluble products present in the paper (Fig. 2). The intensity of the flow varies depending on the slope whose angle may be negative, positive or zero according to the different authors.

Paraprint can also be used for paper cleaning in the same way as damp blotters. This involves placing the paper on several superimposed sheets of wet Paraprint. Soluble compounds migrate into the Paraprint by diffusion and capillarity, and gravity furthers such transport. It is a static system, so it is necessary to change the sheets of Paraprint regularly to maintain diffusion. This method is more efficient than the use of blotters because the pores of the former are larger and allow greater diffusion of the degradation products.

A team from the Istituto centrale per il restauro e la conservazione del patrimonio (ICRCPAL) in Rome has carried out research since 2003 on the use of a rigid gel of gellan gum to clean the paper (Iannuccelli and Sotgiu 2010). A piece of gel, cut to the size and shape required, is applied directly on top of, or underneath, the paper. It releases moisture slowly and evenly, and water can penetrate the paper and dissolve the degradation compounds which are then absorbed into the gel by diffusion and capillary action. According to Mazzuca *et*

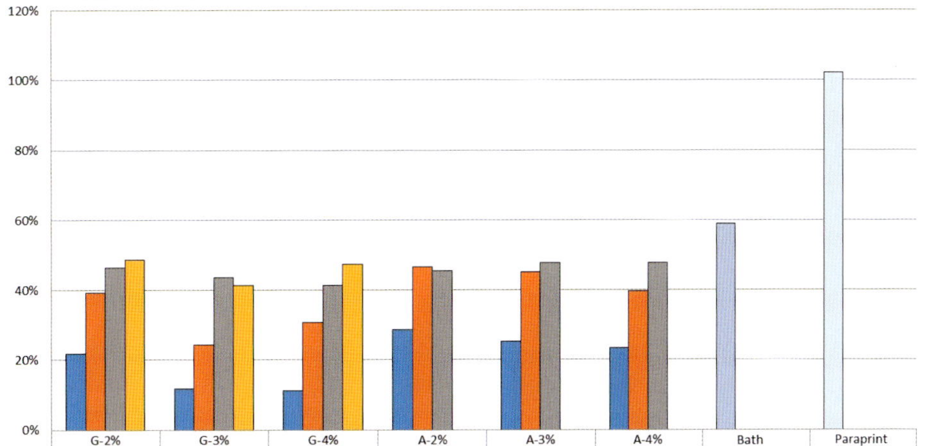

	G-2%	G-3%	G-4%	A-2%	A-3%	A-4%	Bath	Paraprint
5minutes	22%	12%	11%	29%	25%	23%	59%	102%
30 minutes	39%	24%	31%	47%	45%	40%		
1 hour	46%	44%	41%	45%	48%	48%		
2 hours	49%	41%	47%					

FIG. 3 Weight gain of samples treated with different gels.

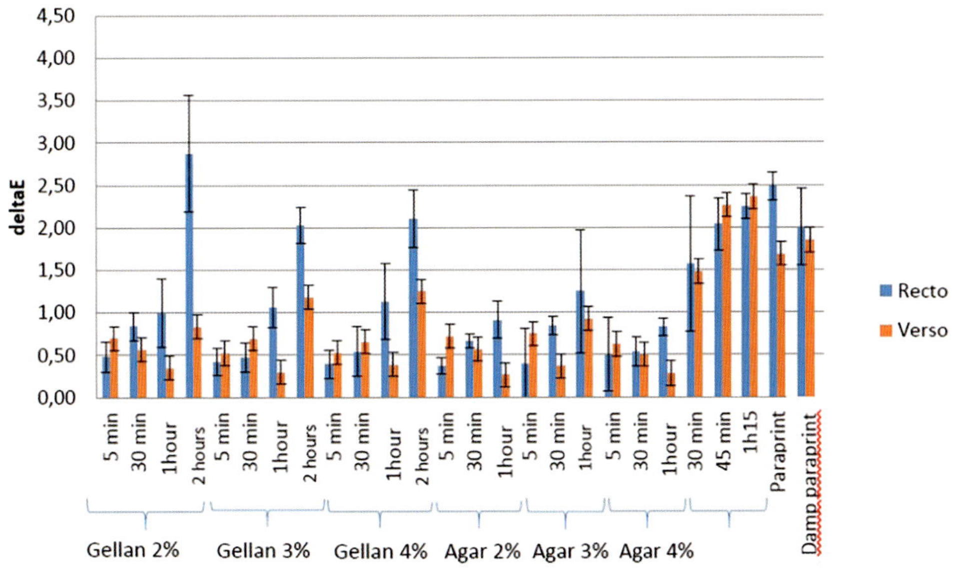

FIG. 4 Colour measurement results.

al. (2014), the method is even more effective than immersion in terms of cleaning. Her team has assessed and compared the amount of degradation products extracted by gellan gum and immersion using liquid chromatography and found greater extraction of pollutants and degradation compounds with the gel. They explain this by the presence of numerous hydroxyl groups which are capable of interacting with acid residues from paper degradation.

Following preliminary tests with agar gels, agarose and gellan gum, and a literature review, we chose to compare gellan gum and agar at three different gel concentrations: 2% 3% 4% w/v. The slabs were placed on top of the paper, followed by a piece of Melinex and a light weight for appropriate treatment times for each gel (Table 1) determined after preliminary testing.

Experiments with treatment methods

Immediately following treatment, the samples were weighed to determine the amount of water introduced into the paper. The comparison bars to the left of Fig. 3 show the percentage weight gain of the samples treated with gels. The release of water was faster with agar than with gellan gum at the same concentration, the latter showing a more progressive release; in both cases water release reached a plateau after an hour of treatment. We therefore chose to focus on gellan gum for further tests. In the case of immersion and Paraprint treatments, the weight gain was equivalent regardless of the processing time: initially very fast and then stable. Samples treated with Paraprint absorb a much greater amount of water than samples treated with other methods. With all gel treatments, even

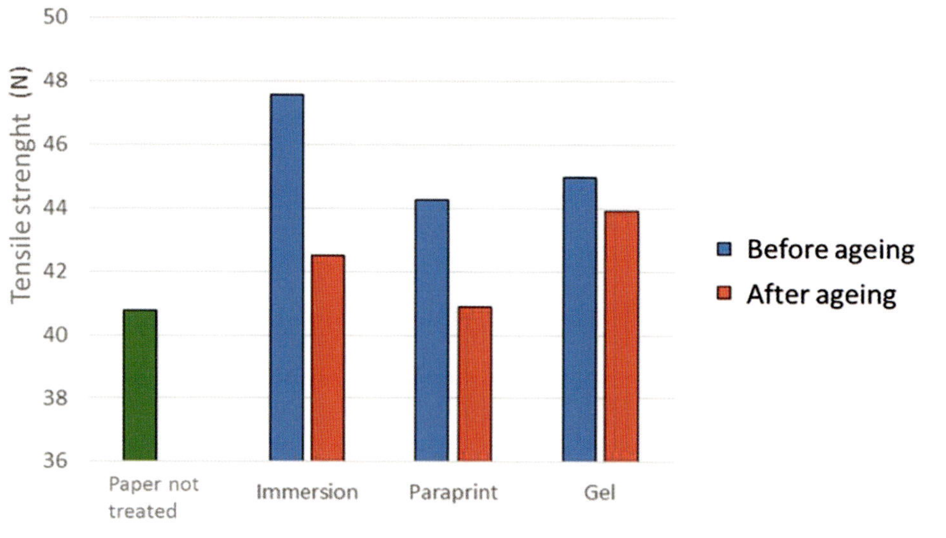

FIG. 5 Ultimate tensile strength for the test paper treated with gellan gum.

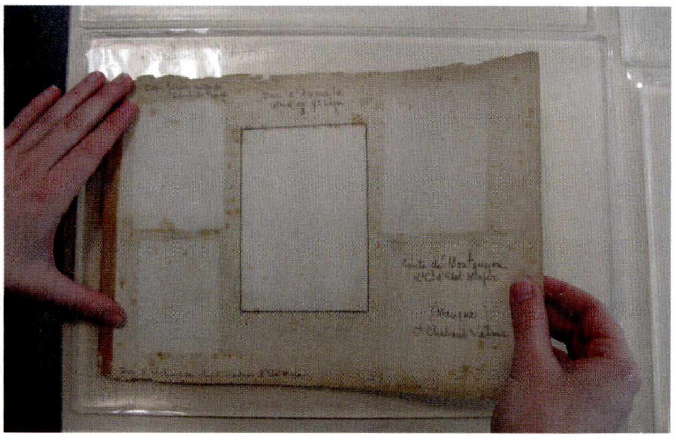

FIG. 6 Placing of paper on the gel slab.

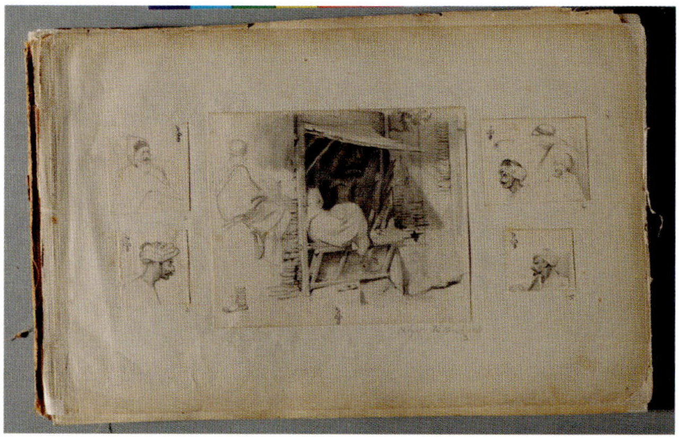

FIG. 8 Page before treatment (© Inp Ghislain Vanneste).

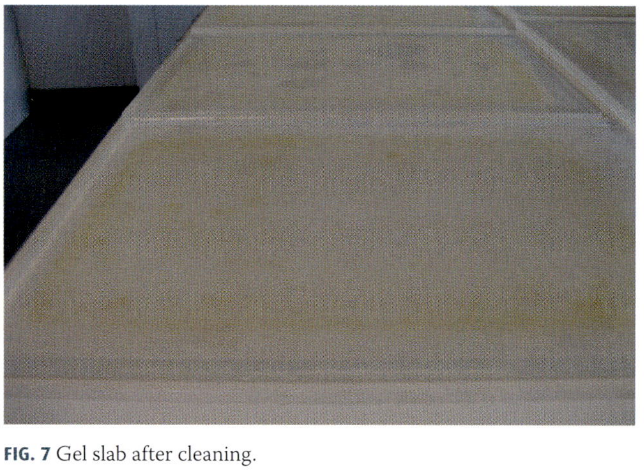

FIG. 7 Gel slab after cleaning.

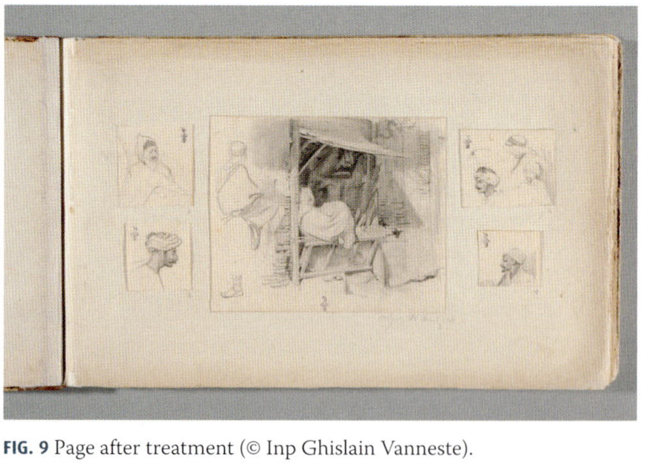

FIG. 9 Page after treatment (© Inp Ghislain Vanneste).

after a long time, the paper absorbs less water than after a short immersion.

Next, a very simple 'water drop' test was performed with a drop of water on the samples to assess the porosity of the paper and check for possible loss of sizing. The paper was heavily sized before treatment: a drop of water applied to the surface did not penetrate for hours in the absence of pressure. After treatment, we found a very strange phenomenon: for the samples treated by immersion or with gels, no change could be seen from the water drop test. On the other hand, all samples treated with Paraprint, both in the capillary unit and with wet Paraprint, exhibited a different behaviour: the drop penetrated into the paper in a few seconds. The paper seemed either to have lost its sizing completely, or at least it was very much reduced. To verify these results, we repeated the treatments on other samples, varying the processing time, but similar results confirmed our initial observations. It seems clear that the Paraprint causes a significant loss of sizing on the paper studied. This loss is also responsible for the high water penetration observed when measuring the weight gain: the paper had become completely porous and waterlogged. The precautionary principle suggested that this type of treatment should not be used for the paper pages of the album, therefore we were left with two choices: immersion or gellan gum. In order to compare them, we decided to conduct pH testing, colorimetry and mechanical testing of the paper. The pH was measured by extraction and the results revealed that all treatments resulted in an equivalent increase of the pH between 0.5 and 1.

Colour measurements taken with a spectrophotometer in several points before and after treatments allowed us to compare the extraction of coloured degradation products in the paper (Fig. 4). They showed that the paper treated with gellan did not exhibit an equal colour change between the side that was in direct contact with the gel, and the other side. However, the colour change was greater than that obtained with an immersion on the side in contact with the gel, but equivalent to that caused by immersion on the other side. At this point, gellan gum seemed to be a good option to treat the paper from the album. We carried out some complementary tests to optimize the treatment, including increasing the time of treatment, and placing the gel slab underneath or above the paper. We concluded that a time of two hours was sufficient, and that results were similar when the gel was applied on top of, or beneath, the paper. To simplify the manipulations of the paper and gain better control of the treatment we chose to place the gel underneath the pages.

We also carried out mechanical testing on paper samples. It is commonly reported that aqueous treatments cause an increase of paper strength; however, some authors contradict this assertion (Carr *et al.* 2006; Moropoulou and Zervos 2003). The paper strength of the treated samples was measured with a dynamometer: half of the samples were tested immediately after the treatments, and the other half following accelerated aging at 80 °C and 65% relative humidity (Fig. 5). All treatments brought about an immediate increase of the tensile strength in the paper. This value is reduced after aging, but less so with the gellan gum treatment.

Treatment of the pages from the album

Our experiments allowed us to rule out the use of Paraprint because the large amount of water intake and loss of sizing were too invasive for the paper type to be treated. In contrast the use of gellan gum seemed to present many advantages: the water intake was lower and more progressive than with the immersion treatment, and there was less swelling of the fibres. In addition, implementation of the gel required less manipulation of the paper, which is particularly advantageous for weak and damaged papers. The gellan gel provided constant control and the treatment could be easily interrupted if necessary, in contrast to immersion. Furthermore, the degradation products were extracted in a directional and homogeneous way, without prior orthogonal diffusion in the paper. We proceeded with the treatment at 3% gellan concentration with the gel slab (Figs 6 and 7) underneath the paper for a two-hour application time. The visual results on the album pages were very satisfying (Figs 8 and 9).

Material and supplier

- Paraprint OL60: GMW - Gabi Kleindorfer, http://gmw-shop.de/

References

Banik, G. and Brückle, I. 2011. *Paper and Water: A Guide for Conservators*. Oxford: Elsevier Butterworth.

Carr, D., Niven, B. and Campbell, L. 2006. 'Effects of selected aqueous treatments on the properties of two papers', *Studies in Conservation* 51: 189–98.

Iannuccelli, S. and Sotgiu, S. 2010. 'Wet treatments of works of art on paper with rigid gellan gels', *The Book and Paper Group Annual* 29: 25–39.

Kirchner, S. 2001. 'Kapillarreinigung. Eine schonende Methode der Feuchtreinigung in der Papierrestaurierung', *Papierrestaurierung* 2 (suppl.): 73–80.

Mazzuca, C., Micheli, L., Carbone, M., Basoli, F., Cervelli, E., Iannuccelli, S., Sotgiu, S. and Palleschi, A. 2014. 'Gellan hydrogel as a powerful tool in paper cleaning process: a detailed study', *Journal of Colloid and Interface Science* 416: 205–11.

Moropoulou, A. and Zervos, S. 2003. 'The immediate impact of aqueous treatments on the strength of paper', *Restaurator* 24: 160–77.

Schalkx, H., Iedema, P., Reissland, B. and van Velzen, B. 2011. 'Aqueous treatment of water-sensitive paper objects – capillary unit, blotter wash or Paraprint wash?', *Journal of Paper Conservation* 12(1): 11–20.

Authors' addresses

- Céline Delattre, Lille, France (celine.delattre.59@gmail.com)
- Stéphane Bouvet, Bibliothèque nationale de France, Scientific Laboratory – Conservation Department , Bussy-Saint-Georges, France
- Emilie Le Bourg, Montpellier, France

Measurement of surface pH of paper using agarose gel plugs: a feasibility study

Amy Hughes

Introduction

The pH value of paper describes the hydrogen ion concentration of paper extracted in purified water. It has long been known that, regardless of storage conditions, acidic paper is more prone to degradation over time and that acid-induced hydrolysis impacts both the optical brightness and mechanical strength of paper (Barrow and Sproull 1959; Banik *et al.* 2012). Therefore, pH is an important marker that conservators use to help interpret the condition of a particular object and it may help to guide treatment decisions.

In the decades since it was first introduced as a diagnostic chemical test for determining the relative permanence of paper, pH has been the subject of energetic study and debate in library, museum, and paper industry laboratories. While pH clearly affects the permanence of paper, it can be difficult to measure accurately. In general, pH measurements commonly suffer from both instrument and user error as outlined in many comparative studies (Joel *et al.* 1972; Kelly 1972; Browning 1977; Saverwyns *et al.* 2002; Strlič *et al.* 2004; Tse 2007). Paper pH can even elude strict definition, as each individual piece of paper possesses its own complex structure and composition of heterogeneous materials (Dwan 1987). Papers, particularly modern papers, are often made with multiple types of fibers, each of which may have separately undergone its own unique chemical and mechanical processing. Prior to sheet formation, these wet fibers are combined with various fillers and treated with internal sizing agents. After formation, coatings and external sizing may or may not be applied to the paper surface. Just to highlight one complicating factor among many, gelatin sizing has been shown to slow degradation processes in acidic papers (Barrett and Mosier 1995). Each of these components and processes may contribute to pH, and it is crucial that conservators bear this in mind when interpreting pH measurements.

There are several methods that may be used to determine the pH of paper but the three methods most commonly used by the paper industry are cold extraction, hot extraction, and surface measurement. In the 1920s and 30s, paper chemists developed methods of pH measurement using pH electrodes to replace time-consuming acid/base titrations (Launer 1939). The Technical Association of the Pulp and Paper Industry (TAPPI), as well as other organizations and bureaus of standards worldwide, developed standard methods for cold and hot extraction of paper after the development of commercial pH meters. Hot and cold extraction methods require large amounts of sample material (up to 2 g depending on the standard). Measurements done in cold water cannot be directly compared with measurements done in hot water. Hot water extract pH values are typically more acidic. Micro-extractions are possible, and correlate well to the standard tests, but even 'micro' samples (~0.01 g) are likely too large when considering individual works of art on paper (Strlič *et al.* 2005).

When sampling is impossible, pH measurements may be taken *in situ*, directly from the surface of a piece of paper. In this type of test, a drop of purified water is deposited on the surface of the paper and after a predetermined amount of time, a flat-head electrode is pressed against the area of wet paper to take the measurement. The surface pH measurement has its own TAPPI standard,[1] which has been found to correlate with cold extraction measurements. However, surface measurements can be inaccurate by as much as an entire pH unit. The discrepancy between surface and cold extraction measurements can be explained by the disproportionate effect of surface sizing and by the heterogeneity of the surface of the paper object. For example, the edges of a naturally aged piece of paper are usually more acidic than the interior of the sheet. Finally, although the surface measurement technique does not require the physical removal of sample material, it still cannot be considered non-destructive. The use of a large water droplet combined with the pressure of the flat-head electrode will often leave a disfiguring tideline or even an impression on the surface of the paper.

In practice, paper conservators working in museums, libraries, and private practice often do not have access to the equipment required for pH measurement of objects under treatment. Electrodes are fragile, expensive, and require a certain degree of maintenance to keep them in working order.

TABLE 1 Comparison of cold extraction, surface, and agarose gel pH measurements for various papers. * indicates pH values and sizing content measured by IPS Testing Services according to TAPPI standards.

	Sample name	Sizing	Cold extraction pH		Surface pH		Agarose pH	
			Mean	Std. Dev.	Mean	Std. Dev.	Mean	Std. Dev.
1	German book paper 1921	Rosin	3.83	0.04	3.48	0.05	4.65	0.29
2	Newspaper 1971	Synthetic (?)	4.28	0.03	4.02	0.05	5.36	0.37
3	Bienfang sketch 1980s	Starch	4.47	0.01	4.41	0.13	5.79	0.15
4	Whatman watercolor 1954	Protein	4.55	0.03	4.68	0.10	5.66	0.12
5	Construction paper (green color) early 20th century	Starch	4.67	0.03	4.20	0.06	4.91	0.21
6	Living book paper 1949	Rosin*	5.10*	n/a	5.36	0.05	5.96	0.10
7	Van Gelder Zonen unicorn 1930s	Protein	5.16	0.00	5.01	0.14	5.35	0.14
8	Album national wove 1895	Starch/Rosin	5.62	0.04	5.88	0.23	6.11	0.09
9	Whatman chromatography 2013	None	5.80*	n/a	4.58	0.51	6.29	0.08
10	Whatman filter paper #1 1920s	None	6.20	0.17	5.97	0.44	6.28	0.12
11	Japanese handmade 1990s	Carbohydrate*	6.40*	n/a	5.46	0.05	6.22	0.08
12	Villedary handmade laid late 18th – early 19th century	Protein	6.58	0.01	6.37	0.12	5.76	0.13
13	BFK Rives 2008	Starch	7.15	0.08	7.07	0.27	6.00	0.10
14	Strathmore 300 2015	Starch/ Synthetic	9.03	0.11	7.86	0.15	6.36	0.04
15	Starwhite printing paper 2013	Starch*	9.20*	n/a	5.82	0.51	6.42	0.04
16	Strathmore 500 2013	Starch*	9.50*	n/a	7.83	0.27	6.46	0.02

So despite the benefits of being able to quantify the acidity or alkalinity of a paper object, pH measurements are not routinely made outside directed research projects.

Recently, conservation scientists Richard Wolbers and Chris Stavroudis, and conservator Daria Keynan have helped to introduce an alternative, more accessible method of pH measurement to paper conservators (Stavroudis 2013; Keynan and Hughes 2013). This agarose gel method was first used to measure the pH and conductivity of sensitive acrylic paint surfaces, and was later adapted to measure the pH and conductivity of paper surfaces. In this method, agarose gel is made into a small 'pellet' or 'plug', using a round punch. The plug is then placed on the surface of the paper object. The gel acts like a sponge, using capillary action to draw soluble materials up and out of the paper. The gel is then removed from the paper and placed into the flat well of a handheld Horiba Laqua Pocket pH Tester along with a drop of water. While the gel plug may leave a small tideline or cleaned area (~4 mm) after it has been lifted from the paper, its trace is typically more discreet than tidelines left after the traditional surface pH measurement (~1 cm). The agarose gel method offers advantages over traditional pH measurement techniques, such as ease of use and minimal damage to the art object. However, the method is as yet critically untested against traditional pH measurement standards.

This paper systematically investigates the relationship between three methods of pH measurement for paper: TAPPI cold extraction, TAPPI surface measurement, and the newly introduced agarose gel method. The three techniques are compared using measurements taken from samples of 16

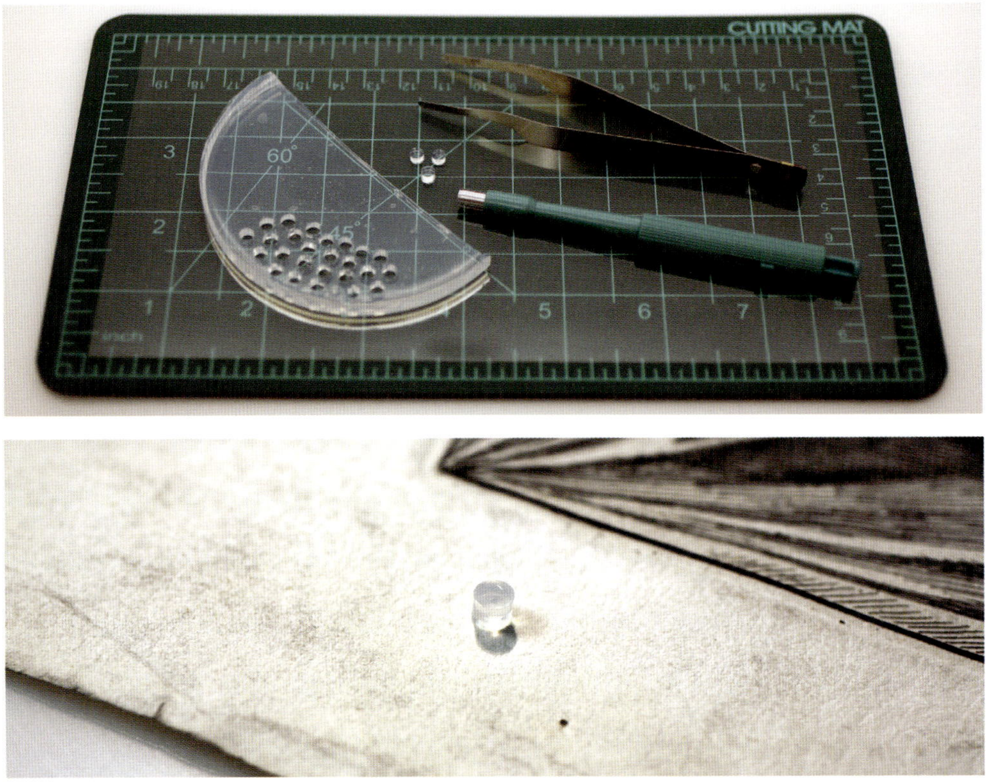

FIG. 1 (*Top*) Agarose gel sheet and punched plugs with Miltex biopsy punch and tweezers; (*bottom*) taking the reading with a 4 mm agarose plug on the surface of a print.

contemporary and historic papers (Table 1). The papers were selected as representative of a broad range of artists' papers available from the 18th through the 21st century. The paper samples were chosen based on availability as well as predicted fiber content and type of sizing. Chemical spot tests were performed on each paper to determine, as much as possible, the type of sizing present, as sizing disproportionately affects surface pH readings.

Experimental

TAPPI cold extraction[2] and TAPPI surface measurements[3] were carried out following these guidelines, with minor modifications as described below. For cold extraction, 1 g of each paper was weighed, cut into small squares, and placed in a glass beaker. A small portion of deionized water (5 ml) was added to each beaker. Samples were then partially macerated and stirred with a glass rod until thoroughly wetted. Deionized water (65 ml) was added to the beaker to bring the total volume of water to 70 ml. The beakers were covered and left undisturbed for two hours, after which time the extract was filtered through plastic mesh and the paper sample discarded. The pH of the extract was measured using a Thermo Orion Model 525Aplus dual channel pH/mV/temperature meter fitted with a Cole-Parmer flat, single junction sealed epoxy electrode. The meter was calibrated against two standard buffer solutions (7.0 and 4.0) and the temperature was manually calibrated. The pH of each sample was measured three times and averaged (Table 1).

For surface measurement, the paper sample was laid down on a thin Teflon film, which was then placed on top of a 1 cm square piece of wool felt. This support provided both flexibility and water resistance. A droplet of deionized water was applied using a glass pipette and left to stand on the surface of the paper for five minutes or until the droplet was completely absorbed. The head of the electrode was then held in contact with the droplet, unless the droplet had been completely absorbed, as occurred with the unsized papers and the degraded papers, in which case the electrode had to be pressed into the moistened area of the paper. The same pH meter and electrode were calibrated as described above. Surface measurements were gathered from three discrete points and the readings were averaged (Table 1).

For agarose gel measurements, the paper sample was placed on top of a sheet of clear polyester film. Using a Miltex biopsy punch, plugs of agarose were punched from a sheet of 4% agarose handled with tweezers (Fig. 1). The cast sheet of 4% agarose had previously been 'rinsed' by immersing the sheet in a large volume of deionized water overnight. The rinsing step is necessary to clear the agarose of residual chemicals introduced during industrial processing. The conductivity of the rinsed agarose measured 33 µS, and the pH measured 6.5. The individual plugs used for taking the pH readings measured 4 mm in diameter and 3 mm in thickness.

For each agarose gel measurement, a plug was placed firmly on the surface of the paper and allowed to rest (Fig. 1). After five minutes, the plug was removed and placed into the well of a Horiba Laqua Twin pH Meter (Model B-713) calibrated against two standard buffers (7.0 and 4.0). One droplet of deionized water (~0.05 ml) was placed inside the well of the meter, directly on top of the plug, and three minutes were allowed to pass before acquiring the pH measurement. During this time, the ionic contents of the plug diffused into the surrounding water droplet. Three minutes allowed sufficient time for absorbed material to release and for the diffusion to reach equilibrium.

The configuration of the Horiba meter's flat sensor is such that to obtain a reading, sufficient moisture must bridge the liquid junction and the reference electrode (Fig. 2). Ions must be able to flow from one part of the electrode to the other, maintaining electrical contact. For this reason, a glass stirring rod was used to draw the droplet across the well, forming a connection between the small grey sensor and the larger black sensor. Adding a second drop of water is not advised, as it will dilute the sample and diminish the accuracy of the pH reading. It is easier to create the bridge using a single droplet if the agarose plug is positioned in the center of the well, resting between the two parts of the sensor. After creating the bridge of water and closing the lid of the Horiba meter to protect the sensor from the influence of light, the pH measurement was recorded. Agarose gel measurements were gathered from three discrete points on each paper sample and the three readings were averaged (Table 1).

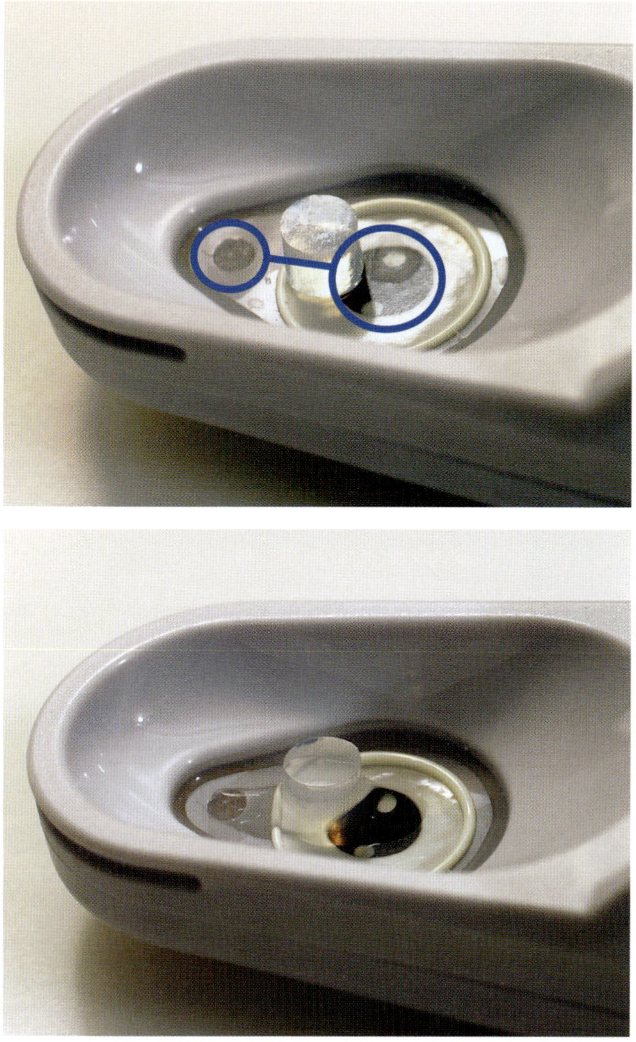

FIG. 2 The well of a Horiba Laqua pH meter with an agarose plug inside. (*Top*) Blue annotations indicate the sensors to be bridged with water. The liquid junction is on the left and the reference electrode is on the right. (*Bottom*) Agarose plug with a single water droplet connecting the sensors.

Discussion

This study confirmed that traditional surface pH measurements exhibit fairly close numerical correlation with cold extraction. Agarose gel measurements also correlate with cold extraction and with surface pH measurements, but this correlation is most meaningful between the pH values of 4.5 and 6.5. Outside this somewhat narrow range, measured pH values gathered with agarose gel do not correspond well with either type of TAPPI measurement. This study indicates that agarose gel has difficulty measuring both acidic and moderately alkaline papers accurately (Table 2). Readings of acidic papers taken with agarose gel are too high (too alkaline by 1–1.5 pH units), while readings from alkaline papers are too low (too acidic by 1–3 pH units). The gel performs somewhat more reliably with acidic papers than alkaline papers, an observation that is consistent with other researchers' reports on the difficulty of measuring the pH of alkaline papers (Strlič *et al.* 2004).

In an attempt to improve agreement between pH measurements taken using the agarose gel method and industry standard surface and cold extraction methods, the procedure of the agarose gel method was experimentally modified and tested on one alkaline paper (Strathmore 300) and one acidic paper (German book). Several potential variables of the agarose gel method were altered, one at a time. The thickness of the gel plug was varied between 1 mm and 5 mm, however, the thickness of the plug proved to have no significant effect on the pH readings of the alkaline or the acidic paper. The diameter of the plug was tested in 0.5 mm increments starting with 2.5 mm through 4.0 mm. Similarly, the diameter of the plug proved insignificant to the pH readings of both papers.

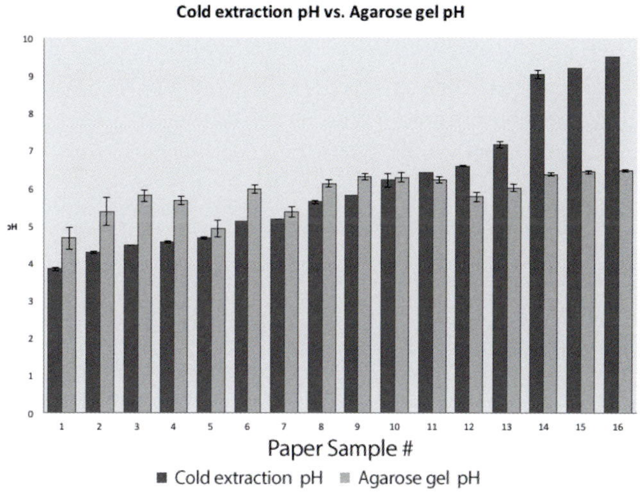

TABLE 2 Comparison of cold extraction pH measurements to agarose gel pH measurements across 16 paper samples.

Next, different concentrations of agarose gel were tested, ranging from 2% through 5%. The dwell time of the plugs on the surface of the paper samples was incrementally increased

TABLE 3 Effects of agarose gel concentration and dwell time on pH measurements. Measurements shown here were gathered from sample #1 (German book paper), an acidic paper with a pH of 3.83 when determined by cold extraction.

		Dwell time			
		5 min	10 min	15 min	20 min
Concentration	2%	4.67	4.46	4.42	4.50
	3%	4.72	4.53	4.68	4.46
	4%	4.65	4.49	4.48	4.30
	5%	4.64	4.70	4.70	4.55

from five minutes up to 20 minutes for each concentration. Table 3 illustrates that neither the concentration nor the dwell time had much effect on the outcome of pH readings, regardless of the acidity level of the paper. It is surprising that the pH does not change significantly over the course of 20 minutes with an acidic, rosin-sized wood pulp paper, considering that the gel continues to imbibe colored degradation products from the paper over the course of that time. The plugs left on the paper surface for 20 minutes were darker in color than the gels left on the paper surface for 5 minutes, but the pH of the plugs did not change significantly.

In practice, a pH value acquired with an agarose plug will provide an accurate indication of the pH of a given piece of paper, unless that paper is very acidic (under 4.5) or neutral to moderately alkaline (7.0 and above). Based on the data gathered during this feasibility study, it is unlikely that the agarose gel method will give acceptable pH measurements outside the 4.5–6.5 pH range. The majority of works of art on paper in public and private collections, and at least those made prior to the 19th century, are apt to be on paper supports with pH in the range measurable with agarose gel. For these works, the agarose gel method is quite useful. The method requires minimal equipment, is relatively simple to execute, and can be a powerful tool when used in tandem with a conductivity meter to measure and record conductivity values. Furthermore, agarose gel is minimally invasive; the small plug is less likely to create a tideline or to locally change the texture of the paper during measurement. For paper conservators, pH is a useful and accessible diagnostic value, which can help to guide preservation strategies and conservation treatment decisions. The agarose gel method is a welcome alternative to other methods of pH measurement currently in use.

Materials and suppliers

- Agarose LE (#A1700): Benchmark Scientific, Inc., P.O Box 709, Edison, NJ 08818, USA, http://www.benchmarkscientific.com/Agarose.html
- Horiba Laqua Pocket Tester, pH (#UX-05754-10): Cole-Parmer, 625 East Bunker, Court Vernon Hills, IL 60061, USA, http://www.coleparmer.com/Product/Horiba_Laqua_Pocket_Tester_pH_Range_2_to_14_pH/UX-05754-10
- Miltex Biopsy Punch with Plunger, 4.0mm (#15110-40): Ted Pella, Inc., PO Box 492477, Redding, CA 96049, USA, http://www.tedpella.com/histo_html/miltex-plunger-punch.htm/

Notes

1. TAPPI T 529 Surface pH measurement of paper.
2. TAPPI T 509 om-15 Hydrogen ion concentration (pH) of paper extracts (cold extraction method).
3. See note 1.

References

Banik, G., Brückle, I. and Daniels, V. 2011. *Paper and Water: A Guide for Conservators*. Amsterdam: Butterworth-Heinemann.

Barrett, T. and Mosier, C. 1995. 'The role of gelatin in paper permanence', *Journal of the American Institute for Conservation* 34(3): 173–86.

Barrow, W.J. and Sproull, R.C. 1959. 'Permanence in book papers', *Science* 129(3356): 1075–84.

Browning, B.L. 1977. *Analysis of Paper*. New York: M. Dekker.

Dwan, A. 1987. 'Paper complexity and the interpretation of conservation research', *Journal of the American Institute for Conservation* 26: 1–18.

Joel, A., Indictor, N., Hanlan, J. and Baer, N. 1972. 'The measurement and significance of pH in paper conservation', *Bulletin of the American Group, International Institute for Conservation of Historic and Artistic Works* 12(2): 119–25.

Kelly Jnr., G.B. 1972. 'Practical aspects of deacidification', *Bulletin of the American Institute for Conservation* 13(1): 16–28.

Keynan, D. and Hughes, A. 2013. 'Testing the waters: new technical applications for the cleaning of acrylic paint films and paper supports', *The Book and Paper Group Annual* 32: 43–51.

Launer, H.F. 1939. *Determination of the pH Value of Papers*. US Government Printing Office.

Saverwyns, S., Sizaire, V. and Wouters, J. 2002. 'The acidity of paper: evaluation of methods to measure the pH of paper samples', in R. Vontobel (ed.), *Preprints of the ICOM-CC 13th Triennial Meeting*. London: James & James, 628–34.

Stavroudis, C. 2013. 'Measuring surface pH and conductivity using water drop and agarose plug methods'. Los Angeles: Getty Conservation Institute. Available at: https://youtube/bOqZEE7Kb8Y/ (accessed 9 January 2017).

Strlič, M., Kolar, J., Kočar, D., Drnovšek, T., Šelih, V.-S., Susič, R. and Pihlar, B. 2004. 'What is the pH of alkaline paper?', *e-Preservation Science* 1: 35–47.

Strlič, M., Pihlar, B., Mauko, L., Kolar, J., Hočevar, S. and Ogorevc, B. 2005. 'A new electrode for micro-determination of paper pH', *Restaurator* 26: 159–71.

Tse, S. 2007. *Guidelines for pH Measurement in Conservation*. Ottawa: Canadian Conservation Institute.

Author's address

Amy Hughes, National Gallery of Art, Paper Conservation Department, Landover, MD, USA (a-hughes@nga.gov)

Cold, warm, warmer: use of precision heat transfer in the optimization of hydrolytic enzyme and hydrogel cleaning systems

Tomas Markevičius, Terje Syversen, Emma Chan, Nina Olsson, Carola Skov Hilby and Rytė Šimaitė

Introduction

In the past decade, conservators have devised multiple methods for the confinement of water, enzymes, chelating agents and detergents for more efficient control over the cleaning process, and reduced impact of water on the artwork. Enzymes suspended in solid gels allow for highly selective cleaning due to their characteristic active site, which 'locks' only to specific substrate molecules. By putting the substrate into a short-lived transitional substrate/enzyme state ES (equation 1) and straining intra-molecular bonds, the enzymes permit reactions under mild conditions.[1] In optimal conditions where the enzymes and substrate are dispersed in water medium the reaction follows Michaelis-Menten dynamics (equation 2) and the reaction rate υ is limited by the amount of the substrate only:

$$E + S \rightleftharpoons ES \longmapsto E + S \rightleftharpoons ES \longmapsto \mathrm{E} + \mathrm{P} \quad (1)$$

$$\upsilon = \frac{V_{\max}[S]}{K_{\mathrm{M}} + [S]}\upsilon = \frac{V_{\max}[S]}{K_{\mathrm{M}} + [S]} \quad (2)$$

where E = enzyme, S = substrate, ES = enzyme/substrate complex, P = product, V_{max} = the maximum rate of the system at (maximum) saturating substrate concentrations, K_M = Michaelis constant, substrate concentration at which the reaction rate is 50% of V_{max}, [S] is the concentration of the substrate S.

However, in the real conditions of a cleaning treatment, which aims to remove undesirable organic materials (glue, starch, oil, mould etc.) the enzymes must react with a solid phase substrate, which includes pigments, hard particulates, dust, soot, metal ions and other materials present on the surface beneath. Such actual treatment conditions are very inefficient, since enzymes, being fairly large molecules (with dimensions over 200 nm), do not diffuse well when placed on complex interfaces. In these challenging conditions, enzyme dynamics are limited by diffusion, convection and enzyme kinetics.

Diffusion, enzyme kinetics and temperature

The performance of enzymes and hydrogels can be optimized by considering several factors: the concentration of the hydrogel, enzyme concentration, pH, conductivity, cofactors (Ca^{2+}, Mg^{2+}, Mn^{2+} ions), chelators, detergents, etc. (Wolbers 2000; Campani *et al.* 2007; Stavroudis and Doherty 2013). Further optimization can be achieved by accurately controlling the temperature, which has a direct impact on the diffusion, reaction rate, capillary action and enzyme kinetics, but has not been exploited in the past due to the absence of sufficiently precise heat transfer instrumentation. A rise in temperature increases the speed at which molecules move, resulting in more frequent and more energetic collisions. Molecular vibration increases with temperature, stretching and bending the intra-molecular bonds, hence enzyme activity increases rapidly, ascending to the optimum temperature (expressed as a curve of activity and temperature) where the enzyme performs at its best. Having surpassed the peak, the enzymatic activity drops rapidly as the vibrational forces denature the shape of the active site. For example, at room temperature (21 °C) the protease Alcalase 2.4L is only 15% active, yet achieves 100% activity at 60 °C.[2]

Yet another important factor in the hydrogel cleaning mechanism is the transport of the active content (water,

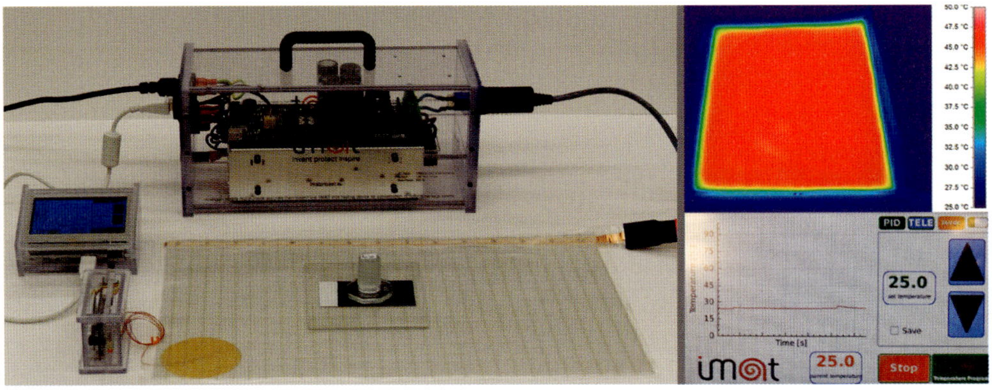

FIG. 1 IMAT precision heat transfer system (*left*), IR thermography showing even heat distribution over the surface (*top right*), and touch screen (*bottom right*) showing a steady heat transfer pattern.

enzymes, ions), which consists of movement of the fluid and within it. The former is determined by the fluid's physical properties, by the physical-chemical properties of the gel and of the substrate. Movement within the fluid depends on thermal motion and, at increased temperature, the diffusion is faster and more even. Hydrodynamic transport of globular enzyme molecules through any porous media could be described by the derivative Stokes-Einstein equation (equation 3), where diffusivity is directly proportional to temperature and inversely proportional to the frictional drag ff, which is a resistance force between the enzyme-carrying fluid and solid-state substrate (Amsden 1998):

$$Do = \frac{kB\ T}{f} Do = \frac{kB\ T}{f} \qquad (3)$$

where Do = diffusivity at infinite dilution; kB = Boltzmann's constant, T = temperature, f = frictional drag coefficient.

From this, it follows that increased adsorption is expected due to an accelerated diffusivity, and that the amount of adsorbed enzymes increases at elevated temperatures (Koutsoukos *et al.* 1983.

New accurate heat transfer method using conductive nanomaterials: IMAT heater

Most heating instruments employed by conservators lack accurate control and even heat distribution across the surface (Markevicius *et al.* 2013a). To control the temperature of hydrogels and enzymes, an innovative flexible transparent heat transfer mat (prototype) based on e-textile containing conductive nanomaterials and developed in the European IMAT project (2011–2014) was used.[3] The IMAT prototype mats employ the thermo-electrical properties of carbon nanotubes and other nanoscale materials to reach up to 85 °C at 36 V with an accuracy of 0.1 °C. The IMAT heat transfer system is composed of diverse lightweight, flexible, thin-profile transparent or breathable heating mats, and an associated main control unit with a touch-screen control console and a wireless temperature sensor (Fig. 1). The IMAT heater has woven-in parallel silver-coated yarns (electrodes) and when voltage is applied, the current is uniformly distributed over the layer of conductive nanomaterials, and heat is instantly generated evenly over the entire surface. If compared to conventional resistive heating elements (wound wire, etched foil) nanotubes and other conductive nanomaterials have an exceptionally low thermal mass, which allows for a rapid thermal response, an important factor to achieve even and steady heat transfer. The IMAT controller analyses the data coming from the wireless T-type thermocouple 40 times per second via a Bluetooth TC board, and regulates the temperature using the classical approach of PID-PWM regulation (proportional integrate derivative pulse width modulation), which permits ultra-steady heating with an accuracy of 0.1 °C. The thin and flexible IMAT profile allows it to be placed where the heat transfer is needed, and it can conform to diverse surface geometries (Markevicius *et al.* 2010, 2011; 2013a).

Coupling enzymatic hydrogels and heat transfer

To verify the feasibility of temperature-based optimization, three hydrogels were tested, with and without the hydrolytic enzymes, on mock-ups with rabbit skin glue and wheat starch substrates, followed by actual treatments.

Mock-ups

Two types of samples with rabbit skin glue and wheat starch were prepared. For the α-amylase gels, strips of Munktell filter paper (batch 1678) and Canson drawing paper (123 g/m²) measuring 5 × 10 cm were sized on both sides with 20% GMW wheat starch. In addition, samples of a 19th-century handmade paper from the Strandmøllen paper mill, Denmark, were used. The samples were coated with 7.5% Fagron Tritici amylum wheat starch and artificially aged according to ISO 5630-3 (14 days at 80 °C, 65% RH in an ESPEC LHU-113 climatic chamber).

Staining

An iodine test was selected to render the cleaning results visible, and was tested on the same paper sheet with and without the starch sizing. For protease testing, 5 × 10 cm strips of commercially primed off-white linen canvas coated with 20% rabbit skin glue (Kremer) with carbon black were used. The samples were examined visually and photographed under normal conditions.

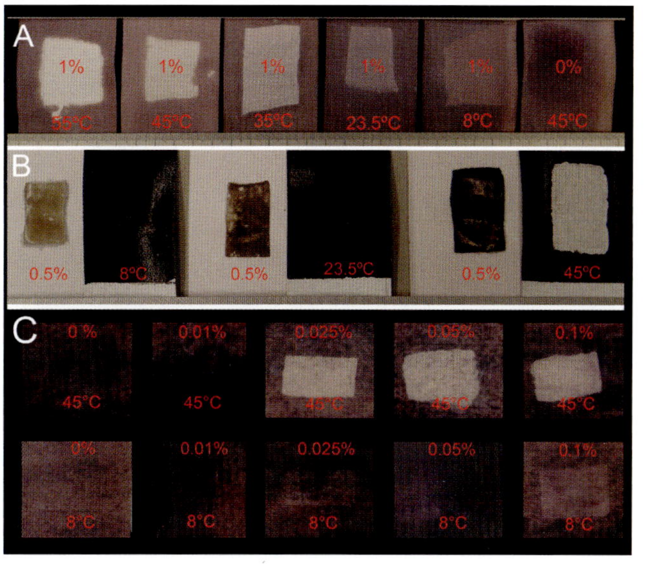

FIG. 2 (A) Ten-minute cleaning treatments at different temperatures using a 1.5 wt% agarose hydrogel containing 1% α-amylase Amplify 12L, or (B) 0.5% Alcalase 2.4L. (C) Agarose (1.5 wt%) gel containing 0–0.1% α-amylase BAN 480 at 8 °C and 45 °C after 5-minute treatments.

Enzymatic hydrogels

Agarose and gellan hydrogels at 1.5 wt% containing α-amylase Amplify 12L,[4] BAN 480L[5] and protease Alcalase 2.4L[6] were prepared. Hydrogels were prepared with 0%, 0.25%, 0.5% and 1% (v/v of stock solution) of protease Alcalase 2.4L (buffered at pH 7, with conductivity adjusted to 2000 μS) and α-amylase Amplify 12L (buffered at pH 6, with conductivity adjusted to 2000 μS).

Testing

An experimental treatment was carried out by placing the enzymatic gel on a mock-up with starch/glue substrate. The samples were treated at: 4 °C, 8 °C, 12 °C, 23.5 °C, 35 °C, 45 °C and 55 °C for 2 , 6, 10 or 24 minutes. To render the result visible, paper samples with starch were rinsed with iodine solution, which stained the starch a purple-blue color, whereas the areas where the enzymes hydrolysed the starch resulted in a lighter stain. The samples coated with the rabbit skin glue containing carbon black pigment were treated following the above procedure. After exposure, the liquid phase residues of hydrolysed glue substrate were absorbed with filter paper and the result could be observed directly.

Amplify 12L and BAN 480L (α-amylase) hydrogel formulations

Both hydrogels at 1.5 wt% without the enzymes swelled the substrate without any identifiable cleaning action at all temperatures tested. In samples containing enzymes no cleaning action was observed at 4 °C, although cleaning action was noticed at 8 °C with 0.5% and 1% α-amylase. At 23.5 °C considerably more starch material was hydrolysed, and the reaction rate increased further at 45–55 °C. It was also observed that the increase of temperature was linearly proportional to the degree of penetration of the cleaning solution into the paper substrate, which can be explained by the increased diffusion both within the gel and paper support. In the test sample, only the front of the paper was cleaned at 8 °C, while at 45–50 °C both sides of the paper were cleaned. BAN 480L at a low concentration (0.025%) at 8 °C showed little to no cleaning action, but iodine staining results revealed that considerably more starch substrate was hydrolysed at 45 °C, hence the reaction rate increased substantially. Further increase in enzyme concentration did not improve efficacy, since the specific activity of enzymes at lower concentration was already sufficient to catalyse the given amount of starch substrate in the sample, and the enzymatic activity did not increase any further following the Michaelis-Menten dynamics.

Alcalase 2.4L hydrogel formulation

Agarose and gellan gels without enzymes produced swelling of the glue substrate, but did not engage in visible cleaning action. Minor cleaning action was recorded at 6 °C and 10 °C. At 23.5 °C the amount of glue material absorbed by the gel increased, and some thinning of the glue substrate was noticed on the sample. At 45 °C, the glue was completely removed (Fig. 2B).

Diffusion and capillary force

The effects of temperature on the diffusivity and capillary action were tested on soiled, off-white, primed deaccessioned canvas from pre-1940, from the Munch Museum, Oslo, Norway. Agarose at 3 wt% and 6 wt% containing Richard Wolbers' iso-solution D[7] were applied in the liquid state at 70 °C, 50 °C, 40 °C and 35 °C for 1 minute. Based on visual observations, the results were twofold: at a higher application temperature, liquid hydrogel diffused further into the paint layer. However, the cleaning of surface soiling increased as the application temperature decreased and the capillary force increased, because of the decrease of agarose pore size.

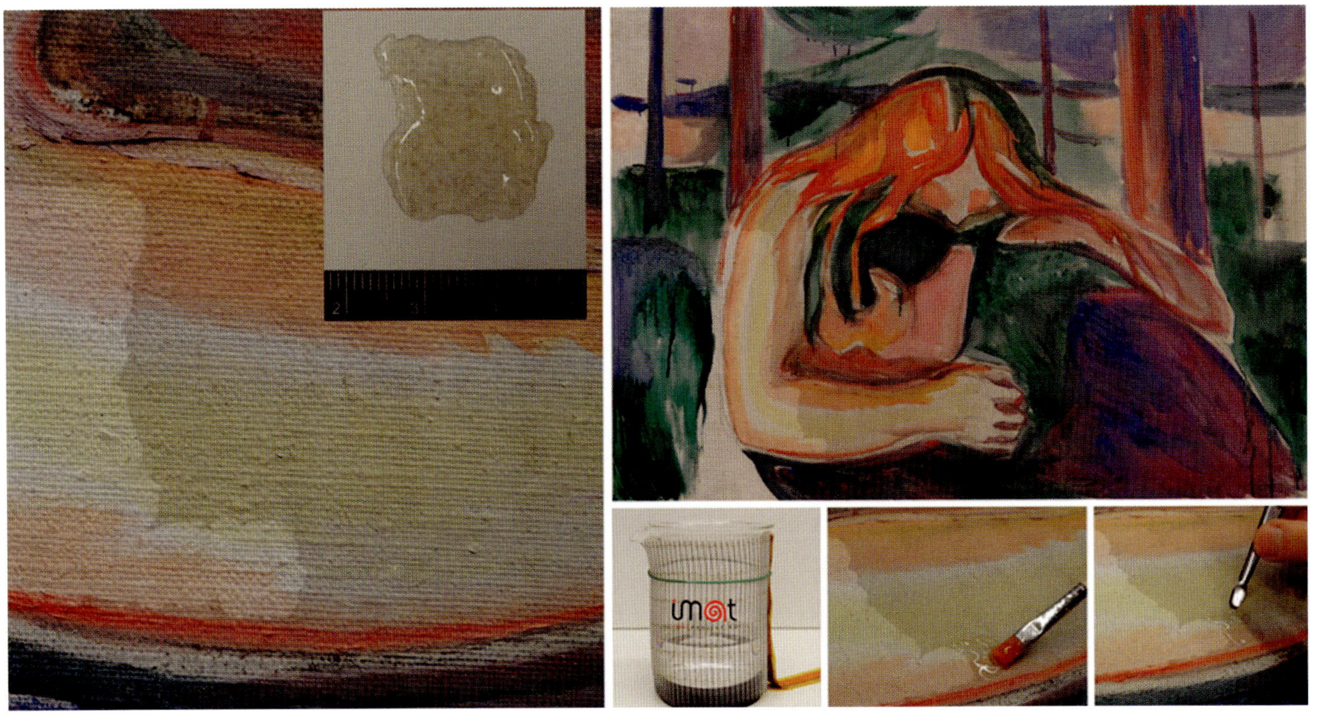

FIG. 3 Edvard Munch, *Vampire*, 1916–18, oil on canvas, 82 × 110 cm, Munch Museum, Oslo, catalogue raisonné no. M1173, and details with surface cleaning in progress. (*Lower centre*): beaker with IMAT sleeve to maintain constant temperature of the liquid-phase agarose gel.

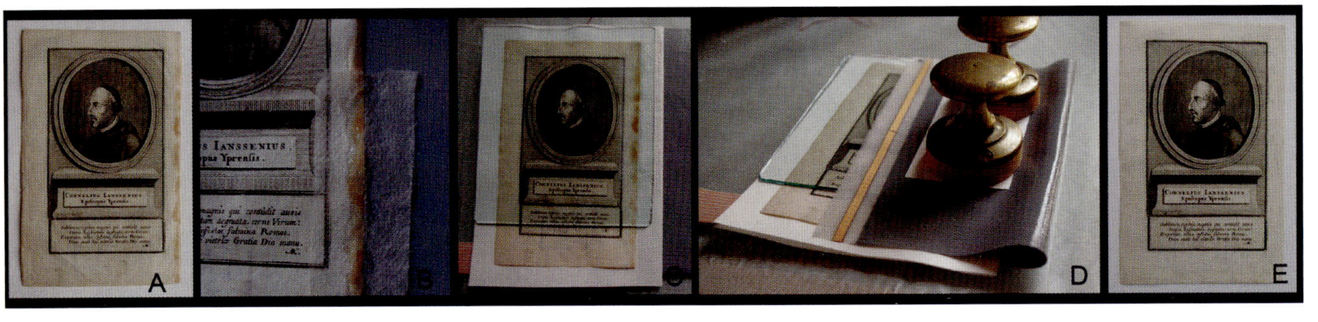

FIG. 4 Gillian van der Gouwen, *Portrait of Cornelius Jansenius*, *c.*1670–*c.*1740, copper engraving, print on paper, 18.8 × 11.2 cm, inv. no. G-11040, Lithuanian Art Museum, Vilnius, Lithuania: showing removal of glue stains using the mild heat transfer with IMAT and a viscous enzymatic gel.

Surface cleaning of an early 20th-century painting by Edward Munch

Temperature-based capillary action of agarose hydrogel (5 wt%) containing Wolbers' iso-solution D with chelator (DTPA) was used to remove a layer of greyish surface dirt from a fragile unvarnished surface on Edvard Munch's *Vampire* (1916–18) (Fig. 3). Agarose gel, when applied as a solid-phase block, produced somewhat incomplete cleaning action due to the lack of contact with the porous surface. A different approach was tested and the agarose was applied in a liquid phase at 39 °C, which changed its phase to solid and after 1 minute, lifted as a film with the entrapped soiling, revealing the intact surface. Precisely controlled temperature was very efficient in controlling the diffusion and capillary action of the gel tailored for this specific case and allowed agarose to remain in a liquid phase during the treatment. Control of the exposure time and temperature was crucial, as minor temperature variations affected the cleaning action and resulted in uneven cleaning. To keep the agarose at a 39 °C (±0.1 °C), the IMAT heater was wrapped around the beaker (Fig. 3).

Removing aged glue stains from a Dutch 17th-century print by Gillian van der Gouwen

Since 2013, the IMAT has been used for diverse enzymatic treatments (e.g. removal of animal glue, starch and oil) of over 80 works on paper at the Lithuanian Art Museum, Vilnius, Lithuania. As an example, aged glue residues were removed from a 17th-century print. Alcalase 2.5L DX at 0.1% v/v was used in 5% Tylose MH300. The gel was applied to both sides of the print, which was covered with polypropylene film, and IMAT mats were positioned on both sides. While at room temperature (19 °C) the enzymes had little effect on the aged proteinaceous stains, treatment at 45 °C removed the stains completely (Fig. 4).

Conclusion

Conventionally, cleaning treatments are carried out in uncontrolled and variable temperature conditions. The transfer of energy to the system determines the liquid/solid phase of the materials, their dynamics, reaction rate, behaviour and the performance of the substance. Thus, heat transfer is a powerful tool in all conservation treatments, yet it cannot be fully exploited without accurate control. In the gel cleaning mechanism, precision heat transfer permits control of the diffusivity, capillary action, reaction rate and enzyme dynamics. In this preliminary study, the precision mild heat transfer system (IMAT) and flexible nanotube mats were shown to establish and regulate temperature conditions accurately; the results show the feasibility and potential of a temperature-controlled approach to optimize cleaning treatments using hydrogels and hydrolytic enzymes.

Initial testing shows that the IMAT provides precision temperature control and that modulating the temperature factor in cleaning treatments is feasible. Research focusing on accurate quantification of enzyme dynamics using fluorescent staining, including other enzymes and modelling methods of application is ongoing. The experimental cleaning tests demonstrate that when greater diffusivity of the liquid phase and enzymes is required, the temperature may be increased to enhance diffusion. Inversely, cooling the surface may reduce the diffusivity, which may be useful when the goal is to limit the cleaning action to the surface only. Capillary force may be controlled by both the concentration of the hydrogel and variation of the temperature. Enzymes are temperature-specific and they catalyse most efficiently in an optimal temperature range, which is frequently above room temperature, and the temperature factor is difficult to engage without precision heat transfer instrumentation. The described temperature-based approach, when coupled with other factors such as concentration of the hydrogel and enzyme, concentration and specific activity (enzyme units U/g, Anson units AU/g, etc.), pH, conductivity, cofactors, chelators, detergents, etc., offers valuable advantages in controlling the dynamics of gels and enzyme kinetics, improving the efficacy of gel cleaning mechanisms to be tailored to specific situations, and contributes to the increased safety of a cleaning treatment for the artwork and better selectivity and nuanced cleaning action.

Acknowledgements

The authors are most grateful to Prof. Richard Wolbers, for his advice and insights during his inspiring workshop at the Munch Museum. We are also grateful for the valuable review by Dr Pedro Gaspar, Head of Conservation and Dr Irina Sandu, Conservation Scientist at Munchmuseet. Special thanks are extended to Dr Julien Coquilat, chemist at Novozymes, for his valuable consultations and advice, as well as Novozymes for the test samples.

Materials and suppliers

- Alcalase 2.4L, DPTA, triethanolamine: www.sigmaaaldrich.com
- GMW wheat starch: GMW, gmw-shop.de
- Amplify 12L, BAN 480 L: Novozymes, www.novozymes.com
- ESPEC LHU-113 climatic chamber: ESPEC, www.espec.com
- Tylose MH300, rabbit skin glue, carbon black: Kremer Pigmente, www.kremer-pigmente.com/de

Notes

1. The classical Michaelis-Menten model includes three principal phases: enzyme [E] transports the substrate molecule [S] and binding to the substrate molecule with the active site (equation 1), forming a temporary enzyme-substrate complex [ES] (equation 2), breaking the intra-molecular bonds and releasing the free reusable enzyme [E] and the product of the reaction [P] (equation 3).
2. The optimum temperature curve for Alcalase 2.4L was provided by Novozymes. Assay prepared using Konelab 30 analyzer (Fisher Scientific).
3. IMAT: prototype/IMAT research project (2011–2014), European Commission's 7th Framework Program for Scientific Research/Coordinated by the University of Florence, Italy.
4. Amplify 12L α-amylase from *Bacillus lichenoformis* (EC: 232-565-6; CAS 900-90-2), hydrolyses (1.4)-alpha-D-glucosidic linkages in starch efficiently in the range 10–50 °Cat pH 6–9, with specific activity 12 SNU-A/g (standard novo units).
5. BAN 480L α-amylase from *Bacillus amyloliquefaciens* (EC 3.2.1.1; CAS 9000-90-2). Optimum pH of 6, temperature optimum 70 °C. Efficient at 10–90 °C.
6. Alcalase 2.4L is a proteolytic enzyme from *Bacillus licheniformis* (EC Number 232-752-2; CAS 9014-01-1). The main enzyme component, Subtilisin A, is an endoproteinase with a specific activity of 2.4 AU/ml (Anson units). Optimum pH is 7–8, optimum temperature is 60 °C.
7. Water, 0.5 % diethylenetriaminepentaacetic acid (DTPA), 0.5% citric acid, buffered to pH 8.0 with triethanolamine (TEA), conductivity adjusted to 2000 μS.

References

Amsden, D. 1998. 'Solute diffusion within hydrogels: mechanisms and models', *Macromolecules* 31: 8382–95.

Campani, E., Casoli, A., Cremonesi, P., Saccani, I. and Signorini, E. 2007. *L'uso di agarosio e agar per la preparazione di 'gel rigidi'/Use of Agarose and Agar for Preparing 'Rigid Gels'*, D. Kunzelman (tr.), *Quaderni del Cesmar7*, no. 4. Padua: Il Prato.

Koutsoukos, P.G., Norde, W. and Lyklema J. 1983. 'Protein adsorption on hematite (a-Fe20) surfaces', *Colloid Interface Science* 95: 385–97.

Markevicius, T. and Olsson, N. 2010. 'Flexible thermal blanket and low pressure envelope system in the structural treatment of modern and traditional paintings on canvas', *Postprints of the American Institute for Conservation of Historic and Artistic Works (AIC) 2010 Annual Meeting*, vol. 23. Washington, DC: AIC, 63–71.

Markevicius, T., Meyer, H., Olsson, N. and Furferi, R. 2011. 'Conductive transparent film heater as alternative to heating table: towards new intelligent mobile accurate thermo-electrical (IMAT) device for structural conservation of paintings', *Postprints of the 10th International Conference on Non-destructive Investigation and Microanalysis for the Diagnostics and Conservation of Cultural and Environmental Heritage*, L'Associazione Italiana Prove non Distruttive Monitoraggio Diagnostica (AIPND), 46–56.

Markevicius, T., Olsson, N., Carfagni, M., Furferi, R., Governi, L. and Puggelli, L. 2013a. 'Imat project: from innovative nanotechnology to best practices in art conservation', *Lecture Notes in Computer Science* 7616: 784–92.

Markevicius, T., Meyer, H., Saborowski, K., Olsson, N. and Furferi, R. 2013b. 'Carbon nanotubes in art conservation', *International Magazine of Conservation Science* 4: 633–46.

Stavroudis, C. and Doherty, T. 2013. 'The modular cleaning program in practice: application to acrylic paintings', in M.F. Mecklenburg, E.A. Charola and R.J. Koestler (eds), *Cleaning 2010: New Insights into the Cleaning of Paintings*. Washington, DC: Smithsonian Institution Scholarly Press, 139–45.

Wolbers, R. 2000. *Cleaning Painted Surfaces: Aqueous Methods*. London: Archetype Publications.

Authors' addresses

- Tomas Markevičius, Conservator of Paintings, The Munch Museum, Oslo, Norway (tmarkevicius@fulbrightmail.org)
- Terje Syversen, Conservator of Paintings, Munchmuseet, Oslo, Norway
- Emma Chan, Paper Conservator, Munchmuseet, Oslo, Norway
- Nina Olsson, Conservator of Paintings, Nina Olsson Art Conservation LLC, Portland, OR, USA
- Carola Skov Hilby, Paper Conservator, Royal Danish Academy of Fine Arts (KADK), Copenhagen, Denmark
- Rytė Šimaitė, Paper Conservator, P. Gudynas Conservation Center/Lithuanian Art Museum, Vilnius, Lithuania

Cleaning the Adolphe Roger murals at the church of Notre Dame de Lorette, Paris

Méliné Miguirditchian, Nicolas Engel, Laetitia Desvois and Anne-Laure Capra

Introduction

Notre Dame de Lorette, designed by Hippolyte Lebas (1782–1867), was the first church built in Paris after the French Revolution. The entire church, consecrated in 1836, was classified as a historic monument in 1984. Taking his inspiration from antiquity, Lebas combined the austere architecture of the building's exterior with a rich interior decor. While the elevation is reminiscent of paleo-Christian basilicas, the iridescent effect of the stucco and gilding framing the murals echoes the opulence of Italian Renaissance churches. The overall decoration, commissioned from several painters by the city of Paris between 1828 and 1836, offers an outstanding overview of the diversity of religious art under the July Monarchy (1830–1848). The corner chapels are consecrated to the four key events in Christian life: baptism, Holy Communion, marriage and funerals. Each one is the work of a different painter: Adolphe Roger, Alphonse Périn, Victor Orsel, and Merry-Joseph Blondel.

Adolphe Roger (1800–1880) was a student of Baron Gros and spent time in Italy, where he studied the Italian Primitives and admired their work. He was also influenced by the German Nazarene painters, an early 19th-century artistic movement. The composition of the décor he painted for the Baptism chapel between 1832 and 1840, organized in small panels, offers a purely logical narrative, built around associations of ideas and parallels. The very rich iconography presents several famous baptisms: Jesus receiving the sacrament from his cousin, John the Baptist, the Roman emperor Constantine, the Merovingian king Clovis, as well as representatives of populations more recently converted to Catholicism in South America (Peru) and Africa (Ethiopia). The main focus was deliberately placed on the sacramental entry into the Christian life and the baptismal liturgy used at the time. One original aspect is the presence of plants, particularly those used to make the chrism, showing the influence of a theologian on the design.

The unusual technical feature of the murals in the Baptism chapel is that they were painted using a new wax technique, developed in Paris in the 1830s. The condition of these paintings has deteriorated over time (due to leaks resulting from poor maintenance of the roof) but not attributable to this new painting technique, which was fully mastered by the artist. The paintings have also been damaged in the past by poorly managed cleaning methods unsuited to this complex technique. More recently, thanks to new cleaning methods, it was possible to develop an emulsion-based process for removing the old varnish used in earlier restorations, while completely preserving the sensitive paint.[1]

Technical study

The wax painting technique originated during the renaissance of mural art in the 19th century and was developed by a series of experiments with two objectives: resisting humidity and giving a matte appearance similar to that of true fresco. The techniques and materials identified[2] in the Roger murals indicated that several artists of the period had these dual aims.

The coloured layer was analysed to determine the type of binder. The compounds clearly identified were beeswax and oil. The presence of amyrins indicated the use of an elemi-type plant resin. Despite the fact that neither spike lavender nor turpentine were identified in the binder, due to their volatility on application, the combination of wax and elemi resin suggested the use of a binder referred to by Paillot de Montabert as 'gluten'[3] and the addition of a lead-based drying oil was indicative of 'mixed gluten'. This binder was truly a new technology developed by the artists of the period, which made it possible to work with the cold wax/resin mixture, without needing to apply heat to the painted surface, thus giving an absolutely matte appearance. Indeed, the slow evaporation of the solvent used to dissolve the wax-resin enabled the artists to work

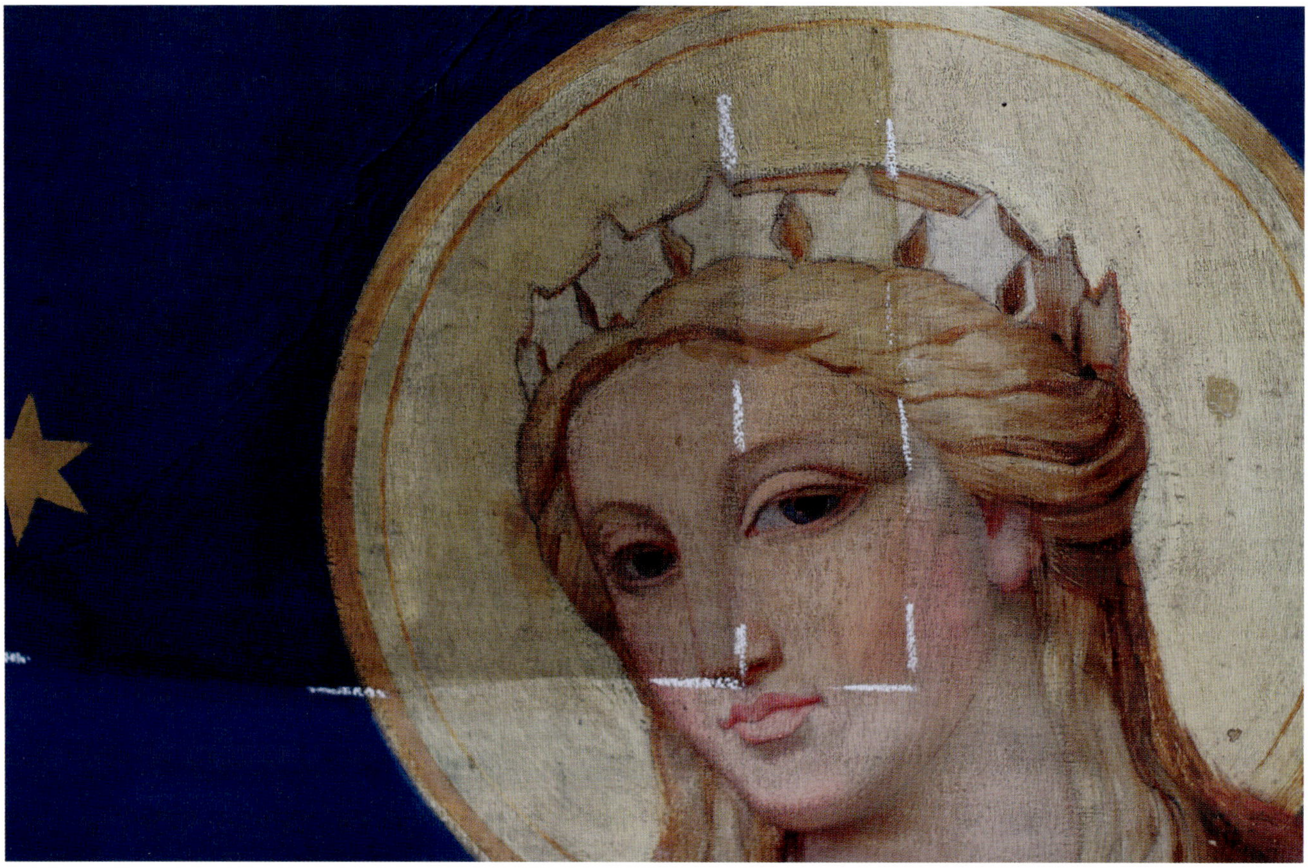

FIG. 1 Details of the removal of the two layers of varnish (photo ©Julien Horon).

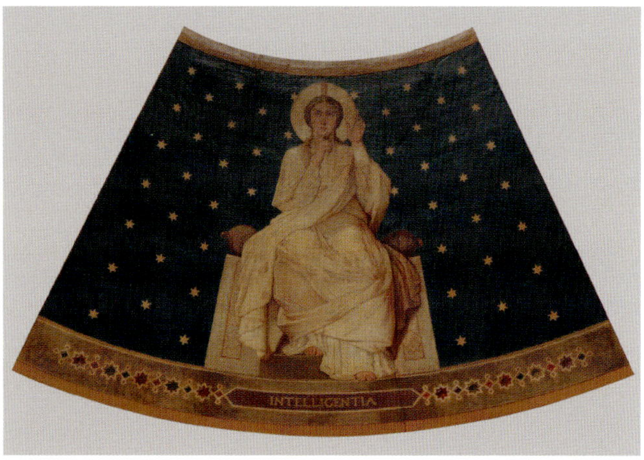

FIG. 2 *Intelligentia*: before cleaning (photo © Julien Horon).

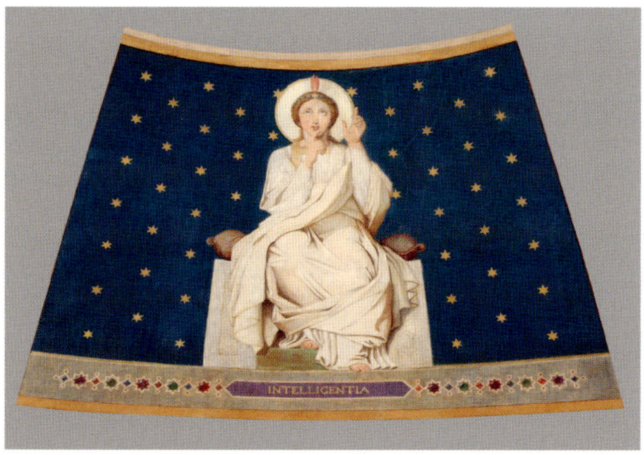

FIG. 3 *Intelligentia*: after treatment (photo © Julien Horon).

longer with the fresh material. The elemi resin added plasticity to the mixture and the drying oil accelerated the drying of the paint layers, making it easier to paint over them.

A more detailed stratigraphic analysis revealed that the artist had applied two preparation coats under the coloured layer, retaining the mixed gluten as a binder and including clay particles as an extender. In the preparation coat immediately below the paint layer, lead white and Naples yellow pigment were used to give this coat a slightly beige colour. Beneath this preparation coat, a wax/oil sealant layer was applied to the limestone surface. This layer, which was not particularly compact and contained shell fragments, lime, and a few grains of quartz, was intended to seal the wall and prevent moisture from seeping through to the painted surface.[4] Stratigraphic cross-sections revealed the deep penetration of this sealant into the stone, indicating that it was applied when hot. Beeswax and lead-based drying oil were identified in this sealant which indicates that it may have been applied following the process developed by chemists Thénard and d'Arcet. The identification of amyrins in this layer may indicate that the elemi resin had migrated from the upper to the lower layer. Another possibility is that the elemi resin was a component of this lower layer, which would indicate that mixed gluten had also been used as a gluten coating. Once this layer had been applied, the surface of the stone was apparently still too uneven because it was found that a

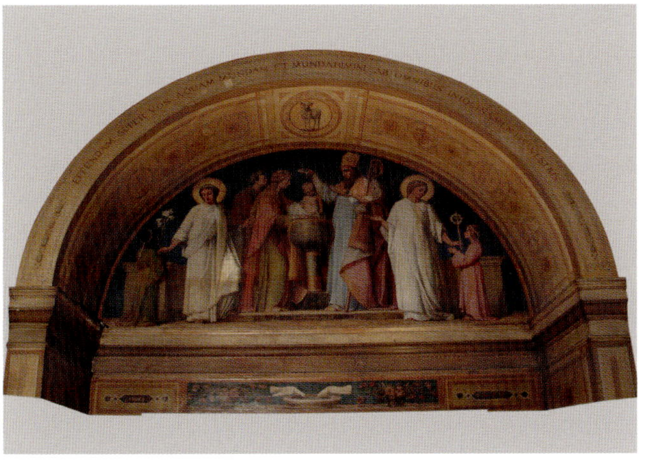

FIG. 4 *Baptimus aquae*: before cleaning (photo © Julien Horon).

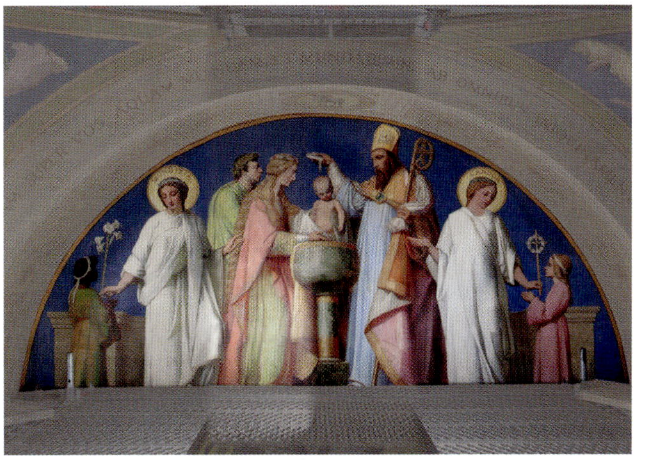

FIG. 5 *Baptimus aquae*: after treatment (photo © Julien Horon).

priming based on white lead and barium sulphate white had also been applied. However, it was not possible to identify the binder used in this layer.

By combining two fundamental principles of this new technique, i.e. a hot base coat and cold wax painting, Roger contributed to the technological research of his period. The end result and the excellent preservation of the paintings show how well this artist had mastered the technique.

Analysis revealed two superimposed coats of varnish. The first coat, in direct contact with the painted layer, was an oil-based varnish containing rosin-type diterpene resin. According to archival documents, two earlier restoration campaigns had taken place – one in 1859, the other in 1933. This oil- and rosin-based varnish was certainly applied during one of these two restorations. The significant thickness of this oil-based varnish was clearly intended to camouflage previous damage to the paint caused by aggressive cleaning methods. This varnish had oxidized and taken on an orange tint over the years. Further analyses then revealed the presence of another, synthetic, acrylic varnish, applied during the most recent restoration in the 1980s. We can assume that the cleaning during this restoration had reduced the matte appearance of the first varnish, so the restorers at that time were obliged to apply more varnish.

The cleaning protocol and its application

Understanding the painting technique used by Roger was the starting point for the entire treatment, including the varnish removal. Once the artist's technique and the various types of varnish had been identified, it was necessary to find a method for removing these two coats of varnish without causing swelling in the mixed gluten compounds. A process was developed using a solvent in a water-based emulsion to remove the two coats of varnish, with the assistance and participation of Richard Wolbers when he visited the site. We chose a 10% benzyl alcohol emulsion in a slightly basic aqueous gel containing 2% xanthan gum, buffered to pH 8. The emulsion was applied with a brush and cleared with water buffered to pH 7. The dwell time was about two minutes.

This emulsion was first tested on several scenes and various colours (Fig. 1). These tests were highly encouraging and it was possible to start removing the varnish. Cleaning started in the dome and progressed to the two lower levels. This particular emulsion made it possible to clean the surface gradually and evenly, without blanching the paint layer (Figs 2–5). The matte finish of the paintings was revealed when the varnish was removed. The low toxicity of the emulsion represented a considerable advantage for the members of the conservation team. The emulsion was easy to prepare and use, making it possible to remove both coats of varnish from a significant total area of 250 square metres.

Conclusion

Those working on the previous campaign in the 1980s showed great caution in deciding not to remove the varnish, as they did not have the appropriate technical solution to do so. Thanks to new, water-based emulsion cleaning techniques, a major obstacle in cleaning varnish from wax paintings has now been overcome. The original colours of Roger's paintings are once again visible. The success of this cleaning process opens up new possibilities for the future treatment of other wax paintings.

Acknowledgements

The restoration of the Baptism chapel at the church of Notre Dame de Lorette, managed by Conservation des œuvres d'art religieuses et civiles (COARC), Paris, received funding from the World Monuments Fund, with the support of American Express. The French state/DRAC Ile-de-France was responsible for scientific and technical management, with the assistance of Prof. Richard Wolbers. I would, therefore, like to thank the following people: Marie Monfort, Nicolas Engel and Pauline Duée, COARC; Dominique Cerclet and Marie-Hélène Didier, DRAC; Prof. Richard Wolbers, University of Delaware, Delaware,

USA; and conservators Sabine Cherki, Anouk Jevtic, Jérôme Dattee, Aude Aussilloux, Maïtena Barret, Cécile Bringuier, Emilie Checroun, Alice Desprat, Marie Dubost, Laurence Fragne, Louise Froidevaux, Florence Labadie, Julie Laubreton, Céline Maujaret, Emmanuelle Paris, Gabriela Szatanik, Pauline Voirin, Bruno Szkotnicki, Lise Bastardos, Diana Da Silva and Alexa Beller; photographer Julien Horon; and graphic designer Raphaël Roig. Samples were taken and analysed by the University of Delaware laboratory, the Laboratoire de Recherche sur les Monuments (LRMH), the LAMOA laboratory, and the University of Avignon laboratory.

Notes

1. The treatment report is available at the COARC resource centre. It presents a precise, detailed account of all the processes carried out between November 2015 and January 2017.
2. The technical examination comprised observations with the naked eye, under a magnifying glass, and using ultraviolet light. These observations guided the sampling and analyses carried out by the laboratories mentioned in the acknowledgements. The analytical techniques used were: scanning electron microscopy coupled with energy dispersive X-ray analysis, Fourier transform infrared spectroscopy, and gas chromatography coupled with mass spectrometry. All the analysis reports are available at the COARC resource centre.
3. Jacques-Nicolas Paillot de Montabert was a French painter who wrote the nine-volume technical treatise on painting *Traité complet de la peinture* in 1829, with an additional volume of illustrations.
4. The sealant layer was applied directly to the stone on every wall in the chapel, but not the dome, where a coat of plaster was applied between the stone and the sealant.

References

Buisson, C. 1983. *La peinture murale à la cire à Paris, dans la première moitié du XIXème siècle*, unpublished dissertation, Institut Français de Restauration des Oeuvres d'Art.

Caffort, M. 2009. *Les nazaréens français: théorie et pratique de la peinture religieuse au XIXème siècle.* Rennes: Presses Universitaires.

Foucart, B. 1987. *Le renouveau de la peinture religieuse en France (1800–1860).* Paris : Arthéna.

Horie, C.V. 2010. *Materials for Conservation: Organic Consolidants, Adhesives and Coatings.* New York: Routledge.

Mecklenburg, M.F., Charola, E.A. and Koestler, R.J. 2013. *Cleaning 2010: New Insights into the Cleaning of Paintings.* Washington, DC: Smithsonian Institution Scholarly Press.

Perego, F. 2005. *Dictionnaire des matériaux du peintre.* Paris: éditions Belin.

Pinna, D., Galeotti, M. and Mazzeo, R. 2009. *Scientific Examination for the Investigation of Painting: A Handbook for Conservator-Restorers.* Florence: Alpi Lito.

Wolbers, R. 2000. *Cleaning Painted Surfaces: Aqueous Methods.* London: Archetype Publications.

Authors' addresses

- Méliné Miguirditchian, Paris, France (meline.miguirditchian @laposte.net)
- Nicolas Engel, Paris, France
- Laetitia Desvois, Paris, France
- Anne-Laure Capra, Paris, France

Investigating the ability of phytate gel systems to treat iron gall ink at the British Library

Zoë Miller, Gayle Whitby and Paul Garside

Introduction

Of the many challenges faced by conservators working in archives and libraries, the self-destructive nature of iron gall ink is one of the greatest (Kolar and Strlič 2006; Brostoff *et al.* 2012). Iron gall ink, developed and valued for its indelibility and permanence for over 1000 years, can also lead to damage: the loss of the inked line and acidity, weakness and discoloration across the substrate. This damage is caused by a combination of acidity (acid hydrolysis) and the oxidative action of iron(II) (Fe^{2+}) ions; both processes require the presence of water. Acid hydrolysis arises from sulphuric acid, a by-product of the reaction that forms the ink; this species is relatively mobile and can migrate and cause damage in the substrate at some distance from the ink. The Fe^{2+} ions are less mobile and tend to only cause damage in the immediate vicinity of inked areas. Ultimately the combination of hydrolytic and oxidative processes can cause the complete loss of the substrate in the region of the ink.

Current treatment options for iron gall ink objects are limited by the need to apply low (or preferably, no) moisture in order to prevent accelerated degradation. At the British Library, we have the additional intervention of aqueous phytate solution as an option for treatment. The traditional immersion technique which allows us to chemically alter and stabilize these inks by carrying out treatment using the combined calcium phytate solution protocol (Neevel 1995) can be intrusive to fragile materials and other historical evidence held within the object. Its application is therefore limited to only those materials and objects whose condition and structure satisfy certain requirements as determined by the conservation decision-making process.

In response to these limitations, and to open up the possibilities for treatment, we have been investigating whether gels can be used locally to introduce calcium phytate to iron gall ink on paper and to determine whether they can be safely applied in their simplest form – made with reverse osmosis (RO) water – as a controlled moisture treatment option without adversely affecting the ink or substrate. Gels have been used successfully with chelating agents for controllable aqueous procedures within other conservation disciplines (Cushman 2006) and can be made up safely and easily within the studio.

This paper is a summary of our initial investigations and observed positive markers for further exploration. Despite commencing with clear and productive objectives, we very quickly discovered the limitations of our procedure and therefore hope to create a platform for sharing problems encountered, and avenues for further research, to encourage collaboration in this interesting area.

Project aims

When questioning our in-house practices in response to current ideas and research, we noted the successful outcome and continuing stability of objects that had been treated using calcium phytate up to 10 years earlier. Fe^{2+} indicator strip tests which turn pink in contact with Fe^{2+}, showed that chemical stabilization had rendered the corrosive capacity of the treated inks inert with little or no detrimental visual or chemical changes (Neevel 2009). The main limitation of this treatment however is that iron gall ink containing folios must be immersed into solutions, therefore bound volumes, parchment and objects with delicate surfaces or other soluble historical features or moisture-sensitive media cannot be treated. An *in-situ* application of calcium phytate solution was therefore sought. In addition to chemical stabilization, this would enable other common treatments to be carried out such as the removal of unwanted stains, adhesive, unsightly or obstructive repairs and linings among many treatments in a book and paper conservator's repertoire.

Recent research carried out in the field suggested that the humidity threshold above which Fe^{2+} migrates is 85–88% relative humidity (RH) before damage occurs (Belhadi *et al.* 2014). This opened up the possibility of applying moisture in controlled amounts using gels as a carrier. Gels have been used as

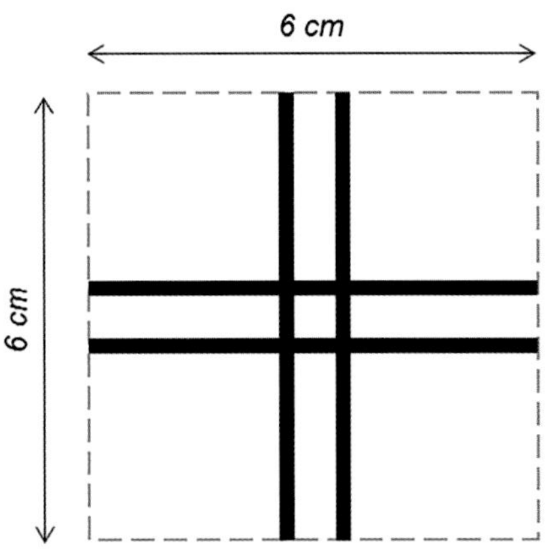

FIG. 1 Pattern for ink application.

a cleaning treatment in paintings conservation for many years (Wolbers 2000; Cremonesi 2013). More recently their use has widened to object, paper and textile conservation for a wider range of treatments (Schönbohm *et al.* 2004; Warda *et al.* 2007; Iannuccelli and Sotgiu 2010; Hawkes 2013). They have been used successfully as carriers for chelating and bleaching agents, enzymes, acids, alkalis, solvents and emulsions and in some cases detergents. Because of their structure, rigid gels can act as molecular sponges to draw out and trap within them impurities, particles and degradation products, leaving little or no residue on the substrate (Khandekar 2004). Their advantages include:

- The ability to control the ingress and diffusion of moisture by altering their concentration.
- Efficiency in drawing out impurities/ softening adhesives etc. in a relatively short timeframe.
- Clean and easy preparation and removal from the substrate leaving little or no residue.
- The ability to produce a localized and potentially *in situ* treatment.

It seemed a natural conclusion to combine gels and calcium phytate. Our aim was to discover whether we could first apply a gel made from RO water to iron gall ink without adversely affecting ink or paper, and secondly make a phytate gel which could be applied directly to the ink to chemically stabilise it *in situ*.

Initial experiment

Our experiment focused on the application of gellan and agarose gels, selected for their safe and easy preparation, formulated with either calcium phytate or RO water. These were applied under controlled conditions to three recipes of iron gall ink applied to Whatman blotting paper using a predesigned pattern which it was hoped would emulate the pockets of degradation created in between handwritten lines (Fig. 1). The recipes included one with a balance of gallic acid and iron sulphate, one with excess iron sulphate and one with excess gallic acid. These were chosen to respond to some commonly found differences in ink quality found in historical examples (Kolar and Strlič 2006).

The samples were divided into three groups: one was left unaged, and the other two subjected to artificial aging (80 °C, 80% RH) for either three or six weeks. A set of control samples from each group was treated using immersion in calcium phytate (reflecting current British Library practice) to ensure a fair comparison. To evaluate our results, Fe^{2+} indicator paper was applied to determine the presence of residual ions following application of the gels. These strips react to the presence of Fe^{2+} ions by turning pink in colour. This enabled us to compare the effectiveness of the phytate gels in treating Fe^{2+} in comparison to those made with RO water. Ultraviolet (UV) imaging was used to assess tideline development and migration of materials within the paper samples and Fourier transform infrared (FTIR) spectroscopy was employed to determine the presence of gel residues as well as the signatures indicative of paper degradation. This revealed no evidence of migration of the gels into the paper; similarly signs of damage in the treated papers were no greater than in the untreated papers.

Initially we experimented with methods for making up the gels involving heating the gel solution in a microwave. We tried 2% gel solutions but these leached too much moisture when placed onto blotting paper. Therefore, 4% gels for both gellan and agarose were selected for the remaining testing (Iannuccelli and Sotgiu 2010). The gels were made up using a 0.0017 molar solution of phytate based on the quantities in the combined calcium phytate solution protocol (Neevel 1995). Gels were made up in round Petri dishes and cut into quarters. These were applied to the substrate through a thin Hollytex barrier to protect media and surface fibres on the samples.

Fresh gel pieces were applied at half-hour intervals until a total, uninterrupted treatment time of six hours had been reached (Fig. 2). This timeframe was selected based on earlier investigations balancing evidence of chemical and physical change with realistic timescales for object treatment. Uptake of water by the paper sample chosen was also considered. Fe^{2+} indicator strips were applied between each application to assess progress. Samples from each group (of inks and aging variables) were also tested at half-hour intervals by immersion in small bags of phytate solution as a comparison. The results were recorded from a single prolonged application of gel for six hours to investigate whether the active component in the gel became exhausted if the gels were not exchanged (Fig. 3).

Observations

The gels were easy and quick to make with few resources. The transparency of both agarose and gellan particularly, allowed us to see what was occurring beneath the gel and the addition of the Hollytex barrier permitted a gentle application of moisture to the surface of the samples. Test strips indicated reduced Fe^{2+} ions in the corresponding ink after applications of both phytate

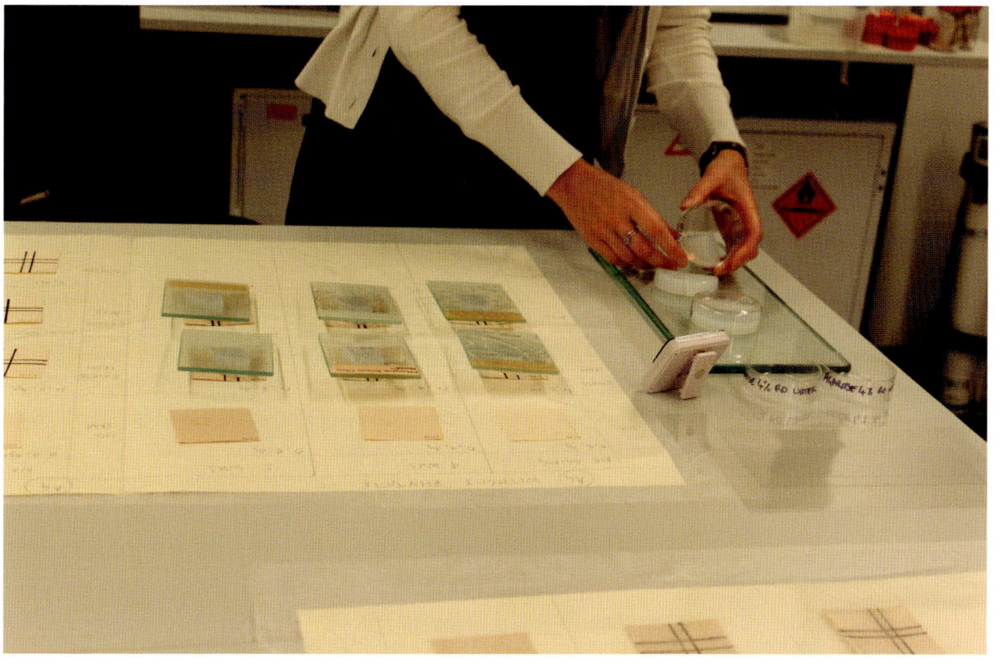

FIG. 2 Gel being applied to samples at 30-minute intervals.

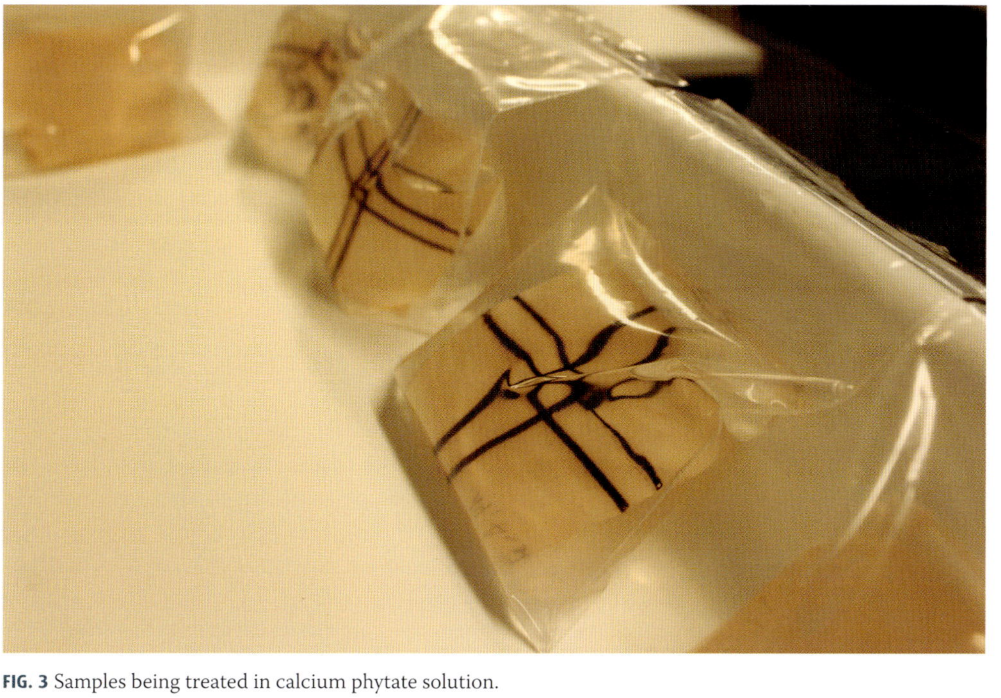

FIG. 3 Samples being treated in calcium phytate solution.

and RO water gels but the changes were slow. A visible reduction in pink colour was noted after the full six-hour treatment time. The phytate gels appeared to be slightly more effective than those made with RO water, and the effects of both were faster on the balanced ink recipes than those which were unbalanced. Some particularly aggressive samples (unbalanced ink recipes aged for six weeks) showed little or no reduction in pink coloration on the indicator strips throughout the experiment, indicating an unsuccessful treatment. This is consistent with our understanding of the unpredictable aging characteristics of historic samples of iron gall ink and reaffirmed the decision to test such various inks (Kolar and Strlič 2006).

Problems encountered

After carrying out these investigations a few difficulties were noted. The first was that the combination of our selected ink recipes and the chosen paper had created particularly aggressive corrosion characteristics in both the unaged and, in particular, the aged samples (Fig. 4). Furthermore the unsized, aged Whatman paper contributed to greater than expected transference of media and surface fibres which detrimentally affected the results. The loose fibre structure, easy wettability and softness made for a poor choice in hindsight as it allowed penetration of the applied ink through to the verso

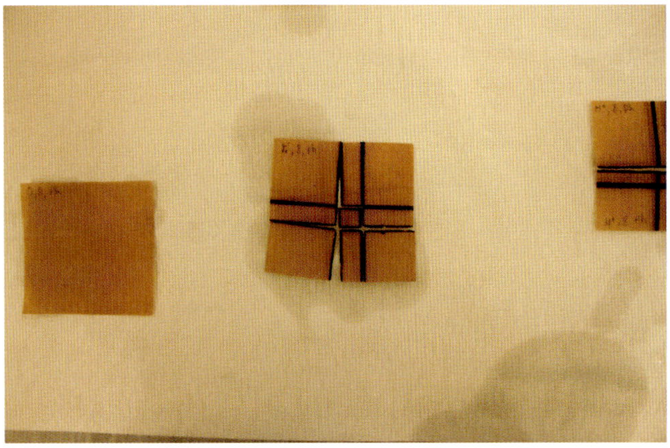

FIG. 4 Aggressive ink sample.

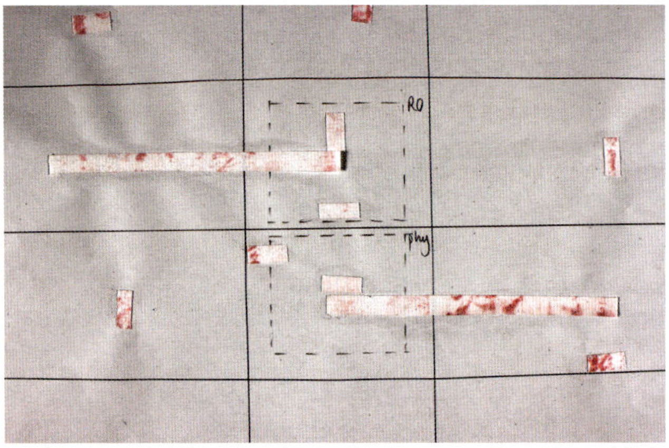

FIG. 5 Pink strip test on historical item showing reduction in Fe^{2+} in areas where gel had been applied.

of the samples, many of which fractured along the inked lines making treatment very difficult (a feature often found in objects showing stage 4 iron gall ink corrosion) (Reißland 2001 and Hofenk de Graaff). A side effect of this was that we were unable to achieve a fully successful result using the traditional immersion treatment for any of the samples, which put into question the results of the equivalent gel testing.

Another outcome of this experiment – which produced negative consequences for the results but positive ones for future use – were the variations in the method of application of the Fe^{2+} indicator strips. We had placed considerable weight on this method to determine the outcomes of our experiment and it became clear early on that variations in our technique of applying the test strip (for example quantity of water used; size and shape of strip and dwell on the ink area) could endanger the collection of clear results. This was an error because the strips are an indication of the presence of free iron and not a reliable indication of its quantity. A wetter indicator strip (for example one allowed to draw up more water before application to the sample) will bring out a deeper pink coloration on the strip than a less wet one. Additionally, a wetter sample of ink (for example one which has had more moisture imparted to it as a result of the concentration of the applied gel) may appear to indicate the presence of more free iron by creating a pinker strip.

We also observed that repeated application of a test strip in the same location mechanically removed particles of ink and/or paper fibres from the sample which adversely altered results. The amount of pink noted on the strip can also be affected by pressure and dwell variation by the conservator. In order to acquire more comparable results, an agreed procedure has been adopted to eliminate some of these problems and applied department-wide for future tests on collection items.

At 4% gel concentration, a defined tideline was created around the applied gel in the substrate where degradation products within the paper and ink were pushed towards the wet–dry boundary. Gellan gel was of particular concern here as it leached water more readily than agarose. Tideline formation was almost immediate in the Whatman paper samples. There were no observed differences here between the RO and phytate gels. We consider tidelines to be an unacceptable side effect of the experiment given its current parameters, particularly as we hoped to investigate it as an *in-situ* treatment option for paper substrates which cannot undergo any other aqueous treatments, such as those within bound volumes. Further testing of higher concentrations of gel could mitigate this in addition to altering treatment times to find an optimal dwell and number of repeated applications. Perhaps the most important observation was that although calcium phytate solution appeared to create a successful gel structure, we have not yet determined whether the heat applied to the solution when making up the gels (or indeed the addition of the gel particles themselves) changed its ability to chelate or rendered its action inert.

Avenues for further work and collaboration

It has been our intention in this paper to present both the aspects of our investigation that failed and those which produced potential for further investigation with an aim of sharing our ideas with fellow professionals and opening a dialogue for collaboration. To address the need for more treatment options for iron gall ink, our logical combination of established treatments has resulted in data that pose more questions than answers. Additional testing was carried out on an historical document which was not part of the British Library collection to mitigate some of the problems that we encountered with our ink samples and to mimic the effects of locally applied moisture on a real object. The results were mixed. The gels effectively reduced Fe^{2+} ions in the ink after two applications of two hours each, which was very encouraging (Fig. 5). We plan to revisit this test to study if the ink which was treated with a phytate gel shows redevelopment of less or no Fe^{2+} over the long term than that treated with reverse osmosis water gel. This would be consistent with theories concerning the efficacy of plain washing to remove corrosive products in iron gall ink. The creation of tidelines as a result of gel application continues to be an area for further work as does the efficacy of the 'phytate gel' as a concept.

We would be interested in collaboration and sharing of this research with similar investigations carried out by the wider community into alternative applications of calcium phytate, use of gel systems with other chelating agents and also with *in-situ* treatment options involving controlled moisture for iron gall ink on paper and parchment.

Materials and suppliers

- Hollytex: Preservation Equipment Ltd, Vinces Road, Diss, Norfolk IP22 4HQ, UK
- Gellan gum: GMW - Gabi Kleindorfer Shop, Gabi Kleindorfer, Aster Str. 9, D-84186 Vilsheim, Germany
- Agarose: VWR, Magna Park, Lutterworth, Leicestershire LE17 4XN, UK
- Indicator strips: Conservation by Design, 2 Wolseley Road, Kempston, Bedford, MK42 7AD, UK
- Phytic acid and other chemicals: Sigma Aldrich Ltd, The Old Brickyard, New Road, Gillingham, Dorset SP8 4XT, UK

References

Belhadi, O., Phan Tan Luu, C., Jacobi, E., Meslet-Struyve, S., Vez, S., Reissland, B. and Rouchon, V. 2014. 'The Dutch Fe-migration mending test', *Journal of Paper Conservation* 15(1): 9-15.

Brostoff, L., Ponce, A., Gaskell, K., Wolbers, R. and Lagalante, A. 2012. 'New research on iron gall ink (11/10/2012 webcast)', Library of Congress. Available at: https://www.loc.gov/today/cyberlc/feature_wdesc.php?rec=5699 (accessed 22 November 2016).

Cremonesi, P. 2013. 'Rigid gels and enzyme cleaning', in M.F. Mecklenburg, A.E. Charola and R.J. Koestler (eds), *Cleaning 2010: New Insights into the Cleaning of Paintings*. Washington, DC: Smithsonian Institution Scholarly Press, 179-84.

Cushman, M. 2006. 'A new method for the treatment of iron-stained architectural marble: in situ reduction of iron III using photovoltaic polymers and the introduction of a new chelating agent for conservation', Association of North American Graduate Programs in Conservation. Available at: https://ford.ischool.utexas.edu/bitstream/2081/3215/2/2006ANAGPIC_Cushman.pdf (accessed 2 March 2017).

Hawkes, R. 2013. 'Gel media in aqueous cleaning methods on paper: a lecture by Professor Wolbers'. Available at: https://thebookandpapergathering.org/2013/11/08/gel-media-in-aqueous-cleaning-methods-on-paper-a-lecture-by-professor-richard-wolbers-university-of-delaware-presented-at-the-wellcome-institute-london-june-19th-2013 (accessed 14 April 2015).

Iannuccelli, S. and Sotgiu, S. 2010. 'Wet treatments of works of art on paper with rigid gellan gels', *The Book and Paper Group Annual* 29: 25-39.

Khandekar, N. 2004. 'Gelled systems: theory and early application' in V. Dorge (ed.), *Solvent Gels for the Cleaning of Works of Art: The Residue Question*. Los Angeles, Getty Conservation Institute, 5–11.

Kolar, J. and Strlič, M. 2006. *Iron Gall Inks: On Manufacture, Characterisation, Degradation and Stabilisation*. Ljubljana: National and University Library of Slovenia.

Neevel, J.G. 1995. 'Phytate: a potential conservation agent for the treatment of ink corrosion caused by irongall inks', *Restaurator* 16(3): 143–60.

Neevel, J.G. 2009. 'Application issues of the bathophenanthroline test for iron(II) ions', *Restaurator* 30(1–2): 3–15.

Reißland, B. and Hofenk de Graaff, J. 2001. 'Condition rating for paper objects with iron-gall ink', *ICN-Information* 1: 1–4. Available at: http://www.cultureelerfgoed.nl/sites/default/files/u6/ICN_info_01_condition_rating_en.pdf (accessed 19 December 2016).

Schönbohm, D., Blüher, A. and Banik, G. 2004. 'Enzymes in solvent conditioned poultices for the removal of starch-based adhesives from iron gall ink corroded manuscripts', *Restaurator* 25(4): 267–80.

Warda, J., Brückle, I., Bezur, A. and Kushel, D. 2007. 'Analysis of agarose, Carbopol, and Laponite gel poultices in paper conservation', *Journal of the American Institute for Conservation* 46(3): 263–79.

Wolbers, R. 2000. *Cleaning Painted Surfaces: Aqueous Methods*. London: Archetype Publications.

Authors' addresses

- Zoë Miller, British Library Centre for Conservation, The British Library, London, UK (Zoe.Miller@bl.uk)
- Gayle Whitby, British Library Centre for Conservation, The British Library, London, UK
- Paul Garside, British Library Centre for Conservation, The British Library, London, UK

Evaluation of leather cleaning with a rigid hydrogel of gellan gum on two composite Amharic shields from the Museo Nazionale Preistorico Etnografico 'Luigi Pigorini', Rome

Flavia Puoti, Anna Valeria Jervis, Roberto Ciabattoni, Egidio Cossa, Antonella Di Giovanni, Maria Rita Giuliani, Giuseppe Guida and Marcella Ioele

Introduction

In recent years, experiments have been carried out on innovative cleaning methods for use in the cultural heritage field. Due to their retention of liquids, and their viscosity, gels significantly reduce the penetration of water and the consequent swelling of underlying surfaces and coatings. Leather is a material that barely tolerates aqueous treatments, and overwetting may cause all sorts of problems including distortion, discoloration, hardening, salt or tannin migrations, and tidelines. During the conservation treatment (Puoti *et al.* 2017) of two Amharic shields (Spring 1993: 99–100) made of thick leather covered by a black coloured layer,[1] we decided to evaluate the use of rigid polysaccharide gels, such as gellan gum, for the cleaning of these objects. The purpose of this research[2] was to provide data on the use of a rigid polysaccharide gel for the cleaning treatment of historic and ethnographic leathers, following some experiments carried out at the Leather Conservation Laboratory of the Italian Istituto Superiore per la Conservazione e il Restauro (ISCR) on artefacts made of gilt and painted leather (Iafrate *et al.* 2014: 265–73).

Gellan gum (Cremonesi 2011) is a polysaccharide produced by *Pseudomonas elodea*, and it is a gelling agent widely used in the food, biomedicine and pharmaceutical industries. It consists of β-D-glucose, α-L-rhamnose and β-D glucuronic acid repeating units. The deacylated type of gellan gum, selected for conservation purposes, forms rigid gels in the presence of cations (especially calcium, Ca^{2+}). The rheological properties highlight a structure with a very high degree of viscoelasticity and is film-forming, homogeneous, transparent and very stable both to high temperatures and to pH variations (Sworn 2009). The sol–gel transition process takes place by cooling a water dispersion of gellan gum from 100 °C to 30–40 °C until it becomes a gel. As well as being directly influenced by temperature, the mechanism of gel formation is related to the presence of mono- or bivalent cations in water.

FIG. 1 Schematic for the application of the gel on leather samples (image: Flavia Puoti).

Surface cleaning with rigid gellan gels

The protocols reported by Iannuccelli and Sotgiu (2010a: 25–39; 2010b: 47–51) were followed for hydrogel choice and preparation.

Materials and methods

Low-acyl gellan gum, Kelcogel CG-LA (C.P. Kelco, Atlanta Georgia, USA)[3] was used to prepare the hydrogel. An aqueous solution of gellan, deionized water and calcium acetate (0.4 g/L), giving a pH of 7, was placed in a 1000 W microwave oven for about two minutes until it boiled and became transparent; it was then left to cool at room temperature, on a flat surface.

The cleaning procedure consisted of the direct application of the gel on the grain (flesh) side of the leather,

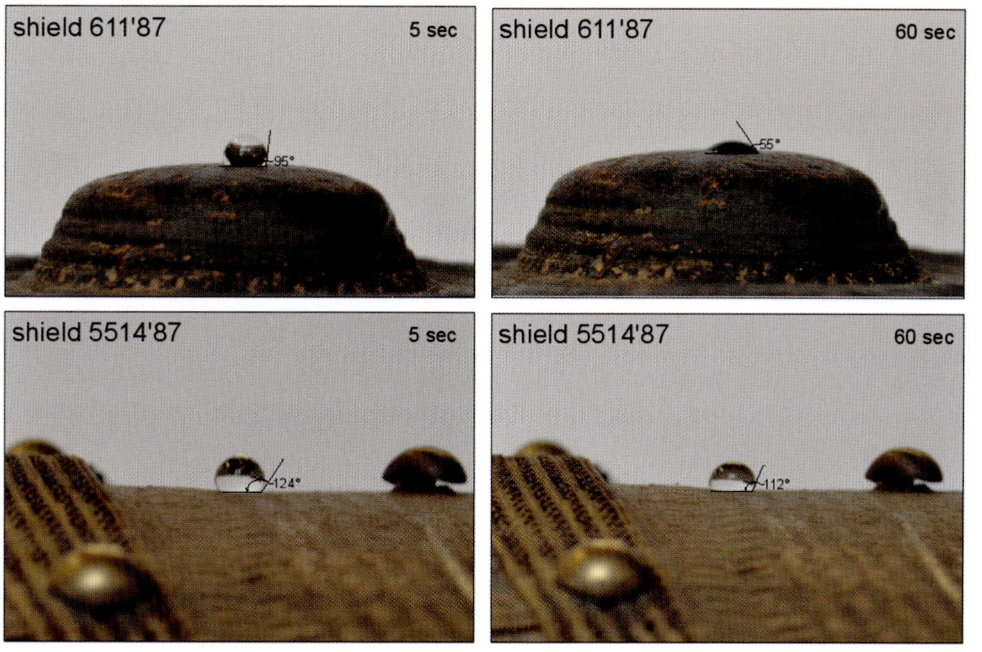

FIG. 2 Contact angle measurements on the leather supports for shields numbered 611'87 and 5514'87, performed with two droplets of water (image: Flavia Puoti).

covered with a polyester film (Melinex), and uniformly pressed with sandbags in order to ensure an even contact between the gel and the leather (Fig. 1). This methodology is based on the capacity of the gel to change the properties of pure water, as its addition increases the wetting power (Shah and Jani 2009: 48–58) and decreases the penetration into the artefact. The release of water to the surface is inversely proportional to the concentration of gellan in the gel: that is, the higher the quantity of gellan gum, the lower the amount of water diffusing into the leather. The gel is able to remove particulate matter, dust, fatty substances that resemble white efflorescence or resinous exudates and water-soluble degradation products.

Preliminary tests: contact angle measurements

To determine the most appropriate concentration of gellan, a preliminary wettability test was carried out on the leather surface of the shields to be treated. The contact angle[4] was measured three times for each type of leather through a drop of distilled water placed on the surface with a micropipette. Images were obtained with a camera located on a tripod positioned perpendicular to the drop, initially 5 seconds after deposition, and again after 60 seconds. The time interval between the two measurements was used as an indicator for the speed with which water penetrates into the leather.[5]

A contact angle less than 90° (low contact angle) usually indicates that the surface can be easily wetted, therefore fluid can spread over a large area. Contact angles greater than 90° (high contact angle) mean that wetting of the surface is unfavourable, so fluids minimize their contact with the surface and form a compact liquid droplet (Piergiovanni and Limbo 2010: 21–9). The leather support for shield no. 611'87 had a low wettability with a high contact angle (average value 96.5°). Furthermore, the speed with which water penetrated into the support was remarkable (the contact angle dropped to 60° after 60 seconds). The red leather lining that covers the inner side of this shield had lower values of contact angle and a lower penetration rate, with homogeneous values around 80° after 5 seconds and 60° after 60 seconds (Fig. 2).

Shield no. 5514'87 had a substantially hydrophobic surface and a significantly lower penetration speed, with contact angles of 98° after 5 seconds and 90° after 60 seconds. In addition to fatty substances added during the tanning, the lower wettability is probably a result of the homogeneity and better condition of the black protective layer which, being made of fatty substances, makes the surface more hydrophobic and occludes the microporosity of the leather.

The contact angle measurements provided data on the wettability of the different kinds of leather used for the shields, and on how fast they absorb water. These tests provided useful information for choosing which leather specimens were to be used in the trials and, later, for the most suitable cleaning methodology.

Preparation of leather specimens

Two series of leather specimens (50 × 50 mm) were prepared, the first with modern leather, not aged artificially, and the second with 18th-century gilt leather samples.[6] The materials used in the first series of samples were chosen according to the characteristics of the leather observed on the shields. Sole leather, obtained by an industrial process of beating and compression of the fibres, vegetable-tanned, and approximately 5–6 mm thick, was chosen as representative of the leather

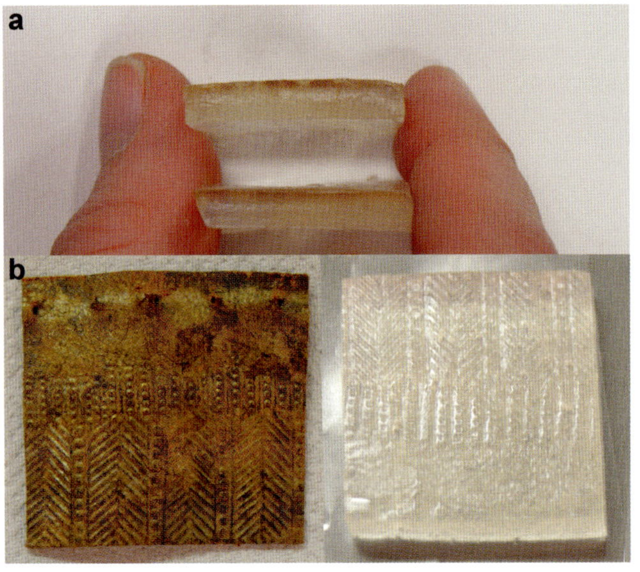

FIG. 3 (a) Yellowed gel after cleaning treatment on shield 611'87. (b) Ancient leather sample on the left, and impression on the gel surface on the right (images: Flavia Puoti).

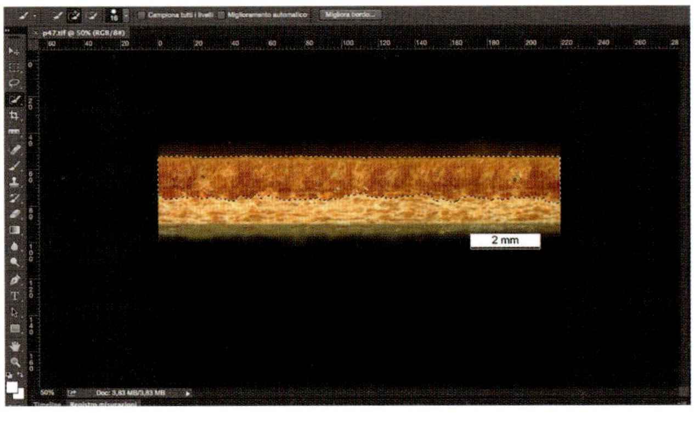

FIG. 4 Screenshot of the wet area measurement in a cross-section (image: Flavia Puoti).

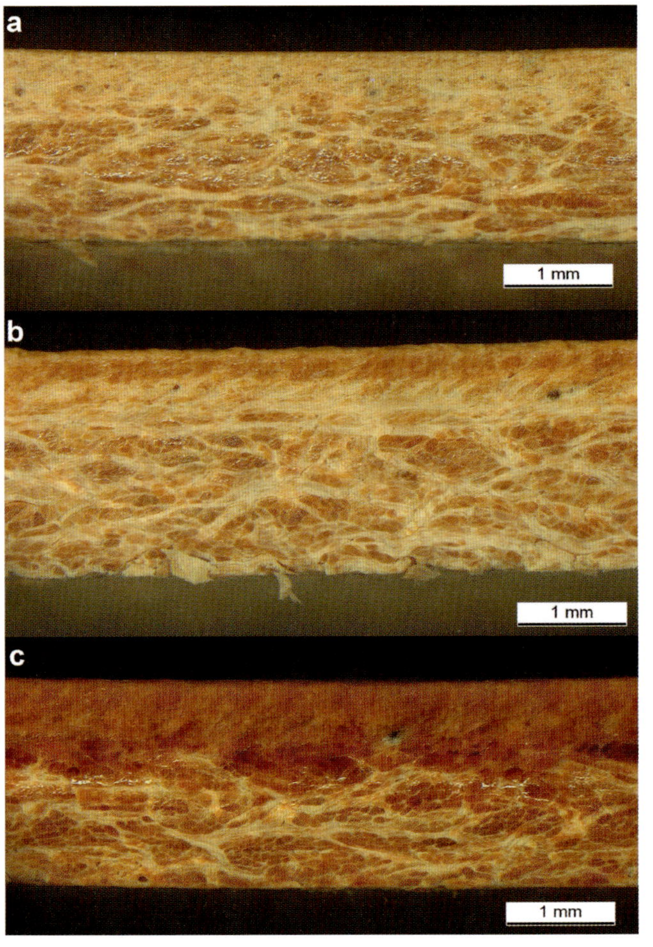

FIG. 5 (a) Untreated leather sample cross-section; (b) leather sample cross-section after 3 minutes of gellan gel treatment; (c) leather sample cross-section after 10 minutes of the same treatment. It is possible to observe the minimal water intake at 2 or 3 minutes contact time (images: Flavia Puoti).

used as a support for shield no. 611'87, which was made of hippopotamus skin. The other type of leather sample was a soft, thin, vegetable-tanned, 3 mm thick leather, comparable to the red leather lining. The historic samples had different types of surface processing (punched, embossed and gilded). These specimens allowed for the verification of the gel action on leather samples similar to those of the shields, as well as for testing the gel on gilt leather, a technique represented in a number of artefacts in European historical leather collections.

Experimental procedures

For the preparation of the rigid gel, we decided to use two concentrations of gellan gum, 3% and 4%, to verify the different absorption capacity and the degree of solvent release. As previously stated, water uptake of the support is inversely proportional to the concentration of the gel: the greater the quantity of powdered gellan, and the more tightly reticular the structure of the gel, the smaller the quantity of water transferred into the leather during the same time interval. We also tested two different thicknesses of gel compresses, 4 and 10 mm, to assess the potential for greater extractive capacity as the thickness increases, and four different application times (2, 3, 5 or 10 minutes). The period of time was varied according to the thickness of the deposit, its nature and its condition. The gel was applied directly on the grain layer of the samples: as it gradually releases water into the leather, the most common degradation agents present on the surface are absorbed. The yellowing of the gel after treatment is due to the migration of water-soluble chromogenic components from the leather support, including organic acids (Fig. 3a).

We tested at least three leather specimens for each concentration, thickness and application time (total 96 samples). The specimens were weighed and imaged before and after the application of the gel; they were then cut and a cross-section was observed under the stereomicroscope. Observation of the cross-sections proved to be the most useful tool for evaluating gel use in the cleaning of leather objects. To obtain repeatable and comparable values of the water penetration in the samples, we used a microscope camera, and calculated the wet area visible in cross-section (Fig. 4).[7]

Results

Data obtained from the weight increase and the wet area in the cross-section gave some useful results:

- The percentage of gellan gum in the gel greatly affects the water penetration: at 3% concentration the gel action is plainly horizontal and at 4% concentration the penetration of water in the leather samples is minimal, with its action limited even to the surface (Fig. 5).
- Contact angle measurement provides useful information on the characteristics of the surface to be treated, and it allows us to anticipate the penetration rate of the solution; we were able to confirm that hydrophobic leather samples were less susceptible to water penetration by increasing the time of application.
- The most appropriate application times for cleaning treatments are 2 and 3 minutes. After this time there are minimum increases in weight, no permanent changes in the surface appearance, and the wet area in the cross-section is minimal.
- The difference in the thicknesses of the gels (4–5 mm and 10 mm) did not significantly affect their extraction capacity. Therefore, the thickness of the gel should be chosen according to the characteristics of the surface of the artefact: when the surface is not perfectly smooth, it is preferable to use minimum thicknesses that allow a greater surface adhesion (Fig. 4b).

Microscopical observation of cross-sections of historic leather was more problematic because these were only 1.5 mm thick, and the degraded fibres made the reading of the colour change caused by water absorption very difficult. Nevertheless, even if the samples had different surface characteristics, they showed comparable values for water penetrability.

Conclusion

Gellan hydrogels may represent a compelling cleaning procedure for cultural leather artefacts as they allow minimal water release and do not cause any mechanical disturbance to the grain layer. This cleaning method was tested on different leather samples, characterized by different aging and properties. Comparison of the results from this study suggest that gellan gel applications of 2–3 minutes do not induce any apparent surface change (e.g. tidelines, colour variations) and that this method is suitable for the cleaning of leather objects made using different techniques.

Acknowledgements

The authors would like to thank ISCR and its scientific laboratories for support, advice and for the use of facilities.

Materials and suppliers

- Kelcogel CG-LA, Melinex: CTS (Italy), Via G. Fantoli n. 26, 00149 Rome, Italy, www.ctseurope.com
- Leather samples: Barberini Andrea – Leather, fur and accessory, via Giovanni Bettolo, 34, 00195 Rome, Italy

Notes

1. Fourier transform infrared (FTIR) spectroscopy with an IN10 IR microscope (Thermo-Fisher) and a diamond cell, or a micro-ATR device applied to a cross-section, and scanning electron microscopy, were used to identify the black layer as mud clay, containing calcium carbonate and partially saponified fat.
2. This paper represents part of Flavia Puoti's Master's project at the ISCR, supervised by Anna Valeria Jervis, Head of the Leather Conservation Laboratory, ISCR.
3. Produced by C.P. Kelco. For product information see the *Kelcogel® Gellan Gum Book*, 5th edn (www.appliedbioscience.com/docs/ Gellan_Book_5th_Edition.pdf).
4. The contact angle is the angle, conventionally measured through the liquid, where a liquid–vapor interface meets a solid surface. It quantifies the wettability of a solid surface by a liquid, by use of the Young equation (Iannuccelli and Sotgiu 2012: 27–8).
5. The measurement was carried out with the 'angular dimension' command in AutoCAD software.
6. These samples, gilt leather small fragments of unknown provenience, were donated by Palazzo Chigi (Ariccia, Rome) for study and research.
7. The area was calculated with Adobe Photoshop CC 2014 for Mac.

References

Cremonesi, P. 2011. *L'ambiente acquoso per la pulitura di opera policrome*. Padua: Il Prato.

Iafrate, S., Moretti, F., Ioele, M., Jervis, A.V., Paris, M., Miracola, P. and Biocca, P. 2014. 'Il Paliotto dei tre martiri. Studio e restauro di un manufatto in cuoio argentato, dorato e dipinto', paper presented at the *Italian Group of the International Institute for Conservation (IG-IIC) 12th meeting, Lo Stato dell'Arte 12, Milan, 23–24 October 2014*.

Iannuccelli, S. and Sotgiu, S. 2010a. 'Wet treatments of works of art on paper with rigid gellan gels', *The Book and Paper Group Annual* 29: 25–39.

Iannuccelli, S. and Sotgiu, S. 2010b. 'A new methodology for wet conservation treatments of graphic art on paper with a rigid polysaccharide gel of gellan gum', in L. Watteeuw (ed.), *Preprints of Choices in Conservation Practice versus Research, 6–8 October 2010*. Copenhagen: Copenhagen Royal Library.

Iannuccelli, S. and Sotgiu, S. (eds) 2012. *La pulitura a umido di opere d'arte su carta con gel rigidi di Gellano: presupposti teorici, metodologia applicativa e verifica analitica*, *Quaderni del Cesmar7*, no. 11. Padua: Il Prato.

Piergiovanni, L. and Limbo, S. 2010. *Food Packaging: Materiali, Tecnologie e Soluzioni*. Milan: Springer Science & Business Media.

Puoti, F., Jervis, A.V., Ciabattoni, R., Cossa, E., Di Giovanni, A., Giuliani, M.R., Guida, G. and Ioele, M. 2017. 'Conservation and study of ethnographic materials: two Amhara shields made of leather and metal from the National Prehistoric and Ethnographic Museum "Luigi Pigorini"', *International Journal of Young Conservators and Restorers of Works of Art* 1 [in press].

Shah, D.P. and Jani, G.K. 2009. 'Modification and characterization of gellan gum', *Pharmaceutical Technology* 33(7): 48–58.

Spring, C. 1993. *African Arms and Armour*. Washington, DC: Smithsonian Institution Press.

Sworn, G. 2009. 'Gellan gum', in G.O. Phillips and P.A. Williams (eds), *Handbook of Hydrocolloids*, 2nd edn. Oxford, Cambridge and New Delhi: Woodhead Publishing Limited, 204–27.

Authors' addresses

- Flavia Puoti, Rome, Italy (flaviapuoti@gmail.com)
- Anna Valeria Jervis, Istituto Superiore per la Conservazione e il Restauro (ISCR), Rome, Italy (annavaleria.jervis@beniculturali.it)
- Antonella Di Giovanni, Istituto Superiore per la Conservazione e il Restauro (ISCR), Rome, Italy (antonella.digiovanni@beniculturali.it)
- Roberto Ciabattoni, Chemistry, Biology and Physics Laboratories, Istituto Superiore per la Conservazione e il Restauro (ISCR), Rome, Italy (roberto.ciabattoni@beniculturali.it)
- Maria Rita Giuliani, Chemistry, Biology and Physics Laboratories, Istituto Superiore per la Conservazione e il Restauro (ISCR), Rome, Italy (mariarita.giuliani@beniculturali.it)
- Giuseppe Guida, Chemistry, Biology and Physics Laboratories, Istituto Superiore per la Conservazione e il Restauro (ISCR), Rome, Italy (giuseppe.guida-01@beniculturali.it)
- Marcella Ioele, Chemistry, Biology and Physics Laboratories, Istituto Superiore per la Conservazione e il Restauro (ISCR), Rome, Italy (marcella.ioele@beniculturali.it)
- Egidio Cossa, National Museum 'Luigi Pigorini', Rome, Italy (egidio.cossa@beniculturali.it)

Challenges of a virgin water-sensitive surface: designing a cast agar gel cleaning system for Patrick Heron's *Still-life against the Sea: 1949*

Alysia Sawicka, Mia O'Toole and Katherine Ara

Introduction

A modern, unvarnished oil painting presents the conservator with a number of unique criteria for consideration during treatment. Establishing and protecting the intended surface aesthetic of the work is essential to preserving the artist's vision. Without a protective interlayer, treatment options are far more limited, and can be further complicated by the now widely recognized phenomenon of water-sensitive oil paints. Issues of water sensitivity in 20th- and 21st-century paintings call into question the suitability of aqueous cleaning systems commonly employed in the surface cleaning of traditional varnished paintings.

In terms of surface character, an unvarnished painting may be likened to the porous surface of wall paintings. In this field in particular, the advantages of polysaccharide-based rigid hydrogels have been commended as a controllable method of delivering water to painted surfaces. The method was originally introduced at a workshop by Richard Wolbers in 2003 and subsequent technical investigations exploring the use of agar gels have been published, although practical application has largely been restricted for use on artefacts other than easel paintings (Wolbers 2004; Campani *et al.* 2007; Anzani *et al.* 2008; Volk and van den Berg 2014).

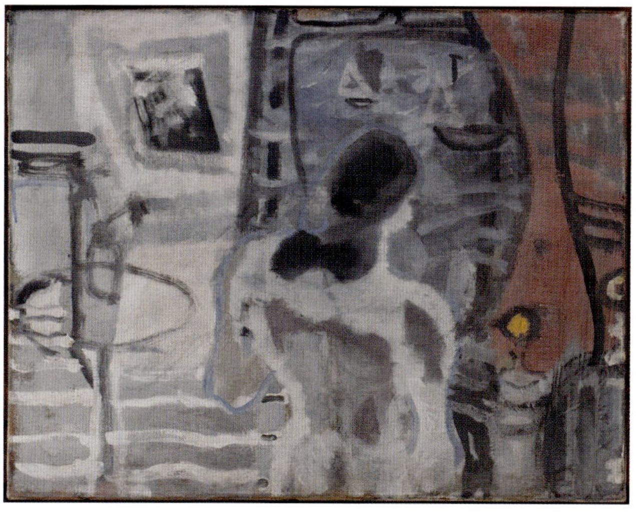

FIG. 1 Patrick Heron, *Still-life against the Sea: 1949*, 1949, oil on canvas, 40.5 × 51.0 cm, private collection: after treatment (© Estate of Patrick Heron 2017).

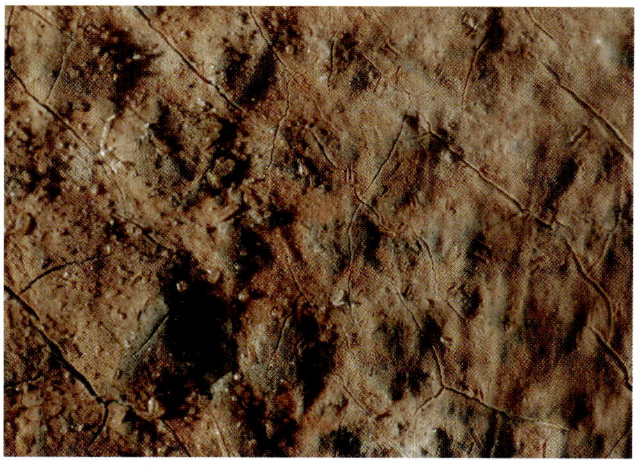

FIG. 2 A partially cleaned area examined in raking light with a stereomicroscope. *Left:* the shards still partially embedded within the paint film. *Right:* sharp divots visible where the crystals have been removed.

Overview

Such characteristics were central in treating *Still-life against the Sea: 1949* by Patrick Heron (1920–1999) (Fig. 1).[1] The artist himself had given the work to the current owner's mother in 1949, and the unvarnished painting had hung untouched ever since. Framed without glass, the paint surface had naturally accumulated a significant layer of loose

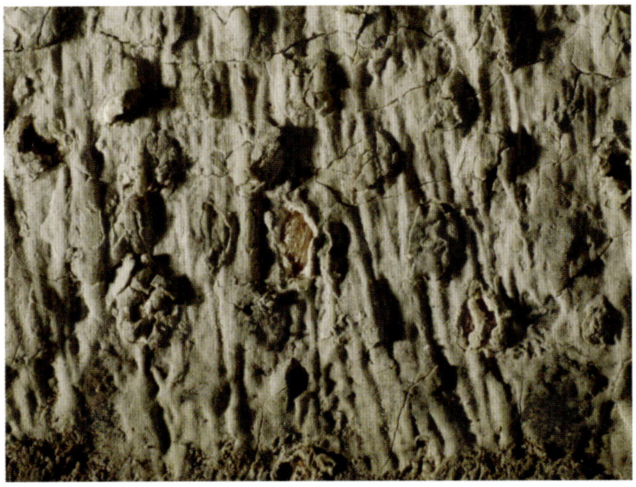

FIG. 3 Ground erupting over peaks of the canvas weave noted during initial aqueous testing.

particulate dirt, comparable to that found on the frame, as well as other superficial splashes and drips. Since this layer was visually disturbing and compromised the subtle surface nuances integral to the work, surface cleaning was undertaken in 2015 at Katherine Ara's studio.

The work, executed in oil on a commercially primed canvas, presented a complicated surface. The surface was lean and desaturated, in large part due to the nature of the water-sensitive ground. Poor adhesion between the paint and ground layers had resulted in microscopic flaking and loss.[2] Many significant *pentimenti* were visible due to the increased transparency of the upper paint layers and the texture of the underlying brushwork. As a result, areas of brittle age craquelure associated with the impasto of the underlying *pentimenti* were present.[3] Microscopical examination of the surface further revealed transparent crystalline shards partially embedded in the paint film, evident across the entire surface, which gave the painting a 'sparkling' appearance (Fig. 2). These crystals were not present on either the frame or the priming where exposed at the tacking margins, therefore it was thought that the source of some components of the crystalline material must originate from within the paint film. The crystals were unaffected by a range of solvents, but were readily soluble in water. It was hypothesized that the crystals may be magnesium sulphate heptahydrate, commonly called epsomite, due to similarity in their behaviour as well as optical appearance. Previously identified in other works using Winsor & Newton paints from this period, the needle-like structures are formed from the reaction of magnesium carbonate with environmental sulphur dioxide at elevated relative humidity (Cooper *et al.* 2014).

Issues of water sensitivity have been noted in other works by Heron; Bustin also observed that the majority of paint swatches in the Heron Paint Archive had become water-sensitive (Bustin 2014: 40). Bustin's studies suggest that Heron used paints produced by several manufacturers – the majority of paints catalogued were from Roberson & Co., although Heron also used Winsor & Newton paints.

Testing

Initial dry cleaning tests proved ineffective, as both brushing the surface and rolling with 'tacky swabs' were largely unsuccessful at dislodging the crystals. Tacky swabs did not affect the dirt or the crystals.[4] Brushing did not remove the surface dirt and the majority of the crystals were unaffected since they were partially embedded in the paint. Further tests quickly ruled out conventional aqueous cleaning systems due to the sensitivity of the ground. Although successful at removing the particulate dirt and crystals, penetration of moisture to the ground from rolled swabs wetted with both saliva and 1% tri-ammonium citrate caused 'volcanic' eruptions to occur over the peaks of the canvas weave due to delamination at the interface between the canvas and the ground (Fig. 3). The presence of craquelure penetrating through both the paint and ground layers probably aided moisture saturating the ground from these free aqueous solutions. Furthermore, the mechanical action of swab rolling caused abrasion to the surface. These aqueous cleaning tests resulted in a similar level of abrasion and increase in gloss. Cleaning with 1% tri-ammonium citrate was initially more successful at wetting the surface, although it required a longer exposure in order to remove the surface dirt. Under the stereomicroscope, the surface of the original paint appeared worn and burnished, resulting in increased gloss. An additional concern was the exposure of depressions where the crystals were removed: the paint structure could potentially be weakened by the removal of the crystals, disrupting the paint film and possibly enabling moisture to penetrate the surface more readily.

Given the ineffectiveness of dry cleaning and the dangers of the combination of mechanical action and free aqueous solutions, it was clear that any aqueous cleaning system must be modified in order to prevent moisture penetration and reduce mechanical action as far as possible. Methods such as increasing the evaporation rate through the addition of acetone were similarly unsuccessful. Gelling agents such as Velvesil, employed to limit the penetration of water into the paint structure, were largely dismissed on account of the mechanical action necessary for clearance. Methods to minimize mechanical action such as applying water through Eltoline tissue with a brush caused paint micro-flaking and were also dismissed. Recent research investigating the use of agar gel for surface cleaning water-sensitive oil paint suggested a potential solution (Volk and van den Berg 2014). Agar gel readily extracts surface dirt and can be easily removed by gently peeling it back. Without the addition of surfactants or chelating agents, further clearance is not necessary, as the gel has not been observed to leave behind residual material. As a result, minimal mechanical action is required during application and removal of the gel.

Development of cast agar gel

An agar gel was prepared according to the recommendations of Anzani *et al.* (2008). A 3% w/v solution was made up in batches by soaking 6 g powdered agar in 200 ml deionized water.[5] In order to obtain a homogeneous gel, the agar-water mixture was twice heated to 95 °C, cooled in between to below its 30 °C gelation point, replenishing the volume of any evaporated water after each cooking phase. In testing carried out by Volk and van den Berg (2014), the optimum results were achieved with a semi-solid agar brushed on before gelation with short, repeated application times as required. Following these findings, initial tests were carried out with solutions of 3% w/v agar brushed on while still warm (approximately 37 °C) and left for a duration of 30 seconds. No tests were made on how any additions to the gel might impact the delivery of the agar or the effectiveness of the cleaning, as the pure water-agar gel was sufficient to remove all surface dirt present. While safe to use over the stable red paint layer, this technique was obviously not suitable for use in areas with micro-flaking, since the gel contracted and pulled up the paint film upon cooling.

As a consequence, a system of pre-casting the gel was developed in order to avoid the gel contracting on the paint surface. Warm agar was poured onto tensioned silicone-coated release film in a thin even layer (*c.*2 mm thick to allow the insertion of a swab stick for removal) and left to cool (Fig. 4). The cooled agar formed a rigid gel, damp to the touch and with a slightly yellowish cast, although sufficiently transparent to observe the surface beneath. Lower concentrations of agar (2% w/v) were markedly floppier and easily released water, quickly wetting a contacted surface, while a stiffer gel formed by higher concentrations (5% w/v) did not conform as well to textured surfaces and was significantly more opaque. For this treatment, the properties of a 3% w/v gel most closely met the necessary criteria as it conformed well to the paint surface, with a more controlled release of water than a lower concentration.

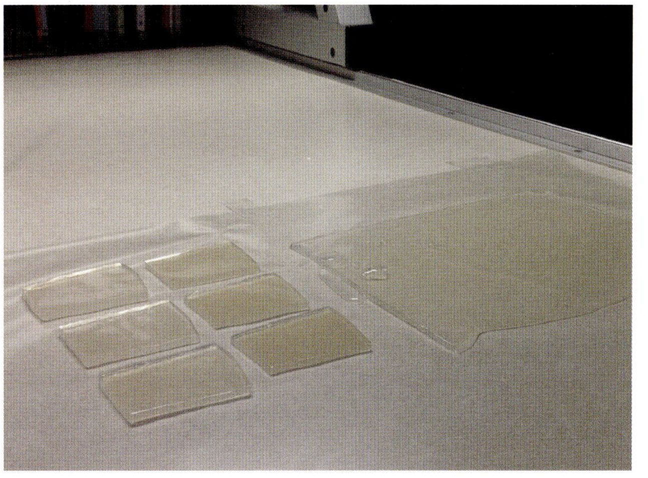

FIG. 4 Sheet of agar cast during testing with prepared shapes cut out.

Cast agar gel in practice

The cast gel was applied to the surface of the painting and lightly pressed down to ensure complete contact with the impasted paint, left for the allotted application time and subsequently peeled back using the swab stick. A single 30-second application removed the water-soluble crystals and some of the dirt, although in more soiled passages the dirt was not entirely removed. After allowing time for residual moisture to fully evaporate from the paint surface, a further 30-second application of freshly cast agar drew up the remaining dirt. Corroborating the observations of Volk and van den Berg (2014), these repeated applications were more effective than a single application of 1 minute, and had the additional advantage of further limiting the penetration of moisture into the porous paint and ground layers, which readily extracted moisture through capillary action. The success of multiple shorter applications may relate to a decline in the agar gel's ability to absorb the dirt as it progresses toward a saturation point.

The success of the treatment was indicated by several factors:

- Upon removal of the gel, the surface texture of the paint could be seen clearly impressed into the agar, demonstrating the ability of the gel to conform to and contact the entire surface.
- Following removal, the gel released greyish water when pressed, indicating that the dirt had been successfully absorbed into the gel. The discoloration of the gel confirmed that the majority of the surface dirt had been extracted during the first application, with a further application removing any residual dirt and a third application of freshly cast gel releasing clear water: the treatment was therefore carried out with only two applications.
- Visual examination confirmed that the dirt had been successfully extracted from the paint surface. Other surface accretions such as drip marks along the left side of the painting were also removed. The agar gel had made a visible difference as the paint surface appeared lighter and cleaner, with no change noted in saturation or gloss.
- By applying the cast gel at room temperature, no surface tension was created by the contraction of the gel upon cooling, avoiding the issue of micro-flaking experienced with brush application of warm agar.
- No delamination of the paint or ground layers occurred as with other aqueous cleaning methods tested, indicating that the moisture penetration was sufficiently minimized by the more controlled release of water achieved with short, repeated applications.
- It was clear from microscopical examination that the surface texture had not been affected, the lack of mechanical action necessary meaning that no abrasion was caused. Crisp, sharp depressions remained as a result of the removal of the crystals.

In practice, a further advantage of casting the agar was that it allowed for tailored shapes to be cut from the cast film of gel, in order to avoid the creation of cleaning lines. A tracing of the composition of the painting was made with a Melinex overlay for use as a template. Shapes were cut from the cast gel to align perfectly with corresponding compositional forms (Fig. 5). However, when checked after application under ultraviolet

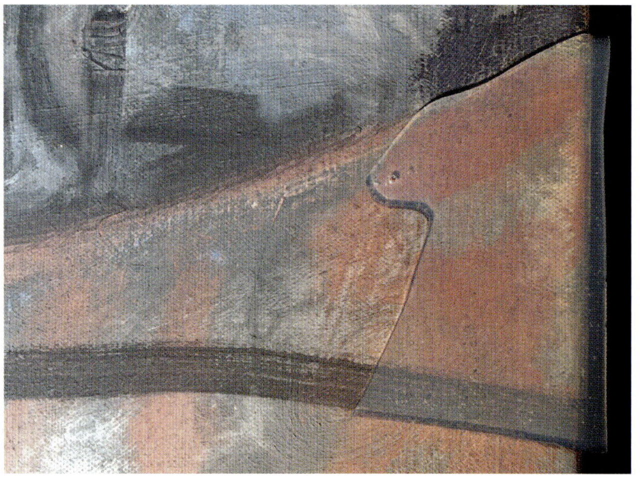

FIG. 5 Example of the cast gel applied to the surface, tailored to fit within elements of the composition.

(UV) light, a slightly fluorescent tideline was visible at the perimeter of one test area. The water retention of the agar seemed to be poorer at the edges, the surface dirt not drawn up into the agar gel, but left deposited on the surface around the perimeter of the cast shape.[6] This proposed mechanism may have been amplified by moisture released while pressing the gel in place. These tidelines were eliminated by slightly overlapping the pre-cut forms, and subsequently no residues were detectable in UV light following treatment.

Conclusion

The development of this cast agar gel cleaning system was designed to address the specific challenges presented by Heron's *Still-life against the Sea: 1949*, an unvarnished oil painting with a virgin surface. The combination of accumulated particulate dirt and superficial crystalline deposits was aesthetically disturbing. The various inherent instabilities within the painting demanded specific parameters for a cleaning system, in particular the water-sensitive ground and the poor adhesion between the paint layers. The combination of these factors dictated that a cleaning system be devised that restricted both mechanical action and moisture penetration as far as possible. Agar gel, an established cleaning tool across various fields in conservation, was selected on the basis of these criteria. The innovation of pre-casting and applying the cast gel at room temperature negated the problems identified with brush application of the warm gel contracting on the paint surface, and also enabled the precise application of tailored shapes. Although the method was time consuming since it required repeated applications, it could be easily controlled and sufficiently reduced the diffusion of water into the ground. As with any cleaning system, the long-term preservation of the painting is paramount. Potential degradation of the paint surface caused by residual cleaning material was avoided because the gel was readily removed by peeling back from the surface, with no further clearance necessary. By careful manipulation of materials in response to observations noted by the conservator, it was possible to achieve a successful, even surface clean of *Still-life against the Sea: 1949*, which is now behind glass. Agar gel proved to be a malleable tool for surface cleaning easel paintings, easily tailored to the surface in question. Various potential practical applications for the gel are suggested by the different adaptations made possible by physical modification such as grating (Cremonesi 2016) or through additions to the cast gel, manipulating the agar's pH, conductivity, and ability to chelate ions.

Acknowledgements

We are grateful for the advice of Bronwyn Ormsby, Mary Bustin, and Stephen Hackney during this treatment and the writing of this paper.

Material and supplier

- Agar E406 (a lower grade agar): Special Ingredients, www.specialingredients.co.uk

Notes

1. A comparative before treatment image has not been included, as the visible change before and after cleaning is quite subtle on this scale.
2. Areas of flaking were consolidated using a 20% w/v solution of Paraloid B-72 in xylene.
3. Surface cleaning would fully reveal both the *pentimenti* and the extent of blanching to the paint surface, further complicating the treatment. To unify the surface, disturbing *pentimenti* were later suppressed through retouching and the blanched paint locally saturated with Paraloid B-72.
4. At this studio, a method for coating swabs with Lascaux 303 HV to create a tacky film once dry, as an alternative method of dry surface cleaning, is currently under development.
5. Although agarose is more refined with greater transparency, agar is more commonly used in conservation studios, as it has a tighter polymer network and releases water more slowly. It is also considerably cheaper (Cremonesi 2013).
6. The exact mechanism of agar gel in terms of its use as a cleaning method is as yet unclear. See Anzani *et al.* 2008 and Volk and van den Berg 2014 for further discussion.

References

Anzani, M., Berzioli, M, Cagna, M., Campani, E., Casoli, A., Cremonesi, P., Fratelli, M., Rabbolini, D. and Riggiardi A. 2008. *Gel rigidi di Agar per il trattamento di pulitura di manufatti in gesso/Use of Rigid Agar Gels for Cleaning Plaster Objects, Quaderni del Cesmar7*, no. 6. Padua: Il Prato.

Bustin, M. 2014. 'Towards interpretation of making, meaning, and change in British twentieth century oil paintings: the relevance of an artist's paint archive', in K.J. van den Berg, M. de Keijzer, A. de Tagle, J. Kreuger, G. Heydenreich, A. Burnstock and T. Learner (eds), *Issues in Contemporary Oil Paint*. Cham: Springer, 33–43.

Campani, E., Casoli, A., Cremonesi, P., Saccani, I. and Signorini, E. 2007. *L'uso di agarosio e agar per la preparazione di 'gel rigidi'/Use of Agarose and Agar for Preparing 'Rigid Gels'*, D. Kunzelman (tr.), *Quaderni del Cesmar7*, no. 4. Padua: Il Prato.

Cooper, A., Burnstock, A., van den Berg, K.J. and Ormsby, B. 2014. 'Water sensitive oil paints in the twentieth century: a study of the distribution of water-soluble degradation products in modern oil paint films', in K.J. van den Berg, M. de Keijzer, A. de Tagle, J. Kreuger, G. Heydenreich, A. Burnstock and T. Learner (eds), *Issues in Contemporary Oil Paint*. Cham: Springer, 295–310.

Cremonesi, P. 2013. 'Rigid gels and enzyme cleaning', in M.F. Mecklenburg, A E. Charola and R.J. Koestler (eds), *Cleaning 2010: New Insights into the Cleaning of Paintings*. Washington, DC: Smithsonian Institution Scholarly Press, 179–84.

Cremonesi, P. 2016. 'Technical note: surface cleaning? Yes, freshly grated agar gel, please', *Studies in Conservation* 61: 362–7.

Volk, A. and van den Berg, K.J. 2014.'Agar – a new tool for the surface cleaning of water sensitive oil paint?', in K.J. van den Berg, M. de Keijzer, A. de Tagle, J. Kreuger, G. Heydenreich, A. Burnstock and T. Learner (eds), *Issues in Contemporary Oil Paint*. Cham: Springer, 389–406.

Wolbers, R. 2004. *Un approccio acquoso alla pulitura dei dipinti*, *Quaderni del Cesmar7*, no. 1. Padua: Il Prato.

Authors' addresses

- Alysia Sawicka, London, UK (alysia.sawicka@gmail.com)
- Mia O'Toole, London, UK (miaotoole@rocketmail.com)
- Katherine Ara, Katherine Ara Ltd, London, UK (studio@katherineara.com)

Gelling predictions: the challenges of taking research into practice

Emma Schmitt

Within the field of textile conservation, wet cleaning processes utilized to improve the aesthetic of an object as well as facilitate preservation through the removal of harmful stains is accepted practice (Tímár-Balázsy and Eastop 2008). The methods for cleaning are many and variable, depending on circumstance. From full immersion cleaning, to the use of suction tables or discs and poultices, the method selected is object, time, and resource-dependent. The introduction of agarose gels into conservation, their subsequent utilization in various sub-fields and successful treatments on textile objects prompted fundamental research into how they respond on various textile substrates (van Dyke 2004; Stockman 2007; Warda *et al.* 2007; Ellis 2009; Sahmel *et al.* 2012; Pouliot *et al.* 2013; Shaeffer and Gardiner 2013; Younger and Duffus 2014; Gulotta *et al.* 2014).

A dissertation project carried out at the Centre for Textile Conservation and Technical Art History at the University of Glasgow, Glasgow, UK, in the summer of 2014 examined the use of agarose gels on textiles (Schmitt 2014), by investigating published treatments and research in the conservation of all materials. An extensive literature review provided information of trends in the use of the gel within conservation. Further research examined the chemical structure and properties of agarose. This information helped to clarify which properties were important for use within conservation. From this information a research plan was developed that, taking these trends into account, attempted to verify the most effective method of application within textile conservation.

The project was developed to examine three fiber types (silk, cotton, and wool), three gel concentrations (1%, 2.5% and 4%), and two depths of gel (3 and 5 mm). These variables were identified as fundamental properties that could be easily controlled for treatment. Examination of fiber type allowed for a better understanding of how the fiber content of an object can change the efficiency of the gel. The range of concentrations chosen covered trends in current practice and sought to offer assessment of the range of concentrations suggested for use within conservation treatments (van Dyke 2004; Stockman 2007; Warda *et al.* 2007; Ellis 2009; Sahmel *et al.* 2012; Pouliot *et al.* 2013; Shaeffer and Gardiner 2013; Younger and Duffus 2014; Gulotta *et al.* 2014). The thickness of the gels was selected because 0.5 cm is the most common gel described in conservation literature and 0.3 cm is the height of the free-form gel that forms when approximately 100 ml 1% agarose is poured on a flat surface (van Dyke 2004; Sahmel *et al.* 2012; Schmitt 2014).

Water-soluble ink was applied to the fabric samples using the gel molds as a template. Agarose gels were placed onto the stained fabric samples; weights were applied and left for one hour. At this point the gels were removed, photographed and assessed, evaluated by their ability to achieve good contact with the textiles, establishing a system of capillary action that effectively drew the ink from the textile without spreading it further into the substrate, forming tidelines.

The results from this research showed that the effectiveness of the gel was reliant on the fiber to which it was applied, as well as the contact that was achieved between the textile and the gel. Gels applied to silk exhibited extensive wicking to the textile due to the hydrophilic nature of silk (Tímár-Balázsy and Eastop 2008). All silk samples exhibited some tidelines and in the case of 1% agarose gels, the textile was wetted out completely. As a result of this, poor capillary action was achieved and very little of the ink was drawn out of the textile (Fig. 1).

Cotton samples gave better results but the agarose gel (1%) treatment still showed distinctive outward movement of the ink and the formation of tidelines. More concentrated agarose gels (2.5%) proved most effective when good contact was achieved between textile and gel. The first two tests indicated that without good contact (Fig. 1), wicking and tidelines occurred. Once light finger pressure was applied, as in tests three and four, capillary action was achieved, wicking was greatly reduced, and the amount of ink entering the gel increased. In 4% gels, less outward movement of the liquid was noted and less ink was drawn into the gel.

Tests on wool were similar to those on cotton. The hydrophobic nature of wool reduced the movement of the ink through the textile overall (Lindberg 1953). Gels (2.5%) exhibited good contact and removal of the ink from the sample and 4% gels showed limited outward movement of ink and limited uptake into the gel. The results were the same for these fabrics using both 3 mm and 5 mm gels. The only difference that could be discerned was a reduction in the movement of the ink in all tests, which we postulate is a consequence of the smaller amount of water in the 3 mm gels.

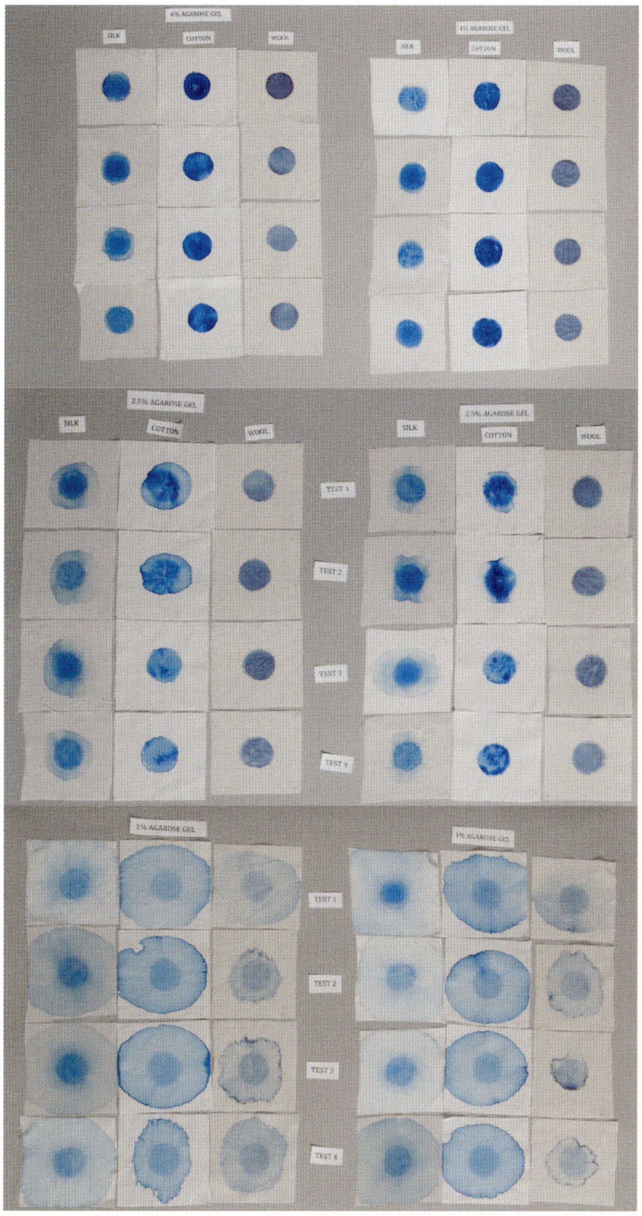

FIG. 1 Movement of ink through fabric samples treated with 4%, 2.5% and 1% agar gels. Gels were either 5 mm thick (*left*) or 3 mm thick (*right*). Fabrics are laid out from left to right in order: silk, cotton, wool.

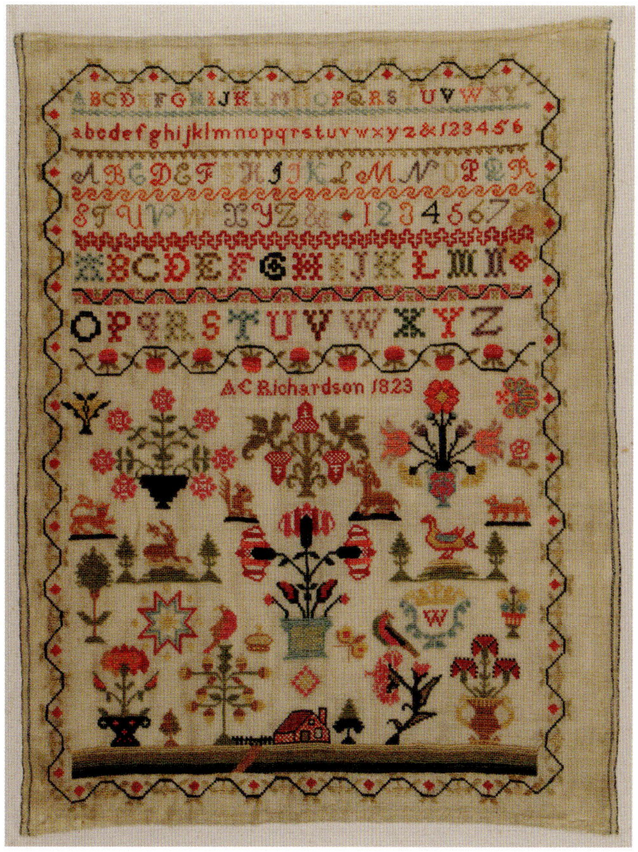

FIG. 2 A.C Richardson, sampler, Denver Art Museum: prior to treatment.

FIG. 3 Water beading on the surface of the sampler showing the effect of the surface treatment.

It was determined that agarose gels pose some risk if used on a silk substrate. The lack of control over the liquid increases the likelihood that staining could be moved through the object while very little soiling is drawn into the gel. The use of these gels on cotton and wool are much safer, especially at a 2.5% agarose concentration and above. At this concentration, the amount of water entering the textile was able to solubilize a stain with little wicking into the surrounding substrate.

In an effort to assess success in practice, an object was identified within the collection of the Denver Art Museum, Denver, USA, that would be well suited for the application of an agarose poultice: an American sampler dating to 1823 (Fig. 2). This multicolored sampler depicts four distinct alphabet styles on the upper half of the object, floral and animal imagery as well as a small house on the lower half, and two figural borders, one dividing the alphabets from the imagery and one enclosing the composition. The sampler was constructed on a loose plain-weave wool ground with silk and wool embroidery. The ground appeared to be coated with some form of stiffener that changed the handling of the wool. The sampler was in good condition with only minor losses on three corners. A brown stain approximately 4 cm across was visible on the proper left side of the piece. The staining was visibly obtrusive, despite the degree of embellishment in the area. It was determined that, if the stain could be reduced or removed, the sampler could be displayed with greater visual effect.

Color fastness testing showed that the embroidery threads offered no risk of bleeding during treatment therefore no barrier layer would be needed. The decision to forgo the use of barrier layers was twofold. The first goal was to follow the

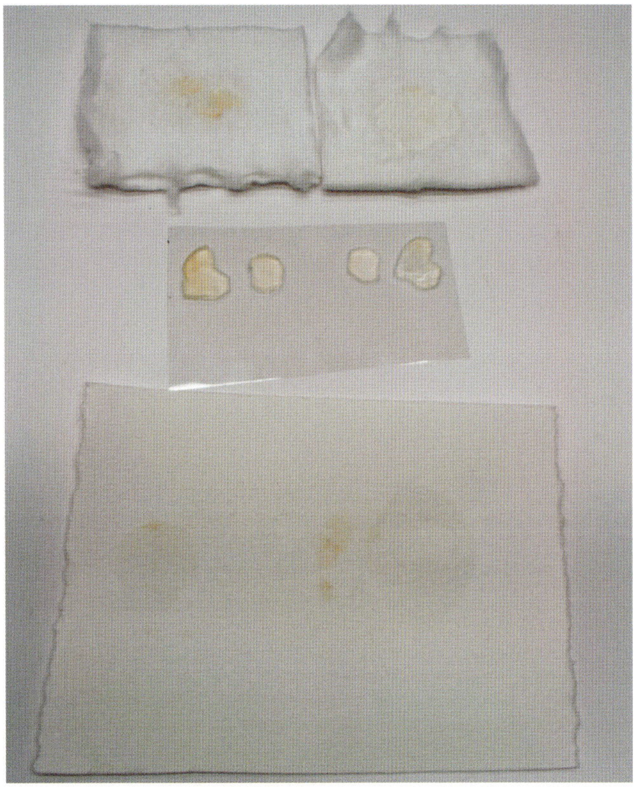

FIG. 4 The gels applied to the surface of the sampler, and the blotter used during suction cleaning, showing how each method was capable of drawing out staining from the textile substrate.

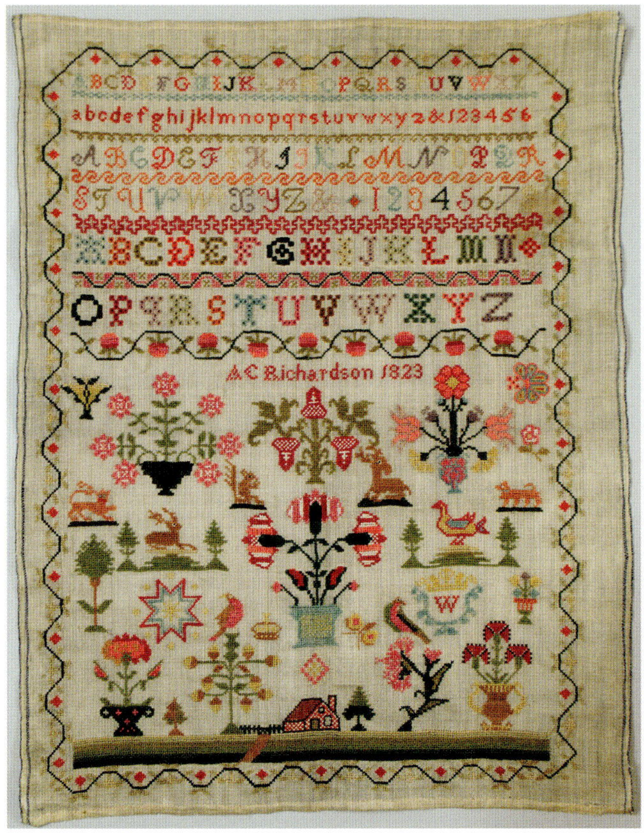

FIG. 5 The sampler after cleaning.

process that had been established in Glasgow; the second was that the introduction of cyclododecane, the barrier layer that has been used in conjunction with agarose in the past (Sahmel *et al.* 2012; Shaeffer and Gardiner 2013) prevents a full understanding of how a liquid might move through a textile.

The research project was crafted to reduce variables that could affect the overall outcome of the treatment. By introducing an object into the same experimental design, control was maintained over the process despite the addition of the multiple variables related to the object. These variables included an aged textile, a coated, possibly starched, open-weave ground, two fiber types, the more three-dimensional contours of the embroidered wool, and multiple colored silk threads. These factors, very common in textile objects, could fundamentally change the outcome of the treatment relative to the control samples in the earlier study.

Based on the wool ground of the textile and the coated nature of the fabric, it was decided to begin the treatment utilizing a 2.5% gel, a concentration that, according to the Glasgow tests, was likely to ensure a successful treatment with little or no movement of water into the ground fabric. A 2.5% gel was prepared and cast at a 5 mm thickness chosen as it was observed that the more hydrophobic wool along with the coating applied to the ground would offer a greater degree of resistance to flushing out the stain (Fig. 3). As the bulk of the staining was located on the wool ground the hydrophilic nature of the silk embroidery threads was discounted.

The stains on the sampler were traced, and the gel cut to the size of the stains. Following the technique developed in Glasgow, the gel was placed on the object and light finger pressure and a weight was introduced to ensure contact between the gel and the textile substrate (Schmitt 2014). Because of the variables introduced, the gel was checked every 15 minutes instead of being left in place for one hour. After half an hour the gel was removed. It was determined that further reduction of the stain was possible and the gel was replaced with a fresh sample (Fig. 4) which was left in place for another 10 minutes. At this point, water wicked out into the textile and a tideline formed. Once this occurred, the gel was removed and the sampler moved to the suction table. The remaining cleaning was carried out with water and a blotter under suction (Fig. 4). After treatment the sampler exhibited reduced staining, although discoloration was still present (Fig. 5).

The formation of the tideline during the treatment of the sampler raised a number of concerns regarding the results of the Glasgow research. Cleaning in practice introduces variables that affect the end results, thus increasing the challenges associated with forming a safe and effective treatment. Earlier treatments utilizing agarose gels used a 1% concentration and barrier layers to prevent and control outward movement of moisture into the textile in an effort to prevent dye bleed (Sahmel *et al.* 2012). As this treatment aimed to spot clean without the introduction of barrier layers, heightened control over the gel was fundamental to the success of the treatment. Based on the results, application of agarose within this context was unsuccessful.

From the research project and this treatment, conclusions can be drawn and inferences made as to the issues with the use of agarose gel poultices on textile substrates. Process and protocols were copied from the Glasgow research, except for the replacement of the gels partially through the treatment. The object introduced a series of variables to the treatment that should also be considered. The coating and open weave of the textile, along with the more textured surface, could also have prevented the gel from achieving the contact needed to establish the capillary action that makes these gels effective and also changed how water moved through the fibers as compared to the scored wool tested in Glasgow. Different choices during the treatment process could have altered the outcome, if not improved it. The second gel applied to the textile could have been exchanged for a blotter. This would have eliminated the introduction of more water, preventing the formation of tidelines, and would have drawn out further staining from the textile ground. The gel could have been reapplied after the textile had dried to establish if further soiling could be drawn from the fibers.

The decision to utilize agarose for this treatment had been taken with the idea that the information gained through research needed to be transferred into practice to fully test the effectiveness of the treatment. The presence of a suction table at the Denver Art Museum and its eventual use for this treatment encourages this faster, safer, and less costly treatment. The suction table offers benefits that a poultice cannot: it allows the object to be fully cleaned, and the area to be rinsed, with numerous applications of a poultice. This treatment could have been finished in an hour, as opposed to the five hours necessary to prepare, treat and triage the object. Because there was no concern over dye bleed for the sampler, the suction table, if not a full immersion wet cleaning, would have been a better option. However, the treatment offered an opportunity to directly assess how research transitions from a controlled setting to active practice. The multiple ways that this object could have been treated allowed this method to be tested safely, in actual practice without the risk of applying the gel to an object that may have responded negatively. It shows that challenges in any textile poultice treatment still exist with agarose gels, and thus risks must still be evaluated. This treatment is also a prime example of pushing the boundaries of effective treatment. Stopping the treatment after the first gel would have prevented the migration of the soiling into the surrounding textile, and while the sampler would not have had the degree of stain reduction that was achieved with the suction table, the results would have been positive and the tidelines that formed with the application of the second gel avoided.

The use of this method within textile conservation remains object-dependent, as it is for the treatment of all other classes of material, and the treatment must be constantly observed by the conservator. Risks may be considered too great for many objects without the implementation of barrier layers or other safeguards. Further testing is needed and care should be taken in assessing all the variables associated with an object before a similar treatment is undertaken.

Materials and suppliers

- IBI molecular biology grade agarose: Universal Medical, 1500 Providence Hwy, Unit 32, Norwood MA 02062, USA, http://www.universalmedicalinc.com/
- Molecular biology grade agarose A9539: Sigma-Aldrich, The Old Brickyard, New Road, Gillingham,Dorset, SP8 4XT, UK, www.sigmaaldrich.com/united-kingdom.html
- Macaroon mold: Lakeland, www.lakeland.co.uk
- Parker Quink:Rymans, www.ryman.co.uk
- Wool delaine, cotton lawn, heavy weight silk *habotai*: Whaleys Bradford Ltd., Harris Court, Great Horton, Bradford, West Yorkshire BD7 4EQ, UK, www.whaleys-bradford.ltd.uk

References

Ellis, S. 2009. 'A passage in the life of a palampore: conservation', *Journal of the Canadian Association for Conservation* 34: 21–8.

Gulotta, D., Saviello, D. Gherard, F. Toniolo, L. Anzani, M. Rabbolini, A. and Goidanich. S. 2014. 'Setup of a sustainable indoor cleaning methodology for the sculpted stone surfaces of the Duomo of Milan', *Heritage Science Journal* 2(6). Available at: http://www.heritagesciencejournal.com/content/2/1/6 (accessed 25 May 2014).

Lindberg, J. 1953. 'Relationship between various surface properties of wool fibers. Part I: Methods for estimating wool fiber modification', *Textile Research Journal* 23(2): 585- 8.

Pouliot, B., Fair, L. and Wolbers. R. 2013. 'Rethinking the approach: techniques explored at Winterthur for the stain reduction of ceramics', in H. Roemich and K. van Lookeren Campagne (eds), *Recent Advances in Glass, Stained Glass, and Ceramics Conservation 2013: ICOM-CC Glass and Ceramics Working Group Interim Meeting and Forum of the International Scientific Committee for the Conservation of Stained Glass.* Amsterdam: ICOM, 211–23.

Sahmel, K., Mina, L., Sutherland, K. and Shibayama, N. 2012. 'Removing dye bleed from a sampler: new methods for an old problem', *Textile Specialty Group Postprints* 22: 78–90.

Schmitt, E. 2014. *An Examination of the Working Properties of Agarose Gels for Textile Conservation,* unpublished Master's dissertation, University of Glasgow, Glasgow, UK.

Shaeffer, E. and Gardiner. J. 2013. 'New and current materials and approaches for localized cleaning in textile conservation', *Textile Specialty Group Postprints* 23: 109–24.

Stockman, D. 2007. 'Treatment options for oil stains on paper', *The Book and Paper Group Annual* 26: 115-26.

Tímár-Balázsy, Á and Eastop, D. 1998. *Chemical Principles of Textile Conservation.* London: Routledge.

van Dyke, Y. 2004. 'Practical application of protease enzymes in paper conservation', *The Book and Paper Group Annual* 23: 93-107.

Warda, J., Brückle, I., Bezur, A. and Kushel, D., 2007. 'Analysis of agarose, Carbopol, and Laponite gel poultices in paper conservation', *Journal of the American Institute for Conservation* 46 (3): 263-79.

Younger, S. and Duffus, P. 2014. 'Trials and tribulations: experiments with agar', paper presented at the *ICON Textile Forum: Joined Up Thinking, London. UK.*

Author's address

Emma Schmitt, Darien Center, NY, USA (easchmitt@gmail.com)

A study of thickened protein glues for the readhesion of absorbent flaking paints with methylcellulose and wheat starch paste

Karolina Soppa, Stefan Zumbühl, Manon Léchenne and Anne Muszynski

Introduction

Since low-viscosity protein glues tend to penetrate into the substrate rather than form a film of adhesive, thus do not provide sufficient adhesion, thickened adhesives are commonly used to readhere absorbent flaking paint or ground. Wheat starch paste (WSP) and methylcellulose (MC) are too viscous to penetrate into the gap caused by delamination, and their adhesion is regarded as too weak: diluted in water, they deliver insufficient adhesive strength. Therefore a modified system of low-viscosity glues is required such as pre-consolidation or isolation with non-polar solvents to prevent the penetration of the adhesive into the adjacent material (Soppa 2016). Another approach is to adjust the viscosity of proteinaceous glues by using gel thickeners. The goal is to create an adhesive system that sufficiently penetrates the gap caused by delamination, while being too viscous to be absorbed into the paint through capillary action. Mixtures of protein glues and WSP or MC as thickeners have been widely used (Bauer and Koller 1965; Heiber 1992;[1] Springob 2001; Laaser *et al.* 2013). It is well known that the properties of mixtures are strongly dependent on the ratios of the individual components. Moreover, the general perception is that mixtures are unstable. The main issues are: what are the requirements for the composition of gel-thickened proteinaceous glue? Do these mixtures form homogeneous or heterogeneous films? Is there a major difference between a protein gel created with MC or WSP?

Generally, gelled adhesives are complex systems. It is important to note that the individual components have different properties. In practice, proteinaceous glues such as gelatine (G) and sturgeon glue (S) are the most frequently used binders. The main difference between the two is the gelation temperature (T_{gel}). A 5% w/w solution of a type A, 180 Bloom[2] gelatine gels at room temperature, whereas a 5% w/w solution of sturgeon glue gels at ~12 °C. MC can be used to create highly viscous gels at lower concentrations than WSP, even though MC has a lower average molecular weight.[3] As a consequence, a 5% w/w concentration MC paste has comparable viscosity to a 10% w/w WSP. Additionally, MC typically exhibits a higher tensile strength than various starch types.[4] Finally, it is important to consider the specific properties of the mixtures. Since the T_{gel} value, molecular weight, shape and polarity of MC and WS influence film formation properties, the fundamental question is: how do these variables affect the miscibility, penetration, performance within a joint, and the adhesion of the mixtures? This paper discusses systematic trials of MC- and WSP-thickened proteinaceous glues on two different absorbent model systems. The aim is to evaluate the performance of the glues as conservation tools that might be used on a real artwork.

Materials and methods

The tests were carried out on two differently absorbing models: delamination between chalk ground on canvas with a WSP isolation layer (model A) and between two absorbent chalk grounds (model B). Glues should be less viscous than a 3% w/w MA A4C solution to penetrate successfully into gaps. Glues and mixtures, as well as their estimated viscosities (all at 40 °C) are described in Table 1. Paint samples and canvas were prepared separately as follows: a washed, ironed, stretched, wetted and re-stretched flax fabric of medium density served as a support. For sizing, a 13% w/w WSP was applied in two layers. The chalk ground plates (1 cm^2) made of gelatine type A, 180 Bloom, 5% w/w and a pigment-filler mixture (chalk, gypsum and iron red, 30:30:1 w/w/w) 5:8, as well as an alkyd paint top layer were attached to the 1 × 15 cm^2 strips of isolated canvas or to each other.

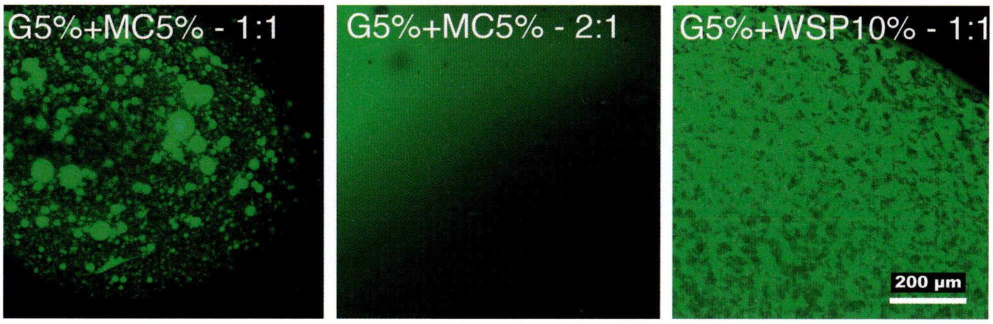

FIG. 1 Mixtures on microscope slides (×5) with green fluorescence for gelatine, using the FITC filterset. (*Left*) 5% gelatine and 5% MC 1:1 is heterogenous; (*middle*) the same 2:1 mixture is homogenous; (*right*) 5% gelatine and 10% WSP 1:1 is heterogenous (images: Karolina Soppa).

					Model A chalk ground on canvas					Model B chalk ground + chalk ground				
	Test series		V	Mixture	A Arithmetic average	F_{max} Arithmetic average	F_{max} Arithmetic average	Failure of joints	Penetration Ø	A Arithmetic average	F_{max} Arithmetic average	F_{max} Arithmetic average	Failure of joints	Penetration Ø
	Adhesive(s)*	Ratio			mm^2 ± SD	N ±SD	N/cm^2 ± SD	–	µm ± SD	mm^2 ± SD	N ± SD	N/cm^2	–	µm ± SD
Protein	S 5% (RT)	–	1	–	58±5	8.4±4.5	14.5	AF/CF	113±30	not visible	0	0	AF/CF	110±43
	G 5%	–	1	–	68±2.3	9.7±3.9	14.3	AF/CF	74±26	not visible	2–3 (invalid)	0	AF/CF	65±21
MC	MC 3% (RT)	–	7	–	39±3.8	9.4±3.4	24.1	AF/CF	92±33	72±4.7	24±6.2	33.3	AF/CF	–
	MC 5% (RT)	–	8	–	45±4.5	12.7±2.2	28.2	AF/CF	39±20	68±12	36±9.1	52.9	SF	–
MC 3%	G 5%	1:1	3	1	52±14.4	12.7±3.5	25.5±8.4	AF/CF	65±20	89±6	20±5.5	22.5	AF/CF	–
	"	1:2	2	2	59±12	15.4±3.2	26.1±4.5	AF/CF	58±23	–	–	–	–	–
	"	2:1	4	3	56±10	11.1±3.5	19.8±4.8	AF/CF	42±13	–	–	–	–	–
	S 5%	1:1	3	4	51±11.7	8.7±5.2	16.3±6.7	AF/CF	–	90±9.3	17.6±7.8	19.6	AF/CF	–
MC 5%	G 5%	1:1	6	5	49±10.3	13.6±4.5	27.6±5.9	AF/CF	28±18	73±12	31±10.6	42.5	AF/CF	11±7
	S 5% (RT)	1:1	6	6	45±9.2	16.8±5.2	36.4±6.1	AF/CF	–	69±10.7	19±8.5	27.5	AF/CF/SF	60±23
	G 5%	1:2	5	7	–	–	–	–	–	58±13.3	15±4.9	25.9	AF/CF	40±11
WSP 10%	G 5%	1:1	4	8	46±10.8	14.8±4.8	30.1±7.8	AF/CF	–	51±10.4	34±18.6	66.7	AF/CF	70±36
	"	2:1	5	9	65±16.8	24.3±7.5	37.2±6.9	AF/CF	–	–	–	–	–	–
	Paygel® + G 5%	1:1	2	10	–	–	–	–	–	55±7.6	33±15.9	60	SF	42±17
	"	2:1	2	11	–	–	–	–	–	62±8.2	28±7.1	45.2	AF/CF/SF	14±7
	Paygel®	–	10	–	–	–	–	–	–	–	–	–	–	–
	S 5% (RT)	1:1	4	12	57±10.6	14.6±7.4	24.9±8.7	AF/CF	–	54±20.6	17±10	31.5	AF/CF	100±35
	–	–	11	–	–	–	–	–	–	57±11	31±16.8	54.4	SF	–

S = Sturgeon glue, G = gelatine, MC = methycellulose, WSP = wheat starch paste
RT = room temperature (22 ± 2 °C)
SD = standard deviation
* General adhesives application temperature = 40 °C
Invalid = No reliable measurement possible
A = lateral glue distrubution
AF/CF = adhesion and cohesion failure in the adhesive, SF = substrate failure
V = estimated viscosity range of the glues (1–10), where 1 represents the lowest viscosity
Red shading indicates undesirable properties
Green shading indicates desirable properties

TABLE 1 Results for adhesives and mixtures showing viscosities, lateral distributions, N, N/cm², failure of joints and penetration into the substrates on models A and B.

The mixtures were stirred for 15 minutes at 40 °C with 350 rotations per minute.[5] The drying time for the adhesive varied from one day to three weeks. On canvas, drying was faster, whereas drying of MC 5% (7 µl) between the chalk ground flakes takes ~15 hours. After pre-testing, the drying times for the mixtures in model B were 2–3 weeks and for model A mostly 48–72 hours. The drying process was accelerated by placing a blotting paper between the samples and the weight (75 g and volume 3 × 2.2 × 1.5 cm); samples were at ambient conditions with relative humidity 50±5% and temperature 22±2 °C.

Penetration and film composition in the joint were investigated by means of fluorochromized gelatine and sturgeon glue, using fluorescein isothiocyanate (FITC) which results in a green fluorescence. In gelatine-MC mixtures only the MC was fluorochomized with dichlorotriazinyl aminofluorescein (DTAF) which results in a green fluorescence, whereas the gelatine was fluorochomized with tetramethyl rhodamine isothiocyanate (TRITC) which results in a red fluorescence. WSP was not fluorochromized but was mixed with fluorochromized gelatine and sturgeon glue (green fluorescence). The mixtures were imaged in thin section (8 µm) using fluorescence microscopy.[6] Some samples were additionally examined with Fourier transform infrared focal plane array (FTIR–FPA) spectroscopy. Adhesion tensile strength tests were carried out according to Comiotto (2009) with a Zwick 1120 (Zwick GmbH &Co – testing speed 50 mm/min) and are described in Soppa *et al.* (2014). The results were recorded in newtons (N) and are presented in N/cm², with the calculated averages and standard deviations. For each test series, 10–15 samples were tested. The different glue mixtures and artificially induced flaking was tricky to evaluate. To maximize the reproducibility of the results of the tensile strength, several individuals conducted the tests.

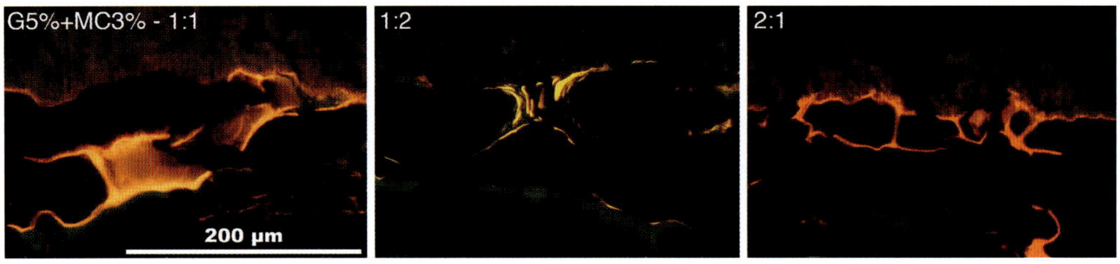

FIG. 2 Typical adhesive bridges shown in cross-section, with red fluorescence for gelatine and green fluorescence for MC. Model A (chalk ground on canvas), showing typical adhesive bridges in a joint made from 5% gelatine and 3% MC, with concentrations indicated; 2:1 had mostly homogenous bridges. The 1:1 mixture surprisingly had two-thirds homogenous bridges while the 1:2 mixture had mostly inhomogenous glue bridges. Imaged at ×20, using the DUAL filterset (images: Karolina Soppa).

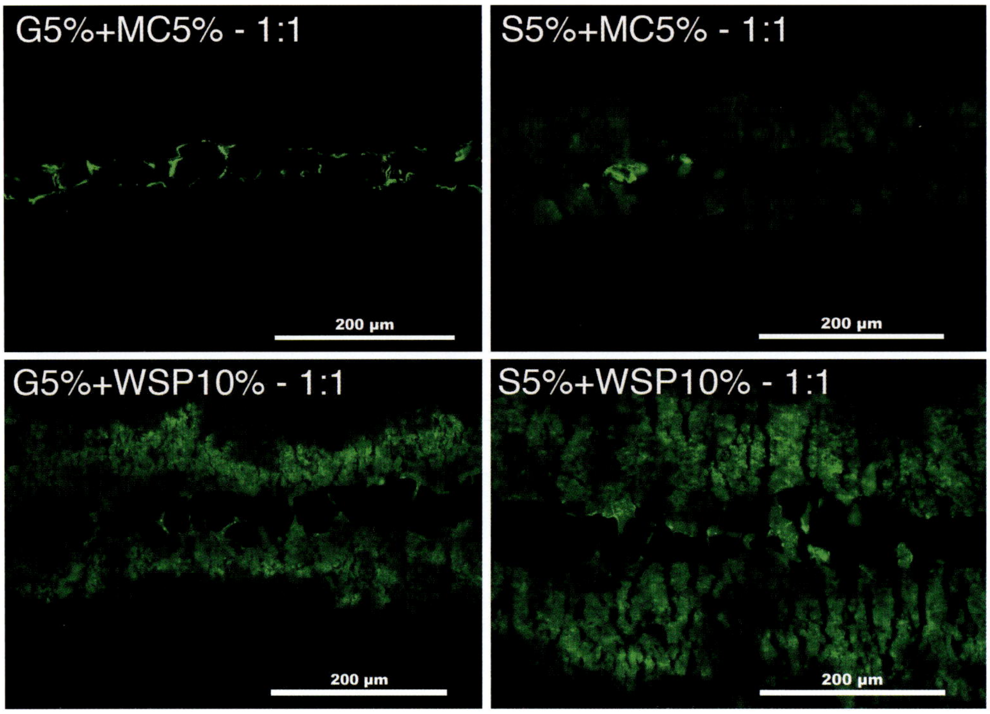

FIG. 3 Penetration and adhesive bridges shown in cross-section, with concentrations and ratios indicated. Model B (chalk ground on chalk ground), showing penetration into the chalk ground and some glue bridges; Gelatine and sturgeon glue have a green fluorescence. The green fluorescence shows the penetration of the proteins into the chalk ground and some glue bridges between the chalk grounds. Imaged at ×20, using the FITC filterset (images: Karolina Soppa).

Results and discussion

All results are summarized in Table 1. First, the solution properties in a beaker were observed. We noted that more rigid gels maintained a homogeneous nature over a longer timespan. Gelatine separated faster than sturgeon glue. To characterize the mixture properties, the pure mixtures were examined under the microscope: both protein types mixed with MC (1:1) formed a dispersion, with MC forming the continuous phase and the protein as the dispersed phase (Fig. 1). In contrast, a mixture of gelatine and methyl cellulose (both 5% w/w, 2:1) was very homogenous, and pockets of MC were very rarely visible. The WSP mixtures were heterogeneous and resembled suspensions to some degree. In mixtures prepared with homemade WSP the starch grains were only partially open, whereas in a pre-cooked WSP (Paygel 290), no starch grains were observed.[7]

Penetration into the substrates

The penetration of glue components into the substrates, bond lines along the interfaces, as well as the glue bridges formed during drying were documented (Fig. 2). As expected, the lower the viscosity of the glue mixture, the higher its penetration into the substrate. Gelatine settled under the flakes on the canvas within 4–5 minutes, whereas sturgeon glue required up to 10 minutes. Gelatine dried faster and penetrated less deeply into the substrate than sturgeon glue, probably due to its higher T_{gel}; this was observed for all mixtures (Fig. 3). In contrast to gelatine, pure 3% w/w MC showed high penetration into the chalk ground and into the isolation layer (Table 1). This can be explained by the long working time before setting and its low molecular weight (41,000 g/mol). When glues were applied, the wheat starch isolating layer on the canvas swelled quite quickly and MC could then penetrate into the isolation layer very easily, whereupon a separation of the mixture was observed (Fig. 3).

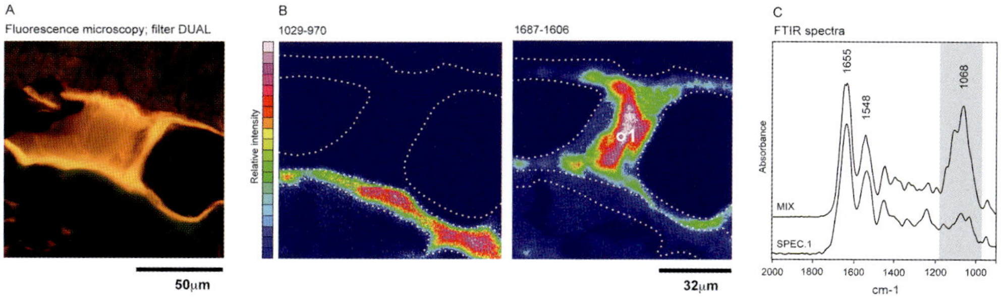

FIG. 4 Adhesive bridge in a joint with 5% gelatine and 3% MC (1:1) on model A. (A) fluorescence image, (B) FTIR–FPA image showing the separation of the binder components: carbohydrates (starch and methyl cellulose, 1029–970cm^{-1}) and protein (1687–1606cm^{-1}), (C) extracted FTIR spectrum (SPEC 1) of the separated protein bridge relative to the film of the original mixture used (images: Stefan Zumbühl).

However, gelatine-MC mixtures penetrated less into the chalk ground (Table 1). Therefore, MC not only thickens the gelatine, but the gelatine also holds the MC in place and the penetration of the adhesive is diminished. In contrast, the WSP-containing protein mixture scarcely prevented penetration into the substrate and exhibited almost the same high penetration of the protein into the chalk ground as for the pure protein glues. This result is probably due to the presence of half-open starch grains in the homemade WSP; when using Paygel 290, the penetration was reduced. It can be concluded that MC at 5% and 3% w/w as a thickening agent worked best, followed by the pre-cooked WSP.

Film formation properties

The experiments showed that the two components of the glue mixtures separate to different degrees during the application, penetration and drying. In the joints, two major decomposition arrangements of the components were visible. In most cases the thickener (WSP or MC) forms glue bridges with small protein inclusions (see Figs 2 and 3), often observed in 1:1 mixtures due to fast drying. This mixture remained heterogeneous while the homogenous 2:1 mixture separated. The second possibility is that the gel carries the protein into the delamination gap and then the separation of the components begins. The separation might be due to the high affinity of one component to the substrate (Fig. 2). For example, the MC in a MC+G-mixture penetrates into the substrates, whereas the protein builds bond lines or adhesion bridges between the interfaces. A two-phase film of MC with gelatine inclusions can often be observed between the protein layers or bridges. This kind of separation can be visualized with different methods. Using fluorescence microscopy the homogenous glue bridges had a red-orange fluorescence, which indicates mainly the presence of protein (red fluorescence). This was confirmed by chemical imaging using FTIR–FPA, which could only detect the protein component. Therefore the MC seemed to penetrate into the WSP isolation layer and/or the chalk ground (see Fig. 4). The following applies: the faster the (1:1) mixtures dried, the more heterogeneous the adhesive film formed. Drying is controlled by the substrate porosity, the location (at the edges of the samples, there were mostly heterogeneous films/bridges, while in the centre, there were more homogenous ones), the T_{gel} (proteins can fuse more quickly when T_{gel} is high; this is equivalent to mixtures in a beaker) and the gel viscosity (thicker mixtures prevent protein mixing).

Adhesion

The adhesion of a gelled glue is strongly dependent on the material interfaces (Table 1). Due to the greater effective surface on model B, the adhesion was higher. On both models the proteins had the lowest, if any, adhesion strength. Gelatine dried faster than sturgeon glue and penetrated less into the substrate. Gelatine generated a higher or equal adhesion with a greater lateral distribution, and it was preferable to most of the mixtures. Considering the substrate, model A had only one highly absorbent side, whereas the WSP isolation layer seemed only to absorb water and MC. Consequently, low-viscosity mixtures performed better on model A. An example can be seen in the mixture of MC and sturgeon glue (both 5% w/w, 1:1). Because of the longer working time, the sturgeon glue could fuse while the MC penetrated into the substrates. In contrast, on model B the same mixture penetrated too far into the chalk ground and resulted in considerably reduced adhesion. The different tests indicated that the mixture of gelatine with WSP thickener (mixture 8) adhered best on model A without substrate failure and showed a wide lateral distribution. In the case of model B, gelatine 5% mixed with MC at 5% as well as with WSP (all 1:1 w/w) adhered best and showed the lowest substrate penetration. The gelatine-MC mixture exhibited adhesion and cohesion failure while the WSP mixtures (1:1) often showed substrate failure and greater substrate penetration, which is particularly undesirable. If the viscosity of a gelatine-MC mixture (both 5% w/w, 1:1) is too high, perhaps a 4% MC could be used as a compromise, given that the 3% variant with gelatine showed a wide lateral distribution but lower adhesion.

Conclusion

All mixtures and the gels adhered better to both models than protein glues alone. Gelatine penetrated less into the substrates and typically adhered better in mixtures than sturgeon glue. The penetration into the substrate of the MC mixtures was always lower than that of their individual components. Although the gelatine-MC mixture (both 5% w/w, 1:1) performed best on model B, penetration into the gap was not satisfactory considering the viscosity. If a wide lateral distribution is required the MC concentration should be lowered or alternatively pre-cooked WS with G (2:1) should be used, since this mixture prevents substrate penetration and has a lower viscosity. The same mixture should be optimal for model A. These results confirm that the adhesion of a mixture of protein glue with different thickeners depends mostly on the capability of the latter to form a stable mixture with the protein component, the T_{gel} of the protein/gel mixture, and the nature of the substrate.

Acknowledgements

The authors would like to thank Christoph Krekel (Stuttgart State Academy of Art and Design), Elke Mentzel (Bern University of Applied Sciences), Brigitte Lienert-Pärli (Bern University of Applied Sciences), Volker Schaible (Stuttgart State Academy of Art and Design), Jörg Scheller (Zurich University of the Arts), Nadim Scherrer (Bern University of Applied Sciences) and Karin von Lerber.

Materials and suppliers

- Sturgeon glue, swim bladders (*Acipenser baerii* BRANDT) purchased and prepared as indicated and recommended (2 hours at 68 °C plus centrifugation): Stoerleim GmbH, www.stoerleim-manufaktur.de.
- Gelatine solution prepared according to Gelita guidelines: 2 hours swelling and 15 min dissolving by warming to 55 °C while stirring.
- MC prepared with the hot-cold method and minimum 12 hours in the fridge. The 3% solution was always gently stirred before usage and allowing to set for a few minutes.
- Wheat starch paste: 15–20 minute cooking until gelation is reached, and 15–20 minute continued heating and stirring, followed by sieving.
- Paygel 290: Archer Daniels Midland Company (ADM), www.pel.eu.

Notes

1. See Springob 2001.
2. The Bloom value indicates gel strength. For proteinaceous glues it is in the range 50–300.
3. The M_n (average molecular weight) is 41,000 g/mol for MC, 160,000–800,000 g/mol for wheat starch as amylose (mainly linear, 25%), and 1,600,000 g/mol for amylopectin (highly branched, 75%) according to BeMiller and Whistler 2009: 457.
4. The tensile strength of MC is 1,000–10,000 N/cm^2 and for starch 160-2,100 N/cm^2, according to Wypych 2016.
5. For materials and application see also Soppa *et al.* 2014 and Soppa 2016.
6. For thin-section production see Soppa *et al.* 2014. Fluorescence microscopy: Leitz DMRB microscope with CoolLED pE-300, 2B:100 (and 3G:5, when using DUAL-filterset), and two filter sets (FITC (green): BP 475-495, LP 510, BP 512-542) and (DUAL (green and red): BP 468 and 553, BP 493 and 574, BP 512 and 630) and a Jenoptik ProgRes camera.
7. Subsequent testing showed that almost all of the grains are opened in 5% w/w WSP prepared using a microwave oven (800 watt, < 60 sec), while a 10% WSP solution showed many closed starch grains, had no remaining water and resulted in a very rigid paste. It is important to view the WSP under a microscope.

References

BeMiller, J. and Whistler, R. (eds) 2009. *Starch: Chemistry and Technology*, 3rd edn. Amsterdam: Elsevier Academic Press.

Bauer, W.P. and Koller, M. 1965. 'Festigung loser Bemalung auf ethnographischen Holzobjekten', *Der Präparator* 4: 235–8.

Comiotto, A. 2009. 'Fine-tuning atmospheric plasma to improve the bondability and coatability of plastics in modern and contemporary art and design', in T. Bechthold (ed.), *Future Talks 009: The Conservation of Modern Materials in Applied Arts and Design*. Munich: Die Neue Sammlung, 33–9.

Laaser, T., Soppa, K. and Krekel, C. 2013. 'Lokalisierung von Konsolidierungsmitteln in Gemälden durch Fluoreszenzmarkierung. Teil II', *Zeitschrift für Kunsttechnologie und Konservierung* 27(2): 218–27.

Soppa, K. 2016. 'Wegweisende apolare Lösungsmittel? Teil 1: Vorabsperrung textiler Bildträger und Kreidegrundierung durch apolare Lösungsmittel zur Malschichtklebung mit Gelatine', *Zeitschrift für Kunsttechnologie und Konservierung* 30(2): 363–78.

Soppa, K., Laaser, T., Krekel, C., Genton, M. and Seidel, T. 2014. 'Adhesion and penetration of sturgeon glue and gelatines with different Bloom grades', in J. Bridgland (ed.), *Preprints of the ICOM-CC 17th Triennial Conference*, paper 1313. Paris: International Council of Museums. [CD only]

Springob, C. 2001. 'Stärkekleister als Verdickungsmittel von Störleim zur Malschichtfestigung', *Zeitschrift für Kunsttechnologie und Konservierung* 15(1): 111–32.

Wypych, G. 2016. *Handbook of Polymers*, 2nd edn. Toronto: ChemTec Publishing.

Authors' addresses

- Karolina Soppa, Bern University of Applied Sciences, Bern, Switzerland (karolina.soppa@hkb.bfh.ch)
- Stefan Zumbühl, Bern University of Applied Sciences, Bern, Switzerland
- Manon Léchenne, Bern University of Applied Sciences, Bern, Switzerland
- Anne Muszynski, Bern University of Applied Sciences, Bern, Switzerland

Agarose-enzyme gels in paper conservation

Yana van Dyke

Introduction

This paper examines the significant benefits of an agarose gel as an enzyme delivery system for use in the conservation of works of art on paper. The estimable value of enzymes has long been recognized in the object-tailored precision treatment of works of art, offering advancements at the physical and chemical levels. Hydrolase-type enzymes are frequently utilized to break down adhesive residues from previous restorations or to facilitate the removal of secondary supports such as linings or mounts. The principal advantages of these enzymes are their specificity and efficiency in catalysing hydrolytic cleavage of polymers such as polysaccharides, proteins, and lipids. Three of the most useful hydrolase-type enzymes to the paper conservator are alpha-amylase, protease, and lipase, which expedite the breakdown of carbohydrates, proteins, and lipids respectively. Enzymes, however, require a moist environment, without which, their unique configuration is lost and their catalytic ability destroyed.

Agarose, a homogenous polysaccharide, is extracted from agar, a more heterogeneous mixture of carbohydrates derived from a species of red algae. Utilized ubiquitously in molecular biology, agarose is a preferred matrix for work with proteins. Electronically neutral and possessing a broad range of physical, chemical, and thermal complexity, it is unlikely to interact with coupled biomolecules, thus preserving the enzymes without altering their effectiveness in conservation treatment. There are numerous advantages for conservators when using an agarose-enzyme gel in appropriate situations. Specificities related to the working properties of both enzymes and agarose will be provided along with fundamentals of selection, cost-effective purchasing, and variables in the preparations. Targeted treatment-specific agarose enzyme gels are particularly pragmatic for works of art possessing water-sensitive media that cannot be immersed, and for localized *in situ* accretion breakdown.

Agarose

Agarose is a purified linear galactan hydrocolloid isolated from agar or agar-bearing marine algae. This gelling polysaccharide is structurally a linear polymer consisting of alternating D-galactose and 3,6-anhydro-L-galactose units. Due to varying combinations of the 1,4 and 1,3 linkages in agarose, helices occur in discreet regions of the chain. The helix-forming regions of agarose are the crosslinking points; polymer chains intertwine at the helices. There is strong intermolecular bonding between polymers at these points as well as hydrogen bonding between spirals in the helix regions of the chain itself. Currently there are at least 12 types of agarose available. Much of the commercial market is driven by the food and drug industry and its universal use in DNA gene sequencing by electrophoresis in biochemical laboratories.

Agarose Type VII or low gelling temperature agaroses are excellent for in-gel enzymatic reactions. Gels exhibit excellent clarity and are available in an appropriate range of gel points (the temperature at which an aqueous agarose solution forms a gel as it cools). Purchased in a powdered state, agarose is added to water heated to 80–100 °C, solvating the polymer chains. Separated, unfurled, and provided with sufficient energy to break bonds, the chains begin to crosslink at helix-forming regions on cooling. The agarose gel on its own has been employed successfully in conservation procedures as a poultice ranging from 1 to 2%.

The concentration of agarose directly affects the extent of crosslinking of polymer chains and in turn the rate of diffusion of the enzymes and water. Because there are greater numbers of chains in solution in gel preparations of higher concentrations, there are more helix-forming regions available for crosslinking, creating a matrix through which water and enzymes diffuse. With greater crosslinking, the matrix is more extensive and dense, resulting in a smaller 'pore size' (Tuvikene *et al.* 2008). This more extensive network slows diffusion because moisture and enzyme has a greater distance to travel to the gel–substrate interface. Therefore, the rate of diffusion is slower in gels of higher concentration. Depending on the desired use, different concentrations can be prepared, described by weight to volume percentages of agarose to water. Generally, a 1.0–1.5% final concentration w/v is useful as a medium for enzymes. Under 1%, the gel may be too wet where planar distortions may be a problem, and tidelines may occur in unwashed objects. Above 1.5%, the gel can be too dense and diffusion of the enzyme may be too sluggish to be effective or useful. Experimentation with preparing the gel is indispensable in order to determine the concentration and consistency most appropriate for a given object or specific conservation problem.

Enzymes

Enzymes are powerful proteins and highly specific catalysts (Wolbers 2000). Present in all living cells, they perform vital functions of metabolic processes whereby nutrients are converted into energy and fresh cell material. Some of the best-known enzymes are those found in digestion and, fortuitously, the most commonly employed in fine art conservation. Pepsin, trypsin, and peptidases break down proteins into amino acids; lipases split fats into glycerol and fatty acids; and amylases break down starch into simple sugars. Long-chain proteins are coiled and folded thousands of times into specific molecular shapes that function to facilitate or catalyse specific chemical reactions, accelerating reaction rates by factors as much as a million times.

Enzymes are specific and efficient. Active sites within the enzyme are related to the physical shape and charge arrangement within the molecule. In the case of lipase (Stockman 2007), protease (van Dyke 2004), and amylase (Ericson 1992), the reactions involve breaking molecular bonds and then adding water to the severed bonds to complete the reaction. As with other catalysts, enzymes are required in relatively small amounts and remain unchanged at the end of the reaction. The molecules upon which the enzyme may act are referred to as substrates, and enzyme-converted molecules are referred to as products.

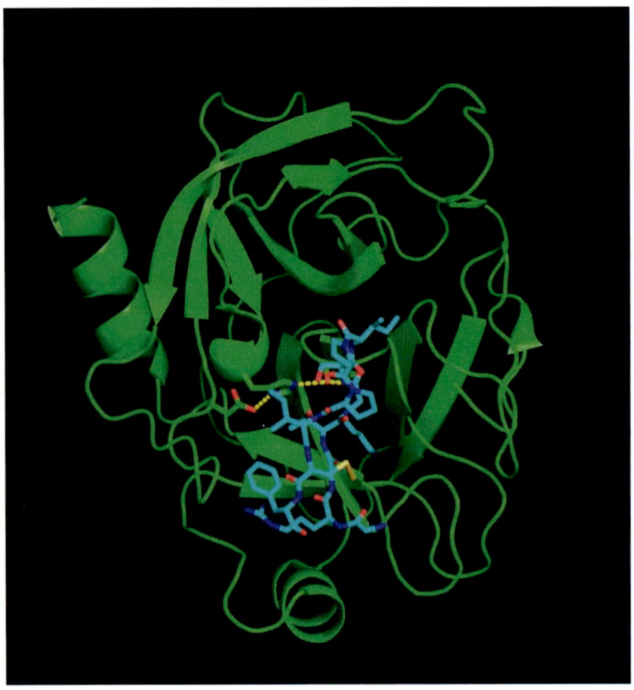

FIG. 1 Computer-generated model of a trypsin enzyme structure. The active site is in the center.

Although enzymes are large molecules with hundreds of amino acids, only a small part of the enzyme participates in the catalysis of biochemical reactions. This is the active site. The three-dimensional structure of the enzyme determines the appearance of the active site, which precisely serves to act upon the substrate to be transformed. Enzyme structures will unfold when heated or exposed to chemical substances disrupting the overall shape and resulting in loss of activity, referred to as denaturation. The catalytic activity of many enzymes depends on the presence of small molecules termed cofactors, and coenzymes that assist in the reactions, often species source dependent (Copeland 2007). Binding loosely to the enzymes, coenzymes are organic molecules often located at the active site, attracting or repulsing a substrate or product (Fig. 1).

Enzyme classification

Categorized by the nature of the bond cleavage reactions that they catalyse, enzymes are assigned numbers by the Enzyme Commission (EC). In 1955, the International Union of Pure and Applied Chemistry formed the Enzyme Commission and began the enzyme nomenclature scheme. Each enzyme is allocated a four-digit EC number: the first three digits define the substrate and types of bonds upon which the enzyme acts more specifically within each; and the fourth is a unique identifier or serial number. The EC number begins with the broadest class, followed by a sequence of subclass numbers.[1]

E.C.3._._._ represents the group of hydrolases. The enzymes most often utilized in conservation belong to the hydrolase group, forming two products from a substrate by hydrolysis.[2] For example:

TRYPSIN 3.4.21.4

3	Hydrolases (enzymes that use water to break up some other molecule)
3.4	Peptidases (hydrolases that act on peptide bonds)
3.4.21	Serine endopeptidases (clipping from within the polypeptide chain)
3.4.21.4	Trypsin (from a multitude of sources: bacterial, various mammalian, etc.)

Enzyme activity units and assays

The measurement for enzyme performance is simply called 'a unit', which is a defined amount of enzyme. A unit can be seen as a measure of speed: the more units, the faster a particular reaction will occur. Activities vary with temperature and pH, and test conditions are usually precisely defined for the assay. When purchasing, look for enzymes with the most units per weight, which will generally give the highest specific activity. They are the purest type of enzyme requiring considerably less mass and lacking unwanted fillers, extenders, and bulking agents. Measuring activity of an enzyme varies among research labs, but according to the Commission on Enzymes for the International Union of Biochemistry: one unit (U) is defined as that amount of enzyme which will catalyse the transformation of one micromole of substrate per minute under defined conditions (Bisswanger 2014).

For investigators in different laboratories to reproduce one another's results, consistency in industrial applications, and a host of other critical reasons, it is imperative that product information be reported in meaningful units and accompanied by sufficient details of the assay used. Standardized reporting of activity measurements is on its way to being adapted universally and, in general, commercial manufacturers provide specifics on each lot of enzyme isolated and sold in the consumer marketplace. Specific buffer systems used in the reaction mixture, the pH and temperature at which the assay was recorded, the time interval over which initial velocity measurements were made, and the detection method used are most commonly provided. Turnover numbers are typically reported in terms of molarity change per unit time per molarity of enzyme, moles substrate lost or product produced per unit time per mole of enzyme or equivalently, molecules of substrate lost or product produced per unit time per molecule of enzyme. The most commonly purchased form of enzyme is a water-soluble lyophilized powder.

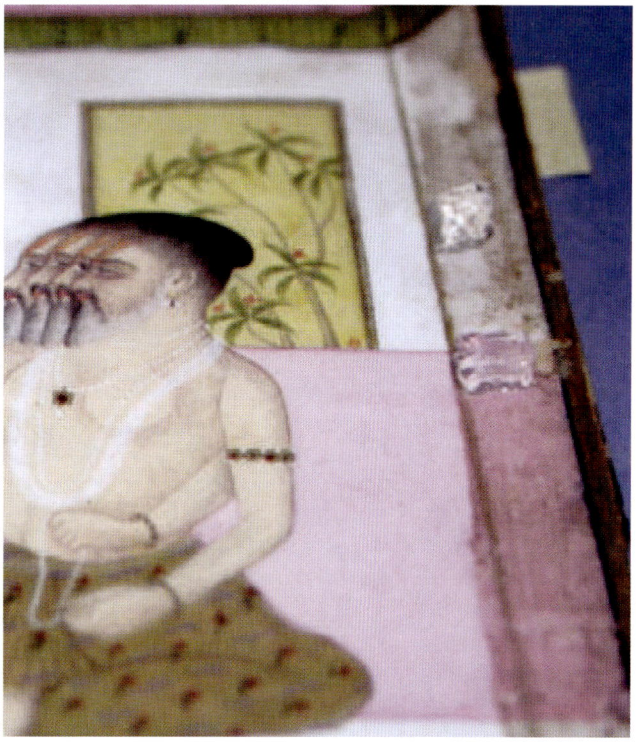

FIG. 2 During treatment of TMMA 2000.321, agarose gel blocks in place along the right edge.

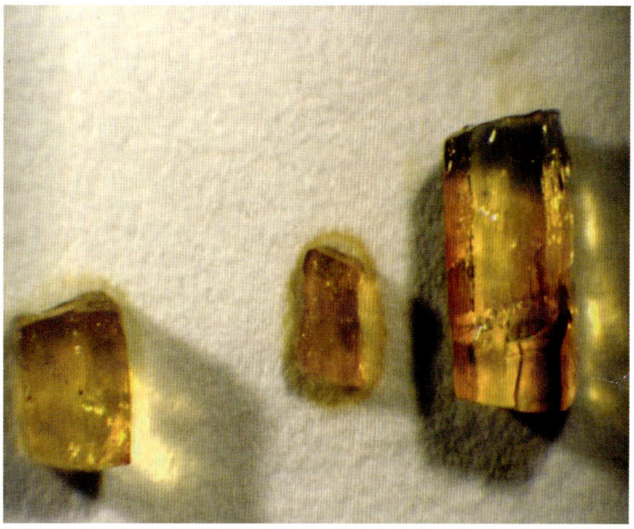

FIG. 3 Detail of used agarose enzyme gel blocks, infused with protein adhesive.

Calculating concentrations

Extensive experimentation has shown that enzyme concentrations of 400 units/ml provide ample digestion activity for successful enzyme agarose gels:

$$\text{Enzyme} = \frac{\text{Concentration (volume)} \times \text{Total Volume}}{\text{Activity Solid}}$$

$$\text{Sample gel: } \frac{400 \text{ units activity/ml} \times 30 \text{ ml}}{17{,}600 \text{ units/mg}} = 0.68 \text{ mg or } 0.00068 \text{ g enzyme}$$

Understanding the unique requirements of the enzyme chosen to carry out desired treatment effects is essential for success. Creating the most favorable environmental conditions for the enzyme to work efficiently and effectively is the goal in conjunction with the wellbeing of the work of art.

Preparing enzyme agarose gels

To use an agarose gel as a vehicular enzymatic delivery system, the conservator should chemically identify the substrate to be acted upon and then select the most suitable enzyme (Odegaard *et al.* 2005). The enzyme will be incorporated into the agarose before it reaches the gel point. When shopping for an agarose it is important to find a low-gelling temperature type in order to maintain the stability and function of the enzyme which will be incorporated as the gel cools at the lowest possible temperature before congealing at its 'gel point'. Gel concentration will directly impact the enzyme's mobility and thus higher concentrations will result in denser matrices slowing enzyme diffusion and efficiency. Based upon numerous experiments, agarose gels above 2% show no advantages over control gels (agarose without enzyme).

A 1.0–1.5% gel is prepared by weighing out and dissolving agarose powder[3] in 90% of the total volume of deionized water until all clumps are broken up.[4] Heat the suspension to 90 °C and allow to boil either in a beaker on a hot plate or in a microwave oven[5] until it is a clear and homogeneous solution. Cool to 15 °C above the gel point. While waiting for the gel to cool, weigh out the enzyme and place in the final container (but see below) and add the remaining 10% of the total volume of 'water' in the form of appropriate buffering

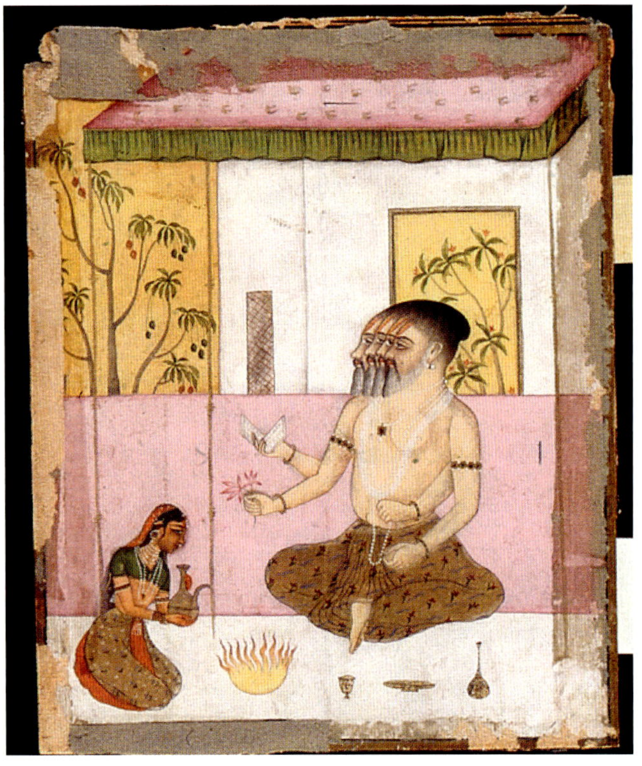

FIG. 4 *Khambhabati Ragini: A Page from a Dispersed Ragamala Series*, *c.*1675, opaque watercolor and gold on paper, 16 × 13 cm, The Metropolitan Museum of Art, 2000.321: before treatment

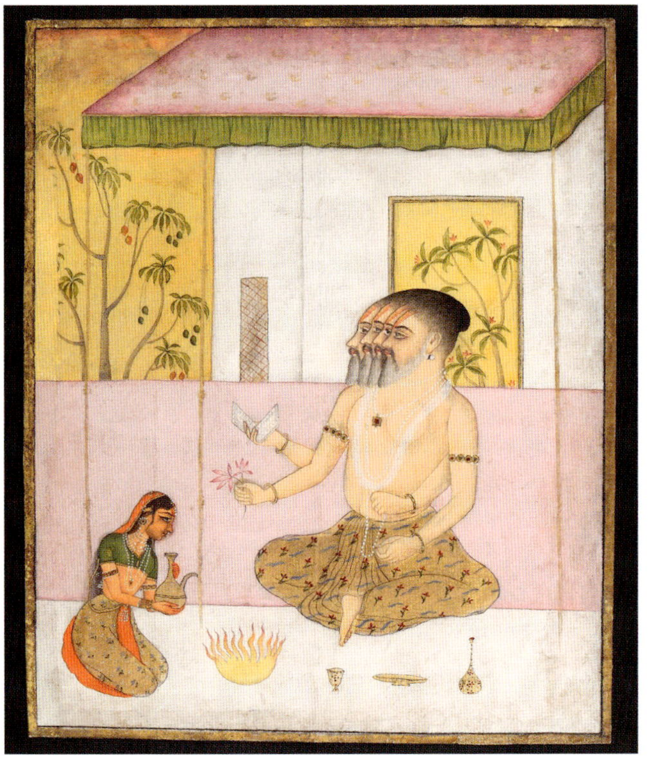

FIG. 5 TMMA 2000.321, after treatment with agarose-encapsulated trypsin gel.

systems[6] and coenzymes if needed. Once the gel has cooled to 15 °C above its gel point, pour the gel into the molding container and gently swirl, avoiding the creation of air bubbles, to incorporate the enzyme mixture evenly into the agarose.

Common problems related to handling issues have been encountered within the research community relating to the loss of enzymatic activity due to protein adsorption onto the surfaces of containers and pipette tips. Proteins bind avidly to glass, quartz, and polystyrene surfaces. Hence, containers made of these materials should not be used for enzyme samples. Containers and transfer devices constructed of low-protein binding materials such as polypropylene and polyethylene, should be used whenever possible: all are available commercially. Maintain high standards of chemical practice when transferring lyophilized solids to the scale; minimize contamination. Store enzymes as recommended by the supplier for longer shelf-life.

Beginning with small pieces of the enzyme agarose gel placed in discreet areas on the work of art, carry out an initial critical evaluation on the working properties of the gel and observe variable responses of the artwork, preferably under magnification. Be alert and plan for interventions related to alterations in surface texture, creation of planar distortion, tideline formation, or disruption of sensitive media. As with any conservation treatment, a host of complex parameters are to be kept in mind while gauging risks against ideal outcome goals; a constant balancing act. Once treatment protocols have been established, cut appropriately sized gel blocks and put in place (Figs 2 and 3). Dwell times per block will vary with the availability of substrate to the enzyme (especially any non-water soluble material). Agarose enzymatic gel treatment advantages include:

- at concentrations of 1–1.2%, the agarose enzyme gel remains soft, flexible, and pliant enough to conform to the surface being treated;
- capillary action coupled with the enzymatic cleavage on specified substrate;
- gel is clear allowing one to see substrate being acted upon;
- gel can be tailored to a specific shape as compared with other viscous poultices;
- minimal physical manipulation is necessary once *in situ*;
- concentration can be slightly varied to alter wetness;
- enzyme gel keeps up to 24 hours with refrigeration, although with reduction in activity;
- isolated treatment area can have extended exposure to enzymatic action;
- treatment is safe, with negligible and minimal residuals (van Dyke 2004; Warda *et al.* 2007);
- green alternative to organic solvents.

Conclusion

Works of art on paper, parchment, and palm leaf that cannot be treated through full aqueous immersion techniques are ideal candidates for in-gel enzymatic procedures. The breakdown of tenacious adhesive residues embedded in the surfaces of palm leaf manuscripts, Indian paintings on paper, European and Islamic manuscript illuminations on paper or parchment, and drawings attached in albums have been successfully treated with agarose-encapsulated enzyme gels. These gels can also serve the rare book conservator in facilitating the removal of adhesive residues along the spine, and address individual pages that are constrained, bound up in the textblock, allowing for treatment *in situ*.

Enzymes and their specificity have been and will continue to be important and powerful tools to the conservator, allowing us to treat objects otherwise untreatable. With an understanding of the technical and physical nature of the objects as well as learning and borrowing from many disciplines, treatment strategies continue to be shaped and refined to fit the needs of these unique and irreplaceable works of art in our care. Analysing and diagnosing the problems as precisely as possible and following through with targeted treatment strategies carries us well into the new millennium exploring vast opportunities for continued refinement. Let the current work inform future experiments in the practical applications of enzyme agarose gels (Figs 4 and 5).

Acknowledgements

I would like to express my eternal gratitude and deepest respect to Richard C. Wolbers, Associate Professor, Coordinator of Science, and Paintings Conservator, The Winterthur/University of Delaware Program in Art Conservation, whose ethos and brilliant mind offers an enduring path of precision and refinement for our profession. His pioneering ideas, so generously shared, invigorate, inspire, and continue to advance enlightened object-tailored conservation treatment procedures, benefitting a multitude of conservation disciplines around the world. Additionally, I am especially grateful to Marjorie Shelley, Conservator in Charge, Sherman Fairchild Center for the Conservation of Works of Art on Paper, The Metropolitan Museum of Art, New York, for her unwavering support and insatiable quest to learn and employ innovative treatment procedures on the multifarious works of art in our care.

Notes

1. The Enzyme Data Bank: http://www.ebi.ac.uk/thornton-srv/databases/enzymes/. This encyclopedic catalog of known enzymes includes extensive information on EC numbers, catalytic activities, specific cleavage sites along with the latest discoveries. Release update: November 30, 2016. ExPASy Molecular Biology Server: http://www.expasy.ch. Home page of the Swiss Institute of Bioinformatics, containing numerous databases and protein analysis tools as well as a host of links to other molecular biochemistry sites.
2. Enzymes sourced from Sigma-Aldrich (http://www.sigmaaldrich.com/):
   - Trypsin EC 3.4.21.4, optimal pH 7.0, Type IX-S lyophilized powder, 13,000–20,000 BAEE units/mg protein, pH 7.6 at 25 °C.
   - α-amylase EC 3.2.1.1, optimal pH 7.5 – 8.5, Type II-A lyophilized powder from *Bacillus* sp. 1,4-alpha,D-Glucan glucanohydrolase, 1,500–3,000 units/mg protein (Biuret). Unit definition: one unit will liberate 1.0 mg of maltose from starch in 3 minutes. pH 6.9 at 20 °C, chloride ion essential (requires a Cl^-); Ca^{2+} and Cl^- necessary for stability.
   - Lipase EC 3.1.1.3, optimal pH 7.2 at 37 °C, Type VII pancreatic, 700–1,500 units/mg solid. Tri-, di-, and monoglycerides are hydrolysed (in decreasing order of rate). Unit definition: one unit will hydrolyse 1.0 micro equivalent of fatty acid from a triglyceride in 1 hour at pH 7.2 at 37 °C (Biuret). Ca^{2+} ion essential, co-lipase counteracts certain inhibitory effects.

   Choose only high purity enzymes usually described in catalogs as crystalline or lyophilized. Avoid using crude forms even if the price is one-tenth of others: the crude ones are full of extenders, fillers, unwanted and sometimes unknown bulking agents. Use of impure enzymes described as 'crude' results in adhesion of contaminant proteins.
3. Suggested sources for purchasing agarose are:
   - Agarose 4018 (Sigma http://www.sigmaaldrich.com/catalog/product/sial/a4018?lang=en&region=US), Type VII low gelling temperature, gel point < 30 °C, exhibits excellent clarity
   - Agarose LE (Benchmark Scientific Inc. http://benchmarkscientific.com/Agarose.html), gel point 36 °C±1.5 °C.
4. If all of the agarose is not dissolved, the result will be non-gelation, and a watery, turbid gel.
5. Microwave for 1–3 min until the agarose is completely dissolved but do not over boil, as this will alter the final percentage of agarose in the gel. Many people prefer to microwave in pulses, swirling the flask occasionally as the solution heats up. A note of caution: wear heat-protecting gloves and be very careful stirring, as eruptive boiling can occur.
6. Suggested sources for buffering agents:
   - Trizma buffers: http://www.sigmaaldrich.com/catalog/substance/trizmabuffersubstance1234598765?lang=en&region=US
   - Bis-Tris buffer: http://www.sigmaaldrich.com/life-science/core-bioreagents/biological-buffers/bis-tris-specification.html

   Ethylenediaminetetraacetic acid (EDTA) is added to trypsin solutions as a chelating agent that neutralizes calcium and magnesium ions that obscure the peptide bonds on which trypsin acts. Removing these ions increases the enzymatic activity. Trypsin solutions in 1 mM HCl are stable for 1 year in aliquots stored at –20 °C. The presence of Ca^{2+} will also diminish the self-autolysis of trypsin and maintain its stability in solution.

References

Bisswanger, H. 2014. 'Enzyme assays', *Perspectives in Science* 1(1–6): 41–55. Available at: http://dx.doi.org/10.1016/j.pisc.2014.02.005 (accessed 25 May 2017).

Copeland, R. 2007. *Enzymes: An Introduction to Structure, Mechanism, and Data Analysis*, 2nd edn. New York/Toronto: Wiley-VCH.

Cremonesi, P. 2013. 'Rigid gels and enzyme cleaning', in M.F. Mecklenburg, A.E. Charola and R.J. Koestler (eds), *Cleaning*

2010: New Insights into the Cleaning of Paintings. Washington, DC: Smithsonian Institution Scholarly Press, 179–84.

Ericson, H. 1992. 'Usage recommendations for alpha-amylases: maximizing enzyme activity while minimizing artifact binding residues', *The Book and Paper Group Annual* 11: 24–33.

Odegaard, N., Scott, C. and Zimmt, W.S. 2005. *Material Characterization Tests for Objects of Art and Archaeology*. London: Archetype Publications.

Stockman, D. 2007. 'Treatment options for oil stains on paper', *The Book and Paper Group Annual* 26: 115–26.

Tuvikene, R., Truss, K., Kollist, A., Volobujeva, O., Mellikov, E. and Pehk, T. 2008. 'Gel-forming structures and stages of red algal galactans of different sulfation levels', *Journal of Applied Phycology* 20: 527–35.

van Dyke, Y. 2004. 'Practical applications of protease enzymes in paper conservation', *The Book and Paper Group Annual* 23: 93–107.

Warda, J., Brückle, I., Bezur, A. and Kushel, D. 2007. 'Analysis of agarose, Carbopol, and Laponite gel poultices in paper conservation', *Journal of the American Institute for Conservation* 46(3): 263–79.

Wolbers, R. 2000. *Cleaning Painted Surfaces: Aqueous Methods*. London: Archetype Publications.

Author's address

Yana van Dyke, The Sherman Fairchild Center for the Conservation of Works of Art on Paper, The Metropolitan Museum of Art, New York, NY, USA (yana.vandyke@metmuseum.org)

Agarose gels with methyl ethyl ketone for cleaning a 19th-century document

Beatriz Aguilar Sánchez and Rodrigo Buentello Martínez

Introduction

The use of gels for cleaning a 19th-century paper document identified as a 'love letter' from the collection of the Historical Archive of Tepapayeca, in the state of Puebla, Mexico, is described. The investigation and treatment of the piece was carried out using methods outlined at a workshop on *Restoration of Documents and Graphic Work in Paper* (*STRDOGP*).

The document consists of a single sheet of cotton rag pulp, with text written in iron gall ink. The aim was to evaluate physical-chemical cleaning using a water-agarose-Canasol NF-1000 rigid gel in combination with a soft gel made from Carbopol 934 and methyl ethyl ketone (MEK, 2-butanone). The gels were selected based on the condition of the letter, and the advantages offered relative to immersion cleaning: control over the wetting of the paper; controlled use of volatile solvents; structural stability of the sample during handling; easy monitoring of action during the treatment time, and localized use of materials. The document was extremely fragile and had undergone a series of alterations due to its use and history. It was folded and fragmented, had significant concretions of dirt and dust on the surface, and generalized abrasion, drying fronts, deformation, and an oily stain covering 50–60% of the area (Aguilar Sánchez and Buentello Martínez 2012). The paper also displayed loss of strength due to extensive degradation and general yellowing due to oxidation of the cellulose (Fig. 1).

Initially, it was washed by immersion in warm water (40 °C) with 0.1% Canasol NF-1000 surfactant. Despite great care being taken during the treatment, the movement of the paper generated new breaks and fragments. Additionally the oil stain could not be removed, so the cleaning methodology was re-evaluated. Priority was given to the following points:

- moistening time for the paper support
- penetration of the solvent into the support, without generating spots or halos
- structural stability of the support
- extraction of yellowing
- removal of the oily substance without abrasion of the support or a loss of a large amount of information.

In addition, penetration of the solvent into the paper could be a disadvantage because solvents applied directly to porous surfaces usually generate halos and may redistribute the stain or solubilized substance without extracting it (Scott 2012).

Rigid gels have been used for the removal of layers of crosslinked polymers from painted surfaces (Campani *et al.* 2007) so we decided to test a rigid agarose gel and Carbopol

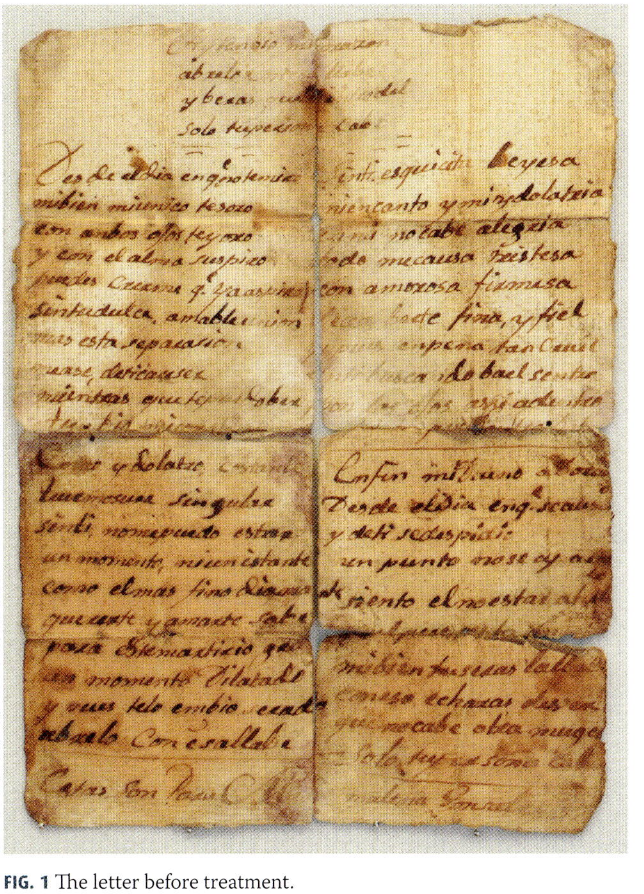

FIG. 1 The letter before treatment.

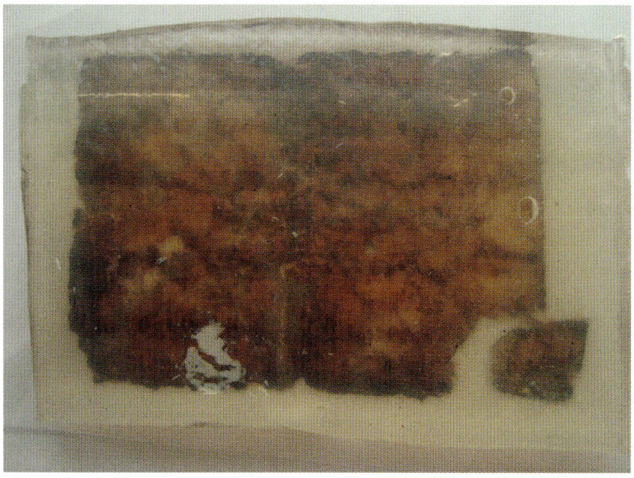

FIG. 2 Application of the rigid gel on a fragment of the letter.

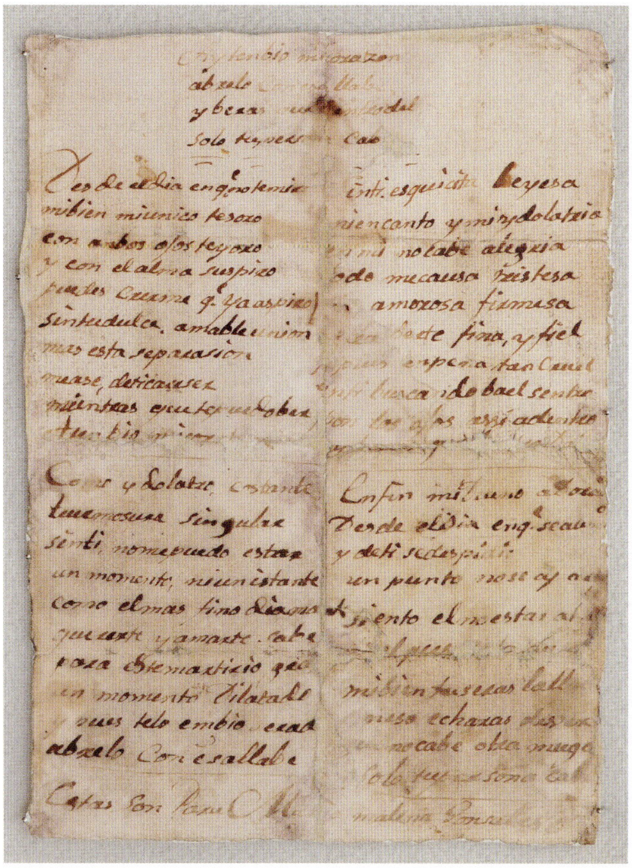

FIG. 3 The letter after treatment.

gels. As a result, the cleaning methodology was divided into two stages: general cleaning with a Canasol-loaded agarose gel and then local removal of deeply anchored oily spots with an MEK-Carbopol gel. Based on several preliminary tests, we chose a 4.4 wt% agarose gel with 0.01% Canasol NF-1000. The application of this rigid hydrogel was appropriate because it offered good contact with the relatively flat and non-textured surface of the paper (Condalab 2016), while its rigidity permitted solvent retention, and made solvent absorption by the paper unlikely. The non-ionic detergent Canasol was chosen because its effectiveness as a cleaning agent has been documented (Wolbers 1988, 2004). Initially, the effectiveness of the rigid gel was verified on a small fragment of the letter. The results were favorable, so a large gel sheet was placed over the entire letter, and left in direct contact for three days, protected by a sheet of polypropylene on the outside, but without using any barrier material between the letter and the gel (Fig. 2).

For the second stage of the treatment – the removal of the oily stain – five suitable solvents were selected based on their solubility, availability, and only moderate toxicity: *p*-xylene, 2-propanone (acetone), MEK, ethyl acetate, water, and ethanol. Gels were prepared from each using Carbopol 934 and Ethomeen C-25 and C-12 surfactants for the polar and non-polar solvents (IIC- Grupo Español de Conservación, 2015), respectively (Aguilar Sánchez and Buentello Martínez 2012). For this treatment, the MEK/Ethomeen/20% Carbopol soft gel was selected, because its solubility parameters are similar to a mixture of xylene in a low ratio with acetone (Barton 1991). After a local application of the MEK Carbopol gel, the letter was subjected to a second immersion washing to ensure removal of residues (Stiber and O'Loughlin 2012). Finally, structural treatment was carried out (Fig. 3).

Results

With the use of a rigid gel it was possible to extract a large amount of oxidation products progressively from the paper. Additionally, the micelle-like action of the Carbopol gel was very successful in removing part of the oily stain from the support. Although an aqueous agarose-Canasol rigid gel proved an ideal solution for the cleaning of the paper whilst maintaining its structural stability, the oil stain was more deeply impregnated and crosslinked and its elimination required the application of a polar solvent and a non-ionic surfactant.

Conclusion

Cleaning is one of the most critical procedures in the conservation of cultural heritage, and is irreversible in nature. Cleaning demands that we have as much information as possible about both the object itself and cleaning materials we bring to it (Ahedo and Cantos 2014). From these discussions and through the results obtained in this case, we can conclude that both control of the penetration into the substrate as well as the removal of dirt, oily material and even oxidation products from the paper were successful, in contrast to immersion washing, which weakened the paper. The agarose-Canasol gel functioned as a synergistic mixture which acted in a gradual and controlled manner (Scott 2012), dissolving the degradation products in the letter by absorbing them from the support, assisted by the surfactant effect of Canasol (Cremonesi 2001, 2002) in combination with the soft MEK gel used for the removal of greasy stains. The effectiveness of the agarose gel is determined by the length of time that it is allowed to act and the cleaning agent incorporated into the gel (Scott 2012).

Due to time constraints, it was not possible to verify by methods such as examination in ultraviolet light, or colorimetry, whether small amounts of gel residues remained on the letter. It is important to ensure that interactions between the treated substrate and any residue will not have detrimental effects on the artwork in the future.

Acknowledgements

We thank the staff of the Restoration and Conservation Department of the National School of Conservation, Restoration and Museology (ENCRyM) for promoting the interest of students in research and interdisciplinary work through the *STRODOGP* workshop, as well as lecturers Rocío Carolusa González Tirado and María del Pilar Tapia López, and conservator Victoria Casado Aguilar, for sharing knowledge.

Materials and suppliers

- Ethyl acetate, 2-propanone: Reasol (Reasol - Morales S. Ma. Arcelia - Analytical and industrial reagents, Mexico City, Mexico, http://mexico.paginasamarillas.com/empresas/reasol-morales-s-ma-arcelia-reactivos-analiticos-e-industriales/mexico-df-16023277
- MEK, ethanol, *p*-xylene, doubly distilled water: Hycel (Zoquipan Ave. 154, Colonia Atemajac del Valle, 45190 Zapopan, Jalisco, Mexico, http://www.hycel.com.mx/empresa/
- Ethomeen C-12 and C-25, Carbopol 934: Talas, 330 Morgan Ave. Brooklyn, NY 11211, USA, http://www.talasonline.com
- Canasol NF-1000, Whatman No.4 filter paper, turpentine: Inquímia Restaura, Talismán 315, Colonia Aragón Inguaran, 07820, México City, Mexico, http://www.inquimia.com.mx/Contacto.html

References

Aguilar Sánchez, B. and Buentello Martínez, R. 2012. *Carta de amor. Metodología de limpienza con gel de agarosa de un documento proviente del archivo histórico de Tepapayeca, Puebla*. Ciudad de México, México: ADABI. Available at: https://www.revistas.inah.gob.mx/index.php/digitales/article/view/4656/4697 (accessed 27 May 2016).

Ahedo, C. and Cantos, O. 2014. *New Polyacrylic Emulsifiers in the Formulation of Aqueous Cleaning Systems: Pemulen® TR-2*. Madrid: Institute of Cultural Heritage of Spain.

Anzani, M., Berzioli, M, Cagna, M., Campani, E., Casoli, A., Cremonesi, P., Fratelli, M., Rabbolini, D. and Riggiardi A. 2008. *Gel rigidi di Agar per il trattamento di pulitura di manufatti in gesso/Use of Rigid Agar Gels for Cleaning Plaster Objects, Quaderni del Cesmar7*, no. 6. Padua: Il Prato.

Barton, A. 1991. *Handbook of Solubility Parameters and Other Cohesion Parameters*, 2nd edn, New York: CRC Press.

Campani, E., Casoli, A., Cremonesi, P., Saccani, I. and Signorini, E. 2007. *L'uso di agarosio e agar per la preparazione di 'gel rigidi'/Use of Agarose and Agar for Preparing 'Rigid Gels'*, D. Kunzelman (tr.), *Quaderni del Cesmar7*, no. 4. Padua: Il Prato.

Condalab, Laboratorios CONDA 2016. Agarosas – *Descripción*. Available at: http://www.condalab.com/es/productos/agaroses/agarosa-descripcion/ (accessed 27 May 2017).

Cremonesi, P. 2001. *L'uso di tensioattivi and chelanti nella polishing di operate policrome*. Padua: Il Prato.

Cremonesi, P. 2002. *L'dei solventi use ORGANICI nella polishing di operate policrome*. Padua: Il Prato.

IIC-Grupo Español de Conservación 2015. *Ethomeen®, fichas técnicas*. Madrid: IIC-Grupo Español de Conservación. Available at: http://ge-iic.com/index.php?option=com_fichast&Itemid=83&tasko=viewo&task=view2&id=106.

Scott, C. 2012. 'The use of agar as a solvent gel in objects conservation', *Objects Specialty Group Postprints* 19: 71–73. Available at: http://resources.conservation-us.org/osg-postprints/wp-content/uploads/sites/8/2014/12/osg019-04.pdf (accessed 27 May 2017).

Stiber, L. and O'Loughlin, E. 1992. *Hinge, Tape and Adhesive Removal: The Paper Conservation Catalog*. Washington, DC: AIC.

Wolbers, R. 1988. *Notes from the Workshop on New Methods in the Cleaning of Paintings*. Los Angeles: Getty Conservation Institute.

Wolbers, R. 2000. *Cleaning Painted Surfaces: Aqueous Methods*, London: Archetype Publications.

Wolbers, R. 2004. 'A alla acquoso approccio polishing dei dipinti', *Quaderni del CESMAR7*, No. 1. Padua: Il Prato.

Authors' addresses

- Beatriz Aguilar Sánchez, National School of Restoration Conservation and Museography (ENCRyM), Naucalpan, Estado de Mexico, Mexico (beatrizag.sanchez@gmail.com)
- Rodrigo Buentello Martínez, National School of Restoration Conservation and Museography (ENCRyM), Mexico City, Mexico (rodrigo_buentello@hotmail.com)

Treating chromatic alterations on barkcloth with agar gel

Chiara Arrighi, Maria Francesca Quarato and Luciana Rossi

Introduction

Stains, particulate deposits and oxidation are the most common chromatic alterations on the surface of *tapas* (barkcloths) (Fig. 1). These beaten bark artefacts from the Pacific Ocean highlands combine the physical and chemical properties of paper with the function of a decorated textile. Unlike paper, however, they undergo extreme procedures to achieve the flexibility and the aesthetic quality of a textile (Smith *et al.*1988; Wright 2001; Daniels 2005). Indigenous tree barks are beaten to obtain sheets of cellulosic fibres, dyed to darken them, and painted with natural pigments in vegetable-based binders (Anati 1997; Neich and Pendergrast 2004).

During this process they acquire a three-dimensional surface as well as deformations, tears and lacunae. Pigments, media and unpainted fibres interact differently in response to changing relative humidity (RH), creating creases and tension across contiguous portions of the same cloth. The background tone of barkcloth naturally darkens with ageing. These characteristics are essential results of the process of creation of such artefacts, and are challenging for the long-term preservation of such material.

Evaluation of production techniques and their consequences is of crucial importance while planning a cleaning treatment. Immersion in water, although widely used, should be performed as a last resort, and only to treat extreme acidic degradation (Johnson 2001). A change in pH may be beneficial for the general preservation of the cellulosic substrate, but can significantly alter the chromatic balance of the painted item, and even compromise the original pigments and dyes. Beaten patterns can flatten or swell irreversibly after prolonged exposure to water.

On the other hand, dry treatments including light vacuum suction (Green *et al.* 1989), soft sponges (Guillemard and Michelet 1987), eraser crumbs, latex sponges, and polyvinyl alcohol sponges (Green and Brown 1986), are not recommended for damaged items, and more generally for a surface made of long uneven fibres. Moreover, they do not provide a means of stain removal.

Agar was chosen in our case for its combination of solubilizing and absorbing properties, high retention power, low interaction with the substrate, and ease of removal. Its use in paper conservation paved the way for trialling it on barkcloth.

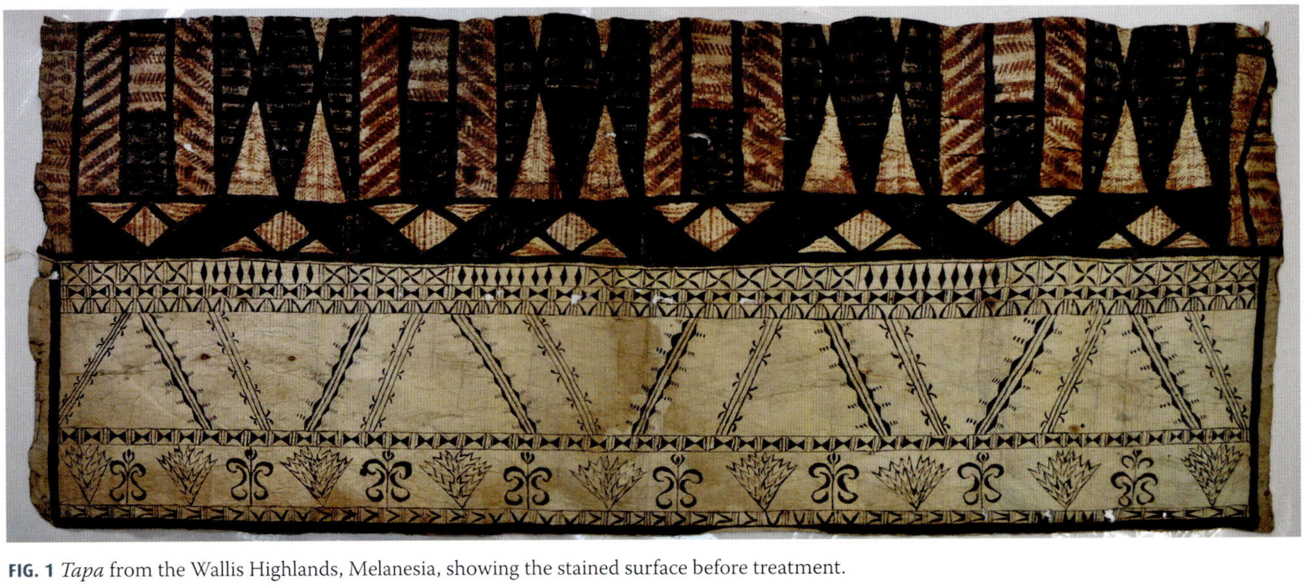

FIG. 1 *Tapa* from the Wallis Highlands, Melanesia, showing the stained surface before treatment.

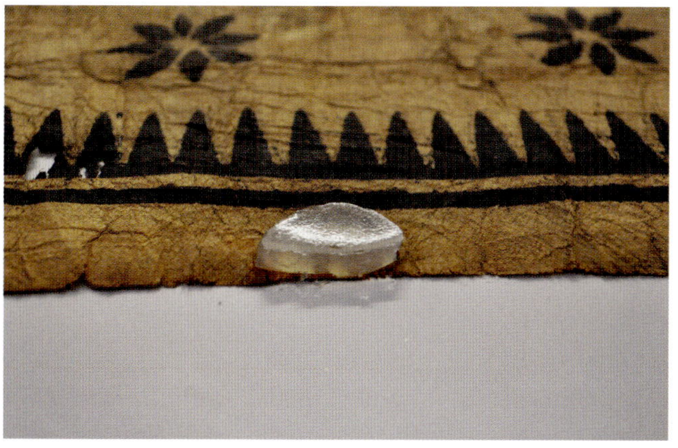

FIG. 2 *Tapa* from the Fiji Highlands, Polynesia, showing agar gel application on an oxidation stain.

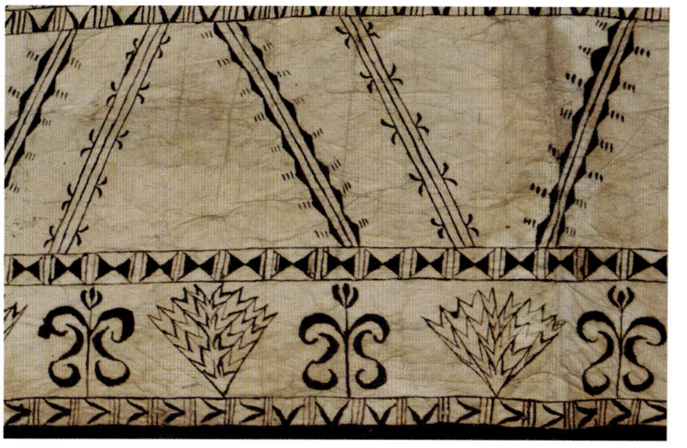

FIG. 3 *Tapa* from the Wallis Highlands, after treatment.

The goal was to perform a localized treatment with limited exposure to moisture and minimum handling. The preservation of a homogeneous background tone was also a priority.

Experimental details

Agar gel was tested on two barkcloths belonging to the Oceania collection of the Museo Nazionale Preistorico Etnografico 'Luigi Pigorini' in Rome. Both *tapas* showed red-brown stains, which altered the chromatic balance of their decoration. The first was a 19th-century *tapa* from the Fiji Highlands, Melanesia, one of the oldest acquisitions of the Pigorini collection. Prior to treatment it was an extremely thin and fragile sheet, fragmented and with tears affecting 80% of its surface. Natural ageing had changed the shade of the background to a golden brown tone. Possibly an increase in acidity had induced the loss of strength: the cloth could not be treated by an overall washing procedure. Any treatment had to be carried out with minimum handling so it was limited exclusively to stain removal. The second *tapa* was a heavy cloth from the Wallis Highlands, Polynesia, whose origins are traceable in the museum archives from 1927. Its decoration consisted of shiny/opaque colours and flat/three-dimensional patterns. Humidity variations made non-painted areas dilate at a higher rate than those protected by paint, and had caused distortion of the three-dimensional design. In this case the goal was to avoid uncontrolled deformation of the cellulosic support while operating directly on stained areas.

The paint of both *tapas* underwent solubility tests by soaking small painted fragments in distilled water for 30 minutes. The fragments were then dried between two sheets of non-acidic absorbent paper. None of the fragments released any trace of colour. A 6% agar solution in distilled water was prepared to provide a rigid gel with adequate characteristics of retention, absorbency and workability. The pH of the gel was adjusted to 8 with 0.5% ammonium hydroxide. The gel sheet had a thickness of 5 mm and was shaped to reproduce the designated area being treated locally (Fig. 2).

Both *tapas* were placed on a smooth flat surface and agar gel was applied directly on the staining for an initial contact time of 5 minutes. After the first application, a small variation in intensity was already noticeable on the stains. Initial results led to a second and a third application, for 5 minutes each. Wet areas were then dried between two sheets of non-acidic absorbent paper.

Results

After each treatment, the red-brown stains increasingly lightened. Their absorption into the gel could be visually monitored from the change in colour of the agar gel sheet. At the end, both *tapas* showed residual oxidation marks, which although still visible, did not interfere with the chromatic balance of the background tones. The localized moisture on small areas did not affect the condition of the entire cellulosic substrate.

Despite its fragility, the first *tapa* could be handled with extreme care during the treatment and its naturally aged tones were preserved before undergoing further operations. Sudden deformation and increased stress in the second *tapa* were avoided by limiting the area treated. Stain removal was not complete in either barkcloth. After treatment, faint traces were still visible on the surface and tidelines that had formed at the borders of stained areas had to be further treated in order to achieve an acceptable balance with the background tones (Fig. 3). The agar gel was mechanically removed from both items at the end of the treatment, without leaving behind any visible residues.

Discussion

Agar proved to be an effective method to treat chromatic alterations on barkcloth when conditions are unsuitable for immersion treatments. The opportunity to place a rigid gel sheet on designated spots without further interaction allows safe treatment despite the low strength of the barkcloth, its very fragile condition and the resulting handling challenges. The localized action of the gel avoids exposing the paint to the unwanted risk of deformation of the three-dimensional patterns and also by avoiding extensive humidification reduces the chance of interference with the natural stresses created by pigment binders. Double-checking variations in colour of both the substrate and the gel allows for constant monitoring of the level of stain removal. Minor collateral effects, such as faint tidelines remaining after drying, can be treated with shorter applications of the same gel. In future interventions, this inconvenience could be inhibited by initially applying a solution of cyclododecane around stain borders.

Traditional cleaning methods were evaluated as either drastic or ineffective for these specific artefacts. Immersion in alkalized water is suggested only to buffer extremely acidic substrates. While certainly washing away grime and oxidation stains, its action can alter original characteristics, resulting in permanent colour change and uncontrolled deformation. Whether 'deacidification and buffering could be ethically or aesthetically acceptable' (Johnson 2001: 82) for *tapas* is still open to debate. Vacuum suction, eraser crumbs, latex sponges and polyvinyl alcohol sponges, on the other hand, remove only surface debris and are only suitable for use on very robust surfaces.

The question is: what can we expect from a conservation treatment on items which have the dual nature of paper and fabric? Original background tones and decoration are not necessarily damaged by ageing, as a general darkening of the surface does not interfere with the aesthetic quality of these particular works – it can be considered their patina that must be preserved. Agar gel provides a satisfactory intermediate solution to alleviate chromatic imbalances where stains and oxidation marks are present. Its localized nature enables treatment to be carried out even when the object is very degraded and fragile, with total respect for the original decoration and patina.

Materials and supplier

- Agarose gel AgarArt and ammonium hydroxide: CTS, Via Gaudenzio Fantoli n. 26, 00149 Rome, Italy, http://www.ctseurope.com

References

Anati, E. 1997. *Tapa tapa: stoffe di corteccia d'albero macerata e battuta di Papua Nuova Guinea*. Capo di Ponte: Edizioni del Centro.

Daniels, V. 2005. 'The characteristics of modern and old barkcloth (*tapa*)', *The Conservator* 29: 95–104.

Green, S.W. and Brown, A.G. 1986. 'Conservation of *tapa* cloth from the Pacific', in *Preprints of the AIC 14th Annual Meeting*. Chicago: AIC, 17–31.

Green, S.W., Zeronian, S.H. and Needles, H.L. 1989. 'Treatment of *tapa* cloth with special reference to the use of the vacuum suction tube', in S. Haig Zeronian and H.L. Needles (eds), *Historic Textile and Paper Materials II: Conservation and Characterization*, ACS Symposium Series 410. Washington, DC: American Chemical Society, 168–87.

Guillemard, D. and Michelet, A. 1987. 'Conservation et restauration des écorces battues de Polynésie', in *Journées sur la conservation, restauration des biens culturels: recherches et techniques actuelles*. Paris: Association des restaurateurs d'art et d'archéologie de formation universitaire, 126–31.

Johnson, E. 2001. 'The deacidification and conservation of a Samoan tapa at the Manchester Museum', in M.M. Wright, *Barkcloth: Aspects of Preparation, Use, Deterioration, Conservation and Display*. London: Archetype Publications, 71–95.

Neich, R. and Pendergrast, M. 2004. *Pacific* Tapa. Auckland: David Bateman.

Smith, L.M., Barclay, R., Gilberg, M., McCawley, J.C. and Stone, T. 1988. 'When textiles are paintings', in R.L. Barclay (ed.), *Proceedings of Symposium 86: The Care and Preservation of Ethnological Materials*. Ottawa: Canadian Conservation Institute.

Wright, M.M. 2001. *Barkcloth: Aspects of Preparation, Use, Deterioration, Conservation and Display*. London: Archetype Publications.

Authors' addresses

- Chiara Arrighi, Rome, Italy (chiara.arrighi2@libero.it)
- Maria Francesca Quarato, Conservation Laboratory, Museo Nazionale Preistorico Etnografico 'Luigi Pigorini', Rome, Italy (mariafrancesca.quarato@beniculturali.it)
- Luciana Rossi, Conservation Laboratory, Museo Nazionale Preistorico Etnografico 'Luigi Pigorini', Rome, Italy (luciana.rossi-01@beniculturali.it)

Local cleaning of tidelines on paper using rigid gels: the influence of pH and conductivity

Sophie Barbisan and Anne-Laurence Dupont

Introduction

In 2015 a project was undertaken at the Institut National du Patrimoine (INP, Paris, France), in collaboration with the Centre de Recherche sur la Conservation des Collections (CRC, Paris, France), as part of a Master's thesis. The research was linked to the conservation treatment of a large-scale charcoal drawing on wove machine-made paper, *Allégorie de Lyon* (Fig. 1) by Louis Janmot (1814–1892). The artwork suffered from severe darkening, as well as extensive tidelines scattered throughout the paper surface. Recent developments in the use of rigid gels in conservation (Iannucelli and Sotgiu 2010; Wolbers 2014) led us to investigate new alternatives for cleaning fragile artworks. Because of the large colour difference between the tidelines and the rest of the paper, local treatment seemed appropriate (Sullivan *et al.* 2014). However, the local use of rigid gels can lead to the creation of additional wet/dry interfaces during treatment, which can eventually develop into new tidelines (Dupont 1996; Souguir *et al.* 2008; Jeong *et al.* 2012, 2014). Parameters such as pH and the conductivity of the water used to prepare the gel are reported to play an important role in cleaning efficiency (Keynan and Hughes 2013; Stavroudis *et al.* 2005) and were investigated to develop the cleaning method.

Pre-testing

To identify the relevant parameters, a series of pre-tests was performed. Cleaning efficiency of agar, agarose, and gellan gum gels were compared by applying small discs of gels for 10 minutes on Whatman paper toned with a methylene blue solution (3 g/L), a water-soluble dye (Couvert and Dupont 2013). Agarose was chosen for its reported cleaning efficiency, its purity and its electric neutrality, which make pH and conductivity adjustment of the gel easier. From the different concentrations of agarose tested, 5% w/v seemed a good compromise between cleaning efficiency and water migration. To limit the latter,

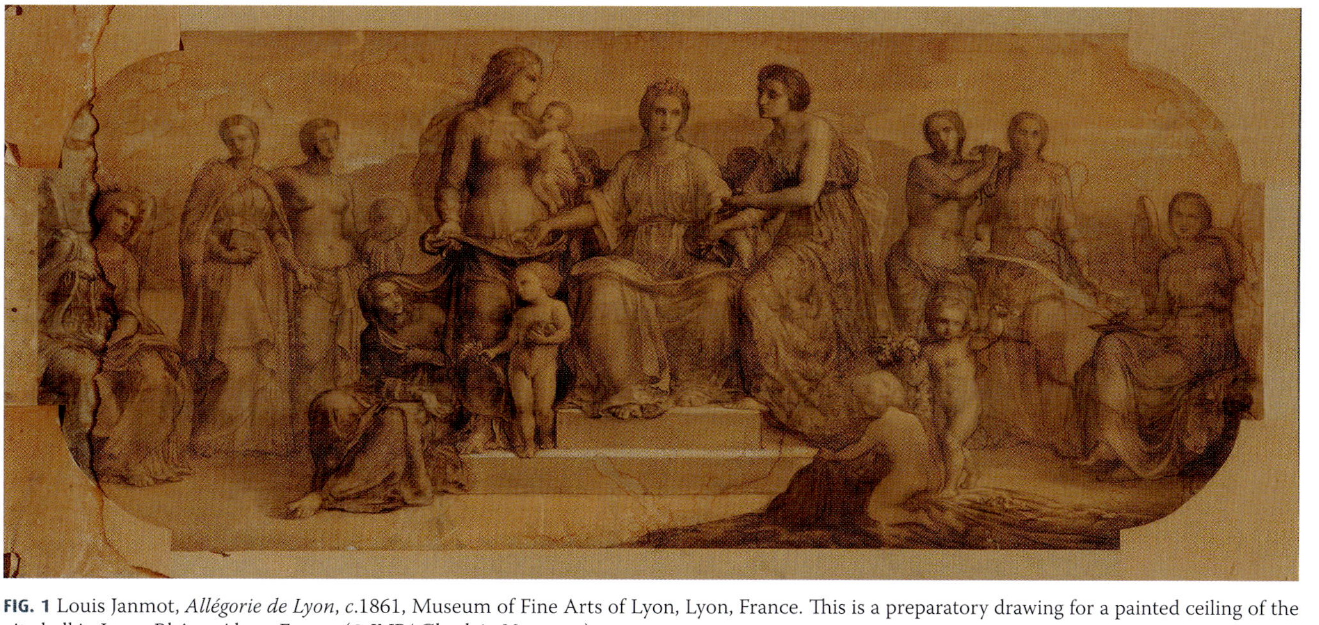

FIG. 1 Louis Janmot, *Allégorie de Lyon*, *c.*1861, Museum of Fine Arts of Lyon, Lyon, France. This is a preparatory drawing for a painted ceiling of the city hall in Lyon, Rhône-Alpes, France (© INP/ Ghyslain Vanneste).

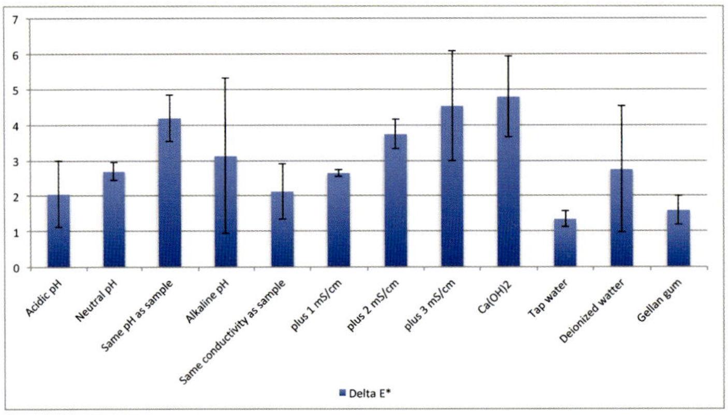

FIG. 2 Colour measurements ΔE^* of the brown line (first phase of testing), corresponding to the colour difference before and after cleaning.

cyclotetrasiloxane (D4) and cyclopentasiloxane (D5), were applied on the paper around the gel. The barrier effect in avoiding the formation of a wet–dry interface was compared with a simple humidification of the samples under Gore-Tex. Both methods seemed effective and pre-humidification was chosen.

Sample preparation

Samples (3 × 10 cm) prepared using Arches paper (pH 8.8) were aged 10 days at 100 °C following ASTM D6819-02e3. Tidelines were created by suspending the aged samples vertically, with a few millimetres of the lower edge dipped into deionized water, for 30 minutes. After air-drying, the samples were aged again for 10 days in the same conditions as before.

First testing phase

In order to follow precisely the contours of the tidelines, the gel application was first tested in a liquid state. Agarose (5% w/v) was applied while hot with a brush through an interleaf of thin Tengujo paper. Agarose gels were prepared directly with dilute aqueous solutions at various values of pH and conductivity, adjusted with glacial acetic acid and ammonium hydroxide. Dilute isotonic solutions, with a conductivity similar to the samples (around 1mS/cm) at pH 6, 7, 8.75 and 9.5, were prepared to evaluate the influence of pH on the cleaning efficiency. Four identical pH solutions with the same pH as the samples (8.75), with stepwise increase in conductivity of 1, 2 and 3 mS/cm from the initial 1 mS/cm, were tested to investigate the influence of conductivity. Alternatively, agarose gels were prepared with tap water, deionized water and a calcium hydroxide solution (pH 10, 12 μS/cm). Agarose powder was dispersed in the solution and the mixture was heated in a microwave for one minute in a beaker with a lid. The different gels were applied for two hours on the treated samples. The impact of cleaning was assessed by colour measurements (using D65 illuminant, 10° observer and CIELAB 1976 colour space) and ultraviolet (UV) and visible light imaging photography using 360 nm for the former lighting.

Results showed that the cleaning efficiency increased with solution conductivity (Fig. 2). With isotonic solutions, ionic exchanges between the paper and the gel are supposed to be limited and thus the cleaning efficiency is low. Conclusions regarding pH were less clear but it seemed that an alkaline pH was favourable. Arches paper is a permanent paper and its pH, even after ageing, is alkaline. It also appeared that similar cleaning results were achieved with pH and conductivity adjusted solutions with high conductivity and alkaline pH, and calcium hydroxide solutions. However, after accelerated aging, unwanted effects were observed on the paper when agarose was applied in liquid phase. This application method created residues and migrations that both aged poorly.

Second testing phase

Previous results made it possible to refine the protocol used in the second phase of testing in which gels were used in rigid form. Agarose (5% w/v) gels prepared with a pH and conductivity adjusted solution (pH of 9.5 and conductivity of 4 mS/cm and 6 mS/cm) and with a calcium hydroxide solution (pH 10) were applied locally. Additionally, a set of samples was sandwiched between two gels made of agarose and gellan gum. In that case, agarose gels (2% and 3% w/v) were applied locally on the tideline, on the recto of the sample, while gellan gum (3% w/v) was applied globally, covering the entire sample surface, on the verso. This was intended to lighten the darkening of the paper and the tidelines at the same time, as well as obtain improved cleaning efficiency while avoiding the formation of additional wet/dry interfaces during treatment. Agarose was prepared with high pH and high conductivity adjusted solutions (pH 9.5 and 6mS/cm) and with a calcium hydroxide solution (pH 10, 12μS/cm).

Although local application of agarose resulted in UV fluorescence of the treated area, this was minimal compared to previous results. The use of the high pH and high conductivity adjusted solutions did not show any particular advantage as similar results were obtained with calcium hydroxide. In some cases, the treatment with these high conductivity adjusted solutions created a darkening of the treated area after aging, which was not observed with the calcium

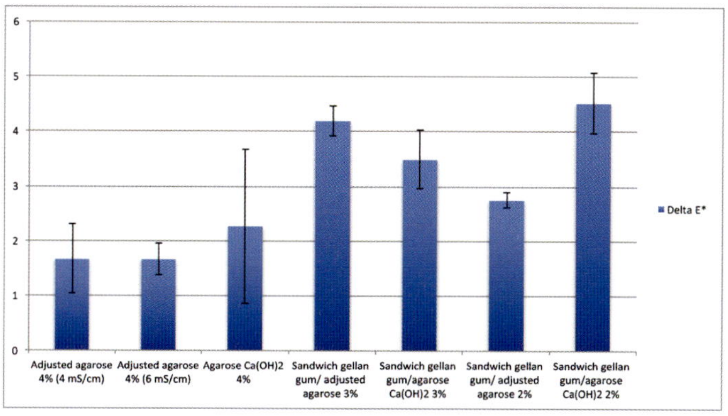

FIG. 3 Colour measurements ΔE^* of the brown line (second phase of testing), corresponding to the colour difference before and after cleaning.

hydroxide prepared gel. The gel sandwiches showed best results in terms of cleaning efficiency (Fig. 3) and behaviour upon accelerated ageing. UV images indicated that no wet–dry interface was created during this cleaning treatment.

Conclusion

Agarose/gellan gum gels used as sandwiches provided the best cleaning results and seem to be a safe method for the cleaning of paper artworks with fragile media. The results showed that high conductivity gels improved the cleaning, confirming earlier publications (Keynan and Hughes 2013). Nevertheless, pH and conductivity adjusted gels did not show any particular advantage over calcium hydroxide gels in the specific case of aged Arches paper. For the cleaning of *Allégorie de Lyon*, several applications of gellan gum (2.5% w/v) on top of an interleaf of Bolloré paper were first tested on the entire drawing surface. This proved sufficient for cleaning both the tidelines and the yellowing. The new cleaning procedure was therefore not used on the artwork, but its efficiency to treat paper locally as well as overall was demonstrated. The cleaning method avoided the formation of new wet/dry interfaces associated with the local application of a rigid gel.

Materials and suppliers

- Agarose and Whatman paper grade 1: Sigma-Aldrich, https://www.sigmaaldrich.com/united-states.html
- Gellan gum/Kelcogel CG-LA: GMW, http://gmw-shop.de/en/
- Arches paper (160 g/m^2): Sennelier, http://www.magasinsennelier.com/
- Bolloré paper (22g/m^2) and Bib Tengujo paper (12g/m^2): Stouls, http://www.stouls.com/index.php
- Methylene blue: Fisher Scientific, https://www.fishersci.com/us/en/home.html

References

Couvert, E. and Dupont, A.-L. 2013. 'Les hydrogels: une nouvelle approche pour réaliser des décollages de papiers sensibles à l'humidité', *Support Tracé* 13: 60–71.

Dupont, A.-L. 1996. 'Degradation of cellulose at the wet/dry interface. I. The effect of some conservation treatments on brown lines', *Restaurator* 17: 1–21.

Iannuccelli, S. and Sotgiu, S. 2010. 'Wet treatments of works of art on paper with rigid gellan gels', *The Book and Paper Group Annual* 29: 25–39.

Jeong, M.J. Dupont, A.-L. and de la Rie, E.R. 2012. 'Degradation of cellulose at the wet–dry interface. I. Study of the depolymerization', *Cellulose* 19(4): 1135–47.

Jeong, M.-J., Dupont, A.-L. and de la Rie, E.R. 2014. 'Degradation of cellulose at the wet–dry interface. II. Study of oxidation reactions and effect of antioxidants', *Carbohydrate Polymers* 101: 671–83.

Keynan, D. and Hughes, A. 2013. 'Testing the waters: new technical applications for the cleaning of acrylic paint films and paper supports', *The Book and Paper Group Annual* 32: 43–51.

Souguir, Z., Dupont, A.-L. and de la Rie, E.R. 2008. 'Formation of brown lines at the wet–dry interface in cellulosic materials', *Biomacromolecules* 9(9): 2546–52.

Stavroudis, C., Doherty, T. and Wolbers, R. 2005. 'A new approach to cleaning. I. Using mixtures of concentrated stock solutions and a database to arrive at an optimal aqueous cleaning system', *WAAC Newsletter* 27(2): 17–28.

Sullivan, M., Brogdon-Grantham, S. and Taira, K. 2014. 'New approaches to cleaning works of art on paper and photographs', paper presented at the *Association of North American Graduate Programs in Conservation conference*. Available at: http://cool.conservation-us.org/anagpic/2014pdf/anagpic2014_sullivan_etal_paper.pdf?bcsi_scan_2687365ababd2c82=0&bcsi_scan_filename=anagpic2014_sullivan_etal_paper.pdf (accessed 6 June 2017)

Wolbers, R. 2014. 'Aqueous methods for paintings', The Winterthur/University of Delaware Program in Art Conservation (WUDPAC), unpublished manuscript.

Authors' addresses

- Sophie Barbisan, Lunder Conservation Center, Smithsonian American Art Museum, Washington, DC, USA (barbisansophie@gmail.com)
- Anne-Laurence Dupont, Centre de recherche sur la conservation des collections – CRCC, USR3224 CRC, Muséum national d'histoire naturelle, Paris, France

Chelating soluble iron(II) from iron gall ink using calcium phytate in agar gel

Avery Bazemore

Introduction

Iron gall ink degradation is a common problem in manuscripts created up to the 20th century. Iron gall ink is made from a source of tannin and a source of iron, usually oak galls and ferrous sulphate, which react to form the characteristic blue-black ink compound. However, historic ink recipes often contain excess iron, which can catalyse oxidative radical chain reactions that break the long chains in the cellulose molecule. The paper becomes weak and brittle and can eventually disintegrate entirely. Ink degradation can be slowed by keeping the object in stable conditions. High or fluctuating relative humidity (RH) transports the water-soluble ferrous ions causing damage in new areas. The ink is often stable enough for the paper to be supported with Japanese tissue where treatment is deemed necessary, but in some cases more invasive treatment may be required.

The current most popular interventive treatment for iron gall ink degradation is the use of phytate, an antioxidant found in plants which chelates the excess ferrous ions and prevents them from participating in radical reactions. However, the phytate treatment is applied in an aqueous bath, which requires volumes to be disbound. Disbinding and washing a book is an extremely invasive treatment, which makes it difficult to justify even if it would prolong the life of the paper. Agar gels may be a solution to this problem. The rigid network structure of the agar gel can contain the phytate solution and allow a controlled application without leaving agar residue, reducing disruption to the structure of the object. The following experiments use an inexpensive and widely available food-grade agar, as it does not appear to produce different results from the purer laboratory-grade agarose when used in conservation applications (Cremonesi 2012).

This project investigates the use of agar gels to apply calcium phytate solution *in situ* to objects that cannot be immersed. The focus of the trials is on chelating iron as tested using bathophenanthroline iron(II) indicator paper. Gels without phytate are used as a control.

Methodology

Samples of Whatman No. 1 filter paper were immersed in a solution of gelatin-alum size as described in a study at the University of Ljubljana, Ljubljana, Slovenia, on the treatment of iron gall ink using phytate and chelators (Kolar *et al.* 2005). Ink was prepared according to the recipe in Neevel (1995), wherein the phytate treatment was proposed, using 0.785 g gum arabic, 1.05 g iron(II) sulphate heptahydrate, and 1.23 g tannic acid (95%), with distilled water added to bring the volume to 25 mL. The ink was applied using a 3 mm broad edge pen in rows of five 15 mm lines. The inked samples were left for three weeks at room temperature to allow the ink to oxidize.

Samples of historic material were taken from early 19th-century letters written on thin, soft lightly sized rag paper purchased from a local bookseller. Solutions of agar (5% w/v) were prepared using tap water which did not react with the iron(II) indicator paper. The solution was boiled and cast into Petri dishes to set. Five percent is a high concentration for an agar gel, and produces a coherent structure which can be easily removed from the paper's surface. A concentrated phytate solution was mixed with the agar before casting.

Concentrated phytate solutions were prepared with 1 mL water. The quantities of the ingredients were calculated to be proportional to the amount of water used to make the gel to ensure that the resulting solution would have the same concentration of phytate as the traditional 0.116% bath (Neevel *et al.* 2007). To produce an 80 mL solution of agar, the concentrated phytate solution was prepared using 1 mL water with 0.23 g of 40% phytic acid and 0.04 g calcium carbonate.

The pH of the phytate solution was increased to 7 with ammonium hydroxide. The resulting gels were approximately pH 7. Phytate treatment usually involves a bath at a lower pH followed by a deacidification bath because the solubility of calcium phytate decreases sharply above approximately pH 5.6 (Neevel 2001). Although the undissolved phytate may have a reduced ability to chelate iron, a preliminary experiment showed no perceptible difference in efficacy between

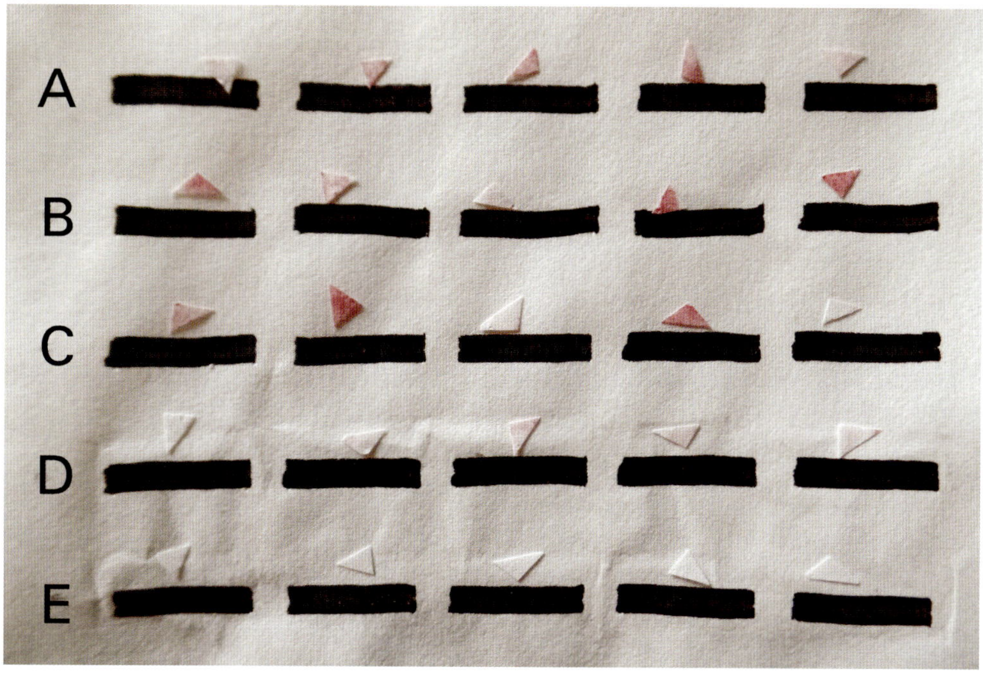

FIG. 1 Residual iron(II) concentration testing using bathophenanthroline iron(II) indicator paper on freshly inked Whatman filter paper samples.

gels containing phytate at pH 7 and phytate below pH 5.6. Because the object would not be exposed to an acidic solution, it would not need to be deacidified, and a one-step process would not expose it to as much moisture.

For tests on treatment duration, 5- and 10-minute exposures were used, employing gels with and without phytate. In trials treating a single line, 5 × 20 mm gels were applied to the samples. To mimic treating a larger area such as an entire page of a book, a gel containing phytate was applied to a 12 × 20 cm letter for 10 minutes. The sample papers were treated on a sheet of Perspex, with a felt and a light weight on top of the gels. After treatment, the samples were dried between felts and under a light weight for 20 minutes before testing for residual iron(II) concentration. Bathophenanthroline iron(II) indicator paper was prepared according to Jacobi *et al.* (2011) by dipping Whatman filter paper in a 0.116% w/v solution of bathophenanthroline in ethanol. Each row on the Whatman paper sample was tested in five locations. On historic material, five areas with high ink concentration were chosen for testing.

Effectiveness relative to treatment duration

The results of iron(II) concentration testing on material prepared for the experiment shown in Fig. 1 indicate that agar gels with phytate were more effective than those without. After a 10-minute exposure (E), iron(II) was no longer detected by this method, and its concentration was greatly reduced after a 5-minute treatment with the phytate-loaded gel (D). Areas of ink exposed to agar gels without phytate also showed a detectable reduction of iron(II) (B and C). However, gels with phytate demonstrated a reduction in the residual iron(II) at a faster rate, indicating that chelation *in situ* is more effective than treatment with a water gel. Gels containing phytate caused noticeable cockling of the paper, which did not improve after drying under weight. This cockling may make the treatment inappropriate for localized use. It is unclear why gels without phytate did not cause cockling to the same extent.

Effectiveness relative to treatment duration on historic material

An agar gel with phytate was most effective in reducing the amount of soluble iron in the historic sample. An agar gel without phytate also produced a significant decrease in soluble iron. The ink on the historic sample did not react with the iron(II) indicator paper as extensively as the ink used on the contemporary mock-ups, so the range of reaction was less pronounced. There was no evident cockling of the paper after treatment, but the treatment created tidelines around treated areas as soluble discoloured material reached the edge of the area wetted by the gel. Much like the cockling in the previous experiment, these effects make the treatment inadvisable for use on discolored objects.

Treating a full page of historic material

After a 10-minute treatment with a phytate-containing agar gel, there was very little detectable soluble iron(II) on the historic sample shown in Fig. 2. However, uneven wetting caused the paper to ripple. The lower areas of the page were not treated as thoroughly and produced a slight reaction with iron(II) indicator paper, shown in the uppermost test area. The cockling was reduced, but not eliminated, after pressing

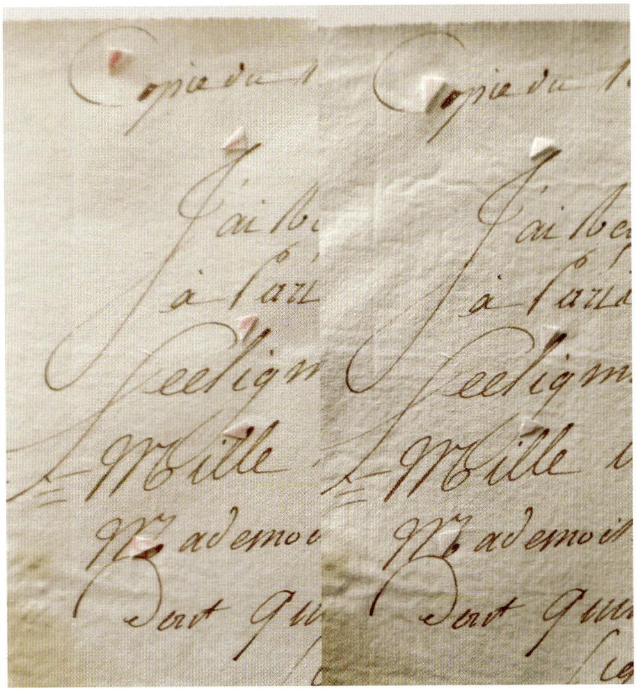

FIG. 2 Historic sample before and after phytate-containing agar gel treatment, showing cockling.

the page between felts to dry. The gel absorbed discoloured degradation products in both experiments on historic material. In the full-page experiment, fragments of ink also adhered to the gel, exposing the oxidized cellulose below. The sample used in this experiment may have been sized more heavily, causing the ink to remain on the surface rather than sinking into the paper.

Conclusion

Agar with phytate appears to be successful in chelating iron on both historic material and the samples prepared for these experiments. A 10-minute application of the gel was sufficient to ensure that no detectable iron(II) remained in new material after the treatment, and to eliminate or dramatically reduce the iron(II) in the historic material. Agar gels without phytate also demonstrated the same effect but to a lesser extent, indicating that some of the success of gels with phytate is due to the agar itself. Some cockling was evident when using agar gels with phytate on the samples produced for the experiments and when treating larger areas of historic material. No cockling was evident when treating small areas of the historic sample, but the treatment caused tidelines and colour change.

Using gels to apply the phytate treatment may prove to be a simple, inexpensive and effective method of treating iron gall ink degradation without immersion. Any conservator using a phytate treatment can easily apply it in a gel. Unfortunately, the treatment is inadvisable due to the cockling and tidelines demonstrated in these experiments, and requires further practical development.

References

Cremonesi, P. 2012. 'Rigid gels and enzyme cleaning', in M.F. Mecklenburg, A.E. Charola and R.J. Koestler (eds), *Cleaning 2010: New Insights into the Cleaning of Paintings*. Washington, DC: Smithsonian Institution Scholarly Press, 179–83.

Jacobi, E., Reißland, B., Phan Tan Luu, C., van Velzen, B. and Ligterink, F. 2011. 'Rendering the invisible visible', *Journal of Paper Conservation* 12: 25–33.

Kolar, J., Šala, M., Strlič, M. and Šelih, V.S. 2005. 'Stabilisation of paper containing iron-gall ink with current aqueous processes', *Restaurator* 26(3): 181–89.

Neevel, J.G. 1995. 'The development of a new conservation treatment for ink corrosion, based on the natural anti-oxidant phytate', in M.S. Koch and J. Palm (eds), *Preprints of the 8th Congress of IADA, Tubingen, 19–23 September 1999*, 93–100.

Neevel, J.G. 2001. '(Im)possibilities of the phytate treatment', in *Postprints from the Iron Gall Ink Meeting*. Newcastle: University of Northumbria, 125–34.

Neevel, H., Reissland, B., Scheper, K. and Fleischer, S. 2007. 'Preparation of calcium phytate', *The Iron Gall Ink Website*. Ljubljana: Instituut Collectie Nederland. Available at: http://inkcorrosion.org/images/file/pdf 9_preparation of calcium phytate_ok.pdf (accessed 27 May 2017).

Author's address

Avery Bazemore, Lambeth Palace Library, London, UK (avery.bazemore@gmail.com)

Biocleaning of wall paintings on uneven surfaces with warm agar gels

Pilar Bosch-Roig, Jose Luis Regidor Ros , Maria Pilar Soriano Sancho,
Rosa Montes Estellés and Pilar Roig Picazo

Introduction

Recent advances in cleaning systems have highlighted the effectiveness of biocleaning methods. Several authors have demonstrated that selected bacterial strains are able to metabolize both organic and inorganic accretions on wall paintings (Ranalli *et al.* 2005; Bosch-Roig *et al.* 2013a,b). However, the development and application of this technique is still at an early stage and its use is limited to specific cases. Currently, the main research directions in this field are focused on: selection of bacterial species capable of metabolizing new substances to be cleaned (Sanmartín *et al.* 2015; Barbabietola *et al.* 2016); solving the application problems of this technique (Bosch-Roig *et al.* 2015); and controlling the safety of the biocleaning system (Bosch-Roig and Ranalli 2014).

There are obstacles to the use of biocleaning systems for wall paintings. The need to maintain optimal temperature and relative humidity for bacteria to metabolize reduces the number of works that can withstand such stress. Therefore a delivery system with a controlled water supply is fundamental to obtain satisfactory results (Bosch-Roig *et al.* 2015). The use of agar gels (agar is a temperature-reversible complex polysaccharide) as a controlled supply of water for the biocleaning of wall paintings has been shown to be very effective (Bosch-Roig *et al.* 2013a, b). The structure and properties of agar allow the formation of rigid gels which in turn allow for internal diffusion of water and act as a sponge that absorbs solubilized material. However, their rigidity poses certain problems for application to rough architectural surfaces.

The rigid agar gel must endure long contact times (hours) and must maintain complete contact with the surface to be cleaned, otherwise air bubbles will be produced between the gel and the surface and unequal cleaning results and/or negative diffusion phenomena of the solubilized materials could appear. These limitations can be diminished by the application of semi-solid warm agar. When close to their gelation point, agar solutions are dense fluids that can be applied by brush to obtain an elastic film well adapted to the surface to which it is applied (Anzani *et al.* 2008; Barbabietola *et al.* 2016.). However, this application system also poses risks and has limitations. On the one hand the application temperature (about 40 °C) must be compatible with the artwork and with the bacteria. On the other hand, the resulting gel layer should be easily removable without the risk of detaching any original particles. Several studies have highlighted the deposition of unacceptable gel residues or even the detaching of the pictorial film when agar is applied warm at certain concentrations (Beltrami *et al.* 2011; Blanco Domínguez and Tortajada Hernando 2013).

Our work is investigating the suitability of semi-solid warm agar as a biocleaning delivery system, with a focus on the application temperature and agar removal after biocleaning. This research has been developed in the framework of the restoration of the large Dionis Vidal wall painting cycle (about 1900 m^2) in the church of St Nicolás, Valencia, Spain.

Materials and methodologies

Site description

The selected site for the cleaning test was a part of lunette 4 of the north area of the central vault of the church. The biocleaning tests were carried out on vertical areas (approximately 60 cm^2 each). The fresco technique was classical Spanish baroque executed with *arriccio* (gesso) and *intonaco* (dry lime), and most of the colours are made from earth pigments. The area was selected because it was optimal for biocleaning – it was detached and had undesired organic (old and hard animal glue from the *stacco*) and inorganic layers on top of the frescoes.

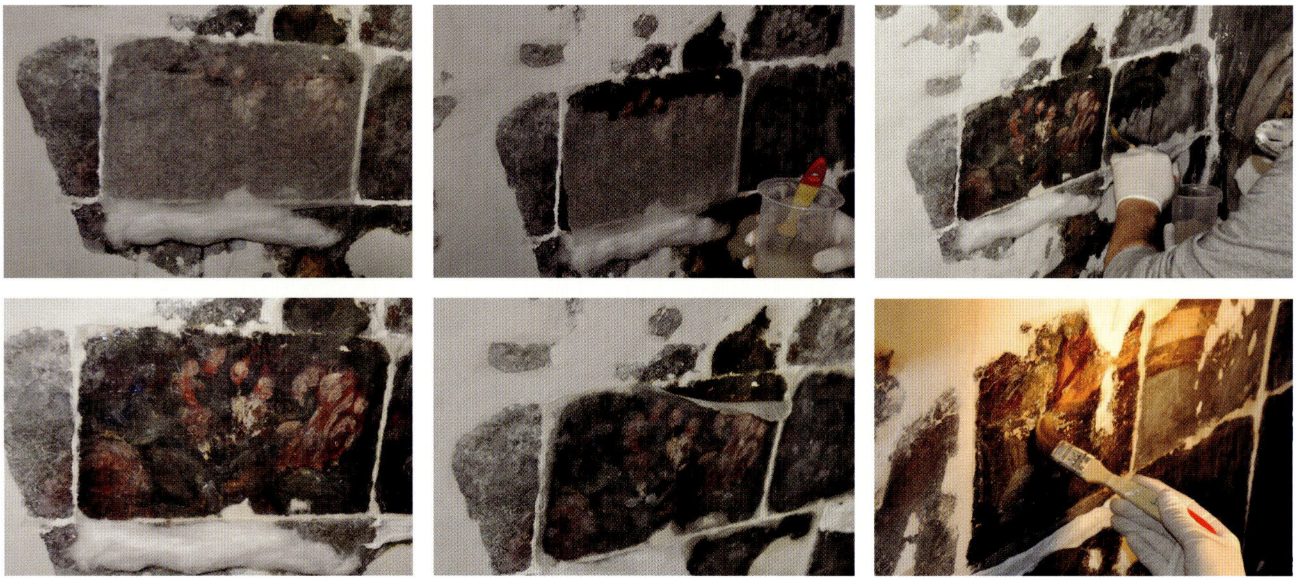

FIG. 1 The biocleaning process for wall paintings. *Upper row, L–R*: the application of Japanese paper, then bacteria, and finally warm agar. *Lower row, L–R*: the agar film formed on the fresco, removal of the agar with cleaning of the surface with sterile water, and brushing after treatment.

Physico-chemical analysis

Microsamples of these accretions were characterized by Fourier transform infrared (FTIR) spectroscopy, optical microscopy (using a Leica DMR microscope) and gas chromatography–mass spectroscopy (GC–MS) (Agilent Technologies, Palo Alto, CA, USA).

Biocleaning treatments

Biocleaning treatment with a warm semi-solid agar (Agarart CTS) delivery system was applied in triplicate and compared with control areas (not cleaned). The biocleaning treatments involved the application of *Pseudomonas stutzeri* DSMZ 5190 over Japanese paper (Bib tenguio. 12 g/m^2, pH 7.1, 100% manila fibre) placed on the fresco surface with a sterile brush. Warm agar was prepared on-site in a Wax Warmer YM-8424 and applied at a temperature of ~ 40 °C with a sterile brush to obtain a layer thickness of 3–5 mm. The biocleaning contact time was 4 hours and the temperature was maintained at 25 °C±4 °C using infrared heat lamps. After treatment, the agar and Japanese paper were removed and the surface bacteria were cleaned using sterile water (Fig. 1). The treated surface was dried at room temperature, and microsamples were collected for microbiological and chemical analysis.

Bacterial growth

Pseudomonas stutzeri DSMZ 5190 was grown in nutrient broth medium and incubated at 28 °C for 48 hours in a shaker (200 rpm). After incubation the samples were centrifuged (3 × 4200 rpm for 10 minutes), washed twice with NaCl 0.8% pH 7.0 and re-suspended in sterile water to a cell concentration of 10^8 CFU ml^{-1}.

Bacterial control

To ensure bacterial absence on the treated surface, contact plate samples were created and analysed in the laboratory. Bacterial viability was measured using a 3M Clean-Trace NG Luminometer (Biotrace-UniLite, Biotrace International plc, Bridgend, UK).

Surface cohesion

The level of surface cohesion and the potential surface resistance to the tensile stress of the wall painting to be treated with warm agar was determined. Calculation of cohesion index by the Scotch tape test was adapted from the ASTM D3359. The test is based on the measurement of the solid surface mass that is fixed on an adhesive tape (mg) in relation to its size (cm^2).

Results and conclusions

GC–MS confirmed the presence of proteins (animal glue) and lipids (wax) on the surface of the frescoes. A 4-hour *P. stutzeri* biocleaning treatment with a warm agar delivery system successfully removed the superficial, undesired organic layer, without leaving any agar residues on the cleaned surface, as confirmed by FTIR and GC–MS.

This study has shown the warm agar delivery system to be an appropriate application support for rough surfaces. Although solid agar has previously been demonstrated to possess good adhesive properties when applied on smooth horizontal, vertical and oblique surfaces, showing no detachments from the wall during treatment (Bosch-Roig *et al.* 2013a, b), the warm agar application was shown to be more appropriate than solid agar on rough surfaces. Optical microscopy of samples of agar and Japanese paper after biocleaning

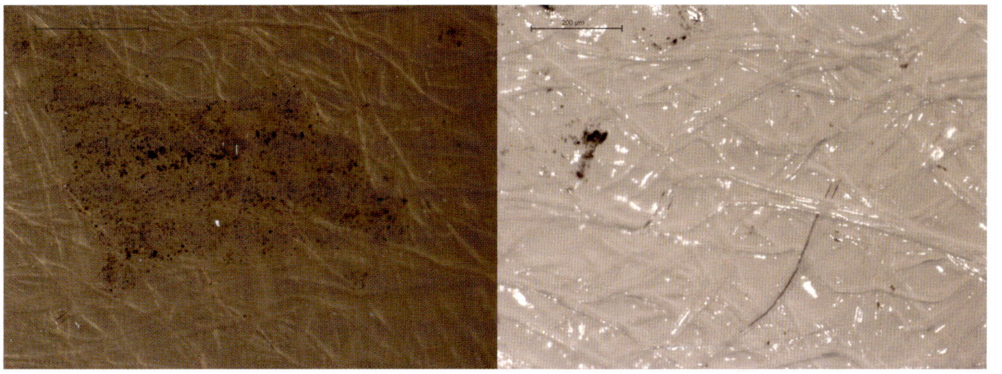

FIG. 2 Micrographs of agar and Japanese paper after biocleaning of the fresco surface, showing fibres from the Japanese paper and dust removed from the samples.

indicated that only small particles of dust could be found on the samples after treatment. No significant detachment of pigment or fragments of the original paint layer was found when the gel was applied to a stable fresco surface (Fig. 2).

The surface cohesion of wall paintings in a good conservation state has values between 0 and 0.3 mg/cm^2. These cohesion indexes have been demonstrated to be sufficient to resist removal of the warm agar applications. However, there may always be areas with larger values, which are difficult to detect, making it harder to ensure that the extensive use of warm agar is safe on the whole surface. Microbiological control of biocleaned surfaces suggested no risk of bacterial regrowth. Very low adenosine triphosphate (ATP) values (144 URL) and few colony forming units (6 CFU) were found on the fresco after the treatment. No significant differences were noted between treated areas before and after treatment.

Acknowledgements

The authors are grateful to: Fundación Hortensa Herrero; Dirección General de Patrimonio de la Generalitat Valenciana; the priest in the church of St Nicolás; and to all the conservators working on the church. This project has been partially supported by the project VALi+d APOSTD/2013/024 from the Generalitat Valenciana, Spain.

References

Anzani, M., Berzioli, M, Cagna, M., Campani, E., Casoli, A., Cremonesi, P., Fratelli, M., Rabbolini, D. and Riggiardi A. 2008. *Gel rigidi di Agar per il trattamento di pulitura di manufatti in gesso/Use of Rigid Agar Gels for Cleaning Plaster Objects, Quaderni del Cesmar7*, no. 6. Padua: Il Prato.

Barbabietola, N., Tasso, F., Alisi, C., Marconi, P., Perito, B., Pasquariello, G. and Sprocati, A.R. 2016. 'A safe microbe-based procedure for a gentle removal of aged animal glues from ancient paper', *International Biodeterioration & Biodegradation*, 109: 53–60.

Beltrami, E., Berzioli, M., Cagna, M., Casoli, A., Cremonesi, P., Cuaz, N. and Selva Bonino, V.E. 2011. 'La pulitura dei dipinti murali: studio comparativo di materiali e metodi', *Progetto Restauro* 58: 2–15.

Blanco Domínguez, M. M. and Tortajada Hernando, S. 2013. 'Cleaning plaster surfaces with agar-agar gels: evaluation of the technique', *Ge-Conservación* 4: 111–26.

Bosch-Roig, P. and Ranalli, G. 2014. 'The safety of biocleaning technologies for cultural heritage', *Frontiers in Microbiology* 5: 111–26.

Bosch-Roig, P., Regidor-Ros, J.L., Soriano-Sancho, P. and Montes-Estelles, R. 2013a. 'Biocleaning of animal glue on wall paintings by *Pseudomonas stutzeri*', *Chimica Oggi/Chemistry Today* 31: 50–53.

Bosch-Roig, P., Regidor-Ros, J.L. and Montes-Estelles, R. 2013b. 'Biocleaning of nitrate alterations on wall paintings by *Pseudomonas stutzeri*', *International Biodeterioration & Biodegradation* 84: 266–74.

Bosch-Roig, P., Lustrato, G., Zanardini, E. and Ranalli G. 2015. 'Biocleaning of cultural heritage stone surfaces and frescoes: which delivery system can be the most appropriate?', *Annals of Microbiology* 65: 1227–41.

Ranalli, G., Alfano, G., Belli, C., Lustrato, G., Bonadduce, I., Colombini, M.P., Zanardini, E., Abbruscato, P., Cappitelli, F. and Sorlini, C. 2005. 'Biotechnology applied to cultural heritage: biorestoration of frescoes using viable bacterial cells and enzymes', *Journal of Applied Microbiology* 98: 73–83.

Sanmartín, P., DeAraujo, A., Vasanthakumar, A. and Mitchell, R. 2015. 'Feasibility study involving the search for natural strains of microorganisms capable of degrading graffiti from heritage materials', *International Biodeterioration & Biodegradation* 103: 186–90.

Authors' addresses

- *Author for correspondence*: Pilar Bosch-Roig, Vitruvian Technologies, Valencia, Spain (pboschroig@gmail.com)
- Jose Luis Regidor Ros, Universitat Politécnica de Valencia, Valencia, Spain (jregidor@crbc.upv.es)
- Maria Pilar Soriano Sancho, Universitat Politécnica de Valencia, Valencia, Spain (pisosan@crbc.upv.es)
- Rosa Montes Estellés, Universitat Politécnica de Valencia, Valencia, Spain (rmontes@btc.upv.es)
- Pilar Roig Picazo, Universitat Politécnica de Valencia, Valencia, Spain (proig@crbc.upv.es)

The removal of lead- and oil-based overpaint from a plaster cast of *Hermes Fastening his Sandal*

Eliza Doherty and Shayne Rivers

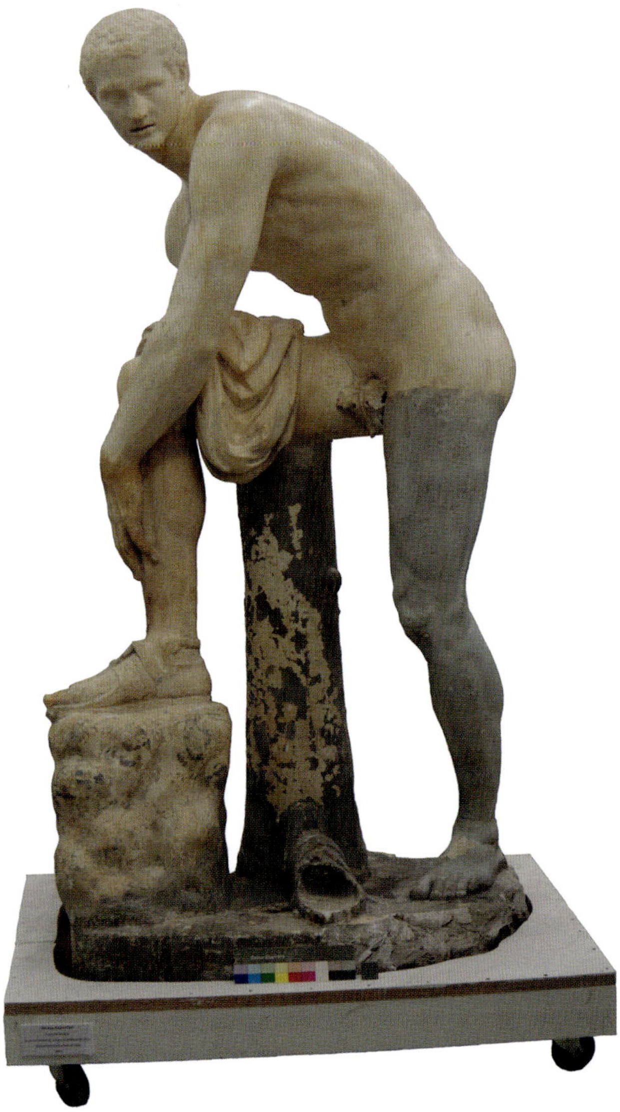

FIG. 1 Plaster cast of *Hermes Fastening his Sandal*, City and Guilds of London Art School, after the previous treatment, showing overpaint on the proper left leg, tree trunk and base.

Introduction

This paper describes the removal of overpaint from a plaster cast of *Hermes Fastening his Sandal*, carried out at City and Guilds of London Art School. Prior analysis of the sculpture had indicated the presence of up to 11 layers of lead-based oil paint over an original shellac coating. These layers had been removed from one half of the cast using dichloromethane-based strippers and a Q:switched Nd:YAG laser (1064 nm, 10 ns). One disadvantage of these strippers is that they are intended to encompass the removal of as wide a range of organic material as possible, which may have adverse effects on a complex, multilayered surface (Fig. 1). A second problem lies in the health concerns surrounding the use of dichloromethane, which is now tightly controlled.[1] It was thought that an alternative methodology, involving a system of solvent and/or aqueous gels, could be specifically formulated for this project. This system was intended to be less hazardous than the previous treatment, while efficiently removing the oil paint and effectively preserving the shellac.

Background

There is no record of when the cast, which closely resembles *Hermes Fastening his Sandal* in the Musée du Louvre (Paris, France) was acquired and little is known about its history, although it was probably sold to the school in the late 19th/early 20th century. It is likely that layers of paint were applied to the object over a number of years, in an effort to hide dirt and defects that may have distracted from the overall form. The cast was subsequently neglected and resided in the women's lavatory for at least 15 years before it was moved to the Conservation Department for examination and treatment.

TABLE 1 Methodology for the removal of the overpaint.

Gel	Dwell time	**Paint layer(s) removed** (numbers in brackets indicate layers that were softened but could not be completely removed)
EDTA at pH 7 with 20% w/w benzyl alcohol	3 hours	1–2 (3)
Deionized water (pH ~5.5) with 20% w/w benzyl alcohol	2 hours	3 (4–5)
Deionized water with 10% w/w benzyl alcohol	1 hour	4
Deionized water with 5% w/w benzyl alcohol	30 minutes	5

The object had been closely examined prior to its initial treatment. Examination of cross-sections of the paint layers under the microscope had revealed 8–11 layers of overpaint over a transparent coating. Staining of these cross-sections had suggested that the layers all contained lead and oil.[2] The transparent coating, assumed to be shellac on the basis of orange fluorescence under ultraviolet (UV) light, appeared to have penetrated the substrate, suggesting it had been applied shortly after fabrication of the cast. A white bloom that was observed on areas of the overpaint had been identified as calcium carbonate through Fourier transform infrared (FTIR) spectroscopy.[3]

Owing to the poor condition of the paint (cracking, flaking and ingrained dirt had been observed), it was decided that all of the overpaint down to the shellac should be removed. This course of action presented an interesting challenge due to the overlapping solubility windows of oil and shellac, and the highly porous nature of the plaster. The previous treatments used to remove the overpaint had included the use of the commercial dichloromethane-based paint strippers Nitromors and Paramose, and a Q:switched Nd:YAG laser.[4]

Treatment

Lead white was considered to be the pigment in the top layer due to the colour of the paint. Since the overpaint was likely to contain a large concentration of Pb^{2+} ions, which could interfere with swelling/solubilization and therefore inhibit its removal, it was thought that an alternative system would benefit from the inclusion of a strong chelator for lead (Wolbers and Little 2004: 166). Wolbers (2000) used an aqueous emulsion gel with the chelator 2-hydroxyethyl ethylenediaminetriacetic acid (HEDTA) to chelate Fe^{2+} ions in highly polymerized oil overpaint containing an iron oxide pigment on an oil painting. A slightly raised pH (9) and a swelling/solubilization agent (xylene) were used to 'open up' the paint film and increase the rate of chelation/solubilization.

The following system was tested on the top layer of overpaint: 5% ethylenediaminetetraacetic acid (EDTA) at pH 7 in xanthan gum, with 5%, 10% and 20% w/w benzyl alcohol. EDTA was chosen because its formation constant (log K_f) for a Pb-EDTA complex (18.3) is larger than the two solubility product constants (pK_{sp}) of lead white (13.13 and 14.93) (Wolbers 2000: 114, 117). EDTA was also likely to be effective in areas of high salt concentration where, as mentioned above, the presence of calcium carbonate had been confirmed through FTIR. The solution was buffered to pH 7 with sodium hydroxide (1M) using a Horiba Laqua Twin Compact pH meter, because EDTA works best above its third highest acid dissociation constant (pK_a) of 6.2 and the estimated conditional stability constant of EDTA for Pb^{2+} peaks at 7 (Rivers and Umney 2003: 545, fig. 11.22). Benzyl alcohol, a high swelling agent for oil, was incorporated into the gel to increase the rate of chelation by swelling the oil medium (Phenix 1998, 2002). The EDTA/benzyl alcohol mixture was gelled using xanthan gum. This gelling agent can tolerate a wide pH range and is compatible with high concentrations of chelating agents and water-miscible solvents (Wolbers and Stavroudis 2012: 512). The gel had the best consistency and working properties at 4% w/v. It was found to be easy to apply and remove, retaining its viscosity even after a dwell time of three hours.

Up to 11 layers of overpaint had been identified in cross-sections. Through solubility testing, five layers were distinguished by colour/shade and their relative thickness established, which aided analysis of the efficacy of the tests:

1. Blue/grey (uppermost layers)
2. Dark brown/red
3. Yellow (thick)
4. Dark brown/yellow
5. Orange/yellow (thick)

With 5% w/w benzyl alcohol, layers 1 and 2 were partially softened after 30 minutes, with increased softening after one and three hours. With 10% w/w benzyl alcohol, layer 1 could be completely removed after 30 minutes and after 3 hours both layers 1 and 2 could be partially removed. With 20% w/w benzyl alcohol, both layers 1 and 2 could be easily removed after 3 hours and layer 3 was partially softened. The surface of the object was closely examined following cleaning tests and no gel residues were observed.

There were two factors to consider following these results. Firstly, EDTA would probably chelate Ca^{2+} ions from a gypsum substrate (Rivers and Umney 2003: 543–4, tables 11.9 and 11.10). Secondly, benzyl alcohol is just within the solubility region of shellac on a Teas Chart and therefore likely to cause swelling/solubilization (Rivers and Umney 2003: 519, fig. 11.10, 520, fig 11.12). With this in mind, a four-stage methodology was established through further testing and observation (see Table 1).

It was judged that the high viscosity of the gel and its separation from the substrate by layers of overpaint would

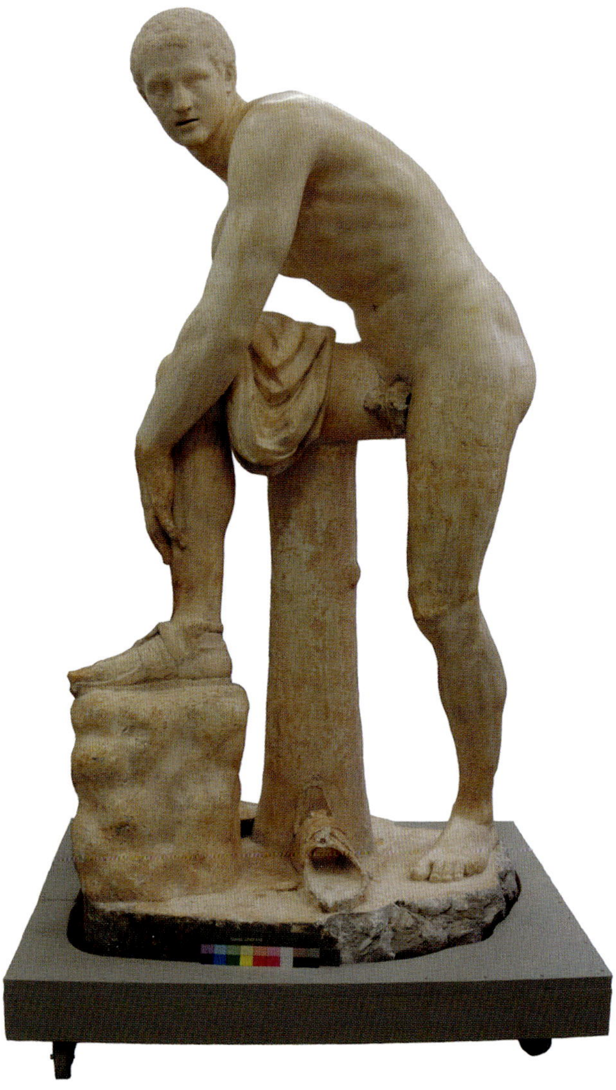

FIG. 2 Plaster cast after treatment.

prevent the chelator in the first gel from coming into contact with the plaster. EDTA was eliminated from the subsequent three gels owing to increasingly closer proximity to the plaster substrate as cleaning proceeded. The gel was found to be effective on underlying layers without the chelator. The proportion of benzyl alcohol was also reduced due to the solubility of the original shellac coating on the plaster. The slightly acidic pH of these gels was also thought to be more compatible with the shellac, since when base readings had been taken of the original coating using a pH meter and agarose gel (4% w/v in deionized water), the average reading was pH 5.0. These indicated that an aqueous solution with a high pH would be more likely to disrupt this coating than an aqueous solution with a low pH.

During the removal of overpaint, measures were taken to reduce the escape of lead residues from the overpaint and fumes from the benzyl alcohol. Appropriate personal protective equipment was worn: nitrile gloves, protective clothing and a solvent vapour mask, and all work was carried out in a well-ventilated workspace. A fume cupboard and portable fume extraction were also employed.

Conclusion

This project is an example of solubility problems encountered in practice. The treatment utilized materials and techniques from related conservation disciplines to address the systematic removal of oil paint layers from a natural resin varnish coating on a plaster substrate (Fig. 2). The materials used were predominantly non-toxic and non-flammable, with evaporation of the solvent as well as exposure to lead reduced via means of a gelling agent. The gel system had good working properties and offered subtlety and control.

Acknowledgements

Thanks are extended to the City and Guilds of London Art School, and specifically Dr Marina Sokhan, Eric Miller, Jennifer Dinsmore, Catarina Ramalho, Hugi Hicyilmaz, Diane Magee, Tim Crawley, and the Conservation Studies class of 2014. Richard Wolbers is also thanked for providing initial advice.

Materials and suppliers

- Benzyl alcohol: VWR International, 201 Rue Carnot, 94120 Fontenay-sous-Bois, France
- Deionized water: Tetrosyl Ltd, Bury, Lancashire BL9 6RE, UK
- EDTA: Sigma-Aldrich Chemie GmbH, Riedstraße 2, 89555, Steinheim am Albuch, Germany
- Sodium hydroxide pellets: BDH Laboratory Supplies, Poole, BH15 1TD, UK
- Xanthan gum: Doves Farm Foods Ltd, Salisbury Road, Hungerford, Berkshire RG17 0RF, UK

Notes

1. A ban on domestic and on-site use of paint strippers containing dichloromethane by the European Parliament came into force in December 2011. The use of dichloromethane is permitted in industrial settings providing safe working practices are followed. See *Decision 455/2009/EC of the European Parliament and of the Council amending Council Directive 76/769/EEC as regards restrictions on the marketing and use of dichloromethane*, Health and Safety Executive. Available at: http://www.hse.gov.uk/aboutus/europe/euronews/dossiers/dichloromethane.htm (accessed 12 March 2016).
2. Dilute nitric acid and a crystal of potassium iodide were used to test for lead: the paint layers turned bright yellow, indicating the presence of lead. A saturated solution of Sudan Black B ((2,2-dimethyl-1,3-dihydroperimidin-6-yl)-(4-phenyldiazenylnaphthalen-1-yl)diazene) in industrial methylated spirits and deionized water was used to test for oil. The paint layers were dyed blue/black, indicating the presence of oil.

3. These samples were taken using the solvent extraction method outlined in Derrick *et al.* (1999: 29–31), using deionized water.
4. Nitromors was used subject to availability. It contains dichloromethane as the active ingredient and a small amount of methanol. The main gelling agent is paraffin wax. Paramose also contains dichloromethane and methanol.

References

Derrick, M., Stulik, D. and Landry, J. 1999. *Infrared Spectroscopy in Conservation Science*. Los Angeles: Getty Conservation Institute.

Phenix, A. 1998. 'Solvent-induced swelling of paint films: some preliminary results', *WACC Newsletter* 20: 3. Available at: http://cool.conservation-us.org/waac/wn/wn20/wn20-3/wn20-308.html (accessed 7 January 2013)

Phenix, A. 2002. 'The swelling of artists' paints in organic solvent and solvent mixtures. Part 2. Comparative swelling powers of selected organic solvents and solvent mixtures', *Journal of the American Institute of Conservation* 41(1): 61–90.

Rivers, S. and Umney, N. 2003. *The Conservation of Furniture*. London: Routledge.

Wolbers, R. 2000. *Cleaning Painted Surfaces: Aqueous Methods*. London: Archetype Publications.

Wolbers, R. and Little, M. 2004. 'The surface revealed: cleaning of two painted plaster sculptures', *Objects Specialty Group Postprints* 11: 154–71.

Wolbers, R. and Stravroudis, C. 2012. 'Aqueous methods for cleaning of paintings', in J.H. Stoner and R. Rushfield (eds), *The Conservation of Easel Paintings*. Abingdon: Routledge.

Authors' addresses

- Eliza Doherty, London, UK (elizadoherty@gmail.com)
- Shayne Rivers, London, UK (shayne.rivers@westdean.org.uk)

Solvent gel versus solvent poultice: evaluating two techniques for the removal of pressure-sensitive tape stains from paper

Cecilia Isaksson

Introduction

Pressure-sensitive tape (PST) is a well-known concern within paper conservation; this paper addresses how to remove the rubber-based tape stains that appear during the crosslinking stage of adhesive degradation, when the carrier has already fallen off, or peels off with ease, and staining is prominent or translucent. Previous removal techniques include using solvent poultices together with local application of solvent. However, Lennig (2010) suggests using rigid solvent gels for these treatments. Advantages included time efficiency, reduction in solvent quantity, and minimized solvent handling, reducing both costs and health risks. This study aims to assess the use of solvent gels for tape stain removal and to compare them with traditional solvent poultice techniques.

Materials and methods

For testing the poultices two artworks with identical paper and identical staining were used, and for testing the solvent gel one artwork with four identical stains was employed. Initial tests for a suitable solvent were made on the suction table and proved acetone efficient. However, the stains needed softening in a solvent chamber first, namely an enclosure with some modern paper soaked in solvent, not in contact with the work. Four poultices were applied to the verso where the most adhesive was located: sepiolite, kaolin, bentonite and Fuller's earth,[1] all on identical tape stains and paper. For the darker coloured bentonite, a barrier layer was used. A circle of dry poultice was made around the stain and a moist mixture of solvent and clay applied to the stain. Using a spatula, the dry poultice was pushed over the moistened area, then covered with polyester sheeting (Melinex), and weighed to keep it airtight. The 'wet pack' was left overnight to dry. The treatment, used for all the poultices, was repeated three times on the verso and once on the recto.

For testing the solvent gels, an ink drawing with four identical tape stains was used. Lennig (2010), found Polysurf 67-CS (cetyl-hydroxyethylcellulose) most effective, however, this was unavailable so Ethulose (ethyl hydroxyethyl cellulose) was substituted. As initial suction table tests indicated that all tape stains were soluble in acetone, Ethulose was only tested in this solvent. A 10 wt% gel (2–3 mm thick) was prepared, transferred to a Melinex tray and then refrigerated overnight. The treatment followed Lennig's method except there was no pre-wetting of the surface with solvent (Lennig 2010: 98); tests were also made both with and without a barrier layer of either blotting paper or smooth non-woven polyester (Bondina). Pieces of gel, slightly smaller than the stains, were placed on blotting paper and saturated with acetone. They were then applied to the stain with different barrier layers: Bondina; blotting paper; Bondina with an extra solvent wetted blotting paper on top of the gel on both recto and verso; or no barrier layer. All solvent gels were covered with Melinex to keep them airtight. After 40 minutes the gel was removed and refilled with solvent. This process was repeated eight times. Following Lenning's example, the gel was replaced after six refills.

Results

Poultices

Observing the stains under normal daylight showed that the most successful poultice was kaolin. Bentonite was least efficacious, which is easily explained as it was the only poultice applied with a barrier layer. Sepiolite proved slightly more effective than Fuller's earth. Under ultraviolet (UV) light, the stains still fluoresced although less than before treatment. The differences were more prominent in daylight. Furthermore, the stain treated with Fuller's earth had developed a dark spot in its centre, a phenomenon that has been reported by Stiber Morenus.[2] The result from the Fuller's earth poultice was considered unsatisfactory, so this treatment was followed up with direct application of solvent on the suction table. To further rinse out adhesive, rehydrate the paper and possibly reduce the dark stain visible under UV, the stain was also treated with water. The combination was effective and much adhesive was removed. Examination under UV showed that the stain fluoresced less, although the dark stain had become larger.

Gels

Firstly, blotting paper proved too thick as a barrier layer. The stain with Bondina as the barrier layer showed results after four refills. The same result was achieved with the double pack (Bondina and dampened blotting papers on recto and verso) although the moisture combined with the gel's thickness caused deformation of the paper. After the first application, improvement was visible where no barrier layer was used. Eight refills produced a good result, but only on the application side.

Discussion

The double pack and the application without a barrier layer were the most efficient techniques but both had pros and cons. The double pack was quicker because it worked from both sides simultaneously, however, it was more difficult to handle and deformed the paper. The non-barrier method was quickest but potentially leaves residues invisible to the naked eye. The results partly agree with Lennig's (2010) conclusion, as she found the double pack the best solution. The residue issue concerning direct application requires further testing, nonetheless, a double treatment with a direct application would improve the technique. Poultices may also leave residues, though typically their colour blends with the paper: it is neither visible to the naked eye nor when examined with UV. As kaolin has long been used in paper manufacturing, it is unlikely to be harmful to paper. Results from the tests, both gels and poultices, could be affected by variations in the amount of solvent, the paper, or the adhesives' degradation process. Fresh, refrigerated gel is most effective – it keeps for 5–6 days after which it shrinks and loses absorbency.

When evaluating poultices versus solvent gels, the latter is preferable as the uncertainty of the poultice drying time is avoided and gels are less messy, therefore facilitating multiple stain removal. Both techniques are best combined with the use of solvents on the suction table, but compared to treatments limited to this, both methods reduce solvent handling with all its benefits for health and reducing costs. Investigations with UV pose the question: how much stain removal is necessary – until satisfactory results are reached in daylight or under UV? The conservator must balance how much solvent the paper can withstand with how much time is available, as well as the costs versus the risks of the stain reappearing, or further degrading the paper. Further research into dark staining using UV examination would also be worthwhile.

To sum up, techniques to remove crosslinked, rubber-based PST still require refining, but this study supports the use of solvent gels as the leading option for this kind of conservation treatment.

Acknowledgements

Special thanks are extended to paper conservator Alison Norton, Modern Museum of Art, Stockholm, Sweden, and to paper conservator Dr Helen Evans, National Museum, Stockholm, Sweden for all their help and support.

Materials and suppliers

- Bentonite: Askania AB, Ågatan 4, 400 20 Mölndal, Sweden, tel. +46(0)31-733 22 00
- Bondina: Preservation Equipment Ltdd, http://www.preservationequipment.com/
- Ethulose: Talas, http://www.talasonline.com/
- Fuller's earth, Kaolin PW 19: Kremer, http://www.kremerpigments.com/
- Hollytex: http://www.preservationequipment.com/Store/Products/Conservation-Materials/Other-Materials/Hollytex-(G), 2013
- Melinex: Preservation Equipment Ltd, http://www.preservationequipment.com/
- Sepiolite: Kremer Pigmente GmbH & Co. KG, http://www.kremer-pigmente.com/en/fillers-und-building-materials/colorless-und-colored-mineral-fillers/siliceous-quartz-granite-and-others/1933/sepiolite, 2016

Notes

1. 'Fuller's earth' is a collective name for a naturally fine-grained clay that easily absorbs oil and fat. The expression derives from the textile industry where the clay was used to clean wool. The workers were called 'fullers', hence giving the name 'Fuller's earth' (*Encyclopedia Britannica* 1994). It is also the name of a poultice/clay often available from conservation suppliers.
2. L. Stiber Morenus, 2001, *Effect of Organic Solvents on Paper*, Washington, DC, unpublished material, handout from the course *Removal of Pressure Sensitive Tapes and Tape Stains* held by E. O'Loughlin, and L. Stiber Morenus at the Royal Danish Academy of Fine Arts Schools of Architecture, Design and Conservation, Copenhagen, 17–21 May 2009.

References

Encyclopaedia Britannica 1994. Available at: http://www.britannica.com/EBchecked/topic/221929/fullers-earth (accessed 14 February 2012); http://www.britannica.com/EBchecked/topic/440361/palygorskite?anchor=ref32863 (accessed 14 February 2012).

Lennig, H. 2010. 'Solvent gels for removing aged pressure-sensitive tape from paper', *Restaurator* 31(2): 92–105.

Author's address

Cecilia Isaksson, Kungliga biblioteket, Stockholm, Sweden (cecilia.isaksson@kb.se)

Using rigid gellan gel for the stratigraphic cleaning of a canvas painting

Giacomo Maranesi

During a treatment which took place at the Istituto di Restauro delle Marche (IRM), Academy of Fine Arts, Macerata, Italy, rigid polysaccharide gels made from gellan proved to be crucial for several aspects of the process (Maranesi 2014, 2017). The previously unknown painting from the School of Vittore Crivelli most probably dates to 1496 and comes from the Monastery of Santa Chiara in Fermo, Italy. The painting was executed using oil and tempera on canvas, on a preparatory layer of calcium sulphate and animal glue, and it portrays the *Madonna delle Grazie (Our Lady of Grace)* also known as *Madonna della Misericordia (Our Lady of Mercy)* (Fig. 1). No previous restoration had been documented, but on the surface there were several layers of overpaint, applied both as taste in style changed and to deal with its advanced deterioration, attributable to the painting technique. The number of overpainted layers ranged from two to five in different areas (Fig. 2).

Preliminary examination included imaging in raking light, infrared reflectography (IRR), false colour infrared (FC-IR) and ultraviolet (UV) imaging, and examination of cross-sections in visible and UV light, to explore the condition of the original paint. We then proceeded with the removal, or in some cases the thinning, of all the non-original layers. This required extreme caution so we carried out a stratigraphic removal of all non-original layers, one by one. The operation proved to be very difficult: these layers were mostly oil paint, and some were interspersed with protein-based and oil-resin varnishes, now deteriorated. Solvent solubility tests all produced negative results therefore it was necessary to investigate aqueous solutions, with slightly basic pH and/or a chelator, in order to swell and remove all the non-original layers. In contrast to the swelling action of the organic solvents, such solutions did not guarantee selective action on the materials to be removed. They presented a high risk to the original paint surface necessitating the use of aqueous solutions with increased viscosity.

To determine which aqueous solutions were more suitable, numerous cleaning tests were carried out on different areas of the background, representative of the various layers to be removed or thinned. Solutions with chelators at both neutral and slightly alkaline pH, as well as chelate-free solutions with a variety of pH levels, were tested, all incorporated into cellulose ether using 4% Klucel G. The second phase of testing dealt with gelling agents. The most suitable aqueous solutions were made more viscous by utilizing three kinds of thickening agents with increasing molecular weight: 4% Klucel G, 2% Carbopol Ultrez 21 and 1–1.5% Phytagel gellan polysaccharide gels were tested. In order to limit the penetration of the aqueous solutions into the original paint, cellulose ether was discarded in favour of polyacrylic acid in some cases. The rigid polysaccharide gel proved decisive in many ways.

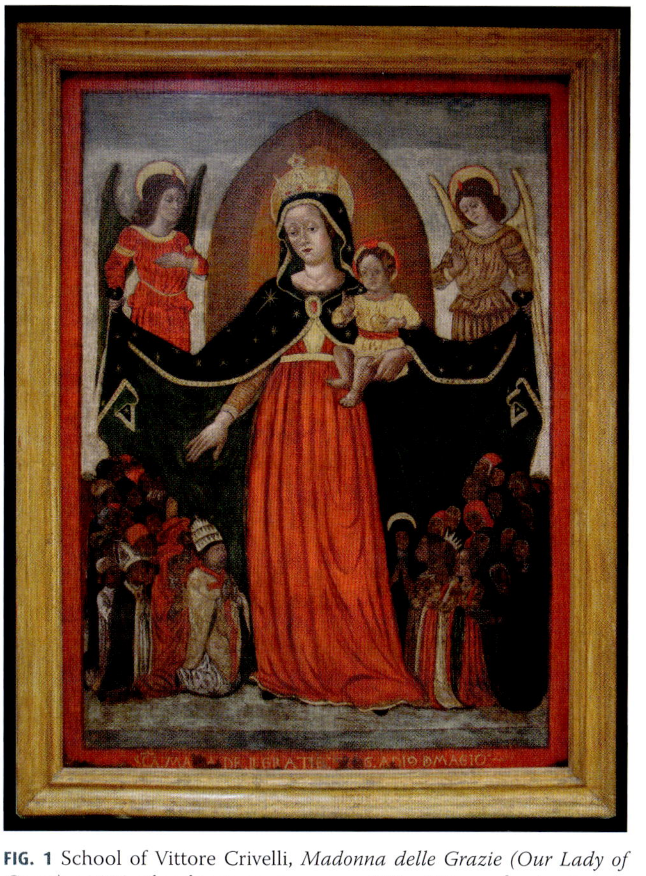

FIG. 1 School of Vittore Crivelli, *Madonna delle Grazie (Our Lady of Grace)*, *c.*1496, oil and tempera on canvas, 180 × 130cm: after treatment.

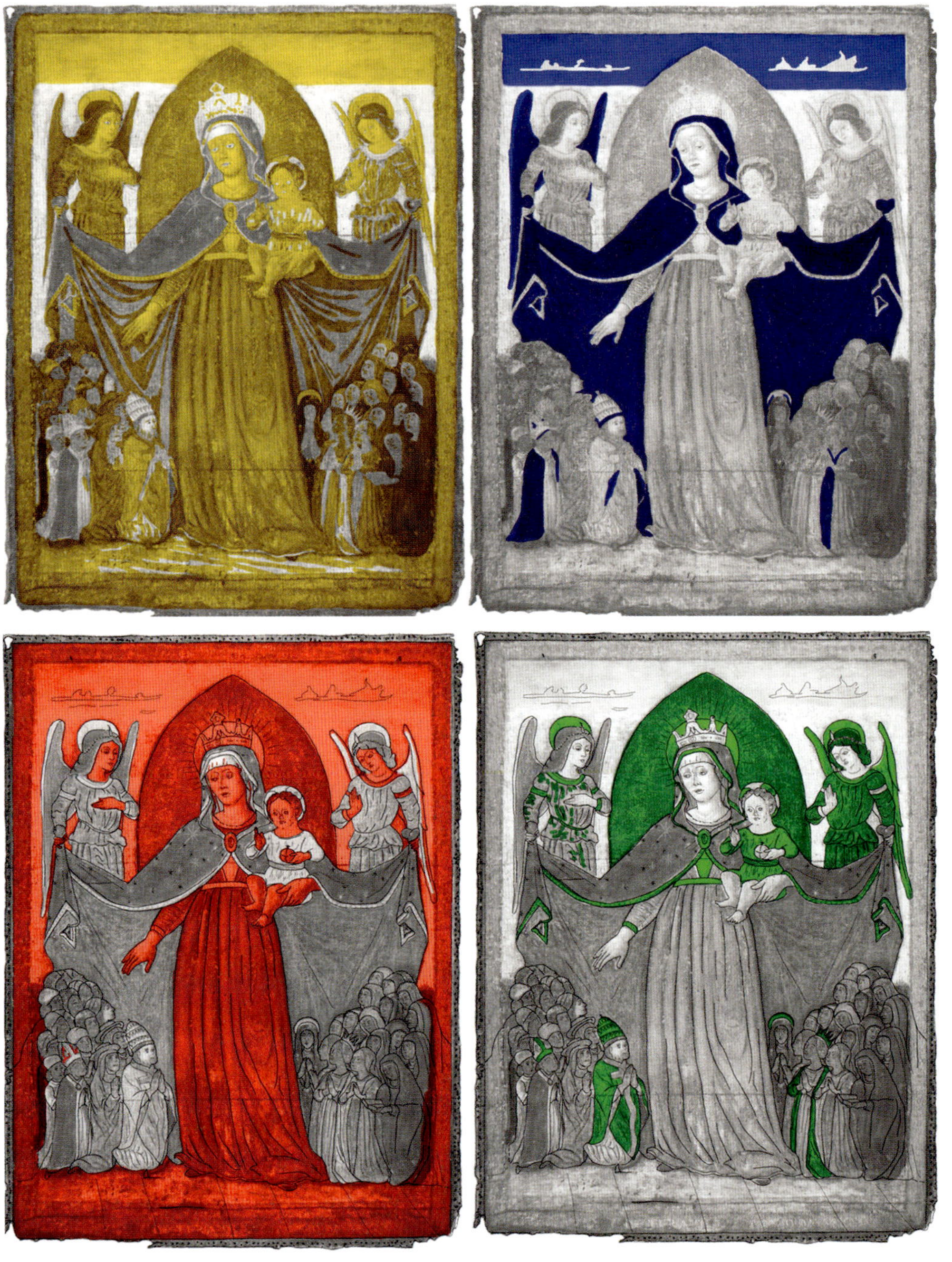

FIG. 2 School of Vittore Crivelli, *Madonna delle Grazie (Our Lady of Grace)*: mapping of some of the layers of overpaint.

Following these tests, four systems were selected: a pH neutral gel composed of water and polyacrylic acid – 2% Carbopol Ultrez 21, neutralized with ammonium hydroxide; a similarly composed gel, but with slightly basic pH; a neutral chelating solution composed of 1% w/v triammonium citrate (TAC) in water supported in gellan; and a buffered chelating solution, consisting of water, citric acid, triethanolamine (TEA) at pH 8, in gellan. These systems all had different effects on the substrate. Clearing for complete removal of the swollen substrate and all the gel residues was carried out using a solution of 1:1 ethanol: deionized water.

The use of gellan made it possible to apply less aggressive solutions by increasing the period of contact and checking the penetration of the solutions. This ensured more selective stratigraphic cleaning, layer by layer (Fig. 3). Using the gellan, the original paint was not touched at all by the solutions and suffered no damage. Moreover, the gel treatment allowed documentation of each layer as it was removed, so that the number of restorations could be established. Gellan requires caution in the formulation and preparation of aqueous solutions, mainly because it is necessary to heat the solution containing the polysaccharide to form a gel. This dictates the selection of materials that are as stable as possible (i.e. non-volatile) for the safety of the operator, as well as the need to determine the optimum concentration of 1–1.5% w/v. On the particularly thin layers of varnish to be removed, we used a gel at 2% w/v. Upon increasing the thickness of the gel a more gradual release of the solution was obtained

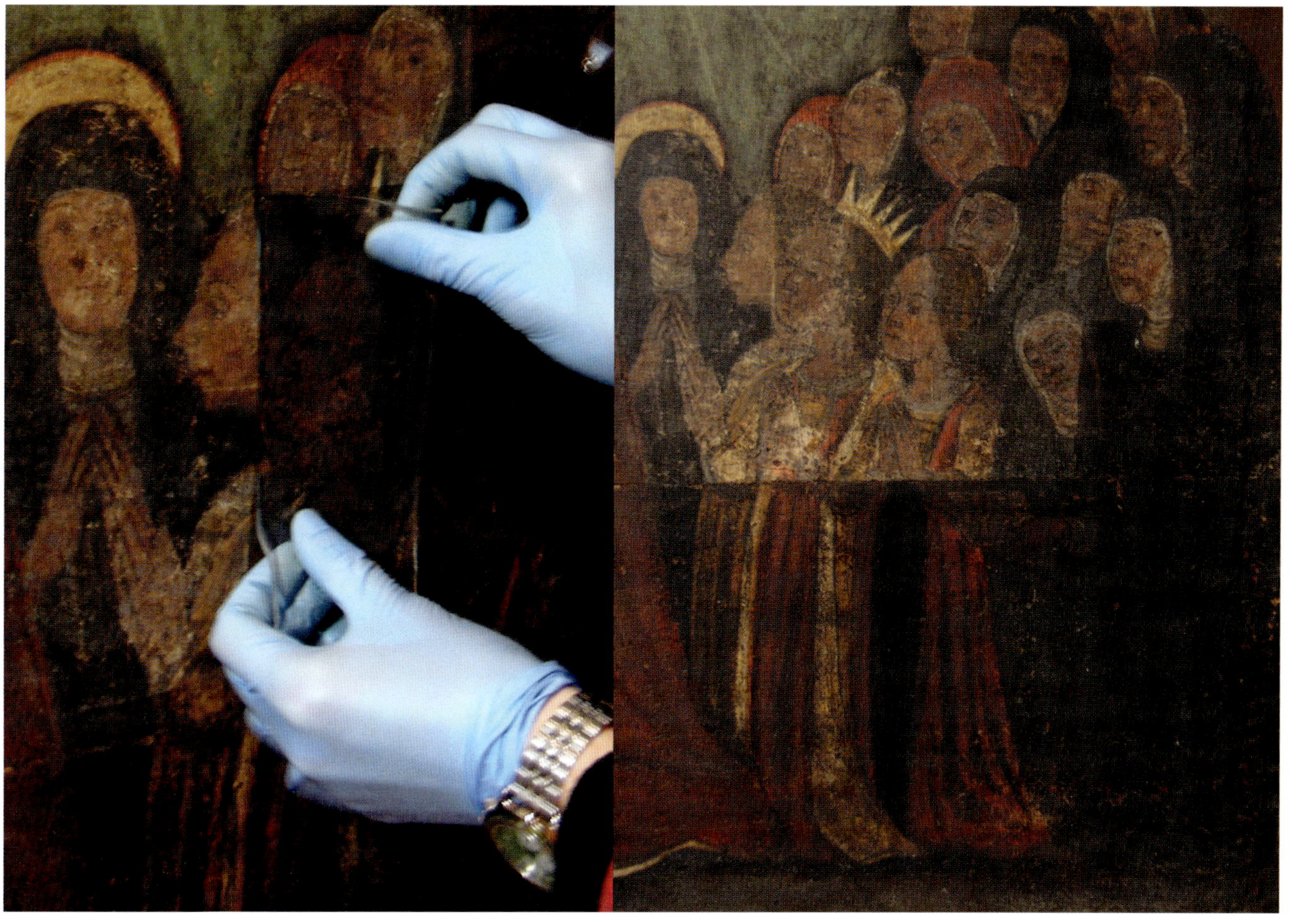

FIG. 3 School of Vittore Crivelli, *Madonna delle Grazie (Our Lady of Grace)*: cleaning using gellan gum.

and the absorption of soluble materials was also augmented. This permitted longer contact periods as well as improved control and efficacy. The gel was prepared by pouring the warm solution onto a smooth surface. Once it had cooled, the gel was cut and shaped mainly according to the shapes of the figures on the painting: it was applied directly to the paint surface with a contact time ranging from 1 to 4 minutes, depending on the nature and thickness of the layer to be removed.

In conclusion, the use of rigid gellan was central to this treatment, making use of:

- its rigidity;
- its high retention of aqueous solutions and solvents;
- its ability to absorb soluble or hydrolysed particles of the layers to be removed;
- its capacity to release aqueous solutions or solvents continuously;
- the lack of residues after removal of the gel.

Materials and suppliers

- Carbopol Ultrez 21,Klucel G, non-ionic hydroxylpropylcellulose: CTS Srl, via Piave 20/22 - 36077, Altavilla Vicentina, Italy
- Citric acid, Phytagel gellan, heteropolysaccharide made from glucoronic acid, glucose and rhamnose, triammonium citrate, triethanolamine: Bresciani Srl, via Breda 142 - 20126, Milan, Italy

References

Maranesi, G. 2014. *Il Monastero di Santa Chiara a Fermo e i sui dipinti. Studi critici, schedatura, analisi conservativa e restauro*, unpublished diploma thesis, Accademia di Belle Arti di Macerata, Macerata, Italy

Maranesi, G. 2017. 'Il Monastero di Santa Chiara a Fermo e i suoi dipinti. Analisi della raccolta e restauro della Madonna delle Grazie', in D. De Luca (ed.), *I giovani e il restauro*. Padua: Il Prato [in press].

Author's address

Giacomo Maranesi, Fermo, Italy (giacomomaranesi87@gmail.com)

The use of agar gel for cleaning and overpaint removal from Romanian ecclesiastical tempera paintings

Andreea Michescu, Daniela Cristina Ilie, Rodica Pavel, Valentina Dudu, Anastasia Floroiu, Dana Postolache and Ioan Darida

Introduction

This paper reports on the use of agar gels on several paintings in egg tempera. Three case studies were selected based on the support: two are 18th-century icons, one with a panel and the other a glass support, and the third is a 16th-century polychrome monastery door. A series of cleaning tests for dirt removal and removal of tempera-based overpaint preceded successful treatment in each case. The results presented illustrate only observations on the behaviour of the original painting layers and the methodological approach in each case.

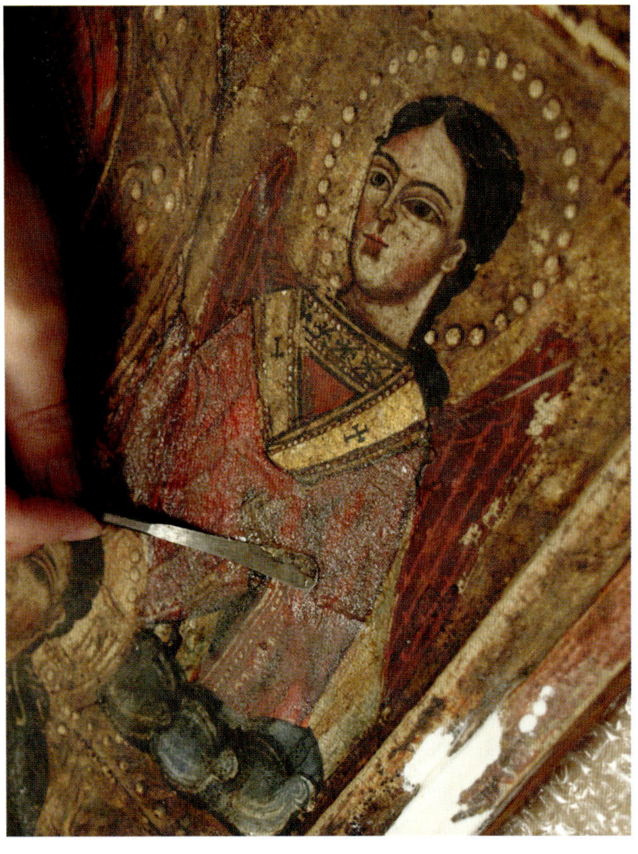

FIG. 1 *Virgin and Child with Archangel*, 18th century, egg tempera on panel, 83.5 ×51.5 × 2.5 cm, Nicula Monastery, Transylvania region, Romania: detail of the icon during removal of the agar gel from the paint.

Virgin and Child with Archangel, tempera on panel

The icon was painted in the traditional tempera technique and before treatment it had typical visual deterioration of the painted surface. The original varnish – identified as a plant gum[1] – was severely damaged, covered by dirt deposits and wax in isolated areas. Preliminary testing showed that the superficial dirt deposits were sensitive to water-based systems; however, the varnish did not respond to aqueous solutions. To prevent the water-based systems from penetrating into the painting structure through cracks and fissures, gelled systems were tested. Synthetic saliva[2] was chosen for testing because of its surfactant and chelating properties as well as its near-neutral pH. The substance was tested at concentrations of 0.1, 0.5 and 1% v/v in both liquid and agar gel systems (Fig. 1).

Preliminary results indicated that synthetic saliva has the same efficiency in the removal of dirt deposits in both liquid and gel states. The optimal concentration of the liquid solution was 0.5%, with a pH of 6 for the final solution. Cleaning tests using saliva-loaded agar[3] gels enhanced the method by reducing the penetration rate of the liquid into the original paint. The concentration of the agar gel (5 wt%) was chosen based on previous tests. The synthetic saliva (1%) was added during the cooling of the agar solution, when it had reached 30 °C. To ensure full contact with the irregular surface of the painting,

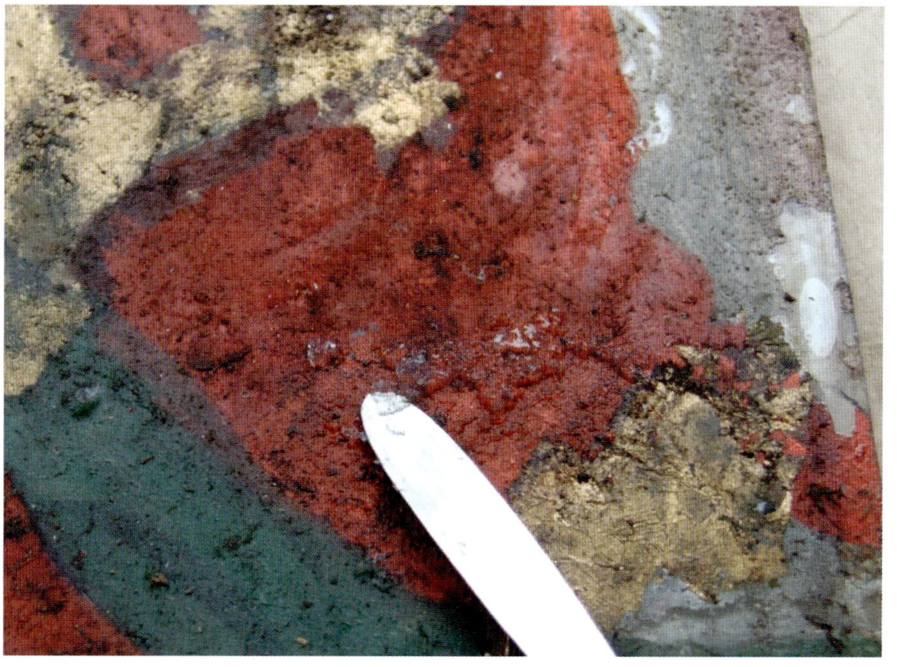

FIG. 2 *Christening of Jesus*, 18th century, egg tempera on glass, 17.5 × 28.5 × 0.5 cm, Transylvania region, private collection: detail of the verso during removal of the agar gel from the paint.

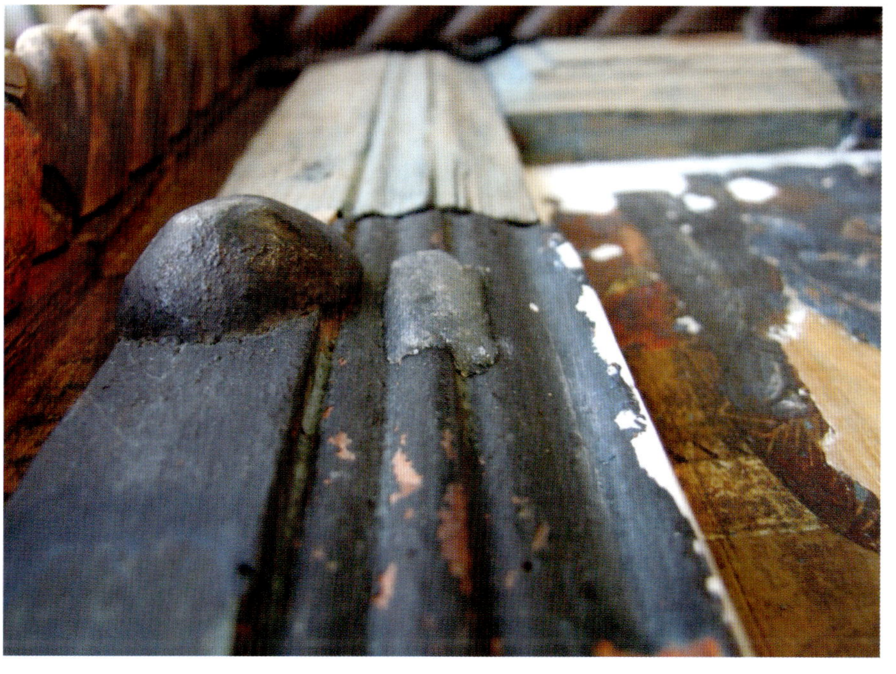

FIG. 3 Door made from polychromed egg tempera on carved wood, Voronet Monastery, Gura Humorului, Suceava district, Romania, 16th century. Detail covered with agar gel that conforms well to the structure. The original red paint layer, which can be seen through losses in the overpaint, was fully recovered by the treatment.

the rigid gel was first mashed, then applied with a spatula. After 1 minute the gel was removed, leaving the superficial dirt layer reduced in thickness; what remained could subsequently be removed by rolling a dry cotton swab. At this stage, the original varnish was ready for the next phase of the cleaning process: partial varnish removal. The advantages of this type of approach lie in the reduction of the exposure time of the paint to liquids, and in the controlled method of using water-based cleaning systems on tempera paintings.

The Christening of Jesus, tempera on glass

Tempera on glass paintings raises difficult problems with respect to the removal of dirt deposits from paint, due to the lack of protective layers. This icon (Fig. 2) had significant flaking of the paint layer, with both highly adherent and poorly adherent dirt deposits, as well as structural damage, namely cracks and deformities. Treatment began with dry cleaning of the surface. The consolidation treatment of the paint layer followed, with the use of egg-yolk emulsion in water (1 part egg yolk, 4 parts demineralized water to which preservatives were added to prevent biodeterioration);[4] the emulsion was sprayed over the flaking paint layer and after 1 minute the emulsion was absorbed by the paint layer. At this point the whole surface was covered with Melinex film to prevent the dislocation of small paint fragments, and the surface was slightly pressed using soft cotton rolls. While successful in fixing the detached fragments of the paint layer, the consolidation process also had the expected effect of causing greater adherence of the dirt to the paint. The soiling removal methodology implied the use of agar gel with a small amount of synthetic saliva (as previously described). Various application times were tested; a 30-second application was found to be sufficient for emulsification of the soiling. The gel was then removed by rolling with a dry cotton swab. The use of an agar gel for the removal of dirt deposits from the surface of the highly water-sensitive tempera painting was a significant improvement on the cleaning methods usually applied to this combination of paint medium and support.

Sixteenth-century polychrome carved door

The entrance to the church of the Voronet Monastery is marked by a wooden door with exquisite polychrome decorations including both painted sculptures and icons, all with complex decorative techniques, attributed to the 16th century when the porch of the church was added to the original building. Long-term exposure to a microclimate outdoors has played a significant role in the deterioration of the polychrome decorations. In addition, the door has undergone at least two interventions including overpainting. Investigation[5] of the original painting materials and the overpaints showed that all were painted using egg tempera.

Tests for the removal of the overpaints were conducted with a focus on water-based systems, using mixtures of water, alcohol, ammonium hydroxide solutions and artificial saliva in varying concentrations, to identify the optimal method for swelling the overpaint layers without affecting the original painting. Three different cleaning methods were used:

- slight mechanical action and rolling with cotton swabs
- cotton poultices
- rigid gels incorporating the mixtures described above.

All systems were applied directly to the paint surface without a barrier tissue. The rigid agar gel was selected because of its rheological properties, which allow it to maintain liquids within its porous structure. The minimum agar concentration necessary to form a gel (1%) provided satisfactory results for this treatment. The best results were obtained from artificial saliva (2%), both in poultices and rigid agar gels. The optimal dwell time was around 1 minute. After this period, the overpaint layers could easily be removed mechanically (Fig. 3).

Conclusion

The aim of this study was to enhance the qualities and versatility of agar gels for the cleaning of varnished tempera paintings on panel, unvarnished tempera paintings on glass, and for the removal of tempera-based overpaint from original tempera paint. Agar gels were successful in these case studies for the elimination of extensive dirt deposits, soiling and overpaint. The main difficulty encountered was controlling the action of aqueous cleaning systems. The results presented here were verified simply by visual observations, and require further research because the use of aqueous cleaning systems for tempera paintings still raises important questions for conservators.

Notes

1. Data collected by Dr Lora Angelova, using a Perkin Elmer Spectrum 100 Fourier transform infrared (FTIR) spectrometer with a ZnSe ATR crystal, using 32 scans, 4 cm^{-1} resolution, 4000–650 cm^{-1} spectral range.
2. Synthetic saliva (CTS Europe).
3. Agar obtained from Fluka, product number 05040, standard quality.
4. The mixture was prepared in the conservation laboratory by Anastasia Floroiu using commercial products. The preservative used is Preventol RI80: CTS Europe.
5. Analysis conducted by Dr Barbu Olimpia-Hinamatsuri and Dr Georgescu Migdonia, senior investigators at the National History Museum, Bucharest, Romania. FTIR and X-ray fluorescence spectrometry were used characterize the origin and nature of both organic (binders) and inorganic materials (pigments).

Authors' addresses

- Andreea Michescu, PhD Candidate, University of Fine Arts, Senior Paintings Conservator, Danart Import – Export. S.R.L., Bucharest, Romania (michescu.andreea@gmail.com)
- Daniela Cristina Ilie, Senior Paintings Conservator, Danart Import – Export. S.R.L., Bucharest, Romania
- Rodica Pavel, Senior Paintings Conservator, S.C. Danart Import – Export. S.R.L., Bucharest, Romania
- Valentina Dudu, Senior Paintings Conservator, S.C. Danart Import – Export. S.R.L., Bucharest, Romania
- Anastasia Floroiu, Senior Paintings Conservator, S.C. Danart Import – Export. S.R.L., Bucharest, Romania
- Dana Postolache, Theological University of Craiova, Department of Conservation of Cultural Heritage, Craiova, Romania
- Ioan Darida, Theological University of Pitesti, Department of Conservation of Cultural Heritage, Pitesti, Romania

Washing works of art on paper using rigid hydrogels containing chelating agents

Ekaterina Pasnak

Introduction

Severely yellowed, coloured works of art on paper pose particular conservation problems. Sensitivity of the graphic media and fragility of the paper support discourages washing with direct application of moisture. Protecting the media with cyclododecane (Rowe and Roseik 2008) or other methods causes distortion of the paper and uneven washing. Conservation treatment can strengthen the paper by washing out degradation products and reducing discoloration or evening it out to a single tone. To retain paper strength and preserve the graphic media, bleaching should be kept to a minimum.

Treatment method

Many of these issues can be resolved by washing with rigid hydrogels, most commonly containing two polysaccharides, gellan gum (Iannuccelli and Sotgiu 2010a) and agar (Casoli *et al.* 2013). With rigid hydrogels there is less swelling of paper fibres because of 'the gradual and controlled release of water molecules from the gel to the paper' (Iannuccelli and Sotgiu 2010b). If pH and conductivity of such gels are adjusted to the parameters of the paper, the swelling of the fibres and the stress caused by osmotic pressure during washing can be minimized. By ensuring that the difference in pH between the aqueous solution and the paper support is no more than 2 pH units, and the conductivity of the solution is not greater than 10 times that of the substrate, this further reduces stress.[1] Areas protected with cyclododecane exhibit less tension at the paper–gel interface during a gel wash. As water is applied from the verso, there is exposure to water of the image, which could lead to movement and solubilization of the graphic media.

Rigid hydrogels can be shaped to specific areas, such as discoloured margins. Deacylated gellan gum (Kelcogel CG-LA) and agarose (agar) have distinct characteristics: agar with a more porous structure can easily fracture during handling while Kelcogel is more stable and manageable. Agar is not as transparent and elastic, but is readily available, cheaper and easier to prepare in large quantities. For larger objects, agar is preferred for use on the verso, while Kelcogel is used on the recto to treat discoloured margins.

Various reagents can be added to gels, such as buffering solutions and chelating agents, either weaker ones such as citric acid salts,[2] or stronger ones, such as salts of ethylenediaminetetraacetic acid (EDTA) or diethylenetriaminepentaacetic acid (DTPA) (Pasnak 2016). For papers with varying degrees of discoloration, gels with stronger reagents can be applied on the darker areas, while the lightest discoloration can be treated with gels containing only water or buffering solutions. Thus, discoloration can be evened out with minimal bleaching.

Case studies

A coloured lithograph on a smooth paper (Fig. 1), *Winter Night in the Mountains* by the Norwegian artist Harald Sohlberg (1869–1935), had been attached to a wood-pulp board; there were tidelines in the margins and the image area, and a burn line in the middle, possibly from a cigarette. After removal of the backing and old in-painting, the print was washed on rigid hydrogels. To cast gels for a large artefact (770 × 720 mm) a wooden frame and a sheet of plate glass was laid on a thick polyethylene sheet and clamped in place, with plastic sheet around the wooden frame to prevent leakage (Fig. 2).

Four gels were cast: one for preliminary washing comprised 4% agar at pH 6.0 with 0.5% calcium propionate and small amounts of citric acid as a buffering agent. The object's pH was *c.*4.5–5 and it was important to raise its pH slowly. Conductivity of the paper was *c.*0.9 mS/cm, while conductivity of the buffering solution was 1.09 mS/cm. Calcium propionate was used as a mild antimicrobial agent. For the second wash, 4% agar gel was mixed with 0.1M sodium and

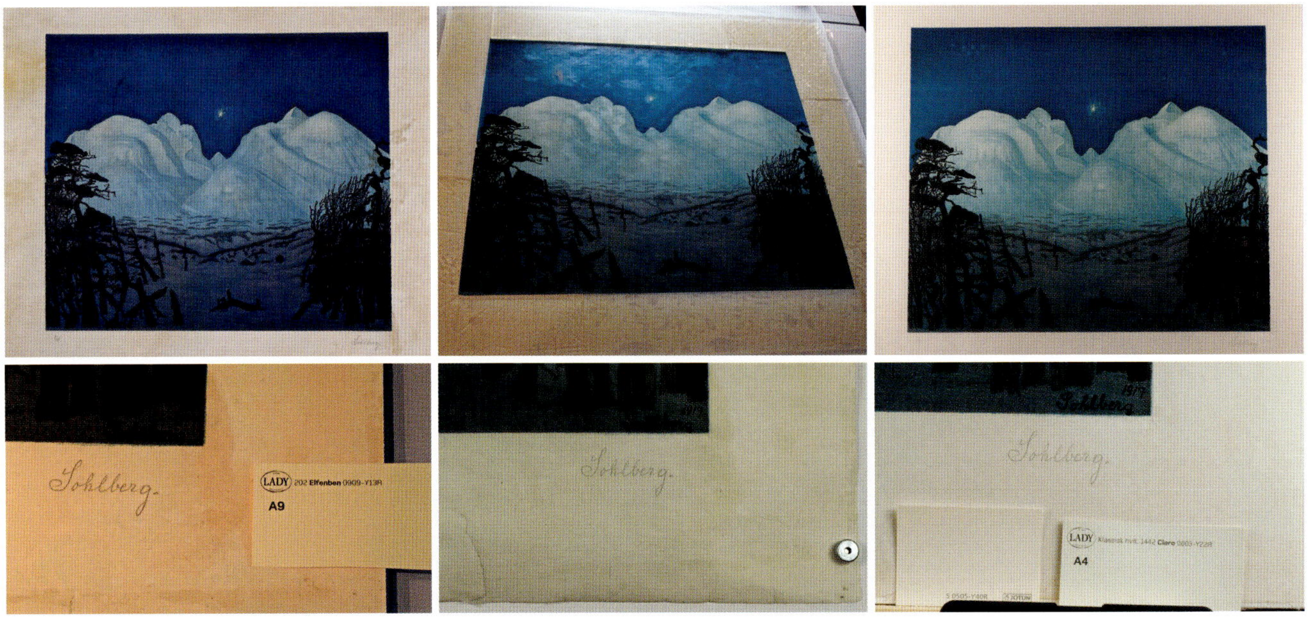

FIG. 1 Harald Sohlberg, *Winter Night in the Mountains*, 1917, colour lithograph, private collection. *Upper row left to right*: overall view before, during and after treatment. The gel on the margins is 2.5% Kelcogel CG-LA in 0.1M sodium and calcium citrate at pH 8.0. *Lower row left to right*: detail of the bottom right corner before, after the gel wash and after final bleaching with 0.5% sodium borohydride in 90% alcohol. The gel is 4% agar in 0.1M sodium and calcium citrate at pH 8.0 (images: E. Pasnak).

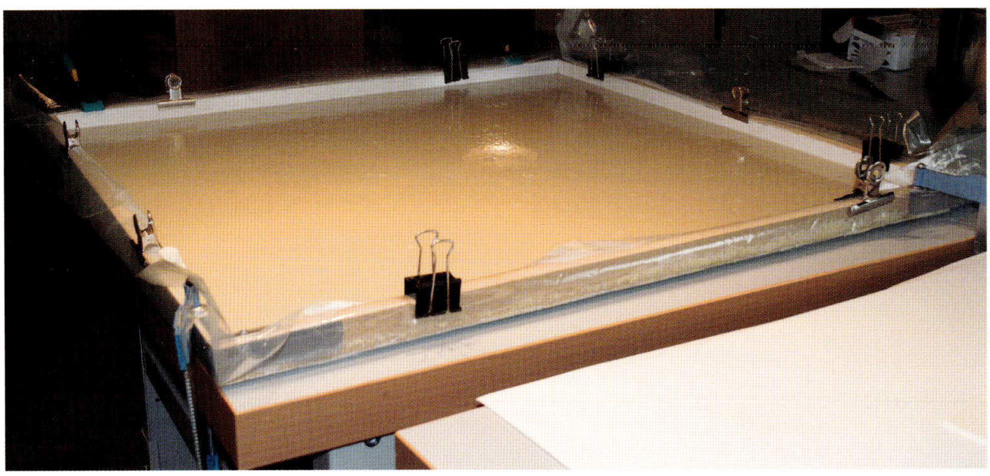

FIG. 2 Casting agar gel for an oversize artwork. Plate glass lies on a heavy polyethylene sheet; on the glass lies a wooden frame the size of the artwork, clamped to the table. The plastic sheet wraps up the sides of the frame and is clipped in place (image: E. Pasnak).

calcium citrate at pH 8 adjusted with sodium hydroxide. For edges during the second wash, 2.5% Kelcogel was prepared with the same citrate solution. Citrate solution was used at pH 8 as it is more effective at reducing discoloration than water but it is not strong enough to remove colorants in the graphic media or paper surface treatment agents; EDTA might be too strong. For rinsing, 4% agar gel was cast at pH 7.5 with 0.5% calcium propionate. For better rinsing, the surface was brushed with purified (reverse osmosis) water with added calcium hydroxide at pH 8 through a thin Viledon non-woven polyester sheet. Finally, remaining darker areas were locally bleached on the suction table with 0.5% sodium borohydride in 90% isopropanol. Based on a comparison with Natural Colour System (NCS) standard coloured chips, the washing treatment caused a decrease of grey and yellow in the paper, from 9–15% grey to 5–6%, while the colour component was reduced from 15% to 3%; the colour shifted slightly from yellow towards red. Therefore the paper appeared whiter and warmer after treatment.

Another example is a hand-coloured etching by Marc Chagall, *Rebecca à la Fontaine* (Fig. 3). Green and yellow were applied as thin washes, while the red on Rebecca's apron was painted with thick impasto. The soft, thick printing paper had turned deep yellow: areas around the platemark were lighter, while the outer edges were more yellow. The colours were consolidated with cyclododecane, and the first wash was carried out on a blotter using calcium bicarbonate solution. To even out the colour on the margins, various gels were prepared; the gel for the verso was 3.5% agar with 0.05M sodium and calcium citrate at pH 8.2. As chelators, citrates are not strong enough to affect colour from the verso. Darker, outer margins were covered with 3% Kelcogel with 0.05M EDTA

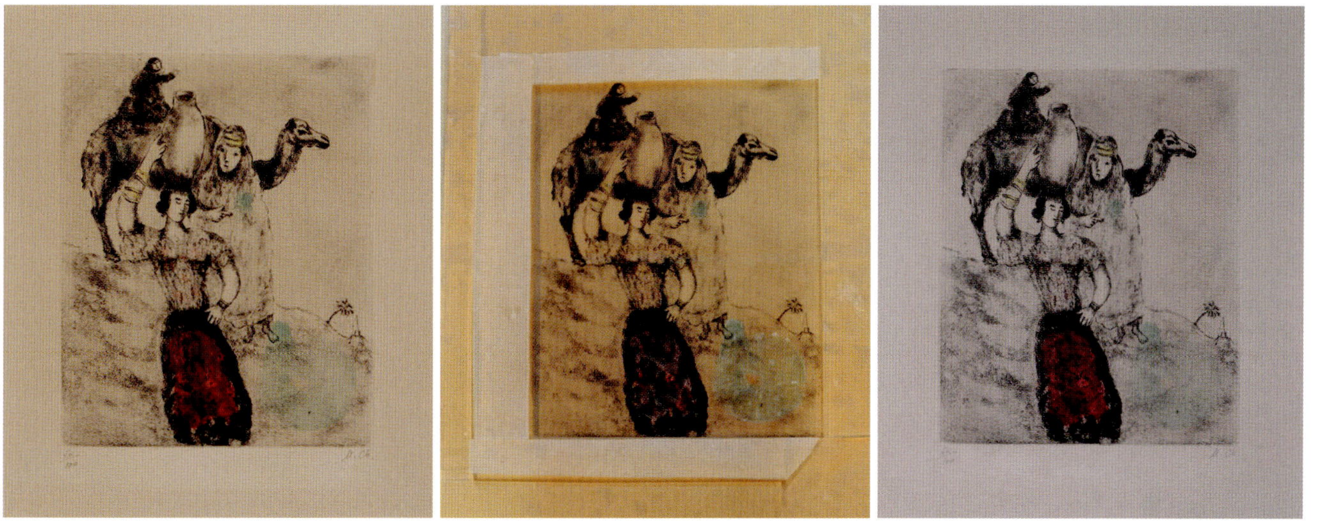

FIG. 3 Marc Chagall, *Rebecca à la Fontaine*, etching with watercolour: (*left*) before, (*centre*) during and (*right*) after treatment. Cyclododecane has been applied to areas of fugitive colorants. The artwork lies on a gel of 3.5% agar with 0.05M sodium and calcium citrate at pH 8.2. On the recto, the gel on the outside margins is 3% Kelcogel with 0.05M EDTA sodium salt at pH 8.5; the gel on the inside margins is 3% Kelcogel with 0.1M sodium and calcium citrate at pH 8.2 (images: E. Pasnak).

at pH 8.5. Another 3% Kelcogel hydrogel with 0.1M calcium and sodium citrate at pH 8.2 was applied on the lighter areas near the image. Finally, the print was rinsed on 3.5% agar gel made with purified water, and calcium hydroxide and calcium bicarbonate to reach pH 8. Although the innermost margin remained lighter, the paper support became much brighter overall while coloured elements were not affected.

Conclusion

As the working properties of gels become better known, it is possible to fine tune washing treatment to specific properties of an artwork. Adding buffering and chelating agents of various strengths into the rigid hydrogels, it is possible to control washing, reducing paper colour in a specific location to a desired level, thus removing products of degradation, improving paper strength and evening the overall colour of the paper.

Materials and suppliers

- Kelcogel CG-LA: Geräte Material Werkzeuge für Papierrestauratoren – Gabi Kleindorfer shop, http://gmw-shop.de/en/tools-and-material/buffering washingcleaning/38/gellan-gum-kelcogel-cg-la?number=41925
- Viledon thin non-woven polyester sheet: Geräte Material Werkzeuge für Papierrestauratoren – Gabi Kleindorfer shop, http://gmw-shop.de/en/tools-and-material/diverse-materialsnon-wovenspolyester-filmmesh/94/viledon-thin?number=37330
- Agar-Agar: Kremer pigments, http://shop.kremerpigments.com/en/search?sSearch=Agar

Notes

1. R. Wolbers and P. Cremonesi, 'Aqueous materials and methods: dos and don'ts. Raising awareness on the possibilities and safe boundaries of surface cleaning', course presented in Valencia, Spain, 30 June 2014.
2. The use of chelators, particularly citric acid and EDTA, is explained in Wolbers 2006: 109–26.

References

Casoli, A., Cremonesi, P., Isca, C., Groppetti, R., Pini, S. and Senin, N. 2013. 'Evaluation of the effect of cleaning on the morphological properties of ancient paper surface', *Cellulose* 20: 2027–43.

Iannuccelli, S. and Sotgiu, S. 2010a. 'Wet treatments of works of art on paper with rigid gellan gels', *The Book and Paper Group Annual* 29: 25–39.

Ianuccelli, S. and Sotgiu, S. 2010b. 'A new methodology for wet conservation treatments of graphic art on paper with a rigid polysaccharide gel of gellan gum', in L. Waatteeuw, M. Vest, J. Palm and D. van der Reyden (eds), *Choices in Conservation: Practice versus Research. Preprints of the ICOM-CC Graphic Documents Working Group Interim Meeting 6–8 October, Copenhagen*, 50.

Pasnak, E. 2016. 'Practical application of an old method: reduction of foxing spots through chelating agents and reducing bleaches', in P. Engel (ed.), *Mould on Books and Graphic Art: A Report on Latest Research Results*. Horn/Vienna: Verlag Berger, 134–58.

Rowe, S. and Rozeik, C. 2008. 'The use of cyclododecane in conservation', *Reviews in Conservation* 9: 17–31.

Wolbers, R. 2000. *Cleaning Painted Surfaces: Aqueous Methods*. London: Archetype Publications.

Author's address

Ekaterina Pasnak, Paper Conservator, Special Collections Library, University of Bergen, Bergen, Norway (Ekaterina.pasnak@uib.no)

Use of a rigid gel for adhesive removal from an early 20th-century poster

María del Pilar Tapia López, Rodrigo Ruiz Herrera and Angélica Vásquez Martínez

Introduction

In 2015, fourth-year students from the BA course in conservation at the Paper Conservation Laboratory in the National School of Conservation, Mexico, treated an early 20th-century poster with rigid gels in order to remove adhesive residues. The treatment was supervised by the laboratory teachers and the Chemistry Education Laboratory, National School of Conservation, Mexico. The poster, titled *Rubolin*, is part of the Museo del Objeto del Objeto (MODO) Museum collection, which focuses on 19th- and 20th-century everyday objects from Mexico. The poster (70 × 1000 cm) is an example of the relationship between art and publishing: it is a chromolithograph with more than 16 colors, made with oil-based inks applied to semi-chemical pulp paper (Peláez *et al.* 2015: 15). A documentary investigation indicated that the poster was in use between 1908 and 1925 to publicize Rubolin from Eugenio Talleri's Brooklyn Varnish Company's (Mexico) brand of paints and glazes. Today the poster is regarded as a collectible museum object (Vásquez 2016). The poster is an allegory of nature, represented by a woman who is painting natural elements with industrial products, and it was influenced by both modernism and art nouveau (Peláez *et al.* 2015).

Evaluation, criteria, and cleaning tests

The main conservation issue was the presence of brown stains on the back of the poster caused by the use of industrially produced cardboard as an auxiliary support, pasted on with an excessive amount of adhesive which had impregnated the paper and migrated into the support fibers, altering the compositional elements and the reading of the image. The stains covered 85% of the poster, as indicated by green fluorescence when illuminated in the ultraviolet (UV) wavelength range 254–365 nm. It was decided to investigate a cleaning method that could eliminate the adhesive stains.

Cleaning tests were carried out with water, ethanol, 1:1 water and ethanol, and acetone, to identify a solvent which could remove the adhesive stains without affecting the poster inks. The results indicated that acetone was the most effective for removing adhesive stains, but because of its volatility, application was difficult as it provided insufficient contact time to swell the adhesive. An alternative method for acetone application was investigated. It was determined that the physical and chemical properties of rigid agar gels were appropriate because of their stability in different environmental conditions (acid or alkaline) and the possibility of adding solvents such as acetone (Lee 2012: 72). The direct application of agar gels does

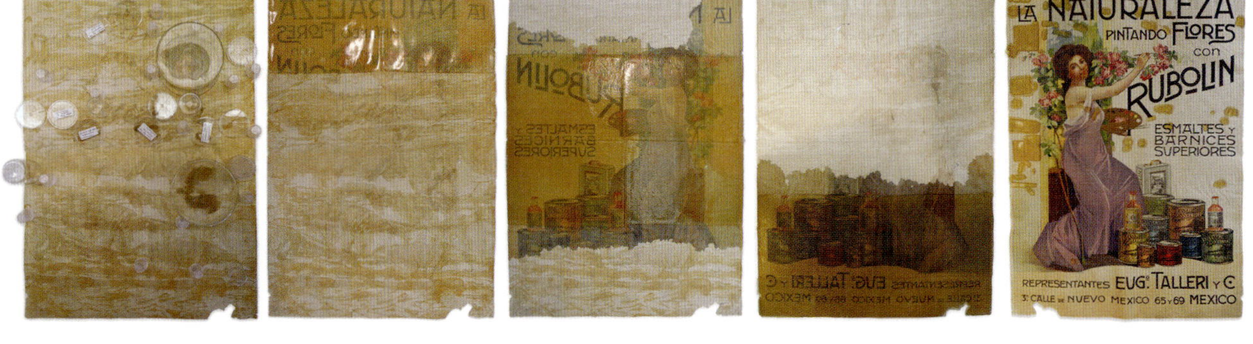

FIG. 1 Gels used in the cleaning procedure.

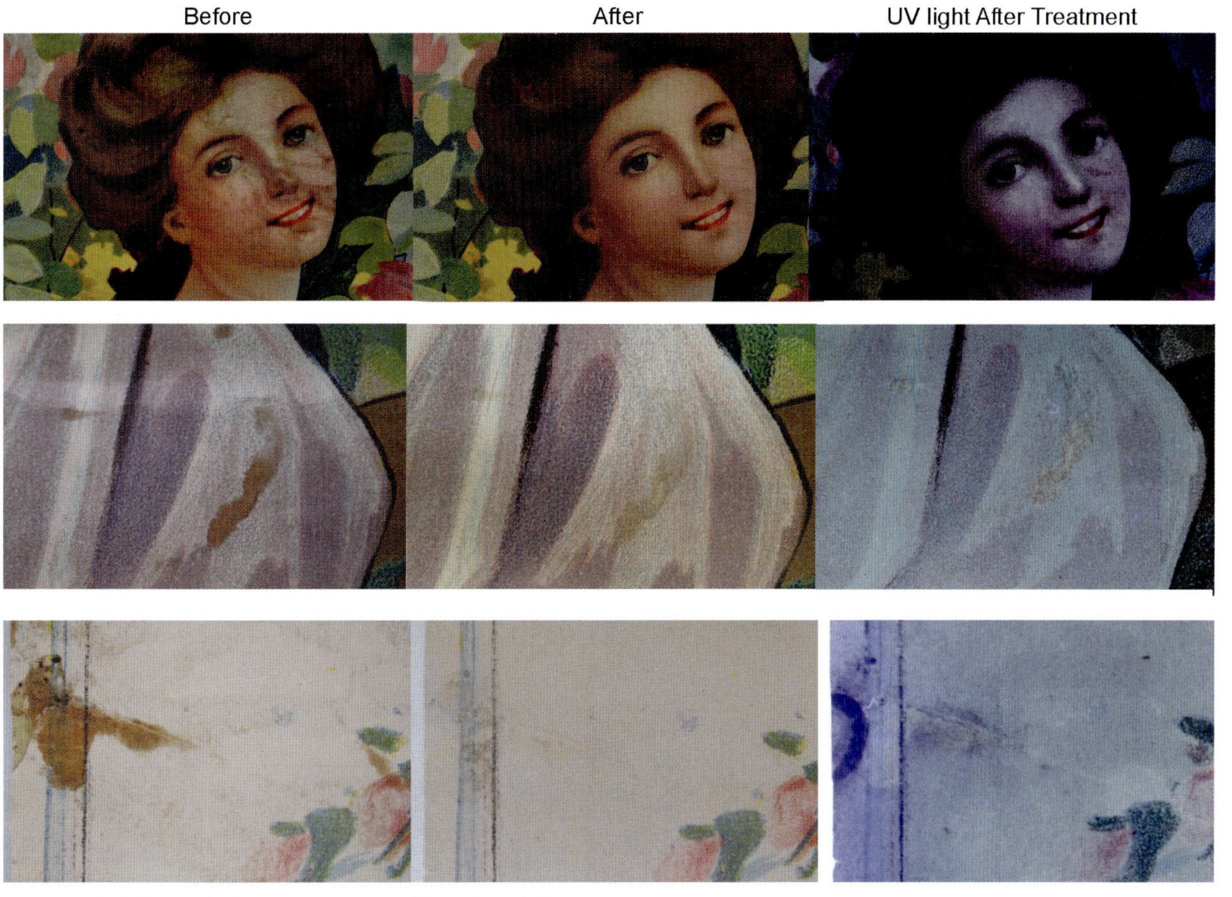

FIG. 2 Details before and after treatment, in visible and UV light.

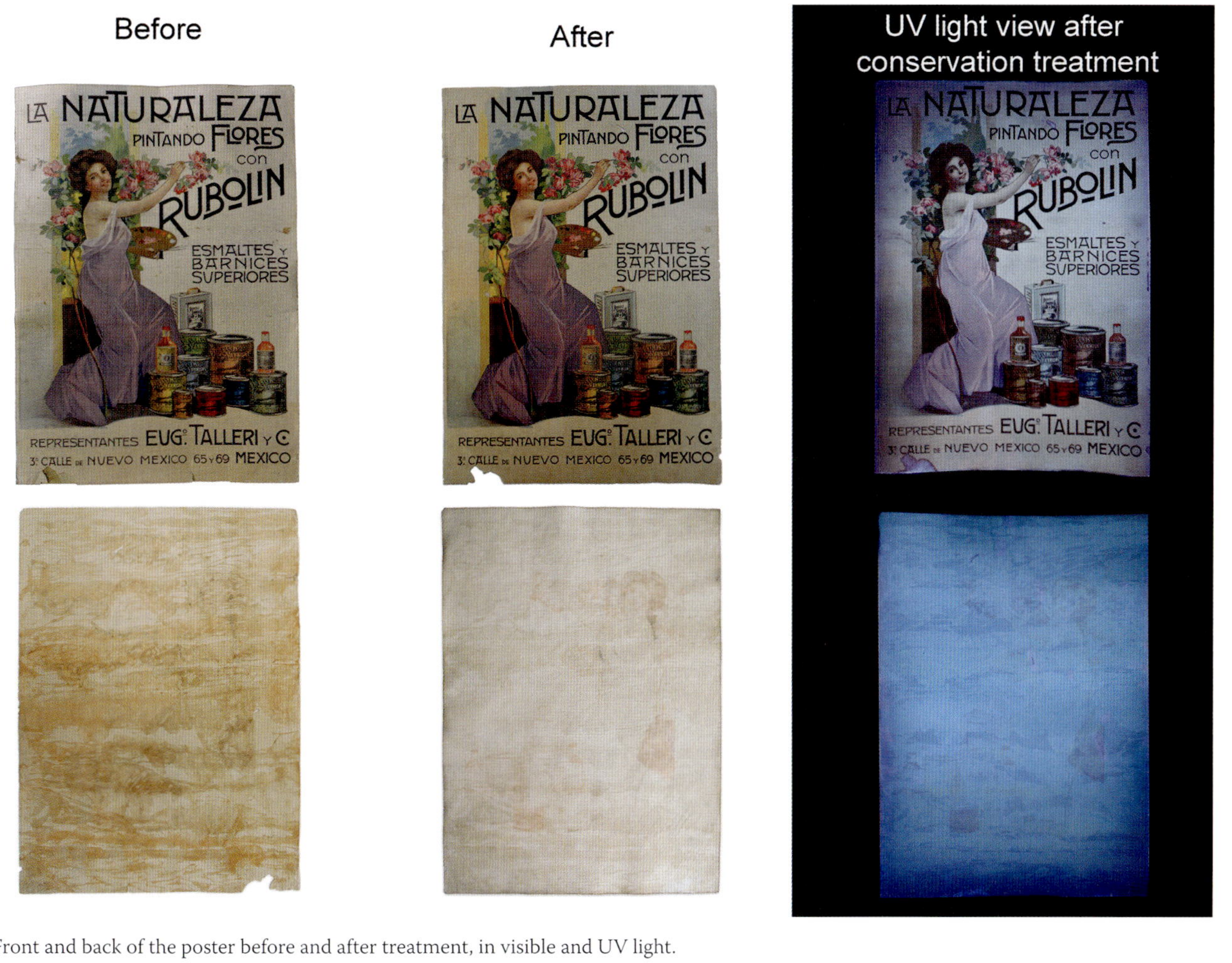

FIG. 3 Front and back of the poster before and after treatment, in visible and UV light.

not modify the color of the paper surface, as shown by Warda *et al.* (2007). In addition, the gels can be removed easily without leaving residues (Sansonetti *et al.* 2012: 3). The application of acetone in an agar gel allowed the solvent to remain in contact for the necessary time, absorbing the adhesive without affecting the surface of the support or the solubility of the oil-based inks, whilst leaving minimum residues.

Experimental procedure

Tests on different areas at various concentrations of acetone were performed: 5%, 7.5% and 10% in rigid, food grade, agar gels, to observe the reaction with the poster inks, support and adhesive. The most suitable concentration was 10% acetone, which was the maximum amount of solvent that could be incorporated into a stable gel; this system was also found to remove the adhesive. The gel was prepared by mixing 900 ml distilled water, 25 g agar and 100 ml acetone.[1] The temperature was kept below 50 °C when the solvent was added (the boiling point of acetone is 56 °C).

The gels were prepared in 35 × 20 × 0.5 cm sheets and their application consisted of the following steps: first, the stains with the greatest concentration of adhesive were treated locally by cutting gel pieces in the shape of the stains, and placing them on the back and front of the poster. During this process the sections being treated were covered with polyethylene film to reduce solvent evaporation and gel dehydration. The second step involved the placement of two gel films across the entire poster, in three applications, to obtain homogeneous cleaning and to prevent drying the fronts (Fig. 1). These applications, to the reverse of the poster, were covered with polyethylene film to reduce evaporation during the process. The gels remained in place for 20 hours; this duration was selected because applications of less than one hour were ineffective, while at extended times of more than 24 hours, the red ink began to react. At the end of the treatment, the remaining stains in the poster image were again treated locally.

After the treatment for adhesive staining, a water wash was performed by immersion in Canasol 1:1000, using delicate brushing with a soft bristle brush to eliminate dirt residues, drying fronts and gel residues. Finally the poster was rinsed thoroughly with water.

Results

At the end of the treatment a radical reduction of the stains had been achieved, with the removal of most of the adhesive. A minimal amount of impregnated adhesive was noted during a follow-up UV examination (Fig. 2). During the cleaning, the rigid gel changed color to a dark yellow, whilst no change in color was noticed in the poster support. Direct contact of the solvent with the inks was minimized by using a gel. Nonetheless, some of the red ink migrated slightly towards the reverse, although this was not visible on the front, at least to the naked eye (Fig. 3).

The cleaning process was controlled effectively because the gel could be shaped to fit the stains. However, it was observed that when using the rigid agar gel in both large and localized sections, the support in the treated area was moistened, and drying fronts were formed. Differential shrinkage was also observed. Therefore, we recommend first treating specific areas and then covering the entire surface of the poster with the rigid gel. In the *Rubolin* poster the acetone gel treatment was effective due to the specific characteristics of the object – an aqueous wash could be carried out to remove some of the residues from the gel, and to even out the support.

Conclusion

This treatment proved the usefulness of a rigid agar gel with a polar solvent for the removal of adhesives from graphic works on industrially produced paper. The technique is quite simple, gel preparation is easy, and it can be executed in any conservation laboratory.

Although the use of rigid gels is not a novelty, each case study contributes new considerations to the application of these methods in a variety of situations. The unique conservation challenges presented by the *Rubolin* poster has led to research into the application of rigid agar gels in the preservation of documents and graphic works with historical and contemporary support in the Paper Conservation Laboratory at the National School of Conservation, Mexico. These gels are being investigated in an effort to find cleaning methods responsive to the nature of the substances that alter these works, and to observe the long-term effects of this type of treatment.

Acknowledgements

The authors thank Javier Vázquez Negrete, chemist, and Luz Esperanza López Méndez, technician, both of the Chemistry Teaching Laboratory, National School of Conservation Mexico; and Antonio Soto, curator, MODO Museum, Mexico.

Note

1. According to a recipe from a Richard Wolbers lecture.

References

Lee C. 2012. 'The use of agar as a solvent in objects conservation', *AIC Objects Specialty Group Postprints C*, 19: 21–83. Available at: http://www.conservation-us.org/docs/default-source/periodicals/objects-specialty-group-postprints-volume-19-(2012).pdf (accessed 23 May 2017).

Peláez A., Ruiz R. and Vásquez A. 2015. *Informe de restauración del cartel publicitario Rubolin*, Museo MODO, México: STRDOGP, ENCRyM-INAH.

Sansonetti A., Casati M., Striova J., Canevali C., Anzani M. and Rabbolini A. 2012. 'A cleaning method based on the use of agar gels: new tests and perspectives', paper presented at the *12th International Congress on the Deterioration and Conservation of Stone, Columbia University, 25 October 2012*. Available at: http://iscs.icomos.org/pdf-files/NewYorkConf/Early%20versions/Sansonetti-August-2014.pdf (accessd 23 May 2017).

Vásquez A. (2016) 'Restauración de un cartel publicitario: Rubolin', *Archivo Churubusco* 1(1). Available at: http://archivochurubusco.encrym.edu.mx/images/uno/rubolin2-g.jpg (accessed 23 May 2017).

Warda, J., Brückle, I., Bezur, A. and Kushel, D. 2007. 'Analysis of agarose, Carbopol, and Laponite gel poultices in paper conservation', *Journal of the American Institute for Conservation* 46(3): 263–79.

Authors' addresses

- María del Pilar Tapia López, National School of Conservation, Mexico (pilar_tapia_l@encrym.edu.mx)
- Rodrigo Ruiz Herrera, National School of Conservation, Mexico
- Angélica Vásquez Martínez. National School of Conservation, Mexico

An initial assessment of local stain reduction using chelate-containing gellan gum gels in paper conservation

Brook Prestowitz

Introduction

Gellan, a fermentation product of the organism *Sphingomonas elodea* (Kelco 2007), has recently been used as a rigid hydrogel for poulticing with water and water-miscible solvents, washing, alkalization, and bleaching treatments by reduction of paper objects (Iannuccelli and Sotgiu 2010; Casoli *et al.* 2014; Maheux 2015; Hughes and Sullivan 2014). Gellan gum's cost advantage over agarose is a significant benefit to conservators as it makes treatment using rigid gels much more accessible. Some initial testing was performed by the author at the Conservation Center for Art and Historic Artifacts (CCAHA) in Philadelphia, USA, to assess the effectiveness and practicality of gellan gels for administering chelating agents to reduce localized discoloration in paper such as light damage, mat burn, and foxing spots. Paper conservators use chelating agents to change the ionic environment and pH of the cleaning system and to complex out metal ions, when addressing staining and discoloration that does not respond to washing or alkalization. By incorporating chelates into gellan gels, conservators would have an additional treatment tool that could increase control over application of solutions for stain reduction. Two chelate salts, triammonium citrate (TAC) and disodium ethylenediaminetetraacetic acid (Na_2EDTA), and two application methods – brush application and gellan gum gel application – were tested to compare stain reduction, lateral movement of chelate solution, and changes in support texture. A chelate-containing gellan gum gel would allow conservators to increase the amount of control during local stain reduction, and to augment the repertoire of cleaning systems that may be delivered by gellan gum gel.

Experimental

Two prints were selected from CCAHA's study collection with staining and discoloration commonly encountered in works of art on paper. Changes in the prints before and after treatment were recorded with digital photography of the prints in normal and raking light.[1] The prints were also examined with ultraviolet A (320–400 nm) (UVA) radiation, noting changes in fluorescence in the supports before and after treatment. Color measurements of selected areas were taken using an X-Rite SP60 portable unit and manually recorded.[2] Microphotographs using a stereomicroscope were taken of the same locations selected for color measurements to monitor any physical changes in the paper fibers.

All chelate-gellan gels were made at a 2% weight per volume (w/v) gellan concentration, with 1 g gellan gum in 50 ml deionized (DI) water. The following chelate concentrations were tested: 0.5% w/v TAC at pH 7.55, 1% w/v TAC at pH 7.58, 0.5% w/v Na_2EDTA, and 1% w/v Na_2EDTA. The pH of both Na_2EDTA solutions was adjusted to pH 6.0 using drops of 1M of sodium hydroxide. A 50 ml, 2% w/v gellan gel with 0.25 g calcium propionate at pH 8.01 was made for an alkaline rinse after chelation.

The prints were surface cleaned, attachments were removed, and the graphic media were found to be stable when tested for solubility in DI water, TAC, Na_2EDTA, and calcium propionate. The chelates were locally applied while the prints were in a blotter wash bath using Tek-Wipe[3] dampened in DI water. Gels were applied to stained areas and allowed to remain in contact with the support for 30 minutes. Chelate solutions were brushed onto stained areas with a size 0 brush three times and allowed to sit for 30 minutes.

The first print, depicting a landscape entitled *Albertani da Venezia* (labeled as 16.01), had red-brown foxing spots with dark centers, and discoloration from light damage where the print was exposed through a window mat (Fig. 1). Staining was treated with 2% gellan/0.5% TAC and 2% gellan/0.5% Na_2EDTA gels. Separate areas of staining were also treated with 0.5% TAC or 0.5% Na_2EDTA chelate solutions applied with a size 0 brush.

The second print, entitled *Still Life* (labeled as 16.03), had acid transfer from an acidic window mat framing the printed

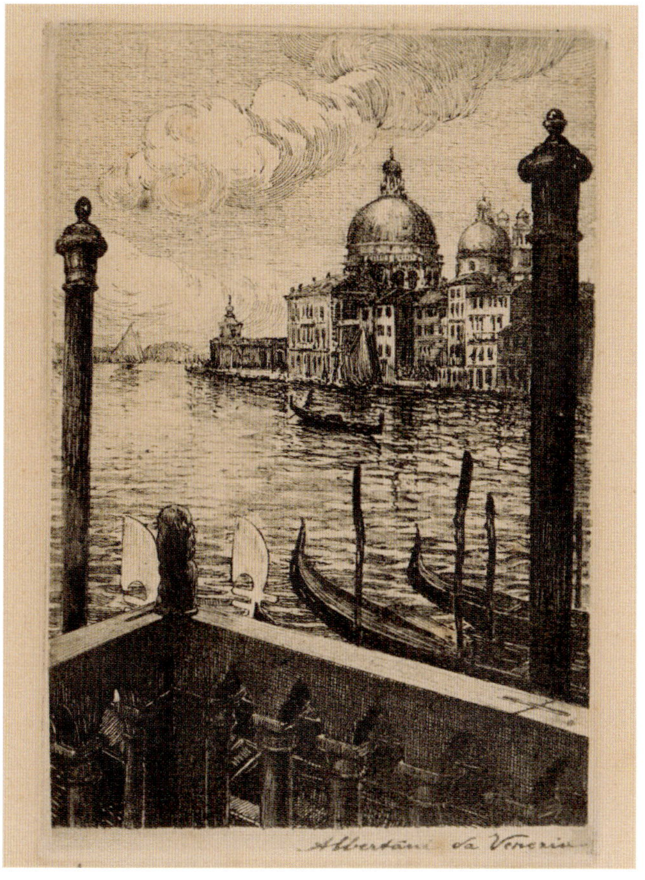

FIG. 1 Recto of 16.01 *Albertani da Venezia* before treatment, in normal light (image: Andrew Pinkham, Manager of Digital Imaging, CCAHA).

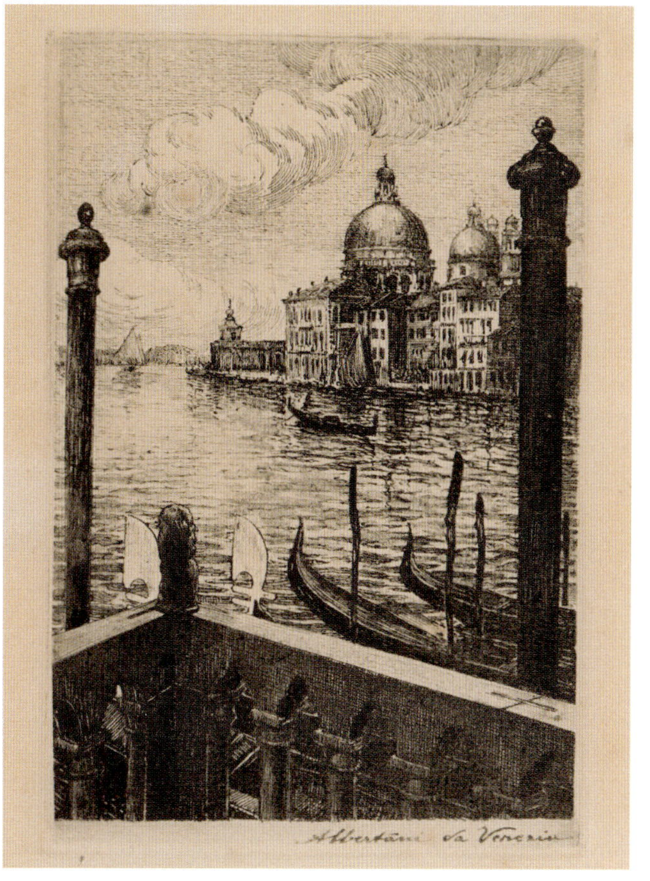

FIG. 2 Recto of 16.01 *Albertani da Venezia* after treatment, in normal light (image: Andrew Pinkham, Manager of Digital Imaging, CCAHA).

image. Staining was treated with 2% gellan/0.5% TAC gel and 2% gellan/0.5% Na_2EDTA gel as well as 2% gellan/1% w/v TAC gel and 2% gellan/1% Na_2EDTA gel. Separate areas of staining were treated with 0.5% TAC, 1% TAC, 0.5% Na_2EDTA, or 1% Na_2EDTA chelate solutions applied with a size 0 brush. The gels were removed and the prints were placed onto clean Tek-Wipe dampened with DI water for an hour to clear chelating agents from the support. Calcium propionate gels were applied to stains treated with chelate-gellan gels and pH 8.01 calcium propionate solution was brush applied to areas treated by brush application of chelate solutions to clear any remaining chelating agents and replenish calcium ions in the supports.

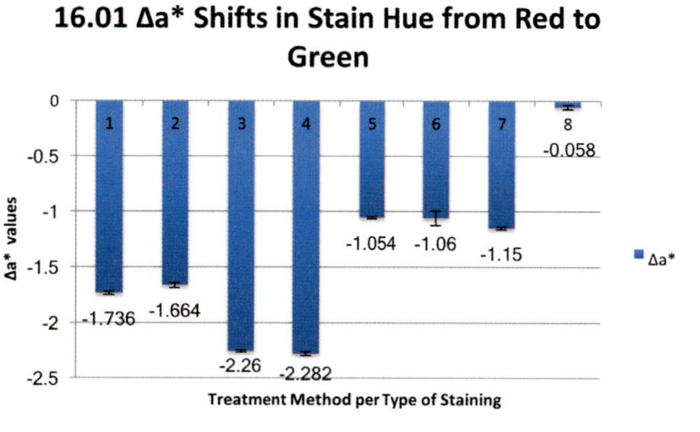

FIG. 3 Color change Δa* of 16.01 following the shifts in stain hue from red to green. 1. Light damage: 2% gellan/0.5% Na_2EDTA; 2. Foxing: 2% gellan/0.5% Na_2EDTA; 3. Light damage: 2% gellan/0.5% TAC; 4. Foxing: 2% gellan/0.5% TAC. 5. Light damage: brushed 0.5% Na_2EDTA; 6. Foxing: brushed 0.5% Na_2EDTA; 7. Light damage: brushed 0.5% TAC; 8. Foxing: brushed 0.5% TAC.

Results

All stains treated with chelating agents were visibly lighter after treatment (Fig. 2). TAC produced better results than Na_2EDTA and worked well on reducing staining caused by light damage (in 16.01) and mat burn (in 16.03). While neither chelate significantly reduced foxing stains in 16.01, examination under magnification showed a notable reduction in foxing stains. Chelate solutions at 1% concentration removed more discoloration than the 0.5% solutions. A change in surface texture occurred in areas that were treated with TAC: the paper fibers were disrupted making the surface look slightly swollen and fuzzy. Examination of the prints with UVA revealed fluorescence in areas where chelating agents had been applied, regardless of solution concentration or application method. The fluorescence was also visible in corresponding areas on the verso. This may be an indication that a chemical change has occurred in the support after treatment with chelating agents.

Greater stain reduction, even to the point of over-cleaning, was achieved in areas that were treated with chelate-gellan gels, particularly those treated with TAC/gellan gels. Color measurements of staining in 16.01 and 16.02 showed notable shifts from red to green and yellow to blue as well as slight shifts from darker to lighter indicating that colored cellulose degradation byproducts were removed from the support during treatment. The change in a* values (Δa*) for print 16.01 shown in Fig. 3 show how much more effective the application of chelate-gellan was compared to brush application. Fig. 3 also shows how much more effective TAC was in reducing stains than EDTA. Cutting and handling the gels was very time consuming, and close attention to the gel application time was necessary to avoid over-cleaning. Results were observed within five minutes of contact. There was a sharp transition noted between areas treated for staining using gels and the non-treated areas. Areas treated by brush application showed some lateral movement of the chelate solutions and resulted in the development of white halos or brightening prominent along the edges of the staining.

Conclusion

Gellan gum gels may be made with chelating solutions and these chelate-sustaining gellan gels have good potential for use in local stain reduction. The chelate-gellan gels increased contact time with the stained support, improving stain reduction and decreasing treatment time. The TAC gels were more successful in reducing foxing and mat burn as well as discoloration from light damage than the Na_2EDTA gels. While it is not desirable, the abrupt visual transition between treated and non-treated areas is an indication of the good control a conservator will have over lateral flow of the liquid phase when using gellan gels for local treatment. Gels must be carefully cut and applied in a manner that will ensure even removal of discoloration and create a soft transition between the treated stain and surrounding areas. Finally, TAC may be too aggressive, as it made a visible change in the surface texture of the prints. Opportunities for future research include identifying alternative chelates that will not cause a change in surface texture, success of local stain reduction with chelate-gellan gels while the object is dry, and the aging characteristics of local stain reduction with chelate-gellan gels.

Acknowledgements

The author would like to extend her appreciation to the CCAHA staff for their advice and encouragement; to Bryn Mawr College for their donations to CCAHA's study collection; Steven P. Wrenn, PhD. Associate Professor of Chemical and Biological Engineering, Drexel University, USA for his expertise; the Kress Foundation; the National Endowment for the Arts for making this research possible; and to Andrew Pinkham, Manager of Digital Imaging, CCAHA for his help and time to conduct before and after treatment photography. A special thanks to Jessica Silverman for her continual support, insight, and encouragement through the course of this research.

Materials and suppliers

- Ammonium citrate tribasic (TAC), Ethylenediaminetetraacetic (EDTA) disodium salt dihydrate: Sigma-Aldrich Ltd, The Old Brickyard, New Road, Gillingham, Dorset, SP8 4XT, UK, http://www.sigmaaldrich.com/catalog/product/sigma/a1332?lang=en&region=US
- Kelcogel LA gellan gum: CP Kelco US Inc., 3100 Cumberland Boulevard, Suit 600, Atlanta, GA 30339, USA, https://www.cpkelco.com/markets-served/household-products/products/kelcogel-gellan-gum/
- Sodium hydroxide pellets: Fisher Scientific Ltd, Bishop Meadow Rd, Loughborough LE11 5RG, UK, https://www.fishersci.co.uk/shop/products/sodium-hydroxide-certified-ar-analysis-pellets-meets-analytical-specification-ph-eur-bp-2/p-8015156#?keyword=NaOh
- Tek-Wipe: William Minter Bookbinding and Conservation Inc., 4364 Woodbury Pike, Woodbury, PA 16695, USA and Polistini Conservation Material: Renate Mesmer, 105 6th Street SE, Washington DC, NY 20003, USA, https://www.polistini.com/product/tek-wipe/

Notes

1. A Nikon D300 camera, Calumet Travelite 750 lights fixed with softboxes were used for photography using a Nikon Camera Control Pro2 and Adobe Lightroom CC to process images.
2. A polyester film template was made for both prints marking the location of the stained areas selected for color measurements to ensure accuracy. Five measurements were made in each location before and after treatment for statistical data. The X-Rite SP60 illumination is D6,/10° and it records data in CIE L*a*b*.
3. Tek-Wipe is a non-woven, hydro-spun cellulose and polyester blend textile used at CCAHA as a sustainable alternative to blotters in a method equivalent to blotter washing.

References

Casoli, A., Isca, C., De Isio, S., Botti, L. Iannuccelli, S., Residori, L., Ruggiero, D. and Sotgiu, S. 2014. 'Analytical evaluation, by GC–MS, of gelatin removal from ancient papers induced by wet cleaning: a comparison between immersion treatment and application of rigid gellan gum gel', *Microchemical Journal* 117: 61–7.

Hughes, A. and Sullivan, M. 2016. 'Targeted cleaning of works on paper: rigid polysaccharide gels and conductivity in aqueous solutions', paper presented at the *Joint 44th Annual American Institute for Conservation Meeting and Canadian Association for Conservation 42nd Annual Conference Montreal, 13 May 2016.*

Iannuccelli, S. and Sotgiu, S. 2010. 'Wet treatments of works of art on paper with rigid gellan gels', *The Book and Paper Group Annual* 29: 25–49.

Kelco 2007. *Kelcogel Gellan Gum Book.* San Diego: CP Kelco US.

Maheux, A. 2015. 'Cross-disciplinary uses for gellan gum in conservation', *The Book and Paper Group Annual* 34: 69–79.

Author's address

Brook Prestowitz, The Conservation Center for Art and Historic Artifacts, Philadelphia, PA, USA (brookprestowitz@gmail.com)

Getting into the (sea)weeds: the use of rigid agar gels for the cleaning of a birch bark canoe

Cindy Lee Scott

Although known primarily for its fine arts, decorative arts, and sculpture, the Detroit Institute of Arts (DIA) is also in possession of an impressive ethnographic collection, highlighting objects from a number of cultures including that of the Anishnaabe First Nations of the Mackinaw/Upper Peninsula region of Michigan. During my time at the DIA as the Andrew W. Mellon Fellow in Objects Conservation, I had the opportunity to work on the conservation treatment of a birch bark and porcupine quill model canoe.

Unfortunately, the history of this particular model canoe is not clear. Until 2012, it resided in two separate museums, with the small prow section having been in the collections of the DIA, while the larger extant section was housed across the street at the Detroit Historical Museum, which does not have conservation services. Furthermore, neither museum has clear records of when either section of the canoe was acquired into their collections, nor the reasons for its separation. The canoe most likely dates to between 1889 and 1908, as it was probably built as a souvenir piece to commemorate the launch of the *North Star*, a Great Lakes freighter, which launched in 1889 and sank in 1908.[1]

The canoe, which measures roughly 104 cm in length, presented a number of conservation problems, above and beyond the fact that it was in two pieces. One of the most significant challenges in the treatment of this piece involved the degree and extent of particulate matter coating all of the surfaces. The soiling was so thick in some areas that it completely obscured the colours of the dyed quills, as well as the underlying grain of the bark. Cleaning tests showed that this particulate matter was not easily removed using dry cleaning methods, including soot sponge. In fact, attempts at removal using dry cleaning methods proved to require too much pressure and risked damaging or abrading the fragile surface of the bark. The wet cleaning of birch bark objects can be highly problematic, however, as many wet-cleaning agents, including water, are known to discolour or darken birch bark, cause staining, tidelines, swelling and/or cracking. Because of these potential dangers, wet cleaning of birch bark can and should be avoided whenever possible (Carrlee 2003; Agrawal 2010).

Solubility testing showed that the soiling could be removed most effectively with water, although ethanol was also effective. Ethanol did not over-wet the surface in the same way as water, but the issue of tidelines, albeit less pronounced, remained. The use of cotton swabs for wet cleaning also proved to be problematic on the sensitive birch bark. Even when rolled, the swab abraded the surface too much. There was also significant darkening of the surface due to wetting as it was quite difficult to control the dispersion of water or ethanol from the swab.

In addition to being encrusted with the same particulate matter as the birch bark surfaces, many of the porcupine quills had also been damaged by insect activity. Unfortunately, the result of this damage meant that there were many splintered and raised edges, as well as displaced quills. Furthermore, the quills had faded and exhibited increased brittleness due to light exposure. The condition of the quills contributed to an increased risk of loss through handling of any kind, which included both traditional wet and dry cleaning methods. The use of cotton swabs to clean the quills was especially problematic, as the splintering and raised edges tended to catch on the fibres of swabs.

Analysis of the colorants used to colour the quillwork was conducted with a Bruker handheld X-ray fluorescence (XRF) spectrometer equipped with a rhodium tube, using a Cu-Ti filter for the detection of heavy metals, and then without any filters (in 'Lab Rat' mode) in order to detect the widest possible range of elements. While copper was detected in the blue colour, no other elements were found in the other colours. It is assumed, based on the provenance of the piece, that all the other dyes are likely to be aniline-based, although Fourier transform infrared (FTIR) spectroscopy has not provided conclusive results as yet.

Given the problematic nature of traditional cleaning methods for the cleaning of the birch bark and quill, a different method of cleaning was selected: rigid agar gels. Agar is a rigid polysaccharide gel derived from the cell walls of a species of red algae of the *Gelidium* or *Gracilaria* genus (Davidson and Jaine 2006). As the gel cools, it transitions

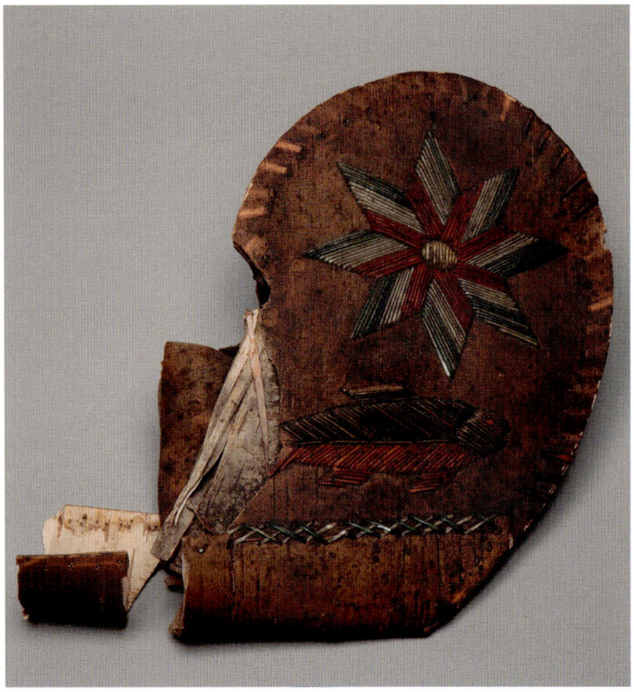

FIG. 1 Prow section before treatment (image: Paul Cooney © Detroit Institute of Arts).

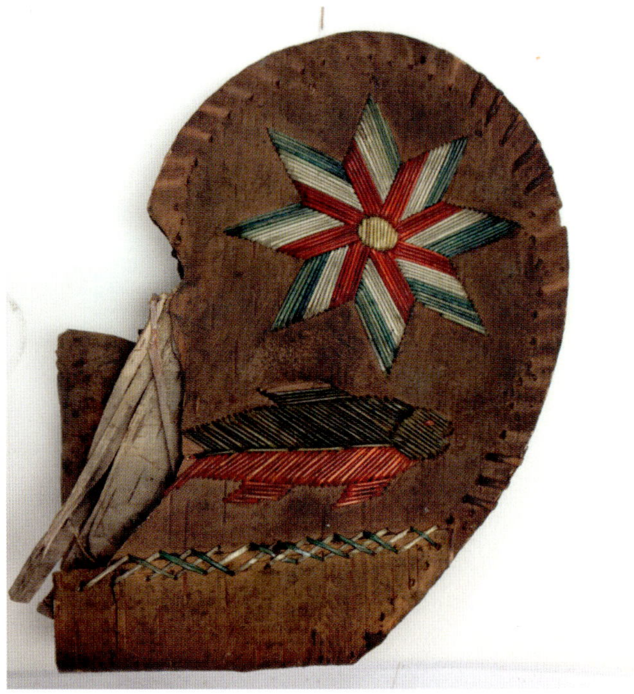

FIG. 3 Prow section after agar cleaning treatment. Only one application was required (Image: Cindy Lee Scott © Detroit Institute of Arts).

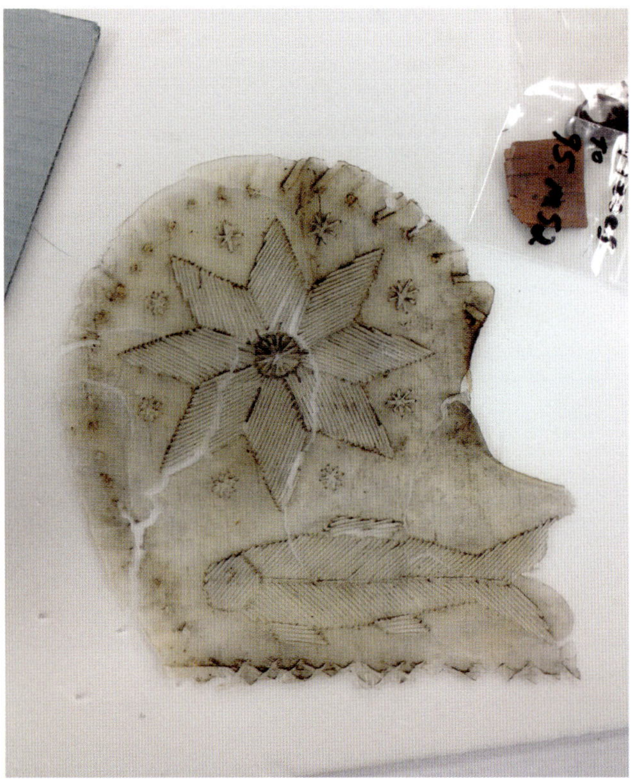

FIG. 2 Agar gel after removal. Application time was approximately 15 minutes (image: Cindy Lee Scott © Detroit Institute of Arts).

from the less viscous sol to a rigid gel phase, allowing for easy removal. The advantages of using agar gels, as opposed to other poulticing or gel materials, lie in its ability to diffuse water and other solvents onto a surface in a controlled fashion, while at the same time, absorbing into its matrix the solubilized materials.

As both the gel and the substrate are cellulosic in nature, the dangers of residues interacting with the surface were minimal. Additional analysis of agar residues revealed that only minimal polysaccharide residues remained after treatment. Both my own previous research (Scott 2012) and that of Anzani *et al.* (2010), have shown that residues remaining after treatment are not detectible using either gas chromatography–mass spectrometry (GC–MS) or FTIR, but rather appear as a slight fluorescence under ultraviolet-induced visible fluorescence at a wavelength of 365 nm. Microchemical spot tests using triphenyltetrazolium chloride (Odegaard *et al.* 2005: 134) confirmed that these residues were polysaccharides. The residues, minimal as they are, are also chemically inert and would not pose a danger to the treated birch bark (Fisheries and Aquaculture Department of the Food and Agriculture Organization of the United Nations 1990).

Cleaning of the birch bark surface was conducted using a 4.5% w/v solution of agar in deionized water. It was applied to the surface at a temperature of roughly 50 °C, in its sol phase, in order to improve surface contact. It cooled to its gel state almost immediately on contact with the surface of the birch bark. The gel was applied at a thickness of around 7 mm to ensure ease of removal. It was allowed to act for 15 minutes before removal, which consisted of lifting an edge with a bamboo skewer and peeling the rigid gel away from the surface in large pieces (Figs 1–3).

The quills were also cleaned in this manner. Solubility testing showed that the dye was slightly mobile with water and ethanol on a cotton swab, but tests with agar indicated that the gel did not affect the dye. As the gel is transparent and colorless, it was possible to monitor for the possible transfer of the dye into the gel. The use of the gel on the quills did cause them to soften temporarily, but this characteristic was found to be advantageous rather than problematic. Instead of breaking or splintering during gel removal, the quills remained flexible; this temporary flexibility also allowed for displaced quills to be placed back into their proper position and consolidated.

Although not typically considered in the cleaning of ethnographic objects, for this highly problematic treatment, agar gels proved to be a very efficacious method for cleaning the bark without leaving tidelines, as well as cleaning and re-humidifying the quillwork without causing further damage.

Material and supplier

- Agar 9002-18-0: Fisher BioReagents, Fisher Scientific, https://www.fishersci.ca/shop/products/agar-80-100-mesh-fisher-bioreagents-3/p-101970#?keyword=9002-18-0

Note

1. *North Star*, American Dive Institute of Central Florida, http://www.adiscuba.com/dive-sites/422/north-star/ (accessed 3 March 2013).

References

Agrawal, O.P. 2010. 'Conservation of birch bark manuscripts: some innovations', *International Preservation News* 50: 36–38.

Anzani, M., Berzioli, M, Cagna, M., Campani, E., Casoli, A., Cremonesi, P., Fratelli, M., Rabbolini, D. and Riggiardi A. 2008. *Gel rigidi di Agar per il trattamento di pulitura di manufatti in gesso/Use of Rigid Agar Gels for Cleaning Plaster Objects*, *Quaderni del Cesmar7*, no. 6. Padua: Il Prato.

Carrlee, S. 2003. 'Canoe Conservation Project', *Alaska State Museums Bulletin* 19: 1–4.

Davidson, A. and Jaine, T. 2006. *The Oxford Companion to Food*. Oxford: Oxford University Press.

Fisheries and Aquaculture Department of the Food and Agriculture Organization of the United Nations 1990. *Training Manual on Gracilaria Culture and Seaweed Processing in China. Chapter III: Properties, manufacture and application of seaweed polysaccharides–agar, carrageenan and algin*. Available at: http://www.fao.org/docrep/field/003/AB730E03.htm (accessed 23 January 2012).

Odegaard, N., Carroll, S. and Zimmt, W.S. 2005. *Material Characterization Tests for Objects of Art and Archaeology*. London: Archetype Publications.

Scott, C.L. 2012. 'The use of agar as a solvent gel in objects conservation', *Objects Specialty Group Postprints* 19: 71–83.

Author's address

Cindy Lee Scott, Objects Conservation, CLS Conservation Services Ltd, Ottawa, Canada (cindylee.scott@gmail.com)

Micro-fragmented agar gels in wall painting conservation: the chapel of Saint Michael in the Royal Monastery of Pedralbes, Barcelona

Rosa Senserrich-Espuñes, Marilena Anzani, Alfiero Rabbolini and Lidia Font-Pagès

Introduction

The frescoes in the chapel of Saint Michael, in the Royal Monastery of St Mary of Pedralbes, in Barcelona, Spain, made in 1346 and attributed to the Catalan painter Ferrer Bassa, are an unique example of artistic quality that combines Barcelona's Gothic style with painting techniques from Tuscany in that century. These magnificent wall paintings, rich in colour and with numerous uses of metal leaf, are executed in a mixed technique that includes fresco and secco using different protein and oil binders. They have survived in good condition in a small cell attached to the cloister of the monastery. However, some alterations due to natural aging of the materials can be observed, and the consequences of different treatments intended to secure the polychrome and 'refresh' the colours. There is documentary evidence of only two interventions, carried out around 1929 and 1951: these have resulted in darkening of the surface and have exacerbated detachment of the paint layer.

One of the key elements in the conservation project was the removal of a soluble fixative (plant gum) applied during a previous intervention, which was causing degradation. In addition there was a need to eliminate less soluble non-original substances, such as animal glues, deposits of different calcium salts and extensive overpainting, which required the use of chelates and solvents. The conservators based in Barcelona wished to use the least toxic methods and the least invasive treatment compatible with the requirements of the frescoes. This led to cooperation and fruitful exchange of knowledge between the head of the conservation team and Italian conservators who had particular experience in the use of agar gels, including micro-fragmented gels which in the Italian and Spanish languages are described as 'destructured gels'.

Choosing the treatment method

Tests with water confirmed that most of the paint layers were not soluble in aqueous media. In addition it was noted that the layers were particularly fragile, and susceptible to abrasion damage from different strata as a result of the combination of mechanical action and the moisture contained in the cotton swab during testing, which decreases the strength of the lime mortar. It was therefore necessary to find a treatment method that only acted on the surface, limiting the movement of water into the porous structure of the painting, and that could gradually remove all the non-original superimposed layers without the need for mechanical treatments at the surface.

After researching various existing treatment options, agar gel was selected as most appropriate to the needs of this work. The use of this cleaning method has spread in recent years (Anzani and Rabbolini 2009; Anzani 2013; Cremonesi 2016; Sansonetti *et al.*, 2012, Senserrich-Espuñes *et al.* 2015; Senserrich-Espuñes 2017; Wolbers 2004; Wolbers 2005). AgarArt is a polysaccharide extracted from red algae, used for the preparation of rigid gel and marketed in the form of dehydrated powder (Bolletino CTS 2009). A new presentation in the micro-fragmented format has been designed by the Milanese company Aconerre Art Conservation Restoration snc, to overcome some limitations in the use of rigid agar gels. It is produced from rigid gelled agar, mechanically reduced to micro-fragments. Recently, it has been introduced as a commercial product, Nevek, that can solve problems related to the preparation of the gel, making available a ready-to-use substance which can either be heated to obtain a fluid gel or left to solidify to create preformed sheets (Bolletino CTS 2016).

This micro-fragmented agar gel has a number of advantages compared to the more common application approaches: it permits cleaning at room temperature, enabling the

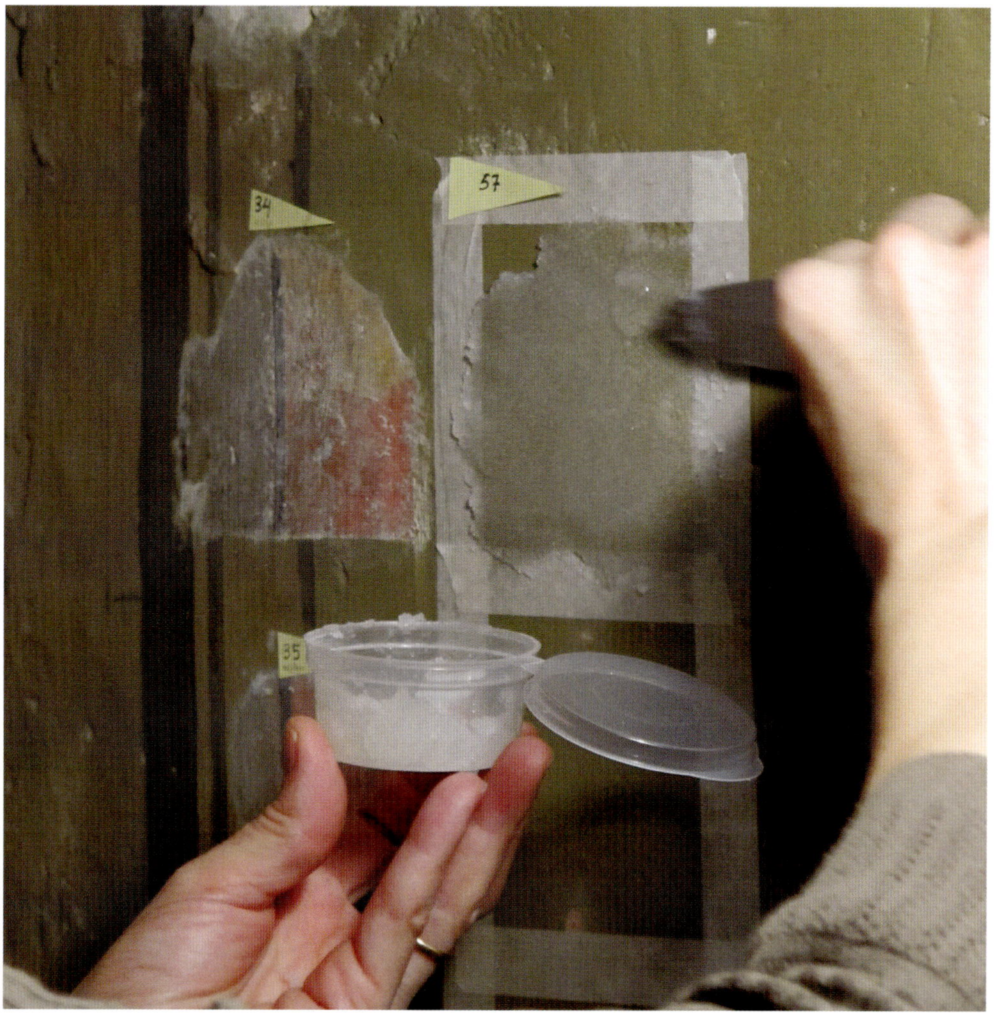

FIG. 1 Overpainting removal test, using ethyl alcohol + acetone with the micro-fragmented agar gel (image © Gonzalo Martí – Royal Monastery of St Mary of Pedralbes, Barcelona City Council).

inclusion of solvents, additives or chelating substances capable of enhancing the cleaning ability of the gel. In addition, its removal is very safe when it is applied through Japanese paper, since this minimizes any risk of flaking of the fragile paint. In case of detachment from the substrate, the paint remains protected by the Japanese paper layer in contact with the wall.

Where surface powdering, flaking or detachment is present – and where even minimum adhesion of gel can be risky for the integrity of the painting – the micro-fragmented form offered the best alternative. Not only is there less adhesion between the poultice and the surface but it also enables the gel to be applied without leaving discontinuities or voids of air which could lead to uneven cleaning, and it provides a more controlled treatment approach.

Agar gel treatment methodology

For cleaning tests, AgarArt agar was used, in concentrations from 3 wt% to 4 wt% in deionized water, with a variable application time between 20 and 60 minutes. The gel was tested in three different states: in preformed sheets, in the fluid state and micro-fragmented. For the elimination of partially or completely insoluble materials present on the painted surface, a chelating agent was added to the poultice formed by water and agar, with the objective of binding the metal ions present in these substances, and making them soluble. The gelling of the solution ensured good surface contact with the painting, minimizing diffusion towards the inner layers. Tests were made with the following chelating agents: triethanolamine citrate, triammonium citrate (TAC) and the trisodium salt of ethylenediaminetetraacetic acid (EDTA), all used in low concentration (0.3–1 wt%) and with application times ranging from 15 minutes to 4 hours, always controlling the pH of the gel to remain within the pH range of the paint layer, which before treatment varied from 6.53 (oil films) to 7.25 (fresco).

The effectiveness of the method was assessed through *in situ* observations with a portable monocular microscope. Control samples were used to verify the efficacy and safety of the proposed method, by analysing the residues of five gel samples after the cleaning process and verifying that all the substances extracted corresponded only to non-original materials. As a result, the treatment method used during the tests was considered effective, with very low or no risk to the painting (Fig. 1).

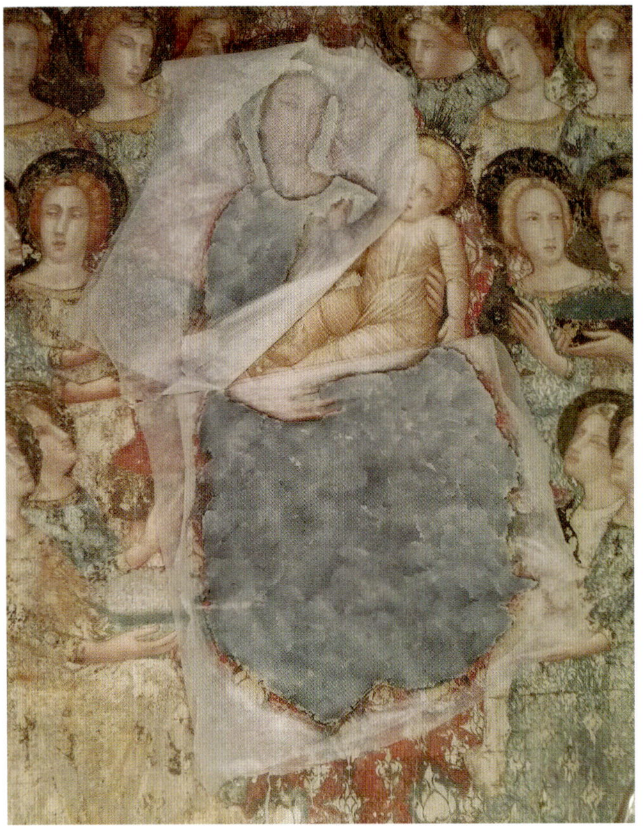

FIG. 2 Micro-fragmented agar gel applied onto two layers of Japanese paper, for paintings with fragile surfaces (image © Rosa Senserrich – Royal Monastery of St Mary of Pedralbes, Barcelona City Council).

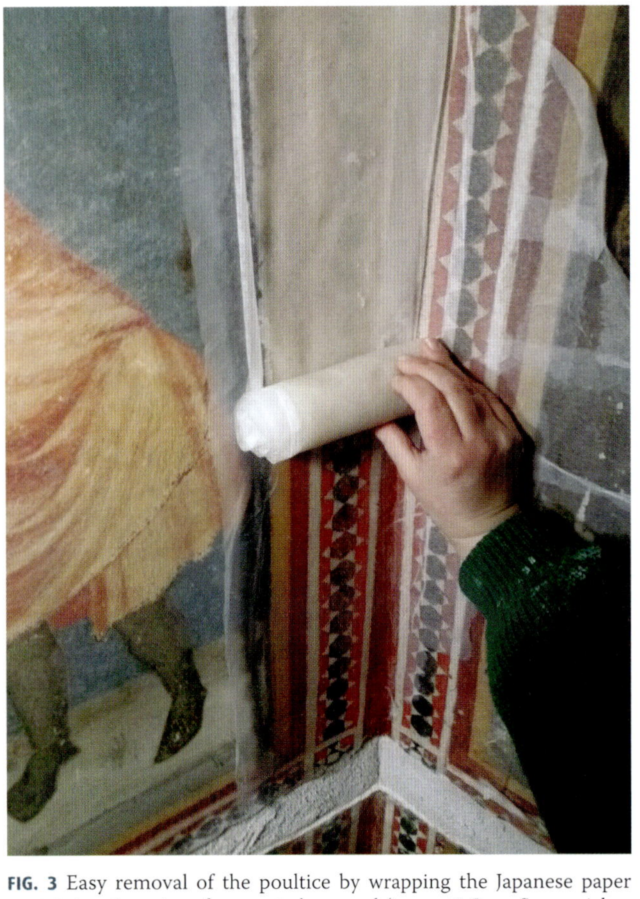

FIG. 3 Easy removal of the poultice by wrapping the Japanese paper containing the micro-fragmented agar gel (image © Rosa Senserrich – Royal Monastery of St Mary of Pedralbes, Barcelona City Council).

Treatment with micro-fragmented agar gel

The treatment began in 2014 on the northeast wall of the chapel, where the central scene of the *Crucifixion of Christ* and the *Triumph of the Virgin* is represented. The micro-fragmented agar gel was used both on the fresco surfaces with pigments that had lost cohesion and on the delicate areas made using a secco technique. For the removal of the non-original vegetable gum, the agar gel was applied at 4%, at a thickness of 0.5–1 cm, onto two pieces of Japanese paper (weighing 6 and 9 g) positioned in advance on the area to be treated, for approximately 30 minutes. In cases of aged animal glues and calcium salt-residues from past biodeterioration, the agar gel was used at 4% following the same methodology, but incorporating TAC at 0.5%. For deposits that were more difficult to remove, consisting of an insoluble whitish veil, it was decided to add 0.5% EDTA trisodium salt to the gel, leaving it for 15–20 minutes.

The ability of the micro-fragmented agar gel to retain organic solvents enabled the removal of oily repaints that were covering the entire base of the faux marble. A 1: 1 mixture of 40% acetone/ethyl alcohol was incorporated, and together with successive applications of the gel with 3% ammonium bicarbonate, produced satisfactory results (Figs 2 and 3).

Conclusion

The application of micro-fragmented agar gel on a 14th-century mural painting is an innovative process that has resolved the main problems concerning the treatment of an area of great technical complexity. The versatility of this system has provided tools to develop various innovative application methodologies adaptable to many problems related to different painting techniques and surface conditions, such as using micro-fragmented agar for the most delicate pictorial layers and to remove oil-based overpaints. The gradual removal of these deposits, the control of water, chelates and other solvents penetration, and the possibility of working without the need for direct mechanical action, were some of the main advantages provided by this method, which allowed the partial recovery of colours obscured by layers of dirt.

Materials and suppliers

- AgarArt, gelling temperature 38–42 °C; citric acid, reagent grade >=99.5%; 2,2',2''nitrilotriethanol triethanolamine 99%: CTS. Srl, Altavilla Vicentina,VI, Italy, http://www.ctseurope.com
- ED3SS Ethylenediaminetetraacetic acid trisodium salt hydrate ≥95%: Sigma-Aldrich Inc., St Louis, MO 63103, USA, http://www.sigmaaldrich.com
- Tribasic ammonium citrate 51997: Bresciani Srl, 20126 Milan, Italy, www.brescianisrl.it

References

Anzani, M. 2013. 'Utilizzo dei gel rigidi ai agar agar per la pulitura del materiale lapideo', in S.F. Musso (ed.), *Tecniche di restauro+Banca dati su DVD,* PLT card 43. Milan: Utet Scienze Tecniche, 526-535/2618.

Anzani, M. and Rabbolini, A. 2009. 'Il restauro del materiale lapideo con i gel di agar agar: il gesso *La bella Pallanza* e il marmo *Busto di donna* di Paolo Troubetzkoy', in D. Rullo (ed.), *Lo Stato dell'Arte 7, Congresso Nazionale IGIIC Napoli*. Florence: Nardini Editore, 661–8.

Bolletino CTS 2009. 'Nuovi Prodotti–AgarArt: l'acqua meno liquida', no. 20.1, 29/10/09. Available at: http://www.ctseurope.com/site/ dettaglio-news.php?id=92 (accessed 23 October 2016).

Bolletino CTS 2016. 'Nuovi prodotti–Il futuro della pulitura si chiama Nevek', no. 43.1, 18/05/16. Available at: http://www.ctseurope.com/dettaglio-news.php?id=355 (accessed 29 September 2016).

Cremonesi, P. 2016. 'Surface cleaning? Yes, freshly grated Agar gel, please', *Studies in Conservation* 61: 362–7. Available at: http://dx.doi.org/10.1179/2047058415Y.0000000026 (accessed 23 May 2017).

Sansonetti, A., Casati, M., Striova, J., Canevali, C., Anzani, M. and Rabbolini, A. 2012. 'A cleaning method based on the use of agar gels: new tests and perspectives', in *Proceedings of the 12th International Congress on the Deterioration and Conservation of Stone, Columbia University, New York City*. New York: ICOM-CC. Available at: http://biblio.iccrom.org/cgi-bin/koha/opac-detail.pl?biblionumber=111327&query_desc=an%3A%224231%22 (accessed 23 May 2017).

Senserrich-Espuñes, R. 2017. 'Il restauro delle pitture murali della capella di Sant Miquel, Monastero de Pedralbes di Barcellona', in *Retrospektive und Perspektive in der Wandmalereirestaurierung – Zwischenbilanz zur Entwicklung ihrer Methoden und Techniken, hrsg. von M.Pfeil, Werkstattgespräch 9. und 10. Mai 2016*, Monaco di Baviera 2017 (Schriftenreihe des Bayerischen Landesamtes für Denkmalpflege, 16) [in press].

Senserrich-Espuñes, R., Anzani, M., Rabbolini, A. and Font-Pagès, L. 2015. 'L'intervento con i gel di agar sulle pitture murali del Trecento nella capella di Sant Miquel, al Monastero Reale di Santa Maria de Pedralbes di Barcellona', in D. Rullo (ed.), *Lo Stato dell'Arte 13, Congresso Nazionale IGIIC Torino*. Florence: Nardini Editore, 161–8.

Wolbers, R. 2004. *Un approccio acquoso alla pulitura dei dipinti, Quaderni del Cesmar7*, no. 1. Padua: Il Prato.

Wolbers, R. 2005. *La pulitura di superfici dipinte. Metodi acquosi*, serie 'I Maestri del Restauro', no.1. Padua: Il Prato.

Authors' addresses

- Rosa Senserrich-Espuñes, Department of Arts and Conservation, Faculty of Fine Arts, Barcelona, Spain (rosa.senserrich@ub.edu)
- Marilena Anzani, Conservation company Aconerre snc, Milan, Italy (aconerre@tin.it)
- Alfiero Rabbolini, Conservation company Aconerre snc, Milan, Italy (aconerre@tin.it)
- Lidia Font-Pagès, Barcelona History Museum, Barcelona, Spain (lfont@bcn.cat)

Towards the sustainable use of agar/agarose in conservation: a case study of the Izu peninsula, Japan

Misa Tamura and Koji Takagi

Introduction

The recent shortage of agar/agarose and the consequent temporary suspension of its supply came as an unwelcome shock to bacteriological research laboratories. However, this news was even less widely known in the conservation community, whose use of agar/agarose appears to be on the increase. This poster focuses on the Izu peninsula in Japan. The region was once renowned internationally as a major producer of *Gelidium* sp., yet has faced a significant decline of its agar fishery. In recent years, a number of initiatives and attempts have been made to restore the *Gelidium*-rich sea beds that surround the region.

Species and available products

Agar can be produced from several different source seaweeds, the three most significant species being *Gelidium* sp., *Pterocladia* sp. and *Gracilaria* sp. Although they all produce agar, the resulting substances have different properties. *Gelidium* is probably the most desired agar seaweed: the agar extracted from it is used for culturing microorganisms, and it is an essential material for bacteriological research. However, this slow-growing seaweed has not been successfully cultivated on a scale that makes it commercially viable. Harvested from natural habitats, its yields can vary from year to year dependent on marine and climatic conditions. It is susceptible to *iso-yake* (rocky-shore denudation), causing devastation to the seaweed habitats by factors such as surging warm currents or infestation by marine life that feeds on the seaweeds (Fujita 2002). The low yield was one of the major reasons for the temporary suspension of agar sales in 2015 (Callaway 2015). *Pterocladia* is similar to *Gelidium* in terms of its habitation patterns but yield tends to be smaller than *Gelidium* (Armisen and Galatas 1987). *Gracilaria*, on the contrary, is a widely and successfully farmed species. A report predicted that the *Gracilaria* industry is perhaps the only agar producer whose market has scope for further growth due to successful farming (McHugh 2003: 15–16). However, the agar produced from this variety of seaweed cannot be used for bacteriological research.

The most desirable property of *Gelidium* for bacteriological research is its low gelling temperature (34–36 °C), in comparison with that of *Gracilaria* (>41 °C). This enables additives to be introduced to the agar without altering their state through heating (McHugh 2003: 21). Both food-grade agar and agarose can be made from *Gelidium*, *Pterocladia* and *Gracilaria*. However, manufacturers and suppliers do not typically state the origins or species of the source material. Another notable difference between *Gelidium* and *Gracilaria* is gel strength. Agar naturally extracted from *Gracilaria* was reported to have less gel strength than *Gelidium*. However, the industrial purification process enables *Gracilaria* agar to increase in gel strength, although the process is costly (Armisen and Galatas 1987). Both of these qualities could also affect the working properties of gels in conservation applications. However, it is very difficult to source provenanced agar/agarose to conduct experiments.

Agar production in Izu

Prior to the post-WW2 economic boom in Japan, the peninsula's thriving agar industry was sustained by diver-harvesters and processors of *Gelidium* and *Pterocladia*. However during the 1960s the region saw an economic transition from the fisheries to the then-lucrative tourism industry which gravely impacted the fisheries. Regular harvesting of *Gelidium* and *Pterocladia* leaves the roots in place, encouraging the growth of new shoots. Moreover, the regular intervention of the harvesters prevents the spread of more invasive species of

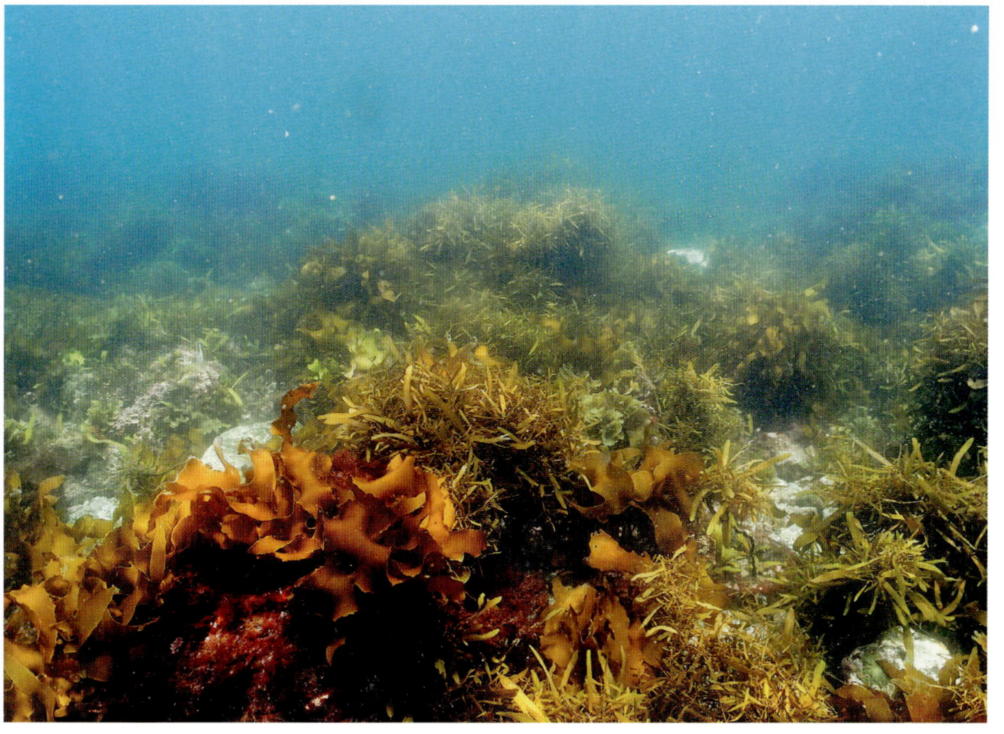

FIG. 1 A sea bed of *Gelidium* and *Pterocladia* (bottom left corner), overtaken by non-agar-producing seaweeds due to the decline of the agar fishery, on Izu Peninsula, Japan (reproduced courtesy of Shizuoka Prefectural Research Institute of Fishery, Izu Branch, 2017).

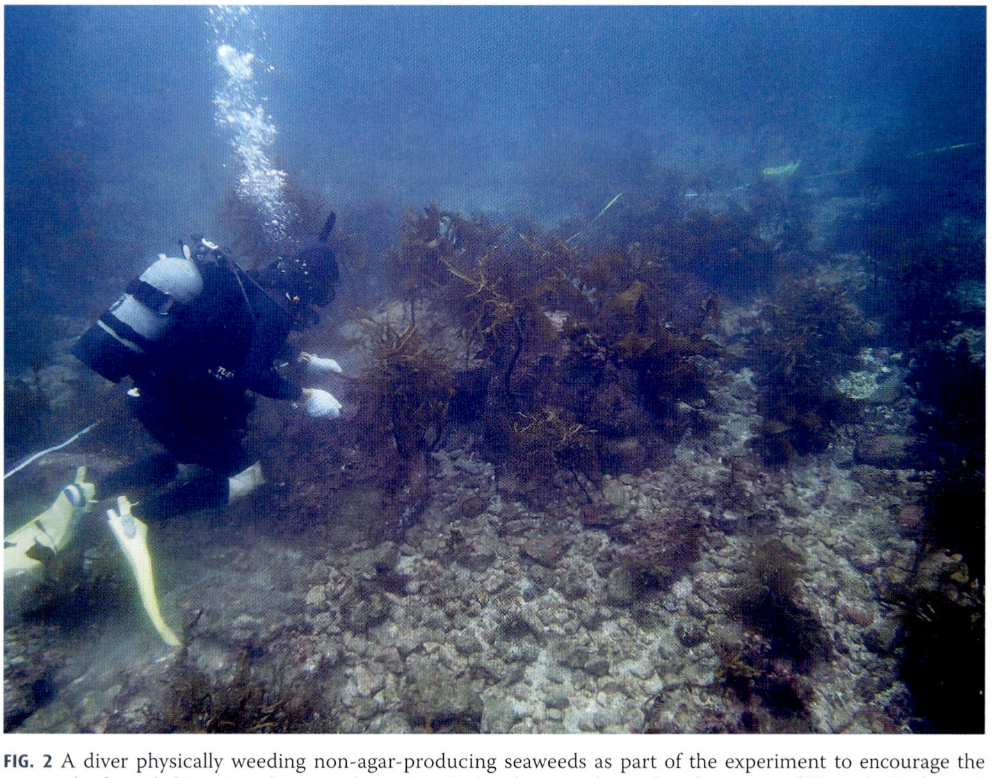

FIG. 2 A diver physically weeding non-agar-producing seaweeds as part of the experiment to encourage the regrowth of agar habitats in a devastated area, Izu Peninsula, Japan (reproduced courtesy of Shizuoka Prefectural Research Institute of Fishery, Izu Branch, 2017).

non-agar seaweeds. As the seaweed farming shrank in scale, many of the agar seaweed habitats became neglected, which resulted in decreasing yields and even the devastation of their habitat (Tengusa Fishery 1998: 72–4, 85–7) (Fig. 1).

The industry faces the challenges of recruiting a younger generation to the occupation. Many aspects of the agar production process have not been automated, making the work very laborious. The rocky sea beds, inhabited by *Gelidium* and *Pterocladia*, need to be harvested physically by hand or using a hoe. The removal of non-agar seaweeds, barnacles and other impurities involves painstaking labour done by hand. The financial gains are not usually sufficient to fully support young families therefore the industry now chiefly relies on the labour of retirees who wish to supplement their income (Tengusa Fishery 1998: 75–6).

In order to revitalize the agar habitats and fishery, the Izu Branch of the Shizuoka Prefectural Research Institute of Fishery has conducted a range of research. For example, the regular physical weeding of invasive non-agar seaweeds by divers in the devastated agar sea beds has been carried out in recent years. The experiment was conducted in an attempt to encourage the growth of *Gelidium* and *Pterocladia*, which had been overtaken by other species. The Institute reported that the experiment has achieved approximately a 20% increase in the agar habitat (Anon. 2016) (Fig. 2). The Institute also records and monitors habitats, the rate of successful rooting of spores, and provides forecasts for the annual yield, in order to inform harvesters, manufacturers and suppliers. It has also been seeking an economically viable labour model for the fishery. One idea suggested would be to pursue a part-time model. For example, local scuba diving instructors would work in the agar fishery outside the tourist season.

Conclusion

In comparison with other industries, the conservation sector's use of agar/agarose is relatively small. However, as long as we have a collective professional interest in our continuing use of the material, it is vital that we strive to promote and raise awareness of responsible and sustainable consumption among our peers. The conundrum we face is in balancing our desire for the agar fishery to revitalize declining production and incentivize the maintenance of its habitats, with the sustainable use of a non-farmable, exhaustible natural resource.

The conservation community could continue the discussion in several ways. First of all, we should proactively communicate with the agar fisheries to make them aware of our professional interests and the existence of demand for their output. Secondly, further investigation should address whether agar/agarose extracted from different seaweeds would be effective in rigid gel systems. This would enable us to gain a much better idea of how much *Gelidium* agar we as a profession actually require. Detailed labelling with more provenance information should perhaps be requested from the manufacturers. Lastly, it is vitally important that individual conservators consider the options for aqueous gel systems by taking into account the considerations discussed above. Could the desired outcome be achieved without using agar/agarose? Are there less environmentally problematic alternative methods of cleaning? Do decisions on conservation treatments also support sustainability?

Acknowledgements

The authors thank Prof. Frances Lennard, head of the Centre for Textile Conservation and Technical Art History, University of Glasgow, Glasgow, UK, and Masatoshi Hasegawa, senior researcher at Shizuoka Prefectural Research Institute of Fishery Izu Branch, Izu, Japan, for their support for the research.

References

Anon. 2016. 'Research into the recovery of agar habitats by weeding'(稲取でテングサ漁場の雑海藻刈り), *Shirahama: News from Shizuoka Prefectural Research Institute of Fishery Izu Branch*(白浜:静岡県水産技術研究所伊豆分場ニュース), November, issue 30, 1.

Armisen, R. and Galatas, F. 1987. 'Production, properties and uses of agar', in J.D. McHugh (ed.), *Production and Utilization of Products from Commercial Seaweeds: FAO Fisheries Technical Paper 288*. Available at: http://www.fao.org/docrep/x5822e/x5822e00.htm (accessed 19 December 2016).

Callaway, E. 2015. 'Lab staple agar hit by seaweed shortage'. Available from: http://www.nature.com/news/lab-staple-agar-hit-by-seaweed-shortage-1.18970 (accessed 19 December 2016).

Fujita, D. 2002. 'Rocky-shore denudation', *Current Issues in Phycology in the Early 21st Century: Japanese Society of Phycology 50th Anniversary Issue*, 102–5. Available at: http://www.sourui.org/publications/phycology21/file-list-21.html (accessed 19 December 2016).

McHugh, J.D. 2003. 'A guide to the seaweed industry', *FAO Fisheries Technical Paper* 441. Available at: ftp://ftp.fao.org/docrep/fao/006/y4765e/y4765e00.pdf (accessed 18 December 2016).

Tengusa Fishery in Izu Compilation Committee (伊豆の天草漁業編纂会) (ed.) 1998. *Tengusa Fishery in Izu*. (伊豆の天草漁業) Tokyo: Seizando Shoten (成山堂書店).

Authors' addresses

- Misa Tamura, Centre for Textile Conservation and Technical Art History, School of Culture and Creative Arts, University of Glasgow, Glasgow, UK (Misa.tamura@glasgow.ac.uk)
- Koji Takagi, Shizuoka Prefectural Research Institute of Fishery Izu Branch, Shimoda City, Shizuoka, Japan (koji1_takagi@pref.shizuoka.lg.jp)

Polyacrylic: Pemulen and Carbopol

Resurrecting a giant: using solvent gels and aqueous systems to restore Villanova University's *Triumph of David*

Kristin deGhetaldi, Brian Baade, Zachary Voras and Emily Wroczynski

Introduction

The conservation treatment of a 17th-century oil painting on canvas, attributed to the workshop of Pietro da Cortona, began in the fall of 2013 within a publically accessible space located in Villanova University's Old Falvey Library situated just outside of Philadelphia, USA. Almost 4 m tall and 6 m wide, the *Triumph of David* arrived at the university in 1950. The composition depicts the celebration immediately following David's victory over Goliath, after which the young shepherd presents the giant's severed head to King Saul. Six years after the arrival of the painting, local restorers[1] were hired by the university to build a new stretcher, re-stretch the canvas, 'clean off surface soil,' apply at least three coats of varnish, and construct a new frame, all of which was carried out over the duration of a single week.[2] In 2001, the painting was again examined by an appraiser who declared that it had been, 'over-cleaned, resulting in major paint loss,' in addition to suffering from darkened overpaint that was determined to be irreversible.[3] These observations led some to assume that it was impossible to successfully restore the painting. The large canvas remained hanging on the south wall in the Reading Room of Old Falvey Library until a team of conservators, scientists, and art historians were asked to examine it in February of 2013.[4] Contributions from many interns, volunteers, scientists, conservators, and other professionals over a period of two years finally led to the successful treatment of the painting (Fig. 1).

General history and provenance

It is surprising that so little is known about the painting's provenance given its prodigious size and monumental scope. Jennie Enfield Berry, a member of American High Society whose family originated from Oak Hill, Georgia, donated the painting along with several other works to Villanova University between 1949 and 1950. Berry married the Italian prince Don Enrico Ruspoli and in 1902 the couple used Berry's inheritance to purchase a 150-room castle[5] from the Orsini family located in the small medieval town of Nemi just outside of Rome (*New York Times* 1901; Bates-Batcheller 1911; Green 2007: 416–18). As there is no mention of this enormous canvas painting in the Orsini family's extensive inventories, it is possible that the princess purchased the Villanova painting in nearby Rome and had it moved subsequently into the castle.[6]

There remains confusion concerning attribution: who originally commissioned the work, and where it was initially housed? Consultations with art historians and conservators have pointed towards Rome as being the likely place of origin, yet other scholars have mentioned possible ties to Bologna, Italy, and even Paris, France. The consensus is that the work was created either by multiple artists associated with Cortona's workshop or by artists who were working in his style.[7] Cortona had a very large workshop throughout the second half of the 17th century, recruiting dozens of easel and fresco painters to meet an ever-increasing number of commissions (Guarino and Masini 1997). It is likely that many assistants helped to create his large-scale canvas paintings, with Cortona himself being involved in the compositional design and/or the rendering of prominent figures rather than the execution of the entire picture. The examination of the Villanova painting strongly supports the notion that more than one painter was involved in the composition; there is more than one distinct style of painting, disparate levels of painting quality, and both subtle and significant changes were identified using non-destructive techniques including infrared reflectography (IR) and X-radiography.[8] The latter also revealed a beautifully executed kneeling male figure in the bottom left corner holding

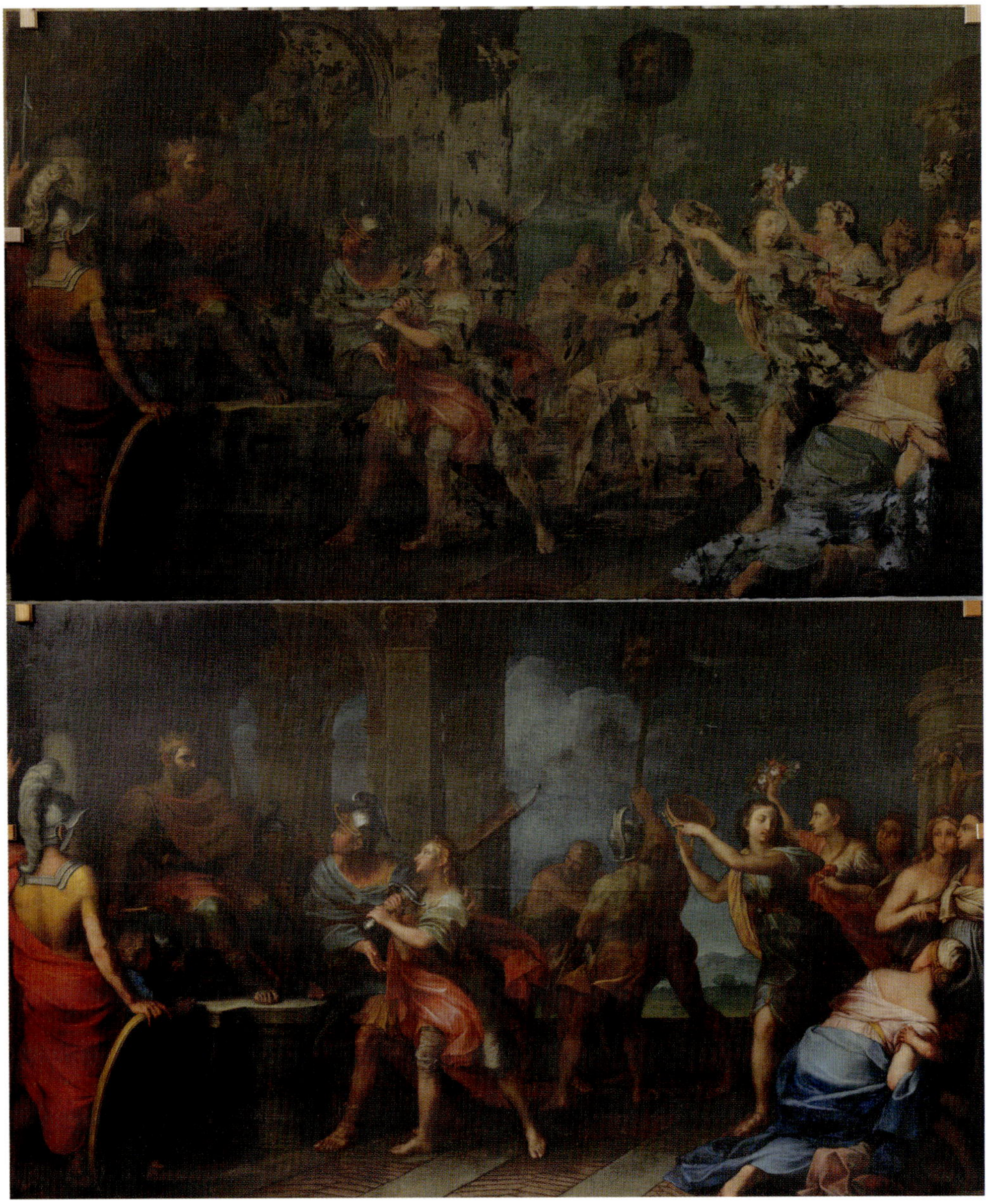

FIG. 1 View of *The Triumph of David* (attributed to the workshop of Pietro da Cortona) before (*top*) and after (*bottom*) conservation treatment. Cleaning was performed using a combination of free solvents, solvent gels, and aqueous emulsion systems (photos courtesy of Villanova University).

a *fasces*, a symbol that was often associated with the power of the Roman magistrate. Further examination of the painting suggests that this figure was painted out many years after the completion of the composition, although the reason for this remains unknown. The discovery of the symbolic *fasces* does suggest that whoever commissioned the picture had close ties with the city of Rome, and may even have held a prominent governmental or diplomatic position.

Construction and materials

Cleaning tests were performed during the conservation team's extensive examination of the painting's condition, and chemical analysis was used to characterize the inorganic and organic components present. Cross-sections (Fig. 2) combined with scanning electron microscopy–energy dispersive (SEM–EDX) spectroscopy revealed that the painting is executed on a red-brown double ground that is rich in iron oxides and alumino-silicates.[9] The original paint layers (containing lead white, vermilion, and Naples yellow) are covered with a layer of overpaint containing zinc white. Some areas of the composition that had been extensively reworked contain multiple layers of paint, while other areas have a relatively straightforward and simple stratigraphy. Nearly all cross-sections collected for preliminary analysis possessed layers of overpaint and/or degraded surface coatings as a substantial portion of the composition was covered with restoration paints and varnish. Scrapings of the superficial varnish layer were analyzed using Fourier-transform infrared (FTIR) spectroscopy and were found to contain mastic resin.[10] Nearly all layers of overpaint were separated from the original paint and ground by a thin, unpigmented layer (most likely containing a

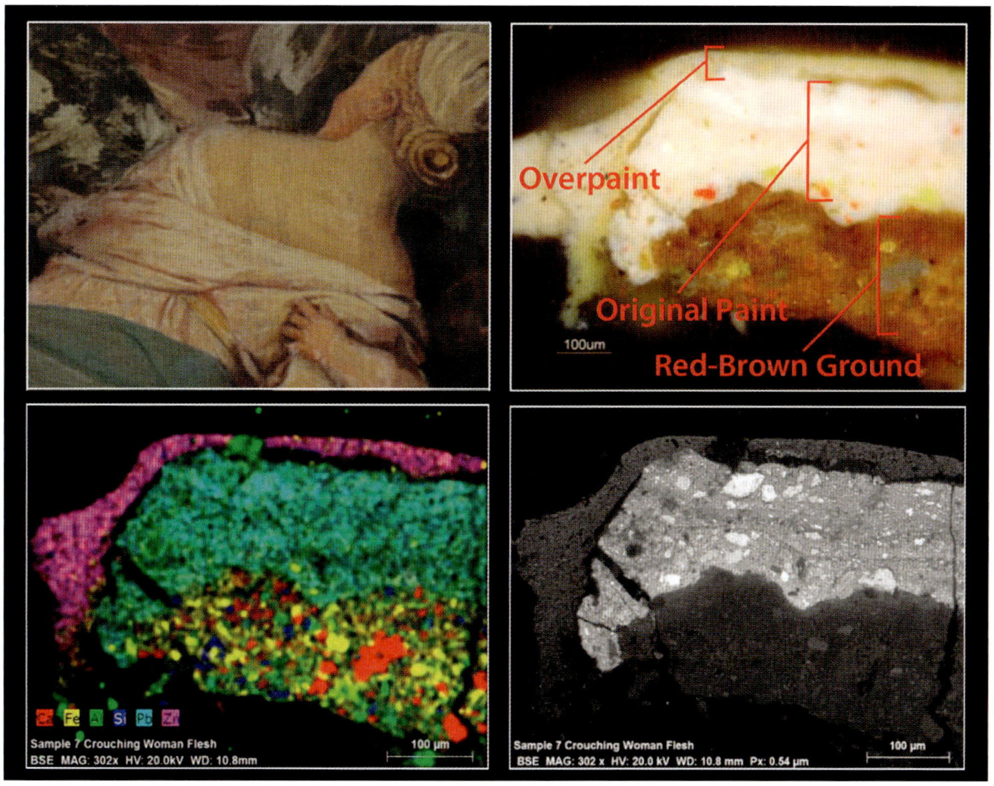

FIG. 2 Detail of the crouching female figure in the lower right corner before treatment (*upper left*) and a cross-section (*upper right*) collected from the figure's flesh. SEM-EDX was used to collect information relating to pigment identification and morphology as well as the characterization of unoriginal and original paint layers (*bottom left and right*).

soft resin such as mastic). Many cross-sections also revealed cracks extending through the original paint and ground layers suggesting that the painting had experienced considerable mechanical stress, possibly from exposure to adverse environmental conditions. Significant damage was found along the outer sections of the picture; these areas had suffered substantial losses to the paint and/or ground layers and possessed multiple campaigns of overpaint. Finally, the entire painting is paste and/or glue-lined to a canvas of similar weight. The type and quality of the lining suggests that this treatment was performed in Italy before the painting was shipped to the USA. The painting is also strip-lined, although it is unclear whether this addition was added during the 1956 restoration, coinciding with the construction of a new stretcher.

Cleaning tests and related analyses

Table 1 summarizes some results of preliminary testing carried out using free solvent mixtures and Carbopol-based solvent gels (Wolbers 2000). In general, a 1:1 mixture of acetone and ethanol was found to be most effective at reducing the varnish layer(s) and areas containing thinly applied overpaint. It is likely that this combination was able to dissolve the oxidized, degraded layer of mastic varnish while softening/swelling the thinly applied layers of overpaint that were applied during the most recent restoration campaign. However, this same mixture of free solvents had little to no effect on areas of thickly applied and/or older overpaint. Additional tests revealed that solvents containing a higher degree of hydrogen bonding (benzyl alcohol and 1-methyl-2-pyrrolidinone) proved more effective in these areas. Ultimately, a 1-methyl-2-pyrrolidinone gel (containing 4 g Carbopol 934, 15 mL Ethomeen C-25, 15–30 mL distilled water, and 200 mL pyrrolidinone solvent) was found to be the most effective at reducing the thickly applied overpaint (Wolbers 2000). The gel was applied locally and allowed to remain on the surface for 1–2 minutes depending on the thickness of the overpaint. This was cleared using a relatively polar mixture of solvents (1:1 acetone/ethanol) and followed by a final rinse with mineral spirits. The viscosity of the gel allowed for successful application on the vertically positioned canvas without the risk of drips. Some areas of thick and older restoration required two to three applications of the gel to effectively reduce the overpaint and reach the original textured surface of the paint and canvas (Fig. 3).

Portions of the lower right section of the composition exhibited a grey-brown surface crust that could not be effectively reduced/removed using solvent systems. A series of Pemulen-based aqueous systems[11] using Lubrizol[12] and Noveon (Ravenel 2010) products was tested. A stock solution was prepared by mixing 100 mL of distilled water with 1 g Pemulen TR-2 powder to which approximately 5 mL of a buffer (triethanolamine) was added until the pH reached 7.5–8. Varying percentages (2, 5, 10, and 20% by volume) of benzyl alcohol were added to the stock gel to determine the most effective proportion. The 2 and 5% emulsions were

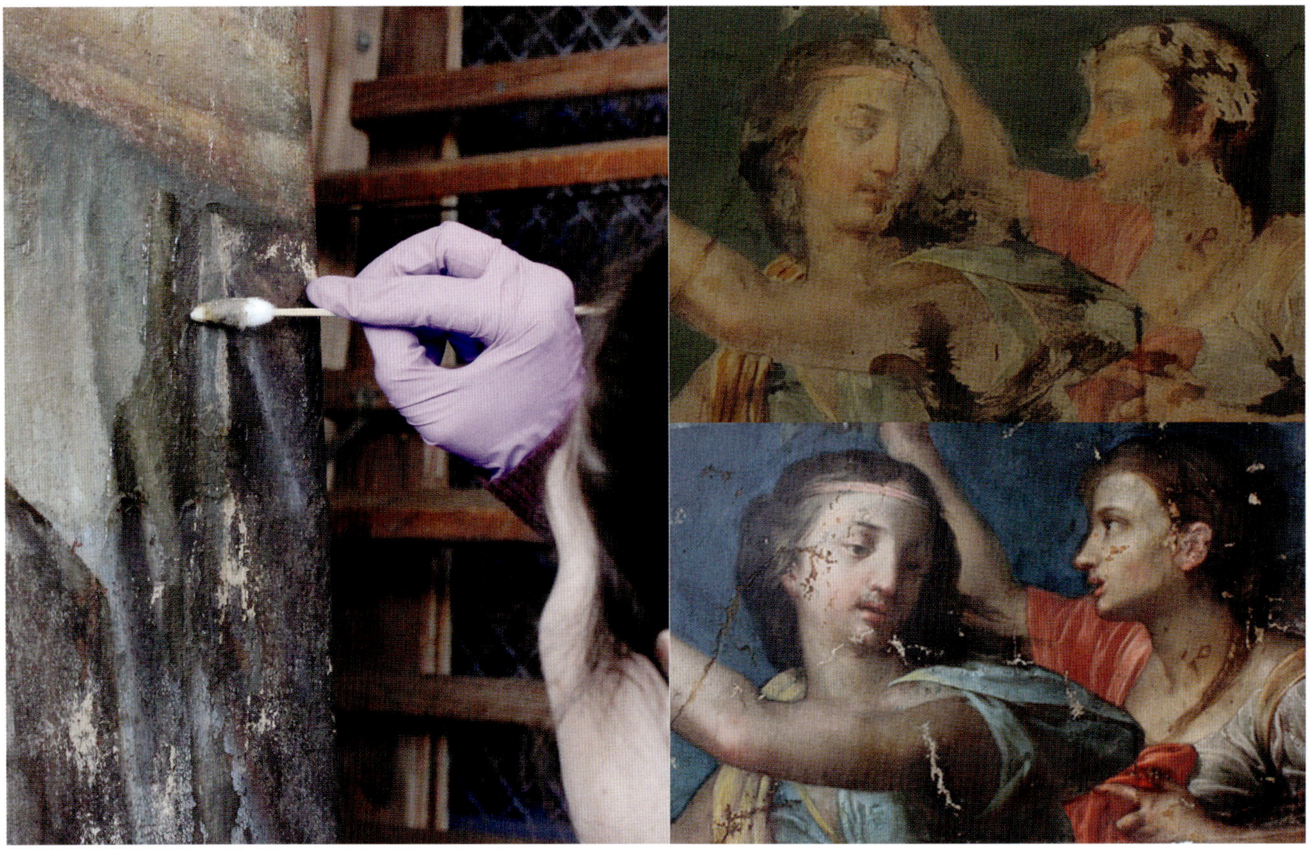

FIG. 3 Intern Maggie Bearden applying a solvent gel to a heavily overpainted area along the left edge (*left*) and a detail of the composition before (*upper right*) and after (*bottom right*) varnish and overpaint removal using free solvents and a solvent gel cleaning system.

TABLE 1 Summary of preliminary testing carried out using free solvent mixtures, Carbopol-based solvent gels, and aqueous cleaning systems.

Testing solution/emulsion/gel	Results and observations
1:1 acetone/mineral spirits	No effect
3:1 acetone/mineral spirits	No effect; slightly breaks through coating
1:1 acetone/ethanol	Breaks through coating; removes most overpaint, but slow and multiple passes required
Pyrrolidinone solvent gel (4 g Carbopol 934, 15 mL Ethomeen C-25, 15–30 mL distilled water, and 200 mL pyrrolidinone solvent)	Very effective; breaks through coating and swells/softens thickly applied overpaint especially if allowed to remain on surface for approximately 30–60 seconds (cleared using 1:1 acetone/ethanol)
Xanthan and benzyl alcohol (20%)	Slightly more effective at 'halos' (tenacious overpaint around old fills) but leaves surface more blanched
Pemulen and benzyl alcohol (2%, 5%, 10%, 20%)	Much more effective than xanthan microemulsion. Cuts through multiple layers of varnish and overpaint in fewer passes; 5% was found to be effective enough without risk of over-blanching surface.
Pemulen and ethanol (5%)	About equally effective as Pemulen and benzyl alcohol microemulsion
Pemulen and pyrrolidinone gel	Not as effective as single solvent gel; harder to work with because of viscosity (sticky consistency)
Pemulen and benzyl alcohol (5%) and EDTA (0.5%)	Slightly more effective than simple microemulsion on tenacious 'halos'; more testing is necessary

found to be no more effective than the pyrrolidinone gel at reducing/removing the degraded surface coating. The 10% emulsion was somewhat more effective but no significant improvement was observed until the 20% benzyl alcohol emulsion was tested. Additional tests using Pemulen TR-2 on areas that did not possess the grey-brown coating (covered only with the degraded mastic varnish) proved effective as well; however, slight blanching was noted after the emulsion was cleared from areas originally painted with thin, dark oil paint. While conservators routinely exercise caution when testing varying cleaning systems, considerable care must be taken when testing aqueous emulsions on

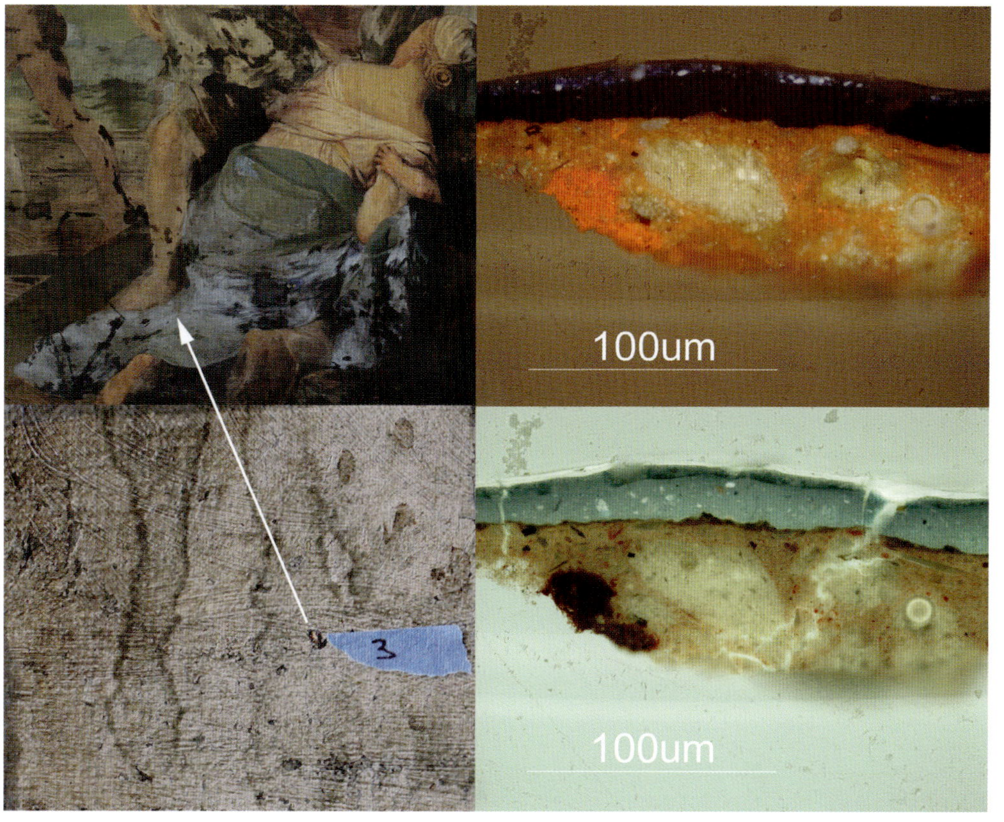

FIG. 4 Detail of lower right corner (*upper left*) and related areas containing a grey-brown, blanched coating (*lower right*). A cross-section (viewed at ×100 magnification) collected from the marked area shown in the lower right image revealed intact paint and ground layers beneath the coating when viewed in visible light (*upper right*) and UV light (*lower right*).

paintings that are potentially sensitive to moisture as was the case with the Villanova painting. The dark red-brown earth ground and its substantially degraded state likely contributed to its moisture sensitivity. In addition, the alumino-silicates present in the ground's iron-oxide matrix suggest that the earth pigments were probably obtained from a clay-rich source, clay being a material that is known for its hygroscopicity. Applying mineral spirits to the surface temporarily reduced the blanching and it was found to completely disappear after the application of an isolation varnish (a reduced ketone varnish, MS2A). More research is needed to better understand the reactions related to the slight blanching; conservators should exercise caution when applying Pemulen TR-2 emulsions to colored grounds that are likely to possess hygroscopic materials and/or are significantly degraded.

Throughout the treatment, the conservation team also considered the potentiality of residues associated with various cleaning systems that were selected for removal of the degraded varnish and/or overpaint. Residues related to Carbopol-Ethomeen gel systems have been extensively studied by an interdisciplinary team of scientists and conservators (Dorge 2004). Radioactive tags were used to monitor the presence of residues before, during, and after clearing with Carbopol-Ethomeen solvent gel systems. On average 0.11% of these gel components were found to remain on the surface; residues associated with the polyacrylate polymers (Carbopol) were found to be negligible while the cationic surfactants Ethomeen C12 and Ethomeen C25 (tertiary cocamine derivatives that are ethoxylated onto an amino group) were calculated to be extremely low, around 60 μg/cm^2 (Dorge 2004: 38). Further testing found that after 12 hours of artificial aging it was difficult to detect the presence of the surfactants using gas chromatography–mass spectrometry (GC–MS), and analysis of paintings that had been cleaned several years earlier using Carbopol-Ethomeen gel systems revealed no detectable residues (Dorge 2004: 128–9). These results supported findings from comprehensive analysis of the gel systems before, during, and after clearance; any surfactant left behind eventually degraded into smaller, more volatile molecules (Dorge 2004: 95).

The topic of potential residues left behind by Pemulen TR-2 systems has not been studied extensively and the subject is beyond the scope of this paper. However, it is worth discussing possible avenues for future research regarding aqueous emulsions that consist of acrylate cross-polymers. An independent review of these polymeric emulsions used in the cosmetic industry noted that they 'share a general lack of chemical reactivity that renders them nearly impervious to degradation. These ingredients are essentially insensitive to solar ultraviolet light (UV) degradation, as the primary UV absorption of acrylics is at a lower wavelength' (CIR 2011: 2). Therefore, the potential for residual solvent and/or residual reagents that might be present in the copolymer after manufacture is really of greater interest when thinking

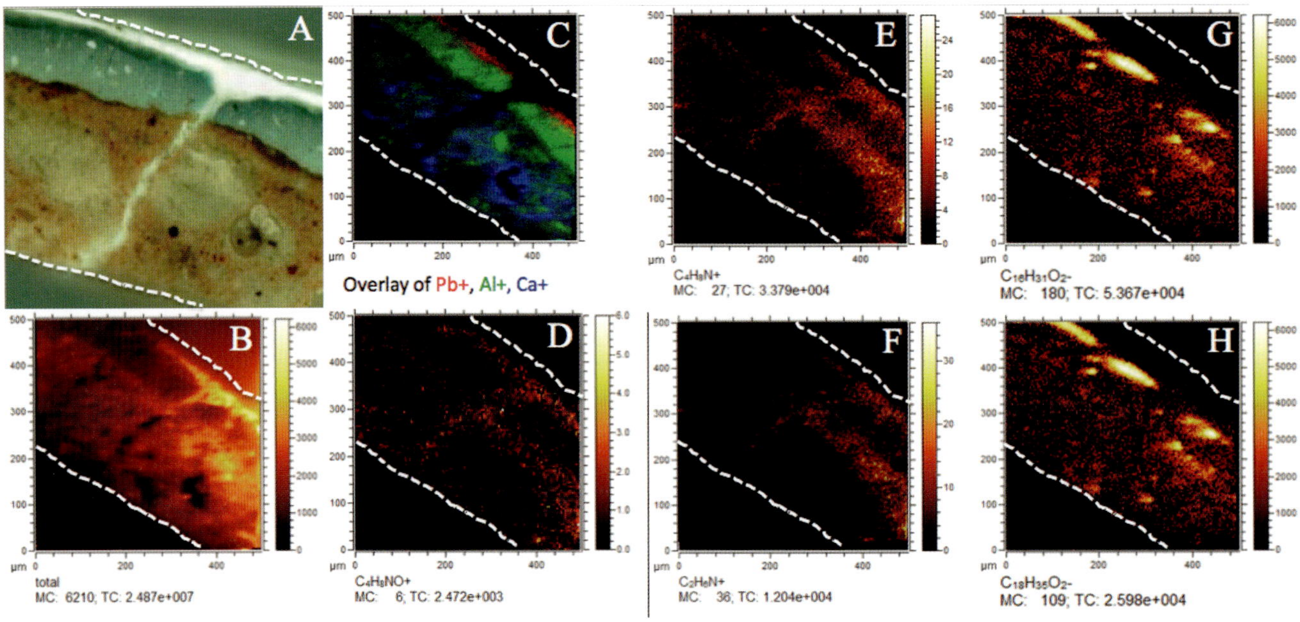

FIG. 5 Comparison of cross-section shown in **FIG. 4**. (A) UV (×100) with images collected using ToF-SIMS. The total ion count collected in positive mode (B) shows all mass fragments (pixel of maximum counts has 6210 counts, total image counts (TIC) is 2.487 × 107). Positive mode revealed the presence of inorganic metal ions (C) corresponding to the presence of lead white (Pb+), lapis lazuli (Al+), and aluminum/calcium-containing particles in the ground layers (Al+/Ca+) in addition to several amino acids including hydroxyproline (5D C_4H_8NO+), glycine (5E; C_4H_8N+), and alanine (5F; C_2H_6N+) suggesting the presence of a glue-based adhesive that had been infused into the paint and ground layers from the surface during a previous restoration campaign. ToF-SIMS in negative mode also confirmed a fairly strong signal for palmitic (5G; $C_{16}H_{31}O_2$-) and stearic (5H; $C_{18}H_{35}O_2$-) acids in the uppermost paint layer with a weaker signal in the ground layers, free fatty acids that are commonly detecting in paints containing drying oil.

about potential residues. A maximum of 0.5% residual solvent in Pemulen TR-2 has been reported and most of the solvents included in the list are volatile at room temperature, making them unlikely to cause problems to a paint film (CIR 2011: 3). Regarding individual monomers and reagents, less than 0.25% (2500 ppm) acrylic acid and less than 0.5% (500 ppm) residual ester have been reported (CIR 2011: 3). Acrylic acid/methacrylic acid is unlikely to react with an oil paint film (particularly an aged paint film) unless in the presence of specific catalysts, as this material is frequently used to create epoxidized oils via acrylation (Dotan 2014: 582–3). The esters listed include two types of allyl ethers: one derived from sucrose and another derived from pentaerythritol (CIR 2011: 8). Minute traces of a sucrose derivative are unlikely to react with an aged paint film and an allyl ester of pentaerythritol would behave similarly to glycerol, a component already present in oil films. While these reports offer a positive outlook, more research is still needed to identify any residues that might be present (comparing various clearing techniques) and to monitor the behavior of such residues if they are identified. Ideal analytical techniques for such research would likely include those used in the Carbopol-Ethomeen study previously described. Radioactive or even fluorochrome tags that are chemically bonded to each type of reagent and/or monomer/polymer would be a suitable method in addition to monitoring residues using mass spectrometry techniques.

Further analysis helped to reveal why an additional cleaning system (aqueous) was necessary to clean a large section of the painting. For areas containing the grey-brown coating (Fig. 4), a Pemulen TR-2 emulsion containing 10% benzyl alcohol was the most effective system and no blanching in the paint or ground layers was observed. A chelator (0.5% ethylenediaminetetraacetic acid (EDTA)) was added to the 10% solution, an addition that was found to increase the efficacy of the cleaning system, likely due to the substantial amount of imbibed grime. Follow-up analysis of the blanched coating from the lower right region of the painting using GC–MS suggested the presence of both fatty acids and protein.[13] As GC–MS does not provide a spatial analysis of organic materials in discrete layers, time-of-flight secondary ion mass spectrometry (ToF–SIMS) was used to further analyze a cross-section (Fig. 5) that had been collected from the crouching woman's blue dress prior to cleaning[14] (Voras *et al.* 2016). The unpigmented surface coating was found to contain a number of amino acids including hydroxyproline, corroborating the results obtained with GC–MS, while fatty acids (palmitic and stearic) were only found in the lower paint and ground layers. The results suggested the presence of an animal glue in the blanched surface coating, a material that was likely used to coat the surface in an attempt to consolidate the paint and ground layer(s) during a previous restoration campaign. The coating could also be seen filling a large crack that extends through both the paint and ground layers as the signals corresponding to the amino acid fragments co-localized along the surface of the paint cross-section and within the interstices of the crack. These findings attest to the difficulty the conservation team encountered during initial cleaning tests using free solvents and solvent gels, and helps to clarify why an aqueous cleaning system (using a Pemulen TR-2 emulsion) proved to be the most successful at removing/reducing the glue layer.

Conclusion

The conservation treatment of Villanova University's *Triumph of David* would not have been possible without the use of solvent gels and aqueous emulsion cleaning systems. The pyrrolidinone solvent gel, the Pemulen TR-2 and benzyl alcohol (5%) emulsion, and free solvents (acetone/ethanol 1:1) were determined to be the most effective cleaning systems depending on the tenacity of the overpaint/varnish and the condition of the original paint beneath. While further research is needed to effectively study potential residues associated with Pemulen TR-2 emulsions, initial findings indicate that starting reagents and polymeric components are unlikely to cause adverse effects to an aged oil paint film. This study demonstrates that multiple cleaning systems are often necessary when faced with the removal of multiple generations of overpaint and varnish as well as proteinaceous residues left behind from previous consolidation attempts.

Acknowledgements

We would like to extend our thanks to project interns Emily Wroczynski, Maggie Bearden, Ellen Nigro, Allison Rabent, Keara Teeter, Serena Vella, Claire Burns, and Katie Rovito; volunteers Sarah Steffan, Tia Polidori, Jenna Borges, Morgan Shankweiler, Emily Immel, Stephanie Carrato, Sarah Beaver, Rachael Modrovsky, Joanna Hurd, and Elizabeth McDermott; and Villanova University students Kristen Watts, Rachel Godat, Menika Dirkson, and Alexandra Stump; for their help and dedication throughout the project. Other individuals at Villanova University who were instrumental throughout the project include David Lacey, Dr Anthony Lagalante, Dr Amanda Norbutus, Brad Thorstensen, Fr Kail Ellis, Dr Timothy McCal, Darren Polley, and Margaret Duffy. Thanks to generous funding provided by the Samuel H. Kress Foundation, the conservation team was able to carry out full-scale X-radiography of the painting (together with GE Energy Management staff James McKinney III and Mark Shurtleff) which revealed multiple significant changes to the composition. Dr Erich S. Uffelman from Washington and Lee University also donated his time and equipment to help the conservation team perform infrared reflectography for the entire painting.

Notes

1. The total cost of the restoration in 1956 was $898.80 (McClees Galleries 1956).
2. McClees Galleries Restoration and Fine Art Gallery, invoice submitted to Villanova, 16 June 1956. D. Falvey, letter to Prince Alexis Droutzkoy, 27 January 1956, in Falvey Memorial Library Archives, Villanova University.
3. A letter in the Falvey Memorial Library Archives, written on 7 December 2001 by Raymond Spiller and Associates Inc. to Fr Richard Cannuli describes the painting: 'Very serious overall large and small areas of darkened overpainting. The entire composition is affected by this black and dark brown staining. It appears to be the overpainting and in-painting from earlier restoration work. Many of the figures were over-cleaned, resulting in major paint loss. None of this is reversible. In my opinion, the value of the painting in its present condition is approximately $5,000 ... However, I do feel that some sections and details could be salvaged, restored and panel-mounted.'
4. In February 2013, Fr Richard Kail Ellis notified the Winterthur/University of Delaware Program in Art Conservation about the poor condition of the painting. Eventually, Dr deGhetaldi was placed in charge of the conservation treatment of the painting, assisted by Brian Baade and several interns and volunteers over the span of two years (2013–15).
5. M. O'Malley, and S. Steffan, pers. comm., 17 January 2015.
6. Orsini family papers, *c.*1150–1950. Box 15, Folder 1. Charles E. Young Research Library Special Collections. University of California Los Angeles, Los Angeles, CA.
7. A number of art historians were consulted throughout this project, including Dr Jorg Martin Merz, Dr David Stone, Dr Anna Lo Bianco, Dr Sergio Guarino, Dr Catherine Puglisi, Dr Richard Spear, Dr Ursula Verena Fischer Pace, Dr Ann Sutherland Harris, and Dr Andrea Bayer.
8. X-radiography was conducted using a Spellman LORAD LPX160 tube in conjunction with high resolution 35 × 41 cm Flex XL Blue digital imaging plates. Over 120 X-radiographic captures were collected and compiled for the final image. For infrared reflectography an IR camera equipped with an indium gallium arsenic (InGaAs) detector was used to collect over 220 images.
9. See Voras *et al.* (2016) for information regarding sample preparation and instrumental/photography parameters used to perform optical microscopy. In preparation for SEM–EDX, excess casting medium was removed from the cross-sections which were subsequently coated with carbon paint. Samples were examined using a Zeiss EVO MA15 SEM with LaB6 electron beam at an accelerating voltage of 20 kV (stage height of approximately 11 mm with a sample tilt of 0°). The EDSX data was collected with the Bruker Nano X-flash detector 6–30 and analyzed with Quantax 200/Esprit 1.9 software. With special thanks to Catherine Matsen for assisting with SEM–EDX at the Scientific Research and Analysis Laboratory (SRAL) at Winterthur Museum.
10. The samples were placed onto a diamond cell (Spectra Tech) and analyzed in transmitted mode using a Nicolet Nexus 670 optical bench equipped with a Continuum microscope (collecting 128 scans at 4 cm^{-1} resolution). With special thanks to Kristen Watts, a chemistry graduate student at Villanova University, for assisting with FTIR analysis. Final interpretation of the spectra was conducted using the IRUG database at SRAL.
11. With many thanks to Professor Richard Wolbers (University of Delaware) for sharing his invaluable expertise on Pemulen cleaning formulations.
12. For Lubrizol see 'Emulsification properties', *Pharmaceutical Bulletin* 8, 2008 at https://www.lubrizol.com/en/Life-Sciences/Products/Pemulen-Polymeric-Emulsifiers/Pemulen-TR-2-NF-Polymer, and for Noveon see 'TDS-114: Introducing Pemulen polymeric emulsifiers', 2010 at https://www.lubrizol.com/en/Life-Sciences/Products/*Pemulen*-Polymeric-Emulsifiers/Pemulen-TR-2-NF-Polymer.
13. With many thanks to Dr Christian W. Petersen and SRAL for assisting with GC–MS. Samples were analyzed using a Hewlett-Packard 6890 gas chromatograph equipped with 5973 mass selective detector (MSD) and 7683 automatic liquid injector. For samples suspected to contain oils,

a MethPrep II reagent (Alltech) was added to convert carboxylic acids and esters to their methyl ester derivatives. Samples were then placed in tightly capped, heavy-walled vials (100–300 μL) and approximately 100 μL (or less for a smaller sample) of 1:2 MethPrep II in benzene was added. The vials were warmed at 60 °C for one hour in the heating block, removed from heat, and allowed to cool. Analysis was carried out using an inlet temperature of 300 °C and transfer line temperature to the MSD (SCAN mode) of 300 °C. A sample volume (splitless) of 1 μL was injected onto a 30 m × 250 μm × 0.25 μm film thickness HP-5MS column (5% phenyl methyl siloxane at a flow rate of 2.3 mL/minute). The oven temperature was held at 55 °C for two minutes, then programmed to increase at 10 °C/minute to 325 °C where it was held for 10.5 minutes for a total run time of 40 minutes. Samples suspected of containing protein were heated for 24 hours at 105 °C with 5.5 M HCl in tightly capped, heavy-walled vials (100–300 μL). The samples were then evaporated to dryness with a stream of air or N_2. Approximately 50–100 μL of the silylating reagent (MTBSTFA + 1 percentTBDMCS, Pierce Chemical Co.) was added and the closed vials were heated at 60 °C for one hour. Analysis was carried out with an inlet temperature of 300 °C and transfer line temperature to the MSD (SCAN mode) of 300 °C and the same column type and injection process. The oven temperature was held at 50 °C for two minutes, then programmed to increase at 10 °C/minute to 325 °C where it was held for 10.5 minutes for a total run time of 40 minutes.

14. With many thanks to Dr Thomas Beebe, Jnr and the Surface Analysis Facility at the University of Delaware for assisting with ToF-SIMS analysis of cross-sections. See Voras *et al.* (2016) for information regarding sample preparation and instrumental parameters used to perform ToF-SIMS analysis.

References

Bates-Batcheller, T. 1911. *Italian Castles and Country Seats*. New York: Longmans, Green, and Co.

CIR 2011. Cosmetic Ingredient Review 2011. 'Crosslinked alkyl acrylates as used in cosmetics.' Draft report presented at CIR Expert Panel Meeting, 3–4 March 2011. Available at: http://www.cir-safety.org/sites/default/files/118_draft_crosslinked_0.pdf (accessed 20 May 2017).

Dorge, V. (ed.) 2004. *Solvent Gels for the Cleaning of Works of Art: The Residue Question*. Los Angeles: Getty Conservation Institute.

Dotan, A. 2014. 'Biobased thermosets', in H. Dodiuk and S. Goodman (eds), *Handbook of Thermoset Plastics*. San Diego: William Andrew and Elsevier Inc., 582–3.

Green, C. 2007. *Roman Religion and the Cult of Diana at Aricia*. Cambridge: Cambridge University Press.

Guarino, S. and Masini, P. 1997. *Pietro da Cortona, il meccanismo della forma: ricerche sulla tecnica pittorica*. Milan: Electa.

New York Times 3 March 1901. Don Enrico Ruspoli Weds.; Is Married in Washington to Mrs. Bruton of Nashville, Tenn. Available at: http://query.nytimes.com/gst/abstract.html?res=990DEED61039E733A25750C0A9659C946097D6CF&url=http:%2F%2Ftimesmachine.nytimes.com%2Ftimesmachine%2F1901%2F03%2F03%2F101069985.html&legacy=true# (accessed 20 May 2017).

Ravenel, N. 2010. 'Pemulen® TR-2: An emulsifying agent with promise,' *WAAC Newsletter* 32(3): 10–12. Available at: http://cool.conservation-us.org/waac/wn/wn32/wn32-3/wn32-304.pdf (accessed 2 May 2017).

Voras, Z., deGhetaldi, K., Baade, B., Gordon, E., Gates, G. and Beebe, T. 2016. 'Comparison of oil and egg tempera paint systems using time-of-flight secondary ion mass spectrometry', *Studies in Conservation* 61: 222–35.

Wolbers, R. 2000. *Cleaning Painted Surfaces: Aqueous Methods*. London: Archetype Publications.

Authors' addresses

- Kristin deGhetaldi, Department of Art Conservation, University of Delaware, Newark, DE, USA (kdeghetal@yahoo.com)
- Brian Baade, Department of Art Conservation, University of Delaware, Newark, DE, USA (bbaade@udel.edu)
- Zachary Voras, Surface Analysis Facility, Lammot DuPont Laboratory, Dept. of Chemistry and Biochemistry, University of Delaware, Newark, DE, USA (voras@udel.edu)
- Emily Wroczynski, Department of Art Conservation, Colonial Williamsburg, Williamsburg, VA, USA (ewroczyn@gmail.com)

The influence of organic and inorganic alkalis on the formulation and properties of Pemulen TR-2 gels

Sofia Hennen, Francisco Mederos-Henry, Cécile de Boulard, María-Fernanda Espinosa and Paolo Cremonesi

Introduction

Pemulen TR-2 is a polyacrylic acid-polyacrylate (C10–C30) ester block copolymer widely used in the field of cultural heritage conservation. The nature of this macromolecule is partly responsible for the thickening, emulsifying, surfactant and chelating properties of Pemulen gels used when cleaning various types of surfaces. The formulation of a Pemulen gel involves the neutralization of the block copolymer with an alkali in an aqueous environment. Generally, triethanolamine (TEA) is the chosen alkali (Wolbers 2013a; Stavroudis 2014). However, TEA has been considered a potential source of residues, possessing long-term deleterious effects (Burnstock and Learner 1992; Bischoff and Erhardt 1994; Tumosa *et al.* 1999).

Although other alkalis have been proposed as substitutes for TEA (Cremonesi 2011; Ahedo Pino and Cantos Martínez 2014), no comparative information regarding their influence on the formulation or working properties of Pemulen hydrogels could be found in the scientific literature (Lubrizol 2009). As a consequence, conservators are lacking crucial scientific

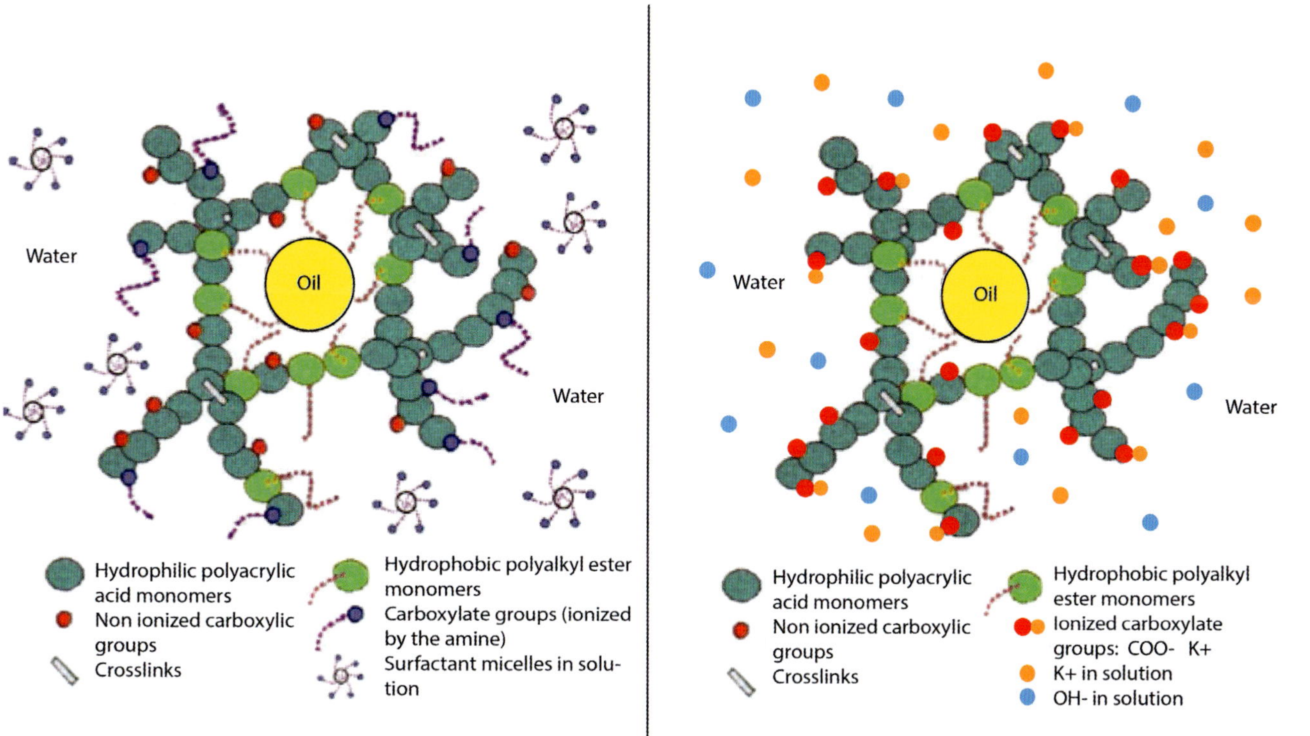

FIG. 1 Neutralization reaction of Pemulen TR-2 with an organic (*left*) and an inorganic (*right*) base (adapted from Wolbers 2013b).

evidence to guide them when choosing alternative alkalis for Pemulen-based preparations. In order to solve this problem, we compared four organic bases (TEA, Ethomeen C-25, Tris and Bis-Tris) and three inorganic bases (potassium hydroxide (KOH), sodium hydroxide (NaOH) and ammonia hydroxide (NH_4OH)) that could be used for the formulation of Pemulen gels. The impact of each base on the preparation, stability, rheological properties and emulsifying capabilities of the cleaning gels was evaluated. The potential risks posed by Pemulen gels for pictorial layers – such as residues or chemical interactions with pigments – were also reviewed.

Physicochemical properties of the selected alkalis

The selected alkalis present different physicochemical properties that could influence Pemulen hydrogel properties and their ensuing cleaning action. Some of these properties include their reactivity, mobility, penetrability and retention. By considering these properties, we can make assumptions on the behaviour of the Pemulen gels and the way these bases could alter pictorial layers during the cleaning process.

Organic alkali

Amines are weak organic alkalis that can accept H^+ ions depending on the availability of their lone N-electron pair. When the polyacrylic acid is neutralized by an amine, the resulting species are partially dissociated in the aqueous environment and exist in both ionic and neutral form. The concentration of each species will be determined by the gel pH and the alkali pK_a values. If present in sufficient amounts, the non-ionized free amine bases form micelle-like structures that might improve the polymer ability to emulsify a non-polar phase (Fig. 1, left) (Wolbers 2013b). The non-ionized molecules can also act as a solvent, being active in the cleaning process (Ravenel 2010), or offer supplementary surfactant and chelating actions during cleaning, especially if there is an amine surplus.

But the use of amines can also pose several risks to the treated surface. They can penetrate and cause swelling and leaching of components from pictorial layers, particularly if they possess a reduced volume. Furthermore, if the amine is viscous (as is TEA) and has a low vapour pressure, it evaporates slowly from the cleaned surface, increasing the risk of leaving residues behind. TEA is a very polar tertiary amine with three primary alcohol end groups. Consequently, it has a similar solvent action to other alcohol solvents and is able to esterify and swell varnish or oil layers (Wolbers 2013a). TEA is also a penetrating molecule with a high retention rate (Tumosa *et al.* 1999). Given its proven affinity for the pictorial layer (Burnstock and Learner 1992), this amine can leave residues that can be absorbed by the oil medium, producing swelling.

Ethomeens C-12 and C-25 are polyethoxylated amines that are generally used to neutralize solvent Carbopol gels. In this study, the Ethomeen molecules have the highest molecular weight of all the alkalis used. Consequently, several size-related effects are to be expected. For example, the alkali molecules might diffuse with greater difficulty into the Pemulen entangled structure, causing a slower neutralization of the polyacrylic acid moieties. In fact, given that both Pemulen and Ethomeen are bulky molecules, their resulting hydrogels are expected to have limited penetration into a porous substrate assuming the pores are of a similar macromolecular size (Cremonesi 2004). Ethomeen degrades fairly rapidly to volatile components and eventually disappears from pictorial layers (Carlson and Petersen 2004). In addition, Miller and Stulik (2004) have proved that residues are satisfactorily removed by appropriately rinsing polyacrylic gels neutralized with Ethomeen C-25.

In our study, tromethamine (Tris or Trizma) is the only primary amine and it possesses the highest pK_a value. Being smaller and more compact than the other alkaline amines used, this molecule might penetrate the pictorial layers faster. This could pose a potential risk given that primary amines are expected to be more reactive than their secondary or tertiary counterparts.[1] Also, for a given metal ion and under similar pH conditions, a stronger chelating action can be expected from primary amines given its greater lone electron pair availability. Nonetheless, no specific information on potential chemical interactions between this molecule and pictorial components could be found.

Bis-Tris methane is a tertiary, less polar amine. Theoretically, since this molecule is larger than TEA or Tris, it should be less mobile and should penetrate the pictorial layer with greater difficulty. This amine appears to have been little used in the formulation of conservation gels and, like Tris, no information on its potential risks is provided by the scientific literature.

Mineral alkali

NaOH and KOH are strong bases. They dissociate totally and exist as highly mobile and reactive ions in an aqueous environment. In a polyacrylic acid system, once the OH^- ions have neutralized the carboxylic groups (COOH), the positively charged K^+ and Na^+ ions are attracted to the neutralized carboxylate ions (COO^-). However, as shown in Fig. 1 (right), a portion of K^+, Na^+ or unreacted OH^- ions might remain freely in the gel structure, becoming potentially problematic for conservation purposes (Wolbers 2013b). Indeed, given their extremely small size, these species have an increased facility to migrate through pictorial layers when water-based conservation operations are performed (e.g. rinsing of a gel). In addition, depending on the system pH, OH^- ions can readily produce hydrolysis reactions in the oil medium and induce certain pigment degradation reactions. Monovalent cations (K^+, Na^+) could associate with ionized acidic groups present on pictorial layers, forming salts.

NH_4OH is an extremely volatile base. For this reason, different authors (Ahedo Pino and Cantos Martínez 2014; Cremonesi 2011) recommend its use. However, it poses

TABLE 1 Volumes and molar concentrations of base needed to reach neutral pH 7 using 1 g Pemulen TR-2 and 100 ml deionized water.

Base	CAS no.	Molecular weight (g/mol)	pK_A	Initial concentration used (%)	Initial molar concentration used (mol/L)	Final neutralization volume (ml)	Final base molar concentration (mol/L)
KOH	1310-58-3	56.08	14.4	18	3.21	3.6	0.11
NaOH	1310-73-2	39.99	14.8	18	4.5	2.1	0.09
NH4OH	1336-21-6	17	9.25	28	14.79	0.75	0.11
TEA	203-049-8	149.19	7.76	99	7.43	1.2	0.09
Ethomeen C25	61791-14-8	789 – 957	7	98	1.18 (average)	6	0.06
Tris	77-86-1	121.14	8.1	40	3.3	4	0.13
Bis-Tris	6976-37-0	209.24	6.46	70	3.35	7.5	0.23

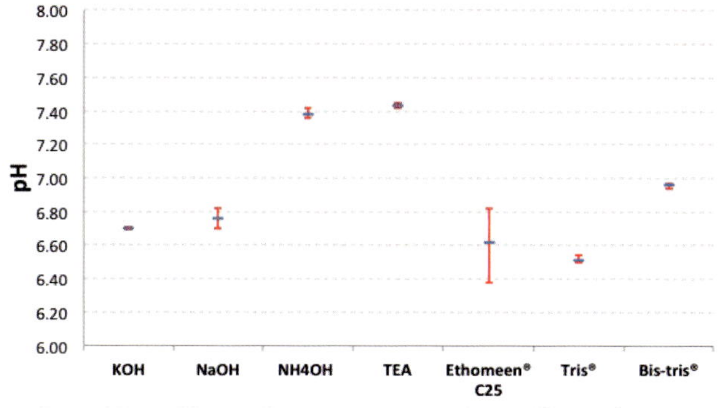

FIG. 2 5-hour average pH (horizontal blue bars) and standard deviation (vertical red bars) for Pemulen hydrogels when neutralized with different alkalis.

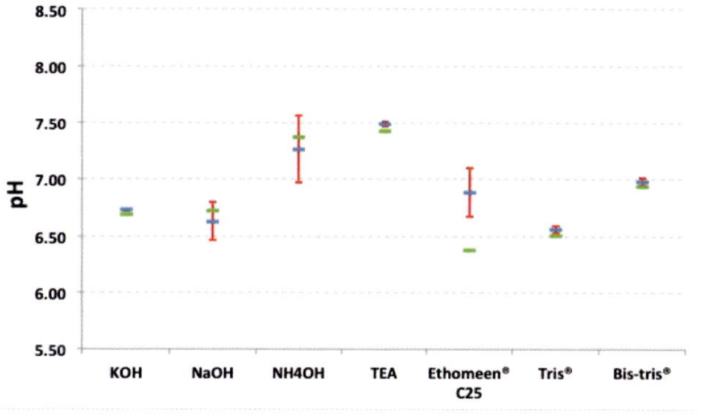

FIG. 3 7-day average pH (horizontal blue bars) and standard deviation (vertical red bars) for Pemulen hydrogels when neutralized with different alkalis. The horizontal green bar indicates the pH after one month.

an acute inhalation risk to the user. Furthermore, given its high volatility, it could compromise the long-term pH stability of a gel as it gradually evaporates. It can also complex some metals such as copper or mercury, and may irreversibly alter pictorial layers containing such metallic pigments (Cremonesi 2004).

Experimental comparison of selected alkalis

In this study, we used five criteria to compare the influence of each alkali on the behaviour of Pemulen gels: the alkali volume needed to neutralize the Pemulen macromolecule; the pH homogeneity and stability in the gel network; the rheological properties of the different Pemulen systems; their emulsifying action; and their tendency to leave residues. Results related to each criterion are discussed below.

Alkali volume

Depending on some of the physicochemical characteristics discussed above, different volumes of alkali are needed to neutralize a given amount of Pemulen TR-2. Using the least possible amount of alkali limits some of the risks posed to the cleaned surface while also reducing potential residues. Nonetheless, these volumes can vary according to different authors' recipes (Wolbers 2013b; Cremonesi 2011; Stavroudis 2014) and technical datasheets, even for a same particular base (Lubrizol 2009; Noveon 1999a). Accordingly, titration curves were obtained for each base to determine the minimal base volume necessary to neutralize an identical quantity of Pemulen TR-2. These curves also make it possible to observe differences in the Pemulen hydrogel buffering behaviour depending on the alkali used. A HI9126 pH meter with a HI1053B contact electrode (Hanna Instruments, USA) adapted to semi-solids and non-polar substances was used, with calibration performed before and after each titration.

As shown in Table 1, the final alkali molar concentration required to produce a neutral Pemulen hydrogel is similar for most bases and ranges from 0.06 M to 0.13 M. The only alkali presenting a significant difference is Bis-Tris (0.23 M). This can be explained by this alkali's pK_a value, which is the lowest of all. The final alkali concentrations are all extremely low (0.1 M average value), which is reassuring in terms of residue and other possible risks for pictorial layers. It is of note that these minimal volumes were found to be generally lower than those reported in the scientific literature (Ravenel 2010; Cremonesi 2011; Cremonesi and Signorini 2012; Wolbers 2013a; Ahedo Pino and Cantos Martínez 2014).

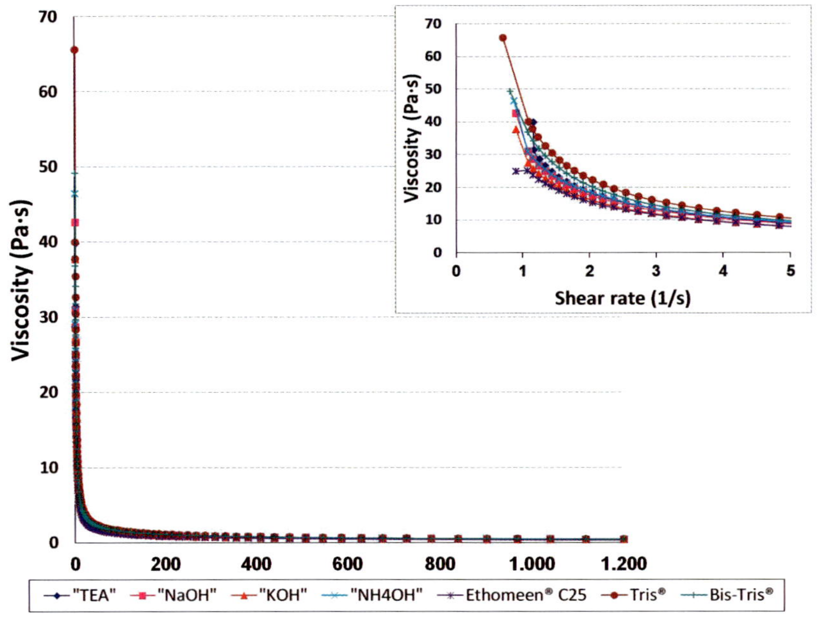

FIG. 4 Viscosity flow curves for the Pemulen hydrogels neutralized with different alkalis. Inset shows their rapid shear thinning behaviour.

pH homogeneity and stability

Periodical pH measurements allow for the evaluation of the system's homogeneity and stability. This gives information on the capacity of Pemulen TR-2 to contain the bases, especially those with high volatility. Two series of hydrogels gels were prepared to gain an initial insight into the reproducibility of the results. Measurements on these 14 gels were performed with the pH meter described above, and systematic calibrations were done between each measurement. Different measurement frequencies were employed to discriminate pH variations due to local heterogeneities in the gels from long-term effects. pH measurements were performed every hour for five hours in a single day, and then once per day for one week. Measurements were repeated after a month.

Using the shortest measurement timeframe (e.g. every hour), local heterogeneities were revealed. As shown in Fig. 2, the gel containing Ethomeen C-25 presented the largest pH variations. This might be due to steric hindrance effects because Ethomeen C-25 is the largest molecule tested. Its large volume might reduce its mobility and diffusion into the polyacrylic acid macromolecular network. Consequently, its homogeneous dispersion in the gel system might be compromised, creating the observed differences.

Our results also show that the different hydrogels are stable over longer time periods (e.g. 7 days and after one month), with pH values varying only slightly (Fig. 3). In the inorganic alkalis, the most important variations were found when using NH_4OH. However, this variability cannot be attributed to the volatilization of the alkali: firstly because the hydrogel pH does not gradually decrease as the daily measurements are performed (results not shown), and secondly because one month later, the pH value is within the range of values obtained during the daily measurements. When comparing the different amines, Ethomeen C-25-based gels again show the greatest variations. Further research is needed to explain the observed trends in NH_4OH and Ethomeen C-25 hydrogels.

Nonetheless, according to these results, it is important to perform pH measurements before using Pemulen gels, especially when cleaning a pH-sensitive substrate or when the maintenance of a particular pH over a lengthy cleaning procedure is necessary. It is recommended to stir the gels before use to promote homogenous re-partition of the base in the polymeric macromolecular network, especially in those hydrogels neutralized with ammonia or Ethomeen C-25.

Rheological behaviour of the gels

Viscosity can influence a Pemulen hydrogel emulsion capacity as well as solvent or water retention (Lubrizol 2009). Also, appropriate rheological behaviour enables localized and vertical gel application while cleaning, avoiding run-offs. In this study, the influence of each alkali on the viscosity of the Pemulen hydrogels was measured with a Carri-Med CSL 100 (Gemini BV, The Netherlands) rotational cylinder rheometer.[2]

As shown in Fig. 4, the rheological behaviour is very similar for all the alkalis. When left undisturbed, they have high viscosity, allowing excellent water retention and vertical and horizontal hold. But when they are increasingly sheared, a comparable action to the movement of a brush or a cotton swab while cleaning, the gels soon exhibit low viscosity. Nevertheless, gels neutralized with Ethomeen C-25 possess the lowest viscosity at rest, meaning that they have the lowest vertical and horizontal hold. This behaviour should be considered in cases where, for example, cleaning is performed vertically and/or long dwell times are needed.

TABLE 2 Selection of base depending on cleaning criteria.

Cleaning with Pemulen TR-2 gels	
Criteria	**Base**
To avoid residues	NH_4OH
Ensure a buffering capacity	NaOH
	KOH
To fabricate a basic gel (pH 7–8.5)	Tris
To fabricate a hydrogel containing an aqueous solution	NaOH
	KOH
	NH_4OH
To fabricate an o/w (oil-in-water) emulsion	NaOH
	KOH
	NH_4OH
	Tris
	Bis-Tris
To fabricate a w/o (water-in-oil) emulsion	Ethomeen C-25
To increase the system's emulsifying capacity	Ethomeen C-25
To fabricate a viscous gel suitable for vertical surfaces, without spreading or sagging, with good water retention	NaOH
	KOH
	NH_4OH
	Tris
	Bis-Tris
To avoid toxicity	Ethomeen C-25
To avoid risks to metallic pigments	Ethomeen C-25

Emulsion capacity

The emulsifying capacity of Pemulen TR-2 is one of its most important properties, since it allows for the creation of oil-in-water emulsions. Thus, non-polar solvents can be added to the Pemulen hydrogels, allowing for useful alternative formulations when seeking to remove non-polar materials. However, no consistent data related to the amount of non-polar solvent that Pemulen is able to emulsify could be found in the literature (Noveon 2009b; Lubrizol 2008; Tegeli 2011; Stavroudis 2012; Ahedo Pino and Cantos Martínez 2014).

We studied the amount of non-polar solvent that is emulsifiable by Pemulen TR-2 when using each alkali. We also studied the effect of pH on this emulsification capacity. Two series of Pemulen (0.5% w/v concentration) hydrogels were prepared – in the first series, a neutral pH was created while in the second the pH value was set to 5. Afterwards, volumes from 1 to 6 mL of white spirit were slowly added. Between each solvent volume addition, the gels were energetically shaken until all the solvent was incorporated into the system.

Stable emulsions were produced with both mineral and organic bases and the emulsifying capacity was very similar for most alkalis. Neutral gels remain stable as emulsions with 45–46% v/v of white spirit while gels at pH 5 emulsify slightly higher amounts (48–50% v/v) of this non-polar solvent. No significant relationship was found between the nature of the alkali (e.g. organic or inorganic) and the emulsion capacity of the hydrogels, indicating that the system's ability to emulsify white spirit is mainly due to the Pemulen polymer. The only significant difference was found with Ethomeen C-25, which showed the best emulsifying capacity, incorporating up to 88% v/v of white spirit. It seems that this amine base considerably enlarges the Pemulen TR-2 emulsion capacity, acting as a co-surfactant.

Residue observations

We observed samples cleaned with gels containing a mineral base (NaOH) and two organic bases (TEA and Ethomeen C-25) with a FESEM Ultra55 (Zeiss, Germany) scanning electron microscope (SEM).[3] For each type of alkali, two Pemulen hydrogel removal methods were tested and observed: dry removal with a cotton swab and dry removal followed by an aqueous rinse. None of the observations allowed us to confirm definitively that gel residues were present on the samples. Several types of deposits and surface features were observed. However, it was impossible to determine if these were a result of a residual substance. Another method worth exploring is the use of time-of-flight secondary-ion mass spectrometry (TOF–SIMS). In theory, this highly sensitive technique could characterize a hydrogel residue on the cleaned surface with micrometre resolution.

Conclusion

Pemulen TR-2 gels have interesting properties that can be exploited to solve complex cultural heritage cleaning issues. However, the formulation of these systems demands the use of an alkali, and choosing an appropriate base is crucial. Our results have shown that several inorganic and amine alkalis are suitable candidates for Pemulen hydrogel formulations. Indeed, they form stable formulations with appropriate pH and rheological properties. Furthermore, as observed with Ethomeen C-25, the hydrogel capacity to form emulsions with organic solvents can be increased.

Our results indicate that other alkalis, which may pose a lesser risk to the treated surfaces, can easily replace TEA in Pemulen formulations. Nonetheless, further studies on the chemical interactions of the alkalis used in this study and the materials found in artworks should be performed. We believe that each alkali could be exploited for a particular cleaning task. As we propose in Table 2, one or several of the tested alkalis could be employed to formulate the Pemulen hydrogels, depending on the requirements of the cleaning intervention.

Materials and suppliers

- Bis-Tris ≥98.0% (titration), Tris® ≥99.9% (titration): Sigma-Aldrich, B-3090, Overijse, Belgium
- Ethomeen C-25, Pemulen TR-2: CTS, 36077, Altavilla Vicentina, Italy
- KOH, NaOH: Merck, B-3090, Overijse, Belgium
- NH_4OH Normapur: VWR (B-3001, Leuven, Belgium
- TEA, white spirit: Kremer Pigmente GmbH & Co. KG, DE-88317 Aichstetten, Germany

Notes

1. See https://www2.chemistry.msu.edu/faculty/reusch/virttxtjml/amine1.htm.
2. All viscosity measurements were made with a disc diameter of 40 mm, an angle between the cone and the plate of 1.56°, a distance between the cone and the plate of 56 μm, and cone-plate geometry, by Prof. Simon Deroover, Coatings Laboratory of the Institut Meurice, Avenue Émile Gryzon 1, B-1070, Brussels, Belgium.
3. All SEM imaging was carried out with a secondary electron (SE2) detector using 1 kV electron acceleration and an average working distance of 5.0 mm, by Prof. Francisco Mederos-Henry, Materials and Process Engineering Institute (IMAP), Place Sainte Barbe 2, Université Catholique de Louvain, B-1348, Louvain-la-Neuve, Belgium.

References

Ahedo Pino, C. and Cantos Martínez, O. 2014. 'Nuevos emulsionantes poliacrílicos en la formulación de sistemas de sistemas acuosos de limpieza: Pemulen™ TR-2', *Informes y trabajos, Ministerio de Educación Cultura y Deporte: Subdirección General de Documentación y Publicaciones* 10: 166–86.

Bischoff, J.J. and Erhardt, D. 1994. 'The roles of various components of resin soaps, bile acid soaps and gels, and their effects on oil paint films', *Studies in Conservation* 39 (1): 3–27.

Burnstock, A. and Learner, T. 1992. 'Changes in the surface characteristics of artificially aged mastic varnishes after cleaning using alkaline reagents', *Studies in Conservation* 37(3): 165–84.

Carlson, J. and Petersen, W.C. 2004. 'Aging characteristics of surfactants', in V. Dorge (ed.), *Solvent Gels for the Cleaning of Works of Art: The Residue Question*. Los Angeles: Getty Conservation Institute, 85–110.

Cremonesi, P. 2004. *L'uso di tensioattivi e chelanti nella pintura di opere policrome*, I Talenti, metodologie, techniche e formazione nel mondo del restauro, no.10. Padua: Il Prato.

Cremonesi, P. 2011. *L' ambiente acquoso per la pulitura di opere policrome*. Padua: Il Prato.

Cremonesi, P. and Signorini, E. 2012. *Un approccio alla pulitura dei dipinti mobili*. Padua: Il Prato.

Lubrizol 2008. *Emulsification Properties*. Technical Bulletin 8. Available at: https://www.lubrizol.com/-/media/Lubrizol/Life-Sciences/Documents/Literature/Bulletin/Bulletin-08---Emulsification-Properties.pdf (accessed 2 September 2015).

Lubrizol 2009. Technical Data Sheet 237. Available at: https://www.lubrizol.com/-/media/Lubrizol/Life-Sciences/Documents/TDS/Neutralizing-Carbopol-and-Pemulen-in-Aqueous-and-Hydroalcoholic-Systems.pdf (accessed 4 June 2017).

Miller, D. and Stulik, D. 2004. 'Research into solvent gel residues', in V. Dorge (ed.), *Solvent Gels for the Cleaning of Works of Art: The Residue Question*. Los Angeles: Getty Conservation Institute, 18–65.

Noveon 1999a. *TDS-114: Introducing Pemulen™ Polymeric Emulsifiers*. Available at: http://pemulentr2.pbworks.com/f/PemulenTR2.pdf (accessed 2 September 2015).

Noveon 2009b. TDS-237: *Neutralizing Carbopol® and Pemulen™ Polymers in Aqueous and Hydroalcoholic Systems*. Available at: https://www.lubrizol.com/Home-Care/Documents/Technical-Data-Sheets/TDS-237-Neutralizing-Carbopol-Pemulen-in-Aqueous-Hydroalcoholic-Systems--HC.pdf (accessed 2 September 2015).

Ravenel, N. 2010. 'Pemulen® TR-2: an emulsifying agent with promise', *WAAC Newsletter* 32(3): 10–12.

Stavroudis, C. 2012. 'Pemulen revised: pHuck the pH meter', *WAAC Newsletter* 34 (2): 19.

Stavroudis, C. 2014. 'Pemulen TR-2: the once and future king (of conservation)', *WAAC Newsletter* 36(2): 10–11.

Tegeli, V.S. 2011. 'Pemulen as a versatile emulsifier', *International Journal of Drug Formulation & Research* 2(1): 52–63. Available at: https://www.ordonearresearchlibrary.org (accessed 3 August 2016).

Tumosa, C.S., Millard, J., Erhardt, D. and Mecklenburg, M.F. 1999. 'Effects of solvent on the physical properties of paint films', in J. Bridgland (ed.), *Preprints of the ICOM-CC 12th Triennial Meeting*. London: James & James, 347–52.

Wolbers, R. 2013a. *Le nettoyage des surfaces peintes: méthodes aqueuses*, S. Mirabaud and L. Desvois (trs). Paris: Eyrolles.

Wolbers, R. 2013b. 'Nuevos métodos de limpieza de superficies. Pintura, papel, textil, escultura', workshop organized by INTI and International Academic Projects in Buenos Aires, 4–8 February 2013.

Authors' addresses

- Sofia Hennen, Conservation et restauration des oeuvres d'art, École Nationale Supérieure des Arts Visuels de la Cambre (ENSAV-La Cambre), Brussels, Belgium (sofia.hennen@gmail.com)
- Francisco Mederos-Henry, Conservation et restauration des oeuvres d'art, École Nationale Supérieure des Arts Visuels de la Cambre (ENSAV-La Cambre), Brussels, Belgium
- Cécile de Boulard, Conservation et restauration des oeuvres d'art, École Nationale Supérieure des Arts Visuels de la Cambre (ENSAV-La Cambre), Brussels, Belgium
- María-Fernanda Espinosa, Laboratorio de Análisis del Centro Nacional de Conservación y Restauración de Chile (CNCR), Santiago, Chile
- Paolo Cremonesi, Lodi, Italy (paolocremonesi57@gmail.com)

Giant sequoia: an extraordinary case study involving Carbopol gel

Chelsea McKibbin, Lu Allington-Jones and Efstratia Verveniotou

Introduction

In 1891 a 'Big Tree' exhibition was commissioned for the American Museum of Natural History (AMNH) and the British Museum (Natural History), today known as the Natural History Museum (NHM). A giant sequoia called 'Mark Twain', was selected from Kings Canyon National Park, California, USA. When felled, it was 1,341 years old, 101 m tall and just over 5 m in diameter. Two sections were cut for display. The bottom and slightly larger one was sent to the AMNH while the top section was split into 12 pieces: one central disc and 11 radial segments. These were shipped to Liverpool, UK, from San Francisco for the British Museum (Natural History), and arrived at the South Kensington, London, site in April 1893. The giant sequoia section went on display the following year, in one of the bays of the central hall. It was moved in 1902 to stand against the wall dividing the north and central halls, and again in 1971 to its current location on the second floor balcony (Tucker 2015) (Fig. 1).

By 2016, a substantial amount of dust had accumulated on the horizontal surfaces of the specimen, the bark suffering the most due to its fibrous nature. The wood surface had

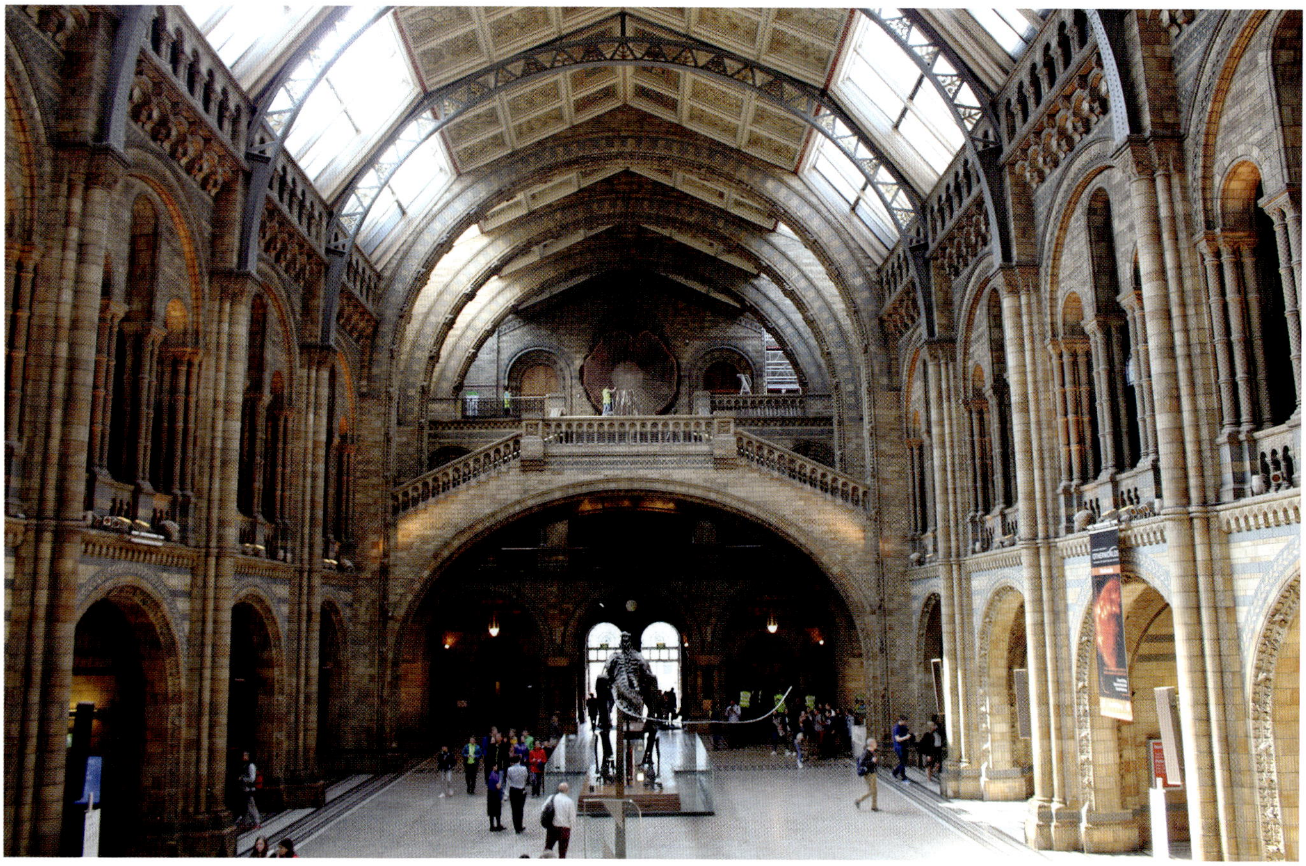

FIG. 1 The section of giant sequoia, on display in the central hall of the NHM, can be seen at the top of the image.

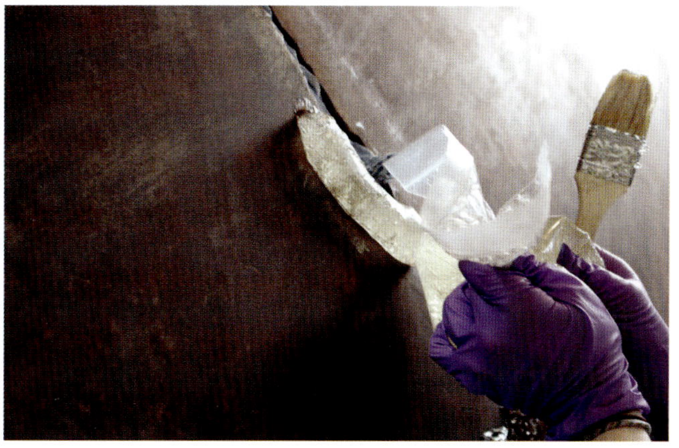

FIG. 2 Applying solvent gel directly from the mixing bag.

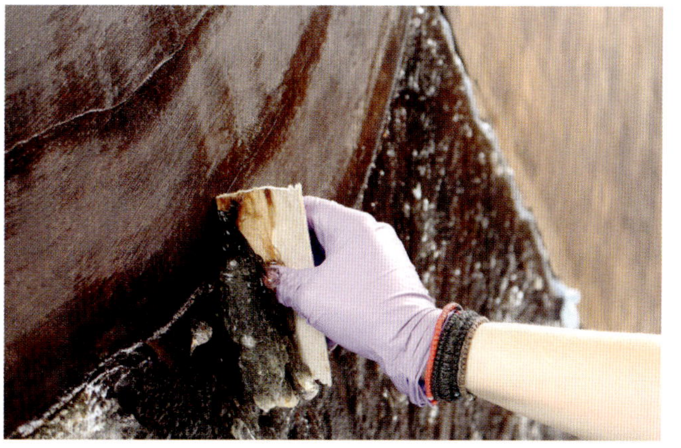

FIG. 3 Removing the solvent gel after treatment. Note the discoloration from solubilized varnish.

previously been treated with a lacquer that had become discoloured, bloomed and cracked, obscuring the growth rings. This was due to darkening with age, exposure to environmental elements and contact with moisture most probably from wiping down with damp cleaning cloths. Liquid and lacquer stains, and drips from adhesives, had greatly discoloured on the lower part of the trunk. Small cracks that appear fresh are evident at the lower part of the trunk possibly due to weight distribution. Nails used to keep the outer bark of the trunk in place are now actively corroding and protruding, and potential brown rot, a wood-decay fungus, has affected discrete areas. These agents of deterioration will need to be monitored in years to come to ensure they are no longer active.

The remit of the project was to remove the dust that had accumulated over the last 35 years, stabilize the friable bark, remove the coatings existing on the surface of the wood and apply a new layer of protective varnish. Before conservation work was undertaken, the curators were consulted and the treatment proposal agreed upon.

Risks to the specimen from conservation treatments included loss of material during dry cleaning, especially on the friable bark surface, abrasion or over-cleaning during the removal of the current varnish layer, loss of integrity to fills during the removal of the current varnish layer, darkening of surfaces during the application of adhesives and consolidants and swelling or distortion of material, in particular the bark during consolidant application. The choice of solvent for removing the old varnish and as a carrier for a consolidant and new varnish resin was therefore a key consideration.

Varnish removal

Wood is a complex mix of components with diverse solubility parameters. Cellulose, hemicellulose, lignin, suberin, cutin, pectin, tannins, wood waxes, resins and oils are adversely affected in different ways by one or more of the following substances: water, ethanol, ether, hexane, benzene, alkaline solutions, acids, enzymes, butanol, methanol, benzyl alcohol, xylene, warm glycerol, sequestering agents and calcium binders (Florian 2012: 22–5; Gilberg 1986: 180; Kronkright 2012: 153,161, 170, 171; Norton 2012a: 85; 2012b: 221).

Petroleum distillates do not swell woody tissues (Kronkright 1981; Norton 2012b: 221) and, of these, xylene appears to pose the least risk to wood, whilst mineral spirits would be the most appropriate for treating bark. The location of the sequoia, within an open gallery, meant that xylene was unsuitable for health and safety reasons (even when fume extraction systems were deployed). Conservators decided that the best solution therefore was to remove the old varnish with a solvent gel, and to discover a suitable consolidation varnish that was soluble in mineral spirits.

Using a gel system reduces the rate of evaporation and diffusion of a solvent, enabling an increased contact period. It also ensures better control over solvent penetration, reduces toxicity, reduces levels of mechanical action in comparison with continuous rolling with free solvents and provides a suitable method of application for a vertical surface. The reduction in volatility 'concentrates' the effect of the solvent, increasing the ability to dissolve the material on which it is working. It allows a solvent with a higher solubility parameter to more readily remove individual layers with potentially less hazardous solvents. The risks posed to the conservator (and the public) would also be limited due to the reduction in evaporation rate of the solvent.

Tests showed that industrial methylated spirit Carbopol gel solubilized the old varnish layer. Carbopol EZ2 (polyacrylic acid polymer) was used as the thickener, Ethomeen C-25 (tertiary amine ethocylate) as neutralizing agent for the polar solution, and low concentrations of distilled water to aid the polyacrylic acid polymer to uncoil and form a stable network (Rivers and Umney 2003). As large quantities of the gel were required, the most effective method for mixing proved to be within polyethylene zip bags, which also minimized wastage and equipment cleaning. The gel was found to achieve optimum smoothness and viscosity after standing for a minimum of 24 hours. However, temperature appeared to have an effect on the resulting gel, particularly noticeable at the time of application. The environment in the gallery fluctuated during the three-month project: whenever the temperature reached 23 °C and above, the gel became 'wetter' and made application much more difficult as it tended to slip off the surface.

The gel was applied directly from the bag by cutting one of the corners and using it to pipe onto the surface (Fig. 2). It was smoothed to a thickness of approximately 20 mm with wide bristle brushes to ensure sufficient contact and even distribution. Solubilization of the varnish became uneven if air pockets were present. The gel was covered with clingfilm to reduce evaporation, and to prevent slippage of the material, and allowed to work for a maximum of one hour. After about 45 minutes the gel became tackier which hindered the removal process. Once the varnish surface solubilized, it was drawn into the gel making it dark brown with saturation (Fig. 3). The spent gel was removed using rudimentary cardboard scrapers, following the grain of the wood. Cardboard was chosen as it was rigid enough to scrape the gel off, but soft enough not to cause abrasion to the wood surface. A second layer of fresh gel was applied to remove the remaining varnish. After the second scraping of spent gel, residue was removed by using IMS-saturated tissue poultices and wipes.

Although every attempt was made to ensure the gel was removed in its entirety, there is an understanding that there may be residual remnants. Attempts at using barrier materials to prevent the risk of contamination proved troublesome and ineffective. Several attempts of various tissue applications were trialed, but maintaining contact on the vertical surface was impossible, resulting in slumping of the material or full

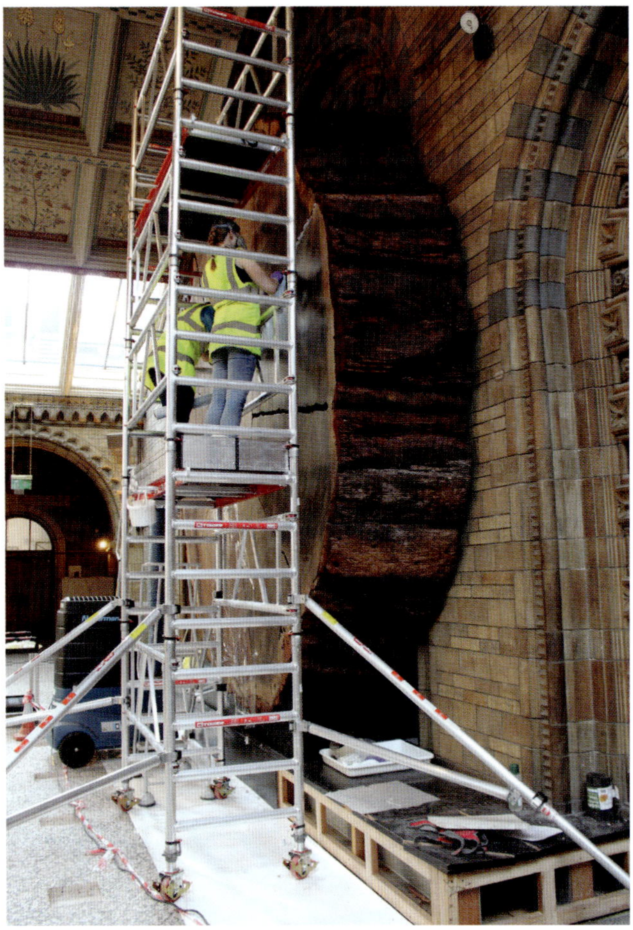

FIG. 4 Working on the scaffolding tower within the gallery.

detachment. Prior wetting with IMS improved gel permeation but contact remained insufficient. In order to remove the varnish, complete contact was necessary, and the success of direct application outweighed the risk of minimal residue.

As the varnish was removed it became apparent that an application of coloured wax had been applied to the top section of the sequoia. This had occurred sometime in the past 40 years as a consequence of various restoration treatments applied to the top and bottom halves of the section. This was removed by applying white spirit to the wax with a stiff brush, followed by a gentle application of soft steel wool following the grain of the wood to agitate the wax. Finally, using IMS-saturated towels, it was wiped until its colour was minimized.

Choice of resin

The next step was to evaluate conservation-grade resins that are soluble in white spirit or Stoddard solvent: Paraloid B-67, ketone resins (polycyclohexanes) and urea-aldehyde resins. Paraloid B-67 is soluble in white spirit with the addition of an aromatic hydrocarbon but it crosslinks under ultraviolet (UV) exposure and may be prone to attracting dust and fracturing (Rivers and Umney 2003: 595). Ketone resins were formerly used in paintings conservation in preference to more stable poly (vinyl acetate) or polymethymethacrylates due to their optical properties (de la Rie and Shedrinsky 1989: 9). Ketone resins, however, are susceptible to photoxidation and lose solubility, requiring more polar solvents for removal (Horie 1987: 116; de la Rie 1988). De la Rie and McGlinchy (1990: 168) suggest using a hindered amine light stabilizer (HALS) such as Tinuvin 292 which acts as an antioxidant by scavenging free radicals (von der Goltz *et al.* 2012: 642). One ketone resin which has been used extensively in paintings conservation (and was formerly a component of BEVA 371) is Ketone N. This was discontinued but replaced by Laropal K80 (Stonor and Rushfield 2012: 623), now also discontinued. Laropal K80 has been replaced in BEVA 371b with an unspecified aldehyde ketone resin (Ploeger *et al.* 2015: 220).

McIntyre (2011) suggests Laropal A81 as an alternative to ketone resins (a low molecular weight urea-aldehyde resin created by a condensation reaction between aliphatic aldehydes and urea). Urea-aldehyde resins have been used as consolidants and adhesives for wood (Horie 1987: 177) and Laropal A81 possesses excellent photochemical stability, resolubility after accelerated light aging, and good wetting capabilities (Leonard *et al.* 2000: 111; von der Goltz *et al.* 2012: 648). It was developed specifically for conservation, in conjunction with conservation scientists (Digney-Peer *et al.* 2012: 624), possesses lower gloss than similar varnishes and produces an even spread due to its low molecular weight (von der Goltz *et al.* 2012: 649). It is also supposed to promote the natural grain of wood (AIC 1998) and gives good colour saturation (von der Goltz *et al.* 2012: 648). Accelerated aging tests of Laropal A81 not only confirmed good photochemical stability but also, to a certain extent, a protective effect from photochemical deterioration of pigments (de la Rie *et al.* 2000). Arslanoglu and Learner (2001: 62; 63) and von

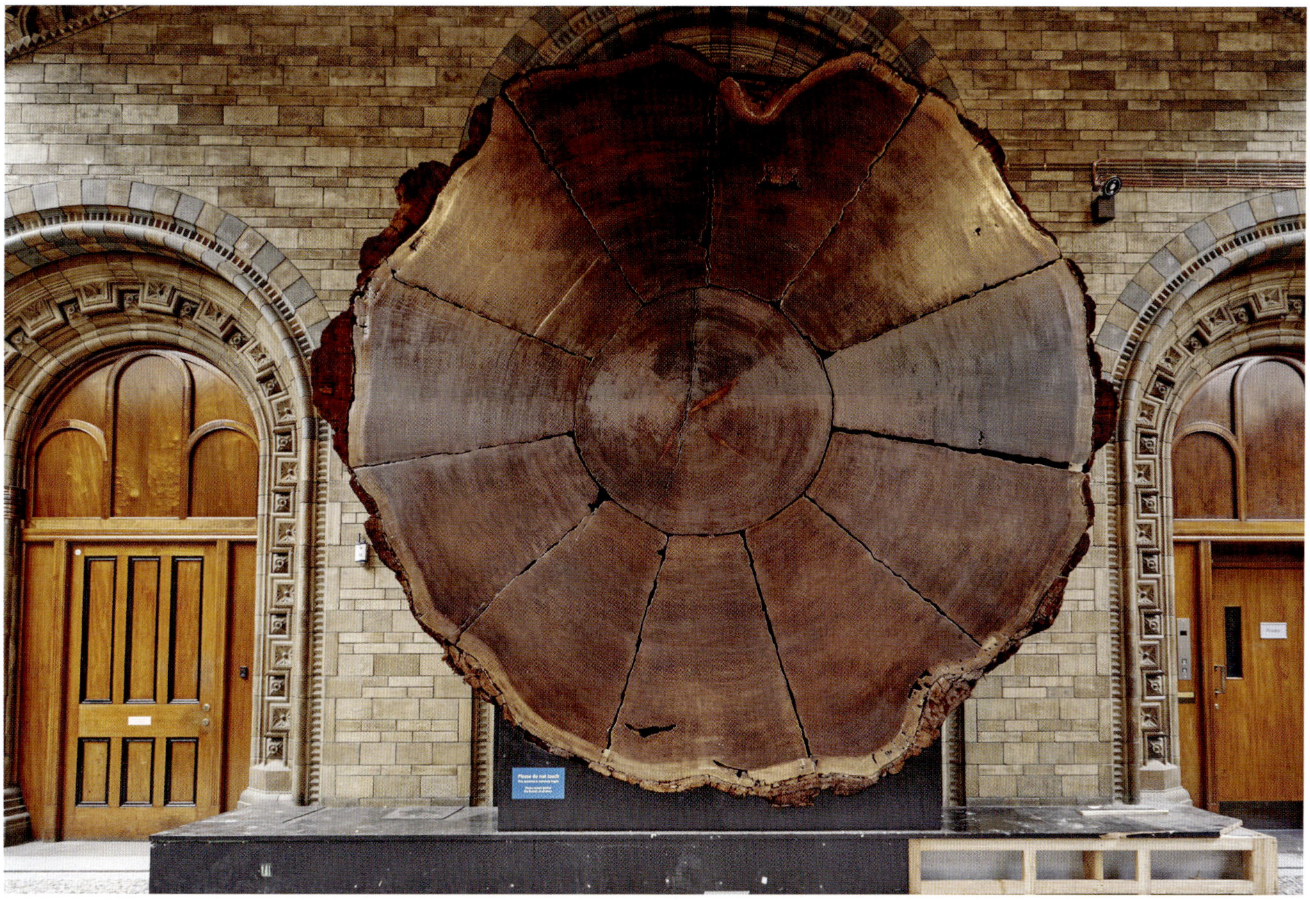

FIG. 5 The section of giant sequoia after conservation treatment.

der Goltz *et al.* (2012: 648) consider Laropal A81 to be the most stable commercially available aldehyde resin and one of the most stable resins known to conservators. Laropal A81's solubility in aliphatic solvents such as white spirit is, however, limited, since it forms a 2-phase gel and solvent mixture (von der Goltz *et al.* 2012: 648). Such solutions can be stabilized by the addition of 2–5% of an aromatic solvent such as Solvesso 100 (BASF 2000). Leonard *et al.* (2000: 113) found a mixture of 1:4 isopropanol in mineral spirits, Arcosolve (propylene glycol monomethyl ether) or equal parts isopropanol, petroleum benzine and Shell TS-28 to be effective solvents. De la Rie *et al.* (2000) used cyclohexane or acetone with toluene as solvent mixes.

Whereas it could be argued that any consolidant is not practically reversible from the bark, some consolidation of the sequoia is essential for continued preservation and stability (especially since the specimen will remain on open display), and choosing the most stable resin is essential. Continued resolubility after aging tests is an indication that crosslinking, chain scission, oxidation and the formation of radicals and other forms of deterioration, which could lead to tackiness, colour change and an alteration in crystallinity (Cruikshank 1985: 55) have not occurred. This is especially important since these could lead to further forms of deterioration such as the attraction of dust, the formation of catalysts for different reactions, and a cracked surface that would trap harmful pollutants, oxygen and moisture (Cruikshank 1985: 57; 63). Therefore 8% Laropal A81 was chosen for consolidation of the bark, dissolved in a 20% white spirit and 1% isopropanol mix.

Paraloid B-72 has high chemical stability and is soluble in xylene, and would therefore be suitable for use for varnishing the wood, but it has high molecular weight which means it exhibits poor wetting on surfaces where old varnishes may have been unevenly removed, and it also may attract dust (Rivers and Umney 2003: 594–5). In addition it must not be layered with a resin in mineral spirits because fogging will occur (Rivers and Umney 2003: 594). The conservation team therefore decided to use Laropal A81 (20%) as a varnish for the wood, which has the added advantage of reducing the number of chemicals introduced to the specimen.

Conclusion

The main challenges faced during this project derived from the sheer size of the sequoia transverse section. The specimen had to be treated *in situ* within the gallery, whilst the lower levels of the hall remained open to the public. The health and safety risks associated with using large quantities of hazardous chemicals, on a scaffold with a 6 m working height (Fig. 4), in a partially open gallery, with conservation of the adjacent mosaic floor occurring below, required careful consideration and management. The volume of solvent gel required was considerable, especially since two applications of a 20 mm layer proved necessary. The authors are very grateful to colleagues who gave up their time to help mix up the copious amounts of gel. The most effective method for making and applying the gel proved to be in zip-lock bags, and the texture of the gel itself improved the longer it was made in advance. The solvent gel was quite labour intensive to remove from the surface of the specimen, especially when it began to dry. An incremental approach to uncovering the solvent gel, scraping away the excess with cardboard spatulas and then cleaning the surface with IMS, before the gel residue dried, proved to be the most viable option. Once the varnish and waxy substance had been removed, the specimen face was sealed and the bark was stabilized using Laropal A81. The final phase was an application of microcrystalline wax using a microfibre cloth, to improve lustre and protection.

Overall the polar solvent gel proved extremely effective at removing the old varnish and revealed the growth rings which had been obscured for decades (Fig. 5). The gel system enabled use of a solvent that posed less risk to the specimen and to staff, and allowed this large-scale treatment to be undertaken within the gallery while it remained open to the public.

Material and supplier

- Laropal A81: Stort Chemicals Ltd, www.stortchemicals.co.uk

References

AIC 1998. *Paintings Conservation Catalogue. Volume 1: Varnishes and Surface Coatings.* Washington, DC: American Institute for Conservation Paintings Specialty Group.

Arsanoglu, J. and Learner, T. 2001. 'The evaluation of Laropal A8: Paraloid B-72 polymer blend varnishes for painted and decorative surfaces: appearance and practical considerations', *The Conservator* 25(1): 62–72.

Cruikshank, W.A. 1985. 'A comparative investigation of two polycyclohexane resins: Ketone Resin- N and Laropal- K80'. A report submitted to the Department of Art in conformity with the requirements for the degree of Master of Art Conservation, Queen's University, Kingston, Ontario, Canada.

BASF 2000. *Laropal A81.* Available at: http://talasonline.com/photos/instructions/Larapol_A81.pdf(accessed 16 September 2015).

de la Rie, E.R. 1988. 'Polymer stabilizers: a survey with reference to possible applications in the conservation field', *Studies in Conservation* 33: 9–22.

de la Rie, E.R and McGlinchy, C.W. 1990. 'New synthetic resins for picture varnishes', in J.S. Mills and P. Smith (eds), *Cleaning Retouching and Coatings.* London: IIC, 168–73.

de la Rie, E.R. and Shedrinsky, A.M. 1989. 'The chemistry of ketone resins and the synthesis of a derivative with increased stability and flexibility', *Studies in Conservation* 34: 9–19.

de la Rie, E.R., Quillen Lomax, S., Palmer, M., Deming Glinsman, L. and Maines, C.A. 2000. 'An investigation of the photochemical stability of urea-aldehyde resin retouching paints: removability tests and colour spectroscopy', in A. Roy and P. Smith (eds), *Tradition and Innovation: Advances in Conservation.* London: IIC, 51–9.

Digney-Peer, S., Thomas, K., Perry, R., Townsend, J.H. and Gritt, S. 2012. 'The imitative retouching of easel paintings', in J.H. Stoner and R. Rushfield (eds), *Conservation of Easel Paintings.* Oxford: Routledge, 607–34.

Florian, M-L. E. 2012. 'Plant anatomy: an illustrated aid to identification', in M-L. E. Florian, D.P. Kronkright and R.E. Norton (eds), *The Conservation of Artifacts made from Plant Materials.* Los Angeles: Getty Conservation Institute, 1–27.

Gilberg, M. 1986. 'Plasticization and forming of misshapen birch-bark artifacts using solvent vapours', *Studies in Conservation* 31: 177–84.

Horie, C.V. 1987. *Materials for Conservation: Organic Consolidants, Adhesives and Coatings.* London: Butterworths.

Kronkright, D.P. 1981. 'New directions in Native American basketry conservation', *Preprints of the American Institute for Conservation 9th Annual Meeting.* Washington, DC: AIC, 95–108.

Kronkright, D.P. 2012. 'Deterioration of artifacts made from plant materials', in M-L. E. Florian, D.P. Kronkright and R.E. Norton (eds),*The Conservation of Artifacts made from Plant Materials.* Los Angeles: Getty Conservation Institute, 139–93.

Leonard, M., Whitten, J., Gamblin, R. and de la Rie, E.R. 2000. 'Development of a new material for retouching', in A. Roy and P. Smith (eds), *Tradition and Innovation: Advances in Conservation.* London: IIC, 29–33.

McIntyre, C. 2011. 'Development of a pigmented wax/resin fill formulation for the conservation of paintings'. Available at: http://www.conservationcolors.com/documents/McIntyre_695_To_Gamblin_2014.pdf (accessed 14 September 2015).

Norton, R.E. 2012a. 'Technology of plant materials used in artefacts', in M-L. E. Florian, D.P. Kronkright and R.E. Norton (eds), *The Conservation of Artifacts made from Plant Materials.* Los Angeles: Getty Conservation Institute, 83–138.

Norton, R.E. 2012b. 'Conservation of artifacts made from plant materials', in M-L. E. Florian, D.P. Kronkright and R.E. Norton (eds), *The Conservation of Artifacts made from Plant Materials.* Los Angeles: Getty Conservation Institute, 195–285.

Ploeger, R., McGlinchey, C.W. and de la Rie, E.R. 2015. 'Original and reformulated BEVA 371: composition and assessment as a consolidant for painted surfaces', *Studies in Conservation* 60: 217–26.

Rivers, S. and Umney, N. 2003. *Conservation of Furniture.* Oxford: Butterworth-Heinemann.

Stonor, J.H. and Rushfield, R. 2012. *Conservation of Easel Paintings.* Oxford: Routledge.

Tucker, P. 2015. *The Giant Sequoia*, Getting to Know the Collections series. London: Natural History Museum.

von der Goltz, M., Proctor, R.G. Jnr., Whitten, J., Mayer, L. and Myers, G. with Hoenigswald, A. and Swicklick, M. 2012. 'Varnishing as part of the conservation treatment of easel paintings', in J.H. Stoner and R. Rushfield (eds), *Conservation of Easel Paintings.* Oxford: Routledge, 635–57.

Authors' addresses

- Chelsea McKibbin, The Conservation Centre, The Natural History Museum, London, UK (c.mckibbin@nhm.ac.uk)
- Lu Allington-Jones, The Conservation Centre, The Natural History Museum, London, UK
- Efstratia Verveniotou, The Conservation Centre, The Natural History Museum, London, UK

Revisiting a shipwrecked felt hat for Chatham Historic Dockyard Trust: a multidisciplinary approach

Deborah Phipps and Jonathan Clark

Introduction

In 2015 Norfolk Museums Conservation Service (NMCS), Norfolk, UK, won a tender to prepare material for a new display at Chatham Historic Dockyards (CHD) in Kent, UK. The aim of the CHD project was to fully record, digitize and exhibit artefacts recovered in the early 1980s from the wreck of the 74-gun flagship *Invincible*, the first of its name, which sank in 1758. As part of the project bid, CHD identified the need for remedial conservation, not only for display but for 'objects deemed to be most at risk which are not going on display but require stabilization treatment'.

A wide variety of objects was identified for conservation including wooden tankards and trenchers, cartridge holders, scrubbing brushes, leather shoes, and textile pieces including stockings, a silk neck tie and a knitted cap. Many of the objects, which had lain buried for centuries in the silt of the eastern Solent, had undergone extensive post-excavation conservation treatments in the 1980s. The results of these treatments varied considerably: some had sympathetically preserved the shipwrecked and waterlogged material, while others had left thick, sticky oil residues, excess adhesive, migrating salts and misshapen objects.

Once the material and documentation had been examined, treatment work was delegated between the NMCS objects and textile conservators, except for one object which is the focus of this paper. The object in question was a soldier's felted wool hat (Fig. 1), one of only two recovered from the wreck. The hat most likely belonged to one of the many troops that the *Invincible* was transporting to Canada. It was among the first of the objects to have been lifted from the wreckage site before excavations had begun, as it had been found 'lying on the seabed surface' (Bingeman 2010:135). CHD had noted that although the hat was not destined for exhibition, they wanted it to be accessible and visually improved, but not necessarily for long-term display. The desire to improve its appearance was not surprising, as the hat had undergone an unsympathetic and distracting treatment, involving bulky gap fills and an adhered lining.

Treatment planning

After an initial assessment, the textile conservator presented this curious treatment to the objects conservator, as it was something she had never come across before. Internally, the hat had been coated with numerous layers of a synthetic paper-like material which had been adhered with a thick application of an undocumented adhesive. The hat had been gap-filled using three pieces of painted cardboard and several sections of blue felt fabric which had been painted brown and adhered with the same adhesive. This previous treatment had left the hat hardened and misshapen. Research conducted by the textile conservator confirmed that the form in which the hat had been 'reconstructed' was also incorrect. Illustrations (Bingeman 2010: 136) confirmed that the hat's correct form had a domed rather than a flat crown, and a flat rather than a curved brim.

After numerous discussions between the objects and textile conservators about how the appearance and stability of

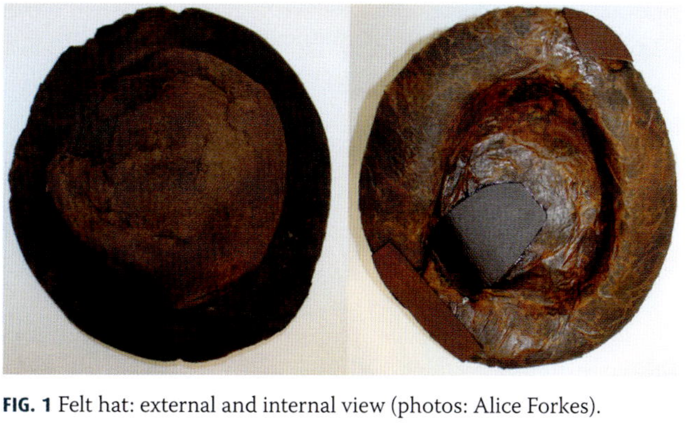

FIG. 1 Felt hat: external and internal view (photos: Alice Forkes).

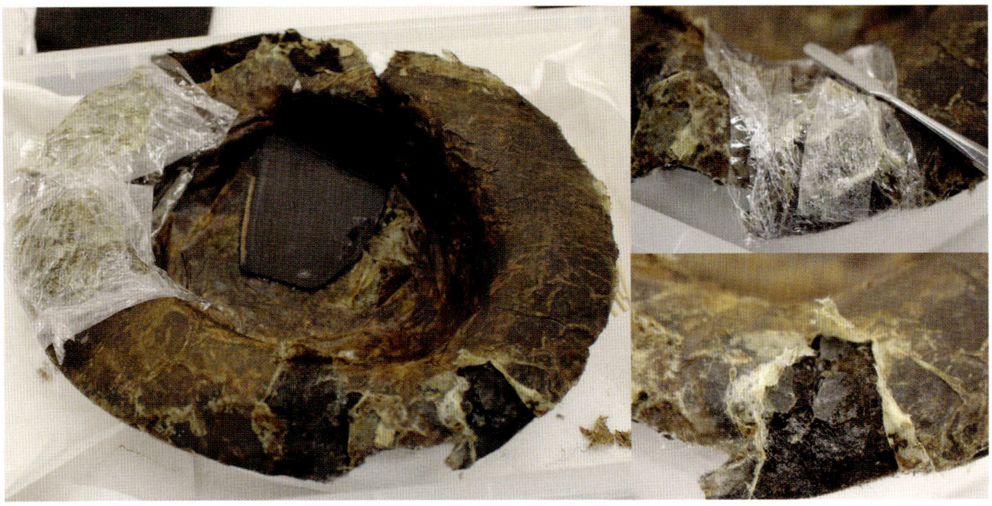

FIG. 2 Removal of the lining using solvent gel, cotton wool swabs and small hand tools (photos: Jonathan Clark).

the hat could be improved, it was agreed that approaching the treatment from both specialisms would be beneficial. Several treatment options were formulated and presented to CHD, which highlighted the potential risk involved in reversing the unsympathetic conservation/restoration treatment. Treatment options ranged from preventive (constructing suitable packaging and returning the hat to storage) to invasive: completely removing the obtrusive and distorting paper lining, felt and cardboard gap fills, and reshaping, gap-filling and mounting the hat. The methodology for this extensive treatment, which was the one favoured by CHD, was formed by an interdisciplinary approach.

The greatest challenge was going to be the removal of the lining. The aim was to minimize damage caused to the hat during this process, and to reduce the movement of the adhesive further into the felt. After numerous tests, using a variety of solvents on cotton wool swabs, acetone proved to be the most effective solvent for softening the adhesive. Initially encouraged by these tests, an estimate of five working days was made, which included the treatment and the manufacture of a basic mount.

After initial attempts to remove a section of the lining, it was soon discovered that the number of layers present and the quantity of adhesive used had been underestimated. The lining appeared to have been applied randomly with a mixture of large and small overlapping pieces, with copious adhesive sandwiched between each layer. Acetone swabs were ineffectual as, unless the swab was soaked with a large quantity of acetone, it evaporated too rapidly, causing the swab to stick to or drag the softened adhesive. There was also concern that applying larger quantities of acetone, via brush or pipette, would allow it to soak into the felt. A polar solvent thus applied to an organic material might have a detrimental effect.

With these issues in mind, the objects conservator introduced the possibility of using a combination of acetone swabs and solvent gels to provide a controlled method of prolonged exposure to acetone while minimizing capillary action and the migration of the adhesive further into the felt, allowing expeditious removal of individual layers of lining. The use of solvent gels for the hat was a method transferred from the objects conservators' previous work of removing aged and discoloured varnish and overpaint from late medieval polychrome wooden altarpiece figures. This process involved the treatment of sometimes friable and porous areas of wooden substrate. In these instances, a barrier layer of close-fibred unbuffered Spider tissue served to allow the solvent vapour to soften the varnish and overpaint, while separating the gel thickeners from the object. It was this methodology that helped inform our treatment of the felt hat.

Treatment of the hat

On small test areas of the hat's lining, an acetone-based solvent gel (2 g Carbopol, 20 ml Ethomeen C-25, 100 ml acetone, 15 ml deionized water) effectively softened the adhesive enough to allow acetone swabs to be rolled under an individual layer, which could then be lifted using small hand tools. Initially, we exposed the adhered lining to the gelled solvent for approximately 5–10 minutes. This exposure was then increased to 15–20 minutes for stubbornly adhered sections. The barrier layer of Spider tissue appeared to be successful in preventing direct contact with the gel for either the lining or the felt, while allowing the active ingredient to act. Although the gel had a satisfactory viscosity, the tissue barrier layer was first wetted down on to the surface of the lining using a brush and a very small amount of industrial methylated spirit, before the gel was applied. This made the application, using a metal spatula, far easier, because the very light tissue was prone to lifting or crumpling during the gel application. The exposed side of the solvent gel was covered with Clingfilm to create a closed microenvironment and enable slow evaporation (Fig. 2).

This methodology was used successfully to remove the cardboard sections that were adhered to the surface of the lining. The blue felt gap fills were more difficult to remove as they had been integrated into the lining. The density of the

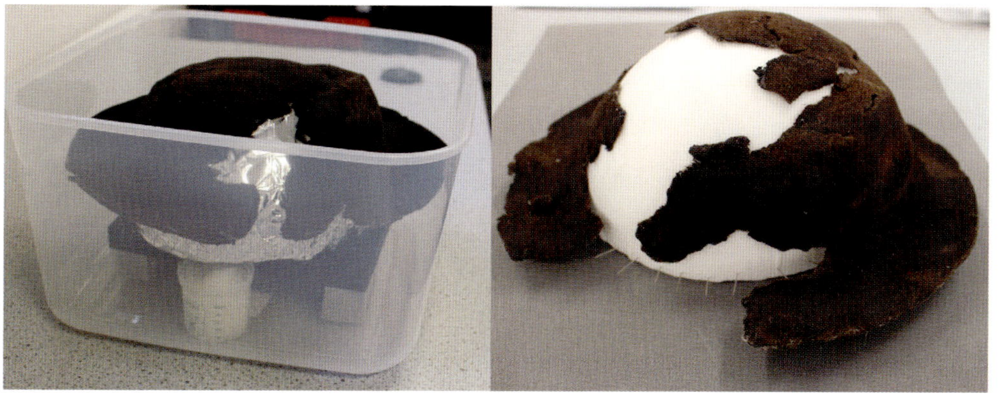

FIG. 3 *Left*: hat in acetone tent before reshaping. *Right*: reshaping and testing the fit of the Fosshape mount (photos: Deborah Phipps).

layered lining was surprising as even after prolonged exposure, the solvent vapour was unable to penetrate more than one or two layers effectively. This may have been the result of the synthetic nature of the added material, or a coating that had been applied to the lining material during manufacture. Even with the solvent gel, the removal of the lining necessitated considerable mechanical intervention using swabs, forceps and metal spatulas, and because the lining material was so difficult to tear, even when wet, scalpels and scissors.

Optimistically, we had hoped that the solidity of the felt hat was a property of the lining rather than a condition of the felt, but as the lining and majority of the visible adhesive residue was removed, just how inflexible and misshapen the hat was became apparent. It appeared that during the application of the lining, large enough quantities of adhesive had been applied to the internal surface to have penetrated the felt significantly. This meant that our plans to reshape the hat would require further intervention and solvent exposure. Solvent gels did not present a practical solution to this dilemma, but as it was deemed that the risk of further migration of the adhesive had been reduced, an enclosed acetone environment (acetone tent) was used (Fig. 3).

The hat was placed on a temporary, aluminium foil-covered internal support made from Fosshape, a shrinkable polyester felt. It was then placed inside a sealed Stewart box with 400 mL of acetone in a glass beaker. After checking at half-hour intervals it was decided that to reach the desired flexibility, the hat needed to be left overnight. This level of exposure made the hat pliable enough to start the reshaping process. Flexible weights and pins were added over a period of two days, to set the hat in the correct form.

For storage, accessibility and potential display, the textile conservator proposed again using Fosshape. This material is increasingly being used by textile conservators to construct bespoke mounts and supports for costumes and accessories. Fosshape can be formed with dry or wet heat, and it keeps its shape when cooled. It can be bonded to itself, stitched or pinned, and depending on the amount of pressure or heat applied during shaping, it can be rendered solid and stiff or remain soft and pliable. It is of archival quality and has passed the Oddy test (Amneus and Miles 2012: 4). It can also be painted with acrylic or watercolour paints.

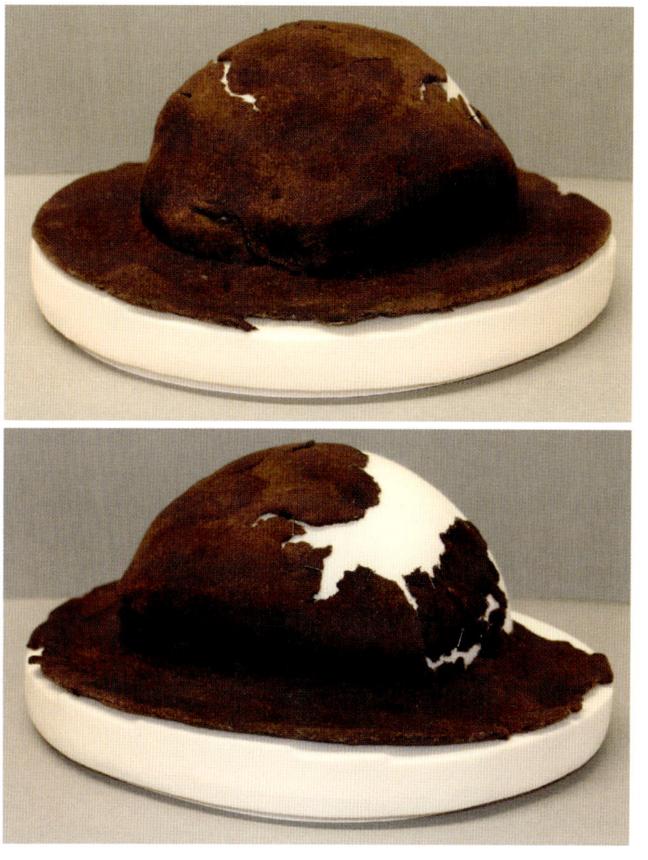

FIG. 4 Finished hat on its Fosshape mount (photos: Alice Forkes).

Using a steamer, the Fosshape was moulded over an existing papier-mâché head to achieve a suitable dome form. The form was then covered with polyester wadding and cotton jersey, to create a cushioned mount. This was set into a base of Plastazote covered with cotton jersey. During this process, some experimentation was undertaken to see if suitable gap fills for the hat could be constructed from Fosshape. Small pieces of the material were shaped, ironed, dyed and painted to determine whether sympathetic fills and colour matches could be achieved. Methods of attaching Fosshape gap fills, using Japanese paper hinges and a variety of adhesives, were also explored. Although CHD had been keen on gap-filling the hat, once it had been placed on its support, it did not seem necessary as the object was stable and readable without the addition of new material that could be confused with the actual object. This opinion was also shared by CHD having seen the mounted hat (Fig. 4).

With the removal of the lining and cardboard and felt fills, separated fragments and loose edges of the hat required securing in place. This was done using bent entomological pins pushed into the Fosshape mount. The small section of the pin visible on the outside of the hat was covered in polyethylene tubing which had been colour matched with acrylic paint.

Conclusion

The treatment of the hat has resulted in an accessible, stable and now historically accurate object, which rather than languishing in a store is now an actively used part of CHD's collection to help tell the story of the *Invincible*. Had more time and analytical resources been available during this project, it would have been interesting to have had both the adhesive and the lining material identified. This project highlighted the benefit of working in a multidisciplinary studio, where different specialisms come together to solve conservation challenges. Although the treatment took longer than originally estimated, without the experience of conservators from different specialisms, the treatment might have been even more time-consuming and far less controlled. In this instance the sharing of knowledge and materials has had a lasting impact as, for the textile conservator, Carbopol solvent gels have become a functional and accessible option for future treatments, such as adhered linings on vulnerable fabrics. Additionally, due to its many diverse applications, Fosshape has been enthusiastically adopted by the objects conservator for numerous recent projects, including the manufacture of a support for a suit of 16th-century armour.

Materials and suppliers

- Acetone: VWR, https://uk.vwr.com
- Acrylic paint: Russell and Chapple, https://www.russellandchapple.co.uk
- Carbopol EZ-2, Ethomeen C-25: Kremer, http://shop.kremerpigments.com
- Cotton jersey: Whaleys, http://www.whaleys-bradford.ltd.uk
- Entomological pins: Watdon, http://www.watdon.co.uk
- Fosshape (polyester felt): Preservation Equipment, http://www.preservationequipment.com
- Industrial methylated spirit: VWR, https://uk.vwr.com
- Plastazote: Polyformes, http://www.polyformes.co.uk
- Polyester thread: MacCulloch-Wallis, https://www.macculloch-wallis.co.uk
- Polyethylene surgical tubing: Adtech, http://www.adtech.co.uk
- Unbuffered Spider Tissue (100% manila): Conservation by Design, http://www.conservation-by-design.com

References

Amneus, C. and Miles, M. 2012. 'A method for invisibly mounting costume using Fosshape™', *Journal of the American Institute of Conservation* 53: 3–14.

Bingeman, J.M. 2010. *The First* HMS Invincible *(1747–58): Her Excavations*. Oxford and Oakville: Oxbow Books.

Authors' addresses

- Deborah Phipps, Conservation Department, Norfolk Museum Service, Norwich, UK (debbie.phipps@norfolk.gov.uk)
- Jonathan Clark, Conservation Department, Norfolk Museum Service, Norwich, UK (jonathan.clark@norfolk.gov.uk)

Carbopol gel as a carrier for oxidizing agents in discoloured lead white conversion treatment

Lana Linda Fiskovic

Introduction

The decision to treat discoloured lead white on early 19th-century Chinese botanical illustrations painted with gouache on paper, from the Kew Archives, Library and Arts department, was made in order to improve the accuracy of their depiction of botanical specimens. Lead white (basic lead carbonate) in gouache had reacted with hydrogen sulphide and changed into lead sulphide (Fig. 1). Carbopol EZ-2 gel was used as the carrier for the oxidizing agents hydrogen peroxide and sodium perborate, to convert the lead sulphide into lead sulphate without exposing the paper support to their action.

The addition of gelling agents such as Carbopol in aqueous treatments, to minimize wetting of the paper support and the graphic media and colorants, is especially useful when it is important to restrict the movement of moisture into papers that are particularly sensitive to water. Carbopol will easily mix with water into smooth, crystal-clear, stable, non-pourable gel solutions. Carbopol is a polymer compatible with many materials including non-ionic and anionic surfactants: it is even compatible with oxidizers. Polymers such as Carbopol are already in their highest oxidation state, so hydrogen peroxide cannot oxidize them further.[1] Carbopol also slows the rate of oxygen release, extending the duration of the reaction (McFarland 1997). In addition, the gel's transparency contributes to visual monitoring of the conversion process.

Sample preparation for gels' pre-treatment testing

Mock-ups were made with paper samples of similar quality, and lead white applications were prepared (following

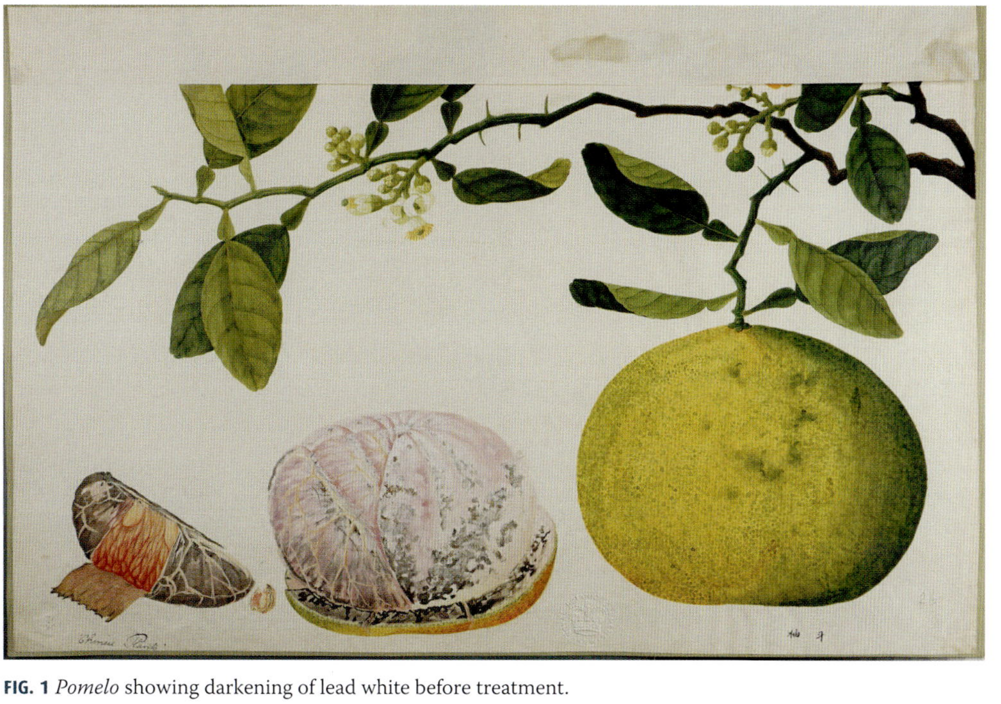

FIG. 1 *Pomelo* showing darkening of lead white before treatment.

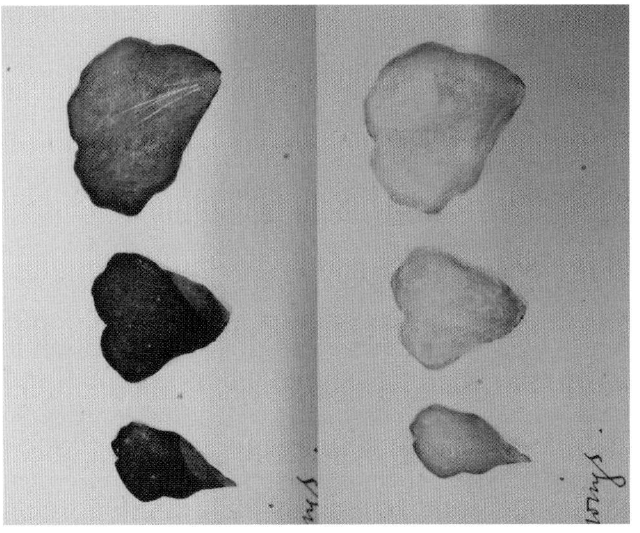

FIG. 2 Detail of another illustration before (left) and after (right) treatment.

risk assessments) for testing to identify any issues such as sensitivity of paper substrates and associated media. Visual examination and ×10–×50 magnifications were used to make a detailed evaluation of the conversion treatment results (Lussier 2006), to ensure that the gels had an efficient yet safe effect on the oxidized lead white pigments, watercolour media, and the paper support, prior to the object's treatment. The tests were also carried out to examine the paper support's absorbency to gel solutions, and to adjust the gel consistency accordingly. The lead white-based watercolour paint was made using a traditional watercolour recipe with gum arabic, applied in different layers to pH-neutral Seawhite watercolour paper (350g/m^2), with a medium cold-pressed surface (internally and externally sized) and placed in a dry oven for three weeks at 80 °C. The samples were then exposed to hydrogen sulphide gas for two hours in a fume chamber; zinc sulphide and hydrochloric acid were mixed to make hydrogen sulphide gas – the released gas from these two compounds darkened the lead white pigment.

The aged paper samples were then treated with cyclomethicone D4, to saturate the paper substrate, before applying the gel. D4 is a volatile silicone solvent which remains in the paper, keeping the gel out and on the surface during treatment; D4 evaporates within an hour at room temperature.[2]

Carbopol gel conversion treatment

Solubility testing of pigment areas with distilled water, industrial methylated spirits and a blotter were conducted to ensure that the paint layer would accept the gel and clearance solution applications. The Carbopol EZ-2 powder was prepared with sample recipes,[3] adjusting the pH to 8 and 9 with sodium hydroxide (NaOH) to create viscous solutions. Hydrogen peroxide 3% and sodium perborate 1M were added and mixed to equal viscosity.

The conversion of darkening was effective but the gel released an unacceptable amount of water to some areas of more absorbent paper substrate, and the gel consistency had to be thickened. A thicker gel solution was prepared and there was less water release, but the treatment time was prolonged and the paper absorbed the liquid from the gel accordingly. Thus, the reagents continued to function after removing the gel, making the treatment not entirely controllable. Taking this into consideration, the reaction has to be stopped before desirable results in visual appearance are achieved.

For an alternative treatment, the 1M sodium perborate was added to a thicker, 1M Carbopol EZ-2 gel, pH 9 of the gel liquid solution when prepared, which with the addition of sodium perborate was pH 10 before readjustment with NaOH back to pH 9. The conversion of darkening was less effective than with hydrogen peroxide, but satisfactory, as the conversion did not occur as quickly. The bleach also continued to react after removal and while drying, causing further lightening of the paint, but as the bleaching was not as thorough this further action achieved the desired result. The time of reaction was slower, as was the liquid absorption.

Both gel solutions were tested for efficiency when applied through lens tissue, to minimize residues in the paint layer or paper substrate. This was not successful because release through the tissue was repeatedly uneven, and left the samples with uneven spotting on the treated areas. Residues were cleared first with a dry swab for gel removal, then with a gentle swab application of distilled water:IMS 1:1, tapping the solution into the substrate. The results were observed under magnification as before but no unacceptable changes were noted on the pigment layer. Ideally, treatment of real artefacts should involve further examination under increased magnification and materials analysis to check for residues of treatment compounds.

Carbopol gel conversion treatment and evaluation

The discoloured areas were slowly converted where Carbopol was used with sodium perborate. With hydrogen peroxide the conversion was quicker and more thorough – the lightening was very intense and the agent continued working even after removal, and showed more intense results after drying. Some areas of discoloration did not react within the same timespan or repeated applications. Taking into consideration possible residues of non-neutral pH, the objects were lightly humidified with alkaline mist (calcium hydroxide pH adjusted 9–10) in a Sympatex chamber before drying in a press between Bondina blotters. The use of Carbopol EZ-2 gel as a carrier for sodium perborate has proved efficient for the conversion reaction; different recipes with the pH adjusted are advisable depending on the object's requirements (Fig. 2).

The timespan of the conversion reaction was considered. Due to its less aggressive action and longer lasting application, which exposes the paper substrate to an increased likelihood of absorption of liquid from the gel, this treatment may be

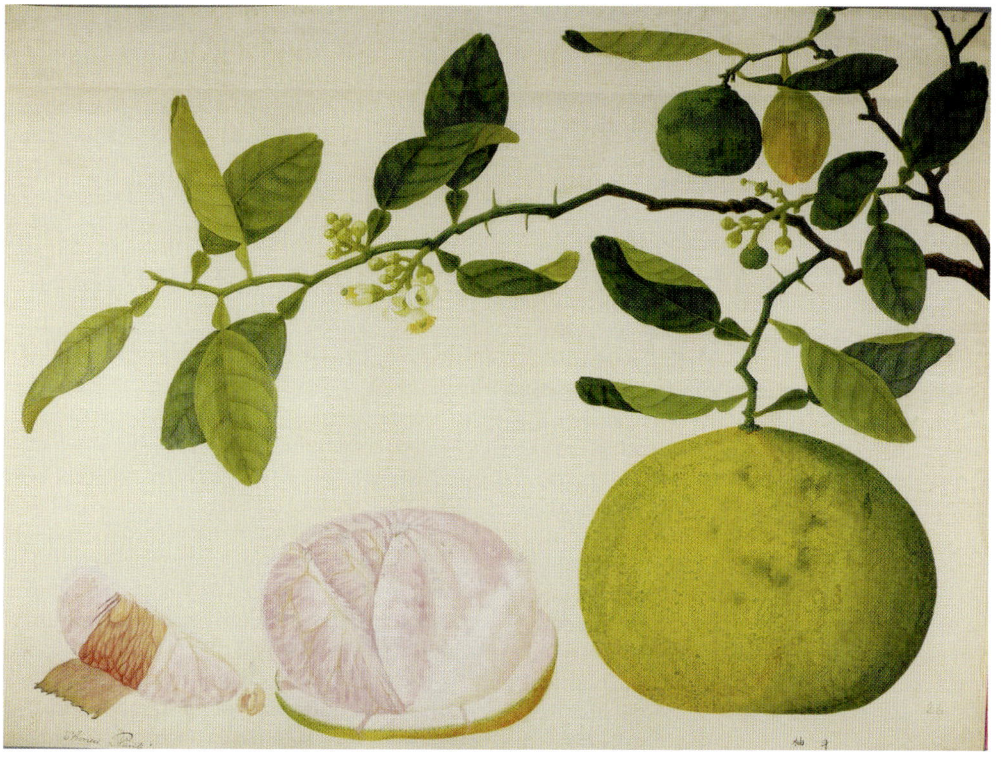

FIG. 3 *Pomelo* after treatment.

inappropriate for discoloured paper substrates where there is a risk of tideline formation and/or unbalanced expansion or cockling. A hydrogen peroxide bleaching agent with a faster action would be more efficient in these cases, but hydrogen peroxide is a powerful bleach and its reaction continues after removing the gels. Close attention to the progress of the reaction is necessary during these applications: it was important to avoid any alteration in the paint and pigment layers. On visible examination, the treatment seemed successful, as there was no undesirable visual imbalance in the graphic media and image after treatment. Discoloration was significantly reduced, making the artworks more realistic in appearance considering their scientific purpose as accurately coloured representations of short-lived botanical specimens (Fig. 3).

Notes

1. Richard Wolbers, pers. comm., March 2016.
2. *Ibid.*
3. *Ibid.*

References

Lussier, S.M. 2006. 'An examination of lead white discoloration and the impact of treatment on paper artifacts: a summary of experimental testing', *The Book and Paper Group Annual* 25: 9–12.

McFarland, M.R. 1997. 'The whitening effects of peroxide gels on darkened lead white paint', *The Book and Paper Group Annual* 16: 56–65. Available at: http://cool.conservation-us.org/coolaic/sg/bpg/annual/v16/bp16-08.html (accessed 27 May 2017).

Author's address

Lana Linda Fiskovic, Zagreb, Croatia and London, UK (lanalindaf@gmail.com)

From brown to bright: removing several varnish and dirt layers from a 17th-century ceiling painted on wood

Roos Keppler, Annefloor Schlotter and Johanneke Verhave

Introduction

The Burgerweeshuis in Zierikzee[1] houses a large room of impressive splendour, combining a 17th-century ceiling painted on wood, 18th-century gilt leather wall hangings and a marble chimney with a chimneypiece by the renowned 18th-century Dordrecht painter Aert Schouman (Fig. 1). The Burgerweeshuis was originally designed as a grand townhouse, and was home to several successive mayors in the 18th century. From the 19th until the first half of the 20th century, the building was used as an orphanage and its current name refers to this function. At the time of the orphanage the room in question functioned as the boardroom, the *Regentenkamer*, for the trustees. Today the Burgerweeshuis and the *Regentenkamer* are cared for by a board of trustees.

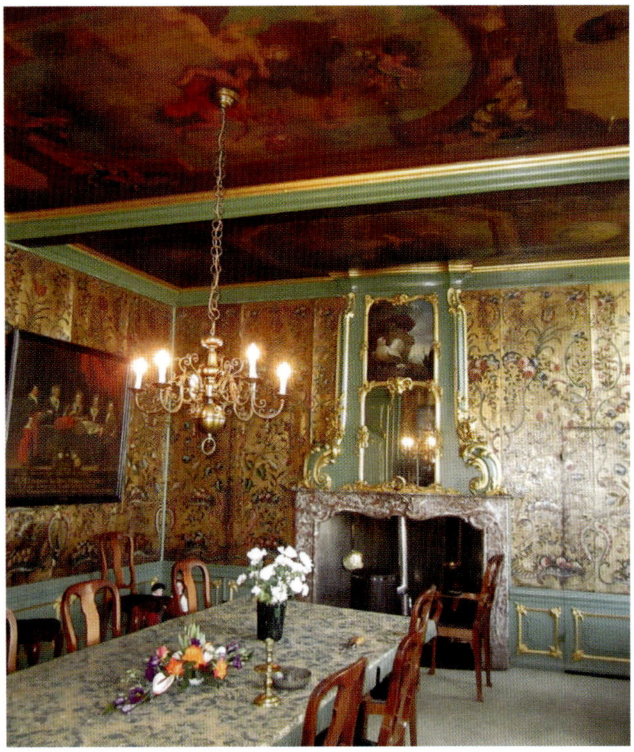

FIG. 1 Overview of the room and the ceiling: before treatment (image: R. Keppler and A. Schlotter).

The splendid boardroom has remained almost intact due to minimal interventions in the past, such as cleaning and structural treatments. However, this state of neglect called for an urgent intervention for the structural preservation of different elements. The leather wall hangings were restored in 2012 and the painted ceiling was treated between 2013 and 2014.[2] An important part of the project focused on the improvement of the suspension of the ceiling planks to the bridging structure above by replacing the fully corroded original nails with stainless steel screws. This paper focuses on the complex cleaning of the painted surface of the ceiling.

The structure of the ceiling

The ceiling is divided in three parts separated by two girders. At an early stage of the building's history, when the floorboards from the first floor, the girders and the bridging structure formed the ceiling, the ceiling had a matt, pale blue finish. At some point the three fields between the girders were covered by approximately 10 rows of pine planks, running parallel to the girders. The planks were attached to the bridging structure using large iron nails. The earliest polychrome finish on the planks is a thick, bright orange oil paint, applied in two layers, and containing the pigments vermilion and lead white. Later, the present decorative scheme was applied by an anonymous artist using oil paint, depicting personifications of gods in clouds, with *tondi* with *putti* representing the four elements in the outer corners. The painting is roughly dated 1610–80.

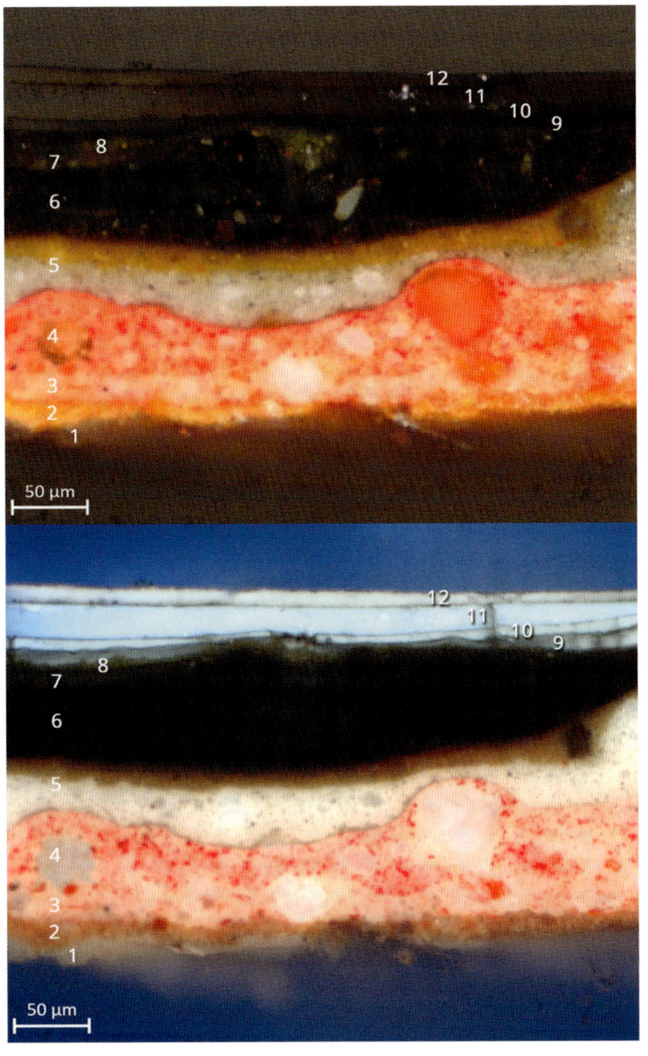

FIG. 2 Cross-section of paint from the green background in the leaves: (*top*) in visible light and (*bottom*) in UV light. On the wood substrate lies a whitish ground and an orange ground layer (1 and 2), followed by two orange red paint layers (3 and 4). These paint layers are the first decorative scheme in this paint sample. Layer 5 is a light grey ground layer on which the current figurative painting is applied (6 and 7). Overpaint (8) and four varnish layers separated by dirt layers lie on top (images: Annelies van Loon).

Layers on the painted surface

The once bright colours of the ceiling painting were disguised by a thick accumulation of dirt, varnish and overpaint. In the pilot project (2013) it was still difficult to understand the build-up of all these layers, which also complicated the cleaning, especially in the darker colours. Therefore samples were taken and analysed (Fig. 2).[3] These analyses indicated that the figurative paintings were not varnished for the first period of their existence. The first intervention was done at the same time as, or shortly after, the gilt leather wall hangings were placed in a new panelling of the room. The bright colours of the paint had already altered due to chemical reactions between the pigments and the paint. The green pigment (probably copper resinate) had darkened and turned brown, and the blue smalt lost its colour and turned grey. To brighten up the ceiling paintings large areas were overpainted at that time, directly on top of the original paint layers. Prussian blue was detected in the blue overpaints on the sky. It is likely that the ceiling was then varnished for the first time. Varnishing has been repeated three times since, each time without an (adequate) surface cleaning. The fluorescence of the four varnish layers seems to indicate that different resins were used each time (Fig. 2).

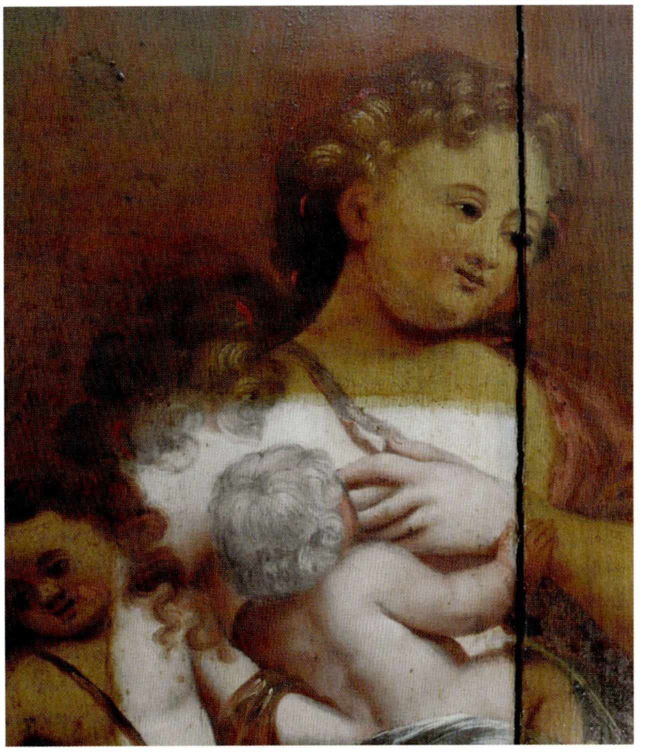

FIG. 3 Detail of the ceiling during cleaning (image: R. Keppler and A. Schlotter).

Cleaning of the ceiling: what to remove and how to achieve this

The combination of the four varnish layers and several dirt layers made the cleaning labour intensive and difficult. Part of the painting consisted of original glaze layers, which proved vulnerable to the solvents required for the varnish removal. It is not difficult to imagine that these thin glazes would have been removed had the painting been cleaned at an earlier stage in its history. Furthermore it was decided to maintain the overpaints directly on the original paint surface. This meant the cleaning could not swell this younger paint layer. Due to the large surface area (30 m^2) and the many different sensitive coloured areas, each with their own pigment and binding medium characteristics, a methodical and precise treatment approach was required.

Some areas, such as the sky, proved resistant to pure ethanol and could be cleaned with a cotton swab and fluid solvent, but the largest part of the ceilings needed to be approached with more caution. A step-by-step cleaning was needed to achieve an even result. After several cleaning tests it was found that removing the later varnish layers was best controlled by using a combination of Carbopol gels and liquid solvents. An acetone-based Carbopol gel[4] was applied to the surface with a brush to swell the top varnish layers. The swelled varnish was then removed with a swab with 3:2 liquid acetone: isopropanol. This step was repeated to dissolve the remaining varnish layers. After this second Carbopol application the surface could be cleaned using a less active mixture of 1:1 isopropanol: Shellsol D40. To ensure that only varnish layers were removed, the surface was constantly checked using handheld ultraviolet (UV) lights (Reskolux UV 365 LED), and multiple paint samples were taken before and after cleaning (Fig. 3).

Conclusion and further considerations

Our extensive series of cleaning tests convinced us that the acetone: isopropanol Carbopol gel was the best mixture to remove all the varnishes and the dirt without touching the first overpaints and the original glazes. As research had revealed that the paint layers were not varnished originally, we tried to keep the saturation of the new varnish (two layers of dammar in Shellsol D40, with the addition of UV stabiliser Tinuvin 292) to a minimum. The cleaning of the ceiling paintings in the *Regentenkamer* of the Burgerweeshuis was as rewarding as it was labour intensive, transforming it from dull brown to colourful and bright. The realization that we were exposing a paint layer that had never been cleaned before was a special experience. We felt that our cleaning method left an intact paint surface.

Materials and suppliers

- Acetone, Carbopol, Ethomeen C-25, Tinuvin 292: Kremer, www.kremer-pigmente.com
- Ethanol, isopropanol (IPA), Shellsol D40: Deffner & Johann, www.deffner-johann.de

Notes

1. The Burgerweeshuis is located on Poststraat 45 in Zierikzee, Zeeland, The Netherlands.
2. Theo Sturge and Lily Dell, UK-based private conservators, treated the wall hangings *in situ*. The commission for the treatment of the ceiling was given to Roos Keppler and Annefloor Schlotter (both conservators of painted finishes in historic interiors). The team was expanded with wood restorer Maurice Steemers and Johanneke Verhave, with the further assistance of Kate Bangerter and Julia van der Burg (paintings conservators).
3. The paint samples were prepared, photographed and a few were analysed by Prof Dr Elisabeth Jägers and Dr Erhars Jägers of Mikroanalytisches Labor, Mülheim, Germany.
4. The acetone gel was a variation on Richard Wolbers' basic Carbopol recipes, as taught in many of his and Chris Stavroudis' cleaning courses 2002–12. It was made from 15 ml acetone, 10 ml isopropanol, 0.5 g Carbopol, 4 g Ethomeen C-25 and 2 g deionized water.

Authors' addresses

- Roos Keppler, Hilversum, The Netherlands (roos@rooskeppler.com)
- Annefloor Schlotter, Haarlem, The Netherlands
- Johanneke Verhave, Rotterdam, The Netherlands

Pemulen emulsions for the removal of overpaint, discoloured remains of lining adhesive, and remnants of older varnish layers on a 17th-century group portrait

Marjan de Visser and Lucy Odlin

Introduction

This paper presents a case study of the application of aqueous gel-based cleaning systems using tailored Pemulen emulsions based on the Modular Cleaning Program (MCP) developed by Chris Stavroudis.[1] Their use is discussed in the context of the full conservation treatment of a large-scale group portrait *The Archers' Guild of Middelburg*, 1650, by Cornelius Jonson van Ceulen (1593–1661).[2] The painting (Fig. 1) depicts the Confraternity of St Sebastian and is displayed in Middelburg town hall in the Netherlands. Notably, *The Archers' Guild* maintains a direct link with the current Confraternity of St Sebastian. Founded in 1217, the Confraternity remains active today, with 24 current members. These members, passionate about their painting, were the driving force behind the conservation project. A key stipulation was that the project should be publicly accessible for the duration of the treatment.

Cornelius Jonson van Ceulen and *The Archers' Guild*

Cornelius Jonson van Ceulen found success in London as court painter to Charles I, but following political unrest and dwindling commissions he returned to his native Netherlands in the 1640s, becoming a member of the Painters' Guild in Middelburg where this painting was executed. Set against a red-brown curtain, the members of the Confraternity surround a table

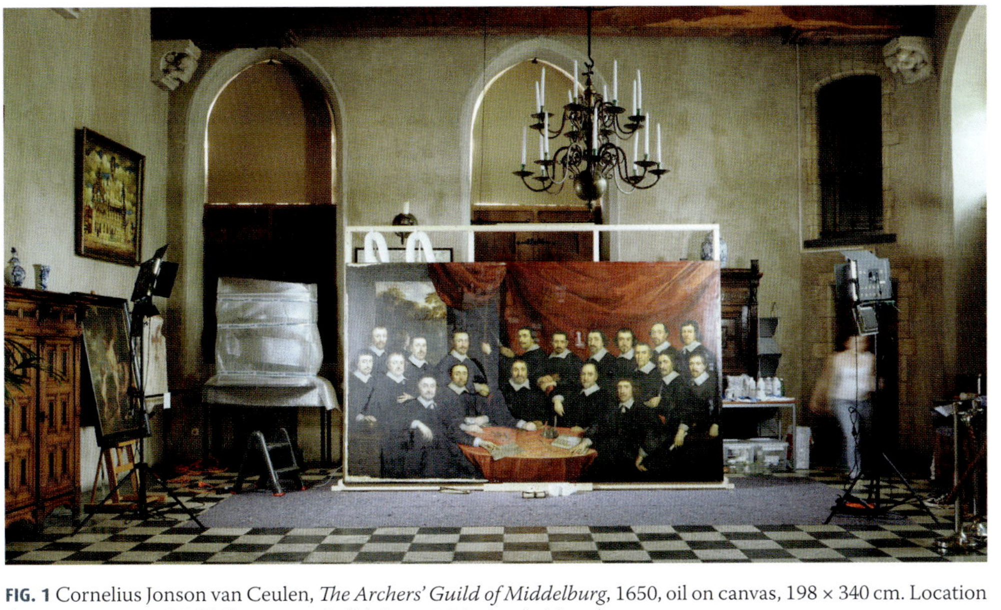

FIG. 1 Cornelius Jonson van Ceulen, *The Archers' Guild of Middelburg*, 1650, oil on canvas, 198 × 340 cm. Location during treatment: Middelburg town hall (photo © Marjan de Visser).

TABLE 1 Pemulen emulsion gels numbered 1–6 as discussed in text.

	1	2	3	4	5	6
Pemulen TR-2						
Buffer (pH)	1 part 6.0	1 part 8.5	1 part 8.5	1 part 8.5	1 part 8.5	1 part 8.5
Conductivity (mS = millisiemens)		1 part *c.*4.0	1 part *c.*4.0	1 part *c.*4.0	1 part *c.*4.0	1 part *c.*4.0
Chelator, ethylenediaminetetraacetic acid (EDTA)			1 part 8.5			1 part 8.5
Surfactant deoxycholic acid (DOA)						1 part 8.5
Demineralized water	1 parts	4 parts	4 parts	4 parts	4 parts	3 parts
Benzyl alcohol	10%	10%	10%	15%	20%	10%

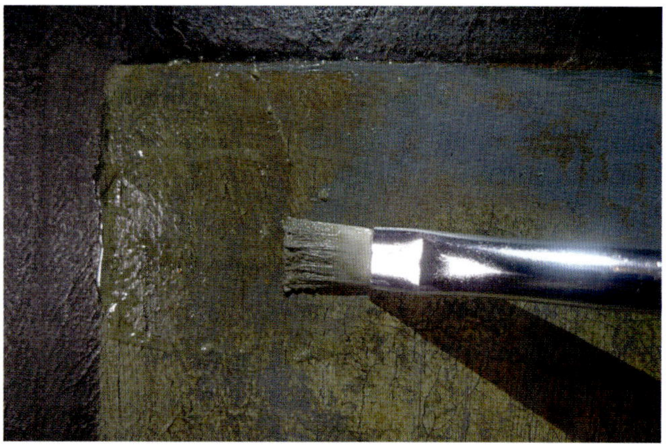

FIG. 2 Cornelius Jonson van Ceulen, *The Archers' Guild of Middelburg*: detail showing the opaque blue layer of overpaint on an ochre-coloured sky during removal of overpaint.

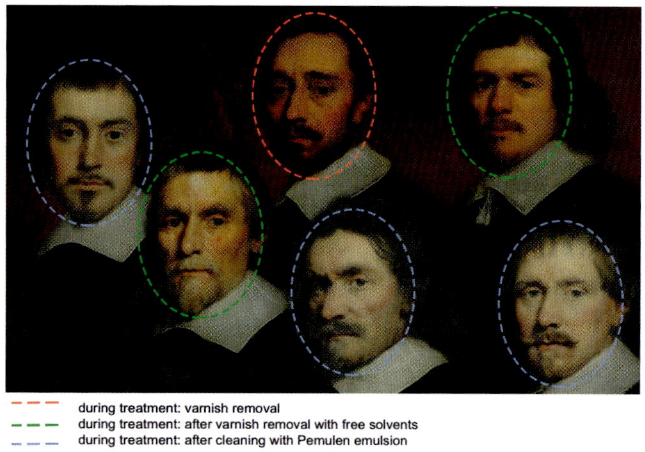

FIG. 3 Cornelius Jonson van Ceulen, *The Archers' Guild of Middelburg*: detail during cleaning, showing three distinct stages of treatment.

covered with a red tablecloth, on which various official items are placed. In the upper left corner there is an aperture with a view to the trees: the turned head of the concierge, his poised bow, and the archer's target are just visible. The curtain displays the coat of arms of the Archers' Guild, and the open book on the table details the names of the board members depicted. In the lower left corner of the tablecloth van Ceulen's invoice can be made out, bearing his signature and the date '1650'.

Condition before treatment

The large composition was painted on a single piece of herringbone weave canvas, and had undergone at least two wax-resin linings, in the 19th and 20th centuries.[3] The original tacking margins were no longer present, and the current tacking margins and front turnover edge were saturated with a thick sticky layer of yellow-brown wax-resin lining adhesive. Two distinct types of filling and retouching provided further evidence of at least two major restorations in the past, increasing the likelihood of varnish and adhesive residues having been deposited on the surface over time. Throughout the composition, the degraded natural resin varnish layer had become opaque and yellowed, dark residues had accumulated in the recesses of the canvas weave and impasto, and together with a dirty nicotine-like layer they were obscuring the image beneath.

Methodology and treatment

The conservation treatment was undertaken in a public space. Free solvent use was minimized and a carbon filter mobile extraction unit was employed for the health and safety of both conservators and the public. Pemulen emulsion gels were a suitable option at this point: they are non-volatile and water-based, can be precisely tailored to the removal of a particular soil, and their rheological properties can prevent penetration into the paint film beyond the targeted layer. However initial dirt and varnish removal was carried out with a buffered solution and free solvents because step-by-step cleaning provided a better understanding of the surface coatings covering the paint film.

The degraded varnish and surface coatings masked widespread discoloured restoration which became more apparent during initial varnish removal using free solvents, particularly throughout the heavily abraded black costumes. The sky was overpainted with an opaque blue layer (Fig. 2), which covered a now ochre-coloured sky. In addition, the discoloured remains of lining adhesive and remnants of older varnish layers remained across the painted surface, emphasized by the herringbone weave (Fig. 3). In lighter passages such as the collars and faces this was considerably disturbing to the image.

The solubility of the overpaint, remains of lining adhesive and remnants of older varnish layers were tested using

Pemulen TR-2 emulsions tailored to different areas following the MCP ingredients and methods. The mixtures were made up of 1 part Pemulen TR-2 to 1 part water-based mixtures (v/v), described in more detail below. Water-based mixtures were made up of 5 parts in total (Table 1). All solutions were cleared using petroleum ether.

- *Pemulen TR-2 stock pH 6.0*: Suspend 4 g Pemulen TR-2 in 100 mL deionized water, stir until uniform and well dispersed. Mix approximately 3.15 g (2.8 mL) triethanolamine (TEA) into 95 mL water. With vigorous stirring, mix the Pemulen suspension with the TEA solution. Adjust to pH 6.0 with additional TEA solution; and bring final volume to 200 mL.
- *Pemulen TR-2 stock pH 8.5*: Suspend 4 g Pemulen TR-2 in 100 mL deionized water, stir until uniform and well dispersed. Mix approximately 10.48 g (10.44 mL) 10% sodium hydroxide (NaOH) solution into 85 mL water. With vigorous stirring, mix the Pemulen suspension with the NaOH solution. Adjust to pH 8.5 with additional NaOH solution, and bring final volume to 200 mL.
- *To prepare the 1% working solution*: Dilute the stock Pemulen gel 1:1 with aqueous preparations.
- *To make the emulsion*: Take the 1% working Pemulen gel; add the benzyl alcohol and shake. It should instantly form a stable emulsion.

Results and evaluation

Cleaning tests were evaluated empirically in visible and ultraviolet (UV) light and using a digital microscope, taking into account the balance between effectiveness, control and length of application of the different emulsion mixtures. Emulsions were applied using white brushes and agitated for up to 30 seconds, until the clear gel showed signs of discoloration, and removed with tissue and cotton swabs. This method was repeated until all residues were removed. Starting with mixtures 1 and 2 and working across Table 1, the two mixtures were then compared and the most effective of these was then compared with the next mixture. All stages of the systematic removal of surface coatings are summarized below:

- *Surface dirt removal*: Nicotine-like surface dirt was successfully removed using an adjusted water solution buffered to pH 8.5 (bicine + NH_4OH), 4.0 mS, cleared with adjusted water pH 8.5, 1.0 mS. A raised pH was chosen as a first step to break down the degraded natural resin varnish.
- *Varnish removal*: The varnish was removed using free solvent mixtures of isopropanol and isooctane, and cotton swabs.
- *Overpaint removal*: Black overpaint in the costumes was removed using Pemulen emulsion gels 4 and 5 depending on the area. Blue overpaint in the sky was removed using Pemulen emulsion gel 6.
- *Discoloured remains of lining adhesive and remnants of older varnish removal*: Residues were removed using Pemulen emulsion gels 2 and 4.

Pemulen emulsions at a raised pH of 8.5 and a 10–20% benzyl alcohol content (emulsions 2, 4 and 5) successfully removed the overpaint and discoloured residual material. The blue overpaint in the sky, most probably a lead white oil-based paint, required emulsion mixture 6, containing the lead chelator ethylenediaminetetraacetic acid (EDTA) and the surfactant deoxycholic acid.

Conclusion

The conservation of *The Archers' Guild of Middelburg* by Cornelius Jonson van Ceulen posed specific challenges related to the public location of the treatment. A Pemulen emulsion gel system based on the MCP proved to be a safe and effective method to remove overpaint, remains of lining adhesive and remnants of older varnish layers from a 17th-century painting.

Notes

1. For an initial bibliography see www.getty.edu/conservation/publications_resources/teaching/caps_modular_cleaning.pdf. MCP software can be downloaded from cool.conservation-us.org/byauth/stavroudis/mcp/. For an introduction to aqueous gel cleaning methods see also Stavroudis and Wolbers 2012.
2. The name 'Cornelius Jonson van Ceulen' is how the artist referred to himself when he moved to the Netherlands, rather than 'Cornelius Johnson' as he is known in the UK.
3. In the 19th century a lining was also carried out to repair tears in both lower corners, and more recently in the mid-20th century, to address a large tear in the centre of the composition.

Reference

Stavroudis, C. and Wolbers, R. 2012. 'Aqueous methods for the cleaning of paintings', in J.H. Stoner and R. Rushfield (eds), *The Conservation of Easel Paintings*. London and New York: Routledge, 500–523.

Authors' addresses

- Marjan de Visser, Restauratieatelier Marjan de Visser, Den Haag, The Netherlands (marjan.de.visser@restauratieatelier.com)
- Lucy Odlin, Paintings Conservator, London, UK (lucy.odlin@gmail.com)

Silicone emulsifiers

Particle-based silicone cleaning emulsions: studies in model paint systems for the cleaning of water-sensitive artworks

Anthony Lagalante and Richard Wolbers

Introduction

In the context of fine art conservation, cleaning is one of the most problematic endeavors given the inherent irreversibility and potential damage that cleaning with solvents can cause. This is no less so with modern water-based acrylic (polymer emulsion) paints, or their water-based or emulsified drying oil counterparts. As unvarnished and unprotected surfaces, both are likely to require cleaning at some point as they age. Both types of paints are laden with surfactants or other chemicals which can contribute to the securing of soils to their surfaces, and can render them extremely sensitive to aqueous and other solvent cleaning methods in terms of physical effects (softening) and extraction of paint film constituents. Although surfactants have been the most often cited extracted materials (Ormsby and Learner 2009), generally all the film-forming materials can be detected in acrylic extraction studies (Smith 2005; Ormsby *et al.* 2006, 2007). Our group has been able to demonstrate, for instance, that in the case of Golden Artist Colors which are heavy-bodied acrylics, non-ionic aromatic surfactant structures (e.g. alkylphenyl ethoxylates (APEOs) resembling the Triton series) can bind carbon-based soils in a manner that is both unique and intractable (Lagalante and Wolbers 2016). Furthermore, we have also demonstrated that the mobility of soils bound in this way follows that of the surfactant's general mobility in an acrylic system. Soils tend to move in or out of acrylic films as a function of both relative humidity (RH) and temperature changes (Ziraldo *et al.* 2016). However, carbon-based soils can strongly bind to surfactants, moving into paint surfaces where they are firmly retained (Fig. 1) (Lagalante and Wolbers 2016).

Our research has pursued the effect of aqueous systems in particular on these films, and has demonstrated that both

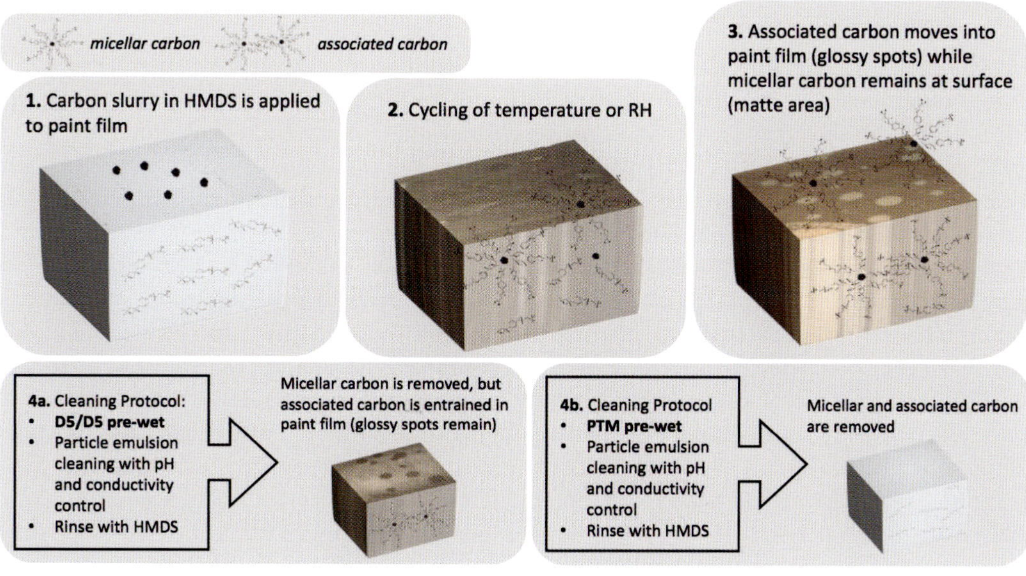

FIG. 1 Model for cleaning carbonaceous dirt from paint surfaces (image: Brynn Sundberg).

pH and ionic effects (conductivity, specific ions) can influence the swelling and extraction of materials (Dillon *et al.* 2014). The observation that aliphatic hydrocarbon-based solvents had at least a minimal swelling/extractive effect on these paints (Jablonski *et al.* 2003), suggests that cleaning systems developed around very low- or non-polar types of solvents (in our case, silicone-based solvents) might pose less risk to these paints, specifically in emulsions of a water-in-oil type (w/o) that predominantly feature continuous phases of non-polar solvents and small or restricted water phases.

An ongoing *Cleaning of Acrylic Paint Surfaces* (*CAPS*) workshop (Stavroudis 2012), has generated successive iterations of w/o microemulsion systems. The first systems introduced at *CAPS 2011* were based on the aliphatic solvent Shellsol D38, a linear alkyl sulfonate and water (30–70%) with a blend of butanol and hexanol as the co-solvent. These were deemed too strong in the general workshop experiences on test films, and a second generation of microemulsions was formulated, again with Shellsol D38, using non-ionic Ecosurf EH-6 (Dow) as the surfactant, a water phase, and mixtures of butanol or hexanol as the co-solvent, as proposed and presented at *CAPS 2012*. By *CAPS 2013*, w/o microemulsions were being prepared and tested in the workshops that included other surfactants (Triton GR-7M; an anionic sulfo-succinate and then the sulfo-succinate in D38), and non-polar silicone solvents (cyclomethicones D4, D5), and silicone-based surfactants (Gransurf 90) to evaluate their effects in w/o microemulsion systems. While the w/o microemulsion preparations are increasingly effective in terms of removing model soils from test films, they are still constructed with relatively large amounts of surfactants.

Surfactants tend to irreversibly intermix, and combined hydrophile-lipophile balance (HLB), critical micelle concentration (CMC) and cloud point temperature (T_{cp}) effects of surfactant combinations are difficult to predict or control. Additionally, surfactants with aromatic chemical groups can add to the problem of carbon soil retention, not remediate it. A typical feature of the adsorption of ionic (anionic/cationic)/non-ionic mixtures is the synergy or anti-synergism (antagonism) at interfaces. For example, the adsorption of one surfactant may either be enhanced or retarded by the addition of a small amount of the other surfactant. In many applications, surfactant mixtures, and in particular anionic–non-ionic mixtures, often give rise to enhanced overall performance over single component systems (Holland and Rubingh 1992). For example, anionic–non-ionic surfactant mixtures might show increased solubilization behavior, exhibit cloud points higher than those of the single non-ionic surfactant alone, and Krafft points lower than those of the single anionic surfactant. Mixed surfactants can often be employed over a wider range of temperature, salinity, and hardness conditions than the individual surfactants (Rosen 1986). Synergy within surfactant mixtures may be one of the fundamental reasons that their use on surfactant-laden surfaces – like artists' acrylic emulsion paints or water-dispersed oils – may be so problematic. Surfactants *do* mix inherently, and along with mixing, both ideal and non-ideal phenomena can occur, which might include non-desirable behaviors with regard to surface tension, CMC, HLB, and cloud point deviations in cleaning mixtures.

Our recent interest has been in developing w/o (or rather w/Si) emulsions that are not based on surfactants, but are in fact surfactant-free, to pre-empt the likelihood of adding to the surfactant burden of these types of paint, and the movement of the surfactant in the dried film (Wolbers *et al.* 2010; Dillon *et al.* 2014; Lagalante and Wolbers 2016; Ziraldo *et al.* 2016). As a result, we have recently introduced an alternative way of constructing w/o type cleaning emulsions using particle-type (Pickering) emulsifiers rather than traditional surfactants, both to avoid surfactant mixing phenomena, and to aid in the clearance of the applied cleaner and emulsifier (Lagalante and Wolbers 2016). Using colloid particles to prepare Pickering emulsions has attracted increasing interest in recent years (Pickering 1907; Aveyard *et al.* 2003; Binks and Horozov 2006; Klapper *et al.* 2008), in contrast to conventional emulsions, which are usually thermodynamically unstable and stabilized by high amounts of surfactants or amphiphilic linear copolymers.

The formation and properties of such surfactant-free particles have been extensively reviewed (Li *et al.* 1997). Several of these polymeric Pickering-type particles are commercially available. Shin-Etsu (Japan) has produced a wide range of crosslinked emulsion-forming particles (the KSG series) based on dimethicone polymer chains that are crosslinked with vinyl (KSG-16), glycerol (KSG-710) or PEG chains (KSG-210, KSG-350z) along with a variety of pendant groups (PEG, aliphatic) to form small particle structures that can be dispersed in aliphatic hydrocarbon solvents (mineral oil), low molecular weight dimethicone fluids, or cyclomethicone D5. Momentive (USA) also produces a commercial polymeric particle/emulsifier trade-named Velvesil Plus which is a vinyl crosslinked dimethicone polymer with pendant PEG chains dispersed in D5. Silicone-based Pickering particles modified with various crosslinking groups, as well as pendant groups, further modify or reinforce particle-to-particle stability.

Following recent successes in utilizing our particle-based w/o emulsifiers to clean water-sensitive surfaces, we surmised that artists' paints containing emulsified drying oil, which are also formulated with dispersion surfactants, are likely to exhibit some of the same soil retention/mobility phenomenon as artists' acrylic paints. As surfactant-laden preparations, surfactant mobility and accumulation at paint surfaces over time is likely to be of concern for paintings with this medium type. Here, we report on additional acrylic cleaning studies, and the extension of particle-based w/o emulsifiers to clean Winsor & Newton Artisan Water Mixable Oil Colors (WNAC). WNAC emulsion paints were examined to determine the nature of the surfactant used in the paint body and their inherent ability to retain model soils, as well as to evaluate strategies for their cleaning with silicone particle-based w/o emulsifiers, and overall clearance procedures from these surfaces post-cleaning.

Experimental

Paint draw-downs

All draw-downs were prepared on borosilicate glass slides (25 × 75 mm) using parallel guide plates set to a uniform wet film thickness of 100 μm when applied to the slide substrate. WNAC paint films were allowed to dry for a minimum of 3–4 months, in a dark, dust-free environment and Golden Artist Colors (GAC) paint films were allowed to dry for a minimum of 18 months, in a dark, dust-free environment.

Acrylics

Fresh paint film draw-downs of GAC heavy-body acrylic paints of the following colors were made using: Naphthol Red Medium (1220) PR5; Titanium White (1380) PW6; Pyrrole Red (1277) PR254; Phthalo Green (Blue Shade) (1270) PG7; Cadmium Yellow Medium Hue (1554) PY83/PY53/PY3; Ultramarine Blue (1400) PB29; and Mars Black (1200) PBk11.

Oils

Draw-downs of WNAC paint films were also made using: Cadmium Yellow Light (113) PY35; French Ultramarine (263) PB29; Lamp Black (337) PBk6; Permanent Alizarin Crimson (468) PV19, PR206; Phthalo Green Blue Shade (522) PG7; and Titanium White (644) PW6/PW4.

Artificial soiling for cleaning studies

For soiling and cleaning studies on both the GAC and WNAC paint films, 20% w/v of HITT fingerprinting powder in hexamethyldisiloxane (HMDS) solvent (at 0.65 centistokes viscosity) was prepared and applied to painted slides. As previously reported (Lagalante and Wolbers 2016), this latent commercial fingerprinting powder is a carbon black derived from combusted lycopodium powder, with red fluorescence nylon particle inclusions (Lagalante and Wolbers 2016). The soiling solution was applied using a polyurethane foam swab to the surface of the painted slide and then washed with a clean foam swab with HMDS until no further soil was removed from the paint surface. The soiled paint slides were allowed to age indoors in the dark for one month at 53% relative humidity (RH) for the acrylics, and five months indoors in open air for the oils.

RH studies on WNAC paints

HITT fingerprinting powder was dusted with a camel hair brush onto the paint films. A custom, sealed microscope stage was fabricated to house the soiled paint film and the RH solution, to allow *in situ* observation under a microscope objective. Saturated salt solutions were used to provide constant RH (Ziraldo *et al.* 2016) moving from low RH to high RH. Fluorescent powder images were captured digitally either on an AmScope epi-fluorescence microscope (at ×200 magnification, using bandpass filter 545/10 with peak transmission at 590 nm) or on a Nikon Eclipse 80i epi-fluorescence microscope, under the same illumination conditions. Image luminosity was quantified over a constant field of view using Adobe Photoshop.

W/Si cleaning emulsions

Acrylic paints

After empirical testing, an 'acrylic cleaning emulsion' was prepared with 10% w/w KSG-350z (Shin-Etsu), 40% of either cyclomethicone D5 or phenyl trimethicone (PTM), and 50% citrate/borate/phosphate buffer to pH 7.

Oil paints

Again, after empirical testing, an 'oil cleaning emulsion' was prepared by weighing the ingredients as 10% KSG-350z, 30% of either D5 or PTM, 60% citrate/borate/phosphate buffer to pH 8. Both cleaning preparations were emulsified by shaking on a SPEX Sample Prep 8000M mixer for 3 minutes.

Cleaning protocol

The soiled test paint films were pre-wetted for 5 minutes with either D5 or PTM. Emulsions were then applied onto the pre-wet surface using cotton swabs and mechanically rolled over the surface until no further lightening of the soiled slide was visibly discernible. At this point, the slide was twice rinsed with D5, and twice with HDMS, wiped dry with a Kimwipe laboratory tissue, and air-dried in under 5 minutes prior to evaluation.

Colorimetry and gloss

Colour measurements (using CIE L*a*b*, C* chroma), and gloss measurements, were made before and after cleaning, using an X-Rite SP62 colorimeter.

Results and discussion

Surfactant in WNAC and GAC dried paint films

The principal surfactant accrued on the *surface* of the acrylic test films was previously determined by us using desorption electrospray ionization mass spectrometry (DESI-MS) to be a 9–10 ethoxylate of nonyl phenol ($NPEO_{9\text{-}10}$) with a structure similar to Triton X-100 (Ziraldo *et al.* 2016). As previously reported (Lagalante and Wolbers 2016), we anticipated that the aromatic structure of the $NPEO_{9-10}$ surfactant may play an important role in retaining carbon-based soils in the additional colors of GAC paints tested.

All of the WNAC test paints studied exhibited identical desorption electrospray ionization–mass spectrometry

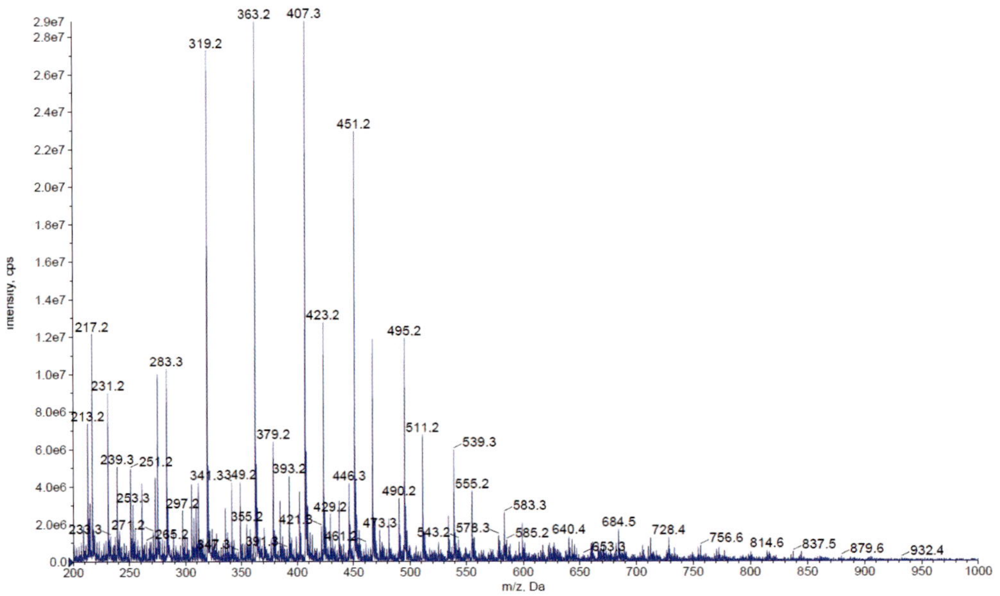

FIG. 2 DESI-MS spectrum of the polyethoxylated surfactant, measured at the surface of a titanium white WNAC paint film.

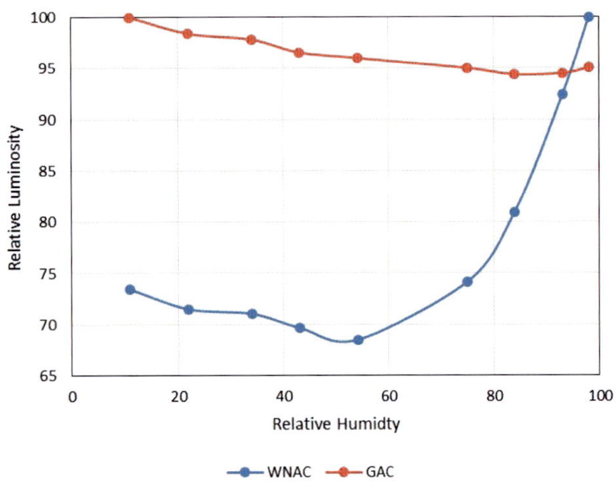

FIG. 3 Artificial soil movement in WNAC and GAC paint films as a function of RH.

(DESI–MS) spectra showing the presence of the protonated $(M+H)^+$ ions of the common commercial surfactant Brij L4, a polyethylene glycol dodecyl ether with an average molecular weight corresponding to $C_{12}H_{25}(OCH_2CH_2)_4OH$, on the test paint surfaces (Fig. 2). The HLB of this alkyl-ethoxylated structure (henceforth, referred to as '$C_{12}EO_4$') is nominally 10, and it can easily form oil-in-water (o/w) dispersions of oils like linseed or other common drying or semi-drying oils. As an alkyl-ethoxylated structure we were expecting that soil mobility and retention would be much less in the WNAC paints than we had found for the GAC colors, given the lack of aromaticity in the surfactant moiety. However, these oil-based binders are additionally 'modified' to enhance their rate of crosslinking, so that they can produce structurally serviceable films in a very short time-frame. Rather than relying on slower auto-oxidation through hydroperoxide production and radical crosslinking, common modifications can include epoxidation of double bonds in the triglyceride structure, followed by derivatization with a more reactive group as an alternate polymerizable functionality. For instance: acrylated, epoxidized soybean oil (AESO) or linseed oil (AELO) can be blended with reactive materials such as styrene (ST), divinyl benzene (DVB), or 4-trimethylsilylstyrene (TMSS) monomers to improve their drying or crosslinking rate (and flame resistance in the latter case of TMSS). AESO-ST or ELO-ST thermosets are, in fact, quite common and useful in various commercial structural applications (Gan and Jiang 2015). The WNAC paints used in this study were all listed by the manufacturer as being 'modified' linseed oil-containing materials, except for a Titanium White, which was prepared with a 'modified' safflower oil. Empirically, it has been our observation that both the acrylic *and* the oil dispersion paints tested here tended to bind carbon-based soil tenaciously. We hypothesize that the binding may occur through different paint film moieties: the $NPEO_{9-10}$ or similar structures present in the acrylic paints, or perhaps crosslinking moieties in the oil emulsion paints tested.

Movement of surfactant and artificial soils with RH

Our previous work (Ziraldo *et al.* 2016) using DESI-MS suggested that the $NPEO_{9-10}$ surfactant migrates *into* the GAC test films with increasing temperature (>25 °C) and RH (>40% RH) and *out* of the films and *onto* the surface of the paint films with decreasing temperature and RH. A similar movement was observed using the fluorescent HITT fingerprint powder as a marker for soil and surfactant mobility on GAC paint films, as well as a hysteresis with the HITT soiling powder (Lagalante and Wolbers 2016). Because of the strong association of aromatic $NPEO_{9-10}$ surfactant (in excess in the paint films) with carbon HITT powder, we reported as the RH and temperature were cycled (in just 4–5 cycles each), that increasing amounts of the HITT powder was irreversibly retained in the GAC films with their reservoir of $NPEO_{9-10}$

TABLE 1 Colorimetry values for the unsoiled, soiled, and cleaned WNAC and GAC paints.

	L*	a*	b*	C*	Gloss
WNAC Titanium White					
unsoiled	96.1	-0.7	5.0	5.1	33.9
soiled	43.9	0.9	4.7	4.8	15.0
30% D5/60% pH 8	94.7	-0.7	5.5	5.6	5.7
30% PTM/60% pH 8	94.5	-0.7	5.7	5.7	5.0
WNAC Lamp Black					
unsoiled	25.3	-0.1	-0.5	0.5	13.5
soiled	20.5	0.6	-1.7	1.8	4.8
30% D5/60% pH 8	25.8	0.1	-0.6	0.6	14.8
30% PTM/60% pH 8	25.9	0.3	-0.5	0.6	9.6
WNAC French Ultramarine Blue					
unsoiled	25.3	2.8	-7.9	8.4	22.5
soiled	28.7	0.2	0.6	0.6	5.8
30% D5/60% pH 8	23.5	3.8	– 10.3	10.9	8.2
30% PTM/60% pH 8	25.7	3.1	-8.6	9.1	19.0
WNAC Permanent Alizarin Crimson					
unsoiled	28.0	15.4	5.7	16.4	40.9
soiled	22.3	1.1	-2.4	2.7	8.2
30% D5/60% pH 8	28.0	14.3	5.2	15.2	5.9
30% PTM/60% pH 8	28.6	13.8	5.0	14.7	11.9
WNAC Phthalo Green Blue Shade					
unsoiled	28.2	– 10.8	0.5	10.8	20.9
soiled	26.3	0.0	-0.6	0.6	4.8
30% D5/60% pH 8	27.8	-10.3	0.3	10.3	8.9
30% PTM/60% pH 8	27.3	-11.1	0.3	11.1	6.3
WNAC Cadmium Yellow Light					
unsoiled	79.2	9.7	35.8	37.1	39.4
soiled	35.6	4.0	10.5	11.3	5.0
30% D5/60% pH 8	61.9	6.8	25.2	26.1	13.25
30% PTM/60% pH 8	76.7	9.3	34.6	35.8	10.6

	L*	a*	b*	C*	Gloss
GAC Titanium White					
unsoiled	97.1	-1.1	0.3	1.1	5.0
soiled	54.7	0.9	3.2	3.3	8.4
40% D5/50% pH 7	72.0	0.3	2.5	2.5	16.4
40% PTM/50% pH 7	97.2	-0.9	0.9	1.3	5.0
GAC Mars Black					
unsoiled	25.9	0.4	0.0	0.4	11.3
soiled	34.8	0.5	2.0	2.1	25.8
40% D5/50% pH 7	32.2	0.2	1.2	1.2	24.3
40% PTM/50% pH 7	30.0	0.2	0.8	0.8	22.2
GAC Ultramarine Blue					
unsoiled	24.7	13.3	– 26.6	29.7	4.6
soiled	30.1	4.1	-8.5	9.4	4.8
40% D5/50% pH 7	31.3	3.7	-7.0	7.9	5.1
40% PTM/50% pH 7	24.9	8.8	-18.7	20.1	4.8
GAC Naphthol Red					
unsoiled	43.0	50.6	32.2	63.6	29.9
soiled	39.8	20.4	11.1	23.3	100.0
40% D5/50% pH 7	41.0	27.0	14.2	30.5	100.0
40% PTM/50% pH 7	43.2	49.2	30.3	57.9	12.7
GAC Phthalo Green Blue Shade					
unsoiled	86.2	-7.1	-3.9	8.1	62.0
soiled	26.3	-2.9	0.5	2.9	100.0
40% D5/50% pH 7	34.6	-3.0	0.1	3.0	100.0
40% PTM/50% pH 7	33.8	-5.5	-3.1	6.3	65.3
GAC Cadmium Yellow					
unsoiled	87.9	-5.8	90.6	90.7	7.6
soiled	55.6	-2.7	34.0	34.1	16.6
40% D5/50% pH 7	65.9	-3.6	50.0	50.1	21.8
40% PTM/50% pH 7	87.0	-4.4	87.5	87.6	13.0
GAC Pyrrole Red					
unsoiled	43.0	54.6	32.2	63.4	29.9
soiled	39.8	20.4	11.1	23.3	100.0
40% D5/50% pH 7	41.0	27.0	14.3	30.5	100.0
40% PTM/50% pH 7	43.2	49.2	30.4	57.8	12.7

surfactant. In comparison to soil and surfactant movement in the GAC paint films, our RH results on the WNAC test films were just the opposite. Fig. 3 shows a comparison of both acrylic and oil test films and the relative luminosity of the fluorescent fingerprinting powder as a function of the RH. As the RH was increased above 60%, the HITT powder (and presumably the $C_{12}EO_4$ surfactant) readily moved *out* of the paint film to the paint film surface. Below 60% RH, the HITT powder moved *into* the paint surface, but this movement appeared to be easily reversed with RH. We speculate that unlike the acrylic paint films tested, there is no inherent aromaticity in the reservoir of surfactant likely to irreversibly hold the soiling powder, as was the case with the Triton surfactant in the acrylic films.

Cleaning of soiled WNAC and GAC paint films

The rationale for the formulation of both the 'oil' and 'acrylic' cleaning emulsions follows from both our empirical experience, and concerted studies of aqueous preparations applied to these types of paints (Dillon *et al.* 2014). The pH and conductivity of aqueous preparations play a distinct role in surface cleaning on paint surfaces. On acrylics, aqueous or aqueous-containing phases in emulsions should ideally be controlled well above the isotonic point of the acrylic paint film (6–10 mS/cm) and at a pH at or below the pK_a of polyacrylic acid (PAA) (pK_a 5.18–7.2) (Hogfeldt *et al.* 1989) to minimize the swelling of the paint film and leaching of extractable materials. On drying oils, the pH could be as high as 8 or 8.5, with the conductivity isotonic or slightly hypertonic to the paint film itself (here, a pH of 8 and a conductivity of approximately 2 mS/cm were used and were the most effective) without adverse swelling effects or saponification of the oil binder. Additionally, in the cleaning emulsions, the tendency was to lower the overall amount of the aqueous phase by mass given the water-sensitivity of acrylic paints, and to employ higher amounts of aqueous phase materials on the young oils, as they are inherently more hydrophobic (that is, water-repelling).

As a general note, pre-wetting with a silicone solvent prior to cleaning was important to the cleaning process as a whole. Neither the $C_{12}EO_4$ found on the WNAC paints, nor the $NPEO_{9-10}$ found on the GAC paints is soluble in the silicone solvents. Gloss measurements following application of PTM or D5 and rinsing with HMDS resulted in no change in the gloss values, suggesting that pre-wetting alone tended not to strip free surfactant from the test paint surfaces. Additionally, pre-wetting may have the advantage of lubricating the surface during the mechanical act of applying and removing the emulsion cleaning systems. Post-cleaning colorimetry results suggest that a high degree of cleaning can be achieved on the WNAC and GAC films (Table 1) by utilizing the proper aqueous pH and conductivity, and silicone emulsion solvent. In WNAC, lightness (L^* in CIE $L^*a^*b^*$ color notation) values following cleaning in either a D5 or PTM based emulsions were commensurate with initial, unsoiled values in all but one paint color. The soiling mixture was especially tenacious on the WNAC Cadmium Yellow Light test film, which was not cleanable without a pre-wetting of the paint surface with PTM prior to the emulsion application. We will continue to investigate this observation utilizing spectroscopic techniques, but we speculate that the WNAC Cadmium Yellow Light paint may contain an aromatic moiety that could be retaining the soil, perhaps depending on the specific type or amount of crosslinking material used to modify the oil drying rate (e.g. the styrene, divinylbenzene, crosslinking moieties).

All of the other WNAC test films were cleanable by pre-wetting with D5 and using the D5-based oil emulsion cleaning system or the equivalent PTM cleaning system. In all cases on the GAC colors, D5-based cleaning systems were inferior to PTM-based cleaning systems, as had been previously observed and attributed to the attractive nature of carbon-based soils and Triton-based surfactants (Lagalante and Wolbers 2016). The present study also confirms that, on GAC paint films, a pre-wetting with PTM, along with the PTM-based acrylic emulsion cleaner, is recommended to remove carbon-based soils effectively, i.e., recover the highest lightness (L^*) values. In the WNAC paints (Table 1), when the films are soiled the gloss *decreases* dramatically and then increases somewhat after cleaning, never recovering to the original value, but definitely recovering some gloss in comparison to the soiled films. In contrast, when soiled, the GAC paint films *increase* in gloss. We propose that the reason for the increase in gloss is the ability of the $NPEO_{9-10}$ to strongly associate with the carbon, effectively mixing with it in a solution sense, and perhaps even concentrating it, at or beneath the paint surface, leading to an increase in specular reflection. This may ultimately represent a very different type of interaction and optical phenomenon than was encountered with the WNAC paints. In the latter, the surfactant/soil relationship is perhaps still micellular, in a traditional surfactant sense. That is, the surfactant *sequesters* the carbon rather than associates with the carbon, creating a more matte surface effect, leaving the paint surfaces more depleted of surfactant on cleaning.

After cleaning and clearance with HDMS, visual gloss changes were measured on both the acrylic and oil paint films. From Table 1 it is evident that while gloss and color values in GAC paint films were recoverable post-cleaning (more so with PTM-based cleaning emulsions), gloss recovery post-cleaning on WNAC with both D5 and PTM-based cleaning emulsions was not possible. To further investigate this loss of gloss, *unsoiled* WNAC paint films that were prewetted, cleaned with the oil cleaning emulsions, and cleared with HDMS did not show a change in gloss. Thus, at this point we can only speculate that the surfactant is removed from the WNAC test films in association with the HITT powder, not by the silicone solvents, but likely through the aqueous phase of the applied cleaning emulsions. However, gloss measurements post-treatment do suggest that the phenomenon may be self-reversing post-cleaning, or by actively intervening with the post-cleaning temperature and RH manipulation. At least some of the gloss can be recovered as fresh surfactant is brought to the paint surface through such manipulation (i.e., on the WNAC films, above a RH of 60% or on the acrylics, below 40% RH, and 25 °C).

Conclusion

In this paper, we report on the further testing and use of a particular silicone particle-based emulsifier, KSG-350z. We have worked with other silicone polymer particle emulsifiers and gelling agents (KSG-16, KSG-210, KSG-710, KSG-850z, and Velvesil Plus) largely with the intention of minimizing, or avoiding altogether, bringing additional surfactant to painted surfaces. However, the KSG-350z material is especially useful, because of its superior ability to emulsify a range of aqueous percentages in a variety of silicone solvents. These emulsions appear to clean a wide range of paint types, including water-dispersed oils. The KSG self-emulsifying gels have the ability to form stable w/o type emulsions with up to approximately 90% of an aqueous component, with very little (3–10%) polymer gel. For cleaning art and artefact surfaces that exhibit some water sensitivity, the w/o, or more specifically the w/Si emulsions that can be created with this emulsifier, have the clear advantage of minimizing and controlling the way in which water can be brought to these surfaces, and are eminently clearable because they avoid the complications of traditional surfactant-based micro- and macro-emulsifiers. Based on the hypothesis that some degree of aromaticity is needed to overcome evidently strong associations between carbon soils and aromatic additives (including aromatic surfactants), PTM in particular can play a critical role as a solvent in helping to overcome strongly bound soils for cleaning and clearance, and has been a revelation to us as well.

We intend to pursue concerted future studies that will seek to further investigate these artists' emulsified drying oil paints, firstly through analytical characterization of the movement of the surface surfactant as a function of temperature and RH analogous to our previous acrylic paint studies, and secondly to characterize analytically the nature of the soil-surfactant association in these emulsified drying oil artists' paints.

Materials and suppliers

- WNAC, GAC: Blick Art Materials, Highland Park, IL, USA
- KSG-350z: Independent Chemical, Glendale, NY, USA
- Fingerprinting powder: HITT Marking Devices, Santa Anna, CA, USA
- HMDS: Clearco Products, Willow Grove, PA, USA
- D4, D5, PTM: Lotioncrafter, Eastsound, WA, USA

References

Aveyard, R., Binks, B.P. and Clint, J.H. 2003. 'Emulsions stabilised solely by colloidal particles', *Advances in Colloid and Interface Science* 100: 503–46.

Binks, B.P. and Horozov, T.S. 2006. *Colloid Particles at Liquid Interfaces*. Cambridge: Cambridge University Press.

Dillon, C.E., Lagalante, A.F. and Wolbers, R.C. 2014. 'Aqueous cleaning of acrylic emulsion paint films: the effect of solution pH, conductivity and ionic strength on film swelling and surfactant removal', *Studies in Conservation* 57(1): 52–62.

Gan, Y. and Jiang, X. 2015. 'Photo-cured materials from vegetable oils', in Z. Liu and G. Kraus (eds), *Green Materials from Plant Oils*. London: Royal Society of Chemistry, 1–27.

Hogfeldt, E., Miyaijima, T., Marinsky, J.A. and Muhammed, E. 1989. 'Application of a simple three parameter model to titration data for some linear polyelectrolytes', *Acta Chemica Scandinavica* 43: 496–9.

Holland, P.M. and Rubingh, D.N. 1992. 'Mixed surfactant systems: an overviewi, *ACS Symposium Series* 501. Washington, DC: American Chemical Society, 2–30.

Jablonski, E., Learner, T., Hayes, J. and Golden, M. 2003. 'Conservation concerns for acrylic emulsion paints: a literature review', *Reviews in Conservation* 4: 3–11.

Klapper, M., Nenov, S., Haschick, R., Muller, K. and Mullen, K. 2008. 'Oil-in-oil emulsions: a unique tool for the formation of polymer nanoparticles', *Accounts of Chemical Research* 41(9): 1190–201.

Lagalante, A.F. and Wolbers, R.C. 2016. 'The cleaning of acrylic paintings: new particle-based water-in-oil emulsifiers', in V.A. Selva Bonino (ed.), *Dall'olio all'acrilico, dall'impressionismo all'arte contemporanea. Studi, ricerche, indagini scientifiche ed interventi conservativ, VII Congresso Internazionale Colore e Conservazione (CESMAR7)*. Padua: Il Prato, 163–76.

Li, M., Jiang, M., Zhu, L. and Wu, C. 1997. 'Novel surfactant-free stable colloidal nanoparticles made of randomly carboxylated polystyrene ionomers', *Macromolecules* 30(7): 2201–3.

Ormsby, B. and Learner, T. 2009. 'The effects of wet surface cleaning treatments on acrylic emulsion artists' paint', *Reviews in Conservation* 54: 29–41.

Ormsby, B.A., Learner, T.J., Foster, G.M., Druzik, J.R. and Schilling, M.R. 2006. *Wet Cleaning Acrylic Emulsion Paint Films: An Evaluation of Physical, Chemical, and Optical Changes*. Los Angeles: Getty Conservation Institute, 189–200.

Ormsby, B., Learner, T., Foster, G., Druzik, J. and Schilling, M. 2007. 'Wet-cleaning acrylic emulsion paint films: an evaluation of physical, chemical and optical changes', in T.J.S. Learner, P, Smithen, J.W. Krueger and M.R. Schilling (eds), *Modern Paints Uncovered*. Los Angeles: Getty Conservation Institute, 187–98.

Pickering, S.U. 1907. 'CXCVI.-Emulsions', *Journal of the Chemical Society, Transactions* 91(0): 2001–21.

Rosen, M.J. 1986 'Molecular interaction and synergism in binary mixtures of surfactants', in J.F. Scamehorn (ed.), *Phenomena in Mixed Surfactant Systems*. Washington, DC: American Chemical Society, 144–62.

Smith, G.D. 2005. 'A "single-shot" separation and identification technique for water extractable additives in acrylic emulsion paints', in I. Verger, (ed.), *Preprints of the ICOM- CC 14th Triennial Meeting*. London: James & James, 824–32.

Stavroudis, C. 2012. 'More from CAPS3: surfactants, silicone-based solvents, and microemulsions', *WAAC Newsletter* 34(3): 24–7.

Wolbers, R.C., Norbutus, A.J. and Lagalante, A.F. 2010. 'Cleaning of acrylic emulsion paints: preliminary extractive studies with two commercial paint systems', in M.F. Mecklenburg, A.E. Charola and R.J. Koestler (eds), *Cleaning 2010: New Insights into the Cleaning of Paintings*. Washington, DC: Smithsonian Institution Scholarly Press, 147–58.

Ziraldo, I., Watts, K., Luk, A., Lagalante, A.F. and Wolbers, R.C. 2016. 'The influence of temperature and humidity on swelling and surfactant migration in acrylic emulsion paint films', *Studies in Conservation* 61(4): 209–21.

Authors' addresses

- Anthony Lagalante, Villanova University, Department of Chemistry, Villanova, PA, USA (anthony.lagalante@villanova.edu)
- Richard Wolbers, University of Delaware, Department of Art Conservation, Newark, DE, USA (doczepp@gmail.com)

Conservation of murals by Eugène Delacroix at Saint Sulpice, Paris

Alina Moskalik-Detalle, Julien Assoun, Françoise Joseph, Marie-Laure Martiny and Marie Monfort

Introduction

The city of Paris owns the Church of Saint Sulpice, as well as 96 other religious buildings. The treatment of mural paintings by Eugène Delacroix (1798–1863) in the Chapel of the Holy Angels, Saint Sulpice began in October 2015 with a four-month study phase (Moskalik-Detalle *et al.* 2016) followed by eight months of conservation interventions. It was completed in November 2016 together with the installation of new lighting. The project was conceived and monitored by Marie Monfort, curator for French heritage; the treatment was entrusted to a team made up of 10 conservators.[1]

The remarkable quality and prestige of the painting, the technical specificities and documentation available (the journals and correspondence of Delacroix and his assistants, which detailed the chronology; sketches and order books)[2] (Hannoosh 2009) was critically evaluated, and each important decision was validated by a scientific committee.[3] This complexity led us to investigate the use of cleaning methods that had been perfected by Richard Wolbers (Wolbers 2000),

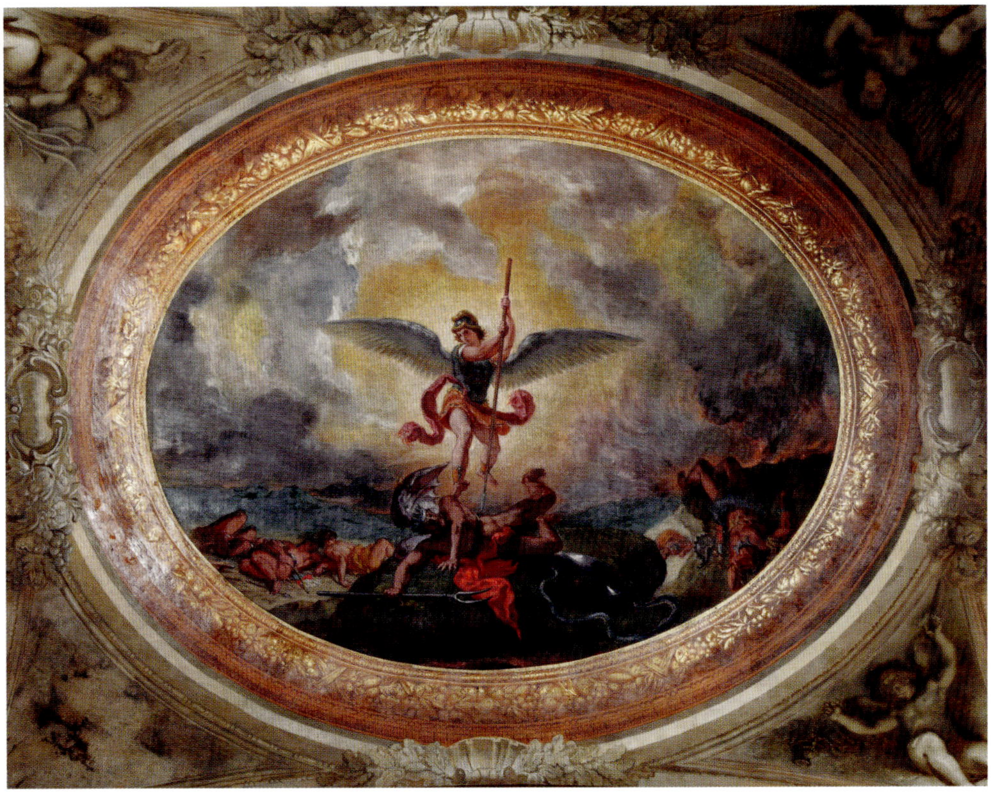

FIG. 1 Eugène Delacroix, *Saint Michael Defeating the Demon*, ceiling, the Chapel of the Holy Angels, Saint Sulpice, Paris: after cleaning (photo © Claire Pignol COARC).

combined with a special approach to mural paintings. The project was therefore an opportunity to use cleaning gels on a monumental scale (a total of 210 m^2).

Saint Sulpice, Chapel of the Holy Angels

Saint Sulpice, the largest church in Paris after Notre Dame, was built in a classical style over a period of 100 years, starting in 1646. During the French Revolution, the church was emptied of its artwork; only part of the work was returned to the building in 1800. Of the 21 chapels that open onto the choir and nave, 17 were decorated with murals during the 19th century and five of those during the period when Delacroix was employed by the commission to paint the Chapel of the Holy Angels; the others were created later. The transepts that Delacroix had expected to decorate as early as 1847 were finally painted by Signol in 1876. Commissions to decorate the chapels were not awarded in a consistent sequence; those of the four chapels nearest Delacroix's on the south side for example, date from 1822, 1845 and 1882.

After composing subjects for the Chapel of the Baptismal Fonts, which had been assigned to him on 28 April 1849, Delacroix was then informed that he should decorate the baptistry in the north tower and not the chapel he had previously been allocated. On 2 October 1849, he received confirmation that he was commissioned to decorate the Chapel of the Holy Angels and recompose all the scenes except for *Saint Michael Defeating the Demon,* inspired by Raphaël's work at the Musée du Louvre. The chapel, like those of the Chapel of Baptismal Fonts, is fairly narrow. For the sake of legibility, the painter had to concentrate the action in the foreground. In his journal and correspondence he mentioned his decorative framing of the scenes with figures. No doubt he had also asked for a plain glass window, since he wanted his paintings to be seen in natural light. The window was changed in 1901 and part of the wall he had decorated with a *trompe l'oeil* tapestry was destroyed.[4]

The mural paintings

Iconography

The surfaces to be decorated by Delacroix were two walls – west and east – whose upper parts are arched, the oval ceiling and four spandrels. The painter chose *Saint Michael Defeating the Demon* (Apocalypse 12:7–9) for the ceiling, surrounded by angels in the corners whose symbols are also linked to the Apocalypse, namely a palm branch, censer, harp and book of life from the Last Judgement (Fig. 1).

On the west wall, Delacroix chose to paint *Heliodorus Chased from the Temple* (Second Book, Maccabees 3:21–28): Seleucus, king of Syria (187–175 BC), having been told untruthfully that the Temple of Jerusalem's treasury contained vast riches, sent his minister Heliodorus to steal them.

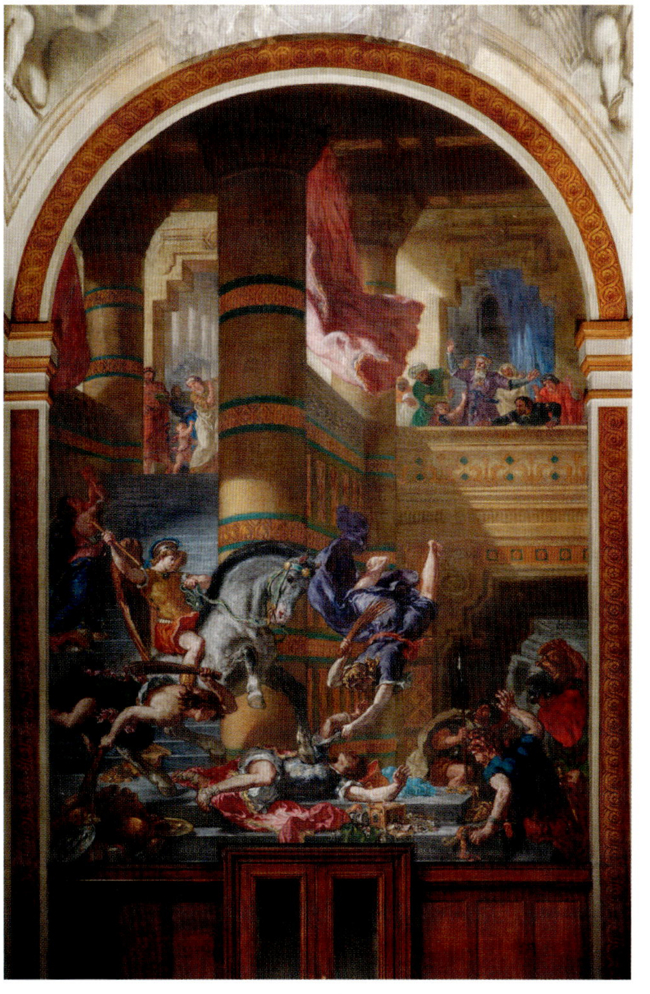

FIG. 2 Eugène Delacroix, *Heliodorus Chased from the Temple*, west wall, the Chapel of the Holy Angels, Saint Sulpice, Paris: after cleaning (photo © Jean-Marc Moser COARC).

Although the high priest Onias, painted in the highest part of the scene, explained that the treasury only contained moderate funds belonging to widows and orphans, and in spite of the supplications of the priests, Heliodorus attacked the temple. But when he arrived, a magnificently caparisoned horse appeared that struck out at him, shown on the ground at the bottom of the composition (Fig. 2).

On the east wall, the painter depicted the best-known painting in the chapel, *Jacob Wrestling with the Angel* (Genesis 32:24–28). Jacob had travelled with a caravan of servants and food to appease the rage of his brother Esau. Having purchased his birthright from Esau, Jacob usurped his father's blessing by passing himself off as his older brother. His brother's anger caused him to flee to Mesopotamia. This scene illustrates Jacob's return to his native land, where he goes to confront his destiny. Left alone during the night after helping his companions cross a river, he is assailed by an angel. Delacroix himself described the scene depicted in the invitation he sent to inaugurate the chapel:

> Jacob travels with the livestock and others present whose aid he hopes will diminish the anger of his brother Esau. A stranger appears and engages him in an obstinate fight that ends when Jacob, wounded on

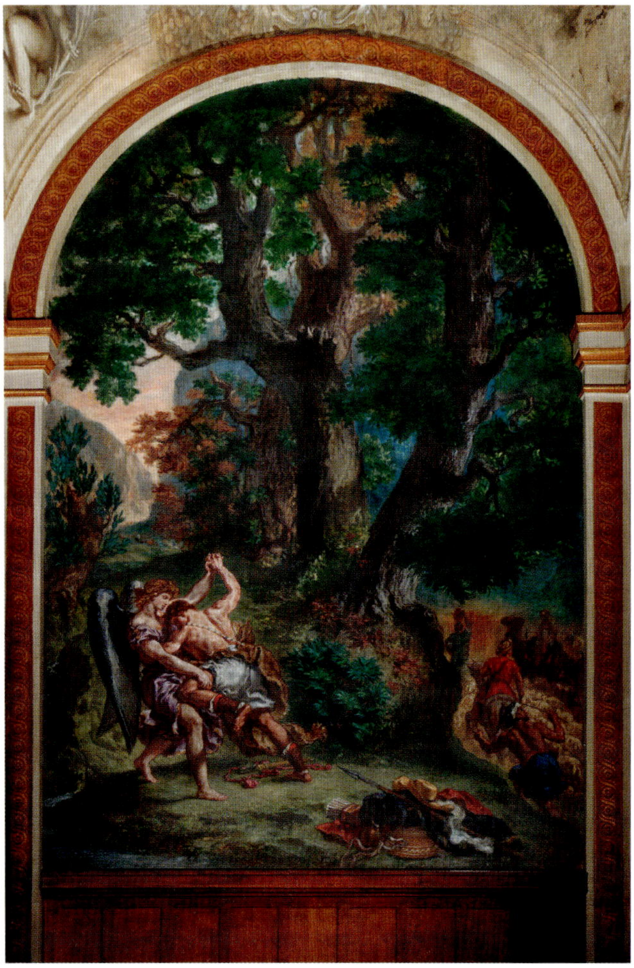

FIG. 3 Eugène Delacroix, *Jacob Wrestling with the Angel*, east wall, the Chapel of the Holy Angels, Saint Sulpice, Paris: after cleaning (photo © Claire Pignol COARC).

> the thigh by his opponent, finds that he is helpless. This struggle is seen in the holy books as representing the trials God sometimes sends to his chosen people.'[5]

Delacroix was the first major painter to undertake this theme of Jacob and the angel. Raphaël had treated the subject in a room within the Vatican, a composition of which Delacroix was aware having seen an engraving by Bartoli (Johnson 1989) (Fig. 3).

The treatment revealed the correspondence between the two scenes and their coherence in terms of the treatment of light. The oak tree in Jacob's fight with the angel echoes the column in the temple of Heliodorus, emphasizing the correspondence between a scene of nature and another in a temple.

Historical data

The chronology for conceiving the paintings and organizing the work was known from several publications (Sérullaz 1963; Spector 1967; Johnson 1989). Approximately 70 sketches created in preparation for this work are known today.[6] At a sale in February 1864, following Delacroix's death in August 1863, 100 preparatory sketch pages were documented for the Chapel of the Holy Angels. The drawings are mainly undated, making the chronology of the sketches difficult to pinpoint.

Delacroix had expected to begin the work in the autumn of 1850 but unfortunately he was hindered by illness, difficult relationships with his assistants and two commissions that interrupted that of Saint Sulpice – the decoration of the Galerie d'Apollon in the Musée du Louvre in 1850 and 1851, and the Salon de la Paix in the Paris City Hall between 1852 and 1854. Despite certain preparatory steps for the walls carried out in advance, the Saint Sulpice site work only seems to have begun in 1854. Delacroix added the final retouches in the summer of 1861.

The stages of the creation are still readable. Delacroix probably did a general drawing first and let his assistants fill in the forms before he intervened for the final touches. He had used this method in the Palais Bourbon and Palais du Luxembourg. While his assistants are frequently mentioned in his letters and journal, close observation of the work confirms the omnipresence of the master's hand, and it remains difficult to distinguish other paint strokes from his own. The painting is interspersed with masterful details reflecting his technique at its height, based on the observation of coloured reflections, accented by arbitrary touches of colour not linked to realistic observations, which reinforce the diverse lighting and texture of the site, themselves accentuated by a sculpted, engraved surface.

Painting technique

The historical context of the mid-19th century, the artist's personality and the especially long period of execution (seven years) resulted in a work with extreme heterogeneity, discernable in all the strata from the deepest preparation layer to the pictorial layer. The cross-sections of several samples revealed the presence of numerous paint layers (up to 15). This technical complexity made it a work that was especially difficult to treat.

The masonry ground and its preparation

This comprised a mixture of limestone bricks with lime plaster grout. The surfaces were chiselled to flatten the levels and even out the stone and grout surfaces. The facing of the stone was soaked with boiling linseed oil and covered with two–four layers of a mixture of white lead, barytes and oil. This mixture was applied heterogeneously – Delacroix sometimes used it again having begun a painting which he would often rub off; he reworked and restructured his impressions by taking into account the motif and distribution of lights in the scene.

The marouflaged canvas

The painting on the central part of the ceiling was on a marouflaged canvas glued down using a blend of oil and lead white. The preparation of the canvas, composed of white lead and oil, is covered with a film of pure wax, possibly applied by cauterization (local intense heating).

The pictorial layer

The painter used a binder made of beeswax and linseed oil, either layered or mixed together. The consistency varied – sometimes thick and sometimes dilute and thin. Standard oil colours were used directly from the highly organized palette, which was composed again every day (Piot 1931). Both full strength pigments and tints can be found. The decorative parts were covered with gold leaf applied on an oil and resin ground layer. While the surfaces have an overall satin finish, some areas have a matte appearance (where the paint layer is thinner), some areas of impasto (thick paint applied sometimes in several layers) and others where the fresh or dried paint has been scratched. There are many premature cracks due to Delacroix's painting technique (Fig. 4). These decorative elements were never varnished by Delacroix.

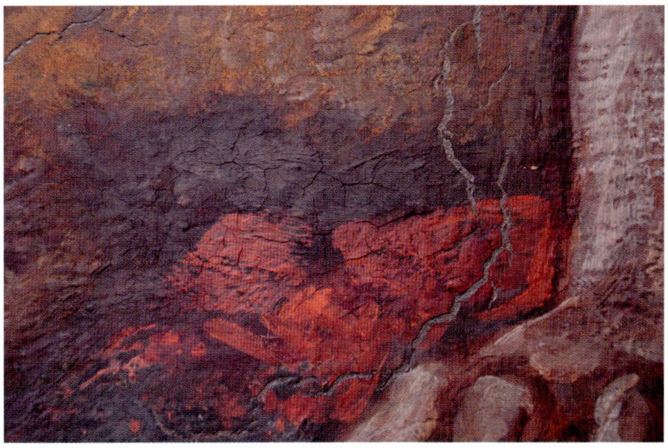

FIG. 4 Eugène Delacroix, *Saint Michael Defeating the Demon*: detail showing premature cracks due to Delacroix's painting technique (photo © Alina Moskalik-Detalle).

Restoration since completion by Delacroix

In addition to the heterogenity of the painting technique, other products were introduced during at least seven fairly invasive restorations carried out over a period of 150 years:

- In 1895 and 1898, the paintings were 'cleaned and washed', the gaps were filled in and the decorative parts were retouched.
- In 1933, 1934 and 1941, periodic interventions 'restored' the paintings, and treatments were carried out due to a loss of adhesion of the paint.
- Two complete restorations of the chapel were performed in the 1950s and the 1970s.

We have no reports describing these earlier treatments: the methods and products used are unknown. The results of the sample taken show the use of products close to the original composition, with the exception of the last restoration in the 1970s when the materials added included a vinyl binder and plaster for loss compensation. Different types of varnish were applied during the various restorations, sometimes very locally: natural resins (dammar or mastic), shellac, acrylic resins (Paraloid B-72) and ketone (polycyclohexanone) resins. Losses of adherence were treated many times, mostly with acrylic resins in recent times.

Current condition

The pictorial layer exhibits multiple flaking following the mortar lines in the masonry. The environmental conditions that caused this are little known, but the heterogeneous stratigraphies imply a moisture sensitivity that had to be considered during the treatment. Under the vault, the decorative parts have some problems linked to rising damp and soluble salts. All surfaces are covered by a thick layer of grime made up of many carbon-based materials from the burning of candles in the building and external pollutants, as well as breakdown products from the masonry and mortar, such as lead sulphate, calcium sulphate and calcium carbonate particles. Preliminary attempts to remove particulate soil clearly indicated that there is an oily, sticky surface.

Removing the soil, varnish and overpaint

Context and restrictions

The complex technology and specific characteristics of the paintings required careful treatment planning, especially in the original paint areas. The presence of a binder made from linseed oil and animal wax, aged and broken up by past cleaning interventions, implied a great sensitivity to organic solvents and water. Mechanical actions had to be limited to avoid polishing the surface. The painting method used also necessitated certain precautions. We found very diverse surfaces with juxtaposed matte or shiny areas, contrasts that had to be preserved. We also had to clean a surface with impasto effects and scratches (Fig. 5), even in the areas carved in relief, without being abrasive. Many glazed areas and transparent layers (possibly with minimal binder because too much diluent had been used when the painting was executed) were observed, which could be even more fragile (Fig. 6).

In addition, some of the previous interventions (carried out chronologically close to Delacroix's original work) had added materials and work methods closely related to the original painting. Finally, since the paintings were executed on a stone surface, any treatment would have to take into account interactions between products that might mobilize soluble salts. We called upon numerous laboratories to help us obtain the maximum amount of data to plan our treatment.[7]

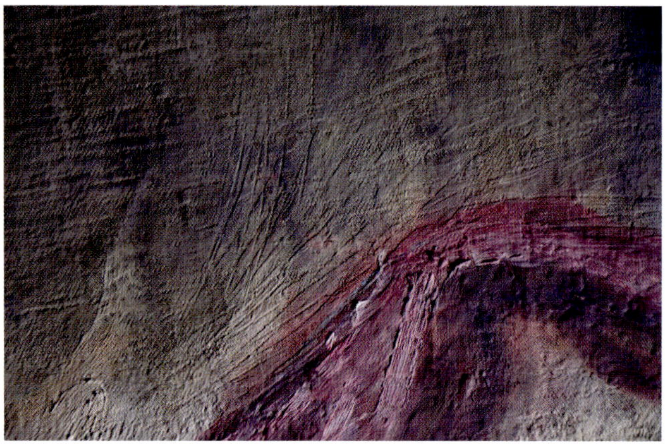

FIG. 5 Eugène Delacroix, *Saint Michael Defeating the Demon*: detail showing scratches and engravings in the fresh paint by Delacroix (photo © Alina Moskalik-Detalle).

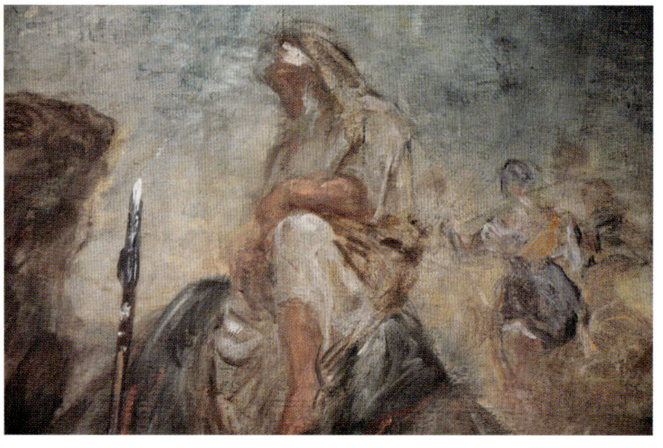

FIG. 6 Eugène Delacroix, *Jacob Wrestling with the Angel*: detail showing glazed areas and transparent layers (photo © Alina Moskalik-Detalle).

Technical examination

Samples

During the study, 37 samples were taken and added to 17 which had previously been obtained. The analyses offered information on the original materials and how they were used, past restoration materials, and observed damage (water content, identification and amount of salts) and surface deposits (soil).

Technical imaging

Ultraviolet (UV) and infrared (IR) imaging was carried out during the investigation.[8] However, the grime that had accumulated on the surfaces compromised observations and interpretation. After the soil removing phase, a range of technical imaging was carried out on the two principal scenes: visible reflectography, UV and IR imaging, IR false colour (IR-FC) imaging, and UV fluorescence imaging.[9] This combined information enabled mapping of the varnishes, glazing and overpainted areas from the various restorations.

Cleaning tests and establishing a treatment protocol

Tests for cleaning soil and eliminating varnishes and overpainting were carried out in the presence of Richard Wolbers, who consulted with us periodically during the project. The mural painting department of Historical Monuments Research Laboratory (LRMH), Paris, France, took samples to evaluate the presence of residue on the painted surfaces, after the cleaning tests.

Wolbers' gels and silicone products

Conscious of the difficulty of working on these paintings, we were convinced of the need to find an existing, safe technique that could be adapted for a monumental surface. As professional conservators, we wanted to work with a specialist to research and develop cleaning techniques, and to improve them by working with a laboratory that could verify our work. Therefore Richard Wolbers actively participated in this process to perfect methods for the work on-site.

Method, principles and their adaptation

Four principles were maintained:

- *Limiting penetration in the deeper layers:* This was effected by advance saturation of the material pores using a penetrating liquid with a retention time which allowed us to work over a specific period of time without acting on the substrata, whilst removing the soil and without leaving residues. We also wanted a product that would evaporate completely in a reasonable amount of time. The silicone solvents tested were cyclic and linear low molecular weight dimethyl silicones that have a chemical skeleton of Si-O-Si-O. Their ability not to cause damage to the paint surface was tested in advance. They did not affect either the natural waxes or the oils. Octamethylcyclotetrasiloxane (D4) was chosen for its volatility, which gave a working time on the painted surface of approximately 30 minutes.
- *Application and contact with a gel:* The cleaning substances (aqueous solutions, organic solvents, chelates) that act on the soil and other materials to be removed were applied using a gel with a paste-like consistency, whose composition was adapted to each need. The gel helped to maintain surface contact without using a surfactant in order to minimize risks created by residues in the substrata. After numerous trials,[10] we chose a gel material without a surfactant but with stable emulsifying properties: Velvesil Plus, based on silicone copolymers.[11] Velvesil Plus is a vinyl cross-polymer of polydimethyl siloxane, substituted with polyethylene glycol pendant groups: it is a particle or Pickering type of emulsifier, and as such does not form emulsions through surfactant chemistry. The gel allows the cleaning products to react only at the interface between the materials to be eliminated and the original substrates to be preserved: only the movements imparted by the tool

lead to surface contact, until the gel is saturated, wiped off and finally cleared. It acts without requiring excessive rubbing over fragile surfaces. Clearance is essential to remove any potentially active residue that would compromise the future stability of the artwork.

- *Control of the action of water:* Since the need to use water has been demonstrated, we sought to tailor its properties and control its action on the soil by buffering and controlling the pH, adjusting the conductivity to near isotonic to the paint surface, and using active substances such as chelates. These variables were adjusted based on the substrates to be cleaned and appropriate proportions in the 'cleaning mixture', because water alone can cause damage and is not efficient enough to reduce the soil.
- *Modifying a mixture of active solvents:* If needed, gels can be prepared with a solvent or mixture of organic solvents with variable polarities in different proportions, to enhance their capacity to solubilize substances from different chemical families: natural resins, synthetic resins, etc.

Protocols for treatment

Once the tests were completed we defined, together with Richard Wolbers, our limitations to modifying the formula (variations and limits for specific characteristics, adaptations to substrates) while respecting the qualities and properties of the original gels.

Removal of soil

The soil analysis and first tests led us to formulate a Pickering-type emulsion that contained both D2 silicone solvent (hexamethyldisiloxane) and cyclomethicone D5 in the solvent phase, and a citrate-diethylenetriaminepentaacetic acid (DTPA) chelate solution in the aqueous phase, having a strong affinity for carbon-based materials (organic soil, combustion residue, etc.) in the gel. The first gel was composed of one part D2; two parts Velvesil Plus; and 10:20:3 w/w of solvent/ Velvesil Plus base gel/ aqueous phase, the last made from 100 ml demineralized water, 0.5 g citric acid and 0.5 g DPTA. The citric acid played the role of a buffer. The pH was adjusted to 6 with a sodium hydroxide (NaOH 1M) solution, and its conductivity was adapted to that of the area to be cleaned.

Elimination of varnish, glazes and overpaint

To eliminate the varnishes, glazes and overpaint, we used a gel with a base of silicone solvent and benzyl alcohol, varying the proportion based on the targeted resins. This second gel was composed of one part D2; two parts Velvesil Plus; and 1:3:20 w/w benzyl alcohol/ Velvesil gel base/D2.

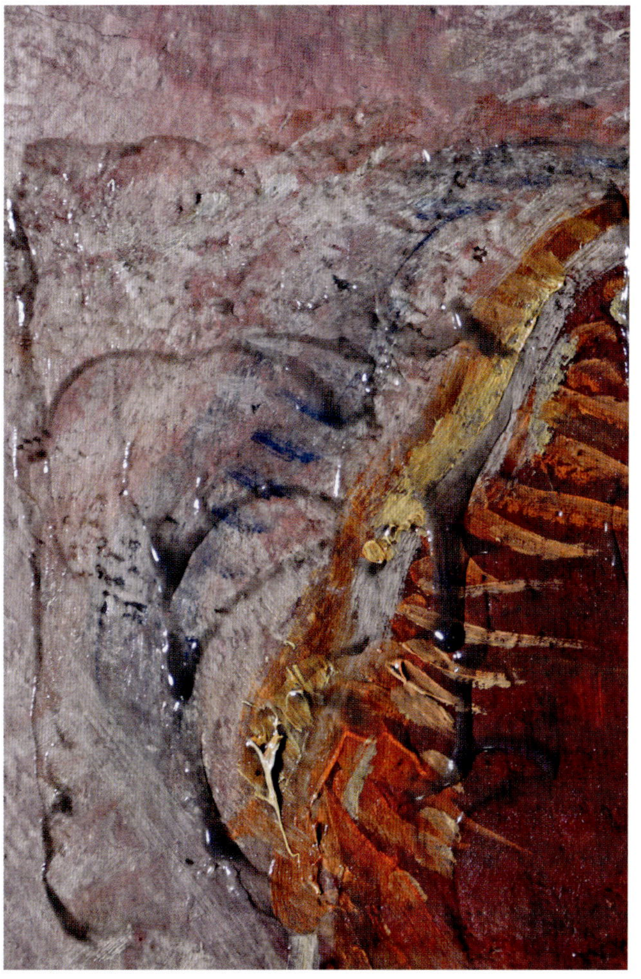

FIG. 7 Eugène Delacroix, *Heliodorus Chased from the Temple*: removing soil with the silicone gel (photo © Alina Moskalik-Detalle).

Elimination of shellac

On the spandrels of the vaulted ceiling, technical imaging revealed the presence of shellac applied in a somewhat random manner, for which the first two gels were found inadequate. Therefore another gel was prepared using a stock gel of 2 g xanthan gum dissolved in 100 ml deionized water. This was futher modified with the addition of 1:20 w/w triethanolamine to the stock gel, and finally 1:20 w/w benzyl alcohol, to form a stable emulsion.

Implementation

The same application was used for all the areas. A small area of approximately 20 × 20 cm was saturated with D4. The gel was applied and spread with a supple brush. The area was then wiped with a soft, absorbent paper and rinsed with D2 on a cotton pad (Figs 7 and 8).

Monitoring and avoiding residues

To evaluate the potential presence of cleaning product residue on the paint surface, a series of samples from the

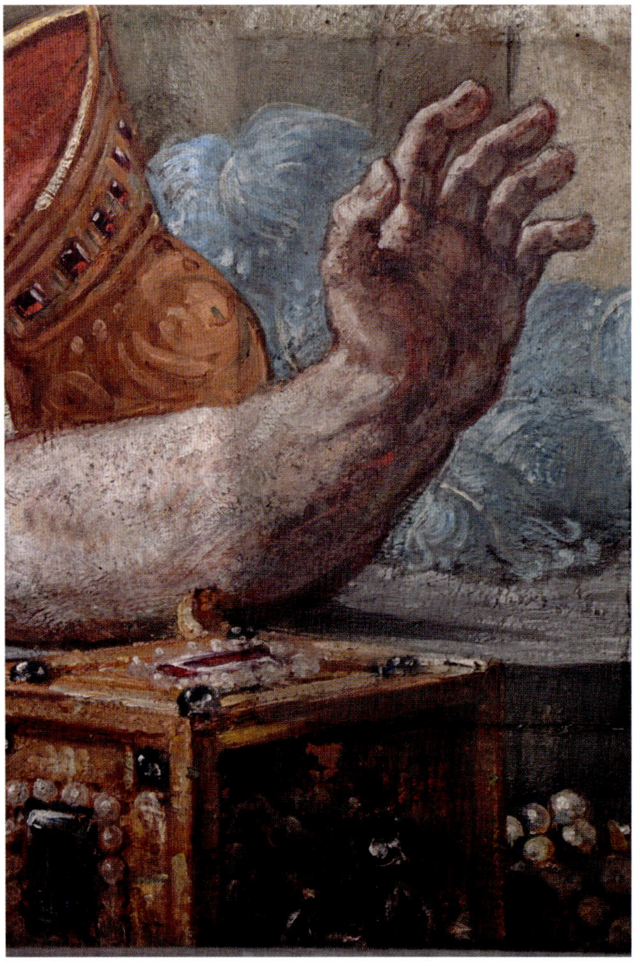

FIG. 8 Eugène Delacroix, *Heliodorus Chased from the Temple*: detail before (right side) and after (left side) removing the soil with the silicone gel (photo © Alina Moskalik-Detalle).

pictorial layers was obtained, before and after treatment, for analysis by Fourier transform infrared (FTIR) spectroscopy.[12] Gel samples were analysed in advance, to serve as a reference. The first series of samples was collected before the consolidation of flaking paint, where sampling was easier. Analysis indicated the presence of a Si-C bond, typical of cleaning product residues: the flaking paint had hindered clearance of the gel. The second series of samples was collected from an area in good condition, where it was possible to carry out a careful clearing procedure. No FTIR signal for the Si-C bond was identified and no trace of residue detected.

Assessment

The cleaning of the entire surface of these monumental mural paintings was performed using silicone products for the first time in France on such a monumental scale. Soil removal was carried out after securing flaking of the pictorial layer, but without prior removal of dust. The result is regular, without any blanching. Returning to the edges of the areas already treated, we found no borders between adjacent cleaned and uncleaned areas. Several gel applications were used to facilitate elimination of soil in the areas carved in relief. There was a low risk of polishing and abrasion since, by using a soft brush impregnated with gel, there was necessarily some degree of mechanical action. The removal of the non-original varnishes, glazes and overpaint proceeded layer by layer, with control over each step made possible by using natural and UV light. The heterogeneous aspects of the original surface and the juxtaposed matte and glossy areas were perfectly respected.

Health and safety

Silicone solvents are used in weak concentrations in the cosmetics industry. For this reason, no toxicity study was carried out on the prolonged exposure to large quantities of these materials during the project to treat an artwork on a monumental scale. However, handling of the waste products is regulated: the cyclomethicone-containing materials were collected and disposed of responsibly through a solid waste method because of the clear and well-established toxicity to aquatic organisms. It is illegal in the EU to dispose of such materials 'down the drain'.

Later work on silicone-treated areas

It was noted that certain readhesion of flaking in the network of fissures and cracks at the edges of the problem areas was difficult to carry out after cleaning with Velvesil Plus. Similarly, the filling of losses with a water-based material was problematic because there was no local adhesion. Did we leave silicone gel behind, after not having rinsed properly? Our examinations of the areas of flaking paint initially tested after cleaning tend to suggest that this is the case: the paint is more retentive of residues and presents more of a challenge for clearance. Vigilance should be maintained when working on such zones.

Specific climatic conditions for the use of D4

The melting point of octamethylcyclotetrasiloxane (D4) is 17.5 °C. When working on-site in winter, the solvent crystallized in the bottle when the temperature fell below 14 °C. We also observed that the solvent crystallized when applied to the cold wall. We acknowledge a potential risk that retained solvent could expand within the wall, affecting the pictorial layer. The addition of a fraction of D2 helps to minimize this inconvenience, but the mixture evaporates more quickly. Impregnation with pure D4 could only begin in April – it was necessary to work at slightly higher temperatures to guarantee the protection properties of the D4 and the gel action on the surface.

Procurement difficulties and costs

The supply chain for large quantities of silicone products adapted to monumental work sites does not yet exist in France. Industrial suppliers offer large quantities, but packaging or stocking them is difficult to manage for independent conservators. Delivery times are extremely restrictive (shipping takes two months from the USA). The long delivery time and restrictions inherent in organizing the project led us to adapt mixtures using more readily obtained products (for example, replacing D4 by a mix of D2 and D5, decamethylcyclopentasiloxane). Suppliers of products for the conservation of cultural heritage should be encouraged to sell the silicone materials in quantities and packaging adapted to their use in conservation.[13] Finally, the high cost can be an obstacle for their use on large-scale projects. For our treatment, we used 205 L of silicone solvents and 30 kg of Velvesil Plus for a total cost of approximately €15,000.

Conclusion

From other projects, we had already gained experience of using gels or microemulsions such as poultices to resolve the problem of surfaces that were difficult to treat by methods that employed organic solvents in very large quantities. This experience with silicone systems offered a solution adapted to the problem of treating Delacroix's paintings, and enabled us to re-evaluate the issues that arise when conserving monumental works on porous materials with a limited ability to control climatic conditions.

It was very reassuring to know that the pictorial layers would be protected by D4 during the cleaning, and that no active agent would penetrate the deep strata or modify the physio-chemical properties of the original layers. Cleaning with solvents, often a very delicate procedure, still does not allow us to distinguish or target layers, particularly when restoration materials whose composition is identical to the original materials have been used at a relatively recent date. We were reassured by the tendency of the silicone systems to eliminate the surface layers progressively, allowing us to arrive at a very homogenous result. Certain non-original glazes in previously abraded areas could be preserved by these selective cleaning methods.

Pre-wetting with neat silicone solvents such as cyclomethicone D4, and the silicone-based Pickering emulsions used to clean the paint surfaces, allowed us to remove the surface soil effectively without stripping wax or oily components from the paint films themselves. The silicone solvents are poor solubilizers of fatty acid and triglyceride materials in general. The emulsions also help to limit the application and amount of aqueous phase brought to the paint surface. In addition, these were surfactant-free emulsions, so clearance of the applied emulsion materials was also easier. The absence of detected residue, confirmed analytically, ensures that the original and the materials added during the present treatment will age well.

The application of the easily manipulated silicone and xanthan gels with a supple brush adds a softness to the conservator's gestures, and eliminates the abrasive action of cotton wool. The adaptability of the composition of gels to the physico-chemical parameters of the surfaces to be treated seems to us important, especially when these cleaning systems are used for mural paintings which may have high water and salt content.

Acknowledgements

The project was made possible thanks to financing by the French Heritage Fondation for the city of Paris and was underwritten by DRAC (Regional Direction of Cultural Affairs) for Ile de France.

Notes

1. The team consisted of Alina Moskalik-Detalle (team leader), Julien Assoun, Diana Bialkiewicz, Claire Dandrel, Diane Henry-Lormelle, Françoise Joseph, Felicja Lamprecht, Tristan Mahéo, Marie-Laure Martiny, Lucia Tranchino. It included the participation of Alexa Beller, intern and student at the Winterthur / University of Delaware program in Art Conservation, Delaware, USA.
2. P. Perrin, *La chapelle des Saints Anges par Delacroix l'église Saint-Sulpice, Esquisses et études préparatoire*, COARC internal document, Paris, 2014. Centre André Chastel, Eugène Delacroix National Museum, Musée du Louvre, *Correspondance d'Eugène Delacroix*, Paris, Centre André Chastel (http://www.correspondance-delacroix.fr) Paris, Musée National Eugène Delacroix, accounts notebook, inventory no. MD 1997-21 (25 June 1854–31 July 1861), expenses for the account of M. Delacroix/St Sulpice church/25 June 1854, accounts notebook, inventory no. MD 1997-23 (24 September 1858–3 August 1861).
3. Composed of: Dominique Cerclet, regional curator for historical monuments, DRAC Ile-de-France; Caroline Piel, general inspector of historical monuments; Sébastien Allard, director of the painting department at the Musée du Louvre; Dominique de Font-Réaulx, director of the Delacroix Museum; Aline Magnien, director of the Historical Monuments Research Laboratory (LRMH); Isabelle Pallot-Frossart, director of the Research and Restoration Centre of the Museums of France (C2RMF); Dominique Martos-Levif, study engineer (LRMH); Isabelle Collet, head curator of the Petit Palais Museum, Paris; Barthélémy Jobert, president of Paris-IV university; Paul Perrin, painting curator at the Musée d'Orsay; and Arlette Sérullaz, honorary curator.
4. The window in a reduced format is visible in an undated engraving signed by Giraudon, annotated 1890 and stamped 'Blondel'. The document is preserved among the civil and religious artworks of the city of Paris.
5. Handwritten note in the collection of the Musée National Eugène Delacroix, a rough draft of the invitation, with a description of the paintings' subjects (MD 2013-1-85), donation of Norbert Ducrot-Granderye.

6. The collection of sketches was established for this project by Paul Perrin, painting curator at the Musée d'Orsay, Paris.
7. LRMH; Scientific Microanalysis of Ancient Materials and Heritage (MSMAP), France; Material Research Study (ERM), France; Richard Wolbers, University of Delaware, Winterthur, Delaware, USA; Institute for Study, Conservation and Restoration of Cultural Heritage at Nicolas Copernicus University, Torun, Poland.
8. By C2RMF.
9. By the Institute for Study, Conservation and Restoration of Cultural Heritage at Nicolas Copernicus University, Torun, Poland.
10. Tests using gels with a base of xanthan gum for targeted actions on overpaint or dammar varnish did not give good results: the action was too rapid, it was difficult to limit the action only to the overpaint while preserving the pictorial layer, and further problems were encountered when clearing while trying to avoid the appearance of blanching.
11. Cyclopentasiloxane/C30-45 alkyl cetearyl dimethicone crosspolymer/PEG/PPG-20/23 dimethicone). Velvesil Plus was then available in France from: SAFIC-ALCAN (www.safic-alcan.fr), their supplier being Momentive Performance, USA. Today it is also available from CTS, 26 Passage Thiéré, 75011 Paris (www.ctseurope.com/fr/). D2 is available in France from VWR International SAS (fr.vwr.com). D4 and D5 were available at that time from Kremer Pigmente GmbH, Germany (www.kremer-pigmente.de) but this supplier offers only D5. D4 is now also available from CTS.
12. Analyses were carried out by LRMH.
13. For this project, the Sisley cosmetics company helped us to find a supplier for D2, for which we are grateful.

References

Hannoosh, M. (ed.) 2009. Eugène Delacroix's *Journal*, 2 vols. Paris: Éditions Corti.

Johnson, L. 1989. *The Paintings of Eugène Delacroix: A Critical Catalogue*. Oxford: Clavendore Press.

Moskalik-Detalle, A., Assoun, J., Joseph, F., Mahéo T. and Martiny M.-L. 2016. *Église Saint Sulpice, chapelle des Saints Anges, peintures murales et toile marouflée d'Eugène Delacroix – study report, March 2016*, unpublished document COARC (Conservation of Religious and Civil Artwork).

Piot, R. 1931. *Les palettes de Delacroix*. Paris: Librairie de France.

Sérullaz, M. 1963. *Les peintures murales de Delacroix*. Paris: Les éditions du Temps.

Spector, J.J. 1967. *The Murals of Eugène Delacroix at Saint Sulpice*. NewYork: College Art Association of America.

Wolbers, R. 2000. *Cleaning Painted Surfaces: Aqueous Methods*. London: Archetype Publications.

Authors' addresses

- Alina Moskalik-Detalle, Igny, France (moskalik-detalle@wanadoo.fr)
- Julien Assoun, Paris, France (jassoun75@gmail.com)
- Françoise Joseph, Paris, France (dcueco.frjoseph@gmail.com)
- Marie-Laure Martiny, Paris, France (marie-laure.martiny@wanadoo.fr)
- Marie Monfort, COARC, Paris, France (marie.monfort@paris.fr)

Gels: evolution in practice

Chris Stavroudis

Introduction

The first time I used a gel cleaning system, as a pre-program apprentice, I had no idea how significant gelled systems would become to my conservation practice. We prepared Ruhemann's ammonia wax paste emulsion, which was gelled by virtue of the high wax content of the solvent/wax continuous phase. The recipe (Ruhemann 1968: 318) was printed as follows:

50 g beeswax
150 mL turpentine or white spirit
15 mL 0.880 [*sic*] ammonium carbonate diluted with
30 mL water

Interestingly, Ruhemann (1968: 206–7) described the paste thus:

> Its special advantages are: it 'opens up' certain films that do not respond to the usual solvents; it 'stays put', does not run or spread, so that the cleaning operation can be confined if desirable to minute spots; its action can be accurately modified by the concentration of the mixture, the thickness of the application and the length of time it is left to act; the strong ingredient evaporates fast but is held *in situ* by the wax as long as desired; when the layer of hard varnish or overpaint is sufficiently softened it can be removed with a mild solvent sometimes with water.

Many of the advantages listed by Ruhemann are still hallmarks of the gelled systems we use today.

Beginning in the 1980s Richard Wolbers at the Winterthur/University of Delaware Program in Art Conservation, at Winterthur, Delaware, USA, introduced a number of 'modern' gelled systems to the conservation field (Wolbers 2000). I have systematized, fine-tuned and incorporated many of these systems into a methodology known as the Modular Cleaning Program (MCP). The MCP is a computer program built in FileMaker Pro that assists with building cleaning systems. Formulae are calculated from physical constants (e.g., pK_a values or Hansen solubility parameters) and specifications such as the target solution concentration or pH. It must be understood that while the MCP provides a systematic approach to building cleaning systems, it *never* makes recommendations to the conservator.

The MCP is both an approach to cleaning and a set of premixed concentrate solutions that can be combined to create test cleaning systems. Because the test cleaning solutions can be mixed rapidly from concentrated stock solutions, the conservator is able to test many cleaning options to arrive quickly and empirically at an optimal, tailored cleaning solution.

Carbopol and solvent gels

Carbopol was the first and perhaps the most significant of the 'new' gelling systems introduced to the conservation profession by Wolbers. Over time it has been followed by a number of other gelled systems, including Pemulen and xanthan gum and others for aqueous systems, and Velvesil Plus and the KSG products from Shin-Etsu for very low polarity solvents. But the history of modern cleaning gels begins with Carbopol.

Carbopol is a high molecular weight polyacrylic acid[1] (Swift *et al.* 2016). It has a carbon backbone with a carboxylic acid group on every other carbon atom in the chain (Fig. 1a). The Carbopol molecule gels a solution by unfurling and adding a three-dimensional matrix of tangled molecules to add structure to the solution – the spaghetti model of gelation. The trick to gelling a solution is modifying the pendant carboxylic acid groups so they are compatible with the solution to be gelled. Fig. 2a shows a model of a Carbopol molecule in its 'unhappy', globular state, in this case as a protonated macromolecule in water. Fig. 2b shows the results of a computer simulation of that molecule after adding sufficient base to deprotonate all the acid groups. This unfurling in a solution is the secret of gelling water or solvents with Carbopol or its cousin Pemulen.

Carbopol was introduced to the conservation world in the 1980s as a vehicle for gelling solvent-based cleaning systems.

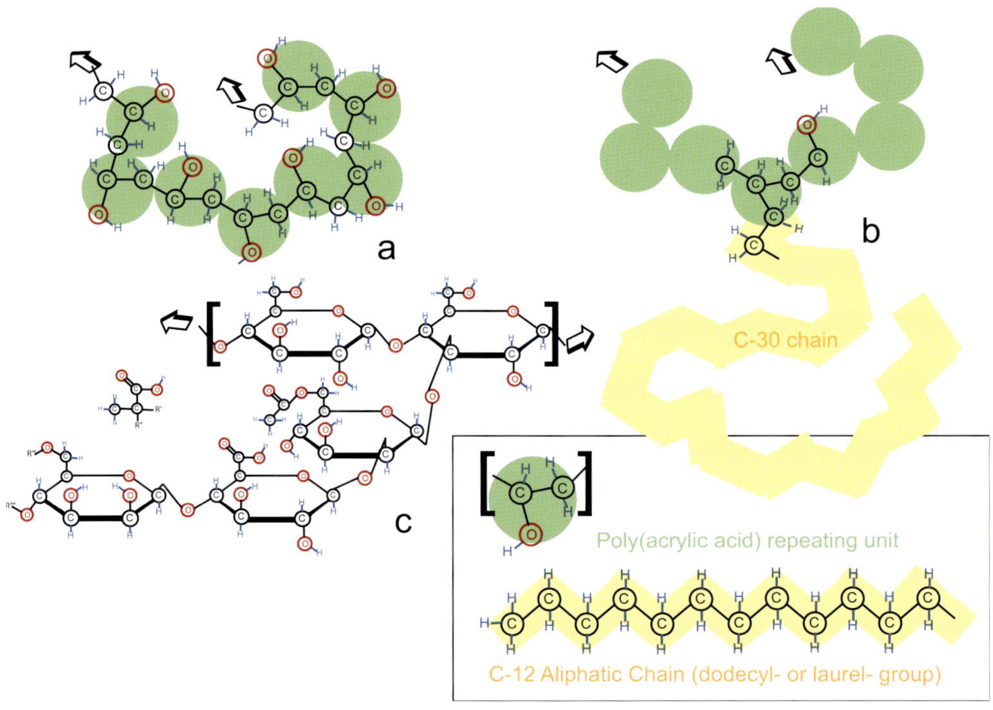

FIG. 1 The chemical structures of Carbopol (a), Pemulen (b) and xanthan gum (c).

Solvent gels are amazing materials – that a polymer at 2% concentration could gel a solvent is astounding. The gels are clear, although sometimes yellow or amber in color due to the discoloration of the amine components, and have good mechanical working properties. They restrict solvent spread and evaporation, are viscous enough to be applied to vertical surfaces and not run, and they work well when applied by brush and cleared with swabs.

Perhaps the most significant feature of Carbopol-based solvent gels is that the utility of the final gel is greater than the sum of the parts. That is to say a Carbopol/amine solvent gel functions differently from those same solvents applied either as a free liquid, a poultice or in a gel form with another gelling system. This is the 'magic' of the Carbopol solvent gel.

As mentioned above, the trick to gelling a liquid with Carbopol is to make the chemical environment of the polyacrylic acid compatible with the solvent to be gelled. Tertiary and secondary amines are used to bridge between the acid groups on the polymer, the amine/carboxylate salts, and the solvent. This increases the solubility range, as well as the range of clearance solvents that can be used. Ethomeen C-25 and Ethomeen C-12 are tertiary amines used to gel high- and mid-polarity solvents, while Armeen 2C is a secondary amine used to gel low polarity solvents (Fig. 3).

While solvent gels can often be replaced with polymeric stabilized emulsions in practical use (which will be discussed later), they still have their place in the 'toolbox' of practicing conservators. The amount of gel residue left behind following clearing has been thoroughly studied (Stulik and Miller 2004) and the amount of residue on a non-porous surface is very small. The residue is a complex salt formed from the polymer and base. However residue remains a significant concern. While polyacrylic acid is reasonably stable – it is the primary compound in Golden Polymer Varnish[2] – the amines are decidedly prone to yellowing. This can be observed in the bulk amine and in older solvent gels. Another concern is that excess amine can form an adduct with carboxylate groups on the artwork surface in the same way they associate with the acrylic acid groups on the polymer.

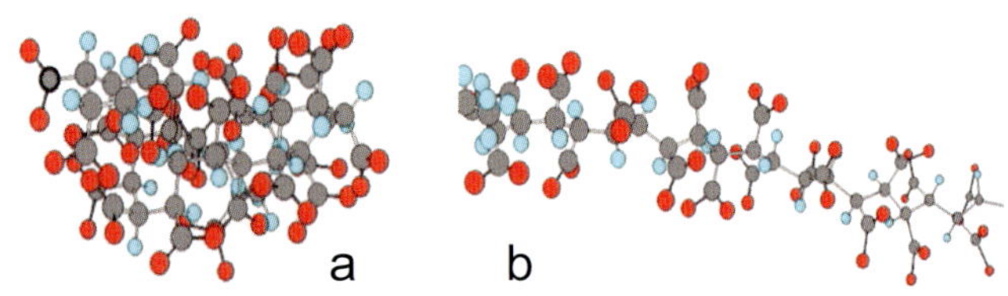

FIG. 2 The Carbopol molecule in water, protonated and 'unhappy' in a globular state (a) and deprotonated in the unfurled state (b) allowing it to gel the water.

Initially, when solvent gels were quite new and recipes for them were new to the practice of many, the standard method to formulate a solvent gel with a new solvent or solvent mixture was to start with Carbopol, add solvent, add the amine and then slowly add water until a gel formed. If the gel did not form or the Carbopol began to precipitate, one fiddled with the components, adding more solvent or more Carbopol or more amine or more water. Eventually, either a gel was formed or the Carbopol/amine precipitated into a 'golf ball' at the bottom of the container.

In those days, a painting arrived in my studio with a very discolored varnish, assumed to be natural resin. Conventional solvents were either too strong, affecting the paint layer below, or too weak to remove the varnish layer. Using the above (less than optimal) system of formulating a solvent gel, I made a xylene-based solvent gel with Carbopol and Ethomeen C-12. To get the gel to form I had to mix in additional Ethomeen and adjust the other components. Ultimately I had a usable

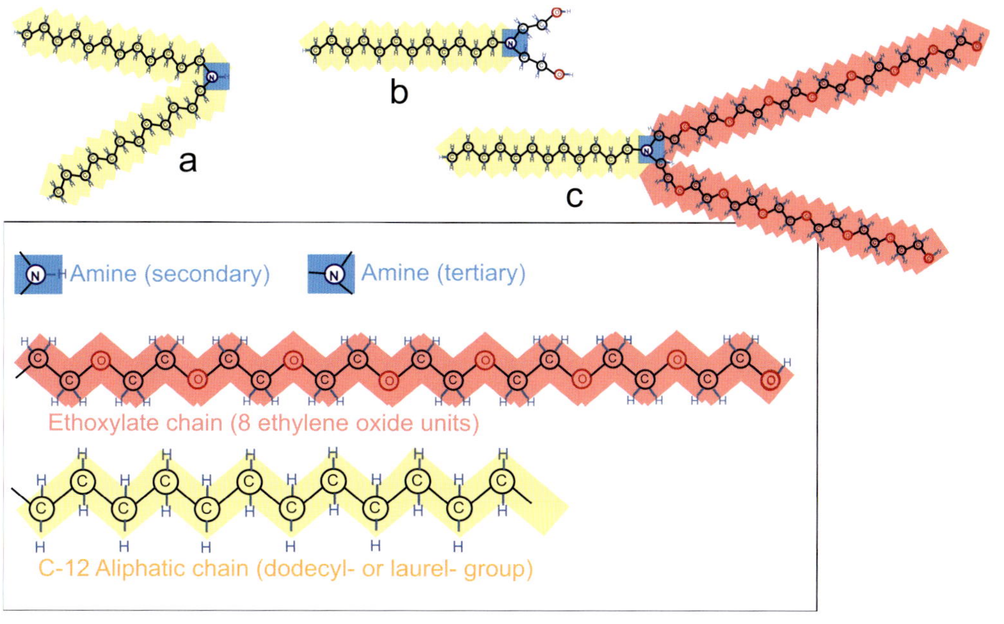

FIG. 3 The chemical structures of Ethomeen C-25 (a), C-12 (b) and Armeen 2C (c).

TABLE 1 Properties of the amines used to create solvent gels with Carbopol.

Amine	MW	Amine type	Molar water capacity	Maximum amount per 2 g Carbopol	Amount water per 2 g Carbopol
Armeen 2C	381	Secondary	2	10.0 g	0.9 mL
Ethomeen C-12	290	Tertiary	4	7.6 g	1.9 mL
Ethomeen C-25	860	Tertiary	17	22.6 g	12 mL

gel that safely removed the varnish layer without affecting the paint below. The solvent gel was applied with a brush, removed with a swab and rinsed repeatedly with xylene-dampened swabs. At the end of the day, the entire surface was rolled with mineral spirits. When the cleaning was complete, the visual outcome was excellent. However, the olfactory results were less satisfying: the painting reeked of Ethomeen C-12. Repeated rinsing with various solvents and water did nothing to reduce the odor. Clearly, excess amine in the gel had combined with carboxylic acid groups on the surface of the painting.

The Ethomeens and Armeen are hybrid surfactants consisting of hydrophobic linear alkanes, an organic base, the hydrophilic moiety, and, in the case of the Ethomeens, polar ethoxylates.[3] When used to make a Carbopol gel, the amine in the solvent environment is in its molecular form acting as a free base. In this form it is not water-soluble but in the presence of a few water molecules is able to interact with the carboxylic acid groups on the Carbopol (or the paint surface). However, if the amine is in an aqueous environment and the pH is lowered, the amine is protonated, becomes water-soluble and is a functional cationic surfactant. Knowing this, I rolled the surface of the painting with a dilute acetic acid solution. The amine became readily water-soluble and rinsed off the surface in a foaming froth. When the foaming stopped, the surface was given an overall rinse first with water, then mineral spirits. When dry, the painting no longer smelled of Ethomeen and the treatment was completed.

This method of working was not acceptable. It led me to think about the chemistry of Carbopol-based solvent gels. Simple stoichiometry allows the calculation of the maximum amount of amine to be added to a given amount of Carbopol. The molecular weight of the repeating unit in Carbopol is 76 AMU. Knowing the molecular weight of the amines allows this maximum 'safe' value to be calculated, as listed in Table 1. While it is perfectly safe to use less amine, more amine than this stoichiometric amount is to be avoided.

Once the Carbopol/amine stoichiometry was established, I made careful measurement of the amount of water required to form a 'beautiful gel' for each amine and appropriate solvents. From the volume (mass) of water added I could calculate the number of moles of water required to form the gel for a given number of acid groups and amine molecules. The empirical testing showed that two water molecules per amine/Carbopol linkage are required to gel a low polarity solvent mixture with Armeen 2C, four water molecules with Ethomeen C-12, and 17 water molecules per amine molecule with Ethomeen C-25. While this does not allow the structure of the carboxylic acid/amine adduct to be determined, we can make reasonable assumptions of where the water molecules are likely located. Fig. 4 shows the likely arrangement for the three amines. Informal testing justifies the assumption that two molecules of water are necessary to establish the bridge between the amine and the carboxylic acid group. With that in mind, we can surmise that the two ethanol groups, each a single ethoxylate addition, on the amine in Ethomeen C-12

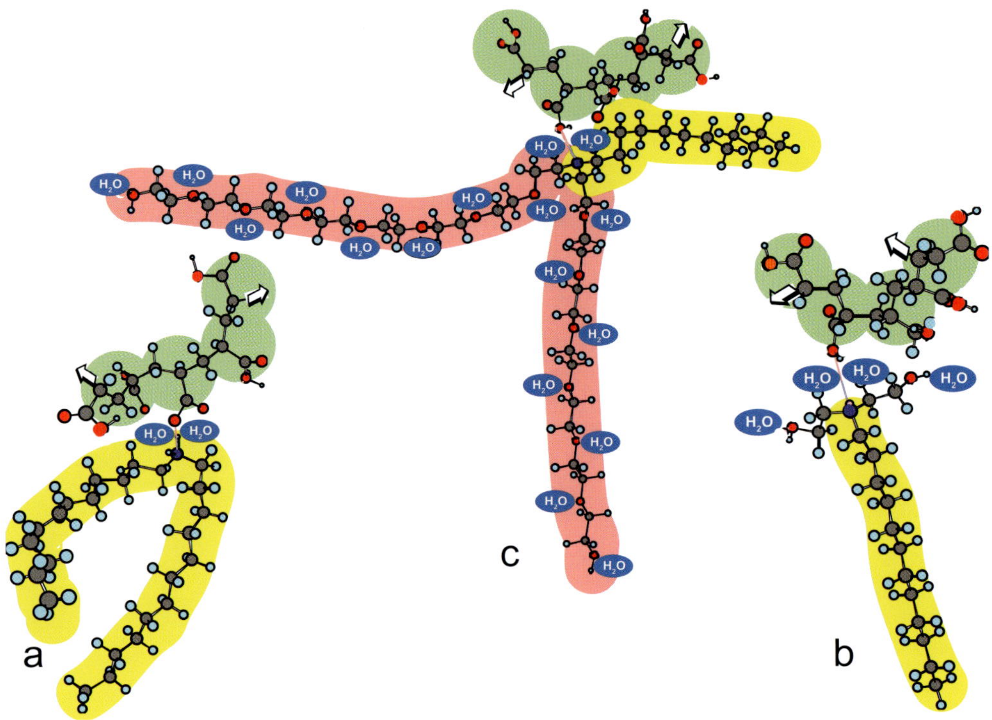

FIG. 4 The proposed adducts of the amines on a polyacrylic acid chain and the location and number of water molecules that allow the gel to form. Again Armeen 2C (a), Ethomeen C-12 (b) and C-25 (c) are shown.

will likely each require an additional molecule of water. While the endpoint of the addition of water to form a 'beautiful gel' with Ethomeen C-25 was impossible to determine precisely, empirical testing indicated a range of 14–20 water molecules, thus a value of 17 seems reasonable, assigning one additional water molecule to each of the 15 oxygen groups on the two ethoxylate chains.

It is worth mentioning that additional water may be necessary to gel solvents that interact strongly with water. Acetone, in particular, absorbs a given quantity of water before any is available to interact with the Carbopol/amine system. My assumption is that the amount of water absorbed by the acetone is related to the acetone/water mixture of 99:1 parts by volume (0.044 mole fraction water). While this is technically not an azeotrope according to most references,[4] some sources report this mixture as an azeotrope or one that forms an azeotrope at higher pressures (Weast 1972: D-3).

As yet, the MCP software cannot match an amine or combination of amines to a specific solvent or solvent mixture to create a successful gel. However, given the amount of Carbopol, it will calculate the amount of amine or mixture of amines as well as the amount of water necessary to form the gel. It also attempts to calculate the additional water necessary to compensate for the water being preferentially absorbed by certain solvents. As for all solvent mixtures, the MCP calculates Hildebrand, Hansen and Teas parameters. These calculations do not, however, take into account the Carbopol and amine present in the gel that gives the gel its 'magic'.

Personally, I use solvent gels less and less frequently, out of concern for residues on more porous surfaces. If a difficult cleaning can be resolved with another system, commonly a gelled emulsion, I will opt for that system. However, there are times when a solvent gel applied to a non-porous surface can be found to give the best cleaning results. One of the most subtle, safe and effective gels for fragile paint below a natural resin varnish is a xylene/benzyl alcohol gel, made as follows:

2 g Carbopol
7.6 g Ethomeen C-12
67 mL xylene
33 mL benzyl alcohol
1.9 mL water

Another very useful gel is Wolbers' gel for the removal of polyurethane varnish, first published as the treatment of a Philadelphia piecrust table (Wolbers 1988: B6), using the following recipe. Note that the amount of Ethomeen C-25 is half the stoichiometric maximum (16.9 g or 16.3 mL) for the amount of Carbopol.

1.5 g Carbopol
8 mL Ethomeen C-25
200 mL acetone
50 mL benzyl alcohol
25 mL water

As with all Carbopol gels, the most expedient method of making a solvent gel is to use a glass jar with a solvent-proof lid, preferably lined with Teflon, both to make and store the gel. The solvent or solvent mixture is measured and set aside in a fume hood. The Carbopol is weighed directly into the jar. The amine is added to the jar very carefully by weight, taking great care not to overshoot the target weight. The amine/Carbopol is briskly mixed into a smooth paste (a small amount of the solvent mixture can be added at this point). Once the Carbopol and amine are well mixed, the balance of the solvent is added to the jar. The jar is capped and given a quick shake. Then the water is added and the solution shaken vigorously until the gel has formed.

Common sense dictates that the rinse solution must possess sufficient polarity to completely thin and dissolve the solvent gel, to avoid leaving large amounts of residue on the surface. Only when the clearance solvent is compatible with the Carbopol/amine system can the conservator obtain the very low level of residue promised by the research cited above. On the other hand, it is often desirable to have the polarity of the rinse solvent somewhat lower than that of the cleaning system to avoid disturbing the difficult-to-have-achieved cleaning result. Thus the selection of the clearance solution is nearly as important as the selection of the solvent gel and must be determined before applying the gel to the artwork's surface.

Unfortunately, the MCP cannot calculate or evaluate potential solvent systems for clearing a given solvent gel. Therefore the following testing procedure should be used before applying a gel on a surface. Mix a small amount of the proposed clearing solution in a small cup. Place a small amount of the solvent gel on a shiny spatula and immerse it in the solvent mixture. Observe the gel using the mirrored surface of the spatula as a reference. If the surface of the gel begins to cloud, the polarity of the gel is not compatible with the solvent mixture. If the gel passes this test, disperse the gel into the solvent mixture and make sure it can be completely thinned which will assure that the gel can be completely removed on a swab dampened with the clearance solution.

Aqueous gels and emulsions

For non-rigid, aqueous gel systems, Wolbers has introduced these gelling agents (in roughly chronological order): hydroxypropylmethylcellulose (HPMC), Carbopol, Pemulen and most recently xanthan gum. While each has its place in the conservator's formulary, xanthan gum has effectively replaced both Carbopol and Pemulen for most aqueous cleaning systems.

In aqueous systems, cellulose ethers like HPMC thicken the solution but, in my experience, do not add any special properties beyond simple gelling. Their compatibility with enzymes gives them special utility in cleaning systems that rely on preserving biological activity. Polyacrylic acids, however, are much more effective in aqueous cleaning systems. In addition to gelling a cleaning solution as effectively as HPMC, they bring a mild chelation-like capability to the cleaning system by virtue of the carboxylate groups along the polymer.

Some years ago, in the pre-Pemulen/xanthan gum days, I formulated a gelled surfactant system in Carbopol to remove recently applied 'rolling writer' ink from a paint surface. Working under a microscope, the thick gel was painted along the ink line as narrowly as was possible. As testing with mock-ups had promised, the ink lifted up off the surface and was suspended in the gel. While I could not compare the effectiveness of the Carbopol system with a similarly formulated cellulosic gel, I feel confident in asserting that the quasi-chelating behavior of the polyacrylic acid groups added that special 'magic' that Carbopol (and Pemulen) do so well.

This extra functionality of polyacrylic acid systems adds a dramatic increase to the efficacy of resin soap gels, allowing deoxycholic acid-based cleaning systems to work on aged natural resin surfaces with a much greater frequency of success than a similarly formulated system gelled with HPMC. A typical 'resin soap' formulation built with the MCP would be:

> 2 mL deoxycholic acid adjusted to pH 8.5 (resin soap)
> 2 mL (0.25M) citric acid adjusted to pH 8.5 (chelator)
> 1 mL (0.25M) bicine adjusted to pH 8.5 (pH buffer)
> 5g 4% Pemulen to pH 8.5
> (Optionally, 2% benzyl alcohol or n-butanol can be added as a co-solvent)
> Note that the final volume is 10 mL, so when the concentrate solutions are diluted the citrate is at 0.05M, the bicine is at 0.025M, the deoxycholic acid is at 10× the critical micelle concentration (CMC), and the Pemulen is at 2%.

As implied in the above recipe, I have switched from using Carbopol to gel aqueous solutions to using Pemulen. Pemulen is a modified polyacrylic acid, a block polymer of both polyacrylic acid and polyactylate esters. The ester side chains are 10-30 carbons in length (see Fig.1b). The aliphatic chains form virtual micelles in the gel analogous to the micelles formed by surfactants. These structures give Pemulen the ability to stabilize non-polar materials, including solvents, into the aqueous gel. Pemulen can suspend non-polar grime into the gel structure without the need for an explicitly added surfactant. This means that upon clearing (rinsing) the gel from the surface, there is no concern about surfactant remaining on the artwork.

An emulsion consists of two phases: the continuous phase and the dispersed phase. The dispersed phase is stabilized into the continuous phase by surfactant micelles in a conventional emulsion or the virtual micelles created by Pemulen. One method of stabilizing an emulsion is to gel the continuous phase to restrict the suspended dispersed phase from coalescing and thereby breaking the emulsion. Pemulen uses both the virtual micelles and the thickening of the continuous phase to create very stable emulsions. This dual nature of Pemulen's ability to form oil-in-water emulsions means that any non-water miscible solvent can be emulsified into a Pemulen gel.

By contrast, conventional emulsions built with a surfactant are very difficult to formulate. Theoretical models of emulsion formulation are not predictive of potential emulsion stability. The proportion of the continuous phase versus dispersed phase and the type and amount of surfactant are generally determined by trial and error. The presence of electrolyte salts in the aqueous phase generally interferes with emulsion formation and adds an additional layer of complexity to the problem.

An emulsion offers a number of benefits to a cleaning system because it can function in two distinctly different cleaning scenarios. When confronted with an intermixture of water-soluble and solvent-soluble materials, the emulsion can simultaneously soften or solvate both materials. The other scenario is to suspend an active agent as the dispersed phase into an inactive continuous phase. In this case, the continuous phase reduces and restricts the action of the dispersed phase on the surface of the artwork. The proportion of the dispersed phase to the continuous phase affords a greater degree of control of the action of the emulsion.

Pemulen belongs to a class of materials referred to as polymeric emulsion stabilizers.[5] These materials, as their name implies, stabilize an emulsion by incorporating surfactant-like structures on a polymeric backbone. The solubility of the polymer determines the continuous phase of the emulsion while the functionality of the surfactant-like structures determine the nature of the dispersed phase. This pseudo-surfactancy of Pemulen was designed to allow the formation of oil-in-water emulsions for the cosmetics industry. Conservation can exploit this designed-in feature of Pemulen for creating solvent-in-water emulsions.

Pemulen's capacity for allowing the incorporation of ionic material makes it a perfect complement for the MCP. The gel can be built with the standard MCP components – concentrated solutions of a pH buffer, and optionally a chelating agent, and/or a surfactant. Once the gel is built and mixed to a smooth, uniform consistency, solvent can be added and stirred to form an emulsion. A recipe for such a system would be similar to the resin soap gel presented above but after mixing, a solvent phase would be introduced and emulsified into the gel.

The disadvantage of using Carbopol or Pemulen is also one of their best features - they are polyacids. The polymers have to be reacted with a base to create an aqueous gel. The pH of the gel must be set when neutralizing the acid groups. In addition, setting the pH of a cleaning system adds control to the cleaning process. The specific polymers used in conservation were chosen to form high viscosity gels at low concentrations. This makes checking the pH of these gels very difficult. pH meters do not give a reliable reading with these gels, possibly due to the low permeability of ions through the gel or because the gel clogs the salt bridge between the measuring and reference electrodes. pH papers can give a reasonable reading of a slightly thinned gel, but the precision of the value is less than ideal. The best way to measure the pH of the gel is with a pH indicating dye. (Obviously, the dye cannot be added to the gel that will be used in a cleaning.) The pH of the gel can be determined by visual comparison with the same dye at the same concentration in a reference solution known to be at the desired pH.[6]

This represented typical professional practice, until Wolbers introduced xanthan gum to art conservators. Xanthan gum is a polysaccharide (Fig. 1c) that is produced by the fermentation of carbohydrates by the bacterium *Xanthomonas campestris*. It is classified as an anionic polymer, due to the presence of carboxylic acid groups in the structure. The acid dissociation constant (pK_a) for xanthan gum is non-linear and a function of other ions present on the polymer, but is generally in the range of 2.4–3.3 (Young *et al.* 1986). This means that within the pH range that conservators typically use, xanthan gum can be considered to be nonionic. The carboxylic acid groups have been deprotonated and, in dry form, are paired with sodium, potassium and/or calcium ions. The net effect is that its addition to a solution will not change the pH of the cleaning system. (This is in stark contrast to Pemulen and Carbopol, which, as discussed above, require the addition of a base to set and adjust the pH of a cleaning system thickened with either of them.)

Before going into detail as to the reasons, xanthan gum has replaced Pemulen in aqueous gels and emulsions in my practice. The fact that xanthan can be added directly to a solution to gel it without having to worry about shifting the pH is a time saver. As it is a polymeric emulsion stabilizer as well, it makes water-in-oil emulsions just as effectively as Pemulen TR-2, and has about the same capacity for solvent in the emulsion. Lastly, it can be made as a 7.5% stock gel and used in the MCP – 1 g or 1 mL being added as one of the five components in a test solution, yielding a 1.5% final xanthan concentration which is a bit thin, but fine for testing purposes and many actual cleanings.

Xanthan gum, being a biopolymer, is much more complex than Pemulen and functions very differently. The molecules have a tendency to form into a double helix, but that helix can unwind and reform at room temperature. In low ionic strength solutions, dilute solutions or at elevated temperatures, the molecules take on a random coil configuration. At cooler temperatures, higher concentrations or with greater ionic content, the chains recombine to form rigid helical rods. At low ionic strengths, this conformational change occurs at 36 °C (Brunchi *et al.* 2016) while at ionic strengths above 0.15M, the helical rod configuration is maintained up to 100 °C.[7] When the molecules reform into their helical configuration, they do not line up perfectly and there are leftover fuzzy, tangled ends on the more rigid helix structure. In turn, the fuzzy ends can entangle with other fuzzy ends to form a semi-rigid, gel-like structure although, technically, it is in actuality a very viscous liquid.

When the fuzzy ends entangle, they form a 3D network of pockets with the sides consisting of the helically wound portions of the xanthan molecules. As the helical sections of the xanthan molecules are held tightly by intermolecular hydrogen bonding, a less polar surface is presented on the inner surface of the pockets. Water molecules can then fill the space inside the boxes, hydrogen bonding with the xanthan and one another. When the gel is stirred, the entanglements of the fuzzy parts are mechanically disrupted and the gel thins – a property known as shear thinning. Once the mixing stops, the pockets reform. This reforming begins as soon as the stirring stops but the gel continues to thicken for a while.[8] It takes perhaps 10 minutes for full thickening to take effect. This can be seen when making a xanthan gel with the MCP. The 1.5% solution seems rather thin initially but becomes much more viscous upon standing.

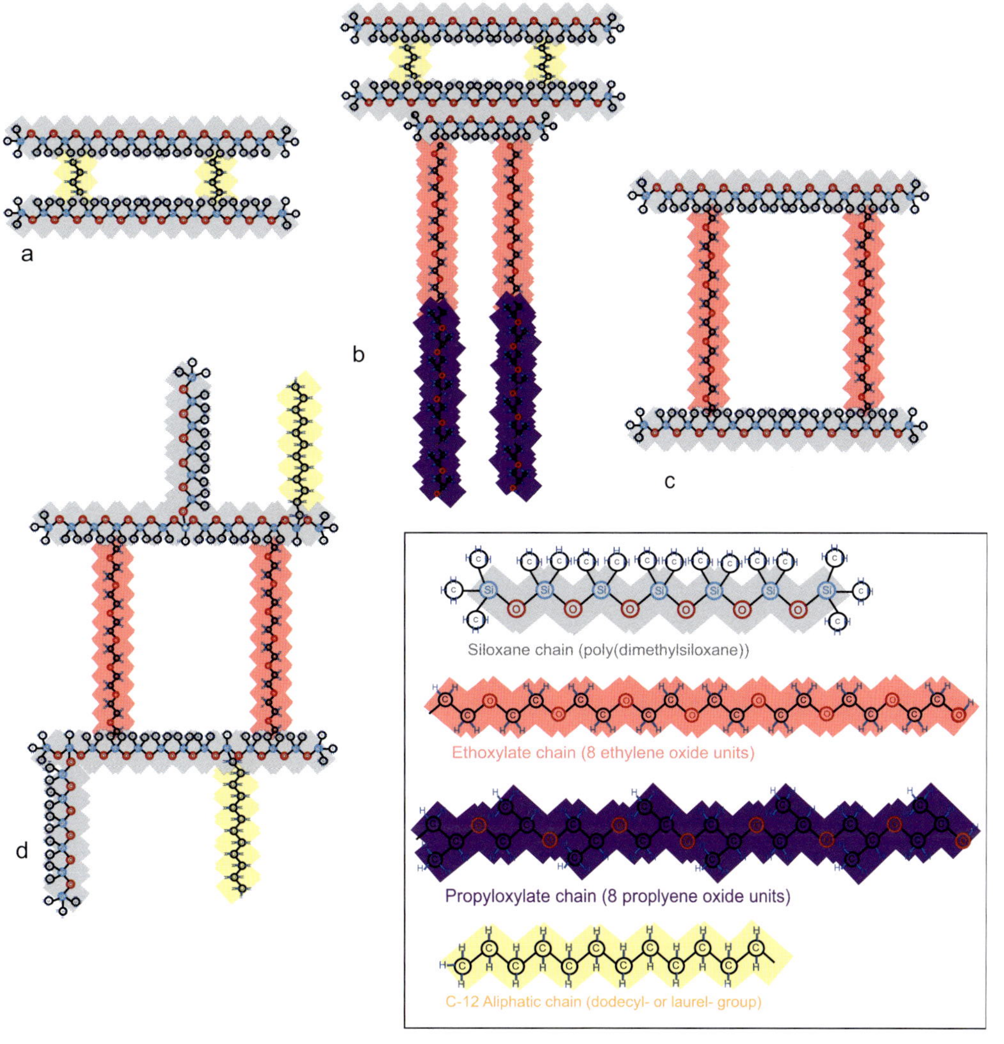

FIG. 5 Schematized structures of Velvesil 125 (a), Velvesil Plus (b), KSG-210 and -240 (c) and KSG-350Z (d).

A typical xanthan gum emulsion built with the MCP would contain:

1 mL Bis-Tris concentrate adjusted to pH 6.5 (pH buffer)
1 mL citric acid concentrate adjusted to pH 6.5 (chelator)
1 mL Brij S-100 concentrate (nonionic surfactant)
1 g 7.5% xanthan gum gel concentrate
1 mL (distilled or deionized) water (diluent, to 5 total mL)
0.5 mL (10%) benzyl alcohol (dispersed phase)

Xanthan gum functions as an emulsion stabilizer as well as a gel thickener. It is, however, not an emulsifier itself. It does not have the non-polar regions present in Pemulen that form virtual micelles. Instead, the pockets that form when stirring is stopped can form around mechanically dispersed droplets of solvent or oil. A peculiar advantage to xanthan emulsions is a by-product of the dispersed phase being held in the 3D network. With agitation, that network is disrupted making the solvent more available to interact with the surface. This is a rare instance where working an area that needs additional attention actually makes a difference in the cleaning efficacy.

The most commonly used solvent in either emulsion system is benzyl alcohol, which has low volatility and very low toxicity. The combination of the aromatic and alcoholic functionalities in benzyl alcohol provides an excellent synergy with the water and other components in the aqueous phase, most commonly a chelating agent. Xylene or mineral (white) spirits can be emulsified into either gel as well. Xylene offers aromaticity without the alcoholic functionality of benzyl alcohol. Mineral spirits can be used in cleanings where neither aromaticity nor alcoholic functionality is desired. Any of these or other non-water-miscible solvents can be combined to provide intermediate activity in the emulsion.

If we can use an emulsion to solvate or swell a material that would otherwise require free solvents or solvent gels to remove, we have built a cleaning system that uses much less solvent. In addition, that solvent emulsified into the gel structure is much less volatile than a free solvent so it is better for the health and safety of the conservator. Being able to use less solvent in a cleaning is also a 'greener' approach to conservation treatments. Finally, using the MCP to build test batches of emulsions with great precision also minimizes the amount of materials needed for testing and the amount of waste.

Clearing an aqueous gel is best done with 'pH adjusted water', i.e., water that has had its pH and conductivity set with a dilute mixture of ammonium hydroxide and acetic acid. Both components of the 'adjusted water' evaporate leaving no residue. With Pemulen it is particularly important to clear with water adjusted to the same pH (or higher) as the gel, but generally it is preferable to clear with 'adjusted water' of the same pH as the cleaning system. It should be noted that 'adjusted water' is not necessarily buffered water. The acetic acid (pK_a 4.76) is a buffer from pH 3.8 to 5.6 and the ammonium hydroxide (pK_a 9.25) buffers from pH 8.4 to 10.2. The pH range from 5.6 to 8.4 is adjusted but is not buffered.

Silicone-based polymeric emulsion stabilizers

As discussed above, Pemulen and xanthan gum can be used to stabilize solvents into a water-based continuous phase. It is often desirable, however, to have a low polarity continuous phase with a dispersed aqueous phase. Complementary to the conventional emulsion system, the reversed phase emulsion allows water to be delivered to an otherwise water-sensitive surface. This class of material was introduced to the broader conservation community by Wolbers at *Cleaning Acrylic Paint Surfaces: Research into Practice* (CAPS-RIP), a colloquium in Los Angeles sponsored by the Getty Conservation Institute in 2009. Along with microemulsions, which will not be discussed here, Wolbers introduced Velvesil Plus,[9] a silicone-based polymeric emulsion stabilizer in cyclomethicone solvent.

Velvesil Plus consists of the solvent cyclomethicone, Velvesil 125, a silicone-based, lightly crosslinked dimethicone polymer gel (Fig. 5a) and a silicone-based surfactant with an ethoxylate and a propoxylate chain grafted to the siloxane backbone (Fig. 5b). In some markets, Velvesil Plus has been replaced with Velvesil 125 and SF1188A, a slightly different silicone, ethoxylate and propoxylate surfactant, that are mixed together by the end user. Although the silicone gel and surfactant components are not covalently linked, they can be considered to be lightly attached by virtue of dispersion forces between the siloxane chains. Velvesil Plus uses the long ethoxylate and propoxylate chains on the surfactant to create micelles that can stabilize water into the silicone solvent continuous phase.

Tiarna Doherty, Smithsonian American Art Museum, was the first to use Shin-Etsu KSG-210 as a substitute for Velvesil Plus (Doherty and Shockey 2013). Wolbers then started looking at other materials in this product line, and found KSG-350Z. KSG-210, -240 and -350Z are true Pickering particle emulsifiers.[10] The polymer in KSG-210 and -240 (Figure 5c) have bridging ethoxylate chains between two silicone chains that provide the higher polarity regions that can hold the emulsified aqueous phase. In KSG-210, the polymer is gelled in a linear silicone solvent while -240 is gelled with cyclomethicone. The most interesting of the family is KSG-350Z (Fig. 5d). The emulsion stabilizer molecule is a modified version of the molecule used in the -210 and -240 products. It has been grafted with additional silicone chains as well as aliphatic chains that give it compatibility with low-polarity hydrocarbon solvents.[11]

In all cases, the silicone-based emulsion stabilizers are as easy to use as xanthan or Pemulen. The thick stock gel is thinned with cyclomethicone to a slightly stiffer consistency than desired for the final cleaning system. The aqueous phase is mixed in and the emulsion is ready: the more efficient the mixing, the smaller the droplets of the dispersed phase. Beyond hand mixing and in the absence of a laboratory shaker, the emulsion can be pushed through a small gauge needle with a syringe to further reduce the size of the emulsified aqueous phase. Usage is equally simple: the milky emulsion is applied, typically with a white-fibered synthetic brush, wiped away with a cotton swab, and rinsed with a silicone solvent. The less expensive, slower-evaporating cyclomethicone D5 (decamethylcyclopentasiloxane) is normally used, but if a faster clearance is needed, particularly during testing, hexamethyldisiloxane will clear equally effectively.

Just as any solvent that is not miscible with water can be emulsified into Pemulen or xanthan gum, any polar solvent that is not miscible with the gel base or cyclomethicone can be emulsified into Velvesil Plus or the KSG gels. So acetone, ethanol, benzyl alcohol to list just a few can be emulsified into the silicone systems. This opens up a whole new range of cleaning options.

Conclusion

When I first began in conservation I remember thinking: 'What can change in the chemistry of cleaning paintings? It's not like there are any new solvents to be invented'. Clearly the world of cleaning *has* changed, giving the conservator more and more tailored tools. With the greater number of cleaning options being researched, developed and tested, it can be difficult for the conservator to weigh the options and develop an intuition about these newer techniques. But the bigger toolbox allows us to clean with precise empirical systems rather than using blind faith while practicing an irreversible process.

It is also important to remember that all the new techniques still rely on the skill and judgment of the conservator. Sometimes the best cleaning can be achieved with isopropanol and mineral (white) spirits. But at other times a better result can be achieved with a cleaning system tailored to the unique creation and history of the artwork. Few things are as gratifying as a cleaning that reveals the artwork as the best reflection of its current state.

Materials and suppliers

- Carbopol, Pemulen TR-2: The Lubrizol Corporation. Obtained from Conservation Support Systems, and also available from Kremer Pigmente and other conservation

suppliers, https://lubrizol.com/Life-Sciences; http://www.conservationsupportsystems.com; http://www.kremer-pigmente.com/
- Ethomeen C-12, C-25 and Armeen 2C: AkzoNobel N.V. C-12 and C-25 were obtained from Conservation Support Systems, and are available from Kremer Pigmente and other conservation suppliers, https://sc.akzonobel.com/en/fa (Ethomeen); https://sc.akzonobel.com/en/personalcare/ (Armeen)
- FileMaker Pro: FileMaker, Inc., http://www.filemaker.com
- Golden Polymer Varnish with UVLS: Golden Artist Colors, Inc., www.goldenpaints.com
- Hydroxypropylmethylcellulose, deoxycholic acid, citric acid, bicine, Bis-tris, cyclomethicone D5, hexamethyldisiloxane: Sigma-Aldrich Co., https://www.sigmaaldrich.com
- KSG-210, -240, -350Z: Shin-Etsu Chemical Co., Ltd. KSG-210, -240 and -350Z were obtained from Shepard Bros., https://www.shinetsusilicone-global.com/; http://www.shepardbros.com/
- Velvesil Plus, Velvesil 125, SF1188A: Momentive Performance Materials, Inc., https://www.momentive.com/#
- Xanthan gum clear, Pemulen TR-2: The Personal Formulator, http://www.personalformulator.com/wvss/

Notes

1. *Polyacrylic acid*, Wikipedia. https://en.wikipedia.org/wiki/Polyacrylic_acid. Last modified 25 December 2016. Accessed 27 January 2017.
2. As Acrysol ASE-60, in this product.
3. For Armeen 2C, see https://sc.akzonobel.com/en/fa/Pages/product-detail.aspx?prodID=8626#specs. For Ethomeen C-25, see https://sc.akzonobel.com/en/fa/Pages/product-detail.aspx?prodID=8654#specs. For Ethomeen C-12, see http://www.sc.akzonobel.com/en/fa/Pages/product-detail.aspx?prodID=8221.
4. *Azeotrope tables*, Wikipedia. https://en.wikipedia.org/wiki/Azeotrope_tables. Last modified 23 October 2016. Accessed 27 January 2017.
5. See *Pemulen TR-2 NF Polymer* data sheet. The Lubrizol Corporation. https://www.lubrizol.com/Life-Sciences/Products/Pemulen-Polymeric-Emulsifiers/Pemulen-TR-2-NF-Polymer.
6. See Stavroudis 2010, 2012 and 2014 for discussions on setting the pH of a Pemulen gel, each advancing a 'better' way to set the gel's pH, and offering recipes based on the proposed method.
7. See Vanzan technical information at http://www.vanderbiltminerals.com/ee_content/Documents/Technical/VANZAN_Brochure_English_Web.pdf.
8. See note 7.
9. Velvesil Plus Gel Technical Data Sheet. 2014. Momentive Performance Materials, Inc., https://www.momentive.com/products/show-technical-datasheet.aspx?id=14438.
10. See Shin-Etsu Personal Care Silicones at https://www.shinetsusilicone-global.com/products/personalcare/products/silicone_gels.shtml.
11. See note 10.

References

Brunchi, C.E., Bercea, M., Morariu, S. and Dascalu, M. 2016. 'Some properties of xanthan gum in aqueous solutions: effect of temperature and pH', *Journal of Polymer Research* 23: 123. Abstract available at: http://link.springer.com/article/10.1007/s10965-016-1015-4 (accessed 26 May 2017).

Doherty, T. and Shockey, L. 2013. 'The use of siloxane-based systems to clean water-sensitive painted surfaces', *WAAC Newsletter* 35(2): 15–17.

Ruhemann, H. 1968. *The Cleaning of Paintings: Problems and Potentialities*. New York, Washington: Frederick A. Praeger.

Stavroudis, C. 2010. 'Using Pemulen with the MCP', *WAAC Newsletter* 32(3): 16.

Stavroudis, C. 2012. 'Pemulen revised: pHuck the pH meter', *WAAC Newsletter* 34(2): 19.

Stavroudis, C. 2014. 'Pemulen TR-2 – the once and future king (of conservation)', *WAAC Newsletter* 36(2): 10–11.

Stulik, D. and Miller, D. 2004. 'Research into solvent gel residues', in V. Dorge (ed.), *Solvent Gels for the Cleaning of Works of Art: The Residue Question*, Los Angeles: Getty Conservation Institute, 18–53.

Swift, T., Swanson, L., Geoghegan, M., and Rimmer, S. 2016. 'The pH-responsive behaviour of poly(acrylic acid) in aqueous solution is dependent on molar mass'. Available at: http://pubs.rsc.org/en/content/articlehtml/2016/sm/c5sm02693h?page=search (accessed 26 May 2017).

Weast, R. 1972. *Handbook of Chemistry and Physics*, 53rd edn. Cleveland, OH: The Chemical Rubber Company.

Wolbers, R. 1988. *Notes for Workshop on New Methods in the Cleaning of Painting*, Getty Conservation Institute Workshop, 15–26 August 1988.

Wolbers, R. 2000. *Cleaning Painted Surfaces: Aqueous Methods*. London: Archetype Publications.

Young, S., Torres, J. and Kienzel-Sterzer, C. 1986. *Potentiometric Titration Studies on Xanthan Gum Solutions*, unpublished Master's thesis, Oregon State University, Oregon, USA, 27–56. Available at: https://ir.library.oregonstate.edu/xmlui/bitstream/handle/1957/27110/YOUNGSAN1986.pdf?sequence=1 (accessed 26 May 2017).

Author's address

Chris Stavroudis, Paintings Conservator in Private Practice, West Hollywood, CA, USA (cstavrou@ix.netcom.com)

Cleaning assessment of a vinyl dispersion paint: comparing Velvesil Plus Gel, microemulsions and aqueous solutions

Emilie Faust

Introduction

Secteur d'Argenteuil by Jürg Kreienbühl (1932–2007), executed in 1965 (in the collection of the Fonds regional d'art contemporain (FRAC), Île-de-France) was painted with a vinyl dispersion binder on hardboard and varnished with the same vinyl dispersion. The painting had a grimy surface with dirt embedded in the varnish, for which aqueous cleaning was required. Since vinyl dispersion paint is known to be water-sensitive, and aqueous cleaning could easily damage both its chemical structure and surface (Pereira *et al.* 2013; Zhijin 2010; Silva 2011), the use of water buffered to a suitable pH and conductivity is advised in order to reduce swelling and loss of material (Dillon *et al.* 2014; Wolbers and Stavroudis 2012). Gels or thickening agents can also be used to reduce the amount of water in contact with/penetrating the surface (Wolbers and Stavroudis 2012).

As traditional aqueous solutions are still used in daily conservation practice, our scientific study[1] focused on assessing the impact of aqueous cleaning on the vinyl dispersion paint, comparing the surface before and after cleaning with both aqueous solutions and two gels. Microtopography was found to be a suitable technique for our investigation which was carried out on mock-ups.

Preliminary study

Identification of materials

Jürg Kreienbühl, a native of Switzerland who moved to France around 1956, employed materials commonly used in the construction of buildings therefore it was assumed that *Secteur d'Argenteuil* was made with such materials. The artist's son and widow had documented the painter's techniques and habits, while paint analyses with Fourier transform infrared (FTIR) spectroscopy, pyrolysis–gas chromatography–mass spectrometry (Py–GC–MS) and examination with optical microscopy was used to identify the binder and varnish composition as well as the painting's stratigraphy. Kreienbühl used pigments ground in a vinyl dispersion, directly applied on a Pavatex hardboard panel. A layer of varnish was later applied to protect the paint. Later inpainting another varnish layer can be seen.

Jürg Kreienbühl used a binder called Elotex 55 W25, sold under the name Dispersionbinder. It was identified as a vinyl homopolymer dispersion with external plasticizers (tricresyl phosphate and dibutyl phtalate) and a poly(vinyl alcohol) colloidal protector. Pigments identified were mainly inorganic, including lithopone white and Prussian blue.

Cleaning tests

The painting had a thin dirt layer embedded in the varnish. Aqueous cleaning was considered because the paint layer was very soft. The pH and conductivity of the paint surface were measured and the aqueous solutions used adjusted and buffered to pH 6 and a conductivity of 2000 μScm^{-1}. Among several cleaning agents tested, a solution using ethylenediaminetetraacetic acid (EDTA) as a chelating agent gave the best result for dirt removal.

Material and methods

Mock-ups

Mock-ups were made with materials from the artist's studio. Hardboard panel pieces were painted with a vinyl dispersion similar to the original one (Liant vinylique 11) which had been used both as a binder and a varnish. Only two pigments were

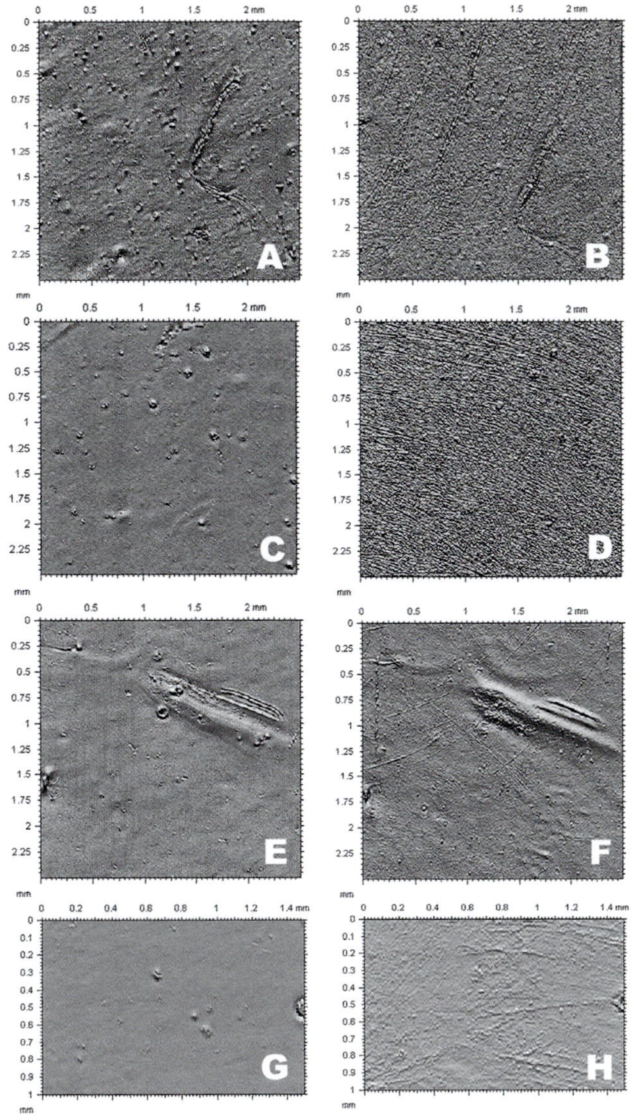

FIG. 1 Two-dimensional images before and after the cleaning of lithopone white mock-ups: A (before) and B (after) cleaned with a Velvesil Plus Gel; C (before) and D (after) cleaned with a microemulsion; E (before) and F (after) cleaned with EDTA; G (before) and H (after) cleaned with citrate solution (© Emilie Faust, 2015).

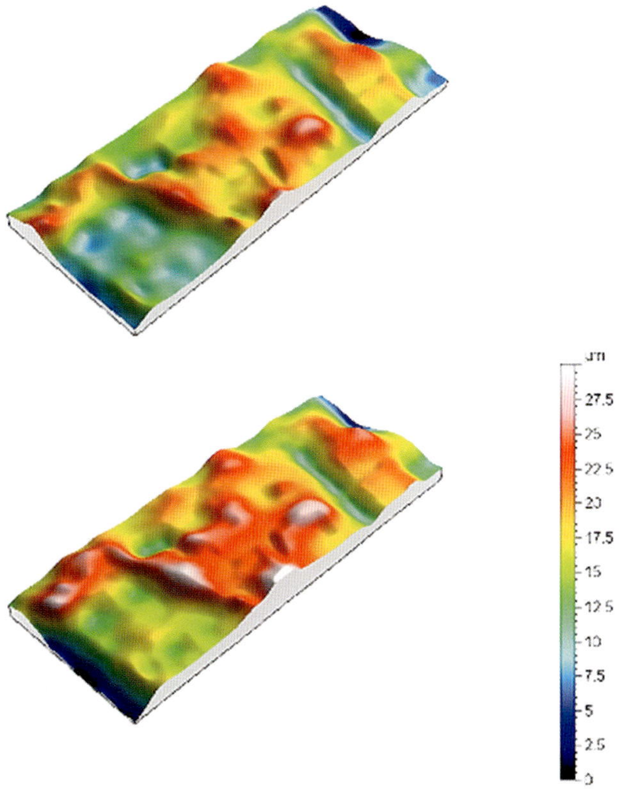

FIG. 2 Three-dimensional images before (*top*) and after (*bottom*) cleaning a white mock-up with Velvesil Plus, highlighting the swelling (© Emilie Faust, 2015).

employed: lithopone white and Prussian blue. Monochrome paint layers were applied on the hardboard panel pieces and dried for one day. The binder was then diluted with water and applied as a thin final varnish.

Twenty-four mock-ups divided into two sets were subjected to accelerated light aging in a Suntest XLS+ Atlas weathering instrument with a xenon arc lamp. Test conditions were 765 W/m^2 for 738 hours with a temperature between 30 and 35 °C.

Aqueous solutions

Two aqueous solutions and two kinds of aqueous gels were made up. EDTA was employed as the main component of the set of aqueous solutions. The pH and conductivity of the water were adjusted according to the measurements of the specimen surface and 0.5 g EDTA was dissolved in 100 mL purified water which was adjusted to a pH 5 with 1 mol/L sodium hydroxide. Purified water was then added to lower the conductivity to 2000 μScm^{-1}. For the second solution, a 1% w/v citrate chelating solution of triammonium citrate (TAC) was prepared in purified water and brought to pH 6 using sodium hydroxide (i.e. this solution was not adjusted to match the pH or conductivity of the surface). Both aqueous solutions were applied with cotton swabs. The EDTA solution was cleared with water which was adjusted to match the pH and conductivity of the surface (pH 5, 2000 μScm^{-1}) while the TAC solution was cleared with tap water.

A Velvesil Plus gel and a microemulsion were selected for their ability to remove materials from water-sensitive surfaces. Velvesil Plus is a ready-to-use dispersion comprising a silicone cross-polymer gel and a silicone gel ethylene oxide/propylene oxide copolymer.[2] Up to 20% of polar and non-polar solvents can be mixed into the dispersion and it will remain very stable. In addition, it has the property of gradually releasing solvents. The EDTA solution (10 wt%) was added to the Velvesil Plus and the mixture applied on specimens with a soft brush, removed with dry cotton swabs, and rinsed with both a non-polar silicone-based solvent octamethylcyclotetrasiloxane (D4) and water (adjusted to pH 5 and 2000 μScm^{-1}).

Microemulsions are emulsions of water in oil (w/o) or of oil in water (o/w) known for their thermodynamic stability. Their main advantage is the formation of very small micelles which increase surface contact and result in improved cleaning efficiency. Their stability allows for safe aqueous cleaning on water-sensitive paints.

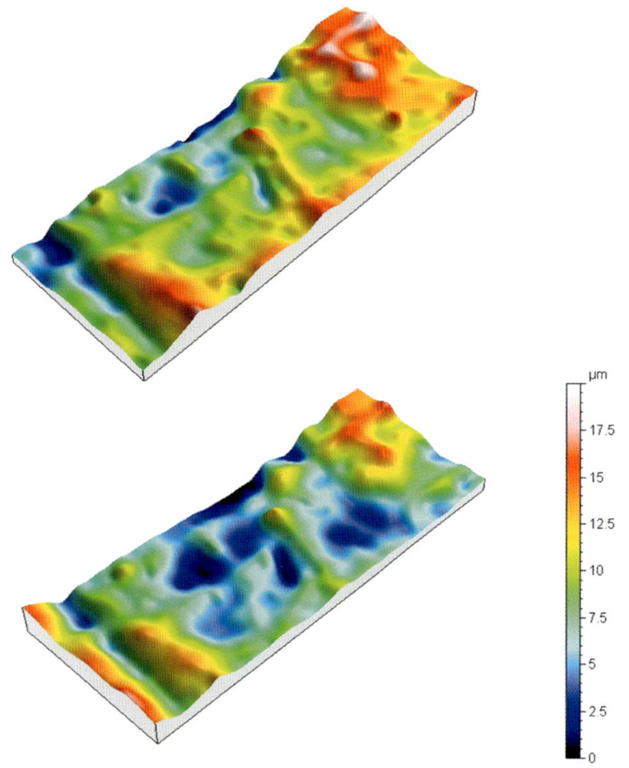

FIG. 3 Three-dimensional images before (*top*) and after (*bottom*) cleaning a white mock-up with microemulsion, showing deeper voids (© Emilie Faust, 2015).

Microemulsions are formed by a balance of an oil phase, an aqueous phase, a surfactant and sometimes a co-solvent. The w/o microemulsion formulated was tailored to include a small aqueous phase and low polarity containing 40% silicone-based solvent (D4), 50% Ecosurf EH3 surfactant and 10% EDTA adjusted aqueous solution based on a phase diagram in Stavroudis (2012). The mixture was applied on specimens with cotton swabs and cleared with adjusted water (pH 5 and 2000 μScm^{-1}).

What is microtopography?

To assess the impact of aqueous cleaning on the vinyl dispersion paint, it was necessary to compare the surface before and after cleaning using both traditional and new aqueous methods. Microtopography is a topographic tool which operates with a confocal technology based on chromatic coding and white light. It allows the measurement of small variations of elevation and surface roughness through chosen roughness indices. In order to understand the resultant modifications scales, we compared the arithmetic surface variations (Sa), the maximum depth of voids (Sv) and the maximum height of bumps (Sp). Three-dimensional images with a Gaussian blur filter gave an indication of swelling, while 2D images allowed visual assessment of surface modifications (Fig. 1). Topographic data were recorded before cleaning on the accelerated light-aged mock-ups and compared with the readings on same cleaned areas.

Results

Microtopographic data indicate that the white mock-ups cleaned with the Velvesil Plus exhibited surface swelling. The height of bumps increased from 1.7 to 9.3 µm and the swelling was obvious when comparing the 3D images before and after cleaning (Fig. 2). Two-dimensional images show deep scratches and abrasion after cleaning (Fig. 1, A and B). Measurements on the blue specimens indicated significant variations with a very important decrease in the height of the bumps from 105 to 42 µm. Two- and three-dimensional images confirmed that serious surface changes had occurred. Some of the original topography had disappeared, resulting in many deep grooves.

The white mock-ups cleaned with the microemulsion show a matte surface just after cleaning. The surface of the blue specimens immediately whitened and clearing caused pigment extraction. Measurements of blue and white mock-ups indicate pronounced surface abrasion highlighted by a decrease of the height of bumps from 36 to 6 µm as well as a reduction in the depth of voids from 40 to 11 µm. Two-dimensional images confirm the measurements showing significant modifications, sometimes leading to a complete surface change (Fig. 1, C and D). Three-dimensional images corroborate those results (Fig. 3).

Cleaning the mock-ups with EDTA solutions led to different phenomena. The blue mock-ups showed a considerable loss of pigment, evident on blue cotton swabs after rinsing, and a visual difference of height level on the cleaned areas. Measurements on the white mock-ups indicated a different result on each one. The first had a slight increase of bump height (~3 µm) and reduction of void depth (~27 µm), suggesting a slight swelling and varnish thinning. The second white mock-up underwent moderate surface abrasion with deeper voids (~6 µm) and decreased bumps (~7 µm); the third mock-up showed significant abrasion with a maximum decrease of bump height of 14 µm. Three-dimensional images both confirm and contradict the measurements, showing a slight swelling for the first and second mock-ups. Two-dimensional images indicate slight surface changes (Fig. 1, E and F): surface texture is slightly reduced while the appearance of scratches and grooves are minimized, resulting in a smoother surface compared to the previous specimens. Measurements on blue mock-ups confirm visual observations showing bump height reduced to 34 µm and void height increased to 46 µm.

Visual observations after mock-up cleaning with the TAC solution are similar to those for EDTA and the matte finish after cleaning is very important. Measurements on white specimens indicate a swelling trend highlighted by the increase in the height of bumps from 0.7 to 15 µm. Three-dimensional images confirm these results showing a swelling after cleaning. However, 2D images indicate significant surface modifications with several scratches and marks while topographic forms disappear (Fig. 1, G and H). Measurements on blue mock-ups show a significant decrease of the bump height from 33 to 7 µm as well as a reduction in void depth from 28 to 18 µm. Surface modifications are visible

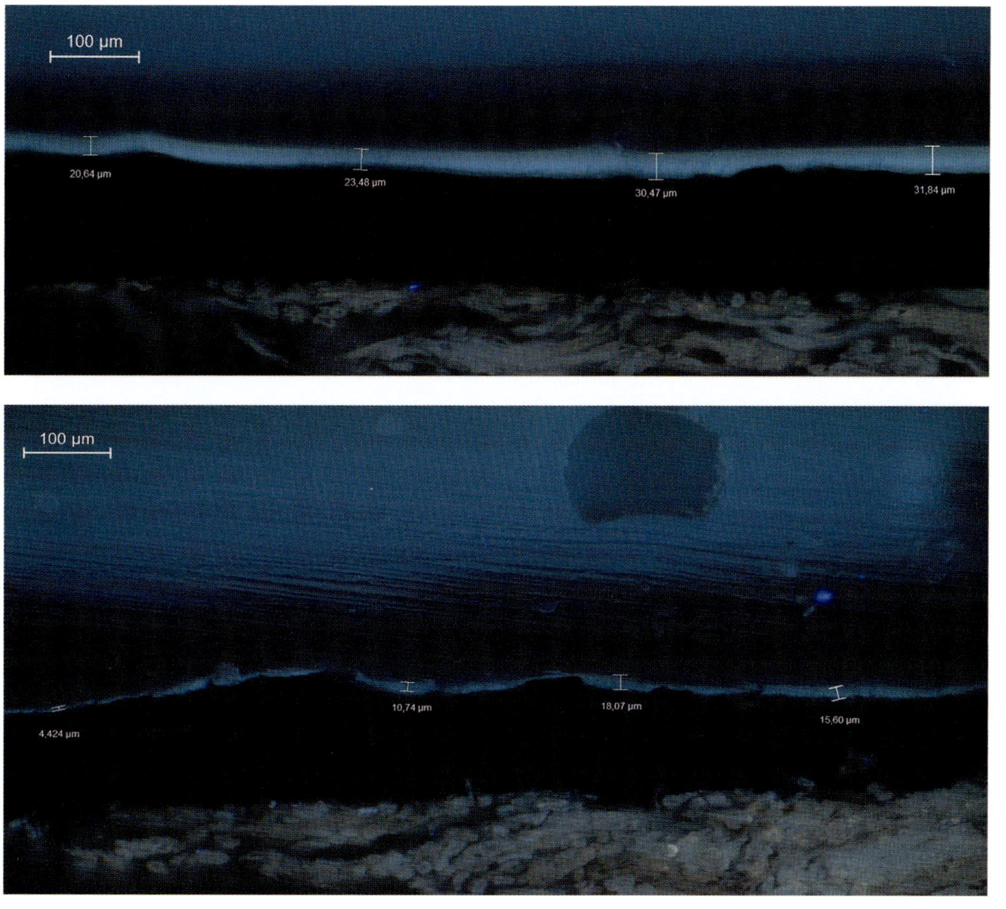

FIG. 4 Prussian blue mock-up samples before (*top*) and after (*bottom*) cleaning with a microemulsion, showing a thinner varnish layer (© C2RMF/ Nathalie Balcar, 2015).

on 2D images and reveal deep marks and scratches as well as a reduction in surface relief.

Discussion

Identifying a general trend from each mock-up set according to microtopographic measurements proved difficult. The results, which were inconsistent, were influenced by several parameters such as the conservator's method of application, the varnish thickness, and the pigments present. Moreover, the high water sensitivity of the mock-ups could be further accentuated by the degradation of the varnish, caused by light aging. The comparison between 2D and 3D images was fundamental for interpreting the results. Cleaning of the white mock-ups with Velvesil Plus provoked swelling and moderate abrasion and the results suggest that the varnish swells before beginning to thin out. Results from the blue mock-ups indicate that varnish thinning and pigment extraction occurred. Corresponding cross-sections were obtained which confirmed that the varnish was thinner after cleaning. It was assumed that gel application and consequent rinses could have led to physical damage and surface abrasion. Additionally, the affinity of EDTA for iron-containing pigments (such as Prussian blue) could cause major surface abrasion.

Blue and white mock-ups cleaned with a microemulsion show significant abrasion, although the swelling is very low. The very large increase in roughness could have resulted from repeated swabbing. Corresponding cross-sections confirm that the varnish layer was significantly thinner (Fig. 4).

The results of white and blue mock-ups cleaned with the EDTA solution are inconsistent and difficult to interpret. However, a general trend emerges from the topographic images. White mock-ups have a high propensity to swell, but surface modifications are minimized compared to other cleaning tests. The abrasion is less pronounced than that following cleaning with Velvesil Plus and microemulsions despite the higher water content of the EDTA solution. The blue specimens were as damaged as the previous blue mock-ups cleaned with microemulsion and gel.

Results for specimens cleaned with the TAC solution are also variable. The roughness parameters of the white mock-ups are of the same order of magnitude as those of specimens cleaned with EDTA solutions. However, the surface exhibited greater topographic changes. This suggests that non-adjusted solutions may favour the softening and consequent abrasion of the paint. Results from the blue mock-ups indicate less significant pigment extraction compared to all other cleaned blue specimens, which may be due to the weaker chelating action of citrate for iron.

Conclusion

Results suggest that in our case, the method of delivering the cleaning solution influences surface modification more than the type of solution itself. Gel or microemulsions require multiple rinses, which is not appropriate when dealing with a painted surface with a very soft and water-sensitive binder. A non-adjusted citrate solution modified the surface more than the EDTA-adjusted solution. However, this may only be applicable to the white mock-ups as it appears that EDTA caused major surface abrasion on the blue samples even in the presence of a varnish, possibly because of the affinity of EDTA for iron-containing pigments.

Ultimately, the surfaces of all the samples were modified after cleaning. However, white samples cleaned with EDTA in adjusted water exhibited less damage – the adjusted solution appeared to give the best results. Nevertheless, the chelating agent needed to be exchanged for one with a lower affinity for iron in order to perform a safe cleaning on the painting. Based on the results of this study, the conservation treatment was tailored using an adjusted solution with a less powerful chelating agent, citrate, at pH 5.5 and 3000 μScm^{-1}. This aqueous solution was applied with cotton swabs and rinsed using a cosmetic sponge to minimize abrasion.

Materials and suppliers

- Pavatex: Pavatex SA, http://www.pavatex.com/home/
- Liant vinylique 11: Marin Beaux Arts, http://www.marinbeauxarts.com/fr/
- EDTA, citric acid, sodium hydroxide, octamethylcyclotetrasiloxane: Sigma-Aldrich Chemie S.a.r.l., http://www.sigmaaldrich.com/france.html
- Triammonium citrate: CTS France, http://www.ctseurope.com/fr/
- Velvesil Plus: Momentive, www.momentive.com
- Ecosurf EH3: DowChemical, http://www.dow.com/

Notes

1. This academic research was undertaken for conservation purposes in collaboration with the French Museum Conservation and Restoration Centre (C2RMF) and the National Institute of Heritage (INP), Paris, France.
2. Introduced by Richard Wolbers for conservation involving highly sensitive paint cleaning.

References

Dillon, C., Lagalante, A. and Wolbers, R. 2014. 'Acrylic emulsion paint films: the effect of solution pH, conductivity, and ionic strength on film swelling and surfactant removal', *Studies in Conservation* 59: 52–62.

Pereira, A., Jão Melo, M., Eaton, P., Schäfer, S. and Learner, T. 2013. 'A preliminary study into the effects of cleaning PVAc paints', in M. Mecklenburg, A.E. Charola, and R.J. Koestler (eds), *Cleaning 2010: New Insights into the Cleaning of Paintings*. Washington, DC: Smithsonian Institution Scholarly Press, 7–78.

Silva, M. 2011. *Analytical Study of Accelerated Light Ageing and Cleaning Effects on Acrylic and PVAc Dispersion Paints Used in Modern and Contemporary Art*, unpublished PhD dissertation, Universitat Politècnica de València and Smithonian Museum Conservation Institute.

Stavroudis, C. 2012. 'More from CAPS 3: surfactants, silicone-based solvents, and microemulsions', *WAAC Newsletter* (34)3: 24–7.

Wolbers, R. and Stavroudis, C. 2012. 'Aqueous methods for the cleaning of paintings', in J. Stoner and R. Rushfield (eds), *The Conservation of Easel Paintings*. New York: Routledge, 500–523.

Zhijin, C. 2010. *Aspects of Water Sensitivity of PVAc Latex Film*, unpublished PhD dissertation, Department of Polymer Science and Engineering, Lehigh University, Bethlehem, PA.

Author's address

Emilie Faust, Paris, France (emilie.faust@yahoo.fr).

Silicone polymers for the removal of grime from a selectively varnished oil, graphite and ink painting

Birgit Straehle, Philip Klausmeyer, Matthew Cushman and Rita Albertson

Introduction

In 2013, a privately owned canvas painting depicting a five-masted schooner, *The Five-Masted Schooner J.B. Prescott* (Fig. 1) was brought to the Worcester Art Museum (Worcester, MA, USA) for conservation treatment. The surface was obscured by grime and discolored varnish, and the unlined canvas had a small tear. The present owner, a direct descendant of the gentleman who commissioned this portrait of his ship, believed the painting had never been treated before. A signature in the lower left corner, 'SFM Badger', identified the artist as Solon Francis Montecello Badger (1873–1919), whose work can be found in the Maritime Museum in Bath (Maine USA) and the National Maritime Museum (London, UK). In the literature, Badger is sometimes referred to as Samuel Finley Morse Badger – perhaps a conflation of his name with that of Boston-born artist and inventor Samuel F.B. Morse (1791–1882).

Born in 1873, S.F.M. Badger was raised in Charlestown, Massachusetts, one of the oldest neighborhoods of Boston located along the banks of the harbor. He likely studied for a short time with William Pierce Stubbs, the son of a shipmaster, who was active as a marine painter in Boston between 1871 and 1894. To achieve precision in his portraits, Badger was thought to have studied blueprints and even sailed Boston Harbor in a small yacht in order to measure vessels. Badger died from tuberculosis in 1919 at the age of 46.

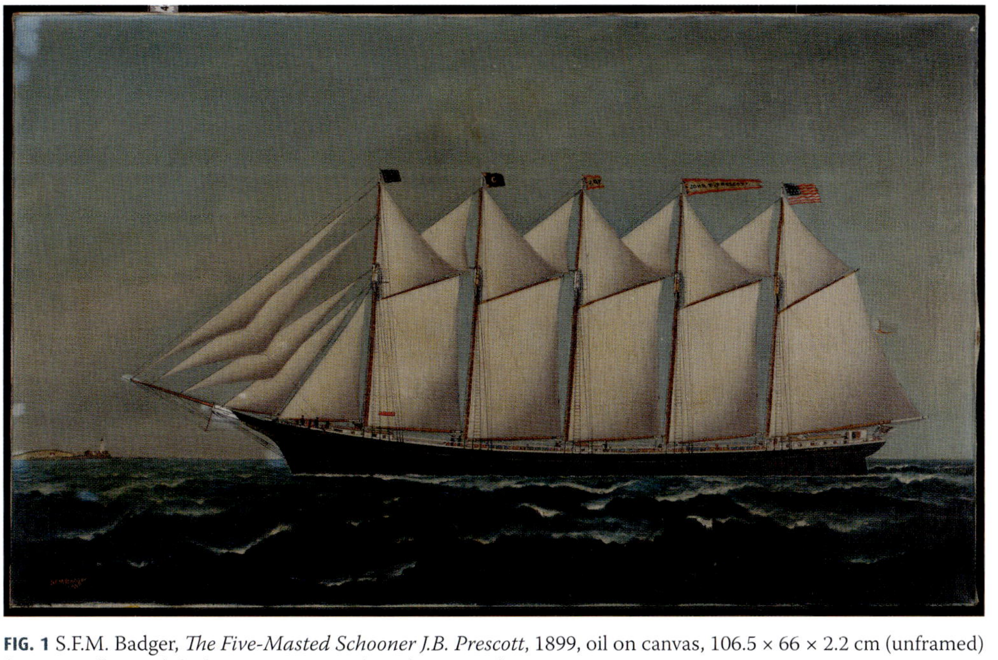

FIG. 1 S.F.M. Badger, *The Five-Masted Schooner J.B. Prescott*, 1899, oil on canvas, 106.5 × 66 × 2.2 cm (unframed) (private collection): before treatment with preliminary cleaning tests.

Examination and condition

The fine, plain-weave, presumed linen canvas support was attached with iron tacks to a wooden, four-member, keyable stretcher. It was prepared commercially with an off-white ground layer that extended onto the tack edges. The reverse of the canvas was dusty and soiled, and dirt and bits of debris were found inside the stretcher pockets. Several different media were used to construct the work, including poor quality oil paint. The paint was applied thinly, especially in the sky and the white sails. The seams of the sail cloth were rendered with thin graphite, and the ship's rigging was executed in different colors of black and brown ink.

Dust, soot and tar residues from tobacco smoke had accumulated and formed a continuous layer of imbedded grime on the paint film. A fine, overall craquelure in the paint was less pronounced in the areas corresponding to the underlying stretcher bars. Aside from minor paint loss associated with the small tear, the paint layers were structurally stable and in good condition. Preliminary aqueous and solvent testing revealed that the painting was largely unvarnished. However, specific design elements, such as the ship's masts and much of the water, had been selectively varnished with a coating that appeared brown.

Challenges

Although aqueous cleaning was suitable for most of the painting, the presence of water-sensitive materials in some design areas necessitated two different cleaning approaches. Challenges to effecting a successful treatment were complicated by the quality and complexity of the artist's materials:

- Poor quality paint was porous due to underbound pigment and a mostly unvarnished surface making it difficult to safely remove grime.
- Selectively varnished areas posed potential complications to cleaning.
- Soil was differentially distributed across the surface making it difficult to clean evenly.
- The coarse grains of the graphite sat atop the high points of the canvas weave and were sensitive to mechanical action.
- Inks were sensitive to aqueous cleaning and mechanical action.

Different strategies for removing soil from each material would be required to achieve a visually coherent cleaning without damaging the picture.

Cleaning tests

The goal was to dismantle the surface grime while minimizing the swelling of the paint. The strategy was to select a mild chelator, add a low-HLB (hydrophile-lipohile balance) nonionic surfactant to help break van der Waals bonding of the grime to itself and the surface, and maintain a low pH to create an environment conducive to removing grime while giving the best chance of maintaining the integrity of the paint film. This aqueous formulation would be used to remove soil from the unvarnished oil painted areas while avoiding contact with the more sensitive ink and graphite elements by cleaning around them. Additional cleaning attempts with traditional and particle emulsions would then be tested for grime removal from the ink and graphite elements. After considerable empirical testing, the following tests were key in devising the final cleaning protocol for removing imbedded soil and achieving uniform cleaning.

Oil painted elements, primarily the sky and much of the water

Cotton swabs were dampened with the following solution (henceforth referred to as 'aqueous solution'): deonized H_2O, 1% sodium citrate (chelator), 1.0% Pluronic L-64 (HLB ~15), 30 mM 4-morpholinoethanesulfonic acid hydrate (MES) buffered to pH 6.16. Although initially the brown coloration of the selectively varnished areas was assumed to be the result of oxidation of a natural resin varnish, solubility tests showed it was due to a preferential attraction and binding of the tobacco smoke and coal-burning residues to the applied resin. These areas were successfully cleaned with the aqueous solution to reveal a perfectly intact clear original surface coating.

Water-sensitive inks and loosely attached graphite linear elements

While working under a stereomicroscope, initial testing with aqueous systems, including the aqueous solution, showed that all colors of ink were adversely affected. In addition, the graphite and inks were negatively impacted by mechanical action due to rolling with a swab. The following testing sequence was conducted:

- *Test 1.* The ink and graphite areas were masked out with Cyclo-2244 cyclomethicone (octamethylcyclotetrasiloxane, or D4) applied with a brush. The D4, which is hydrophobic in nature, protected water-sensitive areas from mechanical abrasion, while also lessening the impact of the aqueous solution used for removing imbedded grime in these areas. In theory, D4 acts to fill cavities in the paint film, such as voids, pores, and microscopic cracks, thereby blocking the aqueous solution from penetrating the paint structure. Because D4 is 100% volatile and evaporates in a short period of time, there were no issues regarding residues and/or clearance. However, the combination of D4 followed by the aqueous solution was not successful, as the D4 did not address the abrasion issue.

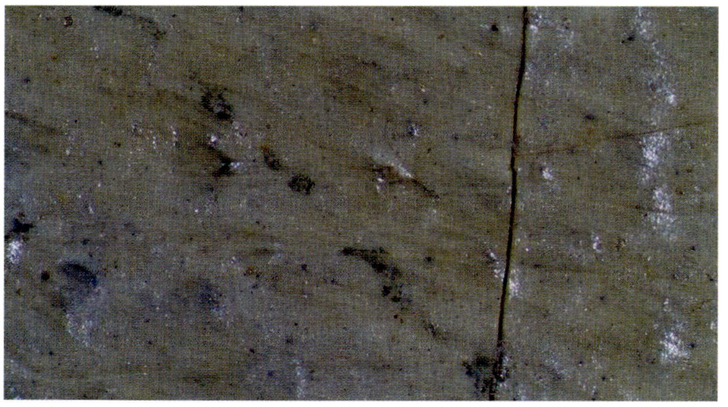

FIG. 2 S.F.M. Badger, *The Five-Masted Schooner J.B. Prescott*: detail showing graphite before cleaning.

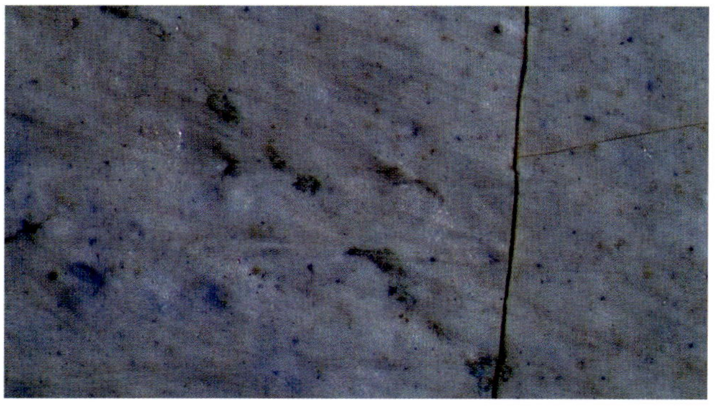

FIG. 3 S.F.M. Badger, *The Five-Masted Schooner J.B. Prescott*: detail showing graphite, after cleaning.

- *Test 2.* To prevent ink and/or graphite abrasion, 30% w/v of the aqueous solution was emulsified with a 70% w/v of Momentive's Velvesil Plus. Velvesil Plus Gel (cyclopentasiloxane ('D5') and C30-45 alkyl cetearyl dimethicone cross-polymer and PEG/PPG-20/23 dimethicone) has many attractive properties: it can be used to create solvent-miscible preparations with very high viscosity, allowing for greater control both spatially across the surface while also limiting diffusion into the paint film. As a 'particle' type emulsifier, it is able to emulsify low percentages of a polar solvent/water phase, without the addition of traditional surfactants. Velvesil (containing D5 as the diluent solvent) also has an extremely low Hanson solubility parameter (5.0), and it can therefore deliver an aqueous phase to a surface in a fairly nonpolar environment, and it can be cleared using silicone solvents with equally low polarity. Although Test 2 gave better results than Test 1, particularly when tested on the graphite, when used with the inks, adverse effects associated with solubility were visible under magnification.
- *Test 3.* A three-step approach was tested by first brushing D4 silicone onto the sensitive ink and graphite areas and then introducing the Velvesil emulsion described above on top of the D4. This had the effect of inhibiting the impact of the Velvesil emulsion on the ink lines and yielding a successful cleaning. No signs of abrasion could be detected under high magnification and the overall evenness of the cleaning was satisfactory. Finally, the Velvesil/aqueous emulsion was cleared with D4 (Figs 2 and 3).

Treatment summary

Dust and debris were removed from the reverse of the canvas and stretcher bars, as well as inside the stretcher pockets. Paint flakes around the tear and small losses throughout the surface were consolidated using 1.5% sturgeon glue. Distortions around the tear were flattened over a long period of time using several applications of a minimally dampened blotter paper and light pressure. As described above, imbedded grime was removed from most oil painted areas, such as the sky and water, with a pH-adjusted aqueous system incorporating a surfactant with a low HLB and a mild chelator. Due to the uneven distribution of grime/tobacco related residue, more than one application was necessary in certain areas (Fig. 4).

As described in Test 3 above, the ink and graphite areas were treated by first masking out small sections of the sensitive areas with D4 silicone. While still wet, the Velvesil emulsion was applied using a small cotton swab. By gently agitating the emulsion on the surface, the grime was dislodged and could be readily picked up. Residues of Velvesil emulsion were then cleared from the surface using a swab dampened with D4. Two applications of this treatment were necessary to remove the soil. Because the ink lines were adversely affected with a longer than necessary dwell time, attention to timing was critically important. Thus, it was crucial for the conservator to limit surface contact to just long enough to remove the soil and to take special care to employ minimal pressure. This multi-step approach yielded excellent results for achieving even cleaning without solubilizing any of the inks and without picking up graphite particles.

Following cleaning, the tear was repaired by harvesting a small piece of canvas from the fold-over edge to make an insert. The threads were aligned and connected using Jade 403, a white aqueous adhesive emulsion containing ethylene vinyl acetate (20/80) copolymer. Losses were filled with a studio-made gesso of calcium carbonate, calcium sulfate and rabbitskin glue, and toned with Schmincke gouache colors. After retouching, a discussion with the owners took place on whether or not to varnish the surface. There was mutual agreement to apply varnish to the partially unvarnished surface using a synthetic resin. A light spray coat of Regalrez 1094, a low molecular weight hydrocarbon resin (25% dissolved in mineral spirits) was applied. Both owner and conservators agreed that this thin, barely visible coating will prevent difficult surface cleanings in the future and offer better protection in the domestic setting. An acid-free backing board with polyester batting was attached to the reverse and minor retouching was carried out using hand-ground pigments bound in MS2A resin (25% dissolved in Shellsol D38). After treatment (Fig. 5) the painting was returned to its original frame.

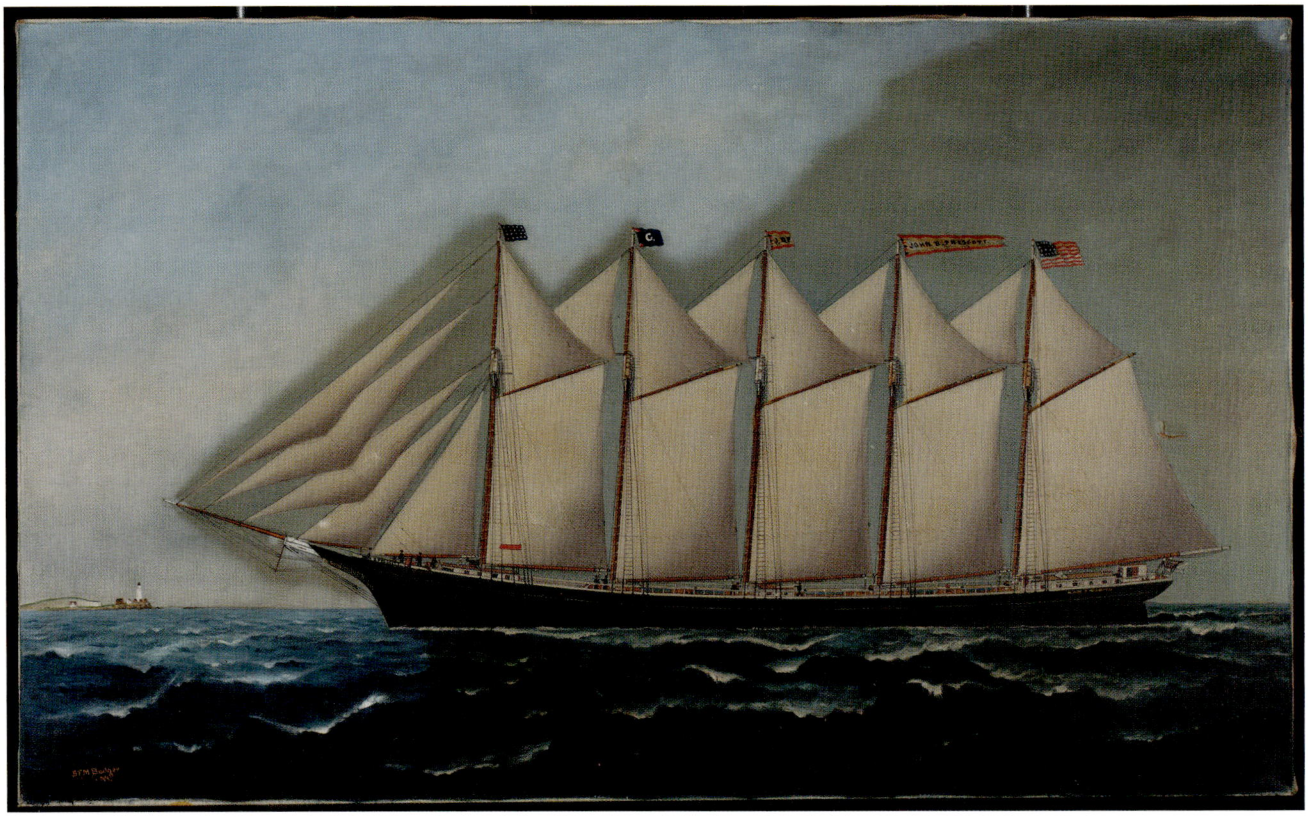

FIG. 4 S.F.M. Badger, *The Five-Masted Schooner J.B. Prescott*: during aqueous cleaning.

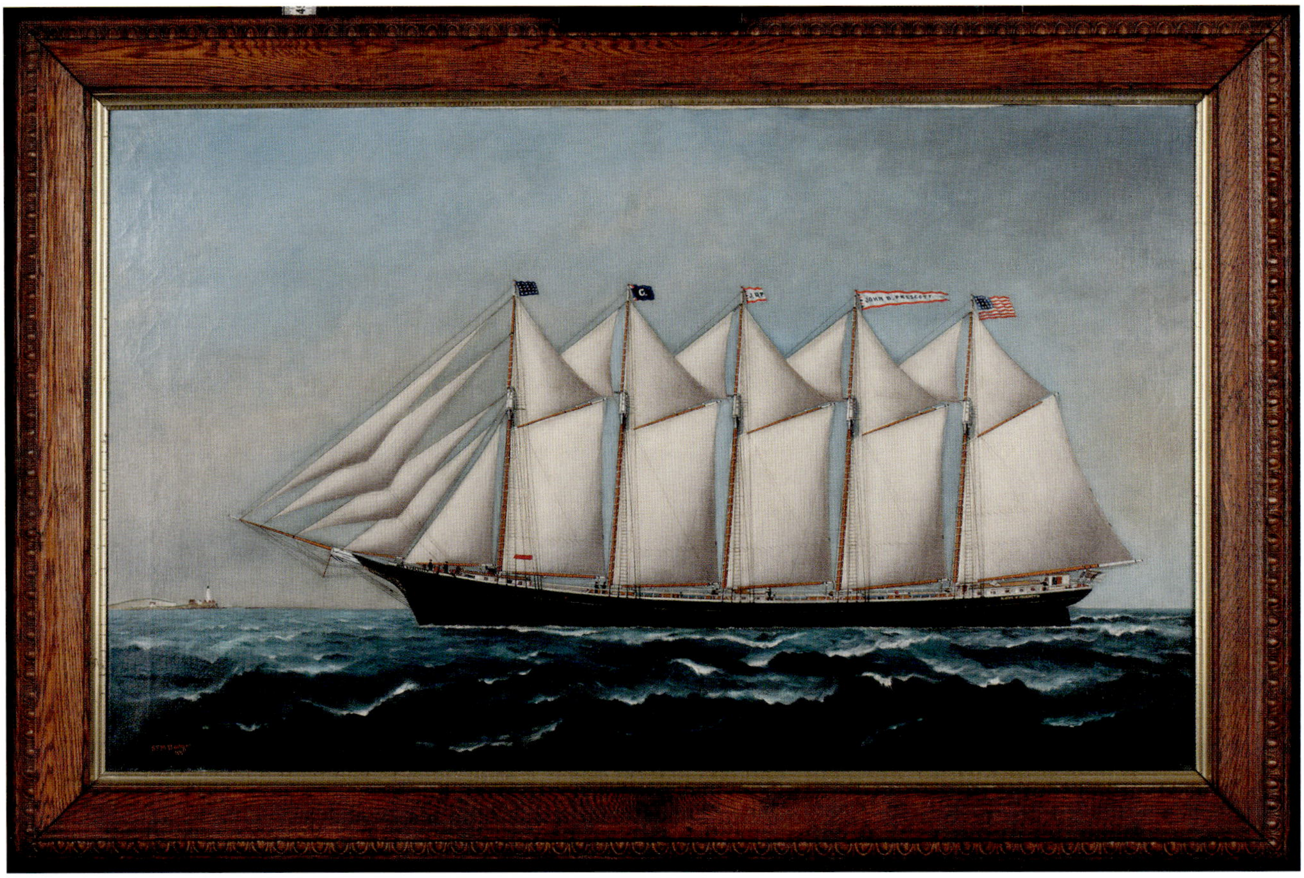

FIG. 5 S.F.M. Badger, *The Five-Masted Schooner J.B. Prescott*: after treatment.

The Five-Masted Schooner J.B. Prescott was transformed by the cleaning treatment. Perceived by the owner as a marine painting with a stormy sky, wispy clouds in a beautiful blue sky were revealed, as were carefully modulated sails and finely delineated riggings. The three-step approach to cleaning this materially complex painting provided an effective and non-toxic method for achieving a sound and aesthetically pleasing result.

Materials and suppliers

- Chalk of Bologna, Chalk of Champagne, rabbitskin glue, Salianski isinglass glue: Kremer Pigments Inc., 247 West 29th Street, New York, NY 10001, USA (212)219.2394, info@kremerpigments.com
- Cyclo-2244: Clearco Products Col, Inc, 3430 G. Progress Drive, Bensalem, PA 19020, USA, www.clearcoproducts.com
- Gouache Colors: Schmincke Co. – GmbH & CoKG, Otto Hahn-Strasse 2, D-40699 Erkrath, Germany, www.schmincke.de
- Jade 403, Regalrez 1094: Talas, 330 Morgan Avenue, Brooklyn, NY 11211, USA, http://talasonline.com
- MES hydrate, 99%: Fischer Scientific, PO Box 3648 Boston, MA 02241, USA
- MS2A: Peregrine Brushes & Tools, 1211 South 60 West Wellsville, Ut 94339, USA
- Pluronic L-64: BASF Corporation100 Campus Drive, Florham Park, NJ 07932, USA
- Sodium citrate: Sigma Chemical Co., PO Box 14508, St. Louis, MO 63178 USA, 314-771-5750
- Sodium hydroxide, 3.0N (3.0M): LabChem, Jackson's Pointe Commerce Park Building 1000, 1010 Jackson's Pointe Court, Zelienopie, PA 16063, USA
- Velvesil Plus 01P: General Electric Company, Waterford, NY 12188, USA,518 237 3330

Dedication

This paper is dedicated to Philip A. Klausmeyer, PhD (1963–2016) who skillfully and patiently guided this treatment. While establishing conservation science at the Worcester Art Museum, he also served as a paintings conservator for nearly 20 years.

Authors' addresses

- Birgit Straehle, Worcester Art Museum, Worcester, MA, USA (birgitstraehle@worcesterart.org)
- Matthew Cushman, Department of Art Conservation, University of Delaware, Newark, DE, USA
- Rita Albertson, Worcester Art Museum, Worcester, MA, USA

Novel methods

Poly(vinyl alcohol)-borax 'gels': a flexible cleaning option

Lora V. Angelova, Emiliano Carretti, Barbara H. Berrie and Richard G. Weiss

Introduction

In an ever-growing realm of thickened and gelated systems available to conservators for surface treatments, the range of physical properties spans from the rigid (e.g. agar, gellan, semi-interpenetrating network gels) to the paste-like thickeners (e.g. xanthan, Carbopol, methylcellulose). Between either ends of the spectrum are semi-soft, viscoelastic materials whose properties continue to be explored. These systems offer the conformability to surfaces afforded by paste-like gels, whilst allowing the comparatively easy residue-free removal often possessed by rigid gels. In some cases, this combination of characteristics is achieved not through the inherent properties of the gel, but by a substrate to which it has been attached; such is the case for methylcellulose-based tissue-gels (Fife *et al.* 2011),[1] and the surface-attached gels currently investigated.[2] An alternative to this approach to creating intermediate handling properties involves materials which are categorized as highly viscous polymeric solutions (HVPS) rather than gels (Fig. 1). The systems of this type discussed here are composed of xPVOH (where x is the monomer % of vinyl alcohol moieties) crosslinked with borate ions (typically derived from sodium tetraborate decahydrate, commonly known as borax). HVPS are not true gels: to understand why, we first need to examine what constitutes a gel.

In the broadest sense, a gel is defined as a material which can be deformed under shear stress, but otherwise behaves like a solid (Nijenhuis 1997; Osada and Khokhlov 2002): for example a lump of Carbopol or a chunk of jelly remain immobile (they do not flow) if left untouched, but if pressure is applied to their surface they will deform to some extent. Gels also have a transition point between the liquid (sol) and gel state, known as the sol-gel transition or the gelation point (Nijenhuis 1997; Osada and Khokhlov 2002). At this point there is a rapid change in the system from behaving as a bulk liquid to something akin to a solid on the macroscale. The change is due to the formation of an infinite cluster network which spans the whole system. In 'physical' gels this transition might occur as a result of changes in temperature (cooling an agar solution) or pH (addition of a base to Carbopol), and it is reversible – the gel can be returned to a liquid state with changes in these conditions. For 'chemical' gels, the transition is non-reversible, and occurs when a covalently crosslinked network is created as in the case of Nanorestore gels – these gels can be used like a sponge with the gel network remaining intact while the liquid it holds can be exchanged for a different one. This last property leads us to the last criterion: gels show heterogeneity in their dynamics on the temporal (time) and length scales. This is because they are composed of more than one component, commonly a less mobile, scaffold-like network (which may be composed of chains of polymers or of small molecules stacked together through physical forces) (Weiss and Terech 2005) and a dynamic liquid portion which has been immobilized on the macroscale within the network, but behaves like a liquid on the microscale within the gel (Nijenhuis 1997; Osada and Khokhlov 2002). A simple example is a cube of jelly – on the macroscale, and if left untouched, jelly looks and behaves much like a solid because its relatively immobile collagen polymer network spans the whole cube; if probed at the microscale, the liquid trapped in the collagen network can behave much more dynamically, as if it is a free liquid.

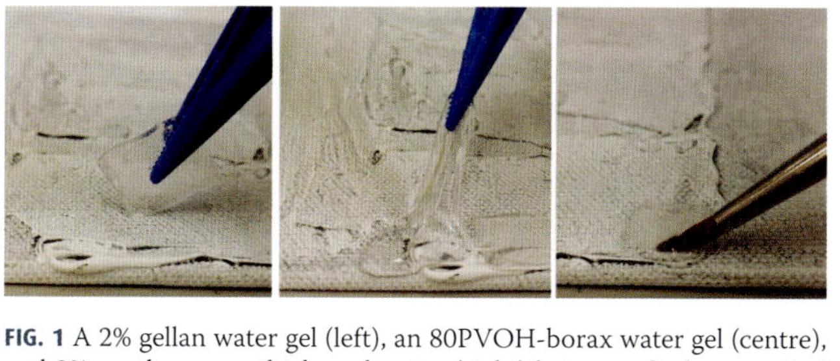

FIG. 1 A 2% gellan water gel (left), an 80PVOH-borax water gel (centre), and 2% xanthan gum thickened water (right) being applied on a soiled acrylic surface. These materials offer different benefits for a treatment: poulticing action from the rigid gellan gel, conformability from the PVOH-borax gel, and continuous manipulation of a desired liquid (or emulsion) on a surface through the xanthan gum.

These physical and chemical properties are what make gels attractive for certain applications such as conservation treatments. On the one hand, the gel can be placed in a specific area of an artwork, where it will remain unless it is moved by application of shear stress, for example, spreading with a brush. On the other hand, the liquid within the gel network behaves, to some extent, as if it were free, so that at the interface between the gel and the artwork surface, dissolution of the desired layer of soil or varnish can occur. In an ideal situation, the solubilized material can be wicked up into the gel, allowing for its safe and efficient removal from the artwork surface.

Although xPVOH-borax systems meet many of the criteria to be gels, they are not true gels for reasons related to their viscoelastic behaviour. In fact, the basic components of the PVOH-borax system are the same as those used to prepare the children's play material, 'Slime', and this term may come closest to what their behaviour resembles. For the sake of sounding less unappealing, we can either refer to the substance as a highly viscous polymer solution or a gel-like material. For simplicity, we will refer to them as gels throughout this paper, although we acknowledge that the term is not completely correct.

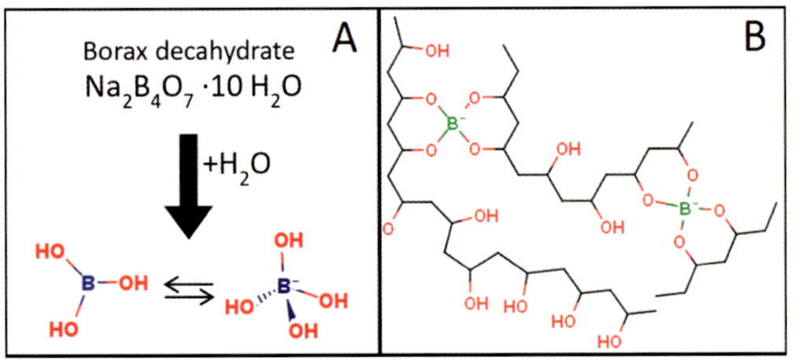

SCHEME 1 Formation of boric acid and borate ion from sodium tetraborate decahydrate on dissolution in water (A), and ester crosslinks formed between poly(vinyl alcohol) and borate through dehydration reactions (B).

xPVOH-borax 'gel' formation and properties

Three ingredients are necessary to form the simplest 100PVOH-borax gel: poly(vinyl alcohol), borax and water. Typically, two aqueous solutions are prepared – one containing the polymer and the other borax – and the two are combined and stirred to form the gel, which occurs immediately on mixing. The borax component provides the crosslinkers for the poly(vinyl alcohol) polymer chains necessary to create the scaffold-like network which holds the water. When dissolved in water, the borax dissociates to form boric acid and borate ions (Scheme 1A) (Sinton 1987; Deuel and Neukom 1949); the resultant solution has a pH of ~9 and acts like a buffer. In the presence of polyols (molecules with many OH, or hydroxyl, groups) such as poly(vinyl alcohol), dehydration reactions between the boron species and four of the hydroxyl groups on the polymer occur (Scheme 1B) (Sinton 1987; van Duin *et al.* 1984). In dilute solutions, a polymer chain will usually have more contact with itself through twisting and rotating than with adjacent polymer chains, and the crosslinks formed on addition of the borax solution will not result in the formation of a gel. However, when we increase the concentration and/or the molecular weight of the polymer, adjacent chains begin to interact and ester crosslinks can now bridge neighbouring polymers (Kulicke and Clasen 2004: 92). When this occurs, the network formed spans the whole system – one of the criteria for forming a gel as outlined in the introduction.

From this simple picture of 100PVOH-borax gel formation, it can already be seen that the concentrations of the polymer and the crosslinker are very important for the creation of a gel with the desired consistency. The polymer molecular weight (i.e. the length of the PVOH chains) also affects the gel properties, with low molecular weight polymers requiring a higher concentration for gel formation (Nemoto *et al.* 1996). In addition, pH plays a central role in the stability of PVOH-borax gels. The equilibrium between boric acid and borate ions (Scheme 1) can be pushed in either direction by altering the pH of the borax solution. Increasing the pH also results in deprotonation of hydroxy groups of the PVOH. Changes in pH equilibrium alter the ability of the gel to form (van Duin *et al.* 1984). Two different mechanisms as to how the crosslinks are formed have been proposed; in both cases, decreasing the pH below 8 hinders gel formation, whilst increasing the pH above 9 creates a more rigid and crosslinked gel. If the pH concentration in the system is too high, the gel may either not form at all, or separate into two phases – a liquid and a rubbery, concentrated polymer portion (Angelova 2013; Bishop *et al.* 2004; Sato *et al.* 1992). This can also occur if the salt concentration in the mixture is too high.

As is the case for all physical gels, large changes in temperature affect PVOH-borax gels (Koga *et al.* 1999; Murakami *et al.* 1980). Although the bonds within the crosslinks (Scheme 1B) may give the impression that these systems are chemical gels, the very low energy required to make and break the crosslinks (reported in the literature to be between *c.*20 and 40 kJ/mol) (Sinton 1987; Koga *et al.* 1999; Schultz and Myers 1969) means that an increase in temperature can cause the gel to return to a liquid state. At room temperature, the weak PVOH-borax crosslinks are advantageous. Because the cross-links are dynamic and short-lived, the gels have a characteristic soft, putty-like consistency which allows them to be moulded and reformed to fit specific topographies on an artwork surface. The labile nature of the crosslinks also gives the gels a plasticity that aids in their removal and reduces the possibility of depositing polymer residues on the artwork surface. PVOH-borax gels have a mechanical property known as dilatancy – they become stiffer when shear strain or pressure is applied – which is postulated to result from the nature of their crosslinks (Robb and Smeulders 1997; Inoue and Osaki 1993; Maerker and Sinton 1986). Thus, even though these gels have a putty-like consistency which allows them to flow slowly under the force of gravity (the main reason for their exclusion from our definition of a gel), they can maintain the elasticity necessary to stay in one piece when peeled from a surface.

Modifications of the classical gel

There is one more variable that can be used when forming xPVOH-borax gels: the ratio of alcohol to acetate pendant groups on the polymer backbone. Poly(vinyl alcohol) is formed by a hydrolysis reaction from poly(vinyl acetate) (Scheme 2). Depending on the extent of the reaction, the polymer produced will have differing ratios of hydroxyl to acetate groups. In 100PVOH, nearly all the acetate groups have been cleaved from poly(vinyl acetate). However, partially hydrolysed xPVOH, with as much as 60% of the acetate groups intact, can be purchased or prepared in the laboratory. In this paper, we have referred to such polymers as xPVOH, where x denotes the per cent hydrolysis; thus, 80PVOH denotes a polymer with 80% hydroxyl groups and 20% acetate groups.

The lower polarity of an acetate pendant group than a hydroxyl introduces a highly useful property to the xPVOH-borax system – the possibility of formulating gels which contain organic solvents (Angelova *et al.* 2011). Whereas 100PVOH is insoluble in many organic liquids, xPVOH can be solubilized in a variety of polar organic solvents such as acetone, methanol, ethanol, 1- and 2-propanol, dimethyl sulfoxide, *N*-methylpyrrolidone, etc., or aqueous mixtures thereof: the lower the degree of hydrolysis, the higher the quantity of the organic component possible in the system. However, when present in a large enough number, the acetate groups may hinder the ability of crosslinks to form since the number of sites for adjacent ester bond formation is decreased; there is a limit to the benefit. Borax is also insoluble in most organic solvents, so some amount of water is always necessary in xPVOH-borax gels. Table 1 describes five polymers which were investigated in detail for their ability to form gels with organic liquids and have subsequently been used in conservation-related investigations and several documented treatments (Angelova 2013). The highly polar nature and slightly basic pH of xPVOH-borax gels makes them suitable for removal of overpaint and varnish that may require aromatic or strongly polar solvents, rather than for surface soil removal from sensitive paint surfaces such as acrylics.

xPVOH-borax gels have been investigated as a tool for conservation treatments for nearly a decade (Carretti *et al.* 2009). In that time, their formulation has been tailored to obtain the best mechanical properties necessary to impart a 'peelable' character to the gels (Natali *et al.* 2011), to incorporate the highest quantity and largest variety of organic liquids (Angelova *et al.* 2011), to ensure that residues of polymer and borate are not deposited to a detectable degree (Angelova *et al.* 2015b), and to understand how the myriad of variables in the system affect its properties. Recently, we also presented a preliminary study which showed that the relatively mobile polymer network of a 80PVOH-borax gel can preferentially orient itself when in contact with a surface, and that this arrangement may have a detectable effect on the liquid at the interface between the gel and the surface (Fig. 2) (Angelova *et al.* 2015a). The purpose of this effort was to elucidate whether the gel network itself adds to the cleaning efficacy of the gel system, which has classically been attributed solely to the gelated liquid. These studies are ongoing and are also being addressed for other types of gels used in conservation (Bertasa *et al.* 2015).

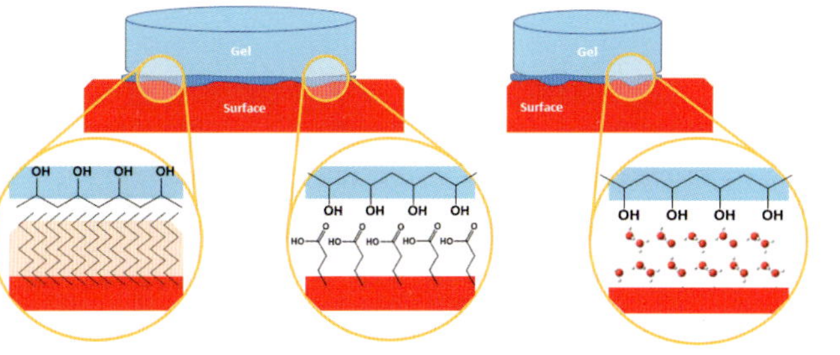

FIG. 2 Schematic representation of the possible arrangement of the PVOH polymer with the non-polar backbone facing a non-polar surface (far left) or the hydroxyl groups facing a polar surface (middle). Illustration of the detected highly ordered water layer between the gel and surface (far right).

As fundamentally aqueous gels, investigations into whether chelators can be incorporated into the 80PVOH-borax network have also been carried out. Although the borate ion itself has chelating capabilities, the stability constants for its complexes with most metal ions are very small (Bousher 1995). The addition of chelating salts into xPVOH-borax gels is a task that demands fine tuning – the change in ionic strength of the liquid portion of the gels can alter the conformation of the polymer chains and disrupt the borate crosslinks, either causing the gel to become excessively rigid or to phase-separate. We have been able to incorporate trisodium citrate (TSC), triammonium citrate (TAC), and ethylenediaminetetraacetic acid (EDTA) into 80PVOH-borax gels at concentrations appropriate for conservation treatments (0.5 wt% or slightly higher in the case of trisodium citrate) (Angelova 2013). Studies of the mechanical properties of gels prepared with increasing concentrations of chelators show that the strength of the gel network tends to increase (Fig. 3A) (Angelova 2013). Nuclear magnetic resonance (NMR) spectroscopy using ^{11}B (^{11}B-NMR) was employed to demonstrate that the increase in gel rigidity is not a consequence of increasing crosslink formation (Fig. 3B); therefore,

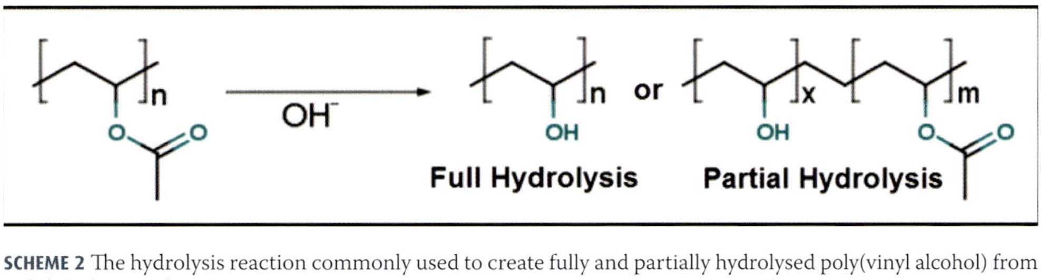

SCHEME 2 The hydrolysis reaction commonly used to create fully and partially hydrolysed poly(vinyl alcohol) from poly(vinyl acetate).

TABLE 1 Summary of xPVOH properties (Angelova 2013).

Kuraray Poval Grade	Exceval HR3010	Poval 48-80	Poval 5-74	Poval LM-20	Poval LM-10HD
Abbreviated name	PVOH	80PVOH	75PVOH	45PVOH	40PVOH
Hydrolysis	>99%	80%	75%	45%	40%
Molecular weight	~31600	~47300	~7300	N/A[a]	N/A[a]
Concentration of polymer necessary for gel formation (wt%)	4	4	6	14	11
Concentration of borax decahydrate necessary for gel formation (wt%)	1.1	1	1.4	1.3	2
Maximum possible organic solvent proportion (wt%)[b]					
Water	100	100	100	100	100
Methanol	30	50	70	75	100
Ethanol	30	50	70	75	75
1-Propanol	30	50	70	75	75
2-Propanol	30	50	70	75	75
Acetone			50	75	75
Methyl acetate	-[d]	-[d]	50	50	75
NMP[c]	30	50	70	75	75

[a] Although Mw data for 45PVAc and 40PVAc were obtained by size exclusion chromatography, they are not reported here due to concerns that agglomeration (from aggregation) of these polymers may have caused erroneous results. [b] The remainder of the liquid is water. [c] N-Methylpyrrolidone. [d] These systems are unstable and undergo syneresis shortly after preparation.

it is probable that it is a consequence of the extension of the polymer chains in the presence of additional ions (Angelova 2013). Testing with chelator-containing xPVOH-borax gels to remove corrosion by-products from metal objects is being carried out by researchers at the Center for Colloids and Surface Science (CSGI) in Florence, Italy.[3]

Because of their particular mechanical characteristics of elasticity and plasticity, the putty-like PVOH-borax materials continue to generate interest and innovative formulations are continuously being created and investigated. Riedo *et al.* introduced an improvement to the mechanical qualities of 86PVOH-borax gels prepared with organic liquids (2-propanol or acetone) by adding a small amount of polyethyleneoxide (PEO) into the system (Riedo *et al.* 2015).[4] Fialová and Kotlík have been incorporating micellar solutions and microemulsions based on sodium dodecyl sulphate or Tween 20.[5] Duncan *et al.* (2016) have been investigating the incorporation of short-chain glycol ethers (water-miscible moieties commonly considered to be very good solvents) into 100PVOH-borax gels.[6] While much of the research into these gels for conservation applications has been carried out using polymers obtained from the Kuraray Company, research by several of the above-listed authors demonstrates that polymers obtained from other chemical suppliers can be substituted in creating xPVOH-borax gels. We have previously published several recipes using polymers detailed in Table 1 (Angelova *et al.* 2015b), and here we include four step-by-step procedures for preparing gels. We encourage conservators (and scientists) to experiment with the formulations: for example, the non-ionic surfactant used in gel 2 may be substituted for a different surfactant at the same weight per cent, whilst the organic liquids in gel 4 can be changed, mixed, or proportionally substituted for some of the water content to produce gels with a larger solvent content.

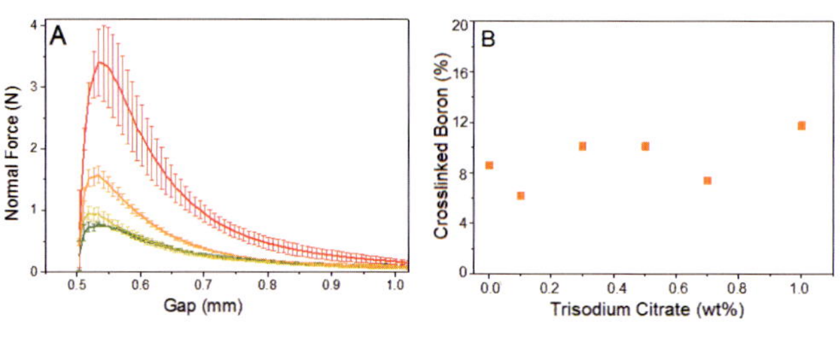

FIG. 3 (A) Extensional rheology (a measurement of the tensile stress the gels can withstand) of samples prepared from 3 wt% 80PVAc and 0.75 wt% borax with 0 wt% (green), 0.1 wt% (yellow), 0.3 wt% (orange), and 0.5 wt% (red) of sodium citrate. (B) Percentage of boron species participating in crosslinks based on integration of the ^{11}B-NMR signals of gels prepared with 4 wt% 80PVOH and 1 wt% borax (Angelova 2013).

Recipes

We used two polymers obtained from Kuraray: 'Poval 48-80', which we refer to as 80PVOH, and 'Poval 5-74' referred to as 75PVOH[7] (for formulations using other PVOH polymers, see the references in this paper).

1. 50 g 80PVOH-borax water gel

In a jar, measure 2 g 80PVOH and 37.5 g water. If a stirrer-hotplate is available, set the mixture to stir at room temperature until the polymer is dissolved. Low heat can be applied to speed up the process. If a stirrer-hotplate is not available, agitate the mixture periodically by hand and allow ~24 hrs for the polymer to dissolve fully. In a second jar, dissolve 0.5 g borax in 10 g water – this should occur without the need for heat, although heating the solution will accelerate the dissolution. Pour the borax solution into the jar containing the polymer and stir the mixture until a putty-like gel is formed.

2. 50 g 80PVOH-borax water gel with 0.5 wt% Ecosurf EH-6

In a jar, measure 2 g 80PVOH, 0.25 g Ecosurf EH-6 and 37.25 g water. If a stirrer-hotplate is available, set the mixture to stir at room temperature until the polymer is dissolved. Low heat can be applied to speed up the process. If a stirrer-hotplate is not available, agitate the mixture periodically by hand and allow ~24 hrs for the polymer to dissolve fully. In a second jar, dissolve 0.5 g borax in 10 g water – this should occur without the need for heat, although heating the solution will help. Pour the borax solution into the jar containing the polymer and stir the mixture until a putty-like gel is formed.

3. 50 g 75PVOH-borax gel with 30:70 isopropanol: water

In a jar, measure 3 g 75PVOH, 14 g isopropanol and 22.5 g water. If a stirrer-hotplate is available, set the mixture to stir at room temperature until the polymer has dissolved. Low heat can be applied to speed up the process. If a stirrer-hotplate is not available, agitate the mixture periodically and allow ~24 hrs for the polymer to dissolve fully. In a second jar, dissolve 0.5 g borax in 10 g water – this should occur without the need for heat, although heating the solution will help. Pour the borax solution into the jar containing the polymer and stir the mixture until a putty-like gel is formed.

4. 50 g 75PVOH-borax gel with 30:70 acetone: water

In a jar, measure 2.5 g 75PVOH, 14 g acetone and 23 g water. If a stirrer-hotplate is available, set the mixture to stir at room temperature until the polymer is dissolved. Low heat can be applied to speed up the process. If a stirrer-hotplate is not available, agitate the mixture periodically by hand and allow ~24 hrs for the polymer to dissolve fully. In a second jar, dissolve 0.5 g borax in 10 g water – this should occur without the need for heat, although heating the solution will help. Pour the borax solution into the jar containing the polymer and stir the mixture until a putty-like gel is formed.

Some practical tips on using xPVOH-borax gels

- Use a spatula, rather than a brush, to manipulate the gel.
- Some conservators have used the gel like a soft, slightly moist eraser that can be rolled over the surface.
- Typical application times vary between 1 and 5 minutes. If leaving the gel for a longer time, cover it with some plastic wrap or Melinex film to reduce evaporation of the liquid component.
- If liquid is observed to leak from the gel when it is in contact with the surface, the cleaning action may be occurring too fast.
- The gel can be removed by lifting with a spatula, tweezers, or by pressing a tissue or swab to which the gel can adhere, and then peeling it from the surface.
- The treated region can be cleared of swelled coatings, solubilized material, and gel remnants with water or a mixture of water and the solvent used in the gel.
- xPVOH-borax gels often soften surface soiling or varnish without lifting it from the surface; the remainder of this material may be removed during the clearance step using traditional swabbing methods.
- The gels are not ideal for certain surfaces, for example porous surfaces such as paper, plaster, and some types of wood; test that the gel can be removed easily from the surface in a discreet area.

Conclusion

Gels, jellies, pastes and putties offer a variety of options to deliver and manipulate a cleaning solution on the surface of an artwork. Some of these materials aid in retaining a solvent at the surface for a longer time in a controlled manner, others offer poulticing capabilities. xPVOH-borax 'gels' provide an ideal vehicle for delivering cleaning solutions to a textured or three-dimensional surface whilst limiting the risk of polymer residue deposition. They are an innovative and useful tool for the conservator's toolkit. The materials necessary for their formulation are inexpensive, readily available and of low toxicity, and their preparation is not complicated, which invites experimentation; there is always room for improvement and further research.

Notes

1. See also G.R. Fife, 'Moving on up: a review of results from SRAL's tissue-gel composite approach', in this volume.
2. See A. Mateescu, S. Freese, P. Frank, U. Jonas and C. Theodorakopoulos, 'Novel surface-attached gels from photo-crosslinkable polyacrylamides for the cleaning of works of art', in this volume.
3. This research has been prepared as a manuscript, not yet published: Angelova, L.V., Matarrese, C., Fratini, E., Weiss, R.G., Dei, L. and Carretti, E. 2017. 'Chelating agents in aqueous, partially-hydrolyzed, poly(vinyl acetate) dispersions crosslinked with borax: physicochemical characterization and an application'.

4. See also C. Riedo, G. Rollo, D. Scalarone and O. Chiantore, 'Improved PVA gels for the cleaning of paint surfaces', in this volume.
5. See also A. Fialová and P. Kotlík, 'Poly(vinyl alcohol)-borate gels as carriers of microemulsions', in this volume.
6. See also T.T. Duncan, B.H. Berrie and R.G. Weiss, 'A comparison between gel and swab cleaning: physical changes to delicate surfaces', in this volume.
7. A sample (typically 500 g) of these polymers can be obtained from the company's website: www.kuraray.com.

References

Angelova, L.V. 2013. *Gels from Borate-crosslinked Partially Hydrolyzed Poly(vinyl acetate)s: Characterization of Physical and Chemical Properties and Applications in Art Conservation*, unpublished PhD dissertation, Georgetown University, Washington, DC.

Angelova, L.V., Terech, P., Natali, I., Dei, L., Carretti, E. and Weiss, R.G. 2011. 'Cosolvent gel-like materials from partially hydrolyzed poly(vinyl acetate)s and borax', *Langmuir* 27(18): 11671–82.

Angelova, L.V., Dhopatkar, N., Berrie, B.H., Ormsby, B., Dhinojwala, A. and Richardson, E. 2015a. 'Gel cleaning: what is happening at the interface?', paper presented at the *SEAHA Conference, London, 14–15 July 2015*. Book of Abstracts available at: http://www.seaha-cdt.ac.uk/wordpress/wp-content/uploads/2015/07/SEAHABoA.pdf (accessed 26 May 2017).

Angelova, L.V., Berrie, B.H., de Ghetaldi, K., Kerr A. and Weiss, R.G. 2015b. 'Partially hydrolyzed poly(vinyl acetate)-borax-based gel-like materials for conservation of art: characterization and applications', *Studies in Conservation* 60: 227–44.

Bertasa, M., Botteon, A., Chiantore, O., Poli, T., Riedo, C., Sansonetti, A. and Scalarone, D. 2015. 'Cleaning materials: a preliminary study of different agar powders', poster presented at *Technart 2015*. Book of Abstracts available at: http://technart2015.lns.infn.it/images/BoA.pdf (accessed 26 May 2017).

Bishop, M., Shahid, N., Yang, J.Z. and Barron, A.R. 2004. 'Determination of the mode and efficacy of the cross-linking of guar by borate using MAS B-11 NMR of borate cross-linked guar in combination with solution B-11 NMR of model systems', *Dalton Transactions* 17: 2621–34.

Bousher, A. 1995. 'Unidentate complexes involving borate', *Journal of Coordination Chemistry* 34(1): 1–11.

Carretti, E., Grassi, S., Cossalter, M., Natali, I., Caminati, G., Weiss, R.G., Baglioni, P. and Dei, L. 2009. 'Poly(vinyl alcohol)-borate hydro/cosolvent gels: viscoelastic properties, solubilizing power, and application to art conservation', *Langmuir* 25(15): 8656–62.

Deuel, H. and Neukom, H. 1949. 'Über die Reaktion von Borsäure und Borax mit Polysacchariden und anderen hochmolekularen Polyoxy-Verbindungen', *Makromolekulare Chemie-Macromolecular Chemistry and Physics* 3: 13–30.

Duncan, T.T., Berrie, B.H. and Weiss, R.G. 2016. 'Colloidal properties of aqueous poly(vinyl acetate)–borate dispersions with short-chain glycol ethers', *ChemPhysChem* 17(16): 2535–44.

Fife, G.R., van Och, J., Stabik, B., Miedema, N., Seymour, K. and Hoppenbrouwers, R. 2011. 'A package deal: the development of tissue gel composite cleaning at SRAL', in J. Bridgland (ed.), *Preprints of the ICOM-CC 16th Triennial Meeting*. Almada: Critério.

Inoue, T. and Osaki, K. 1993. 'Rheological properties of poly(vinyl alcohol)/sodium borate aqueous solutions', *Rheologica Acta* 32: 550–555.

Koga, K., Takada, A. and Nemoto, N. 1999. 'Dynamic light scattering and dynamic viscoelasticity of poly(vinyl alcohol) in aqueous borax solutions. 5. Temperature effects', *Macromolecules* 32(26): 8872–9.

Kulicke, W. and Clasen, C. 2004. *Viscometry of Polymers and Polyelectrolytes*. Berlin: Springer.

Maerker, J.M. and Sinton, S.W. 1986. 'Rheology resulting from shear-induced structure in associating polymer solution', *Journal of Rheology* 30(1): 77–99.

Murakami, I., Fujino, Y., Ishikawa, R. and Ochiai, H. 1980. 'Temperature-dependence of the viscosity of aqueous polyvinyl alcohol)-borax solutions', *Journal of Polymer Science Part B-Polymer Physics* 18(10): 2149–53.

Natali, I., Carretti, E., Angelova, L., Baglioni, P., Weiss, R.G. and Dei, L. 2011. 'Structural and mechanical properties of "peelable" organoaqueous dispersions with partially hydrolyzed poly(vinyl acetate)-borate networks: applications to cleaning painted surfaces', *Langmuir* 27(21): 13226–35.

Nemoto, N., Koike, A. and Osaki, K. 1996. 'Dynamic light scattering and dynamic viscoelasticity of poly(vinyl alcohol) in aqueous borax solutions. 2. Polymer concentration and molecular weight effects', *Macromolecules* 29(5): 1445–51.

Nijenhuis, K.T. 1997. *Thermoreversible Networks: Viscoelastic Properties and Structure of Gels*. Berlin: Springer.

Osada, Y. and Khokhlov, A.R. (eds) 2002. *Polymer Gels and Networks*. New York: Marcel Dekker Inc.

Riedo, C., Caldera, F., Poli, T. and Chiantore, O. 2015. 'Poly(vinyl alcohol)-borate hydrogels with improved features for the cleaning of cultural heritage surfaces', *Heritage Science* 3(1): 23.

Robb, I.D. and Smeulders, J.B.A.F. 1997. 'The rheological properties of weak gels of poly(vinyl alcohol) and sodium borate', *Polymer* 38(9): 2165–9.

Sato, T., Tsujii, Y., Fukuda, T. and Miyamoto, T. 1992. 'Reversible gelation of short-chain O-(2,3-dihydroxypropyl)cellulose/borax solutions. 1. A ^{11}B-NMR study on polymer-ion interactions', *Macromolecules* 25: 3890–95.

Schultz, R.K. and Myers, R.R. 1969. 'The chemo-rheology of poly(vinyl alcohol)-borate gels', *Macromolecules* 2(3): 281–5.

Sinton, S.W. 1987. 'Complexation chemistry of sodium borate with poly(vinyl alcohol) and small diols: a 11B NMR study', *Macromolecules* 20: 2430–41.

van Duin, M., Peters, J.A., Kieboom, A.P.G. and Vanbekkum, H. 1984. 'The pH-dependence of the stability of esters of boric-acid and borate in aqueous-medium as studied by B-11 NMR', *Tetrahedron* 40(15): 2901–11.

Weiss, R.G. and Terech, P. (eds) 2005. *Molecular Gels: Materials with Self-assembling Fibrillar Networks*. Amsterdam: Springer.

Authors' addresses

- Lora V. Angelova, Collection Care Department, The National Archives, Kew, UK (anglovalora@gmail.com)
- Emiliano Carretti, Consorzio Sistemi a Grande Interfase, Department of Chemistry, University of Florence, Florence, Italy
- Barbara H. Berrie, Conservation Division, National Gallery of Art, Washington, DC, USA
- Richard G. Weiss, Institute for Soft Matter Synthesis and Metrology, Georgetown University, Washington, DC, USA

Novel surface-attached gels from photo-crosslinkable polyacrylamides for the cleaning of works of art

Anca Mateescu, Silvia Freese, Petra Frank, Ulrich Jonas and Charis Theodorakopoulos

Introduction

Gel technology for art conservation, introduced in the 1980s (Wolbers 2000), is now established as a means to provide more controlled and better monitored cleaning treatments (Baglioni *et al.* 2015) than liquid cleaning methods. The majority of the gels employed for art conservation are made of deformable materials composed of either a dispersion of a diluted solid network in a continuous liquid or a dispersion of a liquid in a continuous network of a solid phase.

Upon swelling with liquids, the 3D networks of bulk gels can expand freely in all three dimensions, substantially increasing in volume, by a few orders of magnitude (Jonas *et al.* 2012; Mateescu *et al.* 2012). Compared to particulate forms, bulk gels have a comparably smaller surface-to-volume ratio, resulting in a relatively slow rate of the swelling and collapse processes that are limited by the diffusion of the aqueous medium through the interface between the hydrogel and its liquid environment. In contrast, micro- and nanogels possess a large surface-to-volume ratio that allows rapid solvent exchange with the environment, and thus show a faster swelling and collapse behaviour (Pich and Adler 2007). However, upon crosslinking, microgel particles can form 3D bulk gels with similar properties to the latter.

Because of the intrinsic nature of 3D bulk gels, there are limits in the control over the cleaning procedures utilizing such gels. One strategy to improve the cleaning procedure aims mainly at reducing the presence of water on the artwork surface, thereby limiting its diffusion into the bulk of the paint. A unilateral nuclear magnetic resonance (NMR) study of water propagation into acrylic emulsion paints upon cleaning with a number of 3D bulk gels showed that the critical elimination of water diffusion into paints has yet to be accomplished (Angelova *et al.* 2016). Further reduction of water and the limitation of its diffusion is essential to avoid the dissolution of water-sensitive paint (van den Berg *et al.* 2015), to reduce leaching of free fatty acids from oil paints (Stolow 1985), and to reduce the unwanted removal of polyethoxylate (PEO) surfactant compounds from acrylic emulsion paints (Ormsby *et al.* 2009). Depending on the type of hydrogel system employed for the cleaning, there are several concerns related to the post-treatment action of polymer, salts and surfactant residues on the artwork surface (Dorge 2004; Warda *et al.* 2007). Free PEO surfactants in solution can attach via their hydrocarbon segments onto hydrophobic materials on the artwork surface (Wolbers 1990) leading to rapid formation of oxidation-catalyzing species which will eventually accelerate the post-treatment surface degradation, as shown for oil paints (Burnstock and White 1990). The latest physical gels may have addressed the polymer residue issues (Baglioni *et al.* 2015), but elimination of surfactant residues and control of the diffusion of liquids is a difficult task given that the requirement for post-cleaning rinsing (Wolbers 2000) subjects the surface to further wetting, thereby risking liquid diffusion (Casoli *et al.* 2014).

We suggest that these issues can be addressed by utilizing surface-attached hydrogels with customized polymer networks, such as those employed for biosensing (Peppas *et al.* 2006; Mateescu *et al.* 2012), tissue engineering (Okano *et al.* 1993), controlled drug delivery (Bajpai *et al.* 2008) and other biomedical fields (Nicodemus and Bryant 2008; Shoichet 2010; Slaughter *et al.* 2009; Bajpai *et al.* 2008).

Surface-attached gels are based on covalently bonded polymer networks on solid substrates to yield a robust gel film (Jonas *et al.* 2012). By anchoring the polymer network to a substrate, the mechanical stability of these gels is substantially increased and handling is greatly improved. The swelling behaviour is strongly determined by the crosslink density of the network and anchor density of the surface attachment, specifically for very thin films in the micrometre range. Surface-attached hydrogels combine the robust network architecture of 3D bulk gels with the high surface-to-volume ratio of microgels, allowing rapid water and solute exchange between the hydrogel matrix and the surrounding medium. Their thin-film format provides full control over the amount of water that is loaded into the gel matrix upon swelling and that can be released from such a layer system, which is considerably less compared to bulk gels. Moreover, the specific physical properties and application specifications for the gel layer can be tuned by appropriate material selection (e.g. hydrophilicity of the employed monomers) and the synthesis parameters (e.g. crosslinking density and herewith connected swelling capacity and mechanical moduli). One particularly interesting method for the preparation of surface-attached polymer networks relies on photo-crosslinking of a suitable pre-polymer with light-sensitive crosslinker units. Upon irradiation of such a thin pre-polymer layer with light the photo-crosslinker units will be activated to chemically link with neighbouring chains, thus leading to covalently network formation and simultaneous surface attachment. The particular advantage of this strategy lies in the full control over the crosslinking density, which is directly related to the amount of light to which the pre-polymer layer is exposed (Tokarev *et al.* 2008; Matsui *et al.* 2005; Anac *et al.* 2010). This strategy is also applied to the materials used in the work reported below.

This paper presents a pilot study of the development of polymer-backed hydrogel films for use in art conservation. First prototypes of these hydrogel films were prepared by immobilization of novel photo-crosslinkable polyacrylamide (PAM) derivatives on flexible and transparent polyethylene (PE) films. After swelling the network layer with aqueous-based cleaning solutions the efficacy of the resulting hydrogel films to remove soils from accelerated light-aged mastic, dammar, shellac and polycyclohexanone resin coatings, as well as from the varnish of an early 20th-century egg tempera painting, was evaluated.

Attention was directed to evaluation of the cleaning efficacy of plain PAM hydrogel films (no copolymerized surfactant moieties) compared to hydrogel films made from PAM copolymers with covalently incorporated surfactant moieties. These films were either swollen with cleaning solutions containing free surfactants or surfactant-free buffer solution in order to avoid residue issues problems after treatment.

Experimental

Materials

Acrylamide (AM, 98%, Fluka), glacial acetic acid (99.5% pure CH_3COOH, Acros Organics) and aqueous sodium hydroxide (1M NaOH, Acros Organics) were used as received. The polyethoxylate (PEO) surfactants included poly(ethylene glycol) p-(1,1,3,3-tetramethylbutyl)-phenyl ether (Triton X-100, Acros Organics) and poly(oxyethylene)-23 lauryl ether (Brij 35, Aldrich). Acryloyl chloride (97%), trimethylamine (TEA or $N(Et)_3$) (99%), and methacrylic acid (MAA) (99%) (Sigma-Aldrich) were freshly distilled before use. Azobisisobutyronitrile (AIBN, 98%, Acros Organics) was recrystallized from methanol. Organic solvents, including xylene (Riedel & Haen) and ethanol (Merck) were used without further purification. Varnished substrates were made of mastic, dammar, shellac and polycyclohexanone (Laropal K80) resins.

Syntheses

Monomers:

- *TXA*: The acrylic derivative of Triton X-100 was synthesized according to Larraz *et al.* (2003).
- *B35A*: Similarly, the acrylic derivative of Brij 35 was synthesized by reaction of Brij 35 with acryloyl chloride as follows. A solution of acryloyl chloride (1.66 mmol) in dichloromethane was added dropwise to a solution of 1.004 g (0.83 mmol) Brij 35 and 0.12 mL (0.83 mmol) $N(Et)_3$ dissolved in 10 mL dichloromethane, at 0 °C under argon atmosphere and stirred for 55 h at room temperature. Then the reaction mixture was washed five times with 5 wt% aqueous NaOH and the organic phase was dried with magnesium sulphate. After distilling off the solvent the residue was dried overnight under reduced pressure to yield 0.892 g (85.54 %) product. TLC and ^1H NMR (Bruker AMX-300 spectrometer) in deuterated acetonitrile confirmed the purity of the product.
- *BPAAm*: *N*-(4-Benzoylphenyl)acrylamide was synthesized from 4-aminobenzophenone and acryloyl chloride according to the literature (Fadida and Lellouche 2012).

Copolymers:

- *PAM*: The photo-crosslinkable PAM copolymer consisting of 99 mole equiv. of acrylamide (AM) and 1 mole equiv. of BPAAm was synthesized by free radical polymerization in dioxane with AIBN as the radical initiator according to the literature (Beines *et al.* 2007).
- *PAMB*: The copolymer incorporating 5 mole equiv. of the Brij 35 surfactant monomer B35A, 94 mole equiv. of AM and 1 mole equiv. of BPAAm was prepared as PAM above using 44.5 mg AM (6.26 mmol), 400 mg B35A (0.333 mmol), 16 mg BPAAm (0.066 mmol) and 5.42 mg AIBN (0.033 mmol) in 15 mL dioxane. The mixture was purged with argon for 15 min and stirred for 72 hours at 65 °C. The product was precipitated with ethyl acetate (fivefold volume of reaction mixture), separated from the solvent by centrifugation and dried overnight under reduced pressure.
- *PAMX*: The analogous copolymer containing 5 mole equivalent of the Triton X-100 monomer was synthesized in the same way using 29.4 mg AM (4.136 mmol), 150 mg TXA (0.220 mmol), 11.1 mg BPAAm (0.044 mmol) and 3.61 mg AIBN (0.022 mmol) dissolved in 15 mL dioxane. Product isolation was performed as above for PAMB.

Polymer network layers bonded to polyethylene backing

Flexible polyethylene (PE) foils were treated for about 2 seconds per area with a corona discharge device (α-SE, J.A. Woollam Co., Inc.), that generates an air plasma at a distance of 2.5 cm from the PE surface, to increase the hydrophilicity of the PE surface. The copolymer layers for the hydrogel networks were casted onto the activated PE surface with a squeegee blade (1000 μm liquid layer thickness) from 2% w/v solutions of PAM, PAMX or PAMB, respectively in water and dried overnight at room temperature. The dry polymer layer was irradiated with UV light at 365 nm (N_2 atmosphere) to simultaneously induce crosslinking and attachment to the PE substrate surface. In all cases the crosslinking density was varied by changing the irradiation time from 15 to 30 and 60 minutes, generating the following surface-attached copolymer networks: PAM_{15}, $PAMX_{15}$, $PAMB_{15}$; PAM_{30}, $PAMX_{30}$, $PAMB_{30}$ and PAM_{60}, $PAMX_{60}$ and $PAMB_{60}$.

Attenuated total reflectance–Fourier transform infrared (ATR-FTIR) (Nicolet 6700 optical spectrometer) spectra of PAMX and PAMB had an increased absorption at 1110 cm^{-1} due to the presence of the PEO surfactant monomers (Ormsby *et al.* 2009).

Aqueous cleaning solutions and hydrogel film swelling

An aqueous cleaning solution with surfactant was made of 0.6% v/v Triton X-100 in 0.5% v/v aqueous CH_3COOH buffered with 1M NaOH to pH 5.0. A surfactant-free buffer solution was prepared from 0.5% v/v aqueous CH_3COOH buffered with 1M NaOH to pH 5.0. The PE-attached copolymer films were swollen with the appropriate cleaning solutions to generate the active hydrogel layers prior to application to the varnish surfaces. The gel systems were handled by the PE backing using tweezers and remained for 5–10 seconds open to the air to reduce free liquid prior to application on the varnish surface. They were applied with the swollen hydrogel film facing the varnish substrates and left in contact with the surface for 15 seconds, then gently peeled off using tweezers on the PE backing.

Soiled resin substrates

Preliminary tests were carried out on samples made in a previous study (Galatis *et al.* 2012) and on a new set of samples made with the same materials. 30 w/v dammar and mastic resins in xylene (30 w/v) were brought into suspension. Similarly, 15 w/v shellac in ethanol and 30 w/v Laropal K80 in ethanol were also brought in suspension. Once fully blended, the solutions were applied on substrates to form films. The older set of samples had been spin coated at 2780 rpm for 180 seconds on 3 × 3 cm aluminium substrates. The new set of samples was coated with an applicator to produce films of 50 μm liquid layer thickness on glass slides. After drying, the samples were exposed in a xenon-arc fadeometer equipped with an optical borosilicate filter and a window glass filter to imitate sunlight radiation through window glass (315–800 nm), and received a total dose of 11.65 Mluxhrs. Then the samples were soiled with an artificial mixture that was produced according to the literature (Wolbers 1992) and applied on the resin surfaces with a spray-gun. After evaporation and drying of the liquid phase of the soiling mixture, the soiled samples received an additional dose of 582.4 kluxhrs for better incorporation of the soiling particles on the varnishes.

Imaging, optical microscopy and SEM

Typically, a camera was adjusted to f/5.6, a 25 second shutter speed, and a fixed sensitivity of 200 ISO, both for visible and UV fluorescence photography. A linear polarizer from Cinko was employed to reduce glare and enhance picture quality. UV fluorescence was obtained with Philips TL 4W/08 F4T5/BLB lamps (315–408 nm) and a Wratten 2E Kodak cut-off filter on the camera lens. Magnification was determined using an Olympus BX51M Optical Microscope (OM) in reflected visible (VIS) and UV light with a ×10 eyepiece lens and objectives up to ×100. Micrographs were recorded by an attached Olympus DP70 digital camera. Typically, at a sensitivity of 200 ISO, darkfield VIS images were acquired at 1/350 second exposure and UV fluorescence images at 1.0 second. A field emission TESCAN MIRA 3 SEM with a secondary electron (SE) detector was employed. Samples on glass were cut with a glass cutter and sputter-coated with a platinum layer 8 nm thick at 4×10^{-2} mbar, to reduce charging.

Gravimetric studies

Gravimetry was employed with the PE-backed hydrogel films to determine water loading of the cleaning solutions into the hydrogel layer and liquid transfer upon contact with the varnish surfaces. For this purpose, PE-attached hydrogel stripes of 2.5 × 3.5 cm and average weight of 0.05 g were loaded with deionized (DI) water. After swelling, these stripes were left for a few seconds until no further dripping of free water was observed. The swollen PE-hydrogels were brought in contact with non-soiled and dry aged dammar resin substrates and weighted every 5 seconds until no further weight change was detectable.

Results and discussion

PE-backed hydrogel films

To limit the applied liquid volume very thin hydrogel layers would be required, which cannot be handled easily. A solution can be found in surface-attached hydrogel systems. In such a format the hydrogel is firmly attached to a more robust support and after swelling with an appropriate cleaning solution this film is brought into contact with the artwork surface,

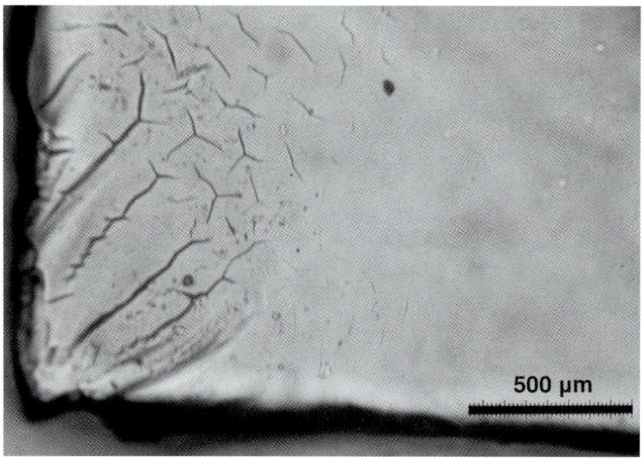

FIG. 1 Stereomicrograph of a PE-attached PAM_{15} film (15 min photo-crosslinked) swollen with 0.5% v/v aqueous acetate buffer system of pH 5.0 with 0.6% v/v Triton X-100.

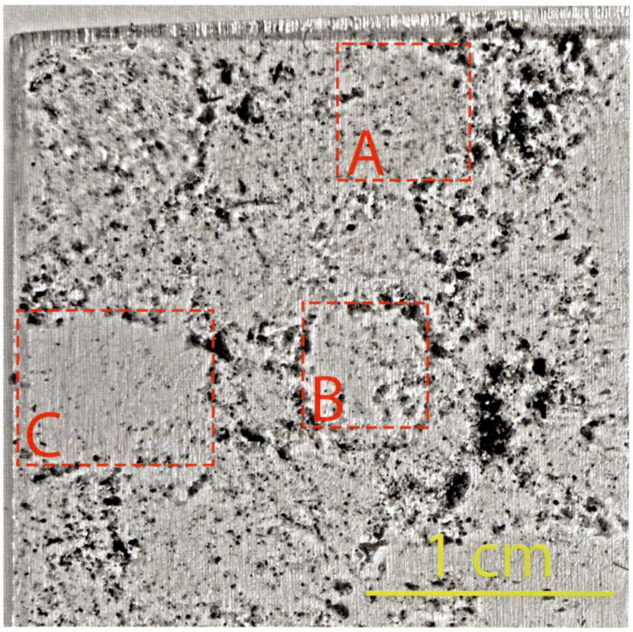

FIG. 2 A soiled mastic resin sample with areas after contact with the PE-PAM swollen by the aqueous Triton X-100–pH 5 buffer solution. Efficacy was improved in inverse proportion to the photo-crosslinking time of the copolymer networks. PAM_{60} (A) was the least effective, PAM_{30} (B) was more effective and PAM_{15} (C) cleaned the surface sufficiently.

like applying a sticky tape. After sufficient contact time to release or dissolve surface contaminations the hydrogel film is detached from the artwork surface and with this, the soil components are removed from the surface.

PE foils were selected as hydrogel supports for their chemical inertness, substantial toughness while being flexible, low cost and wide commercial availability. They also have the advantage of being transparent to allow monitoring the cleaning procedure by simple visual inspection of the artwork surface through the PE-hydrogel film. As these PE surfaces are rather chemically inert and hydrophobic, corona discharge treatment was applied (as outlined in the experimental section), which activated the PE surface by reactive oxygen species and rendered it sufficiently hydrophilic to allow spreading of the aqueous polymer solutions in homogenous liquid films. After drying the polymer layers could be effectively crosslinked and simultaneously bonded to the PE backing by UV irradiation, which activates the polymer-incorporated benzophenone units to undergo photo-induced radical insertion reactions with the surrounding polymer material. From the preparation conditions, a dry layer thickness of around 20 μm for the resulting polymer network can be estimated. Upon exposure to aqueous cleaning solutions the polymer network swells and the layer thickness increases (for the given systems a two–sixfold increase is expected). Depending on the crosslink density (that is controlled by UV exposure time) these hydrogel layers swell more at shorter crosslink times, and while absorbing more liquid, become softer. The particular UV crosslinking times employed (15, 30 and 60 minutes) are indicated in the polymer names by numeral subscripts (e.g. PAM_{15} for a 15 minute crosslinked PAM polymer layer). With the soft hydrogel layers, a good conformal contact with the potentially rough surface of the artwork is expected.

Copolymer systems

Three types of copolymers were synthesized to address the problem of surfactant residues remaining at the artwork surface. The first polymer, PAM, was polymerized from a mixture of acrylamide as the hydrophilic component responsible for hydrogel swelling with the photoreactive benzophenone monomer BPAAm. This polymer system serves as a simple reservoir layer for the cleaning solutions without specific interactions with the cleaning components. In order to increase affinity of lipophilic components of the soils and cleaning solutions, surfactant derivatives were covalently incorporated into the polymer backbone by copolymerization of the acrylamide-BPAAm mixture above with the Triton X-100 monomer TXA or the Brij 35 monomer B35A, respectively. Such surfactant-modified hydrogel networks are expected to absorb free surfactants (particularly in micelle aggregates) and hydrophobic soil components more efficiently by enhanced hydrophobic interactions with the covalently bound hydrocarbon tails of the surfactant co-monomers.

Assessment of cleaning efficacy

For the cleaning procedures two different types of cleaning solutions were tested. The first liquid medium consisted of a 0.5% v/v aqueous acetate buffer system of pH 5.0 without surfactant. The second cleaning solution was made from the same acetate buffer, but containing 0.6% v/v. Triton X-100 as surfactant. Swelling of the photo-crosslinked PE-attached PAM films when loaded with the aqueous cleaning solution could be monitored *in situ* under the stereomicroscope, here shown with a solution containing Triton X-100 (Fig. 1). Swelling of the 15 minute photo-crosslinked network PAM_{15} was more obvious in PAM_{15} than in the more highly crosslinked systems PAM_{30} and PAM_{60}.

For the cleaning method two different approaches were tested. The first cleaning procedure was designed as a

TABLE 1 Visually assessed cleaning efficacy after 15 seconds one-touch contact of the PE-backed hydrogel systems with the varnish test surfaces.

	PAM + TX-100 in pH 5.0 buffer solution			PAM + pH 5.0 buffer solution			PAMX + TX-100 in pH 5.0 buffer solution		
	PAM$_{60}$	PAM$_{30}$	PAM$_{15}$	PAM$_{60}$	PAM$_{30}$	PAM$_{15}$	PAMX$_{60}$	PAMX$_{30}$	PAMX$_{15}$
Shellac	1	3	3	1	2	3	1	4	4
Mastic	1	3	3	1	1	3	1	3	4
Dammar	–	–	–	–	–	–	1	4	4
Laropal K80	–	–	–	–	–	–	–	–	–

	PAMX + pH 5.0 buffer solution			PAMB + TX-100 in pH 5.0 buffer solution			PAMB + pH 5.0 buffer solution		
	PAMX$_{60}$	PAMX$_{30}$	PAMX$_{15}$	PAMB$_{60}$	PAMB$_{30}$	PAMB$_{15}$	PAMB$_{60}$	PAMB$_{30}$	PAMB$_{15}$
Shellac	1	3	3	2	5	5	2	5	5
Mastic	1	3	4	1	4	5	1	4	5
Dammar	1	4	4	–	–	–	1	4	5
Laropal K80	1	3	4	–	–	–	2	3	5

1: not efficient / soils shifted; 2: minor soil removal; 3: moderate soil removal; 4: considerable soil removal; 5 efficient removal

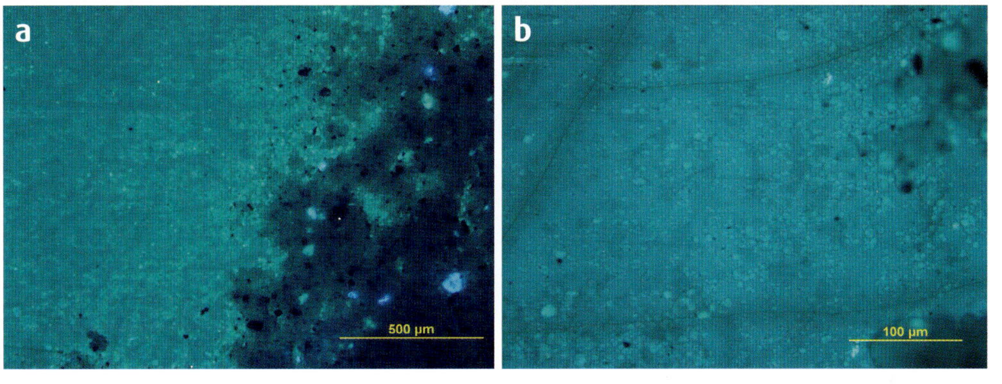

FIG. 3 UV fluorescence micrographs: removal of soils from a light-aged dammar resin coating with the PE-PAMB$_{15}$–pH 5 buffer system: (a) a half cleaned (on the left side) sample and (b) the surface after one-touch application in a higher magnification.

two-step process, with application of the first PE-hydrogel film with surfactant-buffer solution to remove soils as in common cleaning procedures, followed by clearing of desoiled surface from potential surfactant residues by applying a second PE-hydrogel film, this time swollen only with buffer solution (without free surfactant). During this set of experiments, it was observed that the second clearing step alone was already effective by itself to remove soils. Therefore, a second cleaning procedure was designed as a single-step process by using only the surfactant-free buffer solution to swell the hydrogel film in order to avoid contamination of the varnish surface with surfactant.

The results of these tests for the different combinations of hydrogel systems (PAM, PAMX, and PAMB) swollen by buffer solutions – with and without Triton X-100 – are summarized in Table 1 by listing the visually assessed cleaning efficacy. The application of the systems showed that the PE-attached PAM hydrogels interacted rapidly with the soiled surfaces and picked up soils in a fraction of the time compared to methylcellulose thickened solutions. It was also observed that soils were pulled vertically without wiping the surface with the hydrogel films systems. It appeared that the PE-attached PAM hydrogels enhanced the cleaning efficacy of the aqueous solution by allowing soils to be lifted by one-touch application (similar to a sticky tape effect). One important key parameter in this process was found to be the extent of crosslinking, which directly affects the swelling and softness of the hydrogel layer, as pointed out above. The PE-attached PAM$_{60}$ systems showed almost no soil removal, while the PE-attached PAM$_{30}$ removed some soil; the weakest crosslinked system, PE-attached PAM$_{15}$, demonstrated an enhanced efficacy (Fig. 2). Cleaning efficacy improves with reduced crosslink density and consequently increased swelling ratio of the hydrogel layer. This general trend can be found in the comprehensive summary of the visual evaluation of all the systems tested in Table 1.

When the clearing procedure of the two-step cleaning process was carried out by bringing the PE-attached PAMX$_{30}$ systems swollen with the pH 5.0 buffer solution into contact with the cleaned varnish surface for 2 seconds, it was observed that soils were readily lifted from the adjacent soiled surfaces. In line with the cleaning tests with

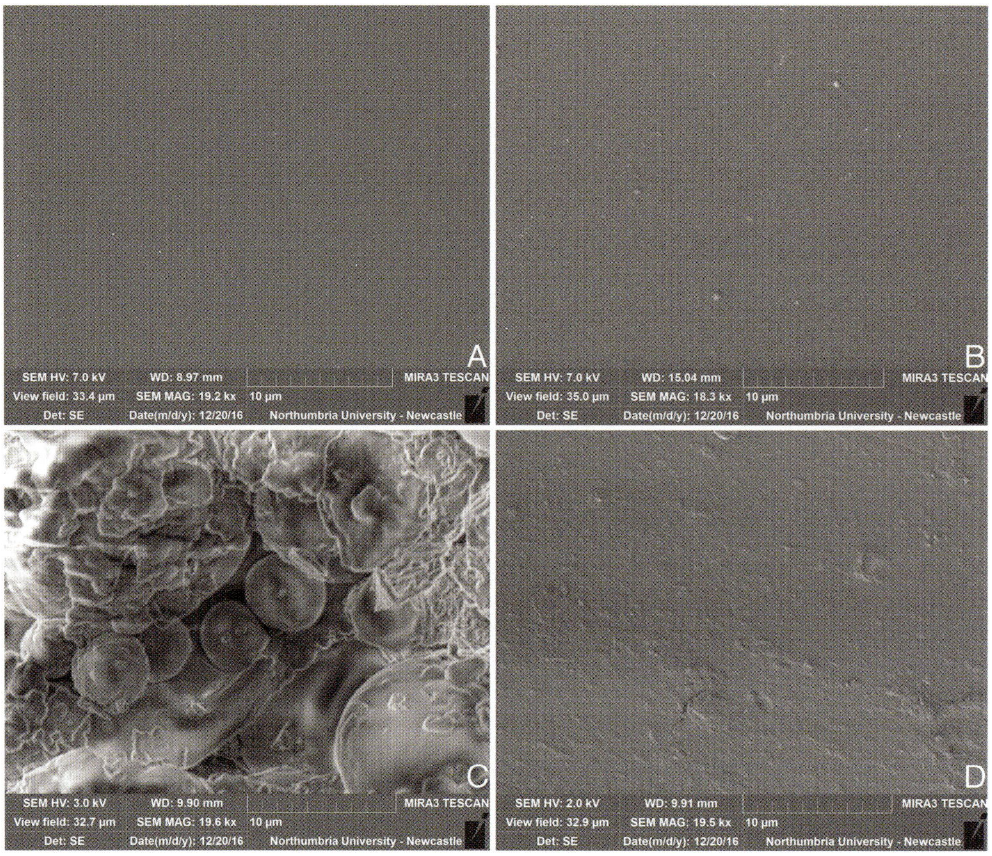

FIG. 4 SEM images of dammar varnish surface of: (A) an aged, unsoiled and untreated control sample, (B) the same sample after contact with the PE-attached $PAMB_{15}$–pH 5 buffer hydrogel, (C) a soiled and aged dammar varnish with a random distribution of 3–10 µm diameter soil particles; (D) the sample from (C) after interaction with $PAMB_{15}$–pH5 buffer hydrogel. Image (D) shows a relatively smooth surface after treatment. Similar results were observed for all the samples tested.

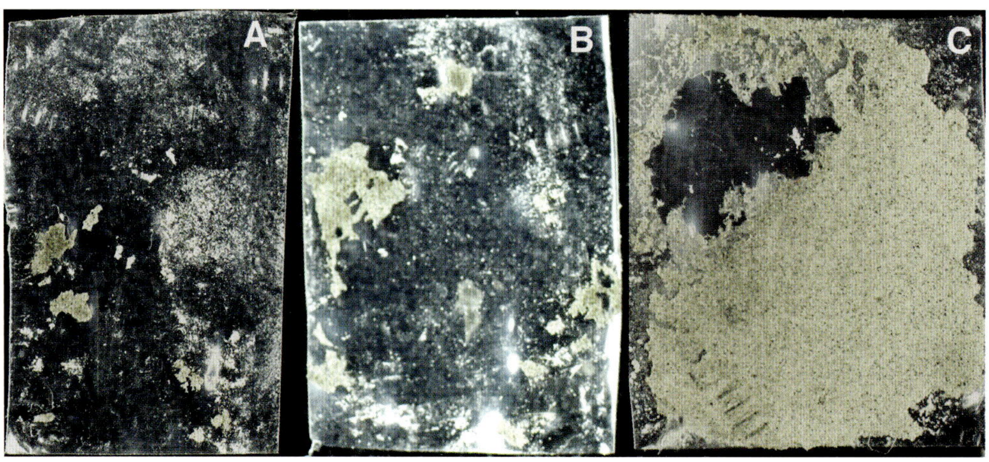

FIG. 5 PE-attached hydrogels based on: $PAMB_{60}$ (A) $PAMB_{30}$ (B) and $PAMB_{15}$ (C) networks after cleaning of soiled varnish surfaces. The efficacy of the less crosslinked hydrogel systems is evident.

PE-PAM above, it was observed that the less crosslinked PE-$PAMX_{15}$–pH5 buffer systems were more efficient than the $PAMX_{30}$ systems. Similar experiments were carried out with PAMB hydrogels containing covalently attached Brij 35, which indicated that the cleaning effects were enhanced when the cleaning solution contained the free surfactant. Again, cleaning efficacy was reduced for the more highly photo-crosslinked PAMB networks and the best results were obtained with the 15 minute photo-crosslinked films. Observations with UV fluorescence microscopy clearly indicate the cleaning potential of the new systems (Fig. 3).

SEM studies confirmed that the interaction of the PE-attached PAMX and PAMB hydrogels systems had only a negligible effect on the plain soil-free resin surfaces. Initially, the systems were tested on light-exposed unsoiled resin films. The gels were left on the surface for about 15 seconds, which was the exact time necessary for lifting soils from the soiled

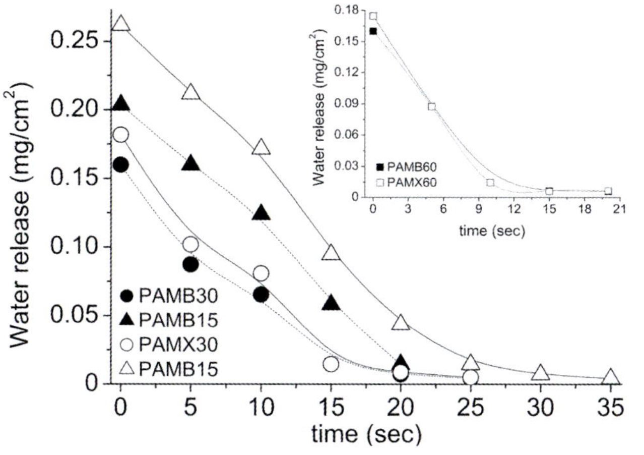

FIG. 6 Gravimetric dehydration curves determined from the interaction of surface-attached PAMX and PAMB systems with an aged, clean dammar varnish. Dehydration of $PAMX_{60}$ and $PAMB_{60}$ (inset) was faster than that of the less photo-crosslinked (and consequently more swollen) films.

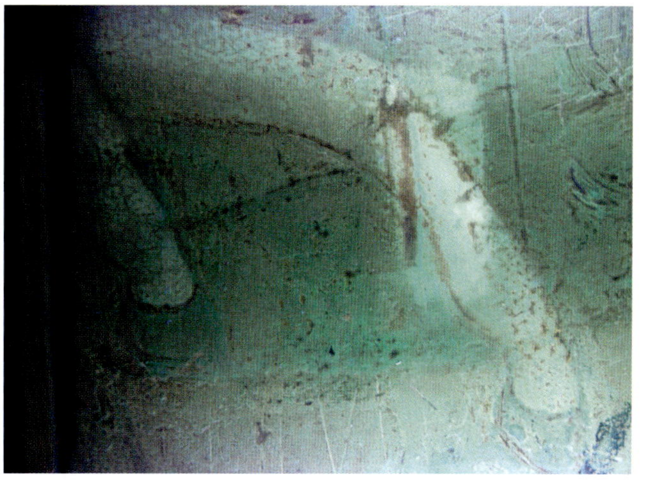

FIG. 7 UV fluorescence of an early 20th-century tempera painting with a mastic varnish after application of PE-attached $PAMX_{15}$ loaded with a pH5 acetate buffer solution. The system efficiently removed originally accumulated surface soils.

samples, and the surface topography was examined by SEM to reveal a barely noticeable change in surface smoothness (Fig. 4A and 4B). In comparison, the soiled surfaces examined before cleaning treatment revealed globular particles with diameters ranging from 2 to 10 μm randomly distributed on the resin surfaces (Fig. 4C). After application for about 15 seconds with the PE-$PAMX_{15}$–pH5 buffer or the PE-$PAMB_{15}$–pH5 buffer systems, respectively, very effectively cleaned and smooth varnish surfaces were obtained, comparable to the control samples (Fig. 4D). The surface roughness of the post-cleaned sample D may also have been created by soiling during the application process rather than interaction with the gel film.

After usage the surface-attached hydrogel systems dried almost instantly. The 30 minute and 60 minute photo-crosslinked films, which were thinner and stiffer than the 15 minute photo crosslinked systems, had a pronounced tendency to warp tangentially towards the hydrogel side of the film. This was a fast process indicating a rapid collapse rate leading to shrinkage that instantly pushed the systems to cup towards the collapsed polymer network side. The systems with the 15 minute photo-crosslinked networks remained flatter for longer but eventually cupped similarly to the 60 minute photo-crosslinked systems. For further examination, the used PE attached films were kept flat by placing them between glass slides immediately after use. Inspection of the dried and flattened PE-attached films showed a varying amount of attached soils from the cleaning process, which served as a direct indicator of the influence of the degree of crosslinking to the cleaning efficacy of the film systems (Fig. 5).

To assess the liquid loading capacity of the films, gravimetric studies were performed during cleaning treatment, which yielded dehydration curves (Fig. 6). These data show that the 15 and 30 minute photo-crosslinked films released water at similar rates, although the PE-$PAMB_{15}$ films contained more water for longer. PE-$PAMB_{15}$ was able to touch the resin surface up to four times prior to dehydration and PE-$PAMB_{30}$, PE-$PAMX_{15}$ and PE-$PAMX_{30}$ were able to touch the surface up to three times prior to dehydration. PE-$PAMB_{60}$ and PE-$PAMX_{60}$ dehydrated between the first and the second contact with the resin substrate.

As the ultimate test, the PE-$PAMX_{15}$–pH 5.0 buffer hydrogels were used on an early 20th-century varnished tempera painting. The systems were placed on the surface for a few seconds and upon lifting the surface was notably cleaner as monitored by UV fluorescence (Fig. 7).

Conclusion

These findings clearly demonstrate that PE-supported hydrogels can be used to remove soils very efficiently from common varnishes. These film systems allow convenient application of cleaning solutions with precise control over the target region on the surface, applying minute liquid volumes in the microlitre per square centimetre range. Covalent incorporation of surfactant moieties in the hydrogel network increased the cleaning efficacy, potentially by increased hydrophobic interactions. Such film systems are also suitable for surface decontamination from cleaning residues by subsequent application of gel films that were swollen with pure, surfactant-free buffer. Optical microscopy and SEM demonstrated that application of the 15 minute photo-crosslinked PE-attached poly(acrylamide-co-ethoxylate surfactant)-buffer systems on aged and soiled mastic, dammar, shellac and polycyclohexanone resin coatings for about 15 seconds led to efficient removal of soil, revealing smooth surfaces that were comparable to unsoiled resin surfaces. Typically for the surface-attached gel technology, water desorption from the systems to the varnish surfaces was rapid and minimal. Given that the films are highly tuneable, further optimization is expected to enable residue-free and water-eliminating cleaning treatments for delicate and water-sensitive surfaces.

References

Anac, I., Aulasevich, A., Junk, M.J.N., Jakubowicz, P., Roskamp, R.F., Menges, B., Jonas, U. and Knoll, W. 2010. 'Temperature dependent swelling behavior of thin poly(N-isopropylacrylamide) copolymer gel layers in ethanol-water mixtures', *Macromolecular Chemistry and Physics* 211(9): 1018–25.

Angelova, L.V., Berrie, B.H., de Ghetaldi, K., Kerr, A. and Weiss, R.G. 2015. 'Partially hydrolyzed poly(vinyl acetate)-borax-based gel-like materials for conservation of art: characterization and applications', *Studies in Conservation* 60: 227–44.

Angelova, L.V., Ormsby, B. and Richardson, E. 2016. 'Diffusion of water from a range of conservation treatment gels into paint films studied by unilateral NMR. Part I: Acrylic emulsion paint', *Microchemical Journal* 124: 311–20.

Baglioni, P., Carretti, E. and Chelazzi, D. 2015. 'Nanomaterials in art conservation', *Nature Nanotechnology* 10: 287–90.

Bajpai, A.K., Shukla, S.K., Bhanu, S. and Kankane, S. 2008. 'Responsive polymers in controlled drug delivery', *Progress in Polymer Science* 33(11): 1088–18.

Beines, P.W., Klosterkamp, I., Menges, B., Jonas, U. and Knoll, W. 2007. 'Responsive thin hydrogel layers from photo-cross-linkable poly(Nisopropylacrylamide) terpolymers', *Langmuir* 23(4): 2231–8.

Burnstock, A. and White, R. 1990. 'The effects of selected solvents and soaps on a simulated canvas painting', in J.S. Mills and P. Smith (eds), *Cleaning, Retouching and Coatings*. London: IIC, 111–18.

Casoli, A., Di Diego, Z. and Isca, C. 2014. 'Cleaning painted surfaces: evaluation of leaching phenomenon induced by solvents applied for the removal of gel residues', *Environmental Science and Pollution Research* 21: 13252–63.

Dorge, V. (ed.) 2004. *Solvent Gels for the Cleaning of Works of Art: The Residue Question*. Los Angeles: Getty Conservation Institute.

Fadida, T. and Lellouche, J.-P. 2012. 'Poly-N-(4-benzoylphenyl) methacrylamide nanoparticles: preparation, characterization, and photoreactivity features', *Journal of Polymer Research* 19(23): 1–12.

Galatis, P., Boyatzis, S. and Theodorakopoulos, C. 2012. 'Removal of a synthetic soiling mixture on mastic, shellac & Laropal® K80 coatings using two hydrogels', *e-Preservation Science* 9: 73–83.

Jonas, U., van den Brom, C.R., Brunsen, A. and Roskamp R.F. 2012. 'Surface attached polymeric hydrogel films', in W. Knoll (ed.), *Handbook of Biofunctional Surfaces*. Singapore: Pan Stanford Publishing, 277.

Larraz, E., Elvira, C. and San Román, J. 2003. 'Novel acrylic macromonomer with amphiphilic character derived from Triton® X-100: radical copolymerization with methyl methacrylate and thermal properties', *Journal of Polymer Science: Part A: Polymer Chemistry* 41(11):1641–9.

Mateescu, A., Wang, Y., Dostalek, J. and Jonas, U. 2012. 'Thin hydrogel films for optical biosensor applications', *Membranes* 2(1): 40–69.

Matsui, J., Akamatsu, K., Hara, N., Miyoshi, D., Nawafune, H., Tamaki, K. and Sugimoto, N. 2005. 'SPR sensor chip for detection of small molecules using molecularly imprinted polymer with embedded gold nanoparticles', *Analytical Chemistry* 77(13): 4282–5.

Nicodemus, G.D. and Bryant, S.J. 2008. 'Cell encapsulation in biodegradeable hydrogels for tissue engineering applications', *Tissue Engineering Part B: Reviews* 14(2): 149–65.

Okano, T., Yamada, N., Sakai, H. and Sakurai, Y. 1993. 'A novel recovery system for cultured-cells using plasma-treated polystyrene dishes grafted with poly(N-isopropylacrylamide)', *Journal of Biomedical Materials Research* 27: 1243–51.

Ormsby, B., Kampasakali, E., Miliani, C. and Learner, T. 2009. 'An FTIR-based exploration of the effects of wet cleaning treatments on artists' acrylic emulsion paint films', *e-Preservation Science* 6: 186–95.

Peppas N.A., Hilt J.Z., Khademhosseini, A. and Langer R. 2006. 'Hydrogels in biology and medicine: from molecular principles to bionanotechnology', *Advanced Materials* 18: 1345–60.

Pich, A.Z. and Adler, H.J.P. 2007. 'Composite aqueous microgels: an overview of recent advances in synthesis, characterisation and applications', *Polymer International* 56: 291–307.

Shoichet, M.S. 2010. 'Polymer scaffolds for biomaterials applications', *Macromolecules* 43(2): 581–91.

Slaughter, B.V., Khurshid, S.S., Fisher, O.Z., Khademhosseini, A. and Peppas, N.A. 2009. 'Hydrogels in regenerative medicine', *Advanced Materials* 21(32–33): 3307–29.

Stolow, N. 1985. 'Part II: Solvent action', in R.L. Feller, N. Stolow and E.H. Jones (eds), *On Picture Varnishes and their Solvents*, rev. edn. Washington DC: National Gallery of Art, 47–116.

Tokarev, I., Tokareva, I. and Minko, S. 2008. 'Gold-nanoparticleenhanced plasmonic effects in a responsive polymer gel', *Advanced Materials* 20(14): 2730–34.

van den Berg, K.J., Burnstock, A., de Keijzer, M., Krueger, J., Learner, T., de Tagle, A. and Heydenreich, G. (eds) 2015. *Issues in Contemporary Oil Paint*. Amsterdam: Springer.

Warda, J., Brückle, I., Bezur, A. and Kushel, D. 2007. 'Analysis of agarose, Carbopol, and Laponite gel poultices in paper conservation', *Journal of the American Institute for Conservation* 46(3): 263–79.

Wolbers, R. 1990. 'A radio-isotopic assay for the direct measurement of residual cleaning materials on paint film', in J.S. Mills and P. Smith (eds), *Cleaning, Retouching and Coatings*. London: IIC, 119–25.

Wolbers, R. 1992. 'The use of a synthetic soiling mixture as a means for evaluating the efficacy of aqueous cleaning materials on painted surfaces', *Conservation-Restauration des Biens Culturels* 4: 22–9.

Wolbers, R. 2000. *Cleaning Painted Surfaces: Aqueous Methods*. London: Archetype Publications.

Authors' addresses

- Anca Mateescu, Continental Automotive Romania, Research and Development, Timisoara, Romania
- Silvia Freese, Department of Chemistry and Biology, University of Siegen, Siegen, Germany (silvia.freese@student.uni-siegen.de)
- Petra Frank, Department of Chemistry and Biology, University of Siegen, Siegen, Germany
- *Author for correspondence*: Ulrich Jonas, Department of Chemistry and Biology, University of Siegen, Siegen, Germany (jonas@chemie.uni-siegen.de)
- *Author for correspondence*: Charis Theodorakopoulos, Department of Arts, Science in Conservation of Fine Art, Northumbria University, Newcastle upon Tyne, UK (charis.theodorakopoulos@northumbria.ac.uk)

Innovative techniques for treating the reverse of paintings: gel systems and Aquazol 500 pre-impregnated facing pads

Enrica Boschetti and Christian Tortato

Introduction

In the last 40 years structural conservation treatments for oil paintings on canvas have changed radically. An established preference for minimal intervention treatments has had a strong impact on conservation practice in most museums and private studios, leading to a critical reassessment of lining and the increasing use of alternative techniques for consolidation. In the past, lining was often applied as a cure-all preventive treatment, intended to last indefinitely and able to cope with a wide range of not entirely predictable issues such as a poorly controlled environment. Nowadays, there is a strong reluctance to re-treat previously lined paintings, especially if their condition is stable. In this case, the main challenge is trying to preserve the existing lining for as long as possible. The physical stress that lined paintings may endure during further structural treatment cannot be justified, if it is not necessary.

Most of the traditional lining techniques permanently alter the surface texture of paintings and additional problems have arisen over time. The lack of control in past environmental conditions, as well as the action of acidic pollutants, has played a role in the deterioration process. The decay of lining materials can sometimes become so advanced that their degradation products may trigger further chemical and mechanical alterations, with negative consequences for both the paint and the ground layers (Ackroyd 1996; Young and Ackroyd 2001; Mecklenburg and Fuster López 2012; Fuster López *et al.* 2016). Such situations often require complex treatment decisions, involving de-lining and extracting old deteriorated adhesives or consolidants when they have completely lost their functionality and instead produce extreme stiffness, and/or no longer fulfil the original function.

The prospect of returning a lined painting to an unlined state has been considered attractive for a long time, but not free of risk (Hackney 2004, 2005). Questions such as whether the 'lining cycle' is meant to last indefinitely or whether it is possible to de-line paintings that have suffered severe damage due to deterioration caused by lining materials are worth investigating. This paper explores and attempts to address these questions.

Principles and methods

The processes of de-lining and extraction of contaminant materials from the original canvas imply a functional correlation between the cleaning system used on the reverse and the facing applied on the front. When the paint and ground layers are brittle and unstable, the application of the facing is strictly necessary to secure them during de-lining and extraction treatments. The choice of facing tissues and adhesive, as well as the application technique, must be tailored to provide an efficient temporary protection with good mechanical stability. In particular, the facing adhesive should build a satisfactory bond in between the protective tissues and the paint surface, at the lowest possible concentration. It should not penetrate into the substrates to avoid interfering with the cleaning treatments to be carried out on the reverse of the painting, and it should be reversible to allow an easy removal by completion of de-lining and extraction processes.

The use of gel systems is an important option to accomplish safe cleaning treatments on textile supports. For the extraction of polar materials (e.g. glue paste) from the reverse of paintings, the application of aqueous-based systems loaded into agar or gellan gum rigid gels has proved to be generally suitable, but with some drawbacks: the implementation of rigid aqueous gels – especially if applied in a fluid state – can often cause movement within the fabric, which risks putting the ground and paint layers under stress.

The extraction of non-polar materials (e.g. wax-resin or synthetic adhesives in hydrocarbon solvents) from the reverse of paintings can be carried out with Carbopol/ Ethomeen gels, combined with the use of aliphatic hydrocarbon solvents,

FIG. 1 Application of the Aquazol 500 pre-impregnated pad, and release from the supporting Melinex sheet.

mixed on occasion with other solvents. These methods allow for the treatment to be performed only on the uppermost surface of the canvas. This is vital, because possible capillary diffusion of the organic solvents could be harmful for both the ground and the paint layers.

This research aims to develop technical solutions with more benefits and fewer disadvantages than existing processes. An innovative facing system using Aquazol 500 pre-impregnated pads on the front (Borgioli *et al.* 2016) has been combined with gel cleaning treatments on the reverse, coupled with the use of silicone-based materials (Boschetti 2016). Since being introduced by Richard Wolbers for the surface cleaning of acrylic paint[1] (Lagalante and Wolbers 2016), silicone solvents and gels are gradually entering into conservation practice as part of a broader range of treatments (Cremonesi 2016). In our research these materials proved to be invaluable tools to enhance the safety of extraction processes for lining adhesives.

In the use of aqueous systems, high volatile silicone solvents such as decamethylcyclopentasiloxane (cyclomethicone D5) and hexamethyldisiloxane (D2) can provide a temporary hydrophobic barrier, which protects the fabric support during the extraction process. As they are extremely non-polar materials, silicone solvents can improve the control over cleaning treatments carried out with fluid or rigid aqueous gels, seemingly without affecting the painting materials (Cremonesi 2016; Stavroudis 2016) and without compromising the stability of the Aquazol 500 facing pads. When organic solvents are necessary for lining removal, they should be contained within a highly stable and non-polar gel. For this purpose, a silicone-based polymeric gel, Velvesil Plus (Momentive) was tested in this study to design innovative cleaning systems, coupled with the use of hexamethyldisiloxane as the volatile solvent used for clearance.

Facing with Aquazol 500 pre-impregnated pads

Among various available adhesives, Aquazol 500 is particularly suitable for designing a facing technique that provides strong and effective protection, a high degree of reversibility, and low impact on artworks. Aquazol 500 (poly-2-ethyl-2-oxazoline) has a molecular weight of 500,000, and is completely soluble in water and in a wide range of polar organic solvents. Due to its high molecular weight it is not expected to seep into the oil paint film, except through cracks. To minimize the risk of diffusion into substrates, the facing method described below avoids any form of direct application and exploits the excellent adhesive power of Aquazol 500 with a simple swelling action. Aquazol 500, dissolved in demineralized water, is used to prepare tissue paper that can be re-moistened as needed, and whose adhesive power is reactivated with a minimal moistening action. In order to manage the whole process safely, the pH and conductivity of the reactivation solution must be carefully adjusted, providing a good match between the pH and isotonicity of the paint surface to avoid swelling the paint during its application.[2]

Preparing the facing pads

Lightweight tissue papers (25 × 25 cm) were placed on a table covered with siliconized Melinex, impregnated with a solution of Aquazol 500 (3% in demineralized water), and dried for 24 hours. Once released, the side of each tissue facing the Melinex appears to be coated by a uniform and compact dry adhesive film of Aquazol 500, whose adhesive property can be reactivated just prior to use by a light micro-nebulization of buffer solution at adjusted pH and conductivity.

Applying the facing pads

The application is carried out by placing each Aquazol 500 pre-impregnated pad onto a slightly oversized, siliconized Melinex sheet with the adhesive side facing up: once reactivated by controlled re-wetting, the pad can be used immediately. Each facing tissue is applied to the paint surface using the Melinex sheet as support, just like an adhesive pad. Light surface pressure, achieved by dabbing the Melinex support with a soft sponge, can help to lay down each pad even on a textured surface. When adhesion has been achieved, the Melinex support can be peeled off (Fig. 1). Once dried the pre-impregnated pads provide stable and efficient protection, permitting de-lining and cleaning on the reverse of the support.

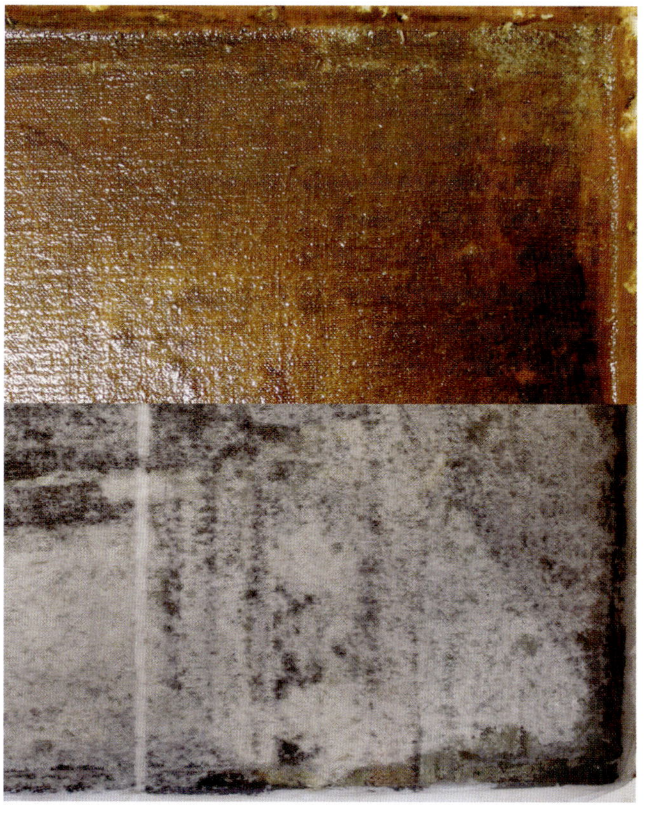

FIG. 2 Hydrophobic pre-treatment with cyclomethicone D5 on the reverse (upper side), and stable adhesion of the Aquazol 500 pre-impregnated pad facing on the front (lower side).

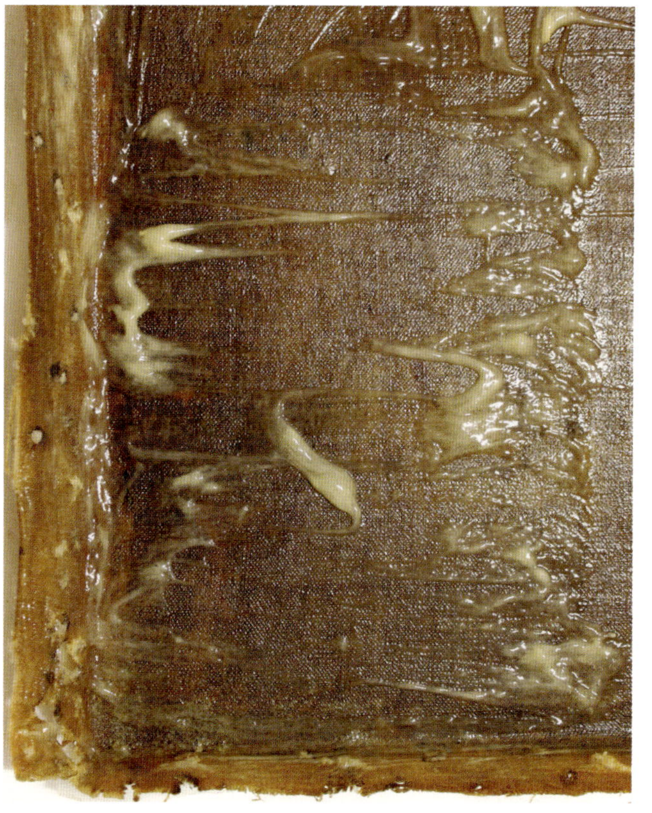

FIG. 3 Extraction process of glue paste residues with xanthan gel applied on the canvas after hydrophobic pre-treatment with cyclomethicone D5.

Removing the facing pads

The facing removal is also carried out using minimal moisture with an aqueous solution similar in pH and conductivity to the application solution. The reactivation solution can be delivered using a micro-nebulizer or with rigid gellan gum gels. With short application times, gellan gels can release the minimum amount of moisture necessary to swell the facing adhesive. Due to their elasticity and transparency, gellan gels provide optimal contact with the surface and permit visual inspection during the release of the facing pads. Through this minimal humidification, the majority of the Aquazol 500 remains within the fibrous structure of the facing pads. If small amounts of residues are left behind, the paint surface can be gently and locally cleared with the reactivation solution.

De-lining and extraction processes: issues and treatments

The facing system has been applied to two case studies.

Painting A

This was dated 1881, measured 76 × 112.5 cm, and was executed in oil on canvas. The painting had been double-lined with glue paste. The extreme stiffness, increasing acidity and growing detachment between the lining canvases had already caused severe distortions and cleavages in the paint layers. These issues led to the decision to de-line.

Following taking pH and conductivity measurements, the painting was faced with Aquazol 500 pre-impregnated pads and removed from its stretcher. Then, both lining canvases were carefully removed. The treatment on the reverse was carried out in two steps: a careful mechanical removal of the thick adhesive residues using scalpels, followed by the extraction of adhesive residues incorporated in the fibres and weaving structure of the canvas (Figs 2 and 3). To ensure safe extraction, the fabric support was saturated with cyclomethicone D5 (Fig. 2) and then treated with an aqueous gel of xanthan gum (4% w/v) in demineralized water. The silicone solvent was used as a hydrophobic barrier: it prevented the water from penetrating into the canvas and then spreading out on the front, thus securing the stability of the facing. Once applied on the fabric saturated with cycomethicone D5, the xanthan gum gel was gently brushed for the time necessary to allow the entrained adhesive to be swollen and extracted into the cleaning system (Fig. 3). The residues were removed first with a fine spatula and then with dry cotton swabs. The treatment was completed with a twofold rinsing process: first using cotton swabs slightly moistened with demineralized water, and then by swab rolling with hexamethyldisiloxane. Where necessary, the extraction treatment was repeated, always taking advantage of the extremely hydrophobic properties of cyclomethicone D5. The xanthan gel is shear sensitive: movement of a brush through the gel allows its viscosity to thin

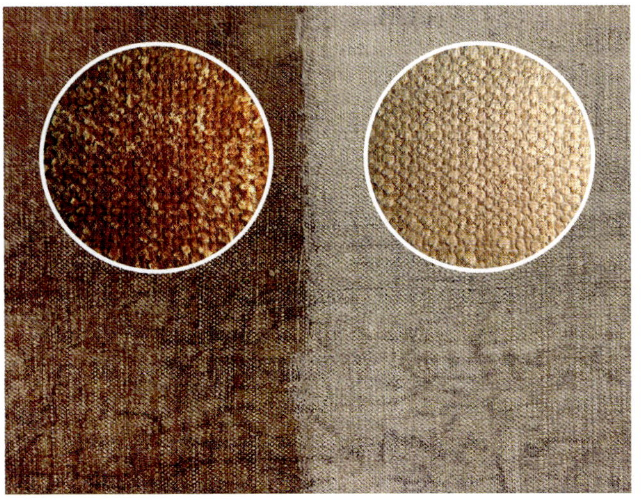

FIG. 4 Painting A: comparative details of the original support before and after the extraction of the glue-paste lining adhesive.

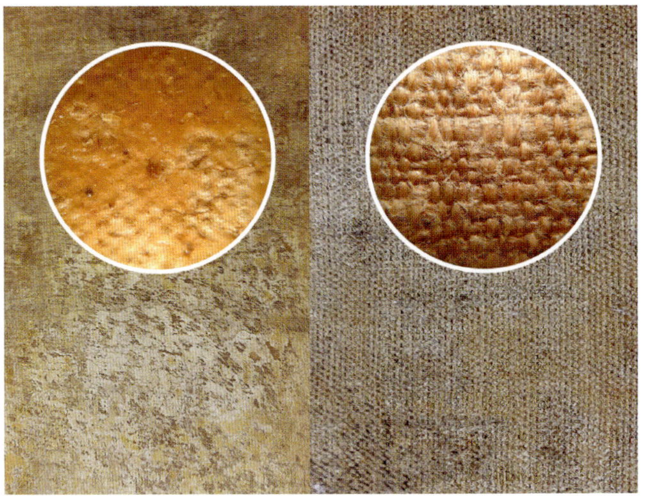

FIG. 5 Painting B: comparative details of the original support before and after the extraction of the wax-resin lining adhesive.

and provides more access to the support threads. The higher density of the aqueous gel with respect to the D5 solvent also allows for better contact with the solvent-laden support. This method has led to a positive outcome (Fig. 4): most of the old lining adhesive was extracted, the applied facing remained stable, no alteration occurred in the paint layers, and the original support lost its stiffness and regained flexibility. The extraction of contaminant materials allowed for the completion of the treatment whilst keeping the painting unlined.

Painting B

This canvas, dated to the beginning of 20th century, measured 126 × 126 cm and was executed in oil using a divisionist technique. The painting was heavily lined with wax resin. Over time, the extreme plasticity of this lining adhesive had caused dramatic distortions to the support. The initial detachment between the canvases produced mechanical defects, leading to cracking, cupping and losses in the paint layers, which led to the decision to remove the lining. By adding extreme weight to the original canvas, wax resin was the main trigger for the failure mechanisms.

The main issue was finding the best extraction system for the wax resin. The painting was removed from the stretcher and the lining fabric was detached using careful mechanical action. After de-lining, the original canvas was found to be coated with a very thick layer of unevenly distributed wax resin. Some areas were hidden by plaques of this material; others, now covered only by thinner layers, had suffered a deeper and invasive penetration of the adhesive. Before proceeding with the removal and extraction of this material, the pH and conductivity were measured and Aquazol 500 facing pads applied on the paint surface. On the reverse, the wax-resin layer was carefully reduced with scalpels. After this first removal, the extraction process was carried out with a miscible solvent mixture added to the Velvesil Plus gel (30% w/w of the stock gel). The solvent mixture added to the gel was a 7:3 mixture of cyclohexane:ethyl acetate (w/w). The silicone gel was brushed gently, changing the duration of contact according to the thickness of the adhesive layer to be extracted. The residues were removed first with a fine spatula and then with dry cotton swabs. Before rinsing with hexamethyldisiloxane, the extraction treatment was repeated on those residues deeply entrained inside the canvas. The painting was in effect returned to an unlined state as a consequence of the extraction process, optimized by repeated cleaning steps with the silicone gel and by clearance steps with a silicone solvent which evaporates completely without any observed interaction with the painting materials (Fig. 5). For a painting lined with a similar technique, reaching such an outcome without the use of silicone materials would have been difficult. In fact extracting such a large amount of lining material would not have been possible without running the risk of significant damage.

Conclusion

Our study has focused on the evaluation and application of combined methods for the safe and efficient removal of old deteriorated linings, in order to avoid the need for a new lining. The results achieved so far are encouraging. The methods are suggested as alternative solutions to expand existing options. We have attempted to evaluate the treatment method critically at each stage, bearing in mind not only the potential advantages, but the overall limitations and possible collateral effects. Analysis would be beneficial to better assess the long-term stability of this treatment.

Materials and suppliers

- Carbopol Ultrez 2: CTS Srl, https://www.ctseurope.com/en/scheda-prodotto.php?id=335
- Ethomeen C12: CTS Srl, https://www.ctseurope.com/en/scheda-prodotto.php?id=355

- Water-resistant Bollorè Japanese paper CTS. Weight: 12.3–22 g/m^2. Composition: Manila hemp, bleached cellulose pulp, polyamine epichlorohydrin: CTS Srl, https://www.ctseurope.com/en/scheda-prodotto.php?id=2354
- Aquazol 500, poly (2-ethyl-2-oxazoline): CTS Srl, http://www.ctseurope.com/en/scheda-prodotto.php?id=131sd
- Vanzan NF-C (xanthan gum): CTS Srl, http://www.ctseurope.com/en/scheda-prodotto.php?id=382
- Velvesil Plus (Momentive Performance Materials, Inc.), cyclomethone D5 (decamethylcyclopentasiloxane), hexamethyldisiloxane: Luigi & Felice Castelli SpA, http://www.lfcastelli.it

Notes

1. R. Wolbers and P. Cremonesi, 'The Cleaning of Modern Paints', unpublished reports from annual workshops held in Lodi, Italy, 2014–2016.
2. Unpublished reports from the lecture given by Richard Wolbers at the CESMAR7 Conference *Nuove prospettive ed approfondimenti per l'uso di poliossazoline Aquazols*, Workshop B, 5 June 2014, Centro Conservazione e Restauro la Venaria Reale, Turin, Italy.

References

Ackroyd, P. 1996. 'Glue-paste lining of paintings: an evaluation of the performance of some additive materials used in glue-paste formulations', in J. Bridgland (ed.), *Preprints of the ICOM-CC 11th Triennal Meeting*. London: James & James, 231-8.

Borgioli, L., Boschetti, E. and Tortato, C. 2016. 'I cerotti di Aquazol® 500. Una procedura alternativa per la velinatura dei dipinti', *Progetto Restauro* 73: 3-13.

Boschetti, E. 2016. 'Utilizzo dei materiali siliconici durante la rimozione di adesivi/consolidanti dai supporti in tela', in P. Cremonesi (ed.), *Proprietà ed esempi di utilizzo di materiali siliconici nel restauro di manufatti artistici – Parte prima*. Padua: Il Prato, 15-21.

Cremonesi, P. 2016. 'Struttura chimica e proprietà dei materiali siliconici', in P. Cremonesi (ed.), *Proprietà ed esempi di utilizzo di materiali siliconici nel restauro di manufatti artistici – Parte prima*. Padua: Il Prato, 5-14.

Fuster López, L., Ottolini, S., De Paola, L. and Yusa Marco, D.J. 2016. 'Meccanismi di deterioramento e studio di patologie su tre dipinti a olio su tela del XX secolo foderati a colla di pasta', in V.E. Selva Bonino (ed.), *Dall'olio all'acrilico, dall'impressionismo all'arte contemporanea*. Padua: Il Prato, 274-9.

Hackney, S. 2004. 'Paintings on canvas: lining and alternatives', *Tate Papers* 2. Available from: http://www.tate.org.uk/research/publications/tate-papers/02/paintings-on-canvas-lining-and-alternatives (accessed 29 March 2017).

Hackney, S. 2005. 'Relining, lining, de-lining', in *Minimo intervento conservativo nel restauro dei dipinti, Quaderni di CESMAR7*. Padua: Il Prato, 34.

Lagalante, A. and Wolbers, R. 2016. 'Cleaning acrylic paintings: new particle based water-in-oil emulsifiers', in V.E. Selva Bonino (ed.), *Dall'olio all'acrilico, dall'impressionismo all'arte contemporanea*. Padua: Il Prato, 107-14.

Mecklenburg, M. and Fuster López, L. 2012. 'Determining the acceptable ranges of relative humidity and temperature in museums and galleries', in C. Lodi and C. Sburlino (eds), *Prima, durante, invece del restauro*. Padua: Il Prato, 165-81.

Stavroudis, C. 2016. 'Silicone-based solvents in conservation: as free solvents and components of gel systems and microemulsions', in V.E. Selva Bonino (ed.), *Dall'olio all'acrilico, dall'impressionismo all'arte contemporanea*. Padua: Il Prato, 176-84.

Young, C. and Ackroyd, P. 2001. 'The mechanical behaviour and environmental response of paintings to three types of lining treatment', in *National Gallery Technical Bulletin* 22: 85–104.

Authors' addresses

- Enrica Boschetti, Milan, Italy (boschetti.enrica@gmail.com)
- Christian Tortato, Milan, Italy (tortato@stratirestauro.it)

A comparison between gel and swab cleaning: physical changes to delicate surfaces

Teresa T. Duncan, Barbara H. Berrie and Richard G. Weiss

Introduction

Gel-like materials form when aqueous solutions of partially hydrolysed poly(vinyl acetate) (xPVAc, where x is the percentage of hydrolysis) are combined with borax ($Na_2B_4O_7{\cdot}10H_2O$) (which dissociates into boric acid and borate anions in water). Di-diol esterification between the polymer and boric acid/borate anions leads to crosslinks that cause an increase in viscosity (Rietjens and Steenbergen 2005; Schultz and Myers 1969). Although the resultant material is really a high viscosity polymeric solution (HVPS), it will be referred to here as a 'gel' to differentiate these materials from formulations used in moistened-swab cleaning. Other types of gels and HVPSs have been examined and tested for cleaning works of art (Wolbers 2000). xPVAc–borate materials can slow the penetration of solvent into subsurface layers, their malleability leads to good contact with topographically complex surfaces, and their physical properties allow for easy removal by 'peeling off' after treatment (Angelova *et al.* 2011, 2015). Components have been added to aqueous xPVAc–borate gels to enhance their cleaning action for specific conservation applications. Examples include: organic solvents for varnish removal (Carretti *et al.* 2009; Natali *et al.* 2011), chelators to aid in removing metal corrosion stains (Cushman and Wolbers 2007; Angelova 2013), poly(ethylene oxide) for Paraloid B-72 removal (Riedo *et al.* 2015), and short-chain glycol ethers for soil removal (Duncan *et al.* 2016). The gels used in this study are aqueous 3 wt% 80PVAc and 0.75 wt% borax gels, with and without the addition of one of two polyethylene glycol (PEG)-based surfactants – nonaethylene glycol monododecyl ether ($C_{12}E_9$) or Brij® S100 ($C_{18}E_{100}$), which we hypothesized would enhance cleaning efficacy through micellar interactions (Fig. 1). Swab and gel cleaning methods, with and without the addition of these PEG-based surfactants, are compared in terms of both cleaning efficacy and physical damage to the surface of an artificial soil-coated acrylic paint and a brass sheet.

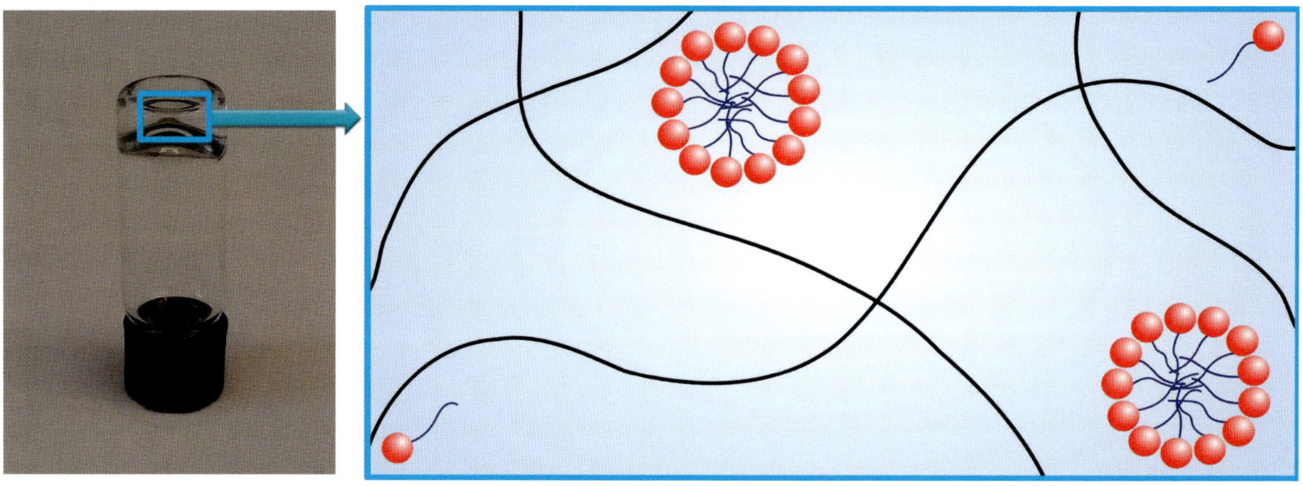

FIG. 1 80PVAc–borate gel formed with aqueous 1 wt% $C_{12}E_9$ solvent system; arrow points to conceptual representation of surfactant distribution and micelle formation in the solvent portion of a xPVAc–borate network.

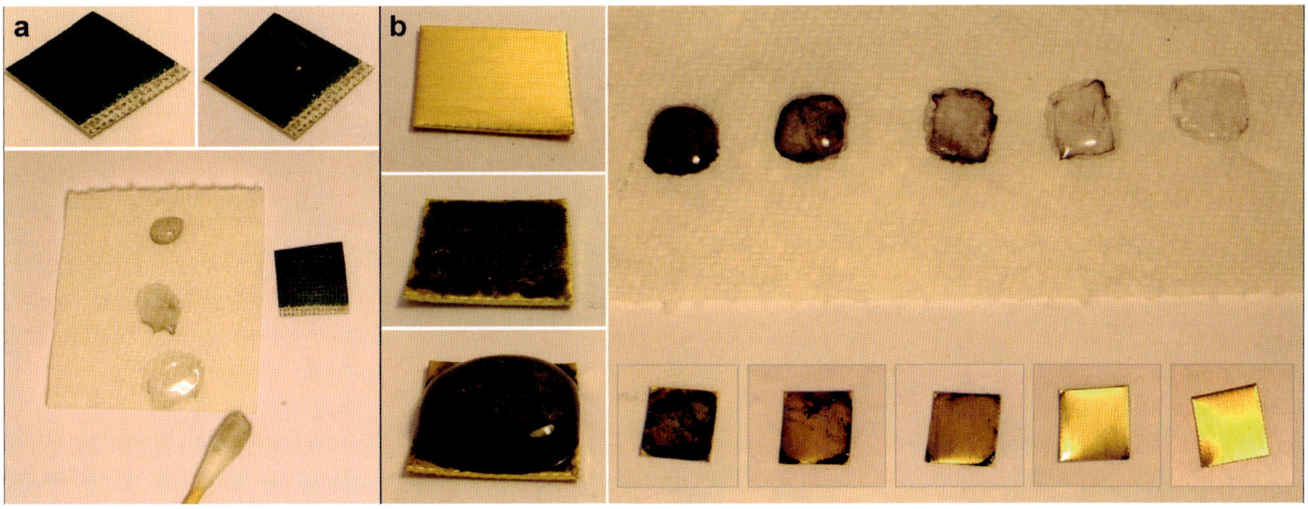

FIG. 2 Illustrations of surfaces cleaned using surfactant-loaded gels: (a) artificial soil-coated acrylic paint on canvas and (b) artificial soil-coated brass coupons, with gels and brass surface after each sequential cleaning displayed left to right.

Experimental

Gel preparation

Gels of 3 wt% 80PVAc and 0.75 wt% borax, with and without the addition of $C_{12}E_9$ or $C_{18}E_{100}$, were prepared as follows. 80PVAc and the appropriate amount of either $C_{12}E_9$ or $C_{18}E_{100}$ were dissolved in water by stirring at room temperature overnight. An aqueous 3 wt% borax solution was added, and the resultant gel was stirred with a spatula for 1 minute. Gel cleaning was performed 1–2 days after gel preparation to allow bubbles to dissipate.

Acrylic paint sample preparation and cleaning experiment

One coat of phthalocyanine green acrylic paint (~0.05 mm thick after drying) was brushed onto a primed canvas support and dried for 5 weeks under ambient conditions. 'Artificial soil' (prepared according to Ormsby *et al.* 2010) was suspended in hexanes and applied with a foam applicator to the paint layer, leaving ~0.8 mg/cm^2 after evaporation of hexanes. After 4 weeks, the prepared canvas sheet was cut into ~4 cm^2 squares, and each sample was cleaned using a swab or gel cleaning (Fig. 2a). For swab cleaning, a cotton swab was dipped in a solution and rolled over the surface for 2 minutes, cleaning a spot of ~1 cm^2 in diameter. For gel cleaning, three aliquots were placed sequentially, for 3 minutes each, on the same area (approximately ~1 cm^2 in diameter) and removed with a spatula. Thereafter, a cotton swab dipped in water was rolled across the surface for 20 seconds to lift any potential residual gel.

Scanning electron microscopy with energy dispersive X-ray analysis (SEM–EDX)

After the cleaning experiments, a small section (~3 mm^2) was cut from the paint sample and imaged with an SEM (Hitachi S-3400N) paired with an EDX spectrometer (Oxford Instruments, Aztec) and X-ray detector (Oxford Instruments X-Max silicon drift detector, 80 mm^2).

Optical microscopy

A Leica DMRX microscope with a PL Fluotar P ×10 objective, paired with a Canon EOS 5D Mark III camera, was used to collect a series of photographs at various focal depths. The batch of photographs was stacked with ImageJ (Schneider *et al.* 2012), and a 'stack focuser" plug-in (Umorin 2002–2006) was employed to create one image with an extended depth of field.

Brass coupon sample preparation and cleaning experiment

A sheet of brass (~70% Cu, ~30% Zn) was cut into ~1 cm^2 squares. Two drops of artificial soil, suspended in hexanes, were placed on the surface and allowed to dry, leaving ~1–2 mg. The procedure was repeated using hexanes alone on one coupon to act as a reference. Four soil-coated brass coupons were cleaned by rubbing a dry swab, rolling a dry swab, rolling an aqueous 0.1 wt% $C_{18}E_{100}$-moistened swab, or using an aqueous 0.1 wt% $C_{18}E_{100}$ gel (Fig. 2b). Swab cleaning was performed until the surface appeared clean to the eye. Gel cleaning was performed using five sequential applications of aliquots. Each aliquot was removed by pressing a paper towel to the surface of the gel and peeling it away from the coupon.

Atomic force microscopy (AFM)

After the cleaning experiments, all brass samples were sonicated in deionized water for 10 minutes, dried with a stream of compressed air, and imaged using an NT-MDT NTEGRA Prima AFM in semi-contact mode.

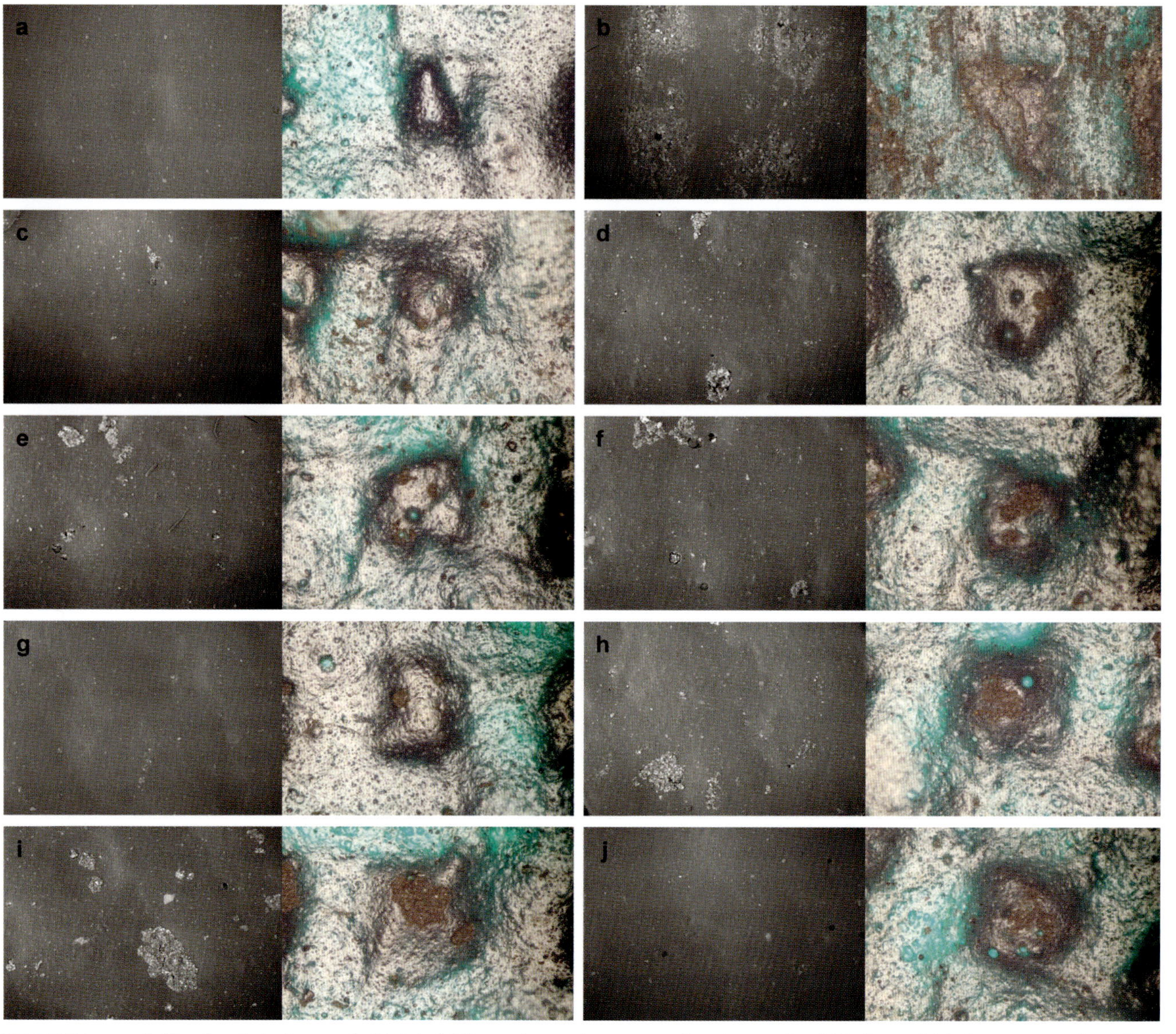

FIG. 3 BSE images (left) and optical micrographs (right) of different sites on the same acrylic paint samples: (a) before and (b) after coating with artificial soil; after cleaning with (c) a water swab, (d) a water gel, (e) an aqueous 0.1 wt% $C_{12}E_9$-moistened swab, (f) an aqueous 0.1 wt% $C_{12}E_9$ gel, (g) an aqueous 1 wt% $C_{12}E_9$-moistened swab, (h) an aqueous 1 wt% $C_{12}E_9$ gel, (i) an aqueous 0.1 wt% $C_{18}E_{100}$-moistened swab, and (j) an aqueous 0.1 wt% $C_{18}E_{100}$ gel. BSE images are 2.1 mm wide and optical micrographs are 1.4 mm wide.

Cleaning artificial soil-coated acrylic paint

Backscattered electron (BSE) and optical micrographs of the acrylic paint before and after coating with soil are shown in Figs 3a and 3b, respectively: the BSE image of the soil-coated paint displays four evenly spaced squares with a higher soil content, and these squares are identified by optical microscopy (of a different location on the same sample) as the interstices formed between overlapping canvas fibers. The soil appears to have collected primarily in these interstices during its application, and a majority of it is removed from the interstices through both swab and gel cleaning methods with all solvent mixtures (Fig. 3 c–j). There were no systematic differences between gel and swab cleaning, nor any trends associated with using different solvent solutions. The presence of surfactant appears to provide no enhancement in the removal of artificial soil from the paint surface; nevertheless, these surfactant-loaded gels may still enhance cleaning for more difficult-to-remove lipophilic soil.

SEM and element mapping (Fig. 4) were performed on the samples in Fig. 3 at a higher magnification in order to assess in more detail any changes to the surfaces caused by cleaning. Areas selected for imaging were of lower soil content, so that the surface was visible. Although the signal from chlorine, a component of the phthalocyanine green acrylic paint, is attenuated when the surface is coated with synthetic soil, it returns upon removal of the soil. Silicon and calcium are present in the sample (in the ground, the paint, or both) before application of artificial soil; their signal intensities increase after application of soil due to their presence in the silica, kaolin, and cement components. A comparison of the elemental profile in the surfaces before soil application and after all cleaning methods indicates no significant changes in content and distribution of the paint; however, small holes (~3–6 μm) appeared in the paint layer after all cleaning experiments. The holes appear to be larger and more numerous in the paint cleaned using swabs than

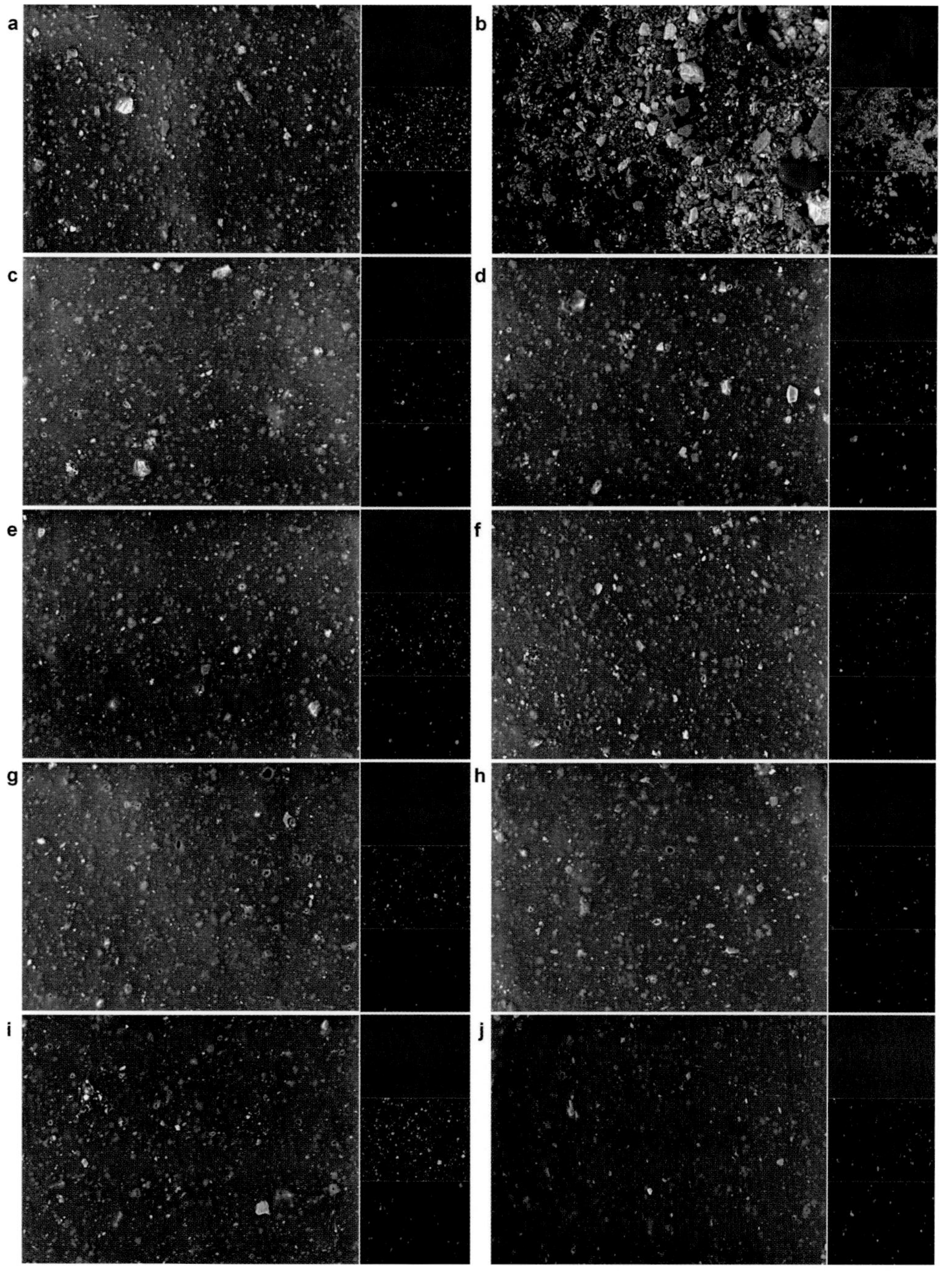

FIG. 4 BSE images of acrylic paint surfaces and element maps of the same areas of chlorine (red), silicon (yellow), and calcium (green): (a) before applying soil, (b) after applying soil, and after various cleaning methods: (c) a water swab, (d) a water gel, (e) an aqueous 0.1 wt% $C_{12}E_9$-moistened swab, (f) an aqueous 0.1 wt% $C_{12}E_9$ gel, (g) an aqueous 1 wt% $C_{12}E_9$-moistened swab, (h) an aqueous 1 wt% $C_{12}E_9$ gel, (i) an aqueous 0.1 wt% $C_{18}E_{100}$-moistened swab, and (j) an aqueous 0.1 wt% $C_{18}E_{100}$ gel (width of all images: 404 μm).

using gels. It is well known that acrylic paint swells when in contact with solvents (Wolbers *et al.* 2013; Dillon *et al.* 2014) and gels (Angelova *et al.* 2016). The swelling may have led to the artificial soil becoming embedded in the paint layer. We conjecture that the holes arise from the extraction of soil embedded in the paint layer. The use of a swab may push soil particles further into the paint layer, causing more damage. To determine the specific reason(s) for the holes, follow-up experiments will be required: both gel cleaning without a swab clearance with and without artificial soil application, and cleaning of older paint films.

Cleaning of brass metal

AFM was employed to look for microscopic scratches on the brass coupons caused by the cleaning methods. Swab cleaning was performed along one diagonal direction so that scratches would be apparent. The soil-coated brass could not be imaged because the soil adhered too weakly to the surface. However, comparisons could be made between a sample with no soil (Fig. 5a) and one after soil-coating and cleaning (Fig. 5 b–e). In the left column of Fig. 5, the AFM images are displayed with the light to dark height scale over the entire height range; in

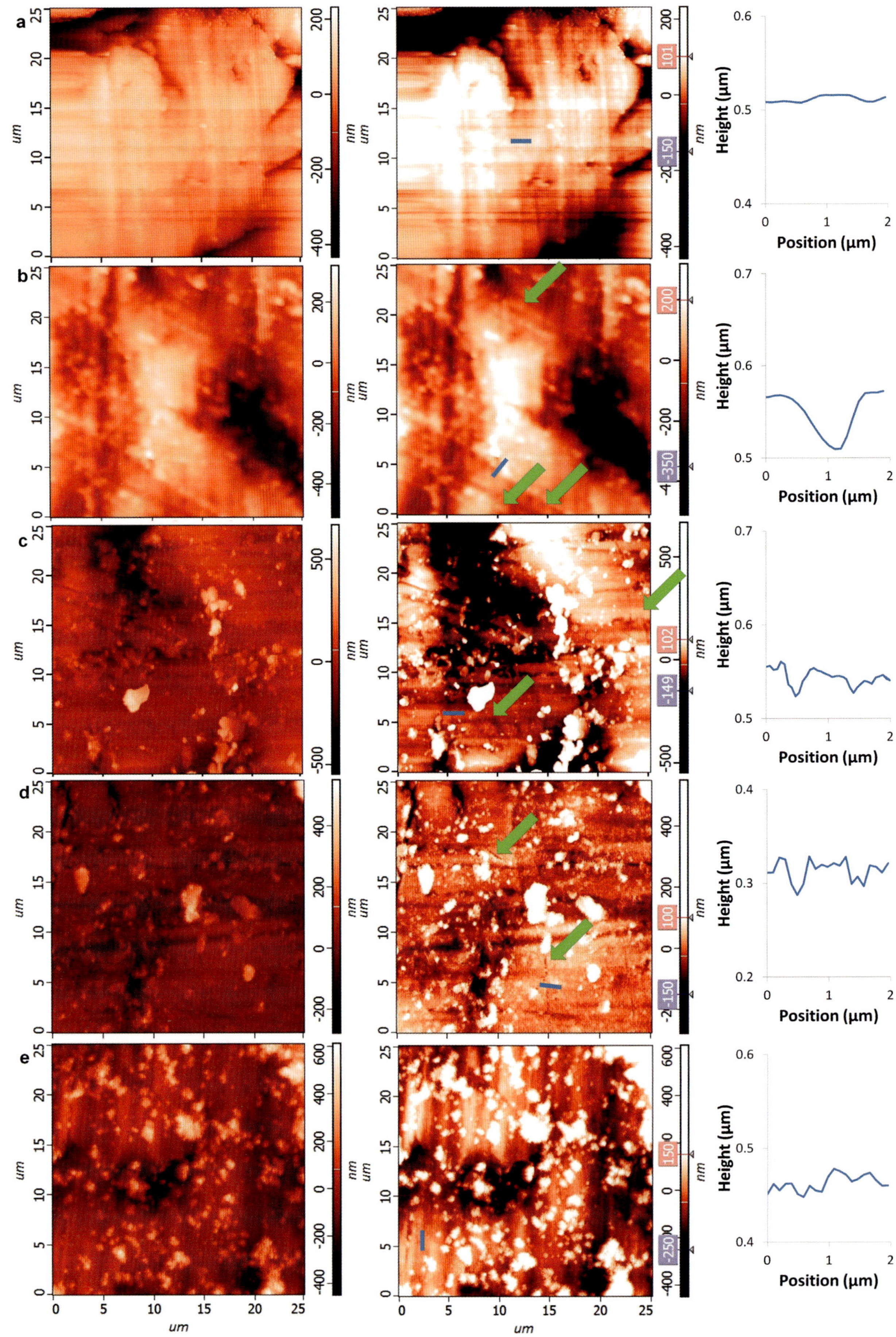

FIG. 5 AFM images (same images shown horizontally, but with different color outputs) of brass coupons: (a) no soil deposited; and after soil-coating and cleaning with (b) dry swab rubbing, (c) dry swab rolling, (d) an aqueous 0.1 wt% $C_{18}E_{100}$-moistened swab rolling, and (e) an aqueous 0.1 wt% $C_{18}E_{100}$ gel cleaning. Green arrows point to some scratches; topography of the blue lines shown at right.

the right column, to emphasize small height changes near the brass surface and better visualize shallow scratches, images are shown with the color output expanded. The untreated surface (Fig. 5a) shows a constructive/destructive interference pattern from the laser due to the reflective surface of the brass, but no scratches. The brass coupon cleaned by rubbing with a dry swab (Fig. 5b) exhibits diagonal scratches (green arrows), indicating that the surface is scratched by the soil particles; the number of scratches was reduced by rolling a swab (Fig. 5c). The directionality of the scratches from rolling is not as consistent as those from rubbing because the swab pushes particles in unintended directions. Gel cleaning (Fig. 5e) was not as efficient as moistened-swab cleaning (Fig. 5d), evident from the larger number of bright spots, indicating soil on the surface; however, gel cleaning is the only method that left no visible scratches on the surface. Five aliquots of gel were used on the brass coupons, and more aliquots likely would have increased cleaning effectiveness.

Conclusion

Two PEG-based surfactants, $C_{12}E_9$ and $C_{18}E_{100}$, have been incorporated into 80PVAc–borate systems with no deleterious effects on the stability, workability, or optical properties of the materials: the materials are clear, peelable, and stable at room temperature (in capped vials to prevent evaporation) for more than two months. Perhaps as a result of its gentler physical action and reduced swelling of the paint, gel cleaning of artificial soil-coated acrylic paint appears to cause less disruption to the acrylic paint surface than swabbing. However, because the paint is relatively fresh, this test case may not be a good model for aged paints. In the cleaning of brass coupons, rolling and rubbing swab methods create scratches on the metal surface; none were detected upon treatment with a peelable gel. In that regard, gels offer a clear advantage. Although incorporation of PEG-based surfactants to the gel did not enhance removal of synthetic soil, insofar as analyses of the mineral and inorganic components indicate, they may be more successful for the removal of lipophilic soil. A conservator must rely on a host of different technologies to confront a wide array of cleaning challenges. As indicated here, gels can provide methods for addressing many of those challenges.

Experimental

Water was deionized (DI) or ultrapure (filtered through an EMD Millipore Direct-Q ultra-pure water system). 80PVAc (Kuraray poly(vinyl acetate); 80% hydrolysed, molecular weight 47,300 g/mol) was rinsed with ice-cold DI water, filtered, and vacuum dried (~125 mm Hg) at room temperature. $C_{12}E_9$ (Sigma-Aldrich, purity unspecified) and $C_{18}E_{100}$ (Sigma-Aldrich, M_n ~4,670, purity unspecified) were used as received; C_nE_m-type surfactants are typically mixtures of homologues (Sharma *et al.* 2011). Borax (sodium tetraborate decahydrate, Fluka, Analytical Ultra, >99.5%) was used as received. 'Artificial soil' was prepared (Ormsby *et al.* 2010) by combining kaolin (1.0 g, Fisher Scientific Laboratory Grade), soluble starch (0.5 g, Mallinckrodt), gelatin (0.5 g, Sigma-Aldrich Type B from bovine skin), iron oxide (0.025 g, raw sienna, Conservation Materials), lamp black (0.1 g, Conservation Materials), silica (0.088 g, Fisher, special fine granular, 80–120 mesh), cement (0.88 g, Sakrete Portland Cement Type I-II), mineral oil (1 mL, Fisher Scientific, light), olive oil (0.5 mL, Crisco pure olive oil), and hexanes (45 mL, Fisher Scientific, HPLC Grade). Brass sheet (C260: 68.5–71.5% Cu, 28.5–31.5% Zn, 0.05% max Fe, and 0.07% max Pb) was purchased from onlinemetals.com. Acrylic paint (Grumbacher, Academy Acrylic, thalo green – a chlorinated copper phthalocyanine pigment), canvas sheet primed with acrylic primer (Fredrix, canvas pad, medium texture real artists canvas), and a boar bristle brush (Winsor & Newton, Rathbone Filbert #3) were used to create painted samples.

Acknowledgements

We thank the US National Science Foundation (Grant CHE-1502856) for supporting part of this research. Teresa Duncan thanks the Achievement Rewards for College Scientists/Metro Washington Chapter for a scholar award. We extend our gratitude to Dr Xinran Zhang in the Physics Department of Georgetown University for help collecting the atomic force microscopy images and Dr Anna Fricker in the Physics Department at Imperial College, London, for discussions on AFM to image scratches on plastics. The Kuraray Co. Ltd is thanked for donating the xPVAc samples employed in this research.

References

Angelova, L.V. 2013. *Gels from Borate-crosslinked Partially Hydrolyzed Poly(vinyl acetate)s: Characterization of Physical and Chemical Properties and Applications in Art Conservation*, unpublished PhD dissertation, Georgetown University, Washington, DC.

Angelova, L.V., Terech, P., Natali, I., Dei, L., Carretti, E. and Weiss, R.G. 2011. 'Cosolvent gel-like materials from partially hydrolyzed poly(vinyl acetate)s and borax', *Langmuir* 27(18): 11671–82.

Angelova, L.V., Berrie, B.H, de Ghetaldi, K., Kerr, A. and Weiss, R.G. 2015. 'Partially hydrolyzed poly(vinyl acetate)-borax-based gel-like materials for conservation of art: characterization and applications', *Studies in Conservation* 60(4): 227–44.

Angelova, L.V., Ormsby, B. and Richardson, E. 2016. 'Diffusion of water from a range of conservation treatment gels into paint films studied by unilateral NMR. Part I: Acrylic emulsion paint', *Microchemical Journal* 124: 311–20.

Carretti, E., Grassi, S., Cossalter, M., Natali, I., Caminati, G., Weiss, R.G., Baglioni, P. and Dei, L. 2009. 'Poly(vinyl alcohol)-borate hydro/cosolvent gels: viscoelastic properties, solubilizing power, and application to art conservation', *Langmuir* 25(15): 8656–62.

Cushman, M. and Wolbers, R.C. 2007. 'A new approach to cleaning iron stained marble surfaces', *WAAC Newsletter* 29(2): 23–8.

Dillon, C., Lagalante, A. and Wolbers, R.C. 2014. 'Acrylic emulsion paint films: the effect of solution pH, conductivity, and ionic strength on film swelling and surfactant removal', *Studies in Conservation* 59(1): 52–62.

Duncan, T.T., Berrie, B.H. and Weiss, R.G. 2016. 'Colloidal properties of aqueous poly(vinyl acetate)–borate dispersions with short-chain glycol ethers', *ChemPhysChem* 17(16): 2535–44.

Natali, I., Carretti, E., Angelova, L., Baglioni, P., Weiss, R.G. and Dei, L. 2011. 'Structural and mechanical properties of "peelable" organoaqueous dispersions with partially hydrolyzed poly(vinyl acetate)-borate networks: applications to cleaning painted surfaces', *Langmuir* 27(21): 13226–35.

Ormsby, B., Soldano, A., Keefe, M.H., Phenix, A. and Learner, T. 2010. 'An empirical evaluation of a range of cleaning agents for removing dirt from artists' acrylic emulsion paints', *AIC Paintings Specialty Group Postprints* 23: 77–87.

Riedo, C., Caldera, F., Poli, T. and Chiantore, O. 2015. 'Poly(vinylalcohol)-borate hydrogels with improved features for the cleaning of cultural heritage surfaces', *Heritage Science* 3(23): 1–11.

Rietjens, M. and Steenbergen, P.A. 2005. 'Crosslinking mechanism of boric acid with diols revisited', *European Journal of Inorganic Chemistry* 2005(6): 1162–74.

Schneider, C.A., Rasband, W.S. and Eliceiri, K.W. 2012. 'NIH Image to ImageJ: 25 years of image analysis', *Nature Methods* 9: 671–5.

Schultz, R.K. and Myers, R.R. 1969. 'The chemo-rheology of poly(vinyl alcohol)-borate gels', *Macromolecules* 2(3): 281–5.

Sharma, K.P., Choudhury, C.K., Srivastava, S., Davis, H., Rajamohanan, P.R., Roy, S. and Kumaraswamy, G. 2011. 'Assembly of polyethyleneimine in the hexagonal mesophase of nonionic surfactant: effect of pH and temperature', *Journal of Physical Chemistry B* 115(29): 9059–69.

Umorin, M. 2002–2006. 'Stack Focuser'. Available at: https://imagej.nih.gov/ij/plugins/stack-focuser.html (accessed 1 June 2017).

Wolbers, R. 2000. *Cleaning Painted Surfaces: Aqueous Methods*. London: Archetype Publications.

Wolbers, R., Norbutus, A. and Lagalante, A. 2013. 'Cleaning of acrylic emulsion paints: preliminary extractive studies with two commercial paint systems', in M.F. Mecklenburg, A.E. Charola and R.J. Koestler (eds), *Cleaning 2010: New Insights into the Cleaning of Paintings*. Washington, DC: Smithsonian Institution Scholarly Press, 147–57.

Authors' addresses

- Teresa T. Duncan, Department of Chemistry, Georgetown University, Washington, DC, USA
- Barbara H. Berrie, Conservation Division, National Gallery of Art, Washington, DC, USA
- Richard G. Weiss, Department of Chemistry and Institute for Soft Matter Synthesis and Metrology, Georgetown University, Washington, DC, USA (weissr@georgetown.edu)

Moving on up: a review of results from SRAL's tissue-gel composite approach

Gwendoline R. Fife

Introduction

When employing organic solvents to remove non-original layers from the surfaces of paintings, conservators are invaluably supported by research findings regarding the nature of paint films and potential solvent effects. Nonetheless the individual nature and history of each painting means that conservators must critically rely on their observations and experience during practice. Cleaning remains a complex procedure for which no generalized solution can exist.

Given the potential for subtle long-term effects which may be difficult to observe at the point of treatment, one logical ideal is to reduce solvent contact with the original materials as far as possible. This idea corresponds with the concept of thickening solvents for cleaning applications since increasing the viscosity of a solvent reduces its movement through cracks via flow and capillary action, thus limiting the opening of new diffusion fronts. In addition, restriction of solvent release at the paint surface due to its retention within the gel matrix has also been indicated (Dorge 2004; Fife *et al.* 2014).

While young paint films may be considered semi-permeable, this does not apply to paintings past a certain age: the typical cracks, losses and porosity of older paint films increases the relevance of residue concerns from non-volatile gelling ingredients. Accordingly, minimizing residue issues while harnessing the advantages of using a gelled solvent was an underpinning concept for the tissue-gel technique developed by Stichting Restauratie Atelier Limburg (SRAL) for treating the paintings in the Oranjezaal in the Royal Palace, Huis ten Bosch in The Hague, the Netherlands.

Results

Beginnings: the Oranjezaal

The cleaning (varnish removal) of these oil on canvas paintings posed many complicating issues:

- The team comprised many conservators with different backgrounds, training and approach.
- Being an entire ensemble the Oranjezaal treatment required a unified result.
- The total surface area of the paintings covered approximately 304 m^2 and there was a strict deadline for completion of the project.
- The paintings were unlined and extremely porous. Any application of swabbed free solvent to the varnished paint film was immediately observable on the reverse of the canvas.

The requirements for a cleaning technique were therefore:

- To be able to regulate the specific solvent mixture used and contact time with the varnished surface in order to diminish the effect of 'different hands', and ensure an even cleaning.
- To remove the varnish as safely and efficiently as possible.
- To prevent penetration of solvent through the painting structure.

Solvent tests had indicated solubility of the varnishes whereby the cellulose ether, Klucel G, could be used in low concentrations as a thickening agent. Aging tests have indicated that this material is relatively stable (Feller and Wilt 1990), and although any residues left on the paint surface may undergo some yellowing, it critically maintains its solubility behaviour, and is inert with regard to any chemical effect it may have on the original paint surface. Testing with a gelled solvent (2% weight Klucel G/volume of solvent) showed no penetration of solvent through to the reverse. The requirement for the application of a regular quantity of gel resulted in the conception of a gel-impregnated tissue, the contact and thus working time of which could be precisely controlled. The use of a secondary dry tissue step – with additional gentle pressure from a roller or spoon – was found to maximize effective absorption and removal of the dissolved residues.

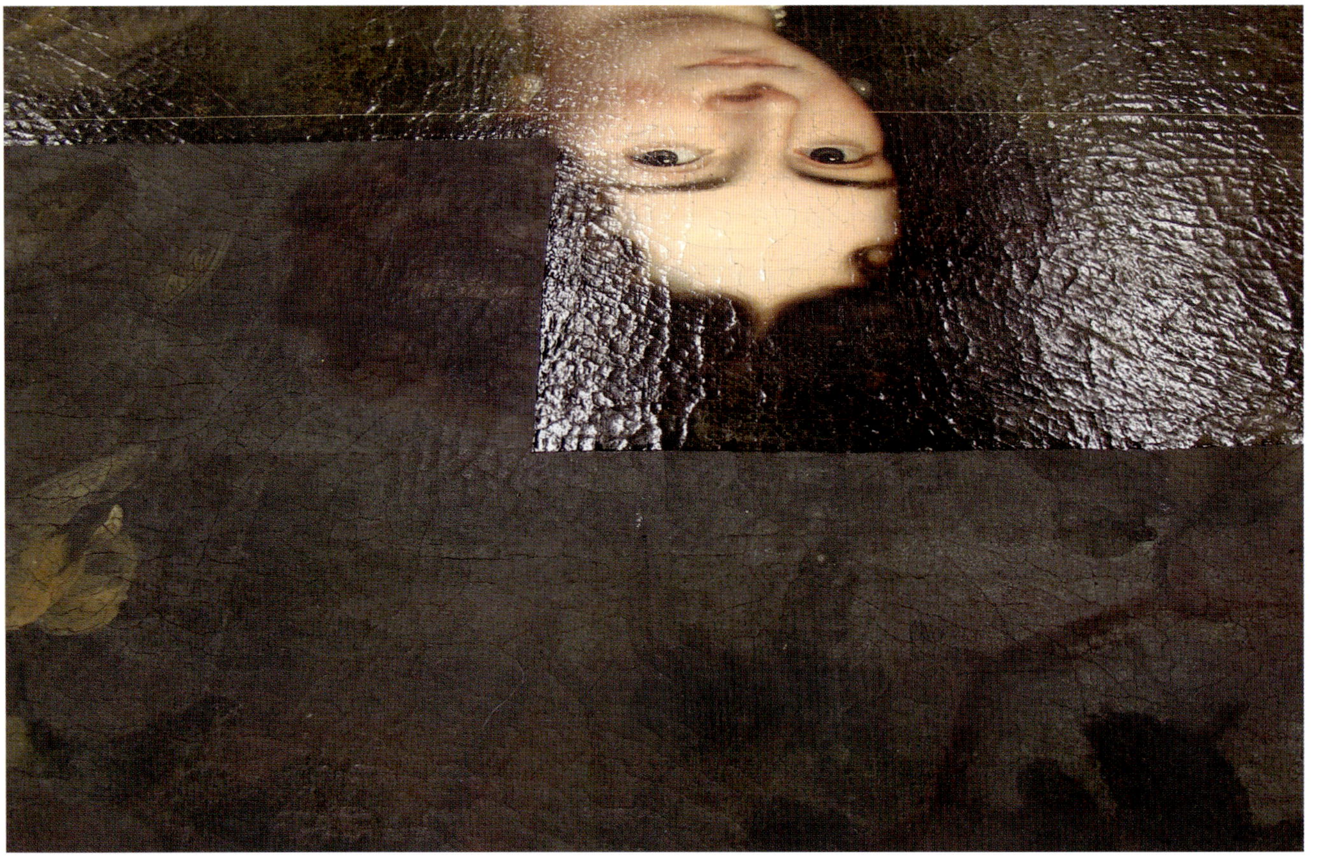

FIG. 1 Detail during varnish removal from the portrait of Ursula Philippota van Raesfelt.

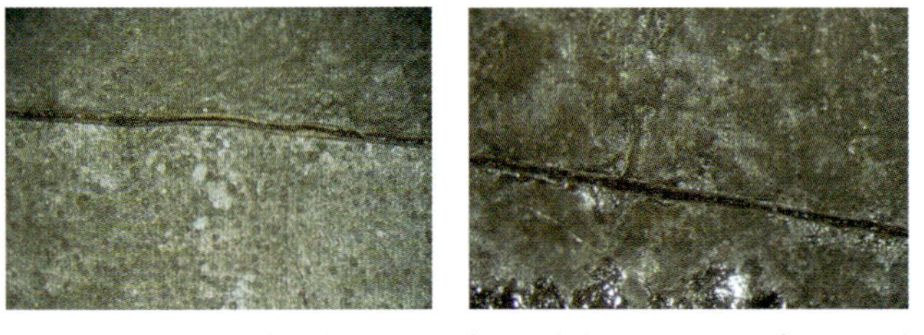

FIG. 2 A painting surface (×15 magnification) during testing of tissue-gel composite (*left*) and free solvent/swab (*right*).

During testing (and ultimately treatment) no detrimental effects on the paint surface were observed using this technique, and neither solvent nor dissolved varnish was visible on the reverse of the paintings. Hence, degradation resulting from acidic resinous varnish residues further penetrating the original canvas supports was prevented – a factor considered especially pertinent for unlined paintings in general (see below). Varnish removal in this manner provided a very revealing intermediate cleaning step, whereby the effects of previous partial cleanings carried out during the paintings' history could be clearly observed. This was of importance for the research and documentation aspect of the project, as well as informing (with the aid of ultraviolet (UV) light) the lead conservators where further cleaning was specifically required. A unified cleaning was safely achieved with a minimum exposure of the paintings to solvent, and to this day the annual controls carried out by SRAL have not indicated any negative side-effects.

Adaptability

Case studies

Selected case studies at SRAL indicate the adaptability and reveal some of the other advantages observed with this type of composite technique. In 2000, SRAL commenced a conservation project at Castle Amerongen, Amerongen, the Netherlands, which included 18 family portraits from the Grote Galerij, dating from the 17th to the 19th century. Many had remained unlined and had extremely cupped paint films. While restoration treatments were excluded from the project mandate, removal of the varnish(es) proved a conservation consideration for the majority of paintings in the collection. Effective consolidation, the reduction of stresses on the paint surface (which exacerbate cupping) induced by the brittle varnish, and as far as possible the removal of acidic resinous varnish materials from the painting surface and structure, were all conservation issues. During testing, some typical problems using swabs and free solvent on such topographies were observed:

- Mechanical damage could be incurred on a macro scale with sharp raised edges vulnerable to breaking.
- It was difficult to achieve an even removal of varnish since the swabs removed varnish from the peaks of the paint film, while leaving more residues in the 'cups'. To remove these latter varnish residues, not only was

more solvent exposure required but also the 'tops' in the paint film could be further abraded.

- The varnish was not effectively removed from the cracks or from under the raised edges. With the paint film remaining cupped after cleaning, and the retention of old restoration material therein, subsequent consolidation could not be effective.

Tests with the tissue-gel composite cleaning showed that macro mechanical action on the vulnerable and cupped paint films was reduced to the extent that no paint losses were incurred. There was no direct rubbing on the paint surface or 'catching' of cotton fibres as occurred with the swab, and less pressure was needed since the surface area of the 'tops' of such cupped paint films were effectively increased by applied tissue layers, and the contact area of the spoon or roller used in the absorbent step was greater than that of a 'typical' swab. More effective removal of varnish from the 'cups' could also be achieved, resulting in a more even removal of varnish from the entire paint surface (Fig. 1).

This 'even' removal has been seen repeatedly with comparative examinations of test areas on various paintings, before and after both swab/solvent and tissue-gel cleaning, at ×15 magnification and samples at ×250 magnification. It has also been observed that the technique appears to swell and dissolve the varnish residues in typical paint cracks and under raised paint edges, in a more gradually homogenous manner. The gradational swelling of these varnish residues (from the top down) may explain the frequent observation that application of free solvent to a recently tissue-gel cleaned location no longer results in solvent penetration through to the reverse of the painting, even when this occurred rapidly prior to cleaning. Although a theoretically negative aspect of this could be relatively longer solvent retention (within this non-original material in the cracks), a positive practical aspect is that, during the absorbent step, there exists the potential for more of these residues to be removed (Fig. 2). This was indeed found to be the case for these paintings, and with the cupping reduced and more of the previous restoration materials removed from the structure, more effective consolidation could subsequently be carried out.

Such advantages were demonstrated using an adaptation of the technique which was tested and applied in 2013 in treating another series of five very large canvas paintings from Kasteel Broekhuizen, Leersum, the Netherlands. Some of the edges of these paintings were so damaged and the instability of surrounding paint areas so extreme that any contact with the surface readily generated further loss. In this case a way was found to consolidate these areas 'temporarily' and enable a safe varnish removal by first applying water-based tissue-gel composites. After complete drying, thickened organic solvent (selected for varnish removal) was then brushed onto the adhered tissues. Consequent absorption and removal of the varnish layers – allowing effective consolidation – was thereby implemented without further loss to the original.

Between May 2006 and May 2007 an ensemble of four very large oil on canvas paintings by Ferdinand Bol (1616–1680) from the Peace Palace in The Hague, the Netherlands, were

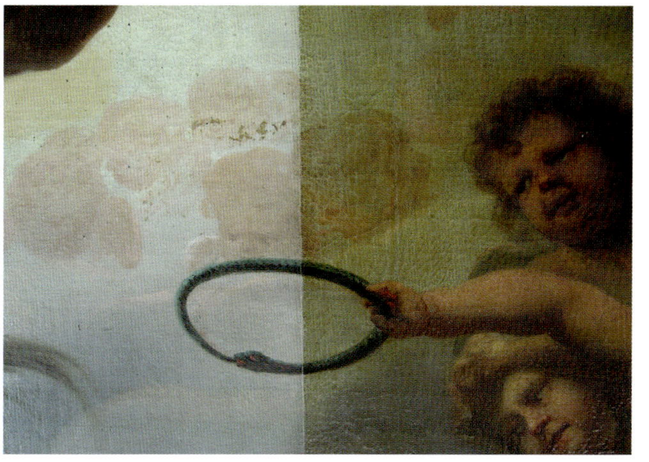

FIG. 3 Detail of *Heer verschijnt aan Joshua* by Ferdinand Bol during varnish removal with tissue-gel composite technique.

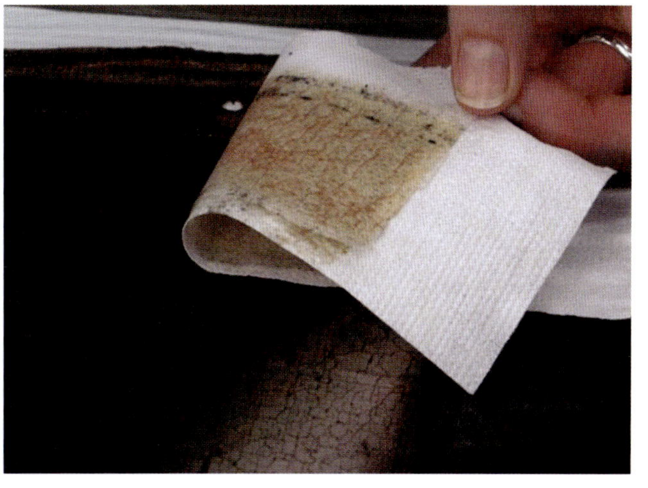

FIG. 4 Detail during cleaning of portrait of Landgraaf Karl van Hessen-Kassel.

technically examined, conserved and restored by SRAL while on public display in the Bonnefanten Museum, Maastricht, the Netherlands (Fife and van Och 2012). Previous partial cleanings, local but extreme cupping in the paint surfaces, and the project requiring a team of conservators of varying background and experience were again considerations. The thickness of the numerous varnish layers established the inherent efficiency of the composite technique. Comparing the quantities of solvent required for applying both the swab/free solvent and tissue-gel composite techniques showed that even when the solvent remaining in the tissues after removal was not subtracted, less solvent was required for varnish removal with the tissue-gel technique. The composite was more efficient since varnish layers could be both dissolved and consequently removed everywhere the solvent was in contact with the surface (as opposed to swab/free solvent cleaning, see Fig. 3).

The effective, 'vertical action' of the technique is best illustrated by the last example to be discussed – the portraits of Landgraaf Karl van Hessen-Kassel and his wife from Castle Amerongen. These two oil paintings on canvas had been previously lined using an unusual adhesive which analysis indicated was composed of a drying oil with both lead and iron oxides.

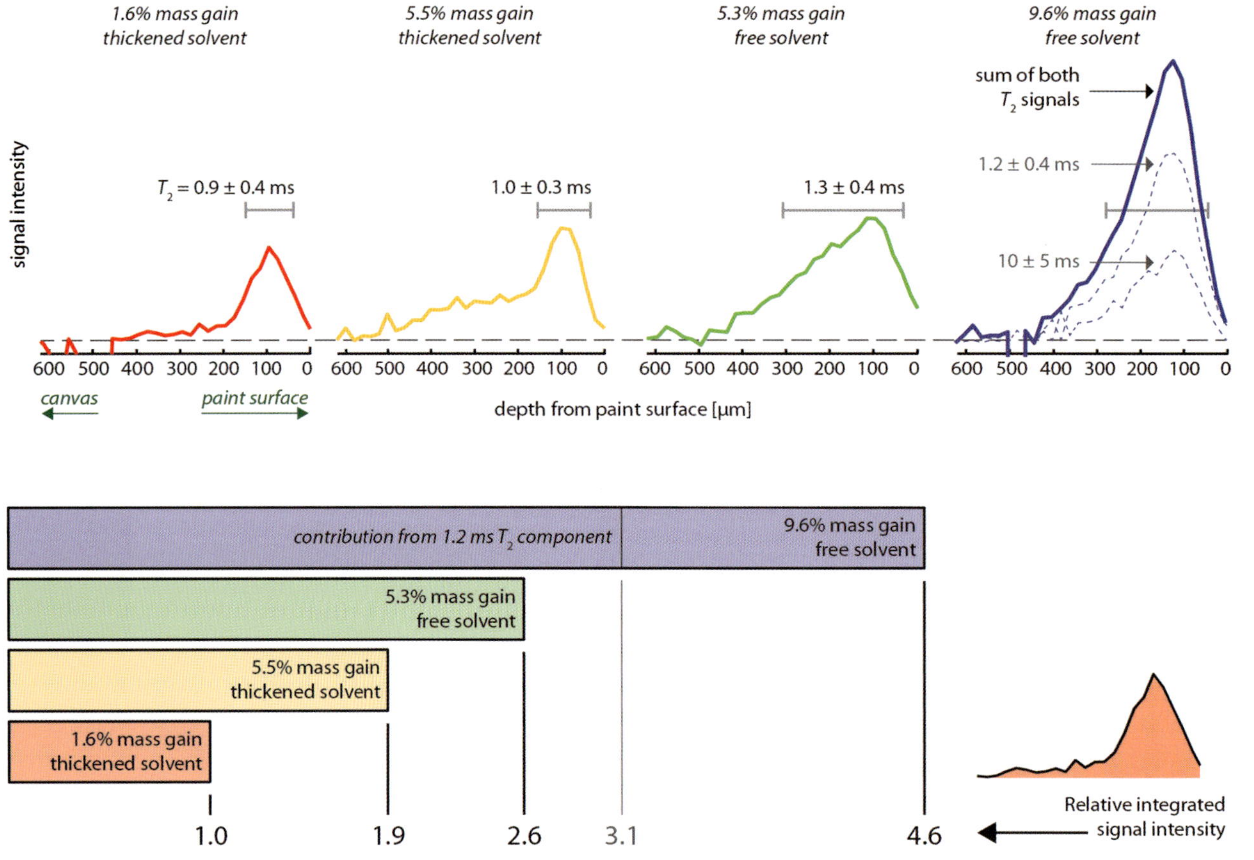

FIG. 5 A comparison of the effects on the model paint film from different treatments with free and thickened solvents. Four treatments were applied to identical pieces of the model, two thickened-solvent treatments (with a 1.6% and a 5.5% mass gain immediately following treatment) and two free-solvent treatments (5.3% and 9.6% mass gain). (*Top*) The thickened-solvent treatments had a narrower range of effect (*c.*100 μm, red and yellow) than the free-solvent treatments (*c.*260 μm, green and blue), determined by the full width at half the maximum value and indicated by the horizontal grey bars. (*Bottom*) Integrating the effect of the different solvent treatments shows that the mass change percentage is directly related to the amount of effect observed. All integrated values are normalized to the thickened-solvent treatment with the 1.6% mass gain.

The adhesive from this historic treatment had gradually flowed from the reverse of the canvas through the cracks in the ground and paint layers and onto the varnished surface. Its intense red colour was particularly visible in the lighter passages and visually disturbed the images in general as it emphasized the pattern of cracks. The possibility of removing the lining adhesive from the surface simultaneously with the varnish layers was explored and, in terms of their solubility behaviour, confirmed during solvent tests. Applications with swabs and free solvent resulted in immediate penetration of the solvent to the canvas reverse, and smearing of the lining adhesive across the surface of the painting. However, tests with the tissue-gel composite technique showed that the system could effectively remove varnish, old overpaints and lining adhesive residues from the surface without solvent penetration to the reverse, and in a vertical manner whereby re-deposition across the painting surface was prevented (Fig. 4).

Discussion

The principles and demonstrations of this method have inspired discussions at SRAL and beyond. The positive advantages outlined above, and the adaptability of the technique with regard to the individual painting's requirements, have led to this type of approach being selected as the most appropriate for their specific use by many different practitioners in many other projects, both at SRAL and elsewhere.

Alongside the potential dangers of naming any approach a particular 'technique', it is considered by the author that the main disadvantages of employing tissue-gel composites, as reviewed here, is the reduced visibility of the paint film surface during treatment, and the delayed ability (as exacerbated with any gelled solvent application) to stop during treatment. Comprehensive pre-testing for solvent sensitivity and response of all areas is required. Accurate timing of the necessary contact time is advised in order to optimize absorption of non-original materials and maximize the benefits of reducing solvent quantities within the layers of paintings.

Various practical issues have also been recounted by others, such as the problem of 'lines' being created by adjacent tissues or from overlapping cleaned areas (especially when no clearance step is undertaken). In the author's experience these are generally resolvable or should not be significant. This type of technique should only be employed if the solvent contact time is not critical, i.e. any indicated solvent sensitivity of an area in the paint film should negate its use.

The main recurring theoretical concern has been the potential for an increased concentration of solvent being held at the paint surface. This was investigated at SRAL to a degree in 2011 when an opportunity arose to examine solvent movement using a novel application of single-sided hydrogen nuclear magnetic resonance (H^1 NMR) (Fife *et al.* 2014, 2015, 2016). Experiments with one paint model type were carried out to comparatively examine the applications of a solvent in both free form (with swabs) and as a thickened gel with tissues (as in the tissue-gel composite approach). The maximum level (i.e. depth) of solvent penetration was found to remain comparable between the two different techniques. But with the tissue-gel composite the quantity of solvent was significantly reduced both at the surface and throughout the entire sample depth (Fig. 5).

While the potential solvent effects associated with varnish applications and removal tend to be explicitly described in research and cleaning reports, the continuing effect of older unstable restoration materials remaining within the entire painting structure does not necessarily arise in such discussions. Although applied to the surface of a painting, these materials can often be observed on the reverse of previously cleaned paintings. The question of how much the removal technique can limit the (further) introduction of these materials deserves explicit discussion in research and in practice.

Conclusion

As highlighted above the tissue-gel composite technique encompasses some advantages – reduced solvent contact and mechanical action, and an 'upward' cleaning action – that have proved highly beneficial in various specific cleaning situations. As an approach conceived to aid what can be a complicated operation, perhaps the most significant aspects are the observations, conceptualizations and reasoning behind its development and its subsequent applications: the way must be carefully related to the why.

References

Dorge, V. (ed.) 2004. *Solvent Gels for the Cleaning of Works of Art: The Residue Question*. Los Angeles: Getty Conservation Institute.

Feller, R.L. and Wilt, M. 1990. *Evaluation of Cellulose Ethers for Conservation*. Los Angeles: Getty Conservation Institute. Available at: http://www.getty.edu/conservation/publications_resources/pdf_publications/cellulose_ethers.html

Fife, G.R. and van Och, J. 2012. 'Bol's chamber paintings in later times', in M. van Eikema Hommes (ed.), *Art and Allegiance in the Dutch Golden Age: The Ambitions of a Wealthy Widow in a Painted Chamber by Ferdinand Bol*. Amsterdam: Amsterdam University Press, 169–87.

Fife, G.R., van Och, J., Seymour, K. and Hoppenbrouwers, R. 2010. 'Tissue gel composite cleaning', in M. Mecklenburg, A.E. Charola and R.J. Koestler (eds), *Cleaning 2010: New Insights into the Cleaning of Paintings*. Washington, DC: Smithsonian Institution Scholarly Press, 197–200.

Fife, G.R., Miedema, N., van Och, J., Seymour, K. and Hoppenbrouwers, R. 2011. 'A package deal: the development of tissue gel composite cleaning at SRAL', in J. Bridgland (ed.), *Preprints of the ICOM-CC 16th Triennial Meeting*. Paris: ICOM-CC, 172.

Fife, G.R., Stabik, B., Blümich, B., Hoppenbrouwers, R. and Meldrum, T. 2014. 'Evidence for the accumulative effect of organic solvent treatments on paintings and what to do about it: a case study of two "identical" 17th-century paintings with single-sided NMR', paper presented at the ICOM-CC Painting and Scientific Research Group Meeting, Washington, DC, February 2014. Available at: http://www.youtube.com/watch?v=qDee6QUc0gg (accessed 26 September 2014).

Fife, G.R., Stabik, B., Kelley, A, King, J.N., Blümich, B., Hoppenbrouwers, R. and Meldrum, T. 2015. 'Characterization of aging and solvent treatments of painted surfaces using single-sided NMR', *Magnetic Resonance in Chemistry*. *Special Issue: NMR in Cultural Heritage* 53(1): 58–63.

Fife, G.R., Stabik, B., Blümich, B., Hoppenbrouwers, R. and Meldrum, T. 2016. 'Evidence for the accumulative effect of organic solvent treatments on paintings and what to do about it: a case study of two "identical" 17th-century paintings with single-sided NMR', in A. Nevin and T. Doherty (eds), *The Non-invasive Analysis of Painted Surfaces: Scientific Impact and Conservation Practice*. Washington, DC: Smithsonian Institution Scholarly Press, 15–24.

Author's address

Gwendoline R. Fife, SRAL, Maastricht, The Netherlands, (gwenfife26@outlook.com)

From biomass to restoration: a new green tool for the cleaning of artworks

Francesca Volpi, Chiara Samorì, Laura Mazzocchetti, Silvia Prati, Paola Galletti, Giorgia Sciutto, Loris Giorgini, Emilio Tagliavini and Rocco Mazzeo

Introduction

The cleaning of traditional works of art, which is usually aimed at removing aged varnish layers, still represents a delicate and potentially hazardous intervention for the safety of the painted surface as well as for the conservator. Traditional approaches have been based on the use of volatile organic solvents, some of which are harmful or historically were toxic to human health and the environment, necessitating specific disposal practices. The more toxic options such as dimethyl formamide are no longer used today as free solvents. In Italy, dimethyl sulfoxide (DMSO) has come into use in the 21st century as a safer alternative solvent. Migration of the solvent into the inner paint layers has to be prevented, to avoid swelling of the underlying layers.

In the last decades, research has focused on the development of new cleaning methods based on gel systems, which are able to contain additional cleaning agents, thereby localizing and controlling both solvent release and the interaction between paint layers and reagents (Carretti *et al.* 2010a,b; Domingues *et al.* 2013; Baglioni *et al.* 2014, 2015). However, in the formulation of alternative methods, little attention has been devoted to the eco-sustainability of conservation procedures. In an attempt to propose innovative green solutions, the present research focused on the formulation of new biodegradable cleaning systems composed of three bio-based constituents: poly-3-hydroxybutyrate (PHB), γ-valerolactone (GVL) and triethyl citrate (TEC). These were employed to obtain a gel-like material for the removal of dammar varnish from oil paintings. PHB belongs to the poly-hydroxyalkanoate family and is produced by bacterial biosynthesis (Sudesh and Iwata 2008). Because of its high crystallinity, its solubility is limited to a small number of solvents, a few of which are environmentally friendly (Fiorese *et al.* 2009; Riedel *et al.* 2013; Samorì *et al.* 2015). GVL is a solvent obtained from lignocellulosic biomass (Alonso *et al.* 2013) which has not previously been used in conservation. Similar to butyrolactone, it is a low-swelling solvent (Phenix 2002). In addition to being green, GVL was selected for its supposed low swelling power. Finally, TEC is a non-toxic and biodegradable component added as a plasticizer to the gel formulation to increase its elasticity and the handling properties of the gel. A detailed characterization of the mechanical properties of the gel has recently been published by Samorì *et al.* (2016).

An ad hoc analytical protocol was applied for the evaluation of the cleaning efficacy of PHB gels on paint mock-ups. The ability of the gels to remove varnish was studied by Fourier transform infrared microscopy in total attenuated reflectance mode (μFTIR–ATR) performed directly on the treated surface before and after the cleaning procedure. This non-invasive investigation permitted the collection of a large amount of data necessary to characterize the PHB-gel action more fully, and to identify the proper sampling area for the preparation of paint cross-sections which were documented by optical microscopy. Finally, solvent retention in the paint layer after cleaning was evaluated by gas chromatography–mass spectrometry (GC–MS) using headspace solid phase micro-extraction (HS–SPME) (Samorì *et al.* 2016).

The gel cleaning action was compared with neat GVL solvent in terms of varnish removal capability and solvent retention. Moreover, the performance of the new systems was compared to traditional solvents such as methanol and acetone. Finally, a cleaning procedure was performed on a17th-century oil painting on canvas.

Materials and methods

The gel preparation is carried out in a single step by dissolving the polymer powder in GVL (9 wt%) and TEC (2 wt%) at 130 °C for 1 minute. Gel formation occurs by cooling to room temperature, when the polymer chains aggregate and the gel becomes slightly opalescent and stiff, allowing for easy handling. Preliminary cleaning tests were performed on paint mock-ups prepared with a complete stratigraphy as found in some Italian paintings of the 17th century, consisting of: a ground layer of gypsum and rabbit skin glue applied over a wood support, a second layer of rabbit skin glue size diluted in water, a paint layer composed of burnt ochre pigment and linseed oil binder, and a varnish layer made of a dammar resin in turpentine. Experiments were carried out on mock-ups naturally aged for 1 year in dark conditions.

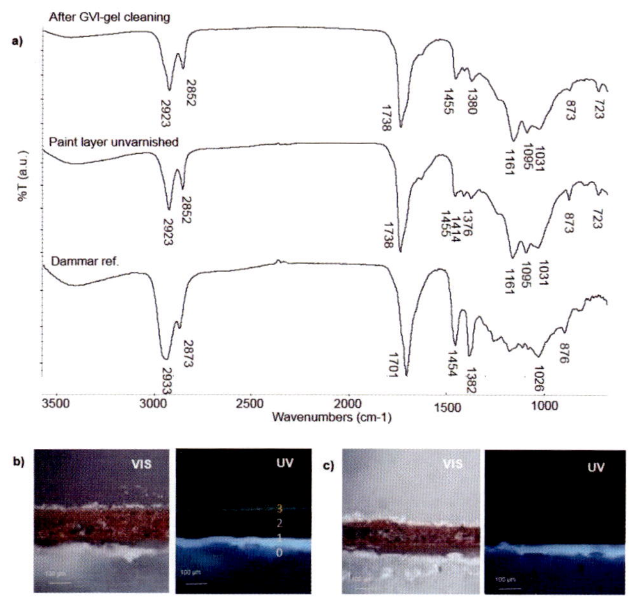

FIG. 1 Cleaning test on dammar varnish and oil medium mock-up. (a) FTIR–ATR spectra from the cleaned area, unvarnished area and dammar resin reference (respectively from top to bottom); (b) cross-sections before cleaning (varnish thickness around 15 μm) and after cleaning with GVL gel; (c), where a reduction of the varnish thickness to <2 μm was observed.

Experimental

Photomicrographs of samples before and after cleaning were obtained with an Olympus (Tokyo, Japan) BX51 optical microscope equipped with a halogen and an ultraviolet (UV) lamp for the acquisition of both visible light and UV images of cross-sections. Analyses were performed by μFTIR (Thermo Fisher Scientific, Waltham, MA, USA) fitted with a mercury-cadmium-telluride detector, using an ATR germanium crystal, operating with a spectral resolution of 4 cm^{-1} in the range 4000–675 cm^{-1}. Surface analyses on mock-ups were acquired on untreated and treated areas with an optical aperture of 200 × 200 μm^2, acquiring six spectra for each area (1 × 1 cm^2). Mapping was performed on a historical sample with an optical aperture of 60 × 60 μm^2, using steps of 12 μm in both directions.

HS–SPME was performed by directly exposing a Carboxen/polydimethylsiloxane fibre in the headspace of sealed vials containing about 1 mg of a fragment collected from treated areas 24 hours from cleaning, using γ-butyrolactone as the internal standard. The SPME fibre was first exposed to the thermally heated vial at 150 °C for 30 minutes (Rombolà *et al.* 2015) and then injected into a gas chromatograph connected to a quadrupole mass spectrometer (Agilent Technologies Inc., Santa Clara, USA). Analytes were thermally desorbed at 250 °C for 10 minutes and separated with a polar column using helium as the carrier gas.

Results and discussion

Evaluation of cleaning efficacy

Following the full characterization of the PHB gel and its physical properties (Samorì *et al.* 2016), the research investigated the cleaning ability of the gel; the confinement of the solvent; and removal from the artwork surface in a manner that avoided leaving any substantial residues. To this end, mock-ups were submitted to preliminary tests by applying the gel (between two sheets of Japanese paper) over the area of interest, using gentle pressure. The application time was fixed at 5 minutes, then the gel was peeled off and a dry cotton swab used to remove the swelled varnish residues. No additional clearance steps were required. Gel efficacy was evaluated in comparison with the cleaning action of non-gelled solvents (GVL, DMSO, acetone, the last two chosen for their high polarity and range of volatility) applied by swab with mechanical action on the surface for about 30 seconds. In Fig. 1a, FTIR–ATR spectra acquired before and after application of the GVL gel on a mock-up revealed the presence of IR absorption bands associated only with the paint layer. In particular, the ester band of oil at 1738 cm^{-1}, bands at 1031 cm^{-1} indicating the presence of silicates and bands at 873 and 712 cm^{-1} ascribable to calcium carbonate, were recognizable. Residues of the triterpene resin after cleaning (which can be identified by the characteristic carbonyl band at 1701 cm^{-1}), were not clearly detected. However, the presence of varnish traces on the paint layer cannot be completely excluded due to the possible overlapping among vibration bands of natural resin with those of the oil binder in the range of carbonyl groups (C=O stretching 1700–1740 cm^{-1}) and aliphatic chains (C-H bending 1455, 1380 cm^{-1}).

Fig. 1b shows the typical paint stratigraphy of a sample collected from the mock-up before the treatment. Over a gypsum ground layer (layer 0) a priming layer of glue was applied (layer 1). Then, a thick paint layer about 150 μm thick was applied (layer 2). The uppermost dammar varnish layer (with a thickness of about 15 μm) is clearly identifiable under UV light by its bluish fluorescence (layer 3). After cleaning, the varnish thickness was considerably reduced (to less than 2 μm), demonstrating the gradual and safe removal obtainable with the gel (Fig. 1c).

As expected, comparable cleaning results were obtained with the swab application of volatile organic solvents, used as references. In particular, the application of acetone or DMSO with a cotton swab for 30 seconds reduced the dammar layer

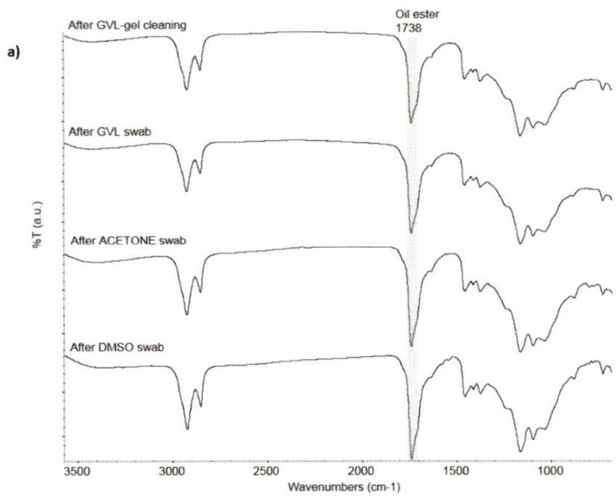

b)

Cleaning system	24h %wt
GVL gel	0,05 ± 0,01
GVL swab	0,15 ± 0,09
Acetone swab	n.d.
DMSO swab	0,38 ± 0,17

FIG. 2 Comparison of cleaning efficacy of the GVL gel and traditional organic solvents applied by swab. (a) FTIR-ATR spectra from areas treated with GVL gel, GVL swab, acetone swabs and DMSO swabs respectively; (b) HS-SPME-GC-MS data on solvent retained by the painted surface after 24 hours.

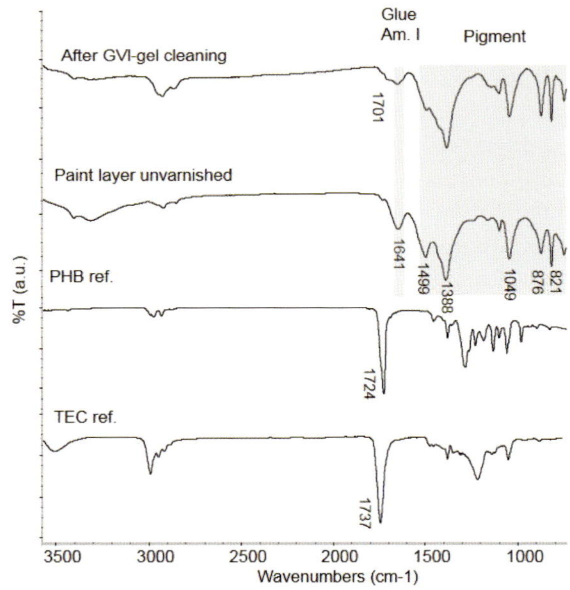

FIG. 3 FTIR-ATR spectra on an area cleaned with GVL gel, unvarnished area, PHB and TEC (respectively from top to bottom). The spectrum after gel cleaning indicates the absence of gel residues.

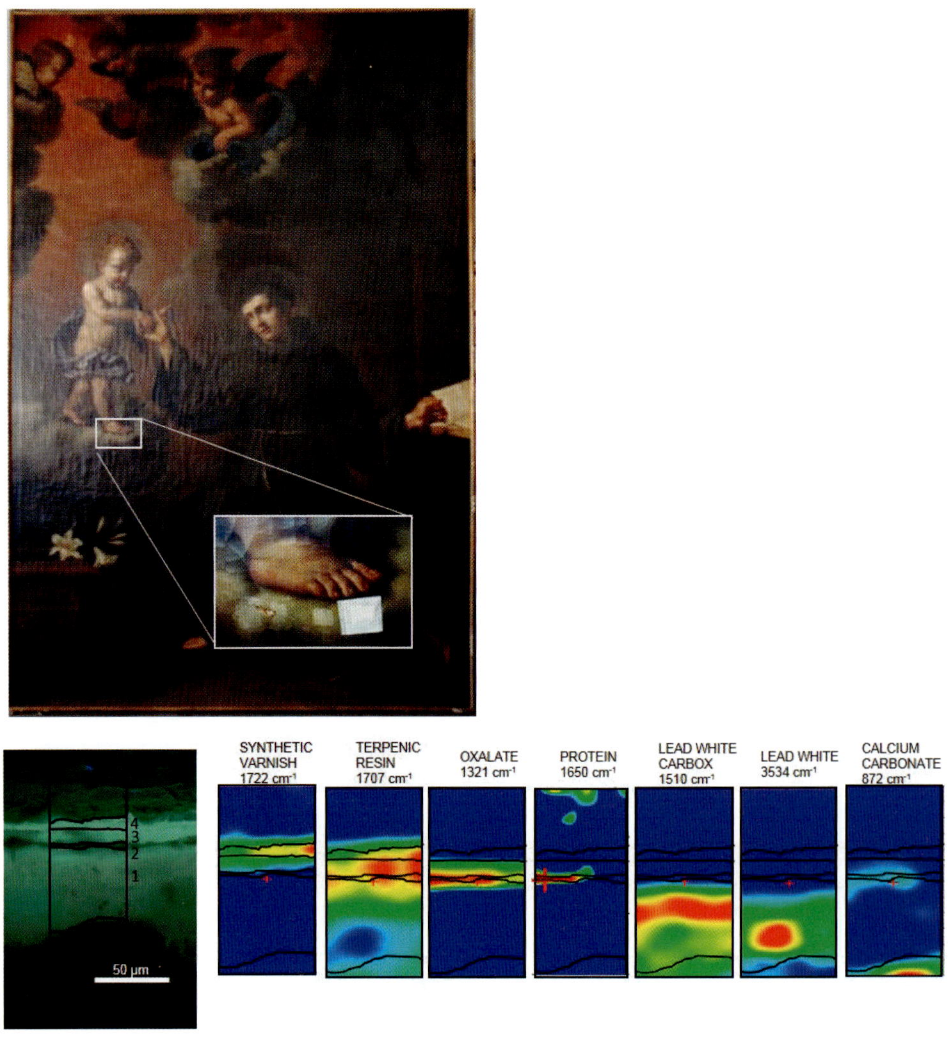

FIG. 4 Benedetto Gennari, *San Antonio*, 17th century, oil on canvas, Church of San Francesco, Ravenna, Italy: FTIR-ATR false colour chemical maps of the identified components.

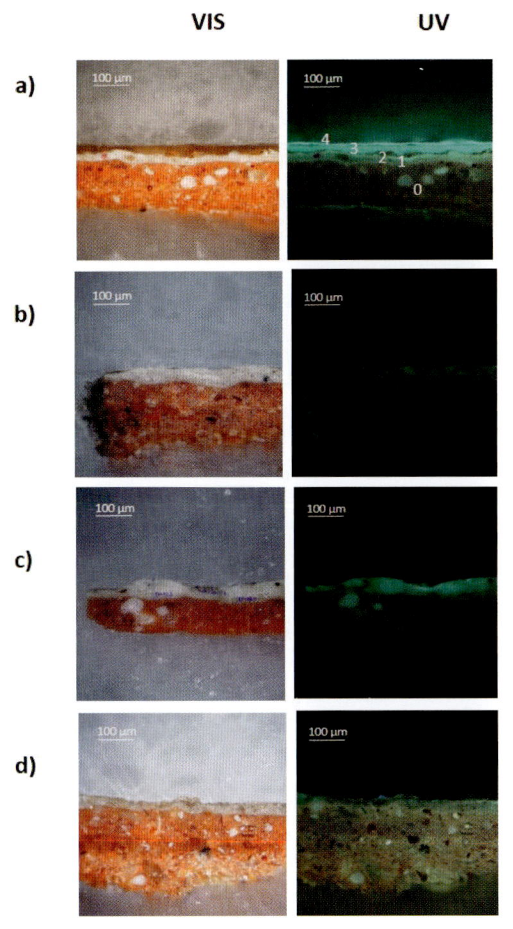

FIG. 5 Benedetto Gennari, *San Antonio*: cross-sections in visible light (*left*) and UV light (*right*) before cleaning (a) and after the application of GVL gel (b), acetone swabs (c) and DMSO swabs (d).

so that it was no longer detectable by FTIR–ATR spectroscopic investigation (Fig. 2a). The proposed gel approach can overcome the problems associated with the use of organic solvents, offering a safer solution for both the artwork and the conservator.

Evaluation of PHB gel residues and solvent retention

To better evaluate the potential presence of PHB gel residues on the painted surface after the cleaning treatment, ad hoc standard samples were prepared, avoiding the use of linseed oil as binder. The ester bands of PHB and TEC are centred at 1724 cm^{-1} and 1737 cm^{-1} respectively. Thus, as previously reported, their presence may be hidden by the absorption band of oil at about 1734 cm^{-1}. For this reason, oil-free mock-ups were prepared with dammar varnish applied over a layer of malachite and animal glue, and submitted to the cleaning procedures described above before single point FTIR–ATR analysis. The collected spectra reveal the typical IR absorption bands of the paint layer (3406, 3312, 1499, 1388, 1049, 876, 821 cm^{-1} for malachite pigment and 1641 cm^{-1} for glue binding medium); residues of PHB or TEC were not detected (Fig. 3).

Particular attention was paid to evaluation of the retention of the solvent by the paint layer after cleaning (which may linked to swelling phenomena), to better evaluate the behaviour of GVL, applied for the first time onto a painted surface. The HS–SPME–GC–MS technique was performed on standard mock-ups prepared with linseed oil binder and red ochre pigment and coated with triterpene resin. The approach avoids chemical extraction methods as the analytes are sampled by a fibre, which is then thermally desorbed in the GC injector (Colombini and Modugno 2009). Results (Fig. 2b) for the amount of solvent retained by the surface 24 hours after cleaning reveal that the amount of GVL within the area treated with the gel is lower than that in the area treated with swab. These evidences suggest that PHB polymer controls the solvent diffusion to the underneath layers, implicating a great advantage in terms of swelling risk. Moreover, GVL has lower retention values compared with the highly retentive DMSO, reported in this case as a known threshold limit for retention risk. Any residue of acetone would be detectable, suggesting that a complete evaporation of the solvent occurred after 24 hours. This result may however be controversial from the point of view of health and safety, because is linked to the risk of inhalation when volatile free solvents are used.

PHB gel case study

Finally, the cleaning efficacy of the GVL gel was validated on a 17th-century oil painting on canvas, *San Antonio* attributed to Benedetto Gennari (Fig. 4). The painting appeared to be obscured by an aged varnish. Cross-sections revealed a complex stratigraphy characterized by several layers (Fig. 5). In particular, there were two distinct varnish layers, 5 and 10 μm thick respectively (layers 4 and 3), between which a thin layer of dirt was present, suggesting consecutive applications; a thin and non-contiguous layer of deteriorated material (layer 2) 3–5 μm thick; the paint layer (layer 1) which was 20–30 μm thick; and the ground layer (layer 0).

Mapping with μFTIR-ATR enabled localization and discrimination between the varnish layers. The outermost layer of synthetic varnish and a lower layer of dammar varnish were recognized by the construction of a chemical map using bands at 1722cm^{-1} and 1707cm^{-1}, respectively. Below the dammar resin a layer containing a proteinaceous material and oxalates, possibly the original varnish, was detected. The paint layer included lead white and a degraded lipid binder, suggested by the presence of lead carboxylates (1510 cm^{-1}).

Cleaning tests with GVL gel were carried out for 5 minutes on the surface and compared with neat DMSO and/or acetone applied by swab. Microsamples collected from areas before and after cleaning were examined with optical microscopy and μFTIR–ATR.

Cross-sections obtained after GVL gel cleaning clearly show satisfactory removal of both the synthetic and the natural varnish, without affecting the paint layer (see Fig. 5b). A similar efficacy was also obtained by the use of acetone and/or DMSO applied by swab, although a possible interaction with the paint layers cannot be excluded. Indeed, cross-sections showed that after DMSO test clean, the paint layer appeared darkened, possibly due to dissolution and migration of dissolved components into the paint layer (see Fig. 5d).

Conclusion

A novel green gel is proposed as a cleaning method for paintings. The complete non-toxicity of its components (polymer, solvent and plasticizer), which are derived from renewable sources, and their biodegradability, represent the true novelty of the PHB-based organogel. Cleaning tests on mock-ups to imitate the stratigraphy of a 17th-century Italian easel painting demonstrate a good capability for removing both acrylic and triterpene varnishes without affecting the underlying paint layer, and without leaving detectable gel residues. The action of the solvent, confined within the gel, ensured a controlled interaction with the paint layers. HS-SPME is proposed as a new method for the evaluation of solvent retention within the cleaned paint surface. The high sensitivity of this technique demonstrated its capability for analysing microsamples, and it can be proposed as part of the analytical protocol for the evaluation of new cleaning systems.

Materials and suppliers

- All solvents and chemicals: Sigma-Aldrich, https://www.sigmaaldrich.com/italy.html
- Dammar resin, turpentine and burnt red ochre pigment (40470): Kremer Pigmente, https://www.kremer-pigmente.com/it
- Linseed oil: Zecchi, http://www.zecchi.it/

References

Alonso, D.M., Wettstein, S.G. and Dumesic, J.A. 2013. 'Gamma-valerolactone: a sustainable platform molecule derived from lignocellulosic biomass', *Green Chemistry* 15(3): 584.

Baglioni, P., Berti, D., Bonini, M., Carretti, E., Dei, L., Fratini, E. and Giorgi, R. 2014. 'Micelle, microemulsions, and gels for the conservation of cultural heritage', *Advances in Colloid and Interface Science* 205: 361–71.

Baglioni, P., Bonelli, N., Chelazzi, D., Chevalier, A., Dei, L., Domingues, J., Fratini, E. and Giorgi, R. 2015. 'Organogel formulations for the cleaning of easel paintings', *Applied Physics A* 121: 857–68.

Carretti, E., Bonini, M., Dei, L., Berrie, B.H., Angelova, L.V., Baglioni, P. and Weiss, R.G. 2010a. 'New frontiers in materials science for art conservation: responsive gels and beyond', *Accounts of Chemical Research* 43(6): 751–60.

Carretti, E., Natali, I., Matarrese, C., Bracco, P., Weiss, R.G., Baglioni, P. and Dei, L. 2010b. 'A new family of high viscosity polymeric dispersions for cleaning easel paintings', *Journal of Cultural Heritage* 11(4): 373–80.

Colombini, M. P. and Modugno, F. (eds). 2009. *Organic Mass Spectrometry in Art and Archaeology*. Chichester: Wiley.

Domingues, J., Bonelli, N., Giorgi, R., Fratini, E. and Baglioni, P. 2013. 'Innovative method for the cleaning of water-sensitive artifacts: synthesis and application of highly retentive chemical hydrogels', *International Journal of Conservation Science* 4: 715–22.

Fiorese, M.L., Freitas, F., Pais, J., Ramos, A.M., De Aragão, G.M.F. and Reis, M.A.M. 2009. 'Recovery of polyhydroxybutyrate (PHB) from Cupriavidusnecator biomass by solvent extraction with 1,2-propylene carbonate', *Engineering in Life Sciences* 9(6): 454–61.

Phenix, A. 2002. 'The swelling of artists' paints in organic solvents. Part 2. Comparative swelling powers of selected organic solvents and solvent mixtures', *Journal of the American Institute for Conservation* 41(1): 61–90.

Riedel, S.L., Brigham, C.J., Budde, C.F., Bader, J., Rha, C., Stahl, U. and Sinskey, A.J. 2013. 'Recovery of poly(3-hydroxybutyrate-co-3-hydroxyhexanoate) from *Ralstoniaeutropha* cultures with non-halogenated solvents', *Biotechnology and Bioengineering* 110(2): 461–70.

Rombolá, A.G., Marisi, G., Torri, C., Fabbri, D., Buscaroli, A., Ghidotti, M. and Hornung, A. 2015. 'Relationships between chemical characteristics and phytotoxicity of biochar from poultry litter pyrolysis', *Journal of Agricultural and Food Chemistry* 63(30): 6660–67.

Samorì, C., Abbondanzi, F., Galletti, P., Giorgini, L., Mazzocchetti, L., Torri, C. and Tagliavini, E. 2015. 'Extraction of polyhydroxyalkanoates from mixed microbial cultures: impact on polymer quality and recovery', *Bioresource Technology* 189: 195–202.

Samorì, C., Galletti, P., Giorgini, L., Mazzeo, R., Mazzocchetti, L., Prati, S., Sciutto, G., Volpi, F. and Tagliavini, E. 2016. 'The green attitude in art conservation: polyhydroxybutyrate-based gels for the cleaning of oil paintings', *ChemistrySelect* 1(15): 4502–508.

Sudesh, K. and Iwata, T. 2008. 'Sustainability of biobased and biodegradable plastics', *CLEAN – Soil, Air, Water* 36(5–6): 433–42.

Authors' addresses

- Francesca Volpi, University of Bologna, Italy (francesca.volpi5@unibo.it)
- Chiara Samorì, University of Bologna, Italy (chiara.samori3@unibo.it)
- Laura Mazzocchettii, University of Bologna, Italy (laura.mazzocchetti@unibo.it)
- Silvia Prati, University of Bologna, Italy (s.prati@unibo.it)
- Paola Galletti, University of Bologna, Italy (paola.galletti@unibo.it)
- Giorgia Sciutto, University of Bologna, Italy (giorgia.sciutto@unibo.it)
- Loris Giorgini, University of Bologna, Italy (loris.giorgini@unibo.it)
- Emilio Tagliavini, University of Bologna, Italy (emilio.tagliavini@unibo.it)
- *Corresponding author*: Rocco Mazzeo, University of Bologna, Italy (rocco.mazzeo@unibo.it)

Testing a potassium polyacrylate aqueous gel for the pretreatment and storage of wet marine iron and organic composite artefacts

Vasilike Argyropoulos, Ageliki Bei and Stamatis C. Boyatzis

Introduction

The use of synthetic clays and gels such as carboxymethylcellulose is quite common in conservation treatments. However, for application to recovered marine artefacts that are waterlogged, it is important that gels ensure moisture retention, are easy to handle (and remove), and can be monitored by conservators. As early as 1987, two types of polyacrylamide superabsorbent polymers (SAPs) (Agrosoke and Erosel) used in horticulture were tested and applied with and without chemical additives as interim storage environments during the transport of waterlogged marine artefacts (Carpenter 1987). Today there are many different types of SAPs on the market and they are commonly used in industry. The most conventional are ionic SAPs, either cationic based on acrylamides (AM) or anionic based on partially neutralized acrylic acid (AA). This paper describes the application of an AA SAP, Aquaperla, as a storage/packing medium to transport marine metal and composite artefacts, as well as for selective treatments.

Applications for anionic SAPS for marine artefacts

Aquaperla, like all SAPs, forms hydrogels by absorbing aqueous solutions through hydrogen bonding with water molecules, and it is commonly used as a water retention agent. These gels swell upon the addition of aqueous solutions,

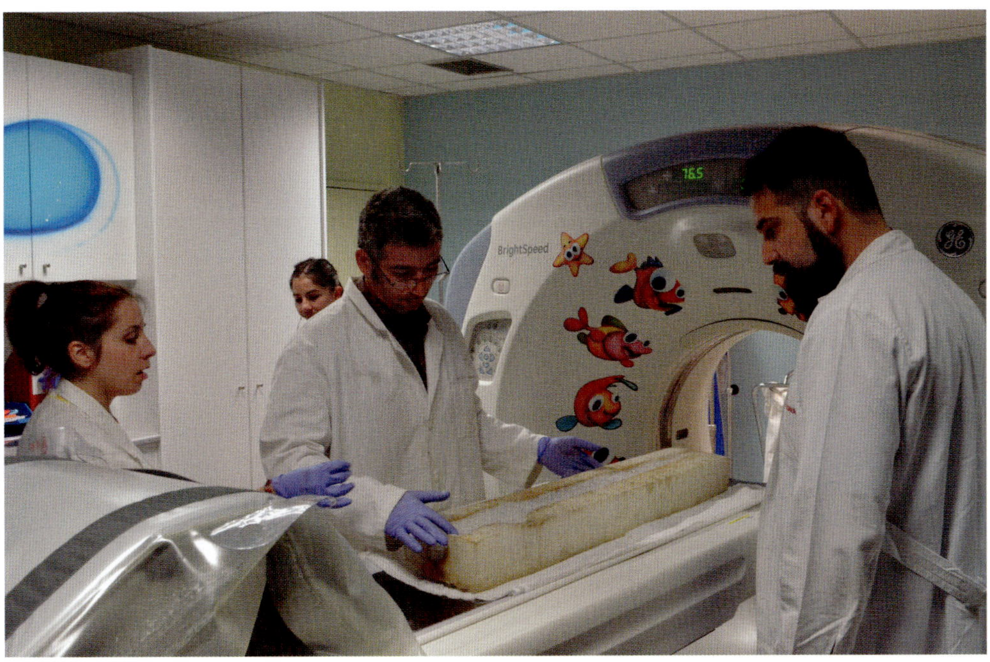

FIG. 1 The use of the water gel as a storage medium for the transport of a marine composite sword to a local hospital, for imaging with tomography.

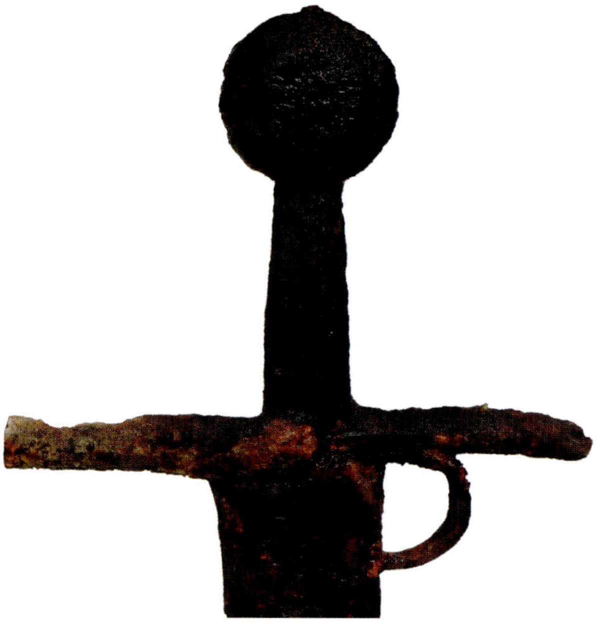

FIG. 2 The stable iron metal surface (pommel and handle) of the medieval sword after the application of Aquaperla with 2% NaOH solution.

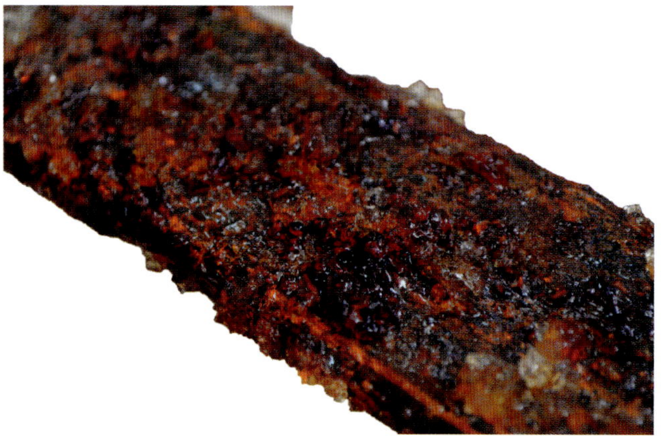

FIG. 3 Areas where the water gel failed to keep the wooden scabbard of the medieval sword moist due to the high concentration of iron in the wood. The dehydration of the surface led to the formation of iron oxyhydroxides.

allowing for the retention of moisture and facilitating wetting to the applied surfaces. The gel was prepared by adding approximately 50 g Aquaperla crystals to 1 L deionized water. The advantages of using this water gel for maintaining the wetness of the artefacts during transport are: its ability to act as a cushion when many objects are stored in the same container; its capability to maintain wetness for over 24 hours; its lighter weight for handling, compared to the use of water baths; it does not need to be removed during the subsequent examination of each artefact (Fig. 1).

Another application involved the use of Aquaperla as a poultice with 2% w/v sodium hydroxide (NaOH) solution for selective treatments of marine iron or composite artefacts with and without electrolysis. The gel crystals were added to the NaOH solution by heating it to *c.*60 °C with continuous stirring; air bubbles were removed under vacuum for a few minutes before the solution was gradually raised to atmospheric pressure.

As an example, local electrolysis was carried out on an iron cannon recovered from the sea that had exhibited signs of active corrosion after its treatment, to remove chlorides. For large marine objects, such as cannons, objects are placed in aqueous baths with alkaline solutions. In our case, it was decided to avoid the use of complete immersion of the object in order to reduce the amount of chemical waste. An Aquaperla gel containing 2% NaOH solution was applied to the affected area. In addition, the gel was covered by an absorbent sponge (containing NaOH solution), and the anode (stainless steel mesh) was placed on top of this matrix. A current was applied to the object, which acted as the cathode. Within 24 hours, the gel had turned from transparent to black in colour, indicating that the active corrosion had been reduced to magnetite. The application of an electrical current to treat marine iron selectively, using absorbent sponges containing alkaline solutions, has already been carried out successfully (Argyropoulos and Batis 2013). However, the gel application was found to be more effective than this previously published method, since it ensured that the surface was more uniformly treated.

Selective treatments using an Aquaperla gel poultice were also used on a 13th-century medieval composite sword recovered from the sea, after its concretions had been removed with a pneumatic air scribe. A gel with 2% w/v NaOH solution (the pH of the gel was 13.5) was applied to cover the iron components, while an aqueous gel was used to cover the organic components. The gel with NaOH maintained the stability of the iron surface for periods in excess of 24 hours, and removed chlorides. Both Ethafoam and wet absorbent sponge was used to sandwich the gels covering the sword, which was then placed in an Escal bag sealed with clips. Each day the gels were checked to ensure there were no colour changes indicative of iron corrosion.

The AA SAP with 2% NaOH solution was successful at maintaining the iron stable and wet for periods from 9 to 14 days, with the pH remaining constant (Fig. 2).

The results varied when the deionized gel water was applied to the organic components of the object. In some areas, it was effective at maintaining the organic components in a wet state, but in other parts of the sword, there was a loss of water retention in the gel (after 4–5 days), with orange staining due to the formation of iron corrosion products. These corrosion products were identified as lepidocrocite and goethite. A possible reason for the failure is that the absorption capacity index (ACI) of AA SAPs is decreased by the crosslinking of the polymer with cations (divalent or trivalent salts for alkali metals and alkaline earth metals ions), meaning that the potassium ions in the Aquaperla exchanged with the Fe(II) ions, reducing the ACI. For this reason, the same areas were treated with a gel made up with ethylenediaminetetraacetic acid (EDTA) solution with and without buffers (pH 5–6). This gel was able to remove iron ions and maintain the area in a wet state (i.e. the ACI was maintained); however, after a few days, the gel poultice completely dissolved (Fig. 3).

Conclusion

AA SAPs such as Aquaperla were found to be effective at maintaining the wetness of composite artefacts recovered from the sea, depending on the type of surface. The gel application was used as a packing medium for the transport of artefacts, treatment of marine iron using local electrolysis, and for selective treatment using chemical additives to a marine composite artefact. Trials on iron surfaces showed that a gel poultice with NaOH could maintain the stability of the metal providing the pH of the gel did not drop below 10. Trials on organic surfaces found the effectiveness of the gel poultice depended on the degree of iron contamination in the organic materials: the water retention of the gel decreased at a faster rate when the organic material contained high amounts of iron. The use of AA SAPs is effective as an interim storage option for marine artefacts, as well for maintaining the surface stability of complex artefacts selectively. However, more research is needed into their application to determine the most appropriate way to remove the chlorides and/or other corrosion products effectively with or without using electrolysis.

Materials and supplier

- DCM Aquaperla water crystals: De Ceuster Meststoffen n.v. (DCM), Bannerlaan 79, B – 2280 Grobbendonk, Belgium

References

Argyropoulos,V. and Batis, G. 2013. 'Saving a marine iron paddle wheel removed from the 1868 steam engine shipwreck *Patris* during an economic crisis in Greece', paper presented at *Big Stuff 2013, Ottawa, Canada*. Available at: https://bigstuff.omeka.net/big-stuff-2013 (accessed 22 May 2017).

Carpenter, J. 1987. 'The uses of soil stabilizing gel media in the conservation of large and small shipwreck artifacts', *International Journal of Nautical Archaeology* 16(2): 95–107.

Authors' addresses

- Vasilike Argyropoulos, Technological Educational Institute of Athens, Department of Conservation of Antiquities and Works of Art, Aegaleo, Greece (bessie@teiath.gr)
- Ageliki Bei, Technological Educational Institute of Athens, Department of Conservation of Antiquities and Works of Art, Aegaleo, Greece
- Stamatis C. Boyatzis, Technological Educational Institute of Athens, Department of Conservation of Antiquities and Works of Art, Aegaleo, Greece

The use of Nanorestore gels in the conservation of lime-based wall paintings

Hanna Eriksson, Ingrid Wedberg, Johanna Nessow and Matilda Bronmark-Thorlund

Introduction

Most lime-based paintings in Scandinavia executed *a secco* with the paint layer on top of almost fully carbonated lime plaster date from the Romanesque period and later. The problem with cleaning such a surface is its uneven texture. The application and composition of the lime plaster gives a rough surface where the tops of the knobbled surface carrying the paint layer tend to lose pigment during traditional, dry, mural cleaning processes that involve abrasive mechanical stress, such as *gomma pane* (a studio-made soft rubber made of copper sulphate, sodium carbonate and flour) or Akapad (Kremer Pigmente). According to Swedish conservation ethics, the preservation of retouched areas is considered important. The work of art is respected in its whole historical context, and additions are seen as a vital part of the artefact's history. The retouching binding medium is often purposely made weaker than the original paint. This renders traditional mechanical cleaning processes too aggressive.

Use of gels in cleaning of lime-based wall paintings

Gels work as vehicles that carry cleaning agents, and aqueous gels are in effect water-based detergent systems. Physical gels made of Carbopol may be loaded with a surfactant, a pH buffer and a variety of solvents. There is a risk when using them of damaging delicate paint, since there is a possibility that the solvent might reach the surface and the gel might leave residues (Khandekar 2004). Due to poor solvent retention, gels based on Carbopol also risk the diffusion of dissolved substances into a porous structure. It is necessary to clear the surface when removing Carbopol gels, usually with organic solvents or water. It is sometimes difficult to avoid loss of paint and to ensure fine control of grime removal.

The paintings in the church of Everlöv, a small village in the south of Sweden, date from 1500. They were lime washed during the 18th century, and partly uncovered in 1850 by a local priest. Since restoration in 1915, no extensive, documented interventions have been carried out. Today, the paintings are heavily contaminated with grime. The paint is very thin and applied to a rough surface, making the cleaning process very challenging. Two earlier cleaning tests had led to significant abrasion of paint. To avoid further damage caused by mechanical abrasion and ensure better control of grime removal, we decided to investigate a more delicate method.

Chemical hydrogels

CSGI (Consorzio Interuniversitario per lo Sviluppo dei Sistemi a Grande Interfase/Research Center for Colloids and Nanoscience) Florence, Italy, has developed a new hydrogel system, Nanorestore gel,[1] for cleaning highly water-sensitive paint such as distemper and tempera as well as lime-based paintings (Baglioni and Chelazzi 2013). The hydrogels are based on semi-interpenetrating chemical poly(2-hydroxyethyl methacrylate) p(HEMA)/poly(vinylpyrrolidone) (PVP) networks with suitable hydrophilicity, water retention properties and mechanical strength, to avoid residues following cleaning treatment (Domingues *et al.* 2013). p(HEMA)-hydrogels have been investigated for medical uses due to their high biocompatibility and PVP is a highly hydrophilic polymer used in the food and cosmetic industries: together they form stable networks with high hydrophilicity. The properties of the gels permit cleaning only at the gel–paint interface. This enables a controlled cleaning action as the gel is simply peeled off the surface after its cleaning action, leaving no detectable residues. This is especially vital on porous lime layers.

TABLE 1 Results of cleaning tests.

Gels	Gel Dry	Extra Dry	Max Dry	PVA1	PVA2	STD	PROH5%
Application period	3 min	3 min	3 min	3 min	3 min	Pressed and rolled over the surface	Pressed and rolled over the surface
Characteristics	Slippery soft	Semi-hard	Hard	Hard but flexible	Hard but flexible	Sticky gel	Sticky gel
Adherence	Very good	Good	Poor	Very food	Very good	–	–
Cleaning ability	Very good. Residues in micro cavities	Very good	Some cleaning effect	Good. Best in coloured area	Very good. Absorbed dirt in white area better than PVA 1. Removed whitish wash/bloom on surface	Fairly good but diminished after a few rolls on surface	Good, better than STD, but diminished after a few rolls on surface
Surface abrasion, obvious pigments on gel	A few pigment particles on gel	Applied gel with more pressure compared to Gel Dry. No pigment on gel	Due to problems with adherence the surface was more exposed to some friction. Pigments on gel	No pigment noted on gel	No pigment noted on gel	Friction due to the application, although less friction compared to Gomma Pane and Akapad sponge	Friction due to the application, although less friction compared to Gomma Pane and Akapad sponge
Evaluation	Best cleaning effect. Best adherence to surface. Gel prone to break into pieces which made second use difficult	Good cleaning effect but adherence is a problem considering the angle variations in a vault	Some cleaning effect but not suitable for use on lime paintings due to the poor adhesion	Good cleaning effect. Gel remained stable. Possible to reuse	Very good cleaning effect. Gel remained stable. Possible to reuse	As the application causes abrasion it is not suitable in areas with critical PVC	As the application causes abrasion it is not suitable in areas with critical PVC

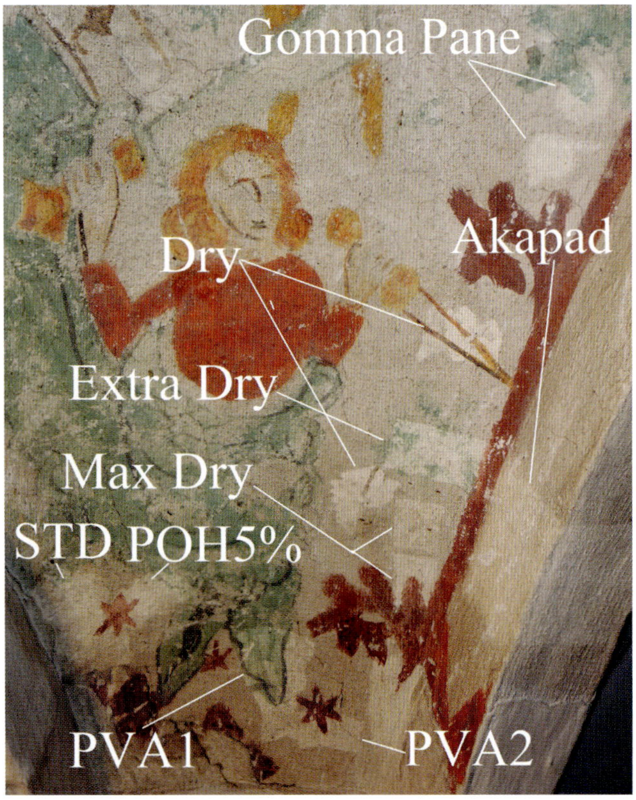

FIG. 1 Results of cleaning tests (photo: Hanna Eriksson).

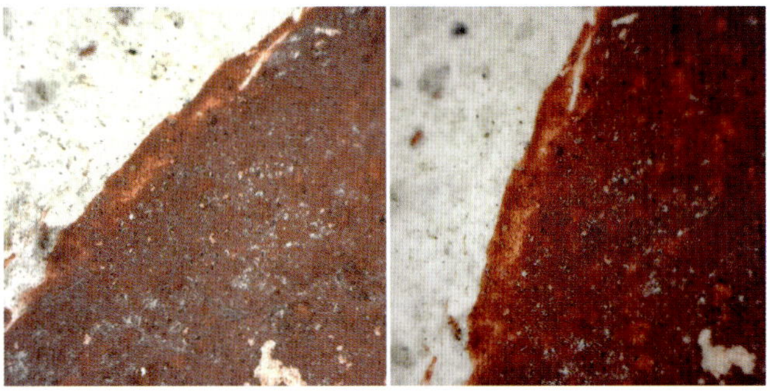

FIG. 2 An area of the lime-based painting at Everlöv, before (*left*) and after (*right*) treatment with PVA2 (USB-microscope ×200) (photo: Ingrid Wedberg).

Tests

This study focused on finding a type of gel that would both work on delicate surfaces and have good adhesion to a vaulted wall surface. This last parameter is crucial for an effective working process on a large scale. The pigment volume concentration (PVC) of such surfaces is high, which mandates a careful cleaning method. The mechanical properties of the gel, such as elasticity and viscosity, are also important to facilitate handling *in situ*. The Nanorestore gel formulations are intended to be reusable 4–5 times, after careful cleaning and soaking between the applications. The gels, which were not loaded with any other polar solvent, were kept in distilled water, making the method easy to work with and non-toxic. The sheets of gel measure between 5 and 10 cm^2 with a width of 2 mm. The different gel types tested varied in adhesion and flexibility.

A series of five types of 'hard' hydrogels was tested: Nanorestore Dry, Extra Dry and Max Dry are based on p(HEMA)/PVP, and PVA1 and PVA2 are based on PVA. Additionally, two so-called sticky gels, STD and PROH were used at 5% concentration; the latter was loaded with ethanol up to 5% w/w. At this stage the exact composition of the PVA series, STD and PROH cannot be disclosed by CSGI. A cleaning test was carried out in the spandrel of the west vault. The harder gels were applied to the surface and left between 2 and 3 minutes before being peeled off. The sticky gels STD and PROH at 5% were applied by hand and used like a rubber. Neither technique required clearance of the surface after treatment.

Conclusion

The results obtained from tests of the hard gels were all satisfactory with regard to their cleaning efficacy. The first test was carried out with Nanorestore Dry. The adhesion of the gel was very good, as was the cleaning effect. Since its mechanical strength was poor, the gel broke into smaller pieces when reused. This is likely to be a problem when the gel is applied on a larger scale to a vertical surface. The gel was cleaned in distilled water for 24 hours before being reused. Both sides of the sheet were used. After the cleaning, residues of dirt could be seen on the gel's contact surface; this may have caused the noted slight loss of cleaning effect when reused. As the rigidity of the gels increased (Extra Dry and Max Dry are successively more rigid) the adhesion, and as a consequence, the cleaning ability decreased. The rigidity of the Max Dry gel was such that it did not remain attached to the surface at all.

PVA1 and PVA2 adherence was as good as Dry, but they were easier to work with due to their greater mechanical strength. PVA2 gave better cleaning results than PVA1. These two gels were also easier to reuse than Dry: they did not break and maintained their cleaning efficiency during repeated use. The sticky gels STD and PROH at 5% w/w, which are applied in a different way, tended to leave grime residues on the surface, although the overall cleaning effect was good. Since their application is similar to traditional cleaning with *gomma pane* the method caused some mechanical stress, which makes it unsuitable in retouched areas. PVA2 and Nanorestore Dry gels gave the most satisfactory results (see Table 1).

Acknowledgements

The authors would like to thank Prof. Piero Baglioni and the staff at Department of Chemistry, CSGI, University of Florence, for their kind assistance in providing the hydrogels.

Note

1. www.csgi.unifi.it/products/gel.html.

References

Baglioni, P. and Chelazzi, D. (eds) 2013. 'Nanoscience for the conservation of works of art', in E. Fratini and E. Carretti (eds), *Cleaning IV: Gels and Polymeric Dispersions*. Cambridge: RSC Publishing, 252–79.

Domingues, J.A.L., Bonelli, N., Giorgi, R., Fratini, E., Gorel, F. and Baglioni, P. 2013. 'Innovative hydrogels based on semi-interpenetrating p(HEMA)/PVP networks for the cleaning of water-sensitive cultural heritage artifacts', *Langmuir* 29(8) : 2746–55.

Khandekar, N. 2004. 'Gelled systems: theory and early application', in V. Dorge (ed.), *Solvent Gels for the Cleaning of Works of Art: The Residue Question*. Los Angeles: Getty Conservation Institute, 7–11.

Authors' addresses

- Hanna Eriksson, Vellinge, Sweden (hanna@skanesmalerikonservatorer.se)
- Ingrid Wedberg, Lund, Sweden (ingrid@skanesmalerikonservatorer.se)
- Johanna Nessow, Malmö, Sweden (johanna@skanesmalerikonservatorer.se)
- Matilda Thorlund Brönmark, Malmö, Sweden (matilda@skanesmalerikonservatorer.se)

New gel formulations for removal of adhesive residues and cleaning of paper and parchment

Marco Fagiolo, Valentina Giunta, Claudia Mazzuca and Silvia Orlanducci

Introduction

In this work, we tested several gels for the removal of protein adhesive from parchment and paper supports. The proposed gels release water in a controlled way on the sample, thereby preventing swelling of fibres and hydrolysis. They do not leave detectable residues on artworks. Their preparation and removal is easy and they are non-toxic for the conservator. The efficacy of these gels is demonstrated with Fourier transform infrared spectroscopy using attenuated total reflectance (FTIR–ATR), Raman spectroscopy and scanning electron microscopy (SEM).

Materials and methods

Nanodiamonds and Proteinase K

We tested two gellan gels mixed with an unusual cleaning agent: Nanodiamonds and Proteinase K. In the field of cultural heritage the use of nanodiamonds (ND) as a cleaning tool is little discussed (Ruobao *et al.* 1992). NDs are a natural material with grains of dimensions 5–10 nm, present as crystals incorporated into meteorites, crystal grains in polycrystalline diamond films, and in powders and suspensions synthesized by detonation (Baildakova and Vul 2007; Reina *et al.* 2014). To prepare the gel, a solution of 4% deacylated gellan and 1% ND powder is placed in a microwave oven until it boils and then cooled in a Petri dish. The gel appears grey because of the ND. The ND gels were compared with a gellan gel (gellan gum 2% and calcium acetate) containing 1 μM of Proteinase K (PK), an enzyme that catalyses the cleavage of peptide bonds. The enzyme was added to the gel while it is cooling. This gel is transparent (Petrella *et al.* 2016).

Gels containing ND and PK were tested on paper samples (7 × 7 mm) from the end leaf of a Persian manuscript. The test sites were first examined using macrophotography and an optical microscope integrated with Raman spectroscopy. The proteinaceous glue layer to be removed was reddish in colour, darker in some areas, indicating trapped lumps of soil components in the glue.

PEOS and PVA

For the cleaning of residues of protein glue from parchment supports we used a hydrogel formed from polyethylene oxide and α-cyclodextrin (PEOS). To prepare it 0.40 g PEO powder (polyethylene oxide) and 0.45 g αCD (cyclodextrin) were dissolved separately in 5 mL of distilled hot water. The solutions were mixed and left to cool (Mazzuca *et al.* 2013). This gel was tested on a fragment of ancient parchment: after a 15-minute treatment, the weight increase of the parchment sample was 33%, which was less than following treatment with non-gelled water or other gels. To prepare PVA gel, 0.7 g poly(vinyl alcohol) 99+ (molecular weight: 85,000 to 124,000), was added to 10 mL distilled water at 100 °C, while stirring. The system was maintained at this temperature with stirring until solubilization. The solution was then cooled on a Petri dish. Subsequently, the gel was covered and subjected to several cycles of freezing/thawing (freezing at −20 °C for 18 hours, and thawing at room temperature for 6 hours). These gels are non-toxic, sufficiently rigid and easily prepared (Peppas and Scott 1992).

Results and discussion

ND and Proteinase K

These gels are able to remove proteinaceous glue residues easily. The areas treated were compared at ×100 magnification. The glue was not homogeneous: in some parts of the paper sample no protein glue was present; in others, there were layers of glue and in still other locations, the reddish colour was darker, indicating dirt and debris trapped in the glue. The gels were applied on the support in contact with the surface and covered with a polyester sheet (Mylar) and a light weight. To assess the efficacy of the gels, we compared FTIR-ATR spectra before and after treatment. The relative intensity of the bands at 1540 cm^{-1} (amide II band, relative to the stretching CN in the peptide bond NH) and 1024 cm^{-1} (relative to cellulose CO stretching) was evaluated. This ratio decreases with the increasing efficacy of the treatment: the less adhesive remaining, the lower the intensity of this band. The normalized values are reported in Table 1, wherein the value of the glued untreated sample was set as 1. The results show that the most effective treatment was with gel ND.

TABLE 1 Normalized ratio for bands at 1540 and 1024 cm^{-1} for the gels.

Sample	Glued	Gellan gel	PK gel	ND gel
I_{1540}/I_{1023}	1	0.92±0.02	0.85±0.02	0.79±0.02

PEOS

PEOS gel was placed in contact with the parchment sample. After about 15 minutes, the gel was removed with a spatula and moistened cotton because it had completely softened the glue. Once the parchment fragments were dry, FTIR-ATR was carried out to ensure the absence of gel residues. To determine long-term effects, untreated and treated samples were artificially aged in an oven at 80 °C, 65% relative humidity (RH) for 30 days, and again examined using FTIR-ATR. As reported in the literature, the ratio between amide I and amide II bands (A_I/A_{II}) and their differences in energy (Δv) are diagnostic of the condition of the parchment (Badea *et al.* 2008). The values reported in Table 2 indicate that treatment with the PEOS gel did not accelerate the sample degradation (measurement error is around 5%).

TABLE 2 Ratios for amide bands (A_I/A_{II}) and their differences in energy (Δv).

Sample	A_I/A_{II}	Δv (cm^{-1})
Parchment untreated, aged	1.11	94
Parchment treated, aged	1.10	92
Parchment untreated, unaged	1.11	88
Parchment untreated, aged	1.10	87
Parchment treated, aged	1.09	89

Based on the results obtained, we worked directly on a 10th-century Greek manuscript parchment guard on areas with gelatin-based adhesive residues. After 15-minute applications of the uncovered gel, the adhesive was removed with a spatula and a cotton bud dipped in a 70:30% water:ethanol solution.

PVA

After the application of PVA gels on new and ancient parchment fragments, we determined that the weight loss of the gel is less than 10%, while the parchment weight increment, due to water uptake, is 25% after one hour of treatment. During this time, the PVA gel changed in colour (yellowing) thus demonstrating an absorption of dirt and degradation products from the parchment. FTIR-ATR examination of the treated areas provided both confirmation of surface cleaning and the absence of gel residues. In fact, the peaks for PVA (at 1146, 922 and 838 cm^{-1}), are not present after cleaning. Moreover, FTIR indicated that the PVA gel does not damage the parchment – indeed it slightly increases the degree of hydration (see Table 3, which has measurement errors of around 5%). SEM confirmed these results.

TABLE 3 Ratios for amide bands and their differences in energy for parchment untreated and treated with PVA.

Sample	A_I/A_{II}	Δv (cm^{-1})
New untreated parchment	1.07	86
New parchment cleaned with PVA1 gel	1.15	86
Old untreated parchment	1.12	88
Old parchment cleaned with PVA1 gel	1.18	93

Conclusion

The development of innovative approaches for cleaning and removing layers of protein adhesive on paper and parchment are presented. For paper, the cleaning action and glue removal using gellan gels with ND or PK is more efficient than using gellan gel alone. The PEOS and PVA gels provide a controlled humidification of the parchment support and efficient cleaning, showing no apparent colour change or halos, and no accelerated degradation afterwards. The results obtained with these gels are a starting point for future improvements to create materials most suited to the conservation of paper and parchment.

References

Badea, E., Miu, L., Budrugeac, P. and Della Gatta, G. 2008. 'Study of deterioration of historical parchments by various thermal analysis techniques complemented by SEM, FTIR, UV–VIS–NIR and unilateral NMR investigations', *Journal of Thermal Analysis and Calorimetry* 91(1): 17–27.

Baidakova, M. and Vul, A. 2007. *New Prospects and Frontiers of Nanodiamond Clusters*. St Petersburg: Ioffe Physico-Technical Institute, the Russian Academy of Sciences. Available at: stacks.iop.org/JPhysD/40/6300 (accessed 27 May 2017).

Mazzuca, C., Bocchinfuso, G., Cacciotti, I. and Micheli, L. 2013. 'Versatile hydrogels: an efficient way to clean paper artworks', *Royal Society of Chemistry Advances* 3: 22896–9.

Peppas, N.A. and Scott, J.E. 1992. 'Controlled release from poly(vinylalcohol) gels prepared by freezing-thawing processes', *Journal of Controlled Release* 18: 95–100.

Petrella, G., Cervelli, E., Mazzuca, C., Micheli, L. 2016. 'A new sustainable and innovative work for paper artworks cleaning process: gellan hydrogel combined with hydrolytic enzymes', *International Journal of Conservation Science* 7: 273–80.

Reina, G., Orlanducci, S., Tamburri, E. and Terranova, M.L. 2014. 'Nanotechnologies for cultural heritage: Nanodiamond for conservation of papers and parchments', in M. Rossi, C. Mariani and M.L. Terranova (eds), *NanoForum 2013. AIP Conference Proceedings* 1603. New York: AIP Publishing, 93–101.

Ruobao, W., Jucheng, L. and Wangdi, W. 1992. 'Diamond: like films as coatings for preservation of culture heritage', in *Proceedings of the EEC China Workshop on Preservation of Cultural Heritage, Xian, China*. Naples: Teti, 19–25.

Authors' addresses

- Marco Fagiolo, Department of Humanities, University of Rome 'Tor Vergata', Rome, Italy (fagiolomar@libero.it)
- Valentina Giunta, Department of Humanities, University of Rome 'Tor Vergata', Rome, Italy
- Claudia Mazzuca, Department of Chemical Sciences and Technologies, University of Rome 'Tor Vergata', Rome, Italy
- Silvia Orlanducci, Department of Chemical Sciences and Technologies, University of Rome 'Tor Vergata', Rome, Italy

Poly(vinyl alcohol)-borate gels as carriers of microemulsions

Anna Fialová and Petr Kotlík

Introduction

Poly(vinyl alcohol)-borate gels can be used as carriers of organic solvents (Carretti *et al.* 2009a; Angelova *et al.* 2011), which are commonly used for removing degraded polymer layers from cultural heritage objects (Grassi *et al.* 2009). Using a gel as a carrier reduces evaporation of solvents and therefore also their environmental impact, limits penetration of cleaning components into the surface of the treated subject and provides easy control of the cleaning process due to transparency of the gel (Carretti *et al.* 2009b). Effective but toxic organic solvents can be replaced with aqueous microemulsions of the same solvents (Grassi *et al.* 2009). Owing to their micellar structure, microemulsions reduce the toxicity of organic solvents to the conservator, as well as lessening the risk of the object being damaged during cleaning (Giorgi and Carretti 2013). Small-angle X-ray scattering proved that microemulsions retain their specific nanostructure while embedded into some polymeric networks (Carretti *et al.* 2009b). A microemulsion carried by a poly(vinyl alcohol)-borate gel should prove to be an ecological and safe cleaning agent (Baglioni *et al.* 2014).

Methods and materials

Four types of poly(vinyl alcohol) (PVAl), Mowiol brand, were used to test how gel viscosity is influenced by the concentration of each component and PVAl properties. Partially hydrolysed PVAl with a degree of hydrolysis of 87.7 ± 1.0 mol % (Mowiol 4-88, molar mass 31,000 $g.mol^{-1}$ or Mowiol 40-88, molar mass 205,000 $g.mol^{-1}$), is water-soluble at room temperature whereas fully hydrolysed PVAl with degree of hydrolysis 98.4±0.4 mol % (Mowiol 4-98, molar mass 27,000 $g.mol^{-1}$ or Mowiol 56-98, molar mass 195,000 $g.mol^{-1}$) must be dissolved in water in hermetically sealed vessels at 90 °C. Sodium tetraborate decahydrate (borax) was added into continuously stirred PVAl solutions. These mixtures were then cured for 24 hours at 20 °C.

The ratio for the starting substances derived from the literature (Carretti *et al.* 2010) was 2 g PVAl and 0.4 g anhydrous sodium tetraborate (i.e. 0.76 g borax) in 100 ml water. Two sets of samples were prepared for each Mowiol type, with linear increase of PVAl or borax content while the volume of the liquid and the other gel component remained constant. The maximum PVAl or borax content was 2.5 times higher than the initial concentrations. To evaluate the impact of the

TABLE 1 Composition of microemulsions (wt%). Organic solvents used: p-xylene (xyl), ethylene glycol (EG), ethylene acetate (EA) and propylene carbonate (PC); 1-pentanol (1-PeOH) was used as co-surfactant.

Microemulsion	ME xyl		ME EGxyl		ME EAPC	
surfactant	SDS	TW	SDS	TW	SDS	TW
	3.9	11.8	4.5	7.6	3.7	11.1
xyl	2.5	2.5	3.0	3.0	–	–
EG	–	–	5.9	5.9	–	–
EA	–	–	–	–	8.0	8.0
PC	–	–	–	–	8.0	8.0
1-PeOH	6.5	–	7.0	–	7.0	–
water	87.1	85.7	79.6	83.5	73.3	72.9

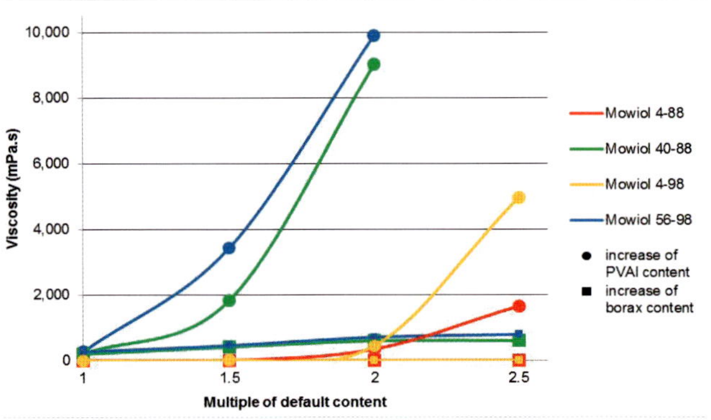

FIG. 1 Viscosity of gels prepared with distilled water.

pH of the aqueous phase on the final gel viscosity, the pH was adjusted to 4–10 with HCl or NaOH. Sample processing remained unchanged and the sample compositions corresponded with the lowest substance ratio.

As a result of the previous experiment, Mowiol 40-88 was chosen as a PVAl suitable for preparation of gels with microemulsion (ME) content. Three pairs of microemulsions were used: each pair contained the same amount of organic solvent but the surfactant was either anionic sodium lauryl sulphate (SDS) with co-surfactant 1-pentanol or non-ionic Tween 20 (TW) (Table 1). We tested seven methods of preparation that differed in the liquid in which the gels were prepared (water, surfactant aqueous solution, microemulsion) and in the order in which the substances were added into the mixture. One method for each microemulsion type was chosen after the result evaluation. The chosen methods were used to prepare two sets of each gel with microemulsion where the PVAl or borax content was increased linearly while the volume of other components remained constant. The maximum PVAl or borax content was twice as high as the starting concentration. Gel characteristics were measured with a vibrational viscometer and pH meter with a combination electrode.

Results and discussion

The viscosity of gels prepared from distilled water increases with molar mass and degree of hydrolysis of the PVAl (Fig. 1). Increasing PVAl content results in exponentially increasing viscosity regardless of the Mowiol type. On the other hand, the degree of hydrolysis of the Mowiol does affect viscosity if the borax concentration is increased (Fig. 1). The increase in viscosity is linear in all cases until the borax content is doubled, at which point, the viscosity rapidly decreases for fully hydrolysed Mowiol and remains unchanged for partially hydrolysed Mowiol.

Changes in pH of the aqueous phase do not have any noticeable influence on dissolving PVAl. The final gel pH is 8.7–8.9 regardless of PVAl or borax content. The use of fully hydrolysed Mowiol results in a mild decrease of the gel pH in comparison to the use of partially hydrolysed Mowiol. Gel viscosity remained constant regardless of the pH of the aqueous phase. A possible explanation is the buffering capacity of borax.

The amount of produced gel with ME content in the research depended first and foremost on the type and amount of organic solvent in the continuous phase. When the microemulsion contained only xylene (immiscible in water) the whole organic phase was trapped in micelles, and a gel could be produced by utilizing the whole sample with most of the tested methods. These gels had a similar consistency and elasticity as gels prepared from distilled water but they differed in transparency: the gels with ME were turbid, which may indicate the changes of microemulsion structure or the changes of polymer solubility, while the gels with water were transparent. The best preparation method seemed to involve dissolving PVAl in a ready-made microemulsion and adding borax afterwards. It was possible to prepare gels from the entire sample using a microemulsion that contained xylene and ethylene glycol (fully miscible in water), but most samples showed a small degree of syneresis within 24 hours and it was not possible to determine whether the microemulsion retained the same composition in the concentrated gel component of such a sample. These gels were less turbid than those with microemulsion xylene but were also less elastic and coherent than gels of distilled water. Gels with an ethylene glycol/xylene microemulsion in later experiments were prepared in the same way as with xylene microemulsions. A microemulsion of ethylene acetate/propylene carbonate contained the largest amount of organic solvents in the continuous phase; both ethylene acetate and propylene carbonate have limited miscibility in water. Gels with a microemulsion of ethylene acetate/propylene carbonate could not be prepared from the whole sample because most of the microemulsion was expelled into the liquid phase. For this reason the ethylene acetate/propylene carbonate microemulsion was not involved in later experiments.

All measurements of the viscosity of gels with microemulsions showed that viscosity depends on the PVAl or borax content just as in gels containing distilled water: the viscosity increases exponentially with linear increase of PVAl content while it increases linearly with linear increase of borax content. If the viscosity measurements of gels with microemulsions are compared, it is apparent that organic solvents in the continuous phase reduce the final viscosity of a gel. However, when the results of mixtures containing the same amount of organic solvents but different types of surfactant are compared, it seems there is no direct relationship between the amount or type of surfactant and viscosity.

Conclusion

Measurements showed that the viscosity of gels increases with rising molecular weight and with the degree of hydrolysis of the chosen poly(vinyl alcohol). These measurements also demonstrated that increasing the amount of poly(vinyl alcohol) causes a non-linear increase of gel-like system viscosity whereas increasing the amount of sodium tetraborate results in a linear increase of viscosity. Different pH values of the initial aqueous phase have no effect on the pH value of the final gels owing to the buffering character of sodium tetraborate; the viscosity of gels is also unaffected by the initial pH of the aqueous phase. The possibility of preparing gels containing microemulsions depends on the amount of organic solvent in the continuous phase, not on the types of surfactants used. The relation between viscosity of gels containing microemulsions and the content of poly(vinyl alcohol) and sodium tetraborate remains the same as suggested above.

Acknowledgements

We would like to express our gratitude to the Czech Ministry of Education, Youth and Sports which supported this project with a grant from the budget of Specific University Research (MSMT No 20-SVV/2016).

References

Angelova, L., Terech, V.P., Natali, I., Dei, L., Carretti, E. and Weiss, R.G. 2011. 'Cosolvent gel-like materials from partially hydrolyzed poly(vinyl acetate)s and borax', *Langmuir* 27(18): 11671–82.

Baglioni, P., Berti, D., Bonini, M., Carretti, E., Dei, L., Fratini, E. and Giorgio, R. 2014. 'Micelle, microemulsions, and gels for the conservation of cultural heritage', *Advances in Colloid and Interface Science* 205: 361–71.

Carretti, E., Grassi, S., Cossalter, M., Natali, I., Caminati, G., Weiss, R.G., Baglioni, P. and Dei, L. 2009a. 'Poly(vinyl alcohol)-borate hydro/cosolvent gels: viscoelastic properties, solubilizing power, and application to art conservation', *Langmuir* 25(15): 8656–62.

Carretti, E., Fratini, E., Berti, D., Dei, L. and Baglioni, P. 2009b. 'Nanoscience for art conservation: oil-in-water microemulsions embedded in a polymeric network for the cleaning of works of art', *AngewandteChemie International Edition* 48(47): 8966–69.

Carretti, E., Natali, I., Matarrese, C., Bracco, P., Weiss, R.G., Baglioni, P., Salvini, A. and Dei, L. 2010. 'A new family of high viscosity polymeric dispersions for cleaning easel paintings', *Journal of Cultural Heritage* 11(4): 373–80.

Giorgi, R. and Carretti, E. 2013. 'Cleaning III: Applications and case studies', in P. Baglioni and D. Chelazzi (eds), *Nanoscience for the Conservation of Works of* Art. London: Royal Society of Chemistry, 225–51.

Grassi, S., Favaro, M., Tomasin, P. and Dei, L. 2009. 'Nanocontainer aqueous systems for removing polymeric materials from marble surfaces: a new and promising tool in cultural heritage conservation', *Journal of Cultural Heritage* 10 (3): 347–55.

Authors' addresses

- Anna Fialová, University of Chemistry and Technology, Prague, Czech Republic (fialovan@vscht.cz)
- Petr Kotlík, University of Chemistry and Technology, Prague, Czech Republic

Using a 'gel' made from nano-lime and nano-silica to repair badly damaged glass mosaic *tesserae*

Monica Martelli Castaldi, Marta Ebbreo, Ilaria Liguori and Simon Elio Leitner

Introduction

The mosaics in the Crypt of the Bishops in the San Gennaro Catacombs of Naples, Italy, one of the most important examples of mosaic art dating back to the sixth century CE, required stabilization (Bisconti 2015). The dire environmental conditions of the catacombs do not favour the conservation of these significant mosaics, so initially the intervention, commissioned by the Pontificia Commissione di Archeologia Sacra, was to be limited to quickly securing the surfaces in anticipation of a full treatment later. However, the emergency intervention itself developed into a complicated task, taking much longer than anticipated and requiring meticulous accuracy. It cannot be considered permanent because of the environmental conditions, which were not addressed. However, it has undoubtedly improved the condition of the surfaces, and further improvement would be possible if treatment were continued. This poster illustrates a particular, and rather surprising, finding obtained by combining nano-silica and nano-lime, which produced a 'gel' that was key to the success of the work, while also enabling the removal of the previous consolidant.

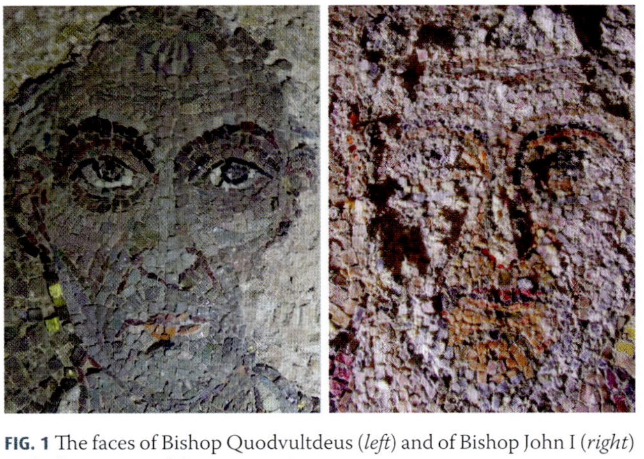

FIG. 1 The faces of Bishop Quodvultdeus (*left*) and of Bishop John I (*right*) at the beginning of the intervention.

Initial condition

The detailed portraits of the bishops in the crypt were created through the careful juxtaposition of tiny glass and marble mosaic *tesserae*, which measure no more than 3–5 mm per side for the face, neck and hands, while for the background and the clothes they measure about 1 cm per side. These nuanced images were created by a careful combination of colours and shades, producing extremely convincing three-dimensionality, as can be seen in the portrait of the Bishop of Carthage, Quodvultdeus (Fig. 1).

Both types of *tesserae* had been severely damaged by efflorescence of salts contained in the tufa support, resulting in detachment from the preparatory layers. In addition, treatments carried out in the past, albeit with the best of intentions, relied heavily on acrylic consolidants, thus saturating the porous preparatory layers and the *tesserae* themselves. These two factors resulted in the disintegration, flaking, pulverization, fracturing and scaling of each *tessera*, preventing the application of a consolidant grouting for this type of surface (Figs 1 and 2).

Testing and treatment

The biggest challenge we faced was to find a way to introduce a new type of consolidant into an acrylic-saturated environment. In order to avoid solubilization and transport of the

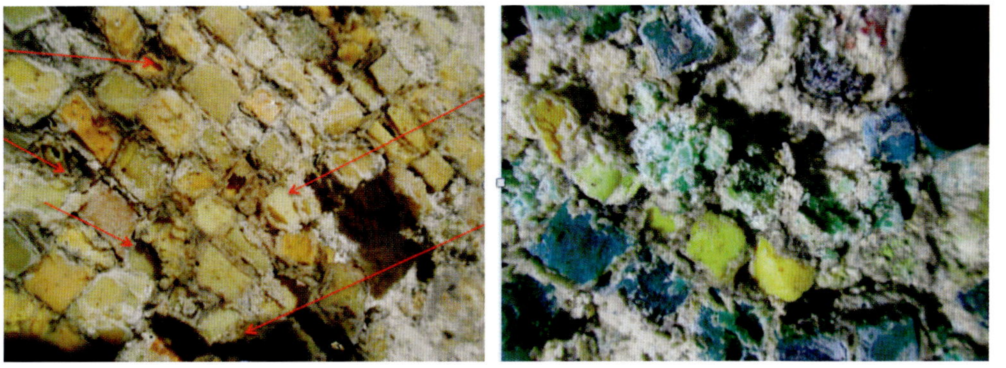

FIG. 2 The surface presenting detachments and various damages of the *tesserae*.

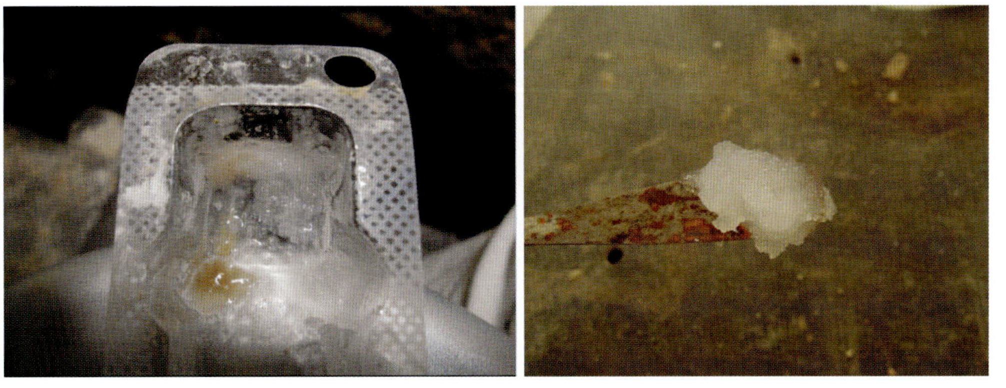

FIG. 3 The 'gel substance' obtained by mixing nano-lime and nano-silica.

acrylic consolidant towards the interior of the mosaic, we attempted to remove it manually, but the *tesserae* were too weak to withstand this. We were therefore forced to consolidate the disintegrated material first, to reinforce the *tesserae*, before proceeding with their re-adhesion and a possible reduction or removal of the acrylic.

Our first attempts were incredibly disappointing. However, we continued, establishing the key points on which we could base a treatment:

- No use of organic polymers as consolidants.
- A requirement to use the least toxic materials, due to the poor air circulation in the catacomb.

Regarding the disintegration and disaggregation problems of both the mortar and the *tesserae*, and the existing damage to the latter, we considered consolidants based on nano-materials, particularly lime[1] for the consolidation of the plaster and the marble *tesserae*, and silica[2] for the consolidation of glass *tesserae*. Not having much time at our disposal, due to the desperate condition of the *tesserae*, the trials and choices were based on personal experience, literature research into similar contexts and verbal communication with fellow conservation professionals.

Results and findings

Our research showed that in recent years, nano-lime has been the subject of many studies, sometimes in subsequent combination with silicic acid esters (Camaiti *et al.* 2008; Borsoi *et al.* 2013). This mixed formulation allows an acceleration of the hydrolysis reaction, resulting in a more dense material, which is able to set even in 'large voids and areas of delamination' (D'Armada and Hirst 2012), where ethyl silicate alone would not be sufficient. The nano-lime[3] also acts as an adhesion promoter ('coupling agent') for silicic acid esters (D'Armada and Hirst 2012). In our case the nano-materials had been used previously or directly mixed together. They have shown reduced rates of reaction, and a rapid evaporation of the solvents, compared to the 2–3 weeks hydrolysis reaction time required by ethyl silicate. The use of colloidal silica in place of ethyl silicate also allowed its use in a humid environment and at very low temperatures.[4] This led to the following treatment decisions:

- for the consolidation of the original mortars and *tesserae*, both nano-lime and nano-silica were used individually;
- for the consolidation of areas of loss, and detached plaster and *tesserae*, various hydraulic mortars were prepared and trialled until we obtained one of distinctly weaker consistency than the original mortar, to mimic the original material in its response to water and salts;
- the same was done for the plasters, to encourage the migration of salts into the replaced mortar.

It was also necessary to find an inorganic material that could be applied at a significant thickness, or to fill voids, to replace areas in the exploded and scaled glass *tesserae*, and to reattach them to the original mortar bed. Products already on the market and tests carried out in Germany were considered, but ultimately it was by directly combining nano-lime and nano-silica that we obtained the most unexpected of results. In addition to the anticipated formation of a nano-particle 'gel' (which occurred particularly fast and proved to have an excellent consistency) (Fig. 3), nano-lime also caused unanticipated swelling of the previously acrylic consolidant (due to both the isopropanol content and the alkaline pH of the nano-lime). The swollen acrylic remained sufficiently flexible and consistent to be removed manually during the consolidation, even from the back of the *tesserae*.

This gel provided the perfect solution as its consistency could be altered according to our needs, and it could either be created *in situ* and on the back of the tiny scales, by injecting both components in the same place, or pre-prepared and applied with a tiny spatula. A light pressure was applied on occasion. Finally, once the disintegrated *tesserae* were consolidated, the gaps at the back and the interstitial spaces were filled by using the hydraulic mortar prepared as above, applied with a spatula or injected.

Conclusion

Considering the initial state of the mosaic surface, the results were far more satisfactory than expected. The 'gel' obtained from the combination of nano-lime and nano-silica proved to be the key to the success for the challenging and delicate conservation treatment of this mosaic. We are still seeking a correct description for the 'gel' substance we obtained and we hope that this poster stimulates interest in studies to improve the methodologies which we adopted for the consolidation of the scales and for the removal of a resin consolidant from very weak and damaged surfaces.

Notes

1. NanoRestore (from CTS), a dispersion of nano-particles of lime in denaturated isopropanol: http://www.ctseurope.com/dettaglio-news.php?id=175.
2. Nanoestel (from CTS), a colloidal dispersion in water of nano-particles of silica: http://www.ctseurope.com/en/dettaglio-news.php?id=276; http://www.ctseurope.com/en/dettaglio-news.php?id=223.
3. Calosil (IBZ): http://www.ibz-freiberg.de/downloads/pdf/produkte/tm/eng/CaLoSiL_E_IP_NP.pdf.
4. http://www.ctseurope.com/en/dettaglio-news.php?id=223.

References

Bisconti, F. 2015. 'Napoli. Catacombe di S. Gennaro. Cripta dei Vescovi. Restauri ultimi', *Rivista di Archeologia Cristiana* 91: 7–34.

Borsoi, G., Veiga, R. and Santos Silva, A. 2013. 'Effect of nano-structured lime-based and silica-based products on the consolidation of historical renders', in paper presented at the *3rd Historic Mortars Conference, 11–14 September 2013, Glasgow*.

Camaiti, M., Dellantonio, G. and Pittertschatscher, M. 2008. 'Restauro dello stemma affrescato del Cardinale Bernardo Cles presso il Castello del Buonconsiglio a Trento: nuove soluzioni per il consolidamento di intonaci dipinti distaccati dal supporto murario', in *Atti del XIV Convegno di Studi Scienza e Beni Culturali "Restaurare i restauri – Metodi, Compatibilità,* Cantieri", Bressanone 24–27 giugno 2008. Marghera: Arcadia Ricerche Ed., 231–41.

D'Armada, P. and Hirst, E. 2012. 'Nanolime for consolidation of plaster and stone', *Journal of Architectural Conservation* 18(1): 63– 80.

Authors' addresses

- Monica Martelli Castaldi, Vietri sul Mare, Italy (m.martelli.c@gmail.com)
- Marta Ebbreo, Palermo, Italy (martaebbreo@gmail.com)
- Ilaria Liguori, Cava de' Tirreni, Italy (liguori.ilaria@gmail.com)
- Simon Elio Leitner, Cava dei Tirreni, Italy (elioleitner@gmail.com)

Improved PVA gels for the cleaning of paint surfaces

Chiara Riedo, Giulia Rollo, Dominique Scalarone and Oscar Chiantore

Introduction

Cleaning techniques typically involve the use of free solvents. This method is not free of inconvenience, due to the penetration of liquids through the surface, potentially causing swelling of pictorial layers and undesired removal of original materials. In addition, solvents can be toxic for users and liquid residues may be retained in the inner layers (Dorge 2004). Polymer gels used as thickeners were introduced by Richard Wolbers in the 1980s for circumventing the drawbacks of free solvents.

A wide selection of gel materials has been studied for application in this field, and poly(vinyl alcohol) (PVA) as a gelling agent is one of the most interesting, due to its capability of forming hydrogel systems, its ready availability, and its biodegradability. In particular PVA-borate gel systems have received much attention because of their interesting chemical and physical properties. PVA in aqueous solution reacts with tetraborate ions giving a dynamic crosslinking known as di-diol complexation (Sinton 1987; Lin *et al.* 2003). The mechanical properties of PVA-borate systems make them suitable for cleaning because they are stable, transparent, and relatively easy to place and remove from the surface (Angelova *et al.* 2011; Natali *et al.* 2011). Organic solvents miscible with water may be added to the formulation (Natali *et al.* 2011). To improve mechanical properties and compatibility with the co-solvent, it has been shown that poly(ethylene oxide) (PEO) can be added to the formulations. PEO is a water-soluble and biocompatible thermoplastic polymer and the literature suggests that it causes an increment of the pores size in PVA/borax hydrogels (Zhang and Ye 2011) while simultaneously altering the mechanical properties of the hydrogel. For these reasons PEO could be a suitable additive for improving the physical characteristics of PVA/borax hydrogels. Furthermore, positive interactions with the miscible organic solvents used in conservation treatments are also possible. PEO was added in concentration from 0 to 3% directly in PVA solution prior to the addition of borax.

The structure of all gels was investigated with rheological techniques and differential scanning calorimetry (DSC) to find correlations between stability behavior and physical properties (Riedo *et al.* 2015). Of primary concern for conservators is the absence of gel residues from the treated surfaces; a number of studies have been devoted to methods for identifying residues (Dorge 2004). For PVA-borax gels, analyses of residues have been performed by Fourier transform infrared (FTIR) spectroscopy (Carretti *et al.* 2010), chemical tagging of the polymers with a fluorophore (Angelova 2013; Angelova *et al.* 2015) and borate detection by inductively coupled plasmas mass spectrometry (ICP–MS). In the present work an analytical method based on FTIR and pyrolysis–gas chromatography–mass spectrometry (Py–GC–MS) was used to identify residues on mock-ups and on works of art after cleaning treatments. These results highlight the importance of clearing the surfaces after the treatment to completely remove gel residues along with traces of the layers being removed.

Materials and methods

Poly(vinyl alcohol) (86–89% hydrolysed, MW 100,000, Fluka), poly(ethyleneoxide) (PEO, MW 35,000, Sigma Aldrich) sodium tetraborate decahydrate (Sigma Aldrich) and acetone (Sigma Aldrich) were used as received. Gels were prepared with different amounts of PVA, borax and PEO, with deionized water and, in selected cases, the addition of acetone. All formulations were prepared by stirring the polymers in deionized water at 70–80 °C until complete dissolution. Sodium tetraborate decahydrate (borax) (6 wt%) was solubilized in deionized water. The organic solvent, when present, was added to the polymer solution, followed by the proper amount of the borax solution.

The hydration in the gels was determined by DSC measurements (Lin *et al.* 2012) carried out to 30 °C with a heating rate of 5 °C/min. The endothermic water melting peaks were

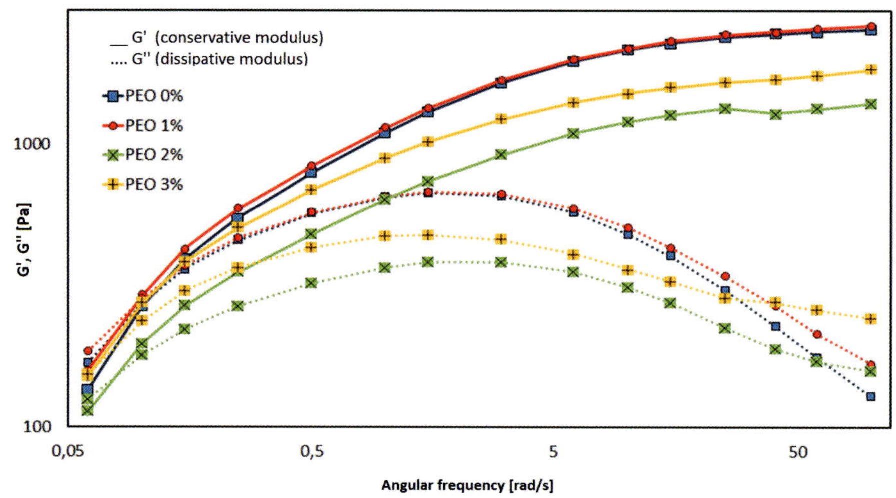

FIG. 1 Frequency-sweep tests performed on PVA-PEO/borax system with 3% PVA, 0.6% borax, 30% acetone and variable amounts of PEO.

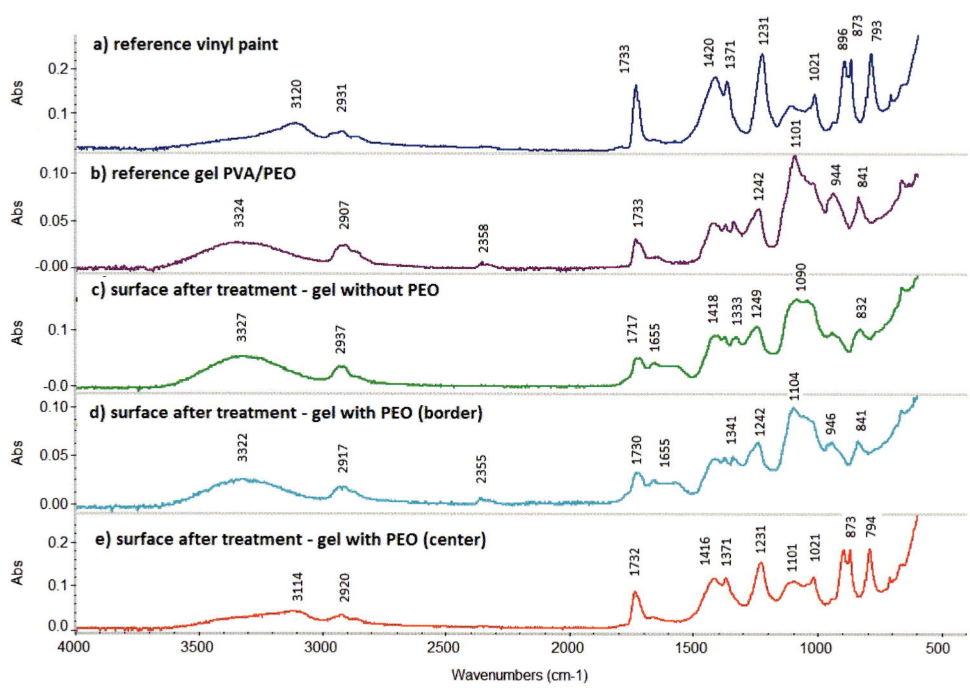

FIG. 2 ATR-FTIR spectra of model samples after cleaning tests performed with different gels (PVA alone and PVA/PEO), and without clearance after gel removal.

integrated to calculate the enthalpy changes necessary to calculate the free water index (FWI) (Carretti *et al.* 2009). A frequency sweep test was carried out on a rheometer equipped with 20 mm plate geometry and Peltier plate temperature control, over the frequency range of 0.005–15 Hz and stress amplitude of 2%. Amplitude sweep tests were carried out to ensure that measures were performed within the linear viscoelastic region.

Mock-ups for tests and evaluation of residues were prepared with a layer of vinyl paint (Maimeri, Italy, with inorganic iron oxide pigment) coated with dammar varnish on a medium density fibreboard (MDF) wood composite support, without priming, and exposed to accelerated light aging for 850 hours. Light aging (Q-Lab Corporation, UK) was carried out with a xenon lamp and Window Q filter, simulating midday summer sunlight streaming through a window. Cleaning tests were performed with and without clearing with cotton swabs, after the gel had been mechanically removed. Cotton swabs were soaked with hexane to prevent mechanical damage of the surface, which could occur if dry cotton swabs are used.

Following the cleaning tests, samples taken from treated areas were analysed with Py–GC–MS. The pyrolizer Pyroprobe 1500 (CDS Analytical, USA) is directly connected with the 6890N Network GC System and 5973 Network MASS Selective Detector (Agilent Technologies, USA). Mass spectra were collected in single ion monitoring mode,

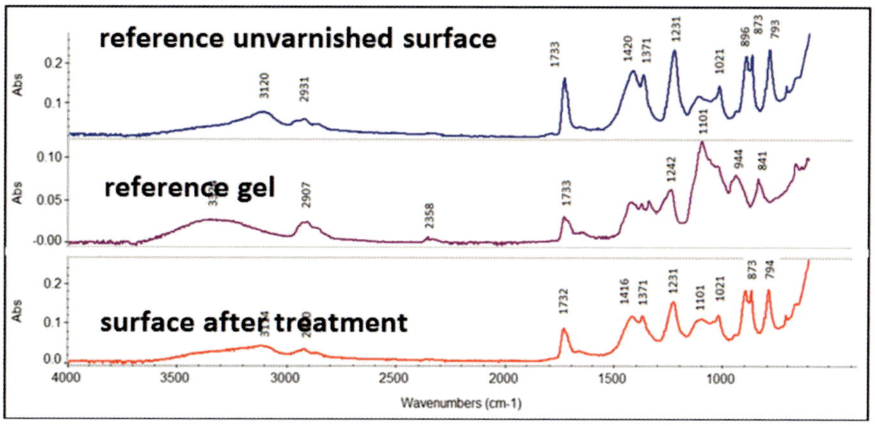

FIG. 3 ATR–FTIR spectra of reference unvarnished surface, reference gel and area treated with gel and cleared with a hexane-soaked cotton swab.

selecting the ions characteristic of the gel polymeric matrix. Attenuated total reflectance (ATR)–FTIR spectra were also collected (FT-IR Nexus Thermo Nicolet, USA).

Results and discussion

The rheological measurements were used to obtain information on structural properties and to study the effect of PEO on the mechanical properties of the gels. The crossover parameters (crossover frequency ω_c and crossover modulus G_c) and the asymptotic value of storage modulus (G'_0) were obtained from the frequency sweep data (Fig. 1) and the apparent relaxation times could be calculated ($\tau_c = 2\pi/\omega_c$). The G'_0 value is directly related to the gel's peelability, and all the formulations with 3% of PVA and 1–3% of PEO showed values above 400 Pa, the limit suggested in the literature (Natali *et al.* 2011) for obtaining peeling.

The increment of PVA concentration gives rise to a drastic increase in G'_0 and relaxation time. The same trend is observed when an organic co-solvent is added. Since G'_0 is related to the entanglement/crosslinking density (Schubert *et al.* 2003), it is suggested that the increased stiffness of the gels containing acetone is due to a larger amount of tetraborate complexation, a phenomenon confirmed by others (Angelova *et al.* 2011) with ^{11}B nuclear magnetic resonance (NMR) studies. The gel with larger amount of acetone is more difficult to apply and mould on the surface to be cleaned, in comparison with the gel containing only water.

When PEO is added, the rheological properties of the gels undergo large changes, as shown in Fig. 1. A reduction of the gel elastic modulus was observed, suggesting that PEO may act as porogen, producing an increase in the mesh size of the network and a partial disentanglement of PVA chains. Simultaneously, an increase of the relaxation times has been observed (Riedo *et al.* 2015). The free water index (FWI) values obtained using DSC were used to check the effect of the PEO addition. The water state is related to the internal microstructure of the polymer network and to the extent of interactions which can develop with the gellant polymer

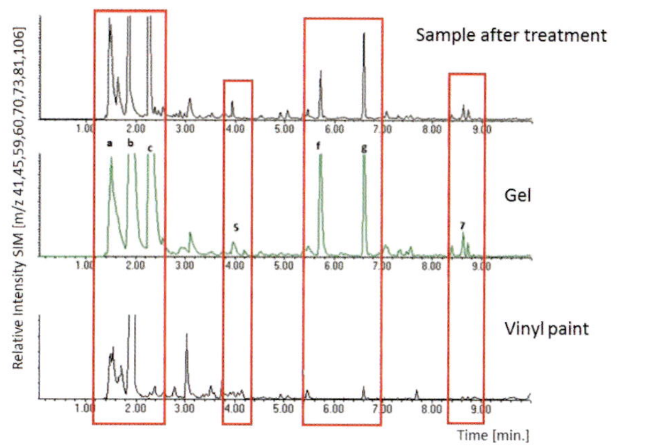

FIG. 4 Py–GC–MS of sample from the mock-up surface after gel removal without clearance, reference gel and vinyl paint.

molecules. When increasing the PEO concentration no changes were observed in the trend of FWI (data not shown), meaning that PEO may not interact with the water dynamics of the PVA-borax system. This observation, together with the rheological data, suggests that PEO is not involved in the formation of the network, but only affects the mesh size and the viscosity of the gel, giving an easily mouldable material.

Application and removal tests performed on mock-ups demonstrate that the addition of PEO makes the workability and application on the surface easier. In contrast to other studies, ATR–FTIR data collected from the treated areas without clearing show spectra superimposable with the spectrum of a reference gel, in particular when the gel without PEO is used (Fig. 2c), indicating that gel remains on the surface after the cleaning action. Gels with PEO show better removibility, in particular from the centre of the treated area (Fig. 2e). When a cotton swab soaked with hexane was used to clear the area after the gel removal, the ATR–FTIR spectrum of the resulting surface (Fig. 3) was comparable to that of the unvarnished surfaces, showing that the gel is capable of efficiently removing the varnish.

Py–GC–MS analysis confirms the presence of residues on our treatment, as shown in Fig. 4, when the pyrogram of a reference dry PVA/PEO gel is compared with that of a sample collected from the surface of the mock-up after removal of the gel and the reference vinyl paint. In the pyrogram of the sample collected from the cleaned area several peaks related to the gel are observed. Only peaks a, b, and g are due to the vinyl paint, therefore we can assume other peaks (c, 5, f, 7) are markers for the gel. Traces of gel were detected by Py–GC–MS both for PVA and PVA/PEO gels. After clearance with cotton swabs only trace amounts of gel markers were revealed, indicating that micro-residues are still present. Since mass spectra of these pyrograms were collected in single ion monitoring mode, the analysis was selective with respect to the pyrolysis products of the constituent polymers of the gel, in particular those of PVA, the majority component.

Conclusion

Rheological and DSC results indicate that in gels containing organic solvents or PEO the stiffness of the gel and the response to external stimuli can be controlled. Addition of PEO allows the gel to be easily moulded and removed from the surface. Cleaning tests performed on mock-ups show that the gel containing PEO is easily removable from the surface, as confirmed by FTIR analysis performed on the treated area. On the other hand, without further clearance with cotton swabs, residues of gel were detected with FTIR analysis of the treated area, in particular when a gel without PEO is used. Py–GC–MS in SIM mode, being selective for the characteristic gel markers, shows the presence of gel residues on the treated surfaces. If a cotton swab is used to clear the treated area after the application of the gel, FTIR is not sensitive enough to detect gel residues, but Py–GC–MS still reveals gel in trace amounts. These results contrast with those reported by others and highlight the need for careful evaluation of the surface to be gel-cleaned – some types of paints and varnishes may be more susceptible to residue adhesion and therefore retention than others.

References

Angelova, L.V., Terech, P., Natali, I., Dei, L., Carretti, E. and Weiss, R.G. 2011. 'Cosolvent gel-like materials from partially hydrolyzed poly(vinyl acetate)s and borax', *Langmuir* 27: 11671–82.

Angelova, L.V. 2013. *Gels from Borate-crosslinked Partially Hydrolized Poly(vinyl acetate)s: Characterization of Physical and Chemical Properties and Application in Art Conservation*, unpublished PhD dissertation, Georgetown University, Washington, DC.

Angelova, L.V., Berrie, B.H, de Ghetaldi, K., Kerr, A. and Weiss, R.G. 2015. 'Partially hydrolyzed poly(vinyl acetate)-borax-based gel-like materials for conservation of art: characterization and application', *Studies in Conservation* 60: 227–44.

Carretti, E., Grassi, S., Cossalter, M., Natali, I., Caminati, G., Weiss, R.G., Baglioni, P. and Dei, L. 2009. 'Poly(vinyl alcohol)-borate hydro/cosolvent gels: viscoelastic properties, solubilizing power, and application to art conservation', *Langmuir* 25: 8656–62.

Carretti, E., Natali, I., Matarrese, C, Bracco, P., Weiss, R.G., Baglioni, P., Salvini, A. and Dei, L. 2010. 'A new family of high viscosity polymeric dispersion for cleaning easel paintings', *Journal of Cultural Heritage* 11: 373–80.

Dorge, V. (ed.). 2004. *Solvent Gels for the Cleaning of Works of Art: The Residue Question*. Los Angeles: Getty Conservation Institute.

Lin, H.L., Liu, W.H., Shen, K.S., LeonYu, T. and Cheng, C.H. 2003. 'Weak gel behaviour of poly(vinyl alcohol)-borax aqueous solutions', *Journal of Polymer Research* 10: 171–9.

Lin, H., Dan, W. and Dan, N. 2012. 'The water state in crosslinked poly(vinyl alcohol)–collagen hydrogel and its swelling behavior', *Journal of Applied Polymer Science* 123: 2753–61.

Natali, I., Carretti, E., Angelova, L., Baglioni, P., Weiss, R.G. and Dei, L. 2011. 'Structural and mechanical properties of "peelable" organoaqueous dispersions with partially hydrolyzed poly(vinyl acetate)-borate networks: applications to cleaning painted surfaces', *Langmuir* 27: 13226–35.

Riedo, C., Caldera, F., Poli, T. and Chiantore, O. 2015. 'Poly(vinylalcohol)-borate hydrogels with improved features for the cleaning of cultural heritage surfaces', *Heritage Science* 3: 23.

Schubert, B.A., Kaler, E.W. and Wagner, N.J. 2003. 'The microstructure and rheology of mixed cationic/anionic wormlike micelle', *Langmuir* 19: 4079–89.

Sinton, S.W. 1987. 'Complexation chemistry of sodium borate with poly(vinyl alcohol) and small diols: a boron-11 NMR study', *Macromolecules* 20: 2430–41.

Zhang, Y. and Ye, L. 2011. 'Improvement of poly(vinyl alcohol) hydrogel by using poly(ethylene glycol) as porogen', *Polymer-Plastics Technology and Engineering* 50: 776–82.

Authors' addresses

- Chiara Riedo, Università degli Studi di Torino, Dipartimento di Chimica, Turin, Italy, and INSTM Consorzio Interuniversitario di Scienza e Tecnologia dei Materiali, Turin, Italy (chiara.riedo@unito.it)
- Giulia Rollo, Koinè Conservazione Beni Culturali,Turin, Italy
- Dominique Scalarone, Università degli Studi di Torino, Dipartimento di Chimica, Turin, Italy, and INSTM Consorzio Interuniversitario di Scienza e Tecnologia dei Materiali, Turin, Italy
- Oscar Chiantore, Università degli Studi di Torino, Dipartimento di Chimica, Turin, Italy, and INSTM Consorzio Interuniversitario di Scienza e Tecnologia dei Materiali, Turin, Italy (oscar.chiantore@unito.it)

Trials of agar gels and task-specific salts for the electrochemical reduction of silver sulphide on silver leaf

Joana São João, Luís C. Branco and Sara Leite Fragoso

Introduction

Silver leaf used for decorative elements can be found in European sculptures, paintings, and religious objects made between the 16th and 18th centuries, with most having been produced during the Baroque period. The technique was frequently used as an alternative to gold leaf, since silver is less expensive. These objects are composed of a complex system of layers, starting with a wood support, followed by the preparation and bole layers, on top of which silver leaf was applied and normally coated with shellac (Thompson 1960; Mactaggart and Mactaggart 2002). The shellac provided a gold colour. In some cases the silver foil was coated with a transparent varnish to preserve its natural colour. Generally, the visible use of gold or silver was associated with very important figures or objects, therefore once the object has lost this visual quality, its value becomes compromised (Dorge and Howlett 1998; Sandu 2010).

A black corrosion layer usually develops on silver over time as a result of chemical reactions with atmospheric pollutants. The most common and worrying corrosion product is silver sulphide (Ag_2S) because it is more stable than metallic silver. When silver becomes contaminated with sulphur it loses its cohesion and becomes powdery (Portel 1991). At present, there is no efficient method for reducing corrosion on silver leaf: treatments consist of the application of a new silver leaf (Portel 1991).

Although corrosion on silver leaf is similar to that on a solid silver object, the treatment approach needs to be rather different. Silver leaf is a particularly thin (1 µm to 500 nm) and delicate surface which is extremely sensitive to abrasion and water, making the existing methods of chemical reduction for solid silver unsuitable for these objects. The oxidation mechanism is the same as that proposed for silver objects in a humid atmosphere with relative humidity (RH) exceeding 60%: a monolayer of water is absorbed onto the surface, which triggers a reaction between silver and oxygen, producing silver oxides and hydroxides (Ag_2O and AgOH), respectively. After these are formed, only a small amount of sulphur-based pollutant, such as hydrogen sulphide (H_2S) is enough to initiate the formation of silver sulphide (Ag_2S) (MacLeod 1979; Ankersmit *et al.* 2000; Lin *et al.* 2013). The electrochemical reduction of silver reduces oxidized silver (Ag+) to its metallic form (Ag) using aluminium as the electron donor, without removing any silver from the surface (Costa 2002; Doménech-Carbó *et al.* 2009):

$$Al\ (s) + 3Ag^{+}(aq) \rightleftharpoons 3Ag\ (s) + Al^{3+}\ (aq)$$

This study focused on objects containing silver leaf as a decorative element, found in Baroque and Mannerist religious objects including paintings, sculptures and altarpieces. The approach involves the application of a methodology already used for silver objects – electrochemical reduction – adjusted for very delicate surfaces which are extremely sensitive both to abrasion and water, which promotes the formation of new corrosion. An agar gel, widely used in conservation treatments for its ability to contain solvents and formulate a rigid gel, was combined with inorganic salts and ionic liquids (ILs), to perform the reduction reaction. ILs are low melting point organic salts, usually composed of an organic cation and an inorganic anion: they have broad applications in electrochemistry due to their high conductivity, wide electrochemical spectrum and large electrochemical window. Additionally, they are environmentally benign since they have a a low vapour pressure which makes them non-volatile, and it is possible to tune some of their physicochemical properties by selective combination of cations and anions (Galiński 2006; Freemantle 2009; Branco 2015).

Experimental procedure

Mock-ups were produced with the same layers found in real objects (Fig. 1), using commercially primed wood panels (PER Belle Arti Gesso primed), which had been silvered

TABLE 1 Results of reduction with agar gels.

Inorganic salts	Chemical structure	
	Cation	Anion
Sodium carbonate (Na_2CO_3)	Na+	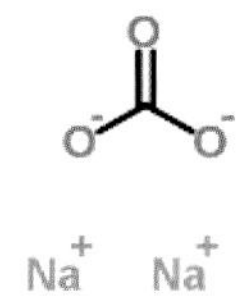
Sodium dicyanamide (NaDCA)	Na+	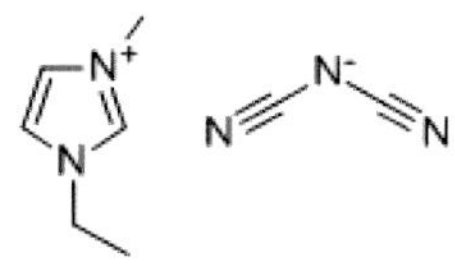
Sodium triflate (NaTfO)	Na+	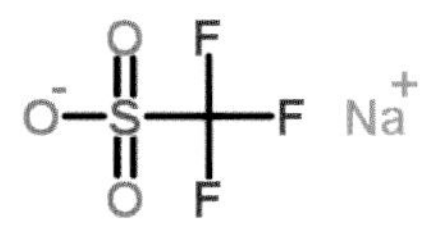
Lithium triflate (LiTfO)	Li+	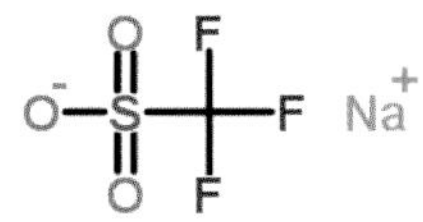
Lithium bis (trifluoromethanesulfonyl) imide ($LiNTf_2$)	Li+ 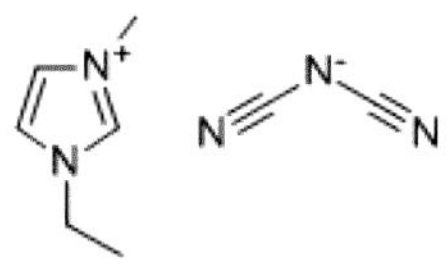	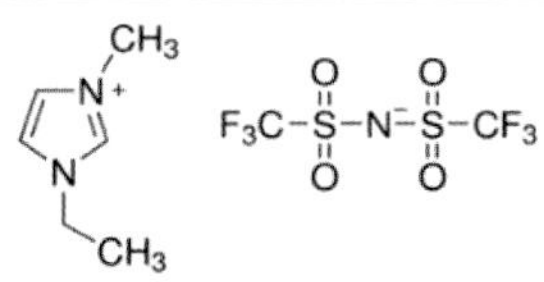
Lithium tetrafluoroborate, $LiBF_4$	Li+	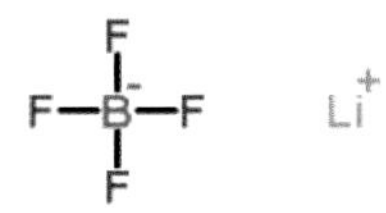
Potassium thiocyanate (KSCN)	K+	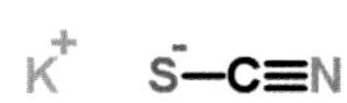
Ionic liquid	**Cation**	**Anion**
1- (2-hydroxy-ethyl) -3-methylimidazolium dianamide [C_2OHMIM][DCA]	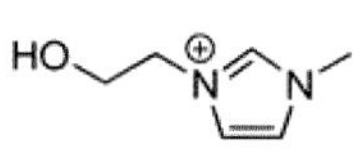	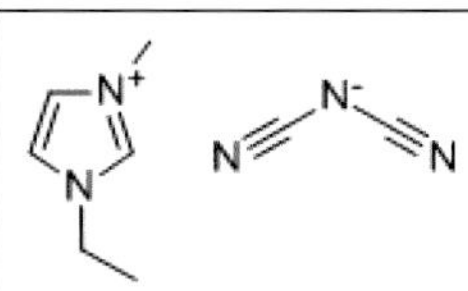
1-ethyl-3-methylimidazolium dianamide [EMIM][DCA]	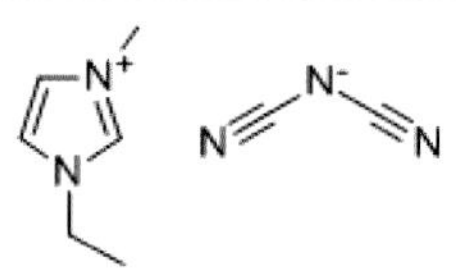	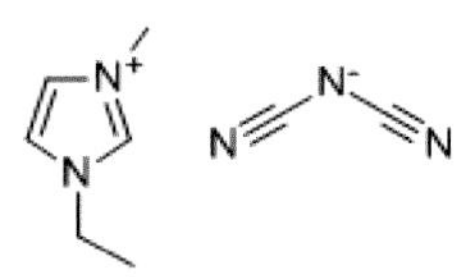
1-butyl-3-methylimidazolium dicyanamide [BDMIM][BF_4]	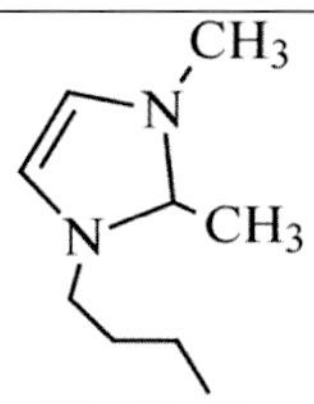	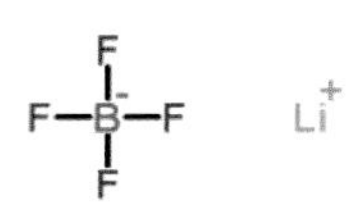
1-ethyl-3-methylimidazolium tetracyanoborate [EMIM][$B(CN)_4$]	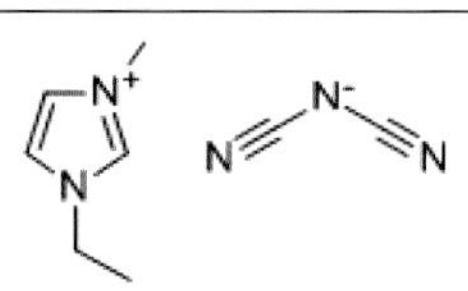	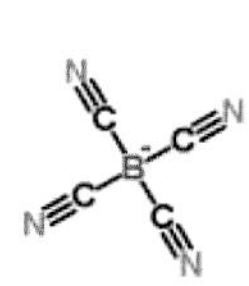

TABLE 2 XPS quantification results.

Gel and inorganic salt	Conductivity (mS/cm) at 25 °C	Application conditions	Results
Agar	0.24	60 s at 3 V	Slight surface reduction, some loss
Agar + Na_2CO_3	12.45	60 s at 3 V	Slight surface reduction (brownish colour)
Agar + NaDCA	28.9	60 s at 3 V	Very intense surface reduction, large amount of losses
Agar + NaTfO	9.78	60 s at 3 V	Very intense surface reduction, no losses
Agar + LiTfO	11	60 s at 3 V	Very intense surface reduction, no losses
Agar + $LiNTF_2$	7.56	60 s at 3 V	Slight surface reduction, no loss
Agar + $LiBF_4$	19	60 s at 3 V	Slight surface reduction, large amount of losses
Agar + KSCN	13	60 s at 3 V	Slight surface reduction, no loss
Gel and IL	**Conductivity (mS/cm) at 25 °C**	**Application conditions**	**Results**
Agar + [C_2OHMIM] [DCA]	25	60 s at 3 V	Good surface reduction, minor losses
Agar + [C_2OHMIM] [DCA] + 10% H_2O	25	60 s at 3 V	Good surface reduction, minor losses
Agar + [EMIM][$B(CN)_4$]	6.67	15 s at 3 V	Good surface reduction, minor losses
Agar + [EMIM][DCA]	27	5 s at 3 V	Very intense surface reduction, no losses

TABLE 3 Additives used on the gels.

Sample	C (%)	O (%)	Ag (%)	S (%)	N (%)
Oxidized sample	72.6	11.0	11.8	4.6	–
Agar with [EMIM][DCA]	71.0	18.2	8.2	–	2.2

(in 2013). At the time of the experiment, these already possessed a layer of natural oxidation. Each plate was divided into several areas, in order to test the methodology. Gels were prepared using a high purity agar, a 6% w/v aqueous solution was prepared consisting of 3% additive and 3% agar, which was heated to 70 °C on a stirrer hotplate until a homogenous solution was obtained. The ILs used (Table 1) had to be in solid form, while the excess of water evaporated. The solution was allowed to cool overnight to form a rigid gel.

Electrochemical reduction was carried out using the agar gels as electrolytes, placing a commercial aluminium foil between the gel and the surface of the silver leaf, to facilitate electrical contact, and applying an electrical current through a power supply using 3 V to accelerate the reaction (Fig. 2).

X-ray photoelectron spectroscopy (XPS)

This was performed on only two samples: one before cleaning and the sample with the best reduction of oxidation. The samples were subjected to sputtering prior to analysis. The data were acquired at 40 eV in 0.5 eV steps, with a residence time of 0.5 s per channel.

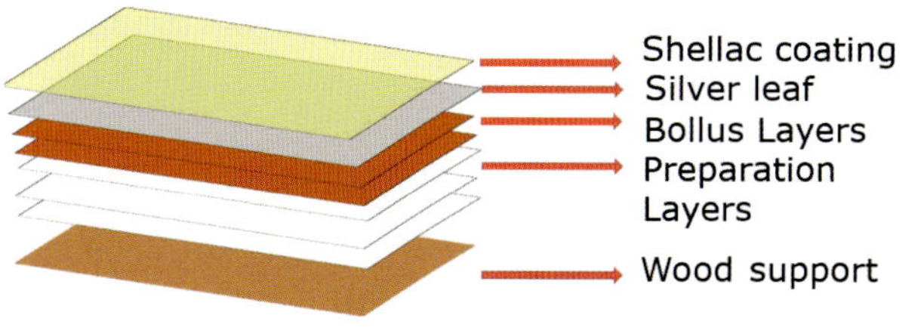

FIG. 1 Layers in the mock-up.

Results and discussion

Twelve agar gels with added inorganic salts or ILs were tested on the oxidized silver leaf mock-ups (Table 2): one gel prepared only with water (reference); seven gels prepared with inorganic salts; and seven gels prepared with ionic liquids. The conductivities of the solutions were measured with an Ion conductivity meter before they were gelled. Application was trialled at 3 V for 1 minute and then modified depending on the additive. Some loss of silver foil occurred during the electrochemical process, especially with gels containing highly conductive additives such as sodium dicyanamide (DCA). It was necessary to adjust the application conditions to use these salts safely.

After optimization of the application conditions, the best reduction results were observed on samples treated with

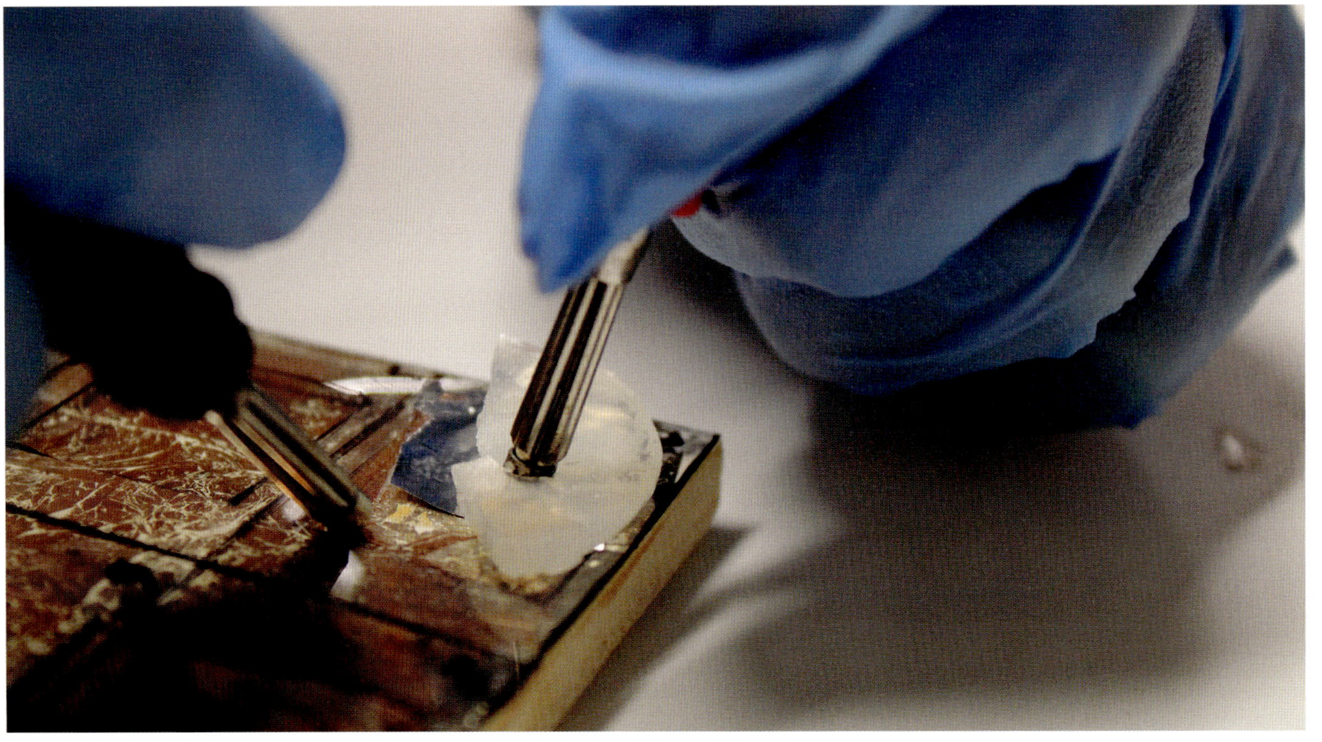

FIG. 2 Application to a mock-up.

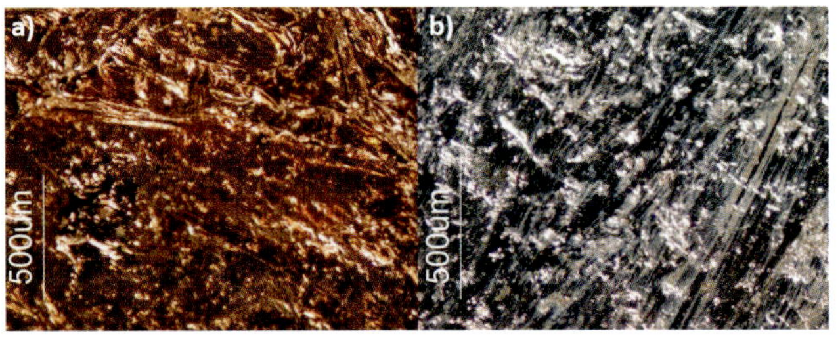

FIG. 3 Reduction of blackened silver using agar gel with sodium triflate (a) before application (b) after application. Dark-field image.

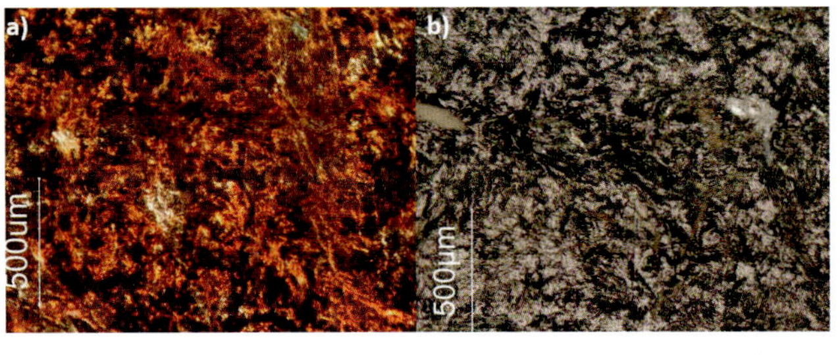

FIG. 4 Reduction of blackened silver using agar gel with EMIM and DCA (a) before application (b) after application. Dark-field image.

agar prepared with sodium triflate (NaTfO), an inorganic salt, and agar prepared with [EMIM][DCA] (an ionic liquid), which reduced the corrosion after only a 5-second application. These examples were examined with optical microscopy using a dark-field objective (Figs 3 and 4) before and after the electrochemical reduction. Both of these treatments proved satisfactory because silver reduction occurred without damaging the foil, providing a very shiny surface, especially on the sample reduced with the [EMIM][DCA]-containing agar gel. The shellac coating was removed during the reduction process, which raises some ethical concerns in real objects, although replacement of the aged shellac layer is a common procedure during the more typical treatment of such objects. The best sample was quantified using XPS to detect sulphur that might still remain after reduction (Table 3). Sulphur was no longer present, however a small amount of sodium was found.

Conclusion

Electrolyte gels prepared with the inorganic salt sodium triflate, or ILs such as [EMIM][DCA], performed best. These additives play a crucial role in electrochemical reduction, and it is necessary to adjust the conditions according to the application. The agar gels were efficient as electrolyte carriers, they are low cost and easy to prepare, but there is a physical manipulation problem due to their low resistance and tendency to crack, which makes the process of applying the electric potential more complicated, especially with non-flat objects. This methodology still needs some optimization for its use on real objects, such as the utilization of a more precise instrument with suitable electrodes for this type of gel, and some optimization to the gel itself, to produce a thinner, more flexible film.

Materials and suppliers

- High purity agar: Sigma Aldrich, Sigma-Aldrich Quimica, S.L.Sintra, Portugal
- ILs and inorganic salts with 98% purity: Solchemar, Rua 5 de outubro, 21c 1°e 7580-128 Alcácer do Sal, Portugal

References

Ankersmit, H., Noble, G., Ridge, L., Stirling, D., Tennent, N.H. and Watts, S. 2000. 'The protection of silver collections from tarnishing', in A. Roy and P. Smith (eds), *Tradition and Innovation: Advances in Conservation*. London: IIC, 7–13.

Branco, L. 2015. 'Líquidos iónicos: aplicações e perspectivas futuras', *Química* 139(5): 15–22.

Costa, V. 2002. 'Electrochemistry as a conservation tool', in J.H. Townsend, K. Eremin and A. Adriaens (eds), *Conservation Science*. London: Archetype Publications, 88–95.

Doménech-Carbó, V., Doménech-Carbó, A. and Costa, M.T. 2009. 'Electrochemical basis of corrosion of cultural objects' in. F. Scholtz (ed.) *Electrochemical Methods in Archaeometry: Conservation and Restoration*. New York: Springer, 123–33.

Dorge, V. and Howlett, F. 1998. *Painted Wood: History and Conservation*. Los Angeles: Getty Conservation Institute.

Freemantle, M. 2009. *An Introduction to Ionic Liquids*. London: RSC Publishing.

Galiński, M. 2006. 'Ionic liquids as electrolytes', *Electrochimical Acta* 51(26): 5567–80.

Lin, H., Frankel, G.S. and Abbott, W.H. 2013. 'Analysis of Ag corrosion products', *Journal of the Electrochemical Society* 160 (8): C345–C355.

MacLeod, I. 1979. 'Conservation of corroded silver', *Studies in Conservation* 24: 123-43.

Mactaggart, P. and Mactaggart, A. 2002. *Practical Gilding*. London: Archetype Publications.

Portel, J. 1991. 'Altered silver gilding', in D. Bigelow (ed.), *Gilded Wood: Conservation and History*. Madison, CT: Sound View Press, 123–43.

Sandu, A.2010. 'Gilding techniques in religious are between East and West, 14th–18th centuries', *International Journal of Conservation Science* 1(1): 28-31.

Thompson, D.V. (tr.). 1960. Cennino Cennini's *The Craftsman's Handbook (Il libro dell' arte)*. New York: Dover Publications.

Authors' addresses

- Joana São João, Department of Conservation and Restauration (DCR), Faculty of Science and Tecnology (FCT), New University of Lisbon, Caparica, Portugal (j.joao@campus.fct.unl)
- Luís C. Branco, Department of Chemistry (DQ), Faculty of Science and Tecnology (FCT), New University of Lisbon, Caparica, Portugal
- Sara Leite Fragoso, Department of Conservation and Restauration (DCR), Faculty of Science and Tecnology (FCT), New University of Lisbon, Caparica, Portugal

Gels for removing varnish and surface stains from Bulgarian icons

Silviya Varadinova-Papadaki

Introduction

A major focus of the educational programme at the Department of Conservation, National Academy of Art, Bulgaria, is the treatment of icons with both canvas and wood supports. To develop the curriculum, the department has recently started exploring the use of gel cleaning systems. The work in this direction was fostered by a workshop on the principles of aqueous and solvent gel-based methods of cleaning painted surfaces, presented by Lora Angelova in 2015. After the workshop, our tests and experiments focused primarily on the use of PVOH-borax gels prepared with a variety of polar solvents. Both canvas paintings and icons have been treated with these systems.

One challenge in the conservation of icons is the partial or complete removal of varnish layers from sensitive paint and gold leaf layers beneath. The proper choice of solvents and the application method is very important for achieving the desired results. All the icons treated in our tests with PVA-borax gels had one or more layers of natural resin-based varnish. With aging, the varnishes had degraded, yellowed, darkened, lost transparency and accumulated soot, dust and other impurities.

Traditional methods used in the department for the removal of varnish and surface cleaning of icons are mechanical or chemical (both aqueous and solvent-based). The mechanical method, which involves a scalpel, is used mainly for preliminary work on very thick layers of varnish and where the varnish is unevenly spread on the surface. The aim of working with a scalpel is to reduce the risk of solvent treatment in those areas with sensitive paint layers and glazes under the varnish, since the quantity of solvents in the next stages of restoration is reduced. However, mechanical removal of the varnish layer is not safe when approaching the paint layers and an alternative, controllable method should be used.

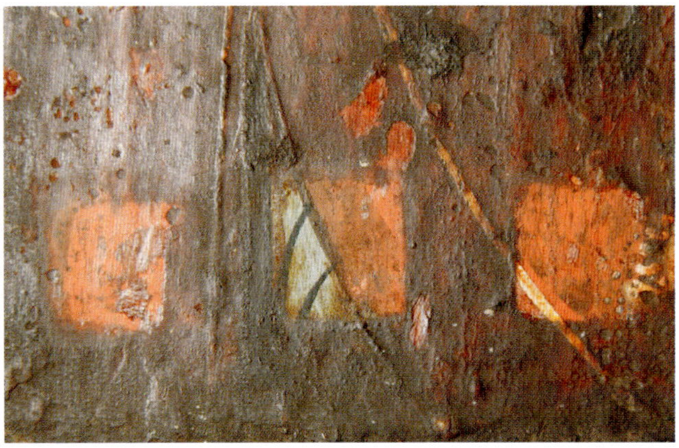

FIG. 1 *Virgin and Child*, triptych, 19th century, tempera on panel, 29 × 22 cm, private collection: detail showing cleaning tests on areas of *c.*1 cm^2.

Aqueous cleaning is usually appropriate for superficial dirt removal. Water can be used pure or mixed with surfactants and other solvents. However, the extended use of water on the surface of icons has been found to soften and damage the gesso layers and is therefore potentially hazardous for the paint. Polar solvents are very effective for removing the residual varnish layers, but are often harmful to the paint layer when used in the traditional manner, applied by cotton swabs or a brush.

Methods

As an alternative to the commonly used aqueous and free-solvent-based methods, we have started to test and employ solvent gels with different compositions, applied to varnish layers of different thicknesses, structure and condition. Three solvents have been used in the main for the preparation of the gels: ethanol, isopropanol and acetone. Gels were prepared with 5 wt% 75% hydrolysed poly(vinyl alcohol) (Kuraray Poval 5-74) and 1 wt% sodium tetraborate decahydrate (Sigma). Solvent to water ratios in the gels tested are typically 10:90, 30:70 or 50:50 (by weight of the liquid component).

Cleaning tests on icons

Tests were carried out on several icons. The first object to be treated was a fragment of a mid-19th-century triptych.

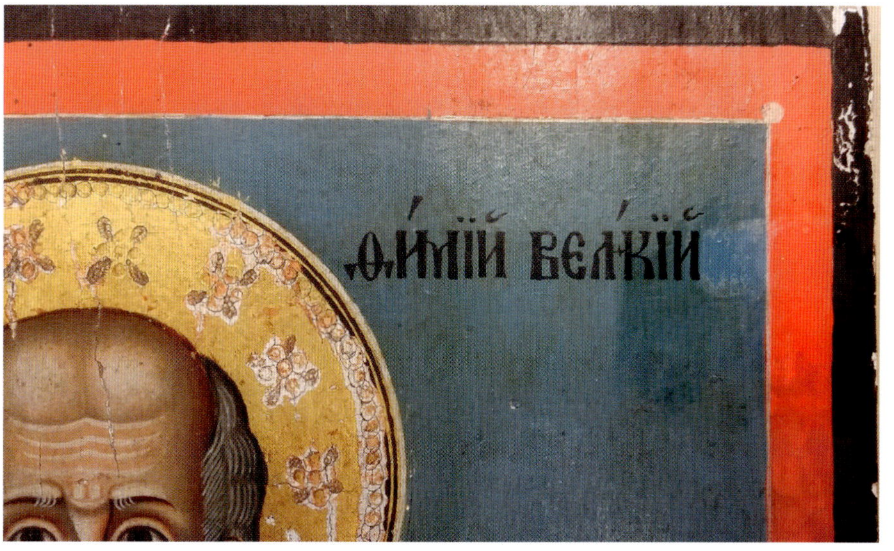

FIG. 2 1 *St Evtimii the Great*, 19th century, tempera on panel, 26 × 19.5 cm, National Gallery, Sofia: detail showing removal of varnish layers with PVOH-borax gel made from 30:70 iospropanol:water.

The natural varnish of the icon was thick, had darkened significantly and accumulated many impurities. As a result, the image was entirely unreadable. Prior to the application of the gel, multiple tests were carried out with the liquid solvent on a cotton swab, as well as applied on a tissue as a compress, for different periods of time. With this technique, the varnish was softened unevenly. Additional mechanical removal with a scalpel of the small spots of varnish remaining was required. We then tested several of the prepared gel cleaning systems and selected a gel of PVOH-borax with isopropanol (50:50 solvent:water) because this solvent had already demonstrated good results in the previous testing phase. The gel dwell time on the surface was determined by multiple trials. Short exposures (1–2 minutes) were not successful, so the time was extended to 3 minutes which proved sufficient for the varnish to be softened evenly. It was possible to remove the dissolved varnish gently with a scalpel without any mechanical force. The gel residues were removed with a cotton swab followed by clearance with white spirit. Only a very thin layer of varnish remained after the treatment. This layer was practically invisible and did not lead to changes in colour. The gel itself turned yellow-brown during the treatment, as did the cotton swabs used for the clearance treatment.

The initial gel tests were carried out on small areas of *c.*1 cm^2; the test site was then gradually enlarged to *c.*8–10 cm^2 (Fig. 1). Working on such an area was effective, allowed monitoring and control over the dwell time, and made possible the safe removal of the dissolved varnish. The positive effects of the gel were recognizable, and the method was an improvement over the previously described time-consuming and potentially harmful scalpel and free-liquid cleaning methods. The time necessary for natural varnish removal with the 30:70 iospropanol:water PVOH-borax gel from icons with a thinner varnish layer is significantly shorter – 30 seconds in our tests. The dissolved varnish was once again absorbed by the gel, but the observed gel colour was paler in this case. This example, *St Evtimii the Great*, is shown in Fig. 2. The gel residues were once again removed with white spirit.

Varnish thinning using a PVOH-borax gel prepared with 50:50 ethanol:water was carried out on the icon of *St Varvara*. Almost half of the painted area is gold leaf on bole, which requires extreme care during varnish removal. The gel was gently moved around on the surface like an eraser, and rarely allowed to remain in the same place until it became slightly yellow. Using this approach, the varnish layer was thinned without damage to the gold leaf layer.

Conclusion

Our cleaning and varnish removal experiments show that the PVA-borax solvent-gel systems have many benefits and positive effects. The gels contain a lower concentration of solvent, and allow for control of the organic solvent evaporation rate. We can control more precisely the dwell time of the liquid during the cleaning process. The cleaning can be kept to a specific and small area, as well as textured areas. They also help to reduce potentially risky treatments, such as the use of scalpels to mechanically reduce varnish, or free solvents which can affect sensitive gesso and paint layers. The PVA-borax gels also allow for a variety of solvent mixtures and ratios which can be tailored to the cleaning problem. However, some limits and drawbacks were observed: in particular, PVA-borax gels cannot be prepared with non-polar organic solvents. During our tests, we also noted that these gels were more difficult to remove from porous surfaces, therefore the potential for residues was greater when treating such objects.

Author's address

Silviya Varadinova-Papadaki, Department of Conservation, National Academy of Art, Sofia, Bulgaria (s.varadinova@yahoo.com)

Responsive bio-based gels for the preservation and treatment of archaeological wooden objects

Zarah Walsh-Korb, Soraya Ruiz-Fourcade and Luc Avérous

Introduction

Wooden objects naturally produce acids during the degradation process, specifically oxalic, glycolic, formic and acetic acid (Fors and Sandström 2006). Due to saltwater-induced degradation of iron bolts and fittings that once maintained the structure of the object, combined with moisture and sulphur from decaying marine organisms, waterlogged wooden objects are often saturated with iron-sulphur salts. These lead to the production of sulphuric acid and/or accelerate the natural production of acids within the timbers through either biological (the presence of sulphur-reducing bacteria) or chemical (redox and Fenton-type) processes (Fors and Sandström 2006). This increased acidity can cause structural instability by further degrading cellulose or producing acidic salts on the surface, resulting in loss and damage of surface layers. Loss of the wood surface not only increases the progression of further degradative processes but, in the case of decorated wooden objects, can lead to damage and loss of surface ornamentation. In addition, reports have shown that off-gassing from wood can result in the formation of acidic salts on other objects in the vicinity (Giorgi *et al.* 2009).

Previous work has shown that surface iron salts can be removed effectively using hydrogels based on functionalized chitosan and guar (Walsh *et al.* 2014). However, the wealth of literature on the use of gels for cleaning painted artworks has demonstrated that solvent or amphiphilic gels are more suited to surface cleaning to remove dirt and grime, leaving surface finishes undamaged (Carretti *et al.* 2009; Bonelli *et al.* 2016). The same must also be true for the removal of surface salts from objects that have already undergone conservation treatment, as the majority of consolidants used today for waterlogged wood are water-soluble, for example, poly(ethylene glycol) (PEG).

Introducing cleaning materials onto wooden objects adds another level of complexity. In addition to protecting any colorants or surface finishes that may be present, there is also the physical condition of the wood to consider, and compatibility between the conservation material and the wooden matrix. Several reports have indicated potential damage to wooden artefacts due to chemical incompatibility (Bardet *et al.* 2007; Walsh *et al.* 2016) or degradation products (Almkvist and Persson 2008; Mortensen *et al.* 2007) of fossil-based consolidants, principally PEG, in contact with archaeological wood. To avoid issues of incompatibility and possible degradative effects, this work focuses on the use of bio-based, renewable polymers, specifically poly(3-hydroxybutyrate) (P3HB) and alginic acid (Fig. 1).

P3HB is a bacterial polyester that is soluble in organic solvents, non-toxic, and can be biologically degraded after use, while alginic acid is a polysaccharide with a similar structure to natural cellulose, which can reduce incompatibility with the matrix, yet is miscible with P3HB in organic solvents. This initial study focuses on the introduction of iron chelating functionality into the polyester and whether it can be applied *in situ*, as a starting point to creating multifaceted adaptable consolidant/treatment gels for application to archaeological wooden objects.

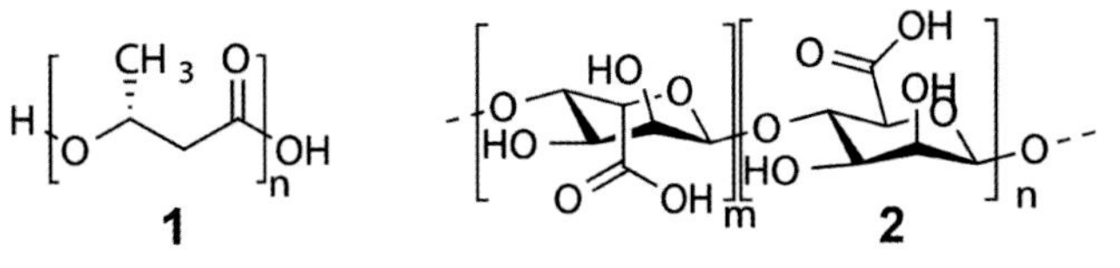

FIG. 1 Structure of poly(3-hydroxybutyrate) (1) and alginic acid (2).

Results and discussion

P3HB is principally bio-produced as a high molar mass polymer (molecular weight approx. 300,000 g/mol) that is sparingly soluble in chlorinated solvents. Despite this, it is of great interest in many application areas due to its

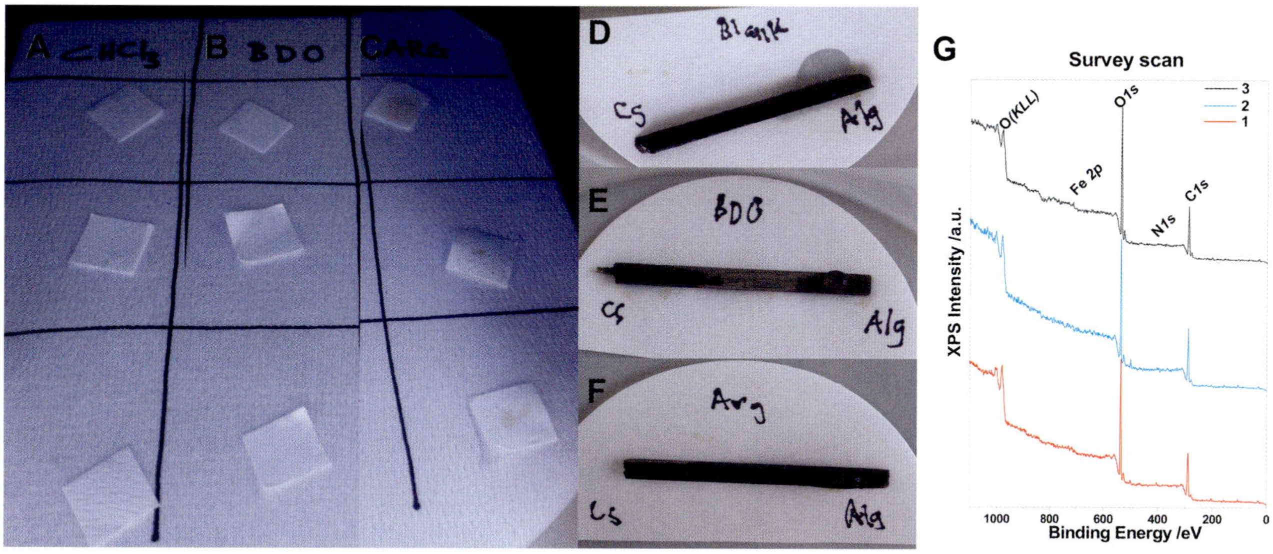

FIG. 2 Attempts to extract iron from wooden test samples using filter papers soaked with (A) $CHCl_3$, (B) P3HB-BDO and (C) P3HB-Arg. Application of (D) $CHCl_3$, (E) P3HB-BDO and (F) P3HB-Arg to iron-soaked oak wood test samples in solutions of alginate to assess film formation on the surface; (G) XPS analysis of the alginate based films containing $CHCl_3$, 1, P3HB-BDO, 2, or P3HB-Arg, 3, after removal from the wood surface.

biodegradability and biocompatibility. In order to create useable materials with improved solubility in other solvents, P3HB must be degraded to lower molar mass polymer chains (oligomers). A popular method is alcoholysis using an acidic catalyst, which produces short P3HB chains with a terminal hydroxyl unit, useful for further modifications. In this work, replacing the diol – here 1,4-butanediol (BDO) was used – with an amino acid, L-arginine, was investigated with the aim of introducing more appropriate chelating units in a one-step process. Due to similarities between amines and hydroxyls in term of reactivity, it was expected that one of the available amines would be covalently attached to the degraded P3HB chains leaving the additional amine groups as well as the carboxylic acid unit free to chelate with available metal ions.

Nuclear magnetic resonance (NMR) studies have shown that the use of L-arginine, while not quite as effective at degrading the polymer chains as BDO, does produce polymer chains of a reasonably short chain length (< 10,000 Da). Disappointingly, small amounts of chloroform were still necessary to introduce the polymer into other solvents.

The ability of the functionalized P3HB chains to remove iron from wooden samples was examined by soaking squares of Whatman No.1 filter paper in 1 wt% solutions of BDO-functionalized P3HB (P3HB-BDO) or arginine-functionalized P3HB (P3HB-Arg) in chloroform ($CHCl_3$) with pure $CHCl_3$ as a control (Fig. 2 A–C). Once dry, the filter papers were investigated under ultraviolet (UV) light to determine if any extraction had occurred. An orange stain was present on the P3HB-Arg tests only, with no change observed on other test papers, indicating that some of the iron from the wood may have been extracted. The need to use chloroform for the extraction is, however, a major limitation for large-scale use in conservation, due to the potential health risks. As such, the application of the P3HBs through another medium was investigated – functional polyesters in small amounts of chloroform were added to solutions of several polysaccharides and the chloroform allowed to evaporate before application to the wood surface (Fig. 2 D–F). The same samples and controls were used as before. However, as can be seen from the images, only alginate mixtures remain as a gel on the surface until a film is formed. Alginate was deemed to be the best carrier for the polymer, resulting in less interaction of aqueous polymers with treatments already in the wood. While the alginate appears to be physically the best carrier for the functional polyesters, X-ray photoelectron spectroscopy revealed that the film of 1 wt% alginate contained 1% Fe after extraction, while alginate with 1wt% P3HB-Arg contained 1.3% Fe (Fig. 2G). While an improvement was evident, it is clear that the alginate alone is also capable of complexing metals through the carboxylic acid groups. Interestingly, the addition of P3HB-BDO resulted in a decrease in uptake of Fe to 0.5% indicating some potential competitive interaction between the alginate and P3HB-BDO.

Conclusion

From these preliminary results it is clear that, while promising, improvements to the preparation of the functional polyesters are necessary, with further investigation of additional appropriate chelating units. Furthermore, grafting of the polyesters to the polysaccharide backbones will be investigated to improve gel formation and create better amphiphilic cleaning materials with enhanced metal chelation properties.

Acknowledgements

Zarah Walsh-Korb and Luc Avérous acknowledge the EU 7th Framework project *Synpol* (Project No. 311815) for financial support, and V. Papaefthymiou for the XPS measurements.

References

Almkvist, G. and Persson, I. 2008. 'Fenton-induced degradation of polyethylene glycol and oak holocelulose: a model experiment in comparison to changes observed in conserved waterlogged wood', *Holzforschung* 62: 704–8.

Bardet, M., Gerbaud, G., Tran, Q.K. and Hediger, S. 2007. 'Study of interactions between polyethylene glycol and archaeological wood components by 13C high-resolution solid-state CP-MAS NMR', *Journal of Archaeological Science* 34: 1670–76.

Bonelli, N., Chelazzi, D., Baglioni, M., Giorgi, R. and Baglioni, P. 2016. 'Confined aqueous media for the cleaning of cultural heritage: innovative gels and amphiphile-based nanofluids', in P. Dillmann, L. Bellot-Gurlet and I. Nenne (eds), *Nanoscience and Cultural Heritage*. Amsterdam: Atlantis Press, 283–311.

Carretti, E., Bonini, M., Dei, L., Berrie, B.H., Angelova, L.V., Baglioni, P. and Weiss, R.G. 2009. 'New frontiers in materials science for art conservation: responsive gels and beyond', *Accounts of Chemical Research* 43: 751–60.

Fors, Y. and Sandström, M. 2006. 'Sulfur and iron in shipwrecks causes concern', *Chemical Society Reviews* 35: 399–415.

Giorgi, R., Chelazzi, D., Fratini, E., Langer, S., Niklasson, A., Rådemar, M., Svensson, J.E. and Baglioni, P. 2009. 'Nanoparticles of calcium hydroxide for wood deacidification: decreasing the emissions of organic acid vapors in church organ environments', *Journal of Cultural Heritage* 10: 206–13.

Mortensen, M.N., Egsgaard, H., Hvilsted, S., Shashoua, Y. and Glastrup, J. 2007. 'Characterisation of the polyethylene glycol impregnation of the Swedish warship *Vasa* and one of the Danish *Skuldelev* Viking ships', *Journal of Archaeological Science* 34: 1211–18.

Walsh, Z., Janeček, E.R., Hodgkinson, J.T., Sedlmair, J., Koutsioubas, A., Spring, D.R., Welch, M., Hirschmugl, C.J., Toprakcioglu, C., Nitschke, J.R., Jones, M. and Scherman, O.A. 2014. 'Multifunctional supramolecular polymer networks as next-generation consolidants for archaeological wood', *Proceedings of the National Academy of Sciences of the USA* 111: 17743–8.

Walsh, Z., Janeček, E.R., Jones, M. and Scherman, O.A. 2015. 'Natural polymers as alternative consolidants for the preservation of waterlogged archaeological wood', *Studies in Conservation* DOI:10.1179/2047058414Y.0000000149.

Authors' addresses

- Zarah Walsh-Korb, BioTeam, ECPM-ICPEES, Université de Strasbourg, Strasbourg, France (walshkorb@unistra.fr)
- Soraya Ruiz-Fourcade, BioTeam, ECPM-ICPEES, Université de Strasbourg, Strasbourg, France
- Luc Avérous, BioTeam, ECPM-ICPEES, Université de Strasbourg, Strasbourg, France

Comparative studies (multi-gel use)

Conservation of a white Louise Nevelson installation: gel systems explored

Sarah Nunberg, Soraya Alcala, Carolyn Tomkiewicz, Cindie Kehlet, Chris McGlinchy and Jens Dittmer

The Nevelson installation

The Chapel of the Good Shepherd Louise Nevelson installation covers the walls of a five-sided room that measures 8.53 m long × 6.4 m at its widest point (Fig. 1). It is a sculptural environment consisting of five three-dimensional sculptures built into the five chapel walls, and one group of three hanging columns at the chapel altar. The sculptures are made in sections from irregular plywood shapes that are nailed and screwed together. The sculptural sections, many comprised of three-sided boxes elaborated with varied elements, are screwed or nailed onto the walls. Originally the sculptures were entirely painted a homogeneous, smooth, yellowish-white except the side panels of the central cross, which are oil gilt with gold leaf. The original paint was applied directly to the wood substrate. Grey-white paint applied during restoration completely obscures the original surface of all the elements, except where the restoration paint has flaked.

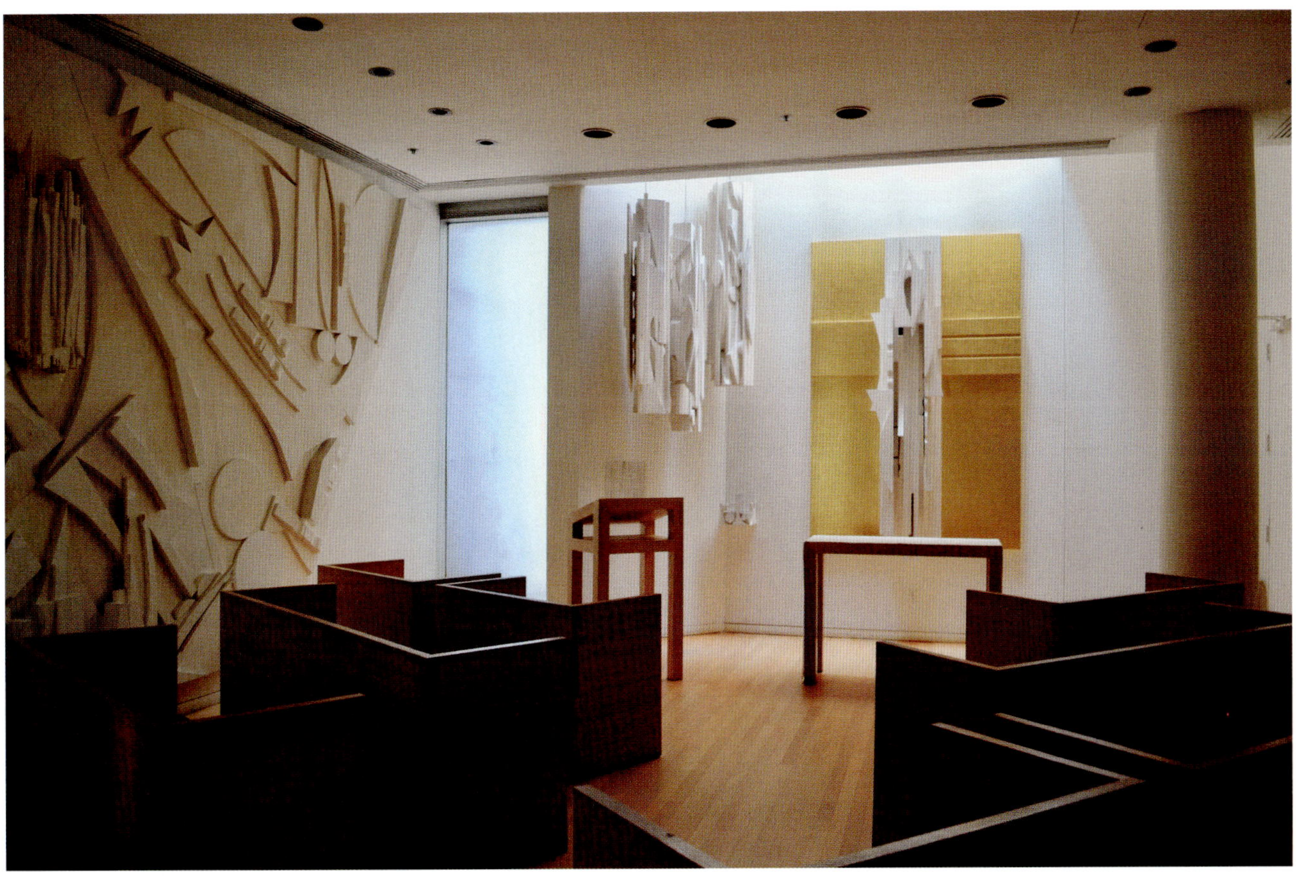

FIG. 1 Chapel of the Good Shepherd, Saint Peter's Church. View of the *Cross of the Good Shepherd*, the *Three Columns*, and *Sky Vestment-Trinity*.

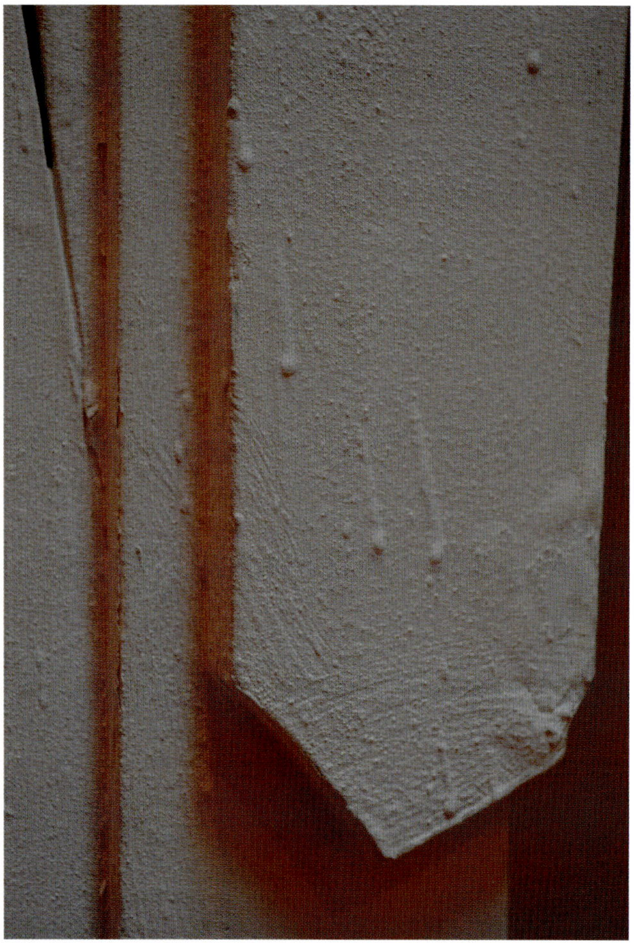

FIG. 2 Detail of *Sky Vestment-Trinity* showing thick, dripping, uneven application of restoration paint.

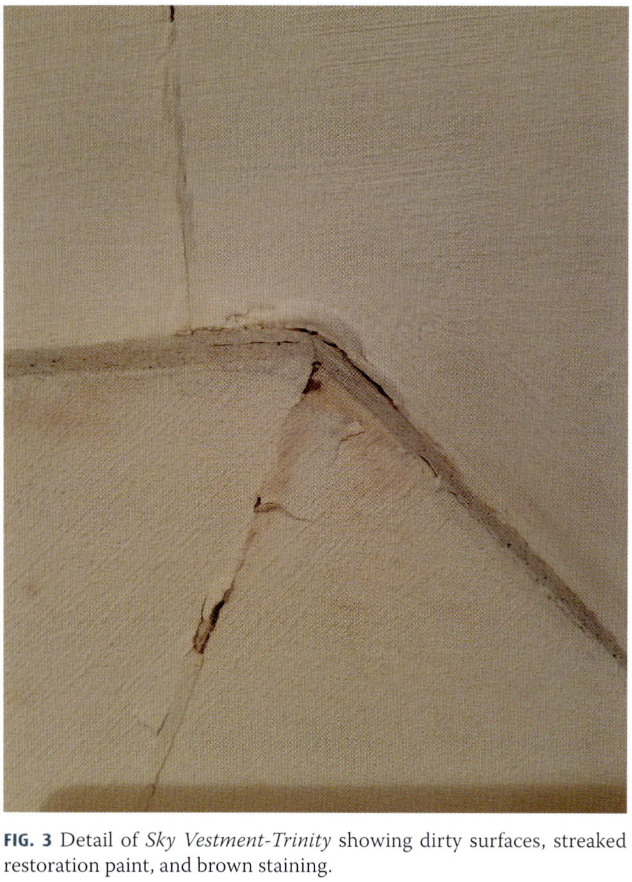

FIG. 3 Detail of *Sky Vestment-Trinity* showing dirty surfaces, streaked restoration paint, and brown staining.

The original paint and restoration paint differ in tone, texture, and application method. The restoration paint was applied by roller and brush, leaving drip marks and uneven layers throughout – sometimes raised and textured, sometimes smoother and thick (Fig. 2). Additionally, the uneven, grey-white restoration paint contrasts with the even, yellowish-white, homogeneous original paint application.

Lack of environmental management and interaction between the two paint layers has resulted in a compromised surface that appears dirty and uneven, with random brown staining throughout. Much of the restoration paint is lifting, flaking and bulging (Fig. 3). In one high-traffic area, the original paint is entirely missing, and the restorer's paint was applied directly onto the unprepared wood.

To identify and confirm the original paint color and Nevelson's working methods, two historically contemporary white Nevelson sculptures were examined. *Bicentennial Dawn*, the largest of the three 1970s Nevelson environments, was installed inside the main entrance to the Philadelphia Courthouse in 1976. Due to its location it was exposed to extreme fluctuations in relative humidity, resulting in the loss or damage of much original paint. The final of at least three restoration campaigns resulted in stripping the sculptures of all original paint and applying a new yellowish-white color matching the tone of original paint layer, similar to the original layer of the Chapel of the Good Shepherd. *White Flame of Six Million* from 1970 in Temple Beth El, Great Neck, New York, was the second installation examined. Areas of the original paint that were inadvertently left behind during restoration also appear yellowish-white.

Paint compositions

The original paint was identified as an oil-based unsaturated alkyd, confirmed by Fourier transform infrared (FTIR) spectroscopy, gas chromatography–mass spectrometry (GC–MS) and nuclear magnetic resonance (NMR). Alkyds are polyesters produced through reactions of fatty acids, polyols, and dibasic acids. The polyol in the original paint composition was identified as pentaerythritol (PE), a widely used alcohol in alkyd paint formulations in the 1970s. In the solid state ^{13}C NMR spectra a strong resonance for carbon-carbon double bonds was observed; the dibasic acid is most likely a phthalic anhydride, which would have made the original paint yellowish-white due to its aromatic structure. Although it cannot be definitively excluded that the yellowish tint is a result of the degradation, it is likely that the current yellowish-white reflects the original paint color; all cross-sections show the yellowish white throughout the original paint layer with no color gradient as would be expected from photo-oxidation.

The restoration paint was identified as a polyvinyl acetate (PVAc) emulsion through ^{13}C NMR, and confirmed with FTIR. X-ray fluorescence (XRF) spectroscopy identified that both paints contained titanium-based white pigments. Signals of molecular PE were found through ^{13}C NMR analysis of samples from both paint layers. In many places PE can be seen macroscopically as a white powder and as brown staining on the overpainted surface, which was confirmed by GC–MS of the powder and scrapings of the brown areas. The free PE was probably added in excess during manufacture rather than originating from degradation of the alkyd paint. The solid state ^{13}C NMR spectra did not show signs of degradation products; in particular, there are no carboxylates that would have resulted from de-esterification of the polymer chain. The migration of the free PE may be a result of the application of the water-based PVAc paint, since PE is a highly soluble molecule.

Solubility tests

Characterizing the original and restoration paint layers through solubility tests played a key role in weighing the treatment options. Initially, preservation and surface cleaning of the restoration paint was explored. A series of tests with cotton swabs dampened with water at pH 6.5 or adjusted with ammonium hydroxide (NH_4OH) to pH 9 gave differing results. Water at pH 6.5 effectively cleaned the surface of dirt and grime with little disturbance to the restoration paint texture, while water at pH 9 or pH 10 removed dirt and grime but resulted in a slightly more compact surface with reduced granularity. Mixtures of chelators, water, buffer and surfactants at pH 5.5, 6.5, 7.5 and 8.5 removed dirt and grime equally well but also altered the surface texture. A modular cleaning mixture (Stavroudis and Doherty 2013) at pH 7.5 altered the physical surface of the restoration paint the least but minimally affected the disfiguring brown staining.

The surface cleaning tests along with materials analyses confirmed that the restoration paint could not be cleaned without changing its surface characteristics and that the alkyd paint was in good, stable condition and probably close to its original color. This understanding led to exploration of a plan to remove the restoration paint, and expose the original alkyd as the presentation surface. The original paint was readily soluble in acetone, but unlike the PVAc, it did not respond to water or 2-propanol, indicating that a system to remove the restoration paint without affecting the original was a viable option. Solvent tests showed that 2-propanol on cotton swabs effectively solubilized the PVAc restoration paint, and that with increased dwell time the efficiency was enhanced. Water at pH 9 with increased dwell time also effectively solubilized the PVAc paint without altering the alkyd. These findings led to investigating the delivery of first the 2-propanol, then the buffered water within a viscous carrier. Due to the large size of the installation and the complexity of its structure, it was crucial to choose a system that could be applied easily, would act quickly, could be cleared entirely without leaving a residue, was non-toxic, and could provide a straightforward cleaning routine. Minimizing impact on human health and lessening environmental toxicity of the treatment, together with avoiding any change to the original alkyd, guided the choice of cleaning system.

Delivery system tests and discussion

Cellulose ether (Klucel M) for solvent delivery

Initial tests for overpaint removal involved a viscous solution of 2.5% Klucel M (hydroxypropylcellulose ether) mixed with 2-propanol, then applied with a thick bristle brush, allowed to dwell for five minutes, agitated, wiped with a dry cotton swab and cleared with a cotton wool damp with 2-propanol. Generally two passes with the Klucel M removed the PVAc paint. If carried out properly, the viscous solution with the solubilized PVAc paint could be removed with minimal surface wetting, allowing for controlled cleaning. However, conflicting discussions concerning possible cellulose ether residues attracting biological growth and inefficient cleaning (Feller and Wilt 1993; Casoli *et al.* 2014; Baglioni *et al.* 2015) led to the exploration of an alternative system. In addition, delivery of the Klucel M with 2-propanol over the large surface resulted in unacceptable solvent exposure, and the adjusted water in Klucel M did not sufficiently remove the PVAc. Given these concerns, the nanostructure chemical gels and the physical hydrogel systems represented an alternative. However, identifying the most efficient application method involved a multi-step process.

Nanostructured colloidal systems for solvent delivery

Many publications describe gel systems and their mechanisms (Baglioni *et al.* 2009; Carretti *et al.* 2009, 2010; Domingues *et al.* 2013). Therefore, this paper will not focus on the chemical compositions or processes involved in making these gels, but rather on the authors' experience in using them. The steps taken to identify the proper colloidal system, including testing of nanostructured systems and their adaptation to separate the PVAc paint from the original alkyd will be discussed. Gels generally are divided into chemical and physical systems. This project found that rigid chemical gels required horizontal surfaces during dwell time, while the physical PVA-borate-based hydrogels based on poly(vinyl alcohol) were more successful on vertical surfaces; they allowed for extended dwell time due to their tenacious quality and easy application to the multicomponent elements.

According to initial on-site testing, the PVAc restoration paint and the original alkyd paint solubility parameters indicated that 2-propanol or water adjusted to pH 9 with ammonium hydroxide would be the most appropriate solvent to load in a gel or dispersion. To test delivery of the 2-propanol and the adjusted water and allow for sufficient dwell time, two high viscosity polymeric PVA-borax gel-like dispersions and three rigid p(HEMA)/PVP hydrogels made by CSGI (Nanorestore Dry, Extra Dry and Max Dry) were explored (see Carretti *et al.* 2014 for a description of the PVA hydrogels and Domingues *et al.* 2013 for a description of the p(HEMA)/PVP hydrogels).

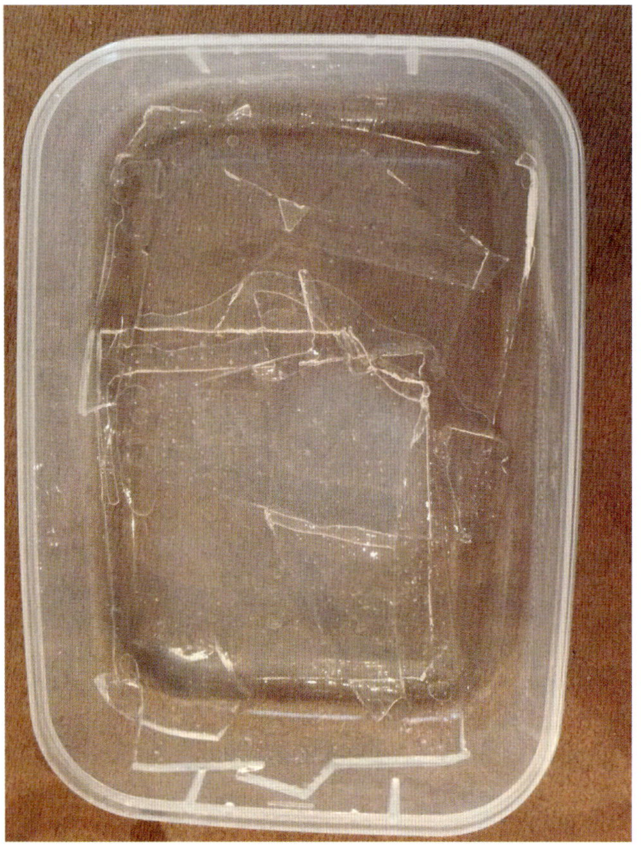

FIG. 4 Open container with Dry gel.

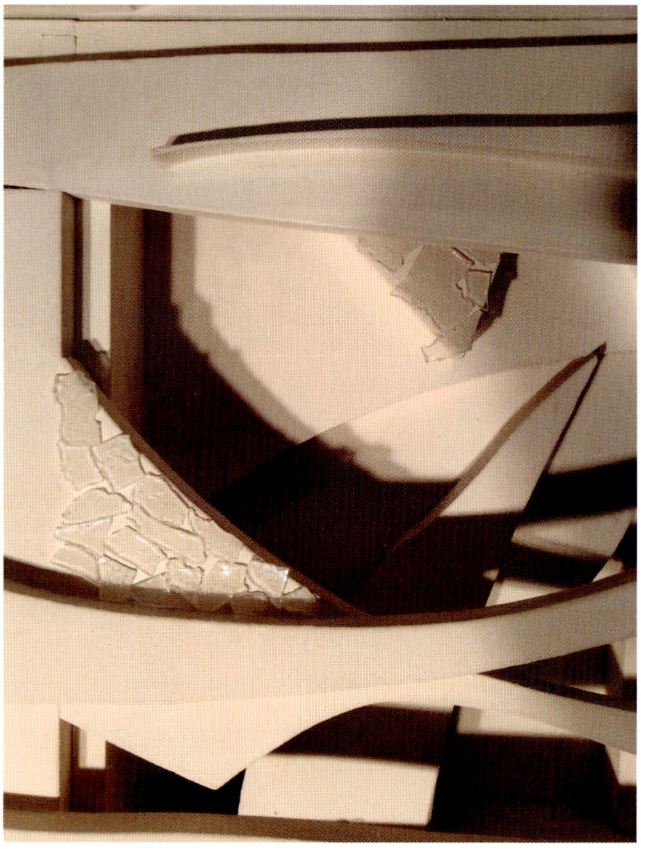

FIG. 6 Gel sheet fragments placed on the *Frieze of the Apostles*. After the second application the sheet broke into many fragments.

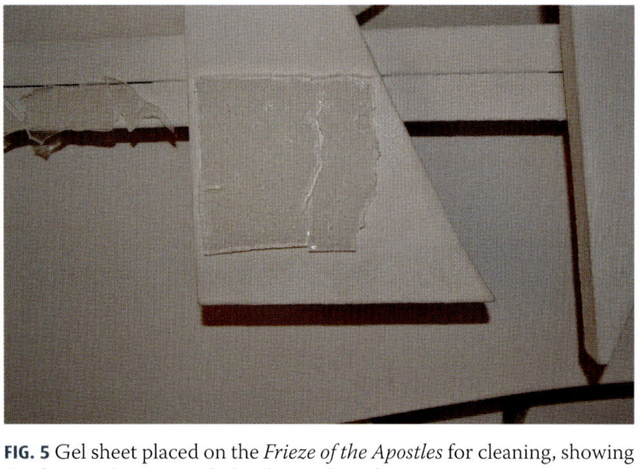

FIG. 5 Gel sheet placed on the *Frieze of the Apostles* for cleaning, showing the first application with the sheet relatively intact.

Rigid chemical gels

Nanorestore Dry, Extra Dry and Max Dry p(HEMA)/PVP chemical hydrogels are characterized by covalent bonds so their structure remains stable regardless of the pH, environment and temperature. Rigid gel preparation is described in depth by Baglioni and Chelazzi (2013), and Domingues *et al.* (2013). These gels are clear and colorless allowing for easy monitoring of the system as the solvent reacts with the surface. The rigid gels are based on an interpenetrating network; they hold the loaded solvent on the painted surface, minimize surface wetting, can be removed as a sheet without leaving a residue, and allow for dry removal of the later paint by rolling off the activated film. This technique avoids abrasion and over-wetting from cotton swab solvent application.

For the Nevelson application, containers with airtight lids similar in size to the gel sheets were filled with 2-propanol. The gels were transferred from the packaging into containers according to preparation instructions (Baglioni *et al.* 2015) (Fig. 4). The solvent-laden gel sheet was cut or torn to the desired size, directly pressed onto the sculpture surface, and allowed to dwell for 20 minutes (Fig. 5). After removing the gel sheet, it was placed back into the container, rinsed by soaking in the 2-propanol for about 10 minutes, and then reused several times. During the first two applications the gels were easily placed onto the sculpture surface and then removed after the dwell time. Following the third application the gels were hard to handle, ripped easily, broke into small pieces, and became more fragile with each additional handling (Fig. 6). Removal of the swollen PVAc paint between each application was achieved using vulcanized rubber sponges, microcellular and/or cotton wool damp with 2-propanol. This procedure was repeated until the PVAc paint was reduced.

The rigid chemical gels removed the restoration paint sufficiently, but they were challenging to use on the large-scale, vertical Nevelson sculpture with its complex construction. The Dry gel was the only gel that stayed on the surface for the desired dwell time – all others fell off after approximately 5 minutes (once the 2-propanol began to evaporate), and required re-immersion in the solvent before reapplication. Cleaning one square meter of the sculpture in this manner took 25 hours and resulted in an uneven, spotty cleaning. Ultimately a different system that would result in more even but controlled cleaning, with faster application and complete removal without leaving a residue, was explored. Max Dry chemical gel in water was tested, but did not stay on the vertical surface long enough to swell the restoration paint, and was not further considered.

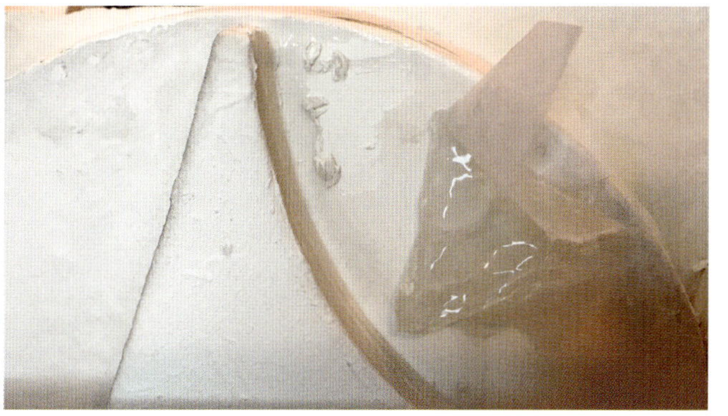

FIG. 7 PVOH-borax dispersion applied on Green's Lens Tissue. The tissue is peeled back and the restoration paint partially scraped off the surface.

High-viscosity polymeric dispersion (physical gel)

The PVA-borax physical viscoelastic systems are described in detail by Carretti *et al* (2009). These dispersions, loaded with either water or 2-propanol, provided a clear, highly viscous material with the clinging properties of the rigid Dry gel, sufficient for vertical surfaces. However, on the Nevelson sculpture the dispersion quickly solubilized the PVAc paint so that the gel slid down the surface, shortening the dwell time. The sudden non-clinging properties were caused by the uptake of solubilized overpaint. PVA-borax hydrogels are susceptible to changes in pH, which alter the equilibrium that maintains their crosslinks; the solubilized and absorbed PVA overpaint probably affected this balance. Although the pH shifted, the dispersion (with sufficient dwell time) still successfully solubilized the restoration paint.

To remedy its lack of retentive properties, a 1 cm diameter ball of dispersion (weighing approximately 0.5 g) was applied to a 3 × 5 cm rectangular piece of Green's Lens Tissue. The dispersion was pressed out to a flat, elastic sheet, molded against the sculpture surface, and covered with tissue to hold it in place. After a few minutes of contact, the hydrogel began to solubilize and absorb the PVAc restoration paint. The gel incrementally decomposed and softened so that it could be further pressed flat. After approximately two minutes the tissue and the bulk of the PVA-borax system were lifted and moved to another area that required cleaning, allowing for rotating surfaces undergoing treatment (Fig. 7). The swelled restoration paint layers were scraped off and then further removed by a final wipe with cotton wool damp with 2-propanol, clearing the restoration and any remaining gel to expose the original surface.

Allowing for efficient removal with minimal surface wetting, the PVA-borax gel-like dispersion sufficiently solubilized and evenly removed the PVAc paint without altering the alkyd. This procedure proved superior to using the rigid gels for a number of reasons: it required smaller amounts of product, produced less than half the waste, took less than one-third the amount of time, and preserved the original structure and quality of the alkyd (Fig. 8).

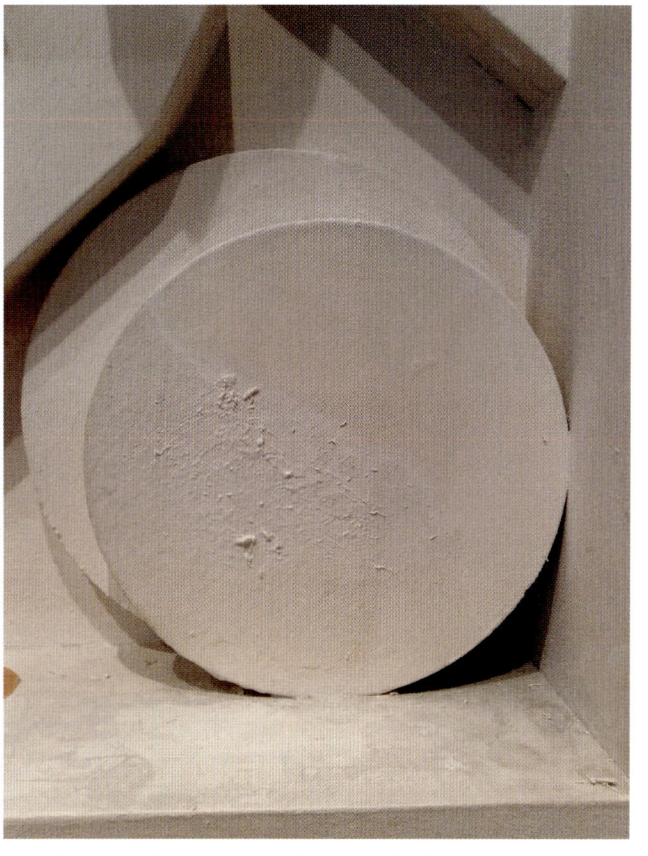

FIG. 8 Removal of the non-original paint on the *Frieze of the Apostles* by rolling it off after contact with the PVOH-borax dispersion. The upper section of the circle is the recovered original paint with its cream-yellow color. The lower section of the circle and all areas surrounding it are still covered with non-original paint.

Discussion

Determining a treatment plan for the Nevelson installation involved multiple tests and close collaboration between all groups involved. Exploration into three solvent application systems led to the most acceptable method for uncovering the original alkyd paint by removing the disfiguring PVAc restoration paint. Klucel M provided a highly viscous solution that allowed for easy brush-on application, appropriate solvent dwell working properties, and dissolution of the PVAc overpaint, although incomplete restoration paint clearing with one application; at least two Klucel M applications were necessary. Also, the possibility of even the slightest residue of cellulose ether, which might foster biological growth, and the insufficient PVAc paint removal prompted further investigation. The application mechanism and solvent exposure of the rigid chemical gels resulted in uneven and spotty cleaning, ruling out this system. The PVA-borax dispersion allowed for relatively fluid application and even cleaning but required wiping with 2-propanol to ensure complete removal. The need to use neat solvents to fully remove the PVA-borax dispersion has led to exploration with CSGI of a new system with similar application methods but removal through peeling (similar to the chemical gels), not wiping with solvent, to eliminate neat solvents and any possibility of residue.

Conclusion

Tests carried out with a highly viscous solution, Nanorestore gels, and a viscoelastic dispersion were effective in leading to an understanding of the best treatment approach, including removing the restoration paint and uncovering the original paint as the presentation surface. Discovering that the later paint could be successfully removed using a highly viscous solution of solvent in Klucel M was a crucial step in identifying a viable system. Extensive communication with CSGI offered a collaborative and productive forum for examining the benefits and limitations of implementing a system for overpaint removal. Adapting the gel for successful delivery of solvent to a vertical surface offered advantageous results for the artwork. It also minimized exposure to toxic solvents for the conservation team. The preferred cleaning method should result in a process that effectively and efficiently uncovers the original alkyd layer without altering its texture or color. Issues have been found related to each gel system and further work will focus on the efficacy of gels for this project. The collaboration between conservators and scientists assisted the process and prompted a better understanding of an optimal cleaning system for this large installation.

Acknowledgements

The authors thank Piero Baglioni and Rodorico Giorgi of CSGI for helpful discussions and extensive collaboration. Sarah Nunberg is grateful to the Stockman Family Foundation for its generous support.

Materials and suppliers

- High-viscosity polymeric dispersion (PVA-borax gel-like dispersion). A clear, colorless high viscosity aqueous polymeric physical gel based on poly(vinyl alcohol) crosslinked by borax. Can load polar solvents such as acetone, alcohol, isopropanol, oil in water microemulsion, etc.: www.csgi.unifi.it/products/gel.html
- Rigid p(HEMA)/PVP chemical hydrogels Dry, Extra Dry, Max Dry are transparent water-based chemical gels, based on a pHEMA/PVP semi-interpenetrated network which has high retention of loaded liquid, and leaves no residues. These gels can be loaded with most polar solvents, surfactant solutions, and oil-in-water microemulsions: www.csgi.unifi.it/products/gel.html
- Green's Lens Tissue, an unbuffered neutral pH tissue: http://www.talasonline.com/Greens-Lens-Tissue
- Klucel M is a hydroxypropylcellulose nonionic water-soluble cellulose ether: http://www.chempoint.com/products/catalog/ashland/ashland-cellulose-ethers/klucel-hydroxypropylcellulose
- TRI-Funori is derived from two species (*Gloiopeltis furcata* and *Gloiopeltis tenax*) of red seaweed (*funori* in Japanese) harvested along the coast of Japan. The key component of the seaweed is a unique polysaccharide starch. A conservation product that is effective in three distinct applications: as a fixative, cleaning agent, and a low-strength adhesive: https://tri-funori.com
- Dry cleaning 'soot removal' sponge made from vulcanized rubber: http://www.talasonline.com/Dry-Cleaning-Sponge-Dirt-Eraser
- Saugwunder poly(vinyl alcohol) microcellular sponge with high-densityopencells:http://www.talasonline.com/Polyvinyl-Alcohol-Sponge

References

Baglioni, P. and Chelazzi, D. 2013. *Nanoscience for Conservation of Works of Art*. Cambridge: Royal Society of Chemistry.

Baglioni, P.L. Dei, E. Carretti, R. and Giorgi, R. 2009. 'Gels for the conservation of cultural heritage', *Langmuir Perspective* 25(15): 8373–4.

Baglioni, P. Chelazzi, D. and Giorgi, R. 2015. *Nanotechnologies in the Conservation of Cultural Heritage:a Compendium of Materials and Techniques*. Dordrecht: Springer.

Carretti, E. Grassi, S. Cossalter, M. Natali, I. Caminati, G. Weiss, R. Baglioni, P. and Luigi, D. 2009. 'Poly(vinyl alcohol)-borate hydro/cosolvent gels: viscoelastic properties, solubilizing power, and application to art conservation', *Langmuir Perspective* 25(15): 8656–62.

Carretti, E., Bonini, M., Dei, L., Berrie, B., Angelova, L.V., Baglioni, P. and Weiss, R. 2010. 'New frontiers in materials science for art conservation: responsive gels and beyond', *Accounts of Chemical Research* 43(6): 751–60.

Carretti, E., Matarrese, C., Fratini, E., Baglioni, P. and Dei, L. 2014. 'Physicochemical characterization of partially hydrolyzed poly(vinyl acetate)- borate aqueous dispersions', *Soft Matter* 10: 4443–50.

Casoli, A., Di Diego, Z. and Isca, C. 2014. 'Cleaning painted surfaces: evaluation of leaching phenomenon induced by solvents applied for the removal of gel residues', *Environmental Science and Pollution Research* 21: 13252–63.

Domingues, J. Bonelli, N., Giorgi, R., Fratini, E., Gorel, F. and Baglioni, P. 2013. 'Innovative hydrogels based on semi-interpenetrating p(HEMA)/PVP networks for the cleaning of water-sensitive cultural heritage artifacts', *Langmuir* 29: 2746–55.

Domingues, J., Bonelli, N. and Giorgi, R. 2014. 'Chemical semi-IPN hydrogels for the removal of adhesives from canvas paintings', *Applied Physics A* 114: 705–10.

Feller, R.L. and Wilt, M. 1993. *Evaluation of Cellulose Ethers for Conservation*. Los Angeles: J. Paul Getty Trust.

Stavroudis, C. and Doherty, T. 2013. 'The Modular Cleaning Program in practice: application to acrylic paintings', in M.F. Mecklenburg, A.E. Charola and R.J. Koestler (eds), *Cleaning 2010: New Insights into the Cleaning of Paintings*. Washington, DC: Smithsonian Institution Scholarly Press, 139–45.

Authors' addresses

- Sarah Nunberg, Department of Mathematics and Science, Pratt Institute, Brooklyn, NY, USA and The Objects Conservation Studio, LLC, Brooklyn, NY, USA (snunberg@aol.com)
- Soraya Alcala, Westchester Art Conservation LLC, Dobbs Ferry, NY, USA
- Carolyn Tomkiewicz, Paintings Conservation, Brooklyn, NY, USA
- Cindie Kehlet, Department of Mathematics and Science, Pratt Institute, Brooklyn, NY, USA
- Chris McGlinchy, The Museum of Modern Art, New York, USA
- Jens Dittmer, Institut des Molécules et des Matériaux du Mans (IMMM), UMR CNRS 6283, LUNAM, Université du Maine, Le Mans, France

Gel formulations coupled with mechanical cleaning techniques to remove old overpaint and pigmented coatings on a large altarpiece by Piero di Cosimo

Irma Passeri, Annika Finne and Anikó Bezur

Introduction

Piero di Cosimo's monumental late altarpiece, *Virgin and Child with Saints Vincent Ferrer and Jerome*, entered the Yale University Art Gallery collection in 1871. The painting has a long and complex conservation history. Originally two sets of angels stood above Sts Vincent Ferrer and Jerome; at some point prior to 1859, the altarpiece was reduced in size and the top profile modified to an arch. The two sets of angels were sold separately on the art market. The spaces in the altarpiece they once occupied were filled with wooden inserts prepared with a ground layer and painted blue to suggest a continuation of the sky above the saints.[1]

A photograph taken at the beginning of the 20th century documents that, alongside these extensive structural interventions, large areas of the composition were repainted and reinforced to revive the previously damaged surface (Fig. 1). In an attempt to remove these restorations, the painting was harshly cleaned between 1957 and 1959. The cleaning, halted at an incomplete stage, left the painting with an uneven surface disrupted by abrasions, losses and residues of old overpaint. The use of aggressively applied solvents and highly alkaline solutions during this treatment has affected the original paint layer (Fig. 2).[2]

In 2013 the conservation department of the Yale Art Gallery began a new examination and conservation treatment[3] of the Piero di Cosimo altarpiece. The objective was to assess the condition of the picture and determine whether it was possible to restore unity to its uneven and partially cleaned appearance. The fragile, compromised paint film and the tenacity of the remaining overpaint challenged traditional cleaning approaches. Extensive testing and critical evaluation of the surface led to the establishment of a cleaning methodology which enabled selective, controlled reduction of pigmented coatings, varnishes, and overpaint (Fig. 3).

This case study discusses several aspects of the Piero di Cosimo treatment. The paper has two main sections: the first describes the restoration history of the altarpiece, and the second, the chosen cleaning methodology – the coupling of mechanical cleaning techniques with carefully tailored gel formulations designed to selectively soften the pigmented coatings and overpaint in need of removal.

Restoration history

Our treatment methodology for the Piero di Cosimo altarpiece follows a path defined by its restoration history. The drastic conservation campaign enacted on the early Italian paintings collection at Yale from the 1950s to the 1970s, which included the Piero di Cosimo, leaves the conservators charged with its care in an unusual position. During this campaign, all past restoration and additions were removed indiscriminately from a large number of paintings, which were then displayed with all damages exposed. This purist archaeological approach must be thoughtfully and critically engaged with when establishing new conservation treatment methodologies for the affected paintings (Aronson *et al.* 2017).

Scattered early descriptions and images of the painting provide some documentation of the physical changes it has experienced. One of the first major alterations was the addition of a star to St Vincent Ferrer's halo in order to transform him to the more well-known St Dominic. This addition occurred between 1583 and 1677: a 1583 inventory describes the painting as a *Virgin and Child between St Vincent and St Jerome*, but by 1677 the tourist guide Giovanni Cinelli describes St Vincent as St Dominic instead (Geronimus 2006: 226). Cinelli also mentions the pairs of standing angels holding books in the sky above the holy figures (Venturini 1994: 61).

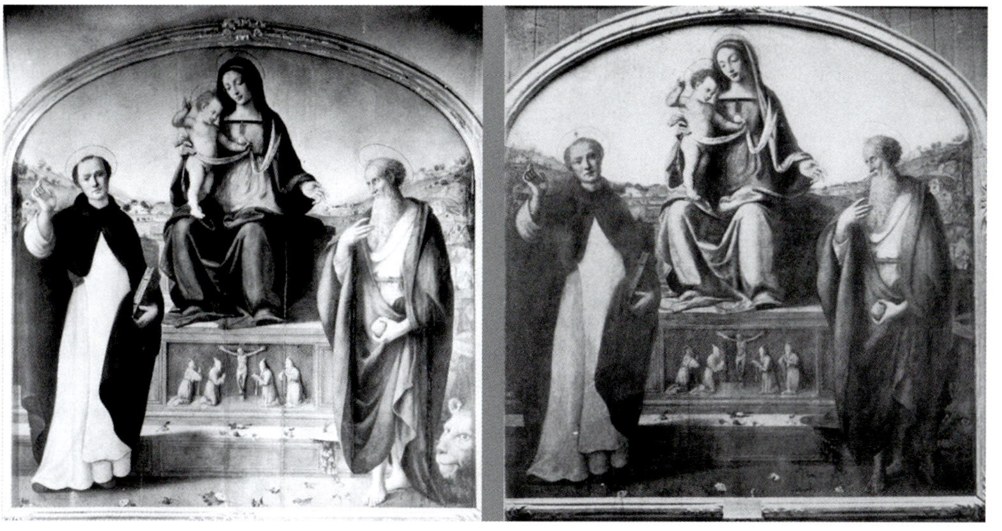

FIG. 1 Early photography of the altarpiece. The left image was taken in the 1920s; the right image was taken in 1895 and published in Rankin 1895.

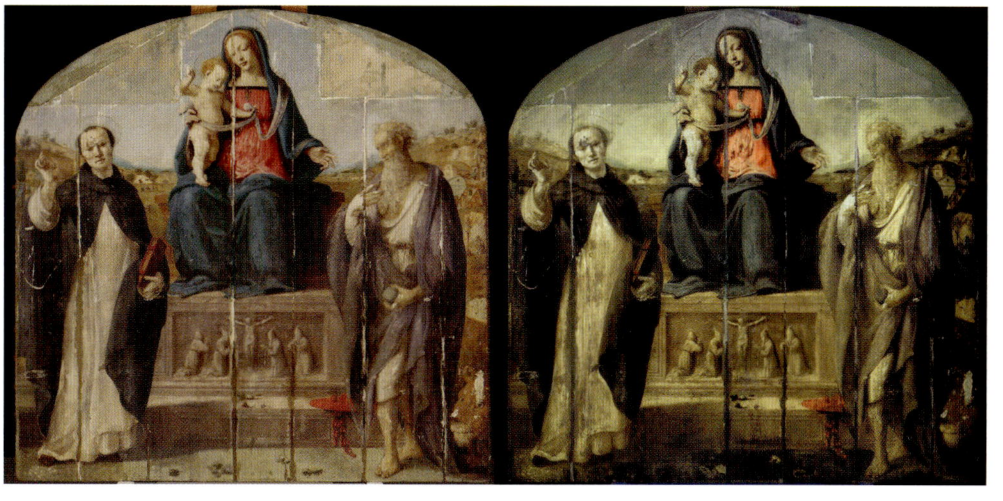

FIG. 2 Photographs taken in 2013 of the altarpiece (*left*) in visible light and (*right*) in UV light, documenting the condition after cleaning in the 1950s and before the present project.

The next documentable alteration is the division of the painting into three parts to separate those pairs of standing angels into two individual panels. At this time the top of the main panel was also rounded into an arch.[4] This process was not altogether insensitive; although the initial saw cuts removed a portion of the Christ Child's proper right elbow, it was then carefully reattached. The division of the painting into sections occurred between Cinelli's 1677 description and James Jackson Jarves' purchase of the painting in the mid-19th century. In our earliest known image of the painting, an 1861 copper-plate engraving published by Jarves (1861: 352) as a Ridolfo Ghirlandaio, the painting is in its smaller, arched state. In this engraving, and 30 years later in an 1895 photograph published by William Rankin (1895: 149), the star in St Vincent Ferrer's halo is still present.

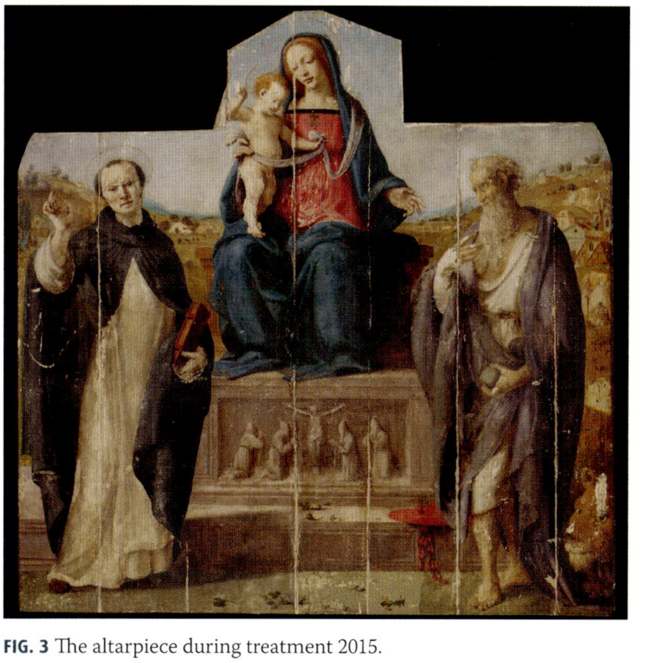

FIG. 3 The altarpiece during treatment 2015.

In 1915 the painting was cleaned by Hammond Smith. A comparison between the 1895 image and the next photograph of the painting, datable to the late 1920s, illustrates the scope of the treatment, which selectively removed the added star and lightened some soot-like darkening along the bottom

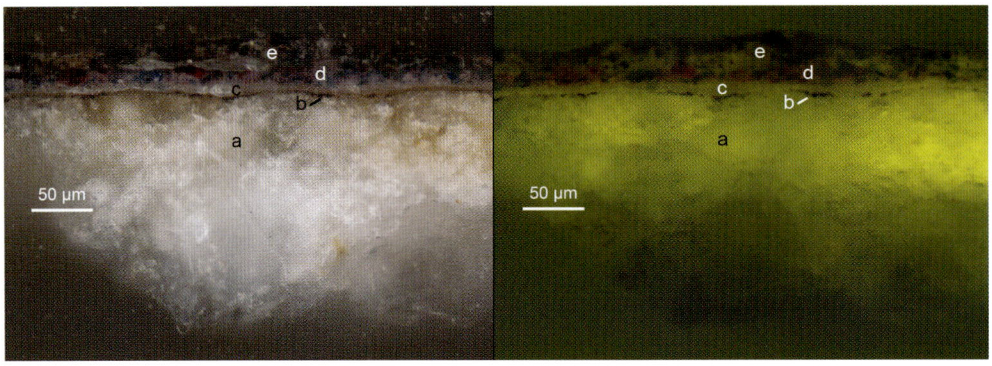

FIG. 4 Cross-section from the purple robe (×7), in visible light (*left*) and UV-induced fluorescence with fluorescence filter set 09 (*right*). Layer sequence: (a) ground layer, (b) underdrawing, (c) medium-rich preparatory layer, (d) original purple paint layer, (e) grey overpaint layer.

of the image (Fig. 1). Under ultraviolet (UV) light, a silhouette of the star remains visible, as the original sky color was exposed in the shape of the star by Smith's cleaning. The surrounding area remains overpainted. With the exception of the star, Smith's cleaning left large areas of past restoration intact. Other documented instances of restoration include Durham's 1931 application of a large mahogany wood cradle and a 1940 treatment by R. Thomas, described only as a 'restoration'. Andrew Petryn's cleaning took place intermittently between 1957 and 1959.

Comparing the early 20th-century images of the painting with its present appearance illustrates both the extent of the old restoration campaigns and the partial, incomplete nature of the 1950s cleaning by Petryn (Fig. 2). When the new treatment began in 2013, the painting was in a state of limbo. The 1950s treatment had partially removed layers of overpaint and pigmented coatings, leaving large remnants of these restoration materials unevenly distributed across the painting surface. These remnants can be considered in two main categories. The first type once aimed to refresh and alter the meanings of portions of the painted composition. Beneath these aesthetic restorations the original paint survives in relatively good condition. The second type of restoration aimed to mask areas of damage.

The first aesthetically motivated restoration we will consider is the dull grey and dark brown overpaint that covered the pedestal of the throne to the left of St Vincent Ferrer. The overpaint obscured the light and clear color of the original paint layer, which resembles a pinkish marble stone, and reinvented the shape of the pedestal by artificially extending the molding in the center of the top step (Figs 1 and 2).

A multilayer paint sample removed from a shaded area of the pedestal to the left of St Jerome and prepared as a cross-section revealed a relatively simple layer structure (Fig. 5). The gypsum-based ground layer is covered by a medium-rich preparatory layer (*imprimatura*), around 15 µm in thickness, which contains a heterogeneous distribution of fine lead white particles: a greater concentration of lead white is present in the upper half of the layer than in the lower half. This preparatory layer, which appears translucent under visible light and is highly autofluorescent under UV, is present in all cross-sections examined (see layer c in Fig. 4 and layer b in Fig. 5a) and appears to have been applied over the underdrawing that is visible as layer b in Fig. 4. Imaging and energy dispersive X-ray (EDX) analysis shows a gradient of lead-containing pigment particles with a higher concentration at the top of the layer and a lower concentration toward the bottom of the layer. Fourier transform infrared attenuated total reflectance (FTIR–ATR) spectroscopy of this layer measured perpendicular to the layer structure of cross-section ×12 (from the shaded area of the step of the pedestal in the foreground), suggests that the preparatory layer is a mixture of a drying oil with lead white. The drying oil-based paint layer, which is about 30 µm thick, consists primarily of a lead white pigment matrix with few red lake pigment particles (on an aluminum-containing mordant, likely hydrated alumina), coarse carbon black, dark iron oxide pigments (iron earth), and a few silica and aluminosilicate particles. The original paint layer is followed by a darkened coating layer (likely aged varnish) which is discontinuous and in some areas is adjacent to a thin layer of discontinuous light brown overpaint. This first layer of overpaint has vermilion and brown iron oxide pigments in a lead white matrix. Another layer of overpaint follows, off-white in color and consisting of lead white, bone black, carbon black and dark iron oxide pigments. A discontinuous varnish layer is present over some portions of the second overpaint layer.

A second significant area of intentional alteration is the reworking of St Jerome's robe. A dull and gloomy opaque grey overpaint was applied to obscure both the original translucent purple color of the robe and alter its folds, increasing the amount of fabric in the robe and stiffening the appearance of the drapery. The full impact of the reworking can be seen in the 1920s photograph (Fig. 1). These alterations make St Jerome seem heavier and more static. The overpaint causes the saint to appear still, whereas in fact he is shown in the act of stepping forward (Figs 1 and 2).

The original paint layer in St Jerome's robe consists primarily of lead white with a mixture of large red lake pigment particles (on an aluminum-containing mordant) and azurite particles (Fig. 4). The presence of angular silicon- and oxygen-rich particles that also contain sodium, magnesium, aluminum, potassium, calcium and manganese may represent the addition of crushed or ground glass. The paint layer appears fluidly brushed but is sufficiently viscous to retain a sense of body. Darker grey overpaint is a combination of drying oil, lead white and ivory or bone black while lighter grey overpaint appears to contain a combination of lead white, Prussian blue and finely ground black pigment particles, likely carbon black (Fig. 4, layer e).

The final example of reinterpreted imagery is in the background. Portions of the buildings and trees on the left side, behind St Vincent Ferrer, were added and revised after the completion of the painting. The pictorial quality of these trees is different from the trees on the right side of the painting. The blue accentuating strokes on the left-side tree branches have a gestural confidence that resembles the original, but they are thicker and more repetitive in direction and color than the right-side tree branches. These may be an old addition or the work of a different hand in the master's studio. Given the apparent age of the background reinterpretations, and the fact that the 1950s cleaning left them intact, they will remain on the painting.

The following instances of reworking were designed primarily to cover areas of damage, although parts of several also seem partially aesthetically motivated. The main types of damage they cover are those left by aggressive past cleaning treatments. These include marks from scraping tools, striations from a rough, sandpaper-like material, and losses with softened, eroded edges typically caused by powerful solvents. It is also possible that areas of loss were originally a result of the painting techniques used to create the altarpiece. Blistering and flaking have been observed in many works attributed to Piero di Cosimo (Walmsley 2015: 81).

The first example of restoration likely intended to repair rather than reinterpret is in the sky, which was entirely reinforced with a blue color slightly warmer than the original. The original paint, where it is exposed, is easily distinguishable by its greenish cast observed under UV light (Fig. 2). It is possible that there is extensive damage to the sky, but it is also possible that the overpaint was meant simply to refresh the color.

The faces of the two saints, the head of the lion, and the flowers in the foreground were reworked and repaired simultaneously. The large losses to St Vincent's face were covered with paint that also accentuated the sockets of his eyes and gave his countenance a sculptural quality. St Jerome's eyelids were covered with deep added shadows to create a similar effect. The restorations applied to the head of the lion cover a large elliptical loss presumably from a burning candle, and, like those applied to St Jerome, solidify the animal's features. The overpaint in these locations was almost entirely removed by the 1950s treatment.

Apart from the restoration of small scattered losses and restoration along the join lines between panels, the most extensive area of overpaint applied to cover past damages is in the foreground. In the images before 1950, this area appears as a relatively flat layer of color that supports 15 scattered flowers, eight of which were added by a later hand, and several of which were reinforced with extra petals and leaves. The 1950s cleaning removed the added flowers and partially removed the reinforced foliage but did not entirely remove the flat, homogenous layer of chalky green overpaint which surrounded them.

Cleaning principles and challenges

A few principles were critical in guiding our methodology. The work of art is a unique creation distinguished by its own history, an object that registers the historical events that have left their marks upon it.[5] Conservators, in collaboration with curators, look through each historical stratum and negotiate its retention or removal, based on a variety of criteria, including the desire to improve the legibility of the image. Subjectivity plays a key role in this process; the aesthetic and cultural background of the conservators and curator impact their visual perceptions of the work of art (De Azevedo Cagiano 1948).

The pursuit of these general intentions in the treatment process for Piero di Cosimo's altarpiece was challenged by the fact that the painting was in a state of visual disorder, with many types of restoration material that concealed parts of the original composition (Fig. 2). Clear, abrupt contrasts in value and tone occurred between the cleaned sections and those untouched or unresolved by prior procedures.

Material properties of the altarpiece presented additional challenges. The original paint film is very fragile. Throughout the foreground the original paint layer does not exhibit the expected resilient and hard quality of a paint film found by analysis to be drying oil. While the characterization of the paint film's mechanical properties with nanoindentation or other instrumental methods was not pursued (see for example Salvant *et al.* 2011 and Vlad-Cristea *et al.* 2012), observations by several conservators involved in the cleaning project converged in finding the layer soft when touched with the tip of a scalpel and in noting its tendency to shave rather than fragment during sampling.

Increase in the softness of dried paint layers may be attributed to saponification caused by the use of alkaline solutions (Hamburg and Morgans 1979: 72, 197; Burnstock and Learner 1992: 167). Saponification specifically refers to the hydrolysis of ester linkages in triglycerides under alkaline conditions, as a result of which 'soaps' or salts of fatty acids form. Hydrolysis of ester linkages may be catalysed by alkaline or acidic conditions. Acid-catalysed hydrolysis leads to the formation of an alcohol and a carboxylic acid. Burnstock and Learner (1992) and Stols-Witlox (2011) discuss multiple examples of historic cleaning protocols that involve alkaline solutions. During its long history, it is likely that the painting was subjected to procedures involving sodium hydroxide or potassium hydroxide solutions that would have led to the saponification of the paint film. Decreased stiffness and strength of paint layers, perceived as softness, can be attributed to hydrolysis, based on mechanical testing of paint systems designed to model the effect of hydrolysis of glycerin ester linkages to fatty acids (Tumosa *et al.* 2005). Mecklenburg *et al.* (2013) found that the use of certain pigments, including earth colors, results in extensive hydrolysis of paint films which relates to losses in stiffness and strength that conservators perceive as softness.

Chain scission during the aging of paint films also leads to the formation of low molecular weight compounds along with hydrolysis and saponification. Mills (1966), van den Berg and van den Berg (1999), and Erhardt *et al.* (2005) discuss chain scission (and subsequent hydrolysis) reactions leading to the formation of dicarboxylic acids such as azelaic acid. These low molecular weight compounds, in turn, act to plasticize the paint film and cause its softening (Tumosa *et al.* 1999: 349; Izzo 2010: 15). Bronken and Boon (2014: 256) point to 'excessive chain scission' as the cause of softness in certain 20th-century oil paints, which may be further exacerbated by hydrolysis of ester bonds. It is worth noting that plasticization of aged oil paint films may also be caused by the introduction of external agents, such as non-volatile solvents (Tumosa *et al.* 1999: 350) and waxes (Hamburg and Morgans 1979: 72).

The discovery of beeswax in blistered areas of the painting indicates that wax was used in a prior treatment, which may have contributed to the softening of the paint film. In several areas of the painting there are many minute blisters containing a white waxy substance which was identified as beeswax using FTIR and Py-GC-MS. A cross-section of multilayer paint removed from the edge of a blister shows the penetration of the wax between the *imprimatura* and ground layer. Although no known records document a consolidation treatment, we believe that since this material penetrated and flowed beneath the paint surface, it is likely a wax that was heated to a very high temperature and pressed down with a tool. IR spectra of several samples of original paint layers, varnish, and overpaint contain a shoulder around 1737–1739 cm^{-1} and weak bands around 730 and 720 cm^{-1}, a good match with features in beeswax reference spectra.

Given the large size of the altarpiece, and the extent of the overpaint remaining on its surface, it was logical for two conservators to work simultaneously on the cleaning process. This introduced a series of concerns regarding the reproducibility of the cleaning results and a need to find ways in which to calibrate individual tool use and perception to achieve consistency with regard to the level of cleaning and removal of overpaint. These constraints structured our chosen treatment framework, which is tailored to accommodate many criteria: our philosophies, the methodologies laid down by the previous unfinished cleaning, and the physical nature of the paint film, its fragility and its size. Since the historical strata of old restorations were already partially removed, it was not possible to selectively retain certain layers. The areas that were partially cleaned had to be brought to completion. Within these parameters, the cleaning would still retain a thin layer of varnish where it remains beneath the coatings and over the original paint layer, and would proceed without causing damage to this paint layer.

After observing the painting in both visible and UV light (Fig. 2), it was decided to selectively focus the cleaning on four areas of the altarpiece: the seven seams, the far left portion of the throne's pedestal behind St Vincent Ferrer, and the foreground. The upper 19th-century additions were removed in order to present the painting as a large fragment of an altarpiece.[6]

A carefully tailored selection of solvent mixtures, gels, and emulsions was developed, and used in conjunction with mechanical cleaning techniques. The resulting methods are able not only to harmonize and unify cleaned and uncleaned areas of the painting, but can also be adapted to the tendencies of different hands, allowing multiple conservators to work on the painting at the same time.

Genesis of the cleaning methodology

The development of the cleaning methodology took place in two phases. In the initial phase, the painting was surface-cleaned and the thickest passages of overpaint, namely those that covered the seams between the panels, were removed. In the second phase, a selection of gel formulations was established in collaboration with Richard Wolbers. These aimed to address the fragmentary overpaint and pigmented coatings that obscured intact original paint.

Analysis of coating or varnish layers as well as overpaint layers was pursued using FTIR as well as Py-GC-MS in order to help understand the difficulties experienced during attempts at their removal. All varnish and overpaint samples were found to contain drying oil, and Py-GC-MS also detected compounds characteristic of *Pinaceae* resins in one varnish sample. The IR spectra of two varnish layers indicated the presence of calcite, in one case together with basic lead carbonate. Overpaint layers contained lead white and in one instance bone black as well. Broad bands around 1650–1630 cm^{-1} and 1535–1520 cm^{-1} in all spectra indicate the presence of metal salts of carboxylic acids, such as lead carboxylates and calcium carboxylates, which likely formed in the varnish and overpaint layers as a result of reactions between the pigments and the drying oils. The low solubility of these compounds in organic solvents typically employed for varnish and overpaint removal added to the challenge of separating oil-based coatings from oil-based original paint layers (see Sawicka *et al.* 2014 for a discussion of strategies tested for the removal of lead soap efflorescence, which comprises lead carboxylates).

The cleaning process began with careful observation of the paint surface, the X-radiograph, the digital IR reflected image and the cross-sections. The first step was the removal of superficial dust and grime. For this operation, the following mixture was prepared: water 100 ml, mucin 0.2 g, triammonium citrate 0.2 g (Cremonesi 2001: 48–9). The solution was applied using cotton swabs and subsequently cleared with deionized water. To introduce new hands to the cleaning and ensure that the various conservators involved achieved similar cleaning levels, the leading conservator first demonstrated the process by connecting the microscope camera to a computer monitor. The new conservator then repeated the actions they had observed under supervision.

Over the centuries, six of the seven seams present in the panel have opened, causing cracks and losses in the ground and paint layer. The planks, no longer aligned correctly, had small steps between them. To compensate for both the damages and the misalignments, fills and retouching that aimed to level the surface had been applied beyond the losses and over large portions of the original paint layer (Fig. 2). We decided to initiate the cleaning in these sections because the discolored and chromatically altered restorations were particularly disturbing. They had the thickest accumulation of added material and divided the surface in vertical sections that interrupted the reading and flow of the composition.

Five to six layers of different discolored retouching and fills were found over the seams and removed, exposing large areas of original material that required some consolidation with a

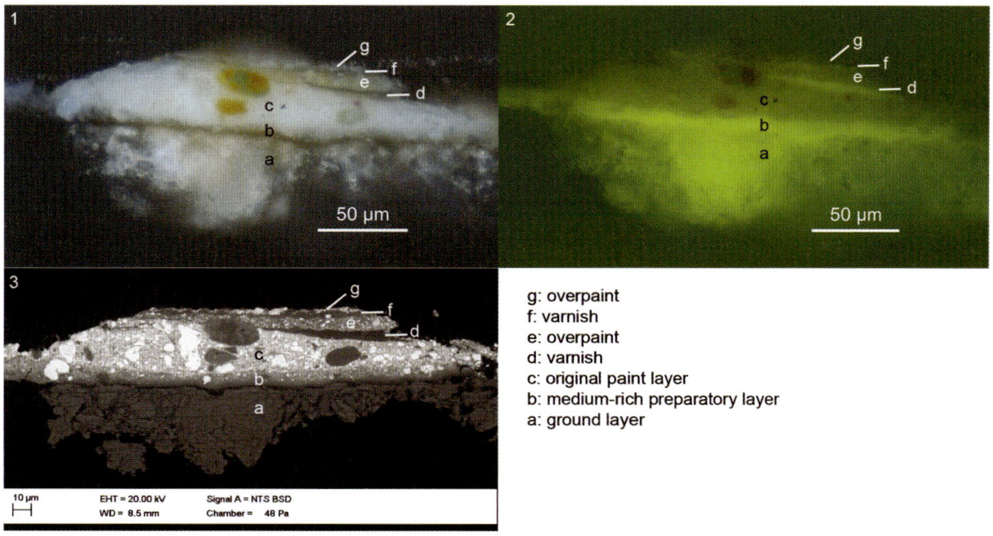

FIG. 5 Cross-section from the foreground (×6): (1) visible light; (2) UV-induced fluorescence with fluorescence filter set 09; (3) backscattered electron image.

3% to 5% solution of sturgeon glue. Initial solubility testing indicated that the oil-based restoration coatings were partially soluble in mixtures of polar solvents. A number of polar solvents, both in free solvent solutions and in gels, were tested. The most successful solvent was an acetone-benzyl-alcohol gel.[7] The gel was applied on the surface with a number 00 sable brush and left to work for roughly one minute. The gel was then removed with a dry swab and the area cleared with a 1:1 solution of isopropanol and Shellsol odorless mineral spirits. Two to three applications were necessary to clear the majority of the overpaint; the remaining residues were removed mechanically. The fill materials, made most likely from a mixture of gypsum and animal glue would soften with a water in oil emulsion (Cremonesi 2001: 52–6).[8]

The gel system developed for removal of the overpaint along the seams was not as effective for certain areas of overpaint in other locations. In the winter of 2014 the conservation department of the Yale Art Gallery consulted Richard Wolbers on the development of a cleaning system for the Piero di Cosimo altarpiece. During the workshop a variety of xanthan gum emulsions, specifically tailored for the requirements of the altarpiece, were prepared. Xanthan gum is a polysaccharide produced by the fermentation of a pure culture of bacteria, the *Xanthomonas campestris*. Inexpensive, safe, and accessible, it has been used for decades in both the cosmetic and food industry for gelling, for emulsions, and as a thickener and stabilizer for suspensions in water-based formulations. Xanthan gum is soluble in water and its viscosity is not affected by changes in pH values or temperatures. Xanthan viscosity can be increased by the addition of a wide range of solvents in small concentrations. The fact that xanthan can carry a wide range of solvents, chelating materials and remain in a gel form at a wide range of pH values makes it an ideal solvent carrier. From a practical point of view, a xanthan gel at a concentration of 1–2% forms a gel that can easily be picked up with a sable brush, applied on the surface and delicately agitated; the gel can be removed with a dry swab and the area cleared first with a solution of water and then with mineral spirits (see also Stavroudis and Wolbers 2012).

Cleaning of the left portion of the throne's pedestal behind St Vincent Ferrer

In this area, two layers of overpaint were present. In the top section of the pedestal, a thick, light grey overpaint lay directly on the pink color of the original paint layer. In the lower section, a dark brown layer was found below the grey overpaint. Beneath the dark brown layer was a thin layer of old but not original varnish. The cleaning progressed from the top edge of the pedestal to the lower edge.

A variety of polar solvents was tested on the pedestal, in the same manner as for the cleaning of the overpaint present along the seams. The mixture most successful for softening and solubilizing the top layer of grey overpaint was the acetone-benzyl-alcohol gel discussed above. The gel was applied to the overpaint either once or twice, with residues removed mechanically. The lower dark brown layer remained mostly intact. To reduce the lower brown layer, different emulsions were prepared from a 2% stock gel of xanthan gum in deionized water. By varying three key elements inside the gels – adjusting the pH, changing the chelating agent and altering the concentration of the emulsified solvent – it was possible to tailor them to the solubility parameters of the brown layer.

The adjustment to the gels was conducted during cleaning tests and the results were monitored under a microscope. Five different emulsions were prepared from a pH of 6 to a pH of 10 and used in response to the thickness and composition of the materials intended to be solubilized. For each prepared gel at a specific pH value, different concentrations of benzyl alcohol were added to the emulsion, namely 5, 10 and 15%. The immiscible solvent is suspended in the emulsion in minute droplets. The emulsified solvent was applied to the surface with a modified 00 sable brush, shaved to an angle to allow for a better traction with the restoration layer. The emulsion was then gently agitated with the brush in order to allow for the small droplets of solvent suspended in the gel to directly contact the surface. The time of contact changed depending on the areas, nevertheless it never exceeded 50

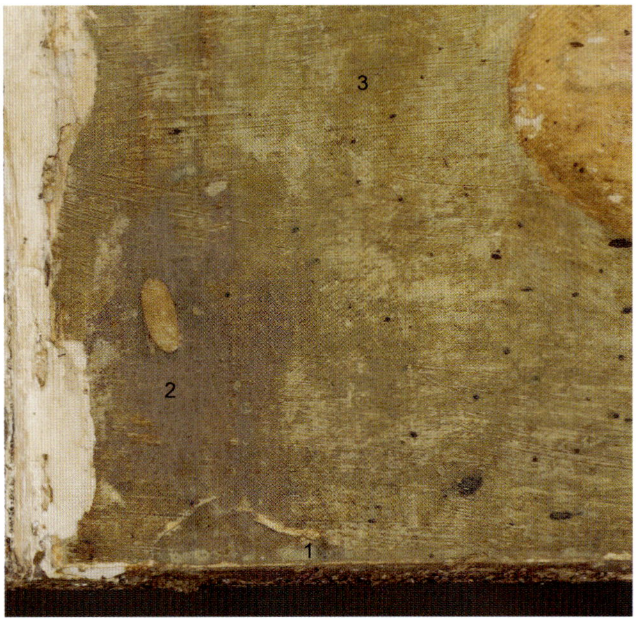

FIG. 6 Lower right foreground during current treatment showing (1) residue of the chalky green layer; (2) thick unevenly distributed grey overpaint and (3) a pigmented layer with high oil-resin content.

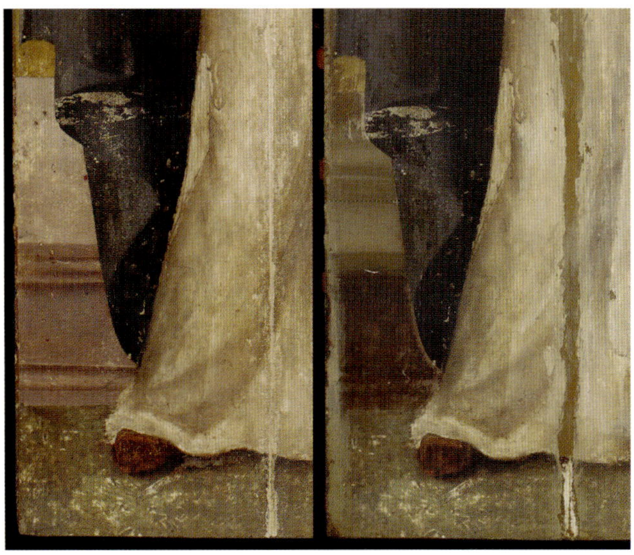

FIG. 7 During treatment (*left*) and before treatment (*right*) of the lower left corner, showing removal of overpaint applied to the pedestal with retention of old, thin varnish layer.

seconds. During the process the gel also becomes cloudier and assumes the color of the material that is softened. All the operations were monitored under a microscope, primarily at a magnification between ×16 and ×40.

The material best suited to reduce the aged and degraded oleo-based brown overpaint was a xanthan gum emulsion at a pH of 10 and containing 15% benzyl alcohol. As we progressed closer to the original paint surface, on top of which an old layer of varnish was present, we opted to use milder xanthan gum emulsions at a pH of 7 in order to avoid damaging the original paint film.[9] Mechanical cleaning was used at the layer closest to the paint surface to preserve the thin and old varnish while still removing the residues of the overpaint. The scalpel blades were shaped and sharpened in order to conform the tip of the tool to the paint topography. The sliding of the blade onto the surface is improved by wetting it with Shellsol odorless mineral spirits, which reduces the traction between the restoration materials and the sharp metal. The mineral spirits also saturates the surface and allows a better understanding of the different physical and optical qualities of the original and restoration materials.

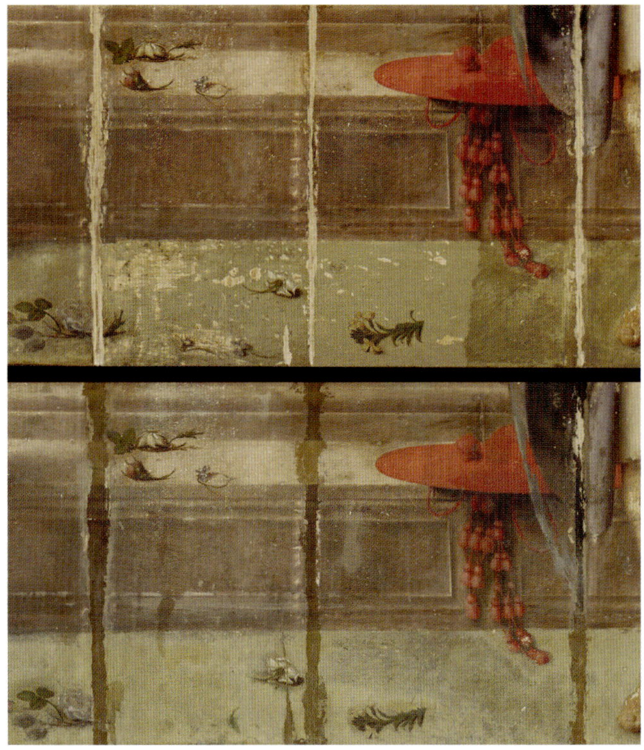

FIG. 8 During treatment (*top*) and before treatment (*bottom*) of the lower center foreground, showing the removal of the chalky green layer, grey overpaint, and pigmented oil-resin coating with retention of the old, thin varnish layer.

Cleaning of the foreground

The foreground presents a particularly complex stratification of different layers of overpaint and pigmented coatings. These layers are not as discrete as those observed in the pedestal; they physically and visually impact each other. As in other areas, the primary challenges of the cleaning methodology are the fragility of the paint layer and its uneven state of preservation (Fig. 5).

The initial solubility test identified the acetone-benzyl-alcohol gel and the xanthan gum gels mentioned previously as the most effective cleaning materials. The combination of these gels with mechanical cleaning enabled selective and highly controllable results. Although similar actions and tools are used across the foreground, the process itself varies in correspondence with the different state of preservation of the paint film and the physical and mechanical qualities of the layers of pigmented and unpigmented coatings.

The first layer of chalky light green overpaint, also present in the left portion of the throne's pedestal behind

St Vincent Ferrer, was removed without difficulty with the acetone-benzyl-alcohol gel following the same procedures described previously. Below this chalky green layer are additional unevenly distributed pigmented coatings: a thick grey overpaint, which is thickest in the lower left-hand corner and becomes more fragmentary towards the center, and a pigmented layer with a high oil-resin content. In some areas the grey overpaint is directly on top of the original paint; in other areas, it is on top of the pigmented layer. An old, also uneven layer of yellowed varnish is present directly over the original paint layer (Fig. 6).

The appearance and behavior of the original paint layer in the foreground offers indications of the various types of damage it has experienced. Where the thin yellow varnish layer is mostly intact, the sensation of the original paint layer is more stable; where the grey overpaint rests directly on the original it fragments more easily. In locations where the various layers of coatings were removed by the 1950s cleaning, the old light yellow varnish layer is entirely absent. The paint layer itself appears slightly thinned, as it is quite smooth, without any trace of the pronounced brushstroke texture visible in areas of the foreground left uncleaned. To cause this erasure of texture, a strong solvent was likely applied and the surface then rubbed. Other locations not touched by the 1950s cleaning appear to have been affected by older aggressive attempts to remove overpaint. Beneath the opaque grey layer, the original paint layer occasionally shows striated damages that suggest the use of an abrasive, sandpaper-like material.

The cleaning process for these two lower layers involves four main steps. The xanthan gum gel with 15% benzyl alcohol, and citric acid as the chelating agent at a pH of 7, is applied to the restoration coatings in the same manner as described before. Mechanical cleaning becomes a greater part of the process in this area due to the sensitivity of the original paint layer. These steps vary in response to the different characters of the painting surface. In the areas of thick grey overpaint with large pigment particles, the gel is worked into the overpaint surface with the small modified brush for a longer amount of time. The binding medium is softened and solubilized and the emulsion gathers up the pigment particles. The physical action of working the gel into the surface allows the suspended solvent droplets to establish a continuous contact time that may vary in each application according to the choice of the conservator. The action of the brush is a key element that influences the effectiveness of the gel. Each physical motion with the brush refreshes the action of the emulsion by bringing new micelles to contact the surface. Particularly thick areas of overpaint can require multiple gel applications, confirming that the gel acts only superficially; it does not penetrate through the paint film.

Where the overpaint is thick the cleaning becomes more tactile. The weight of the scalpel handle, the shaped blade, the controlled circular motions, and different levels of pressure of the hand allow the softened overpaint, which remains slightly resistant after the application of the gel, to be shaved. All cleaning takes place through the microscope, again at a magnification between ×16 and ×40.

Since the lower, thinner coatings are unevenly distributed, several cleaning passes can be made to bring their appearance to an even level. For these levelling stages, the scalpel blade is the primary tool used for cleaning, and the gel is not usually reapplied although this varies from area to area. The exposed original paint is extremely fragile and is left protected by a thin layer of uneven varnish. The multiple layers of overpaint coating St Vincent Ferrer's proper right shoe were also addressed as part of the foreground cleaning. The overpaint layers, although different in color from the foreground, followed a similar layering structure, with a thicker brown overpaint covering several pigmented layers. The cleaning process for the shoe followed the same procedures as described above.

Conclusion

To this point, the ongoing cleaning of the Piero di Cosimo altarpiece has unveiled extensive areas of original, intact paint (Figs 7 and 8). While some revealed areas are better preserved than others, the cleaning is raising the legibility of the image to new levels of comprehension, and enabling new insights into the process of the painting's manufacture.

The restoration of Piero di Cosimo's altarpiece has a pedagogical component. Several interns and fellows have had the opportunity to participate in the treatment. Within the minds of the conservators involved, visual recordings of the restoration coatings and of the original painted surface coexist with the practical muscle memories of the cleaning actions. These recordings influence motor learning. Through practice of varied but similar motions and adaptations, an archive of tactile skills is recorded. Those skills will crucially inform the present and future choices of materials and methods. All collaborators have developed the ability to coordinate and compensate for different hands, cleaning approaches and inherent subjectivities in order to obtain a balanced solution. When restoration is understood as a continuous revealing process of an active critical work, the work of art becomes the source of our knowledge.

Analytical methods

Optical microscopy

Scraped powder samples and un-mounted multilayer samples were examined and documented using a Zeiss SteREO DiscoveryV20 stereomicroscope equipped with a Zeiss AxioCam M5Rc digital color camera. Polished cross-sections were examined and imaged with reflected visible light in dark-field illumination mode using a Zeiss Axio Imager M2m microscope equipped with a Zeiss AxioCam M5Rc digital color camera. The same was used to document UV-induced fluorescence of the cross-sections with fluorescence filter set 09 (excitation: bandpass 450–490 nm, emission: longpass 515 nm). Image acquisition and processing for both microscopes was performed using Zeiss Zen 2012 software.

Scanning electron microscopy–energy dispersive X-ray spectroscopy (SEM–EDX)

Powder samples were placed onto double-sided carbon-filled adhesive discs affixed to 12.7 mm diameter aluminum pin stub mounts. Cross-sections were held in an aluminum multipurpose variable holder. A Zeiss EVO MA 15 variable-pressure SEM equipped with a tungsten filament emission system was used for imaging at chamber pressures between 40 and 60 Pa, achieved by bleeding nitrogen gas into the chamber. Backscattered electron images were acquired with 4 quadrants of a lens-mounted 5-channel diode-based detector. The SEM was operated using SmartSEM software. Elemental analysis was performed using an EDX spectrometer with an Octane silicon drift detector operated using TEAM software. The analysis was carried out at an 8.5 mm working distance, using 20 kV or 22 kV accelerating potential. The beam current was adjusted to keep dead time between 10 and 30%.

Fourier transform infrared (FTIR) spectroscopy

Samples were flattened between two diamond windows in a diamond micro-compression cell. The cell was then disassembled and spectra were measured in transmission while the sample was supported on a single diamond window using a ×15 infrared objective on a Thermo Continuum infrared microscope with an MCT-A detector, using a wavelength range of 4000–650 cm^{-1} and 4 cm^{-1} resolution, coupled to a Nicolet 6700 bench-top spectrometer. A series of infrared spectra along a linear path (line scans) at 1 μm intervals was also collected from cross-section ×12 using a slide-on ATR objective with a germanium (Ge) conical shaped crystal (25° medium angle of incidence, 100 μm contact area, sampling area restricted to 30 × 30 μm using aperture blades). Spectra were examined and compared to reference spectra using Omnic 8.3 software (Thermo Scientific).

Pyrolysis gas chromatography–mass spectrometry (Py–GC–MS)

Samples or portion of samples were transferred to a small pyrolysis cup to which 3 μl tetramethylammonium hydroxide (TMAH) derivatizing agent in methanol (25 wt%) was added. Samples were then introduced into a Frontier Lab pyrolysis unit's (PY-3030D-110) furnace at 550 °C. The volatilized components of the sample were separated on an Ultra-Alloy-5 capillary column (30 m × 0.25 mm, 0.25 μm phase coating, 5% diphenyl, 95% dimethylsiloaxane phase) after passing through a split-splitless inlet (ratio 25:1, split flow 29 mL per minute) heated to 320 °C. The Agilent 7890A GC oven heated the column from an initial temperature of 40 °C (2 minutes) to 320 °C at a ramp rate of 6 °C per minute and maintained the final temperature for 10 minutes. The mass spectrum of the separated components was collected using an Agilent 5975C mass selective detector. Mass spectra were compared to NIST library standards.

Notes

1. For a thorough description of the known history of the painting, see Geronimus *et al.* 2015.
2. A brief conservation report compiled by Andrew Petryn after the treatment lists the solutions and solvents he used: 'naptha, triple-distilled turpentine, xylene, acetone, morpholine and dimethylformamide'. See conservation archive, Yale University Art Gallery.
3. Carlos Moya contributed significantly to the treatment of the painting from 2013 to 2014, through a fellowship funded by the National Endowment for the Arts.
4. The added sections of wood in the upper portion of the altarpiece, datable as a 19th-century restoration since they compensated for the removal of the standing angels, both confused its silhouette and emphasized the fragmentary shape of the original painting. Since they had been partially cleaned, with traces of blue paint remaining and a creamy white preparatory layer left exposed, they no longer approximated the sky color of the altarpiece interior. The sections of wood used to complete the arch were taken from another painting. A portion of what seems to be a hand, painted in a style that appears somewhat later, can be seen in the section added above and to the right of the Madonna.
5. See Baldini 1991–92; Brandi 2005; Philippot 2004.
6. After extensive discussion, it was decided to remove the 19th-century additions to the upper portion of the altarpiece. These additions consisted of 10 wooden inserts, glued together to create an oval profile. The wood used to make the inserts was reused from other old panels as clearly evident from a triangular-shaped insert at the top which includes a set of painted fingers that do not belong to Piero di Cosimo's altarpiece. Seven inserts were placed with the wood grain running vertically, while the two top side ones were glued across the grain direction. The small triangular insert, which includes the elbow of the Christ Child, is original and was not removed. The process started by first clearing the preparation layers applied over the seams where the additions met the original support. The exposed seams guided and facilitated the removal phase. A few of the seams that connected the 10 inserts were already weakened and open. The first step was to slide out the two top horizontal members of the cradle. Then, beginning from the top, the two outer inserts were disconnected from the cradle and the other inserts with a flexible, small Japanese saw. Afterward, perpendicular cuts using a larger Japanese saw were made to the remaining seven inserts. They were carefully detached from the cradle and the original margins of the altarpiece. The protruding vertical and horizontal members of the cradle were then trimmed to follow the new profile of the altarpiece.
7. The acetone-benzyl-alcohol gel included in the mixture: water 25 ml, Carbopol C934 1.5 g, benzyl alcohol 6.5 ml, acetone 25 ml, triethanolamine 5 ml. Good results were also obtainable with *N*-methyl-2-pyrrolidone Carbopol gel, when the following mixture was used: Ethomeen C-25 20 ml, Carbopol C934 2 g, *N*-methyl-2-pyrrolidone 100 ml, water 12 ml. These gels, particularly suitable for oil-based overpaint and oleoresinous coatings, were discussed at a 2007 workshop organized by Paolo Cremonesi at the Opificio delle Pietre Dure, Florence, Italy. See also Wolbers 2000.
8. This *emulsione grassa* described by Paolo Cremonesi was applied with a brush and then cleared with a dry swab and Shellsol odorless mineral spirits (OMS). The emulsion

contains a mixture of: Brij 35 2 g, water 10 ml, Tween 20 2 ml, OMS 90 ml (see also Cremonesi 2001). For more insoluble gypsum-based fills the following mixture worked well: water 100 ml, citric acid 2 g, Triton X-100 0.5 ml, methyl cellulose 1.5 g. For this recipe and its specific use see Wolbers 2000: 119–20.

9. The emulsion comprised: 100 ml xanthan gum gel stock/30 ml 1% citrate solution buffered to pH 7 with diethanolamine and 15% benzyl alcohol.

References

Aronson, M., McClure, I. and Passeri, I. 2017. 'The art of conservation IX: the history of painting conservation at the Yale University Art Gallery', *The Burlington Magazine* 1367(159): 122–31.

Baldini, U. 1991–92. *Teoria del restauro e unita di metodologia*. Florence: Nardini.

Brandi, C. 2005. *Theory of Restoration*, C. Rockwell (tr.). Rome: Istituto centrale per il restauro and Florence: Nardini.

Bronken, I.A.T. and Boon, J.J. 2014. '*Hard Dry Paint, Softening Tacky Paint, and Exuding Drips on Composition* (1952) by Jean-Paul Riopelle', in K.J. van den Berg, A. Burnstock, M. de Keijzer, J. Krueger, T. Learner, A. de Tagle and G. Heydenreich (eds), *Issues in Contemporary Oil Paint*. Cambridge: Springer International Publishing, 247–62.

Burnstock, A. and Learner, T. 1992. 'Changes in the surface characteristics of artificially aged mastic varnishes after cleaning using alkaline reagents', *Studies in Conservation* 37: 165–84.

Cremonesi, P. 2001. *L'uso di tensioattivi e chelanti nella pulitura di opere policrome*. Padua: Il Prato.

De Azevedo Cagiano, M. 1948. *Il gusto nel restauro delle opere antiche*. London: Olympus Publishers.

Erhardt, D., Tumosa, C.S. and Mecklenburg, M.F. 2005. 'Long-term chemical and physical processes in oil paint films', *Studies in Conservation* 50: 143–50.

Geronimus, D. 2006. *Piero di Cosimo: Visions Beautiful and Strange*. New Haven: Yale University Press.

Geronimus, D., Brilliant, V. and Franklin, D. 2015. 'Catalogue' in G. Hirschauer, A. Luchs, S. Padovani (eds), *Piero Di Cosimo: The Poetry of Painting in Renaissance Florence*. London: Lund Humphries.

Hamburg, H.R. and Morgans, W.M. (eds) 1979. *Hess's Paint Film Defects*. London: Chapman and Hall.

Izzo, F.C. 2010. *20th-Century Artists' Oil Paints: A Chemical-physical Survey*, unpublished dissertation, Department of Chemistry, Università Ca' Foscari Venezia, Venice, Italy.

Jarves, J.J. 1861. *Art Studies: The 'Old Masters' of Italy*. New York: Derby and Jackson.

Mecklenburg, M.F., Tumosa, C.S. and Vicenzi, E.P. 2013. 'The influence of pigment and ion migration on the durability of drying oil and alkyd paints', in M.F. Mecklenburg, A.E. Charola, and R.J. Koestler (eds), *Cleaning 2010: New Insights into the Cleaning of Paintings*. Washington, DC: Smithsonian Institution Scholarly Press, 59–67.

Mills, J.S. 1966. 'The gas chromatographic examination of paint media. Part I. Fatty acid composition and identification of dried oil films', *Studies in Conservation* 11: 92–108.

Philippot, P. 2004. 'La notion de Patine et Le Nettoyage des Peintures', G. White (tr.) in D. Bomford and M. Leonard (eds), *Issues in the Conservation of Paintings*. Los Angeles: The Getty Conservation Institute, 391–95.

Rankin, W. 1895. 'Some early Italian pictures in the Jarves Collection of the Yale School of Fine Arts at New Haven', *American Journal of Archaeology and the History of the Fine Arts* 10 (2): 137–51.

Salvant, J., Barthel, E. and Menu, M. 2011. 'Nanoindentation and the micromechanics of Van Gogh oil paints', *Applied Physics A* 104(2): 509–15.

Sawicka, A., Burnstock, A., Izzo, F. C., Keune, K., Boon, J.J., Kirsch, K. and van den Berg, K.J. 2014. 'An investigation into the viability of removal of lead soap efflorescence from contemporary oil paintings', in K.J. van den Berg, A. Burnstock, M. de Keijzer, J. Krueger, T. Learner, A. de Tagle and G. Heydenreich (eds), *Issues in Contemporary Oil Paint*. Cambridge: Springer International Publishing, 311–32.

Stavroudis, C. and Wolbers, R. 2012. 'Aqueous methods for the cleaning of paintings,' in J.H. Stoner and R. Rushfield (eds), *Conservation of Easel Paintings*. New York: Routledge, 500–524.

Stols-Witlox, M. 2011. 'Historical restoration recipes: the cleaning of oil paintings 1600–1900', in J. Bridgland (ed.), *Preprints of the ICOM-CC 16th Triennial Conference*. Almada: Critério. [CD only]

Tumosa, C.S., Millard, J., Erhardt, D. and Mecklenburg, M.F. 1999. 'Effects of solvents on the physical properties of paint films', in J. Bridgland (ed.), *Preprints of the ICOM-CC 12th Triennial Meeting*. London: James & James, 347–52.

Tumosa, C.S., Erhardt, D., Mecklenburg, M.F. and Su, X. 2005. 'Linseed oil paint as ionomer: synthesis and characterization', in P.B. Vandiver, J.L. Mass and A. Murray (eds), *Materials Issues in Art and Archaeology VII: Materials Research Society Symposium Proceedings*, vol. 852. Warrendale, PA: Materials Research Society, 25–31.

van den Berg, J.D. J. and van den Berg, K.J. 1999. 'Chemical changes in curing and ageing oil paints', in J. Bridgland (ed.), *Preprints of the ICOM-CC 12th Triennial Meeting*. London: James & James, 248–53.

Venturini, L. 1994. *Piero di Cosimo e la Pala della Compagnia di San Vincenzo*. Florence: Sansoni.

Vlad-Cristea, M., Riedl, B., Blanchet, P. and Jimenez-Pique, E. 2012. 'Nanocharacterization techniques for investigating the durability of wood coatings', *European Polymer Journal* 48(3): 441–53.

Walmsley, E. 2015. '"A very rich and beautiful effect": Piero's painting technique', in G. Hirschauer, A. Luchs and S. Padovani (eds), *Piero Di Cosimo: The Poetry of Painting in Renaissance Florence*. London: Lund Humphries, 72–81.

Wolbers, R. 2000. *Cleaning Painted Surfaces: Aqueous Methods*. London: Archetype Publications.

Authors' addresses

- Irma Passeri, Yale University Art Gallery, New Haven, CT, USA (irma.passeri@yale.edu)
- Annika Finne, Yale University Art Gallery, New Haven, CT, USA
- **Anikó Bezur**, Institute for the Preservation of Cultural Heritage, Yale University, West Haven, CT, USA

Time-dependent ATR–FTIR studies on the release of solvents from cleaning gels into model systems of oil paint binding media

Lambert Baij, Katrien Keune, Joen Hermans, Petria Noble and Piet Iedema

Introduction

Recently, the focus of cleaning science has shifted towards the confinement of solvents into emulsions and solvent gels for the controlled application of solvents (Wolbers 2000; Domingues *et al.* 2013; Samorì *et al.* 2016; Angelova *et al.* 2016). These thickened solvents and gel systems not only reduce the mechanical action necessary for the cleaning of paint surfaces, but aim to provide the conservator with superior kinetic control of the cleaning process (Baglioni *et al.* 2014, 2015). Unilateral nuclear magnetic resonance experiments on acrylic emulsion paints have shown that the amount of water delivered into the paint by agar gels is on a par with swab cleaning with water, while other gels have shown higher water release (Angelova *et al.* 2016). Combined mass analysis and NMR experiments on oil paint mockups have demonstrated a decreased uptake of solvent by the paint when using gelled systems (Fife *et al.* 2015). Domingues *et al.* (2014) studied the water released by gels on paper by mass analysis and found that the water released by the newly developed rigid gels is significantly lower than that of the polysaccharide gels agar and gellan. In this light, we studied the kinetics of solvent delivery of three cleaning gels on our model systems for aged oil paint binding mediums. These model systems are polymers based on a polymerized linseed oil matrix with Zn^{2+} and Pb^{2+} ions as an integral part of the network, similar to aged binding mediums in paintings (Hermans *et al.* 2015, 2016).

For this study we developed a method based on time-dependent attenuated total reflection Fourier transform infrared (ATR–FTIR) spectroscopy that gives detailed chemical information over time and designed a sample cell that enables careful control of the experimental conditions. Spectral processing and the application of a diffusion model allow for the determination of diffusion coefficients (Fieldson and Barbari 1993). Using this method, the diffusion of a wide range of solvents commonly used for the cleaning of paintings is studied in binding medium models systems containing zinc (Znpol) and lead (Pbpol) ions. In addition, the diffusion of water and ethanol from three conservation cleaning gels – Nanorestore Max Dry, agar and gellan – is studied and compared to free solvent diffusion from a solvent reservoir above the surface of the binding medium model. For free solvents, diffusion and swelling are governed by the equilibrium conditions at the interface of the solvent and polymer

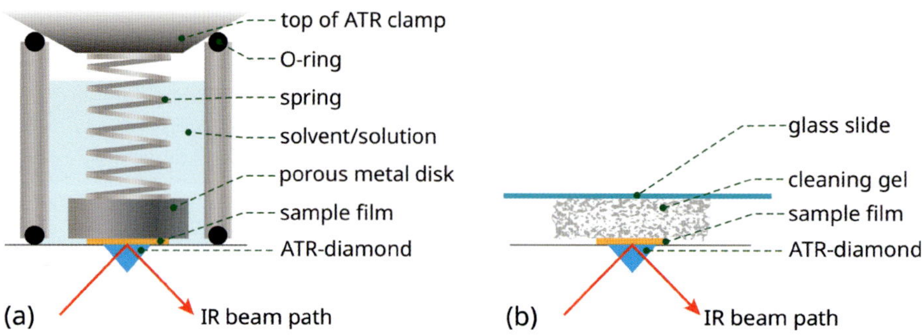

FIG. 1 Illustration of (a) the sample cell used for time-dependent ATR-FTIR measurements of polymer films in contact with solvents or solutions and (b) the setup used for the release of solvent from cleaning gels. For setup (b) the glass sheet lowers evaporation of solvent from the top of the gel.

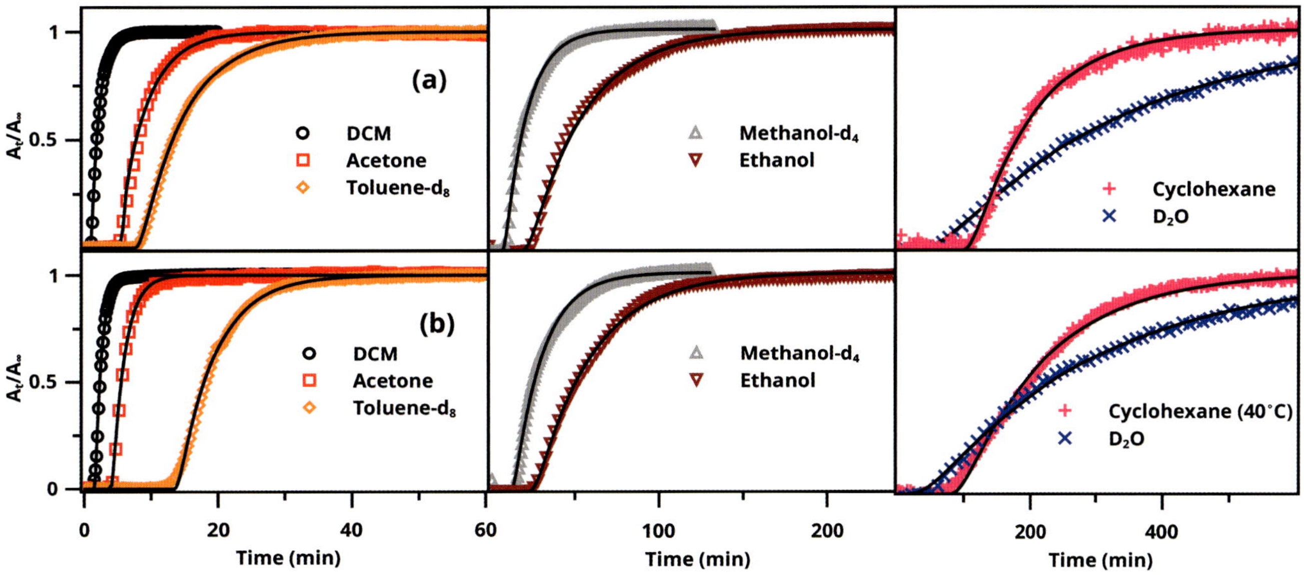

FIG. 2 Concentration profiles of several solvents in (a) zinc ionomer and (b) lead ionomer films of 135–160 µm thickness. Solid black lines represent best fits of the adapted Fickian diffusion model with a lag time τ as extra parameter. For D_2O, τ was estimated. Note the different timescales in each graph. Only a fraction of the collected data points are shown for clarity.

and inside the polymer. In contrast, the observed diffusion for liquids from gelled systems is the result of a combination of equilibrium conditions inside the gel, at the interface and inside the polymer. If the process inside the gel or at the interface is rate-determining, this will be observed as 'retention'.

Experimental

Sample preparation

Binding medium model systems were made according to previously published procedures by Hermans *et al.* (2016) with a metal ion concentration in the polymer films of roughly 160 mM (or TAG : Pb/Zn = 1:0.44 in the initial uncured sample mixture). Agar was purchased from Sigma Aldrich and Gellan Kelcogel CG-LA (LOT NO 3H7072A) was purchased from Azelis and used as received. Gels were made by mixing 4 wt% (agar) and 3 wt% (gellan) in deionized water at 100 °C while stirring for 15 min and subsequently cast on a flat surface. Nanorestore Max Dry was used as received from CSGI.[1] Gels were kept in a sealed container loaded with ethanol or D_2O for at least 12 hours before use. Complete exchange of solvents inside the gels upon loading was not verified analytically. For agar and gellan, no detectable saccharide residues were found in the D_2O supernatant as analysed by 1HNMR while saccharide residues were found in the ethanol supernatant. All gels were blotted with paper tissue and flushed with a flow of N_2 until dry on the surface before application.

Experimental setup

A Perkin-Elmer Frontier FT-IR spectrometer fitted with a Pike GladiATR module that included a heated top plate and a diamond ATR-crystal (ø = 3 mm) was used. In order to measure spectra of polymer samples while they were exposed to solvents or solutions, a custom-built stainless steel cylinder was designed as illustrated in Fig. 1a. As indicated in Fig. 1b, a glass sheet was used to slow down solvent evaporation from the top of the gel.

Data processing

Diffusion in polymer films was measured for cyclohexane (904 cm^{-1}), ethanol (879 cm^{-1}), acetone (529 cm^{-1}), D_2O (2510 cm^{-1}), methanol-d_4 (2485 cm^{-1}), DCM (1265 cm^{-1}), and toluene-d_8 (541 cm^{-1}). The wavenumbers in parentheses for each solvent refer to the positions of a characteristic solvent band sufficiently isolated for accurate band integration. Time-dependent IR absorption band areas were calculated with Perkin-Elmer TimeBase software and used for model fitting without further processing.

Results and discussion

Diffusion of pure solvents in ionomer films

Solvent diffusion in model oil paint binding media was investigated for a range of solvents commonly used for the cleaning of oil paintings. Fig. 2 shows the normalized concentration profiles over time of various solvents in zinc and lead ionomer systems. Values of diffusion coefficients were obtained by fitting the data to Fick's law for diffusion adapted for ATR–FTIR (Fieldson and Barbari 1993). To account for the time before solvent reaches the detector, a lag time parameter τ was employed as an extra parameter to define the onset of concentration increase after the initial swelling period. Fits

TABLE 1 Diffusion coefficients (D) and lag times corrected for thickness (L^2/τ) resulting from a fit Ficks law to the concentration profiles in **FIG. 2**. Swelling factors ($\Delta A/A_{min}$) were calculated by taking the ratio of the highest and lowest intensity of the ester carbonyl band during swelling.

Sample	Solvent	D	L^2/τ	$\Delta A/A$min
		(10^{-8} cm²/s)	(10^{-8} cm²/s)	
Znpol	DCM			
		138	320	1.21
	acetone	61	105	0.73
	toluene–d^8	25	56	0.71
	methanol–d^4	10	41	1.1
	ethanol	4.3	21	0.89
	cyclohexane	1.4	3.7	0.24
	D^{2}O	0.61	8.2	0.05
Pbpol	DCM			
		140	270	1.13
	acetone	61	84	0.89
	toluene–d^8	31	29	0.81
	methanol–d^4	7.2	31	0.56
	ethanol	4.0	16	0.87
	cyclohexane	1.1	5.5	0.45
	D_2O	0.63	8.2	0.13

to the diffusion data were obtained with the band area A_∞, lag time τ and diffusion coefficient D as independent parameters.

Fig. 2 shows that the adapted Fickian diffusion model provides excellent fits for the diffusion of all investigated solvents. Therefore, once the solvent front reaches the detector (at $t = \tau$), diffusion becomes Fickian, implying that solvent–polymer interactions remain constant (van der Wel and Adan 1999). Possible non-Fickian behaviour during solvent swelling (at $t < \tau$) is accounted for by the lag time parameter τ. The reproducibility of the concentration profiles was investigated by repeating the acetone diffusion experiment eight times on Znpol films with a thickness that varies between 124 and 159 μm. The standard deviation in the diffusion coefficient in this set of measurements was 10%.

In general, diffusion coefficients for solvents in zinc and lead ionomer systems were found to be similar. However, a significant variation in the rate of diffusion between different solvents was detected, D_{DCM} being roughly 200 times larger than D_{D_2O}. These values are similar to those determined by Stolow (1957a,b) for swelling experiments on thermally polymerized stand-oil films. The trend in diffusion rates for different solvents is correlated to the kinematic viscosity, with the more viscous solvents having the lowest diffusion coefficients. Similar observations were made by Stolow (1957a,b). The diffusion of D_2O is much slower than what is expected from kinematic viscosity, but is comparable to values reported for water vapour diffusion in pigmented alkyd paints (van der Zanden and Goossens, 2004) and vapour diffusion in historical samples of Prussian blue and basic lead carbonate (Chan and Odlyha 1995).

The lag time τ is a function of film thickness L and the quantity L^2/τ (lag time corrected for film thickness) is known to be proportional to the diffusion coefficient (Vergnaud 1991). Table 1 confirms this, although again water deviates from the trend. The unexpectedly small τ for D_2O is clearly seen in Fig. 2, where the water concentration profile starts before that of cyclohexane, while it is subsequently much slower to reach saturation (for a detailed discussion see Hermans 2017).

The swelling capacity for each solvent was investigated by monitoring IR bands corresponding to the polymer sample. Taking the ratio of the difference in absorption of the ester carbonyl band (1740 cm^{-1}) at saturation and at $t = 0$ ($\Delta A/A_{min}$), a measure is obtained for the degree of swelling. Table 1 shows that DCM swells the investigated model systems more than 110%, while water hardly swells the polymer system at all (5–13% for zinc and lead, respectively). Monitoring the decrease of IR absorption bands of the polymer confirms the low swelling capacity of water and reproduces trends in solvent swelling power reported in the literature (Browne 1956; Zumbühl 2014).

Diffusion of solvents from conservation cleaning gels

We have evaluated the diffusion of water and ethanol from three rigid gels that are used for the cleaning of paintings. This series consists of two polysaccharide gels, agar and gellan, and a type of conservation cleaning gel consisting of a polymer blend commercialized under the name Nanorestore (Domingues *et al.* 2013, 2014; Lahaye and Rochas 1991). While the overall structure remains intact (Scott 2012; Pavitt 2012), minor mechanical and optical changes (decreased opaqueness, increased rigidity) were

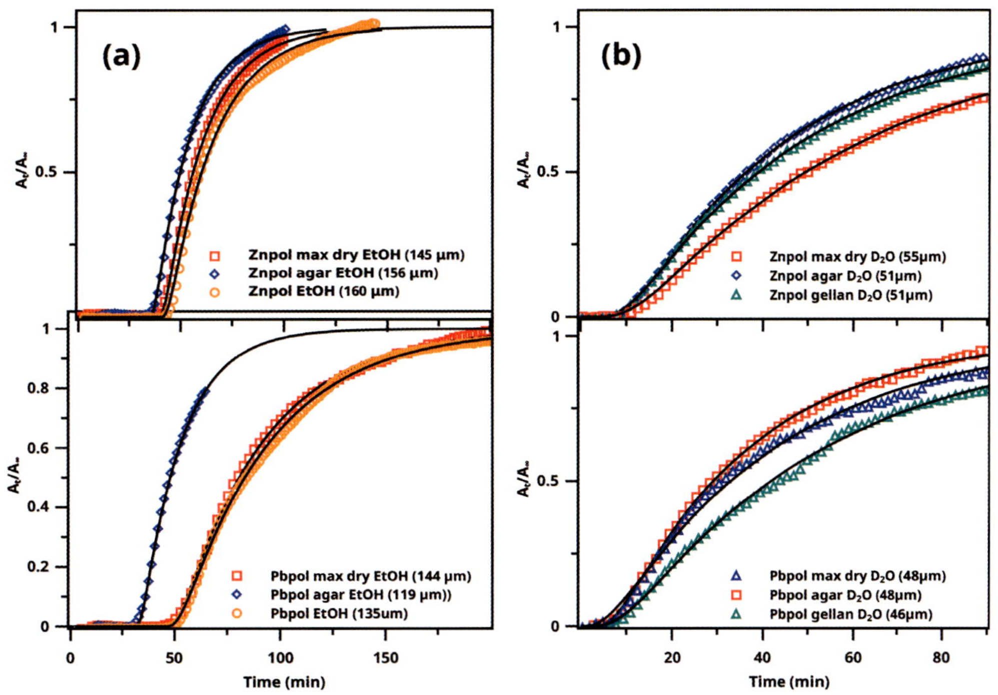

FIG. 3 Concentration profiles of ethanol (a) and (b) D_2O in zinc and lead ionomer films. Solid black lines represent best fits of the Fickian diffusion model with a lag time τ as extra parameter. For D_2O, τ was estimated. Only a fraction of all collected data points and are shown for clarity.

observed upon loading the agar gel (4 wt% in water) with ethanol. The gellan gel (3 wt% in water) was found to be incompatible with the loading of ethanol due to strongly increased rigidity and shrinking and could not be tested at this concentration.

Diffusion of ethanol from gels

Fig. 3a shows the concentration profiles of ethanol in the binding medium model for free ethanol, Nanorestore Max Dry gels and 4% agar gels loaded with ethanol. The difference of the parameter τ between free solvent and solvent released from gels is used as a measure of solvent retention by gels. From Table 2, it is evident that the gelled systems do not show significant retention (for Znpol: $L^2/\tau = 15 \pm 4$ for free ethanol and $L^2/\tau \simeq 10$ for gelled systems). In general, the diffusion coefficient of solvents released from gels in the swollen polymer (at $t > \tau$) is similar (i.e. within the experimental error) to the diffusion of free solvents. For example for Znpol, free ethanol diffusion has a value of $D = 5.6 \pm 1.8 \times 10^{-8}$ cm²/s and ethanol released from gels has a value of $D = 7.6 \pm 0.5 \times 10^{-8}$ cm²/s and $D = 6.6 \pm 2.4 \times 10^{-8}$ cm²/s, for Nanorestore and agar 4%, respectively. Likewise, similar diffusion coefficients for free ethanol and ethanol released from gels were found for Pbpol: $D = 3.2 \pm 0.6 \times 10^{-8}$ cm²/s for free ethanol, $D = 3.40 \pm 0.06 \times 10^{-8}$ cm²/s for Nanorestore and $D = 5.4 \pm 0.3 \times 10^{-8}$ cm²/s for agar 4%. Just as for Znpol, the experimental error in L^2/τ for Pbpol prevented us from concluding on significant solvent retention by gels (Pbpol: $L^2/\tau = 7.6 \pm 2.1$ for free ethanol, $L^2/\tau \simeq 8$ for gelled systems). It should be noted that for the gels, evaporation of ethanol from the gel causes loss of

TABLE 2 Diffusion coefficients D and lag times corrected for thickness L^2/τ resulting from a fit of Fick's law to the concentration profiles in Fig. 2. Values are averaged over two or more measurements. Standard deviations are indicated with ±.

Sample	Solvent	D (10^{-8} cm²/s)	L^2/τ (10^{-8} cm²/s)
Znpol	ethanol	5.1±1.8	15±4
Znpol + Max Dry	ethanol	7.6±0.5	9.7±0.6
Znpol + agar 4%	ethanol	6.6±2.4	12.0±0.3
Znpol	D_2O	0.56±0.08	8.4±0.3
Znpol + Max Dry	D_2O	0.39±0.06	5.4±0.7
Znpol + agar 4%	D_2O	0.44±0.07	6.2±0.2
Znpol + gellan 3%	D_2O	0.39±0.06	6.5±0.4
Pbpol	ethanol	3.2±0.6	7.6±2.1
Pbpol + Max Dry	ethanol	3.40±0.06	8.1±0.0
Pbpol + agar 4%	ethanol	5.4±0.3	9.1±1.0
Pbpol	D_2O	0.58±0.06	7.0±0.5
Pbpol + Max Dry	D_2O	0.42±0.14	6.8±0.3
Pbpol + agar 4%	D_2O	0.42±0.13	6.4±1.4
Pbpol + gellan 3%	D_2O	0.32±0.01	6.7±0.9

contact between the sample and the detector (after ± 100 min for Znpol and ± 60 min for Pbpol with agar) and complete solvent saturation is not reached. In contrast to the diffusion rates, this contact loss prohibited us from determining the exact amount of solvent (e.g. expressed as swelling) delivered by the gels quantitatively. Qualitatively though, the solvent delivery from gels appeared to be on a par with the application of pure solvents at long timescales.

Diffusion of water from gels

It was observed that gels are capable of accumulating water on the surface. If the gels are not properly dried before application, this surface-adhered water seems to be forced into the polymer. Consequently, τ is close to zero and D is up to three times faster than for surface-dried gels. To correct for the error introduced by this hard to control effect, all values were averaged over two or more measurements. Considering the values of D_{D_2O} for all measurements, some gels (gellan and Max Dry for Znpol) show a small but significant decrease in D compared to free water (Table 2). We have no mechanistic explanation for this small decrease in D and it is at present unclear if our dataset is large enough to make a solid statistical argument in favour of this difference. Comparing the retention (τ) of free water and water released by gels in Table 2, no significant increase in τ was found when water was loaded into gels for Pbpol, while for Znpol there seems to be very limited retention (L^2/τ for Znpol is $\simeq 8 \times 10^{-8}$ cm^2/s for free water and $L^2/\tau \simeq 6.5 \times 10^{-8}$ cm^2/s for gels). Clear confirmation of this effect would require a larger dataset.

The fact that there is minimal or no effect on τ when water or ethanol is loaded into gels suggests that diffusion is determined more by the nature (e.g. porosity) of the paint surface than by the method of solvent application (free solvent or gels). In order to inhibit solvent retention from the gels, the diffusion of solvent into the paint must be slower than the diffusion of solvent out of the gel. In that case, small amounts of solvent accumulate on the interface of the gel and the surface. This is a key observation for cleaning with gels: it is the paint surface that largely determines the rate of solvent uptake. This argument holds only for non-porous paints: powdery or cracked paints are expected to show faster diffusion by capillary action. Preliminary experiments on a 40-year-old porous paint sample confirmed this fast diffusion but posed new analytical challenges that will be the topic of forthcoming research.

Conclusion

We have described a time-dependent ATR–FTIR method using a custom sample cell to study solvent diffusion and film swelling in linseed oil-based binding medium models containing lead or zinc (lead or zinc ionomers). The method was used successfully to study the diffusion of a range of seven solvents. Similar diffusion behaviour in zinc and lead ionomers was found. After an initial short non-Fickian phase, all solvents fitted a Fickian concentration profile, with diffusion coefficients similar to literature values. The method was extended to study the diffusion of solvent from Nanorestore Max Dry, agar and gellan gels. We conclude that within this dataset, solvent retention caused by the use of cleaning gels cannot be qualified as significant.

Acknowledgements

Bronwyn Ormsby is thanked for kindly providing the gellan, David Chelazzi for feedback and Piero Baglioni for providing the Nanorestore gels. This research is carried out within the framework of the NANORESTART project funded by the European Union's Horizon 2020 research and innovation program under agreement No 646063.

Note

1. http://www.csgi.unifi.it/.

References

Angelova, L.V., Ormsby, B. and Richardson, E. 2016. 'Diffusion of water from a range of conservation treatment gels into paint films studied by unilateral NMR. Part I: Acrylic emulsion paint', *Microchemical Journal* 124: 311–20.

Baglioni, P., Chelazzi, D. and Giorgi, R. 2014. *Nanotechnologies in the Conservation of Cultural Heritage: A Compendium of Materials and Techniques*. Dordrecht: Springer Netherlands.

Baglioni, P., Carretti, E. and Chelazzi, D. 2015. 'Nanomaterials in art conservation', *Nature Nanotechnology* 10(4): 287–90.

Browne, F. 1956. 'Swelling of paint films in water VIII: swelling of linseed oil paints in water and organic liquids', *Forest Products Journal* 6: 312–18.

Chan, T.Y.A. and Odlyha, M. 1995. 'The effect of relative humidity and pigment type on paint films', *Thermochimica Acta* 269–270(C): 755–67.

Domingues, J.A.L., Bonelli, N., Giorgi, R., Fratini, E., Gorel, F. and Baglioni, P. 2013. 'Innovative hydrogels based on semi-interpenetrating p(HEMA)/PVP networks for the cleaning of water-sensitive cultural heritage artifacts', *Langmuir* 29(8): 2746–55.

Domingues, J., Bonelli, N., Giorgi, R. and Baglioni, P. 2014. 'Chemical semi-IPN hydrogels for the removal of adhesives from canvas paintings', *Applied Physics A* 114(3): 705–10.

Fieldson, G. and Barbari, T. 1993. 'The use of FTIT-ATR spectroscopy to characterize penetrant diffusion in polymers', *Polymer* 34(6): 1146–53.

Fife, G.R., Stabik, B., Kelley, A.E., King, J.N., Blümich, B., Hoppenbrouwers, R. and Meldrum, T. 2015. 'Characterization of aging and solvent treatments of painted surfaces using single-sided NMR', *Magnetic Resonance in Chemistry* 53(1): 58–63.

Hermans, J. 2017. *Metal Soaps in Oil Paints*, PhD thesis, University of Amsterdam.

Hermans, J.J., Keune, K., van Loon, A. and Iedema, P.D. 2015. 'An infrared spectroscopic study of the nature of zinc carboxylates in oil paintings', *Journal of Analytical Atomic Spectrometry* 30(7): 1600–608.

Hermans, J.J., Keune, K., van Loon, A., Corkery, R.W. and Iedema, P. D. 2016. 'Ionomer-like structure in mature oil paint binding media', *RSC Advances* 6(96): 93363–9.

Lahaye, M. and Rochas, C. 1991. 'Chemical structure and physico-chemical properties of agar', *Hydrobiologia* 221(1): 137–8.

Pavitt, R. 2012. 'Cleaning of painted surfaces – Wolbers strikes again! A workshop review by Rebecca Pavitt', *News in*

Conservation 33. Available at: https://www.iiconservation.org/node/3216 (accessed 20 May 2017).

Samorì, C., Galletti, P., Giorgini, L., Mazzeo, R., Mazzocchetti, L., Prati, S., Sciutto, G., Volpi, F. and Tagliavini, E. 2016. 'The green attitude in art conservation: polyhydroxybutyrate-based gels for the cleaning of oil paintings', *ChemistrySelect* 1(15): 4502–8.

Scott, C.L. 2012. 'The use of agar as a solvent gel in objects conservation', *Objects Specialty Group Postprints* 19: 71–83.

Stolow, N. 1957a. 'Action of solvents on dried linseed oil films', *Nature* 179(4559): 579– 80.

Stolow, N. 1957b. 'The action of solvents on drying-oil films: parts I and II', *Journal of the Oil and Colour Chemists' Association* 40(5–6): 377–402.

van der Wel, G.K. and Adan, O.C.G. 1999. 'Moisture in organic coatings: a review', *Progress in Organic Coatings* 37(1): 1–14.

van der Zanden, A. and Goossens, E. 2004. 'The measurement of the sorption isotherm of water in paint films', *Chemical Engineering and Processing: Process Intensification* 43(6): 739–43.

Vergnaud, J.M. 1991. *Liquid Transport Processes in Polymeric Materials: Modeling and Industrial Applications*. London: Prentice-Hall, Inc.

Wolbers, R. 2000. *Cleaning Painted Surfaces: Aqueous Methods*. London: Archetype Publications.

Zumbühl, S. 2014. 'Parametrization of the solvent action on modern artists' paint systems', *Studies in Conservation* 59(1): 24–37.

Authors' addresses

- Lambert Baij, Van 't Hoff Institute for Molecular Sciences, University of Amsterdam, Amsterdam, The Netherlands (C.L.M.Baij@uva.nl)
- Katrien Keune, Rijksmuseum Amsterdam, Conservation and Restoration, Amsterdam, The Netherlands (K.Keune@rijksmuseum.nl)
- Joen Hermans, Van 't Hoff Institute for Molecular Sciences, University of Amsterdam, Amsterdam, The Netherlands (J.J.Hermans@uva.nl)
- Petria Noble, Rijksmuseum Amsterdam, Conservation and Restoration, Amsterdam, The Netherlands (P.Noble@rijksmuseum.nl)
- Piet Iedema, Van 't Hoff Institute for Molecular Sciences, University of Amsterdam, Amsterdam, The Netherlands (P.D.Iedema@uva.nl)

Gels at the Winterthur/University of Delaware Program in Art Conservation: a retrospective review of student projects, 1978–2016

Matthew Cushman

Introduction

In December 1983, a paintings conservation student at the Winterthur/University of Delaware Program in Art Conservation (WUDPAC) wrote about her difficulties in achieving an even cleaning on an easel painting:

> An unsuccessful attempt was made to reduce the splotchy appearance of the remaining varnish in the sky using diacetone alcohol followed by acetone. A dilute ammonia solution (1%) was also unsuccessful. Reduction of the appearance was achieved by using DMF [dimethyl formamide] followed by acetone ... A decision was made to leave this substance on the surface.[1]

Less than two years later, significant shifts in philosophy and practice took hold, coinciding with the appointment of Richard Wolbers to Winterthur Museum's conservation department. Tailoring of aqueous cleaning systems, and exploiting the benefits of gelled solutions, emulsions and solvent mixtures soon became commonplace in the training of WUDPAC students. The innovative treatment methods proposed by Wolbers, coupled with his skill in identifying materials of potential use to the conservator, resulted in an expanding array of tools affording greater specificity for targeted cleaning. This study aims to plot the introduction of new cleaning systems along with the seemingly mundane use of gels at WUDPAC since 1978, tracking experimentation, full adoption, and eventual replacement in favor of materials offering greater specificity as well as safer formulations for the conservator and the environment.

Methodology

Because the present study is concerned with tracking the use of all gel systems over four decades rather than the grand sweep of new innovations, relying on oral history is not ideal. While oral histories have proven useful in recording responses to landmark moments and key figures, they fall short when trying to recount precise details (Hoffman 1996). Treatment documentation captures the circumstances surrounding particular moments in the history of an object, providing data points that, when aggregated from several hundred similar reports, represent practice over a long period of time.

At WUDPAC, student-produced documentation is archived in multiple locations. Reports on Winterthur collection objects are maintained in Winterthur Museum's conservation records. Projects from external sources are given an Art Conservation Program (ACP) number, and ACP object folders are maintained in Winterthur Library's archives. The scope of this study is limited to the ACP-numbered files because it covers student projects since WUDPAC's inception. Aggregating a full record of student work on Winterthur collection objects is a considerably more difficult task; extending this study to include Winterthur object files, student research projects and student publications is under consideration.

Wherever possible, a semi-automated process was used to extract data from individual reports. Reports were scanned in PDF/A format and eventually run through Tesseract,[2] an open-source optical character recognition (OCR) engine sponsored by Google. A parsing script was written in Perl to attempt an extraction of key data from the resulting .txt files: report date, type of object, condition problem, type of gel used, and gel formulation. As the source reports were written in natural language using different formats, the extracted

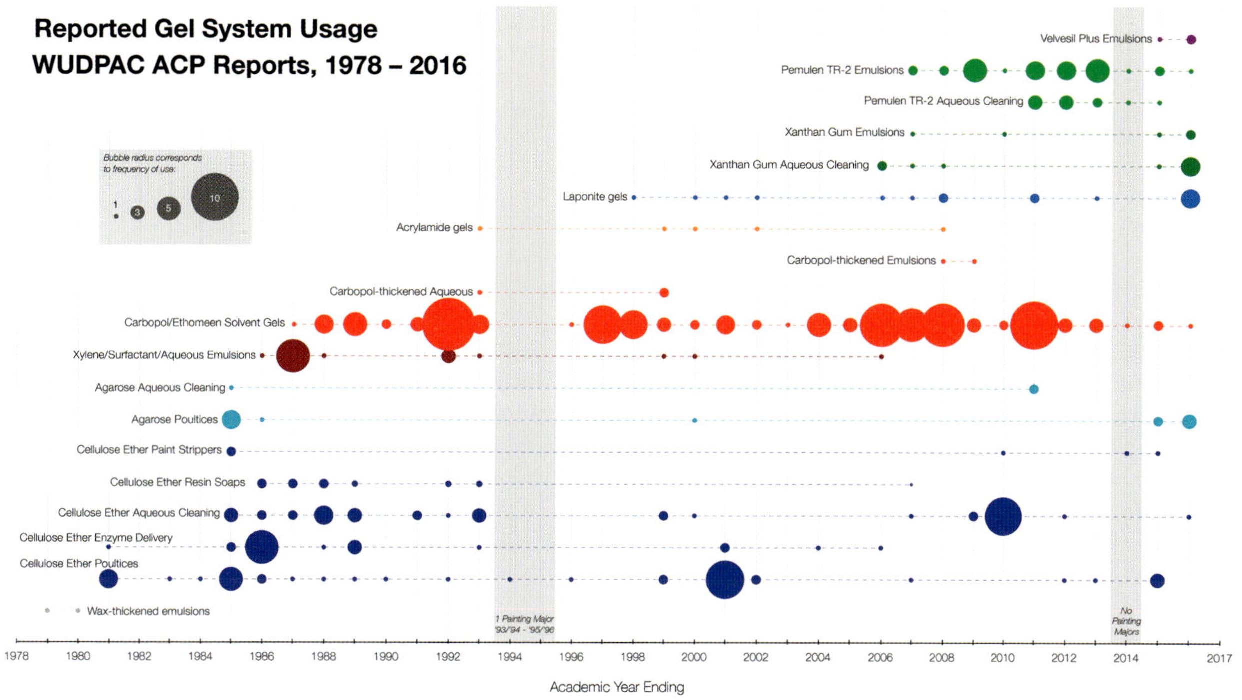

FIG. 1 Gel usage by general formulation.

dataset was unstructured, containing variations in terminology, typographical errors and OCR errors. To achieve some degree of consistency, the dataset was modified using OpenRefine,[3] an open-source data clean-up and transformation application. Despite these efforts, only a small proportion of the files proved appropriate for automated data extraction. Handwritten reports were especially problematic. As a result, almost all of the reports were checked by hand.

Limitations

As noted above, the present study omits student projects on non-ACP objects, namely Winterthur collection objects, some independent study projects, and scientific research projects; also missing are milestones such as the first documented use of xanthan gum as a gelling agent in a cleaning system (Cushman and Wolbers 2007), an important study of gel applications in textile conservation (Shaeffer and Gardiner 2013) and recent research on the use of silicone-based particle emulsions for the cleaning of acrylic paint films (Beller *et al.* 2016). However, given that the ACP files have averaged approximately 50 files per year, this dataset can be considered to be a fairly representative sampling of documented treatment activity at WUDPAC.

Disclaimer

In the sections below, typical formulations are described for several gel systems. These formulations cover a wide range of applications from over 38 years' worth of reports. As such, the formulations presented may not represent current best practice for a given conservation application.

Summary of findings

Approximately 1700 object folders pertaining to 1900 individual objects were surveyed. Explicit descriptions of gel formulations were identified in 238 object folders, with 341 individual gels documented. For the purposes of this study, gels usage was organized according to documented date of use, object type, general use categories (overpaint removal, for example), and type of cleaning system. For hydrogels, a distinction has been made between traditional poultices (that is, a thickener in water) versus cleansing agents (formulations meant to deliver chelating agents, surfactants, pH-buffered solutions, for example). Because the recording of dates is inconsistent across the dataset, gel usage was grouped according to academic year (for example, all treatments performed between 1 July 1999 and 30 June 2000 were assigned the date '2000'). A timeline of reported gel usage is presented in Fig. 1.

Before 1984–1985, gelled formulations were used rarely, although some wax-thickened emulsions and cellulose ether poultices were reported. The earliest case of an enzymatic preparation was during the year ending 1981. The 1984–85 academic year, coinciding with Richard Wolbers' appointment to Winterthur Museum's conservation department, saw a marked uptick in documented gel usage. Most curious are several mentions of agarose gels used in paper conservation at Wolbers' suggestion; these reported but unpublished cases predate by several years the earliest references for agarose listed in AATA (van Dyke 2004) and BCIN (Valentin *et al.* 1996). Cellulose ether preparations are documented in the broadest range of applications, and are the materials that are mentioned over the longest time period.

For coating removal, thickened resin soaps and thick emulsions of xylene, an aqueous phase, and a significant proportion of surfactants, were in use beginning in the 1985–86 academic year, giving way to Carbopol/Ethomeen solvent gels beginning in the 1986–87 academic year. Carbopol gels exhibited a consistency of use unmatched by other gel systems, with decreased usage over the past five years. With the advent of self-emulsifying systems of polyacrylic acid (Pemulen TR-2) and polysaccharides (xanthan gum), and silicone-based particle emulsions, and the adoption of these systems over the past decade, Carbopol/Ethomeen gels have (in general) fallen out of favor, as surfactant clearance and environmental safety continue to be of great concern to conservators.

One trend of note is that paintings treatments represent two-thirds of all gel use at WUDPAC, which should come as no surprise given that Richard Wolbers is a trained paintings conservator. Gel usage by specialty, limited to paintings, works of art on paper and manuscripts, and objects, is illustrated in Fig. 2.

The greater frequency of gel use in paintings helps to explain two gaps in the data as shown in Fig. 1. From 1993 to 1996, there was only one paintings major at WUDPAC, and in the year ending 2014, there were no paintings majors. Additionally, the training and graduation year of the person holding the position of conservator of paintings at Winterthur has some influence over the use of materials, which can explain the recurrence of thickened resin soap preparations in the year ending 2007; this was the year when Mary McGinn (MS, WUDPAC '94) was appointed to this position. The previous documented use of a thickened resin soap was the year ending 1993, when McGinn was a second-year paintings conservation major at WUDPAC.

Observations on specific gel cleaning systems

Cellulose ethers

- *Dates reported*: Year ending 1981–present.
- *Reported uses*: Poultices, enzyme delivery, aqueous cleansers, thickened resin soaps, commercial paint strippers.
- *Typical formulations (all percentages weight-to-volume unless noted otherwise)*: 74.6% of reported cellulose ethers were methyl cellulose; 21.1% hydroxypropyl methyl cellulose (HPMC, 4000cps grade at 2% from Sigma Aldrich); 2.6% ethulose; 1.7% hydroxypropyl cellulose. For poultices, up to 5% in water, typically deionized or distilled. For cleaning solutions, typical concentrations range from 1–2%, up to 2% of a chelating agent added. Often includes approximately 1% of a nonionic surfactant. Reported pH was most often between 5.5 and 8.5, with some outliers as low as pH 4 and as high as pH 9.5. Typical resin soap recipes include 1.3% HPMC, 2–4% of a resin acid (primarily abietic acid), and 3–5% triethanolamine (TEA). Typical enzymatic preparations include 1–2% of an enzyme, buffers, up to 1% of a nonionic surfactant, and 1–5% of a cellulose ether.
- *Notes*: While it is somewhat useful to report percentages, what may be of greater importance is the grade of cellulose ether used or the measured viscosity of the prepared solution.

Agarose

- *Dates reported:* Introduced in year ending 1985, one use in 2000, more frequently used in recent years.
- *Reported uses:* Poultices, enzyme delivery, stain reduction, corrosion reduction, overpaint removal.
- *Typical formulations*: 1–4% in deionized water, typically including buffers, chelators, and sometimes nonionic surfactants.
- *Notes*: Neutral gel, not reliant on ion-facilitated crosslinking (can carry chelators), pore size controlled by concentration.

Xylene/Triton/water emulsions

- *Dates reported:* Year ending 1986–year ending 2006.
- *Reported uses:* Varnish removal, overpaint removal, tenacious grime removal.
- *Typical formulations*: 30–65% xylene, 20–40% aqueous phase including 1–5% triethanolamine, 10–30% nonionic surfactant (Triton X-100 or Triton XL-80N).
- *Notes*: The first reference to 'solvent gels', these preparations are thick, clear emulsions. Supplanted by Carbopol/Ethomeen solvent gels.

Carbopol/Ethomeen solvent gels

- *Dates reported:* Year ending 1987–present.
- *Reported uses:* Coating removal, overpaint removal, adhesive reversal, grime/varnish removal.
- *Typical formulations:* 1–5% Carbopol, 5–25% Ethomeen (C-12 or C-25 depending on solvent mixture), amount of water added depending on Ethomeen used, dryness of Carbopol, solvent mixture. Formulations trending toward 2% Carbopol, 15% Ethomeen.
- *Notes:* High molecular weight, crosslinked polyacrylic acid polymers. Use tapering off post-2010 with the advent of self-emulsifying systems. Some atypical uses include thickened aqueous solutions (1993; 1999) and thickened emulsions (2008–2009).

Acrylamide gels

- *Dates reported*: Year ending 1993–year ending 2008.
- *Reported uses:* Corrosion reduction, grime removal on matte paint, stain removal on ceramics.
- *Typical formulations:* 19% acrylamide, 1% bisacrylamide in 100 mL deionized water, 50 μL

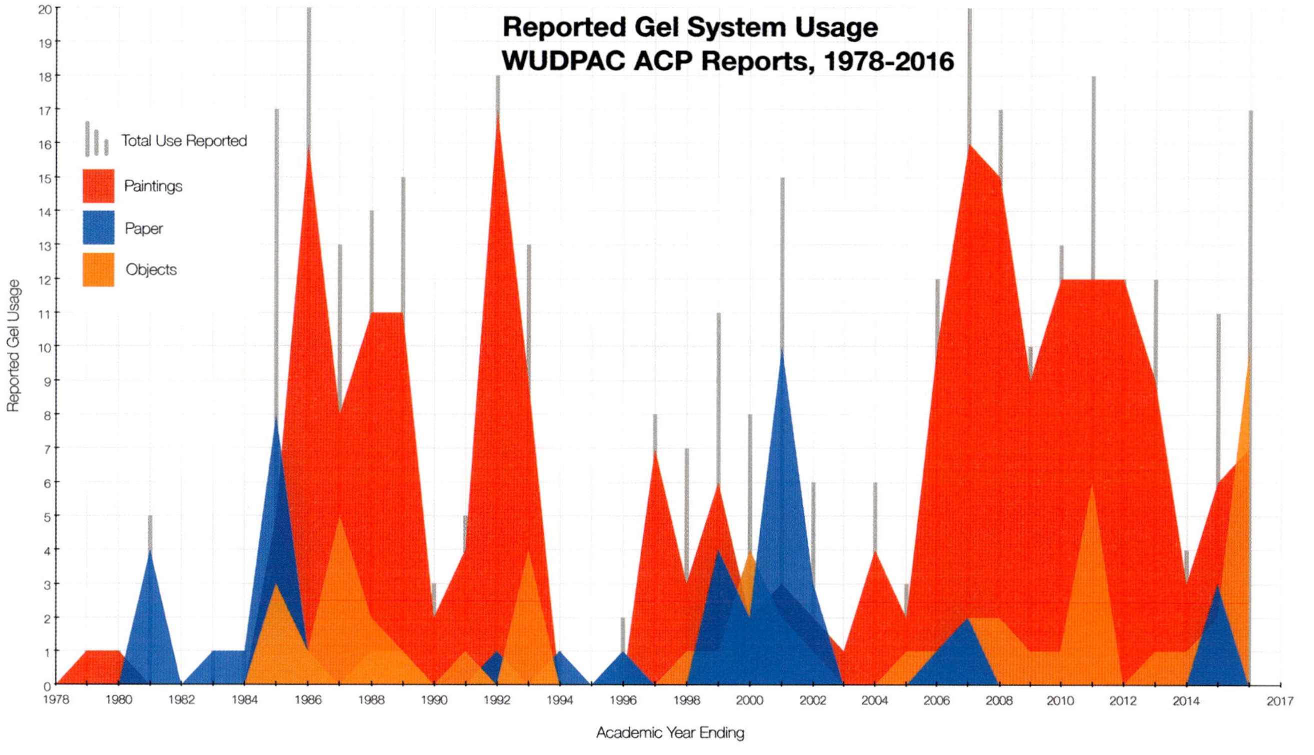

FIG. 2 Gel usage by specialty: paintings, paper, and objects.

tetramethylethylenediamine (TEMED) (stabilizer, improves polymerization), 50 µg ammonium persulfate (gelation initiator); can include surfactants and other chemical agents.

- *Notes:* Stiff gel, relatively inert, with pore sizes dependent on acrylamide concentration and concentration of the crosslinking agent bisacrylamide. Can be grated and used as one would eraser crumbs.

Laponite gels

- *Dates reported*: Year ending 1998–present.
- *Reported uses:* Poultice, stain reduction, iron staining reduction, bleaching of ceramic staining.
- *Typical formulations:* General use: 5% in deionized water; up to 10% reported.
- *Notes:* Most often used in objects conservation. A synthetic smectic clay, creating thick, clear gels in water and other polar solvents.

Xanthan gum

- *Dates reported*: Year ending 2006-present.
- *Reported uses:* Grime removal, coating removal, aqueous cleanser, adhesive removal, stain reduction, tarnish removal, overpaint removal.
- *Typical formulations:* 1–2% xanthan gum, buffers, chelators, <1% nonionic surfactant, up to 40% solvent v/v depending on the solvent (typical: up to 20%).
- *Notes:* Anionic polysaccharide, forms gel-like preparations at concentrations above 0.5%. Stable in nearly all pH ranges, high and low conductivity; able to stabilize high percentages of solvents without the addition of surfactants.

Pemulen TR-2 (Lubrizol)

- *Dates reported*: Year ending 2007–present.
- *Reported uses:* Grime/varnish removal, overpaint removal.
- *Typical formulations:* 1% in deionized water, up to 5–10% TEA, pH 5.5–8.5, 1% chelating agent, up to 15% solvent phase.
- *Notes:* Polyacrylic acid block polymer. Self-emulsifying, requires no surfactant to form stable emulsions at 0.3% concentration and above. Capable of emulsifying up to 45% oil phase by weight. Stable in high and low conductivity solutions.

Velvesil Plus (Momentive)

- *Dates reported*: Year ending 2015–present.
- *Reported uses:* Grime, adhesive and coating removal from water-sensitive surfaces.
- *Typical formulations:* Up to 20% aqueous phase in Velvesil Plus gel. Viscosity sometimes modified using cyclomethicone D4/D5.
- *Notes:* High viscosity silicone cross-polymer gel. Very low Hansen solubility parameter. Compatible with aqueous phases and polar solvents for use on

water-sensitive surfaces. Underrepresented in the data; was in experimental use as early as 2007. Cleared with volatile silicone solvents. A similar line of modified silicone gels called KSG (Shin-Etsu) has been used as well.

Commentary

In general, the history of gel use at WUDPAC has coincided with the adoption of materials allowing for greater specificity and control in critical cleaning applications. Notable trends include the consistent use of cellulose ether gels, early adoption of agarose gels, the prevalence of Carbopol/Ethomeen solvent gels from the late 1980s to 2010, and their eventual replacement with surfactant-free polymer-emulsion systems. With greater attention being given to environmental concerns, the adoption of surfactant-free cleaning systems is a key trend. The ability to deliver aqueous phases to water-sensitive surfaces within a nonpolar gel is a significant development, as is the ability to create water-based, surfactant-free systems capable of incorporating organic solvents for coating and overpaint removal.

Conclusions and future work

By aggregating data gleaned from over 1700 object files, a history was reconstructed showing gel use at WUDPAC. This study serves as an example of the type of information that can be gleaned from bulk analysis of conservation documentation. Through careful extraction of data, histories of material use and conservation practice can be reconstructed. Further work is needed with this dataset to identify trends and perhaps to identify materials worth revisiting – the early use of agarose gels, which were then abandoned for decades, is a prime example. Advancements in natural language processing and cognitive computing, coupled with collaborations between computer scientists and the cultural heritage sector, present great opportunities for research in the history of conservation materials and practices, particularly if conservation institutions are willing to grant collaborative access to archived documentation.

Acknowledgements

The author is grateful for the assistance of the staff of Winterthur Museum's Library and Archives, and for the careful work of Katie Rovito, a pre-program intern at Winterthur.

Notes

1. Carole Abercauph, unpublished treatment report, ACP 585, 12/23/1983.
2. Github.com/tesseract-ocr.
3. Openrefine.org. It should be noted that the author is not, and does not pretend to be, a computer scientist.

References

Beller, A., Summer, J. and Albertson, G. 2016. 'Liquitex acrylic emulsion cleaning study: testing siloxane-based cleaning materials', in A. Nevin, J.H. Townsend, J. Kirby, D. Saunders and A. Brokerhof (eds), *Saving the Now: Crossing Boundaries to Conserve Contemporary Artworks*. London: IIC.

Cushman, M. and Wolbers, R. 2007. 'A new approach to cleaning iron-stained marble surfaces', *WAAC Newsletter* 29(2): 23–8.

Hoffman, A. 1996. 'Reliability and validity in oral history', in D.K. Runaway and W.K. Baum (eds), *Oral History: An Interdisciplinary Anthology*, American Association for State and Local History Book Series. Walnut Creek, CA: AltaMira Press, 87–93.

Shaeffer, E. and Gardiner, J. 2013. 'New and current materials and approaches for localized cleaning in textile conservation', in A. Holden, R. Summerour, E. Scheutz, J. Carlson and G. Petersen (eds), *AIC Textile Specialty Group Postprints* 23: 109–24.

Valentin, N., Sánchez, A. and Harraez, I. 1996. 'Analyses of deteriorated Spanish glass windows: cleaning methods using gel systems', in J. Bridgland (ed), *Preprints of the ICOM-CC 11th Triennial Meeting*. London: James & James, 851–6.

van Dyke, Y. 2004. 'Practical applications of protease enzymes in paper conservation', *The Book and Paper Group Annual* 23: 93–107.

Author's address

Matthew Cushman, Winterthur Museum, Winterthur, DE, USA (mcushm@winterthur.org)

A study of water mobility in hydrogels suitable for biocleaning mural paintings: a unilateral NMR study

Valeria Di Tullio, Eleonora Gioventù, Noemi Proietti, Loredana Luvidi, Anna Maria Mecchi and Donatella Capitani

Introduction

In recent years, the use of *Desulfovibrio vulgaris* bacteria has become a common method for the removal of sulphates from stone. Through metabolic processes the sulphate-reducing bacteria *D. vulgaris* are able to reduce SO_4^{2-} ions in H_2S according to the following reaction (Gauri and Bandyopadhyay 1999):

$$6CaSO_4 + 4H_2O + 6CO_2 \rightarrow 6CaCO_3 + 4H_2S + 2S + 11O_2$$

Compared with other more traditional methods, this biological process results in a more homogeneous removal of the surface deposits and preserves the layers under salt efflorescence and black crusts (Gioventù *et al.* 2011, 2015; Troiano *et al.* 2013). On wall paintings, the method is typically used by applying sulphate-reducing bacteria entrapped in hydrogel poultices. Hydrogel systems offer several advantages for the cleaning of artefacts in terms of selectivity and a controllable and gradable penetration rate at the interface between the gel and the surface of the artefact (Burnstock and White 1993). As a consequence, the study of water mobility and diffusion into the gel system and in the porous matrix as a function of the application time is an important step for the assessment of hydrogel performance in cleaning treatments (Caggioni *et al.* 2007).

In this study hydrogels and sulphate-reducing bacteria formulations applied on wall painting specimens affected by sulphate efflorescence were investigated by unilateral nuclear magnetic resonance (NMR). This technique combines an open magnet and a radio-frequency (RF) coil to generate a sensitive volume external to the sensor where the NMR signal is generated and detected. The magnetic field is applied to one side of the object non-destructively, making this tool particularly valuable for the analysis of cultural heritage (Blümich *et al.* 2010; Del Federico *et al.* 2012; Di Tullio *et al.* 2016; Angelova *et al.* 2016). Unilateral NMR measurements can be used to obtain detailed information on the interaction between water molecules, the gel network and the porous matrix. To assess which layers of the artefact are in contact with water during application of the cleaning system, ^1H depth profiles of water absorbed by the mural painting were used to evaluate the water content and its penetration depth within the porous network. Relaxation times (T_1, T_2) and self-diffusion coefficients were measured to study the molecular motions of water inside the gels.

Materials and methods

Carbogel, gellan gum and Laponite RD gels were prepared at different concentrations. A 2% gellan gum (Kelcogel, CTS srl) gel, a 1% Carbogel (CTS srl) and a 5% Laponite RD (Bresciani) gel was prepared by dissolving the powder in water. The gellan gum formulation was heated in a microwave oven at 750 W for 3 minutes and cooled to room temperature to obtain a rigid gel about 5 cm thick. The cleaning formulations were prepared by dissolving 2.5 g lyophilized bacteria in 2 L water; 50 mL of this suspension was added in 100 mL gel. In the case of gellan gum, the suspension was added during the gel cooling at about 40 °C. Gels with bacteria were applied on the surface of the samples (Fig. 1) for 1 or 4 hours. Before and after the absorption procedure, the wall painting specimens and gels were analysed using unilateral NMR. All measurements were performed at 13.62 MHz on a portable NMR instrument from Bruker Biospin interfaced with a purpose-built single-sided sensor by RWTH Aachen University, Aachen, Germany (Perlo *et al.* 2005). Longitudinal relaxation times T_1 were measured with a Saturation Recovery pulse sequence followed by a CPMG-train in the detection period to increase the sensitivity (Perlo *et al.*

TABLE 1 Longitudinal relaxation times (T_1), transverse relaxation times (T_2) and self-diffusion coefficients on water molecules in gels measured by unilateral NMR.

Sample	T_1 (s)	W_A (%)	T_2A (ms)	W_B (%)	T_{2B} (ms)	D *10^{-9} m^2/s
Carbogel 2%	1.7±0.1	10	6.4±0.8	90	52.2±0.4	2.05±0.05
Laponite RD 5%	0.7±0.1	9	7.1±0.4	91	42.7±0.1	2.03±0.02
Gellan 2%	0.7±0.1	5	5.6±0.5	95	39.2±0.1	1.97±0.02

Note: W_A and W_B are the percent amplitudes of T_{2A} and T_{2B}, respectively

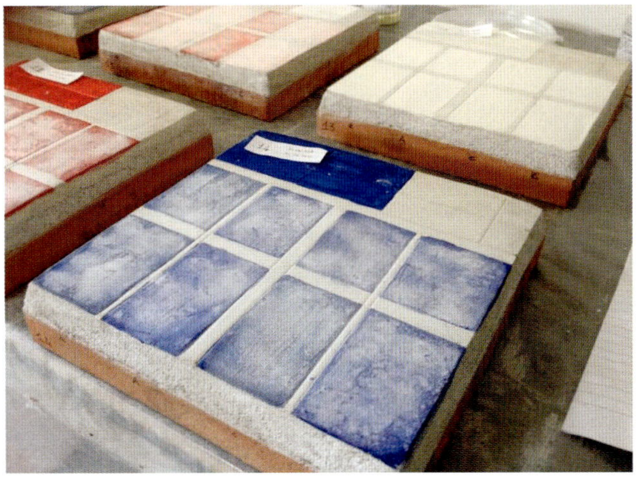

FIG. 1 Wall painting specimens.

2011). Effective transverse relaxation times T_{2eff} (Goelman *et al.* 1995; Hurlimann and Griffin 2000) were measured with a CPMG pulse sequence, 4096 echoes were recorded inside the gel with an echo time 2τ of 71.2 μs. Due to the long dead time of the probe, the strong steady magnetic field gradient and the long echo time used in the pulse sequence, the transverse relaxation decay of non-exchangeable proton of the gel networks are typically too short to detect (usually 10–20 μs) (Belton 1997). The inhomogeneous magnetic field of the portable NMR is a further source of relaxation which shortens the measured T_{2eff} values of the gels, making them definitively shorter than those which would be measured in a homogeneous field.

^{1}H depth profiles collected on specimens after the application of gels were obtained as the addition of the first four echoes acquired with a CPMG pulse sequence with an echo time of 86 μs and a nominal resolution of 23 μm. To optimize the duration of measurements, profiles were acquired by repositioning the single-sided sensor in steps of 250 μm to scan the desired spatial range, from the surface of the specimen to a depth of about 5 mm. Self diffusion coefficients D were measured at 0.36 T (13.62 MHz) with a steady gradient (SG) of 14.28 T/m using a stimulated echo (SGSTE) pulse sequence followed by a CPMG echo train to improve the SNR (Perlo *et al.* 2011). The uncertainty associated with T_1, T_{2eff} and D was obtained by repeating the measurement three times on each sample.

Longitudinal and effective transverse relaxation times were obtained fitting the magnetization decay to the following equation:

$$Y(\tau)=\sum_{i=1}^{n} W_i e^{\frac{-2\tau}{T_{1,2i}}}$$

where n is the number of components of the decay, W_i is the weight of ith component, and $T_{1,2i}$ is the relaxation time of ith component. Before fitting, the sum of weights was normalized to 100%.

Tests were performed on wall painting specimens (30 × 30 × 2 cm^3) prepared ad hoc with 8 mm *arriccio* (1 lime putty + 3 sand, grain size 0.1–1 mm) and 4 mm fine plaster (1 lime putty + 2 sand, grain size 0.5 mm max.). After a 3-hour drying time, pigments were applied using the fresco technique. After 7 months, a spray of gypsum (30% aqueous solution) to simulate a salt efflorescence was applied on the specimens. Analyses by portable NMR were performed on the hydrogels/bacteria and on wall painting samples before and after the treatments.

Results and discussion

Water transport in Carbogel, Laponite RD and gellan gels was investigated by measuring the longitudinal relaxation time (T_1), the effective transverse relaxation time (T_{2eff}) and the self-diffusion coefficient. Relaxation times (T_1, T_2) of the gel polymers depend on the molecular weight, degree of branching, crosslink density, size of the side groups and temperature (Challagan 1991). Relaxation measurements are affected by both rotational and translational molecular motions, whereas diffusion measurements are related only to the translational molecular displacement. The values of longitudinal (T_1) and transverse relaxation time (T_2) found in Carbogel, Laponite RD and gellan gels are shown in Table 1.

As reported in the literature (Homans 1992), the transverse relaxation time (T_2) of a nucleus is sensitive to both slow and fast motions, while the longitudinal relaxation time (T_1) is affected by fast motions only. The observation that the transverse relaxation time T_2 was much smaller than T_1 indicated that water in Carbogel, Laponite RD and gellan gels had a slowed motion due to the presence of the polymeric network of gels. To explain the two observed transverse relaxation times in gels, two proton species must be taken into account: one with a short transverse relaxation time, T_{2A}, related to 'bound' water, and another with a long transverse relaxation time, T_{2B}, related to 'free' water. Proton nuclei in the 'bound' water domain may be affected by hydrogen bonding interactions and/or chemical exchange with polymeric network of the gels.

To study the translational motion of water molecules in the gel/bacteria systems, the self-diffusion coefficient was measured. With unilateral NMR, the amplitude of spin echo

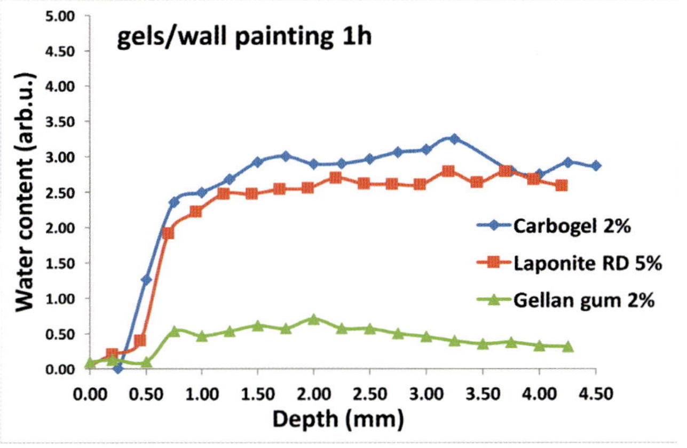

FIG. 2 ^{1}H depth profiles plot and table of the values of wall painting specimens after absorbing water from the three gels tested after an application time of 1 hour.

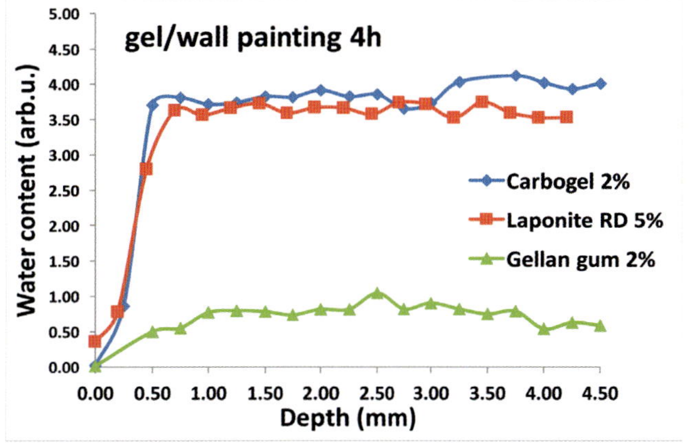

FIG. 3 ^{1}H depth profiles plot and table of the values of wall painting specimens after absorbing water from the three gels tested with an application time of 4 hours.

generated under a strong, uniform and static magnetic field gradient can be used to detect the self-diffusion coefficient, a parameter which can reveal molecular dynamics and sample microstructure (Kimmich 1997; Rata *et al.* 2006).

The normalized echo attenuation $\frac{A_i}{A_0}$ measured by NMR was fitted, and the self-diffusion coefficient, *D*, of the water molecules was obtained from the slope of the plot of $\ln\left(\frac{A_i}{A_0}\right)$ vs $\gamma^2 G^2 \tau_i^2 \left(\Delta + \frac{2}{3}\tau_i\right)$. The self-diffusion coefficients found in the Carbogel, gellan and Laponite gels are reported in Table 1. These were found to be smaller than *D* of bulk water ($2.1*10^{-9}$ m^2/s at room temperature) indicating that the translational motion of water molecules is more hindered in gels than in bulk water.

The mean square displacement $\langle r^2 \rangle$ of water molecules with a self-diffusion coefficient, *D*, during a diffusion time Δ can be calculated from the relationship (Farhat *et al.* 1997):

$$\langle r^2 \rangle = \sqrt{2D\Delta}$$

In a diffusion time Δ of 10 ms, bulk water molecules should undergo a mean square displacement of about 6.5 μm (diffusion length). During this diffusion time, the translational motion of water molecules in a gel is increasingly hindered as the average probability of collision between water molecules and the gel network increases. Comparing the values of *D*, the translational motion of water molecules was most hindered in a gellan gel. In the case of thixotropic gels the pore sizes were not very easy to measure; Mao *et al.* (2001) have measured by cryo-SEM the pore size of gellan gum finding two discrete pore-size distributions on the order of 0.1 and 1 μm. In gellan gels, water displacement may also be hindered by the degree of interconnection between the pores.

Unilateral NMR ^{1}H depth profile measurements were performed to evaluate the water content absorbed by the wall painting specimen during application of the gels. The amplitude of the ^{1}H depth profile is directly proportional to the amount of absorbed water as a function of the depth scanned. Therefore, each experimental point represents the amount of water at the corresponding depth of measurement. Depth profiles showed the trends dependent on the capability of the gel to release water into the porous media. Carbogel and Laponite RD show a very similar trend with the largest amount of water release after a 1-hour application. The same trend was found after a 4-hour application (Figs 2 and 3, Table 1). With respect to Carbogel and Laponite, after 4 hours of application, the gellan gum gel shows the weakest water release capability.

Conclusion

Unilateral NMR is a powerful tool to evaluate water mobility and diffusion in gels used in a fully non-invasive and non-destructive way. Different types of gel formulations to be used as cleaning systems for wall painting were investigated. Transverse relaxation times measured in gel formulations indicate the presence of water in two environments. The water molecules with a short relaxation time may be involved in hydrogen bonding interactions and/or proton exchange with gel macromolecules, while the molecules with a longer relaxation time are free to move through the network, although self-diffusion measurements demonstrate that their translational motion is increasingly hindered as the average probability of collisions with the gel network increases. The release and diffusion of water at the gel–wall painting interface were investigated. Gellan concentration and application time of the gel may be tuned to modulate the amount and the depth of penetration of the liquid phase through the porous network. Water absorption, distribution and penetration in the support depend not only on the physical-chemical and rheological properties of the gel but also on the properties of the support to which the gel is applied. This study is part of a more extensive research effort aimed at obtaining an analytical protocol for establishing *in situ* the performance of cleaning systems.

References

Angelova, L., Ormsby, B. and Richardson, E. 2016. 'Diffusion of water from a range of conservation treatment gels into paint films studied by unilateral NMR. Part I. Acrylic emulsion paint', *Microchemical Journal* 124: 311–20.

Belton, P.S. 1997. 'NMR and the mobility of water in polysaccharide gels', *International Journal of Biological Macromolecules* 21(1–2): 81–8.

Blümich, B., Casanova, F., Perlo, J., Presciutti, F., Anselmi, C. and Doherty, D. 2010. 'Non-invasive testing of art and cultural heritage by mobile NMR', *Accounts of Chemical Research* 43(6): 761–70.

Burnstock, A. and White, R. 1993. 'Cleaning gels: further studies', in N.H. Tennent (ed.), *Conservation Science in the UK*. London: James & James, 36–9.

Caggioni, M., Spicer, P.T., Blair, D.L., Lindberg, S.E. and Weitz, D.A. 2007. 'Rheology and micro-rheology of a microstructured fluid', *Journal of Rheology* 81(5): 851–65.

Challagan, P.T. 1991. *Principles of Nuclear Magnetic Resonance Microscopy*. Oxford: Clarendon Press.

Del Federico, E., Centeno, S.A., Kehlet, C., Ulrich, K., Yamazaki-Kleps, A. and Jerschow, A. 2012. '*In situ* unilateral 1H-NMR studies of the interaction between lead white pigments and collagen-based binders', *Applied Magnetic Resonance* 42(3): 363–76.

Di Tullio, V., Capitani, D., Atrei, A., Benetti, F., Perra, G., Presciutt,i F., Proietti, N. and Marchettini, N. 2016. 'Advanced NMR methodologies and micro-analytical techniques to investigate the stratigraphy and materials of 14th century Sienese wooden paintings', *Microchemical Journal* 125: 208–18.

Farhat, I.A., Loisel, E., Saez, P., Derbyshire, W. and Blanshard, J.M.V. 1997. 'The effect of sugars on the diffusion of water in starch gels: a pulsed field gradient NMR study', *International Journal of Food Science & Technology* 32(5): 377–87.

Gauri, K.L. and Bandyopadhyay, J.K. 1999. *Carbonate Stone: Chemical Behaviour, Durability and Conservation*. New York: John Wiley & Sons.

Gioventù, E., Lorenzi, P.F., Villa, F., Sorlini, C., Rizzi, M., Cagnini, A., Griffo, A. and Cappitelli, F. 2011. 'Comparing the bioremoval of black crusts on colored artistic lithotypes of the Cathedral of Florence with chemical and laser treatment', *International Biodeterioration and Biodegradation* 65: 832–9.

Gioventù, E., Luvidi, L. Mecchi, A.M., Di Tullio, V. and Balloi, A. 2015. 'Biocleaning method on painting layer: preliminary tests for monitoring the effect of sulfate-reducing bacteria on pigments', international workshop *Green Conservation of Cultural Heritage* 27–28 October. Rome Open Access Abstract e-Book. Available at: http://www.yococu.com/news/abstract-book/ (accessed 24 May 2017).

Goelman, G. and Prammer, M.G. 1995. 'The CPMG pulse sequence in strong magnetic field gradients with applications to oil-well logging', *Journal of Magnetic Resonance* 113(1): 11–18.

Homans, S.W. 1992. *A Dictionary of Concepts in NMR*. Oxford and New York: Clarendon Press.

Hurlimann, M.D. and Griffin, D.D. 2000. 'Spin dynamics of Carr-Purcell-Meiboom-Gill-like sequences in grossly inhomogeneous B(0) and B(1) fields and application to NMR well logging', *Journal of Magnetic Resonance* 143(1): 120–25.

Kimmich, R. 1997. *NMR: Tomography, Diffusiometry, Relaxometry*. Berlin: Springer.

Mao, R., Tang, J. and Swanson, B.J. 2001. 'Water holding capacity and microstructure of gellan gels', *Carbohydrate Polymers* 46(4): 365–71.

Perlo, J., Casanova, F. and Blümich, B. 2005. 'Profiles with microscopic resolution by single-sided NMR', *Journal of Magnetic Resonance* 176: 64–70.

Perlo, J., Casanova, F. and Blümich, B. 2011. *Single-sided NMR*. Berlin: Springer-Verlag.

Rata, D.G., Casanova, F., Perlo, J., Demco, D.E. and Blümich, B. 2006. 'Self-diffusion measurements by a mobile single-sided NMR sensor with improved magnetic field gradient', *Journal of Magnetic Resonance* 180: 229–35.

Troiano, F., Gulotta, D., Balloi, A., Polo, A., Toniolo, L., Lombardi, E., Daffonchio, D., Sorlin, C. and Cappitelli, F. 2013. 'Successful combination of chemical and biological treatments for the cleaning of stone artworks', *International Biodeterioration & Biodegradation* 85: 294–304.

Authors' addresses

- Valeria Di Tullio, Laboratorio di Risonanza Magnetica 'AnnaLaura Segre', Institute of Chemical Methodologies (IMC) CNR, Monterotondo (RM), Italy (valeria.ditullio@cnr.it)
- Eleonora Gioventù, conservator-restorer, Rome, Italy
- Noemi Proietti, Laboratorio di Risonanza Magnetica 'AnnaLaura Segre', Institute of Chemical Methodologies (IMC) CNR, Monterotondo (RM), Italy
- Loreda Luvidi, Institute for the Conservation and Valorization of Cultural Heritage (ICVBC), CNR, Monterotondo (RM), Italy
- Anna Maria Mecchi, Institute for the Conservation and Valorization of Cultural Heritage (ICVBC), CNR, Monterotondo (RM), Italy
- Donatella Capitani, Laboratorio di Risonanza Magnetica 'AnnaLaura Segre', Institute of Chemical Methodologies (IMC) CNR, Monterotondo (RM), Italy

The effect of gelling agents and solvents on poly(methyl methacrylate) surfaces: a comparative study

Stefani Kavda, Stavroula Golfomitsou and Emma Richardson

Introduction

Gels were introduced to the field of conservation to optimize cleaning treatments through their assumed ability to control solvent diffusion and limit mechanical stresses. We are currently undertaking a systematic study into the impact of solvent gel combinations to increase our understanding of their effect on poly(methyl methacrylate) (PMMA) artefacts. The isotropic, amorphous nature of PMMA renders it vulnerable to low molecular weight liquid penetration (Shashoua 2008). Their diffusion into PMMA results in changes to mechanical behaviour; a localized reduction in yield strength renders the material highly susceptible to scratches, crazes and brittle fracture (Michler 1989: 378). Based on the hypothesis that solvent-based gels permit solvent penetration into the PMMA surface, while minimizing the risk of material dissolution (Khandekar 2004), this study assesses the damage potential of the individual materials employed in gel systems. Experimental work examined the effect, safety and time-dependent action of solvents, hydrogel matrices and solvent-gel mixtures.

Experimental details

Mechanically cut transparent, colourless PMMA samples (Goodfellow UK, ME303001, 0.38 × 25 × 25 mm) were used. Agar (Sigma-Aldrich) was dissolved in deionized water (conductivity 1 μS) and heated twice to boiling conditions (*c.*90 °C) to form a 2% (all by weight) solution (pH 7.0). Carbopol EZ2/Ethomeen C-25 and Pemulen TR-2 gels (1% Pemulen:5% triethanolamine (TEA)) were prepared according to Stavroudis and Blank (1989) and Stavroudis (2012), and 80PVAc-borax (5%:1%) was prepared according to Angelova *et al.* (2016). All gels, except agar, were left to stand overnight to fully disperse. The same procedures were followed for the solvent gels, loaded with 20:80% solvent:deionized water mixtures, with deionized (DI) water, ethanol (EtOH), isopropanol (IPA) or petroleum ether (PET).

General full-factorial experiments examined the effect of four hydrogels and four free solvents, at two application times (5 and 60 minutes) to understand their interaction with PMMA. Treatments, carried out in triplicate for hydrogels and free solvents, and duplicate for solvent gels, were assigned to samples in a randomized order, to eliminate systematic errors. The solvents were applied as three drops via micropipette directly on the sample surface. All gels were placed on samples with a Teflon spatula. Agar and PVAc-borax gel required a glass weight to increase the contact with the PMMA surface. Gels weighed approximately 2 g and were applied as 2 mm thick layers. During treatment, samples were protected with a cover to limit the evaporation of solvents. Rigid agar and elastic PVAc gels were easily removed from the samples, while gel pastes made from Carbopol/Ethomeen and Pemulen were wiped with a Teflon spatula. Treatments were followed by three separate cotton swabs moistened with DI water, linearly rolled towards one direction.

Methods

Evaluation of treatment outcome was carried out using a Leica M205A stereomicroscope with raking light. Elimination of the scattering effect of light under the stereomicroscope was achieved in a darkroom, against a black background, resulting in images appearing as though grey-scaled. A JEOL JSM-6610LV scanning electron microscope (SEM) was used for imaging gold-coated samples (10 kV accelerating voltage in high vacuum mode). Melinex was employed as masking to standardize the documentation process of a surface area before and after treatment. Treated samples were left at ambient temperature for 24 hours prior to weighing with

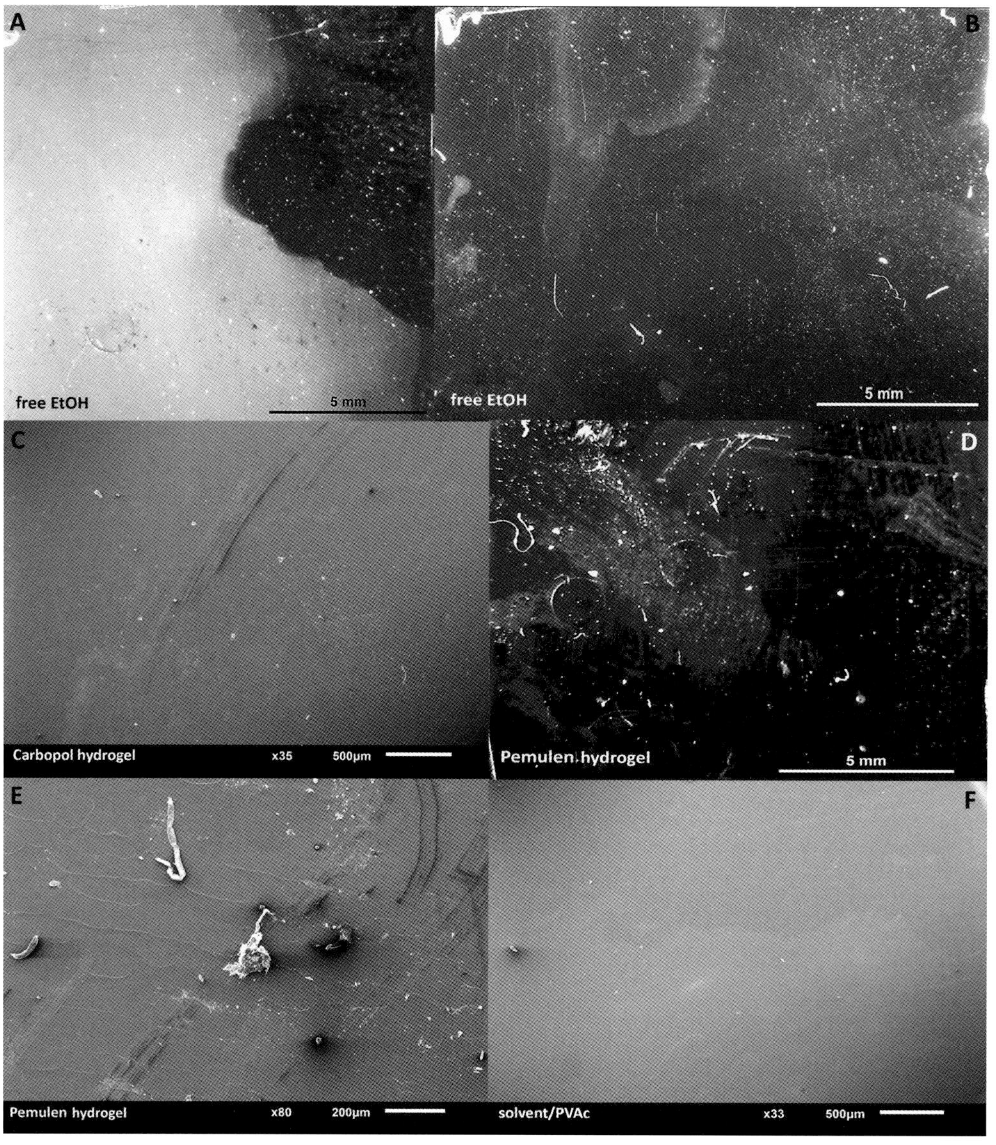

FIG. 1 (A) Raking light image showing clear and dull surfaces upon EtOH. (B) Raking light image showing dull areas upon ethanol (EtOH). (C) SEM image reveals scratching after Carbopol/Ethomeen hydrogels. (D) Raking light image showing greasy surfaces after Pemulen hydrogels and solvent/Pemulen; (E) SEM image reveals Pemulen residues; (F) SEM image reveals discoloration after solvent/PVAc (images: Stefani Kavda).

a Sartorius MSE225S-000-DU Cubis Semi-Micro Balance (accuracy 0.01 mg). Gloss measurements were performed with a Konica Minolta Multi Gloss 268 Plus, and Fourier transform infrared (FTIR) spectroscopy with a diamond Universal Sampling Accessory for attenuated total reflectance (ATR) on a Perkin Elmer Frontier spectrometer from 4000 to 400 cm^{-1}, processed with Perkin Elmer Spectrum software.

Results

Visual and microscopical observation

Free DI water and PET induced negligible blemishes. Application of free EtOH resulted in two distinct behaviours: clear/transparent surfaces after 5 minutes and non-uniform opaque areas and blemishes after 60 minutes (Fig. 1 A and B). Samples cleaned with IPA performed similarly to EtOH cleaned samples. Agar hydrogels caused limited scratching only visible with high magnification under the SEM. Opaqueness caused by free EtOH and IPA was significantly reduced when the solvents were introduced in agar gels. Agar left negligible residues visible under the SEM. Sample behaviour after PET gel cleaning resembled that of samples after free solvent application. Carbopol/Ethomeen hydrogels caused intense scratching visible under the SEM, and left considerable amounts of residues (Fig. 1C). Samples treated with Carbopol/EtOH exhibited non-uniform opaque areas characteristic of the action of the free solvent. All solvent/Carbopol gels left residues rendering surface appearance dull. All Pemulen gels left surfaces scratched and greasy (Fig. 1D), with residues which rendered surfaces stained (Fig. 1E). The samples were visually inferior to those treated with other solvent gels. Surfaces treated with water-based PVAc-borax were minimally blemished. This condition was aggravated upon

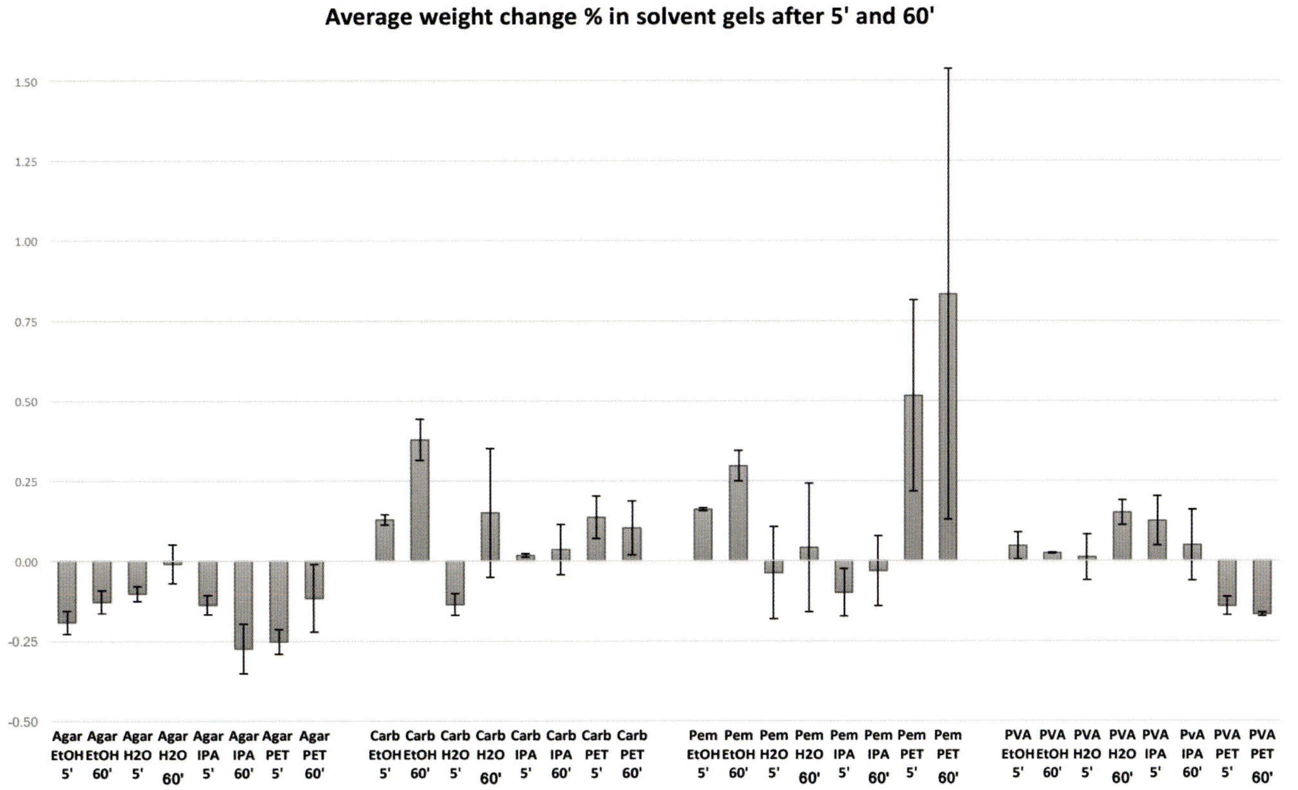

FIG. 2 Average weight change % of samples after solvent gels for 5 and 60 minutes.

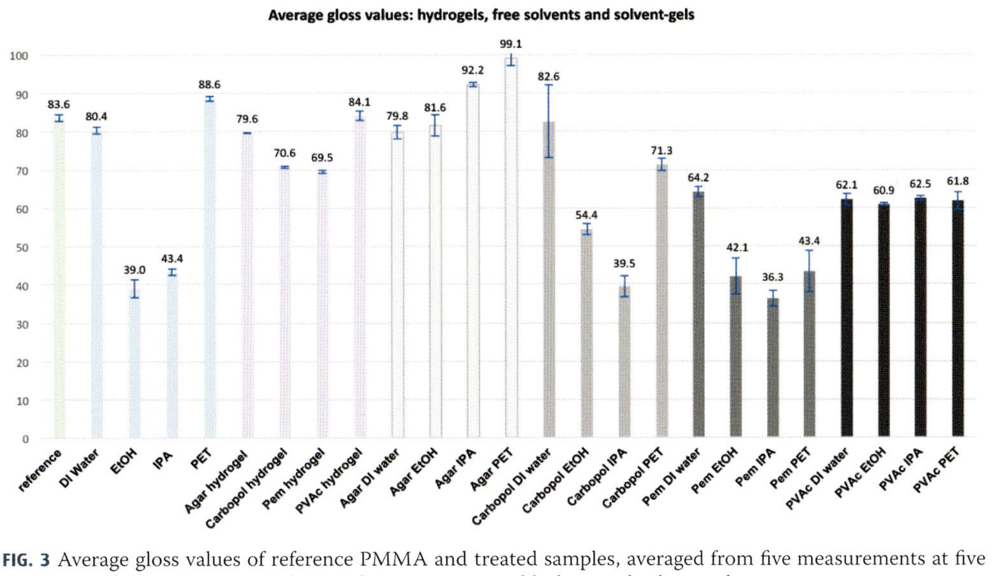

FIG. 3 Average gloss values of reference PMMA and treated samples, averaged from five measurements at five different surface spots at an incident angle at 20°, against a black matte background.

addition of solvents, along with scratching. Occasionally, solvent/PVAc gels caused areas of discoloration on PMMA (Fig. 1F). Minimal gel residues were observed after solvent-PVAc gel treatment. No swelling of samples was visually detected on any of the treated samples.

Weight change

There was no clear pattern of weight change with free DI water and PET. Application of PET for 5 minutes resulted in the smallest weight fluctuations (Fig. 2). Samples showed an overall trend of weight loss after hydro- and solvent/agar gel treatments, which was more pronounced than with free solvents. Carbopol/Ethomeen and Pemulen hydrogels appeared more consistent upon addition of solvents. PVAc-borax hydrogels induced mostly weight gain while solvent/PVAc exhibited the least alterations. EtOH was identified as the solvent having the greatest effect on the treatments; gel type was shown to be the most significant factor in the solvent-gel systems while application time did not impact the outcome (Kavda *et al.* 2017).

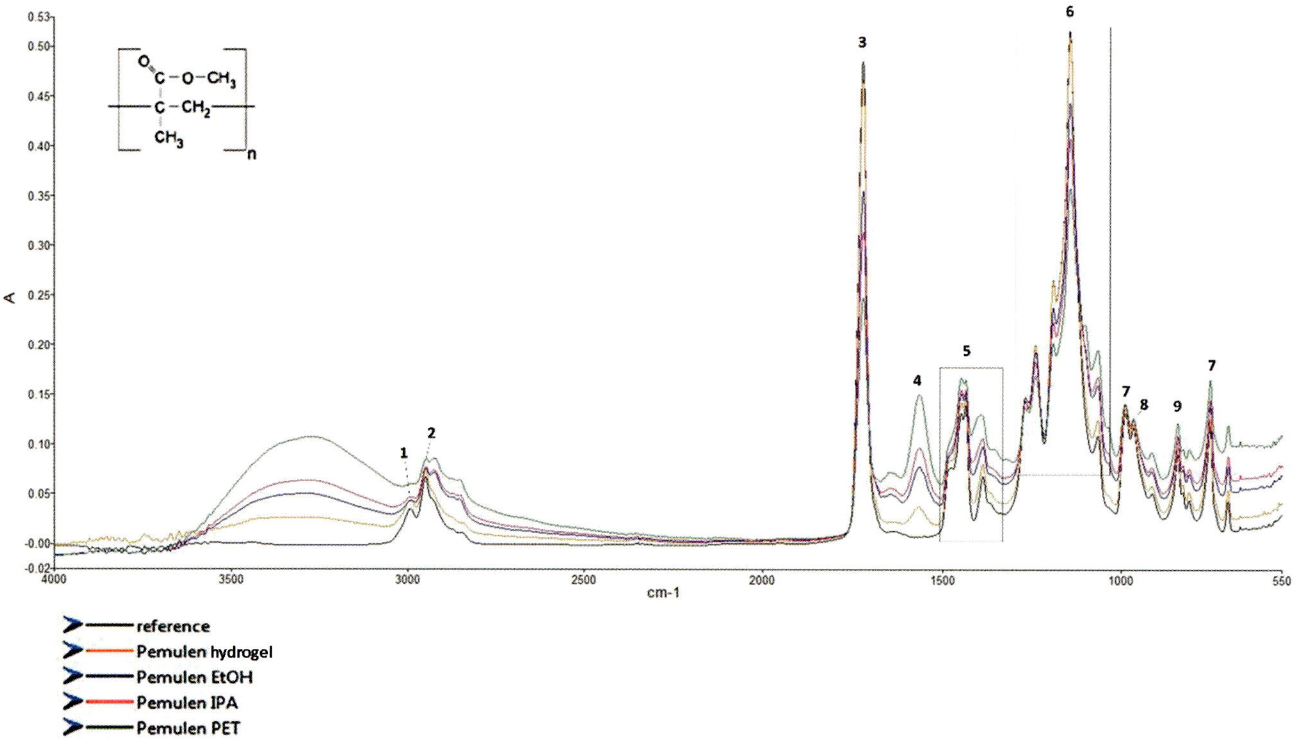

FIG. 4 Spectra of reference PMMA and samples after solvent/Pemulen with 32 scans at ambient temperature and pressure, with a resolution of 4 cm^{-1} equating to a penetration depth of 1.66 μm. Labelled primary absorption bands correspond to assignments in Table 1. The methyl methacrylate (MMA) monomer unit is inset.

TABLE 1 Assignments of diagnostic IR absorptions of reference PMMA (Haris *et al.* 2010).

Labels	Wavenumber (cm−1)	Assignment	Functional group
1	2993	C–H symmetric and asymmetric stretch	CH_3
2	2951	C–H symmetric and asymmetric stretch	CH_2
3	1723	C=O stretch	$COOCH_3$
4	1556	C–O asymmetric stretch	COO^-
5	1500–1300	C–H deformation	CH_3 and CH_2
6	1400–1100	C–O stretch	$COOCH_3$
7	987/750	C–H out of plane bend	CH_3
8	965	C–C	CH_3
9	840	C–H rock	CH_2

Gloss measurements

The use of free DI water did not significantly alter the PMMA surface gloss. The visibly dulled surfaces after treatment with free EtOH showed the greatest variation in specular reflectance. Free IPA cleaning increased opacity while treatment with free PET showed a marginal increase in surface gloss (Fig. 3). DI water/agar and EtOH/agar offered the closest gloss values to the reference PMMA, while IPA/agar and PET/agar increased the gloss. In Carbopol hydrogel treated samples the large deviations of the measurements indicated that replicates were significantly affected. With the addition of solvents, Carbopol exhibited a fluctuation in gloss. Even though Pemulen hydrogels caused small losses, with the presence of solvents, samples behaved poorly with the lowest gloss values among solvent gels tested. PVAc-borax hydrogels resulted in insignificant changes, but once solvents were added, a consistent drop in gloss occurred.

FTIR–ATR

Fig. 4 shows the FTIR spectra acquired from the PMMA control sample and 72 hours after the samples were treated with solvent-Pemulen mixtures. The structure of the MMA monomer is also illustrated in Fig. 4 and the peak assignments are described in Table 1. Agar and Carbopol with EtOH, IPA and PET, as well as Pemulen and PVAc with all four solvents, caused alterations to the PMMA surfaces, demonstrated by changes to the IR absorptions of treated samples. The principal

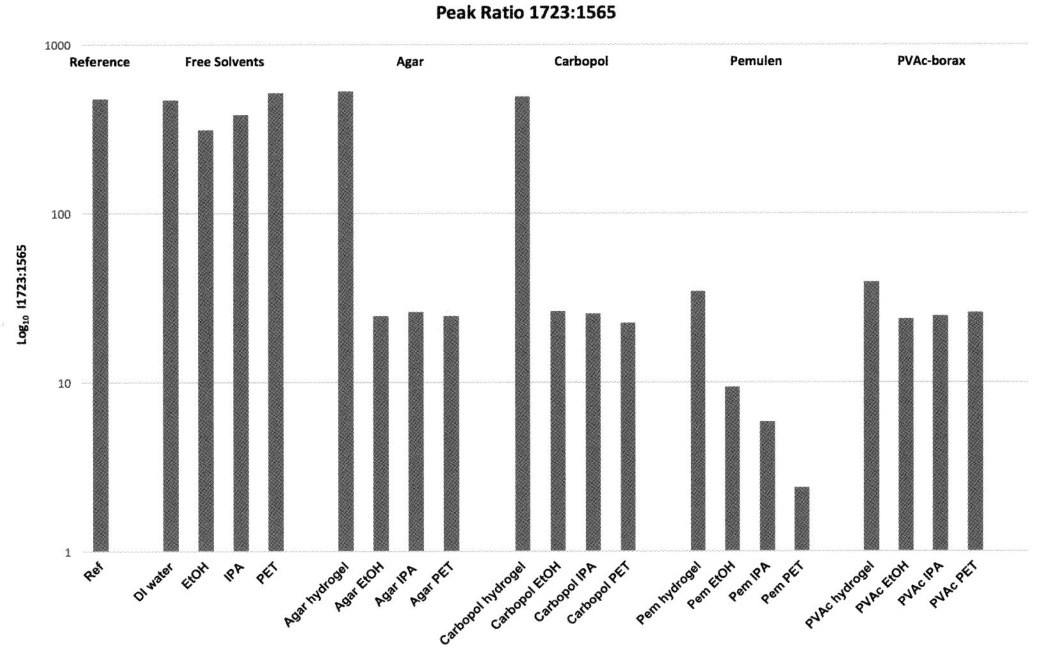

FIG. 5 Ratio between the peak heights at 1723 cm^{-1} and 1565 cm^{-1} of reference PMMA and treated samples.

changes were the formation of absorption bands at 3335 cm^{-1} and 1565 cm^{-1}, also seen after treatment with Pemulen and PVAc-borax hydrogels. The broad bands centred around 3400 cm^{-1} indicate the presence of oxidized structures, in particular hydroxyl groups (–OH) (Kaczmarek *et al.* 2010: 1600), and the absorption at 1560 cm^{-1} is assigned to the presence of a carboxylate ion (Grassie and Fortune 1973: 15). It is apparent from the spectra that an increase in the 1560 cm^{-1} band is accompanied by a decrease in the principal carbonyl band at 1723 cm^{-1} (peak 3),suggesting ester side-group elimination (Kaczmarek and Chaberska 2006: 8189). Decay of absorption at 1239 cm^{-1} and 1144 cm^{-1} (peak 5) is also associated with the cleavage of ester groups. Their decomposition has been shown by Chiantore *et al.* (2000: 1660) to relate to loss of polymer, especially to volatile compounds. They demonstrated that most of the changes in carbonyl band intensity relate to the production of methyl formate, while other gaseous products are methane, hydrogen and carbon oxides (Çaykara and Güven 1999: 226). In addition, an increase in the intensity of the shoulder at 2920 cm^{-1} and 1645 cm^{-1}, indicates the formation of C=C bonds associated with main chain scission reactions (Ayre *et al.* 2014: 83).

The ratio between the peak height at 1723 cm^{-1} and 1565 cm^{-1} (Fig. 5) is used as a measure of change in the samples, showing the progression between the relative decrease of carbonyl groups and increase in carboxylate ion. However, it needs to be borne in mind that the vibrational band at 1565 cm^{-1} also corresponds to functional moieties present in all of the synthetic gel formulations and resulting from solvent cleavage of the glycosidic link in agarose, the gelling fraction in agar. Therefore, the increase in carboxylate ion may be in part due to surface residues. The free solvents of EtOH and IPA were seen to cause a slight decrease in the peak ratio, suggesting ester decomposition, exacerbated by the reduced evaporation rate in the presence of the gels. Further work is under way to establish the depth of the residue layers and the relationship between solvent type and gel deposition.

Discussion

Elastic PVAc gels, together with rigid agar gels, proved the easiest to prepare. PVAc was harder to apply, but easy to remove, and agar gels were the easiest to place and remove due to their rigidity. Agar was not particularly compatible with PET, as exhibited by the gel dissociation, however its introduction did not prevent gellation. In the preparation of Carbopol/Ethomeen gels, difficulties were encountered due to dissociation of the system. Carbopol and Pemulen gels are in paste form that was easy to apply, but prone to leaving residues.

The poor appearance of samples after Carbopol and Pemulen gel cleaning appears comparable. Gloss measurements of the samples after application of the hydrogels presented similar values. Samples after the solvent-gel treatments followed the same pattern of hydrogels having the highest gloss, followed by PET, IPA and lastly EtOH. All gloss values were slightly lower in the Pemulen-based gels. Samples treated with solvent-based Carbopol and Pemulen gels showed a similar pattern of weight fluctuation. The results of poor surface appearance of Pemulen-based gels are in agreement with the gloss and weight measurements, which presented the most deviating values. Samples treated with PVAc gels offered similar visual observations throughout the use of the different solvents, which were comparable to other solvent gels. Interestingly, solvent-PVAc rendered PMMA surfaces equally glossy in a consistent manner. This might suggest that treatment with PVAc gels offers more consistency, and that their action might be quite controlled.

IPA/Carbopol and EtOH/PVAc gels showed the least weight change. In theory, this appears to be a positive result, as it indicates that the solvent-gel action has no net effect on the treated material. Further investigations will determine whether more than one process might be occurring concurrently, as gravimetric analysis is completely insensitive to simultaneous extraction and deposition of material, if this occurs.

Weight changes in samples induced by free solvents were occasionally intensified or moderated by gellation. This suggests that gels influenced the performance of the solvents possibly by slowing their evaporation, while increasing control over their action. This, however, is not always a desirable characteristic: in the case of EtOH, which appears to be aggressive on these PMMA samples, this mechanism would increase solvent diffusion and contact time, aggravating the condition of the sample. Based on the FTIR results, this solvent interaction with PMMA can lead to extraction and/or dissolution of volatile components.

EtOH, followed by IPA, caused the highest visual and physical alterations, supported by gloss and weight measurements. Nonetheless, FTIR showed a similar degradation between PMMA treated with most solvent gels. When coupled with Pemulen, PET instigated the greatest change, followed by IPA and then EtOH (Fig. 5). This might indicate that when added in Pemulen, PET's action is intensified, an assumption also supported by gravimetric data (Fig. 2).

Conclusion

Visual observations, gravimetric and gloss measurements, and chemical analysis showed that hydrogels did not cause significant damage to PMMA surfaces. EtOH influenced the surfaces to the largest degree, appearing unsafe for use on PMMA. The application time did not affect the result of treatments at this stage of the study. FTIR–ATR showed that gels with EtOH, IPA and PET mainly caused cleavage of ester groups in PMMA, increasing its degradation. Overall, solvent-agar gels appeared to be the most effective and safe for use on PMMA, while solvent-based Carbopol and Pemulen performed poorly. Samples after treatment with solvent/PVAc-borax gels generally displayed more consistent behaviours with the least variations.

References

Angelova, L., Ormsby, B.A. and Richardson, E. 2016. 'Diffusion of water from a range of conservation treatment gels into paint films studied by unilateral NMR. Part I. Acrylic emulsion paint', *Microchemical Journal* 124: 311–20.

Ayre, W.N., Denyer, S.P. and Evans, S.L. 2014. 'Ageing and moisture uptake in polymethyl methacrylate (PMMA) bone cements', *Journal of the Mechanical Behaviour of Biomedical Materials* 32: 76–88.

Çaykara, T. and Güven, O. 1999. 'UV degradation of poly(methyl methacrylate) and its vinyltriethoxysilane containing copolymers', *Polymer Degradation and Stability* 65(2): 225–9.

Chiantore, O., Trossarelli, L. and Lazzari, M. 2000. 'Photooxidative degradation of acrylic and methacrylic polymers', *Polymer* 41: 1657–68.

Grassie, N. and Fortune, J D. 1973. 'Thermal degradation of copolymers of methyl methacrylate and butyl acrylate, 3. Residual polymer', *Die Makromolekulare Chemie* 168(1): 13–18.

Haris, M.R.H.M., Kathiresan, S. and Mohan, S. 2010. 'FT-IR and FT-Raman spectra and normal coordinate analysis of poly methyl methacrylate', *Der Pharma Chemica* 2(4): 316–23.

Kaczmarek, H. and Chaberska, H. 2006. 'The influence of UV-irradiation and support type on surface properties of poly(methyl methacrylate) thin films', *Applied Surface Science* 252(23): 8185–92.

Kaczmarek, H., Gałka, P. and Kowalonek, J. 2010. 'Influence of a photoinitiator on the photochemical stability of poly(methyl methacrylate) studied with Fourier transform infrared spectroscopy', *Journal of Applied Polymer Science* 115(3): 1598–607.

Kavda, S., Richardson, E. and Golfomitsou, S. 2017. 'The use of solvent-gel systems for the cleaning of PMMA', *MRS Advances*. DOI 10.1557/adv.2017.249.

Khandekar, N. 2004. 'Gelled systems: theory and early applications', in V. Dorge (ed.), *Solvent Gels for the Cleaning of Works of Art: The Residue Question*. Los Angeles: Getty Conservation Institute, 5–11.

Michler, G.H. 1989. 'Crazes in amorphous polymers I. Variety of the structure of crazes and classification of different types of crazes', *Colloid and Polymer Science* 267(5): 377–88.

Shashoua, Y. 2008. *Conservation of Plastics: Materials Science, Degradation and Preservation*. Oxford: Butterworth Heinemann.

Stavroudis, C. 2012. 'Pemulen® revised: pHuck the pH Meter', *WAAC Newsletter* 34(2): 19.

Stavroudis, C. and Blank, S. 1989. 'Solvents and sensibility', *WAAC Newsletter* 11(2): 2–10.

Authors' addresses

- Stefani Kavda, UCL Qatar, University College London, UK/Doha, Qatar (stefani.kavda.12@ucl.ac.uk)
- Stavroula Golfomitsou, UCL Qatar, University College London, UK/Doha, Qatar
- Emma Richardson, Department of History of Art, University College London, London, UK

The case of Capogrossi in Rome: trials and tests using gels for the cleaning of a contemporary wall painting

Paola Mezzadri, Giancarlo Sidoti and Maria Carolina Gaetani

Introduction

This paper focuses on gel treatments applied to a contemporary mural painting of 1954, by the Italian artist Giuseppe Capogrossi (Capogrossi 1957; Aloi 1958: 354 ff.; Bolpagni 2013; Standeven 2013). This forgotten masterpiece is located on the ceilings of the double staircase at the main entrance of the Airone, built as a cinema and theatre in Rome, Italy.

The building, situated inside a central court enclosed by five blocks of flats, was commissioned by a public corporation and designed and planned during the 1950s by the architects Adalberto Libera and Eugenio Montuori and the engineer Leo Calini (Calandra 1956; Zevi 1956; Aloi 1958: 354 ff.; Montuori 1981: 56–7; Garofalo and Veresane 1989: 159; Bonavita 2003; Pellicano 2007). The Airone cinema is situated 7 m below street level. The abstract geometric style of the painting on the ceilings, comprising graphic and repetitive fields of brilliant 'pop' colours, was obtained utilizing modular elements applied onto the wall. Overall, in line with the architectural plan, the painting recalls the imagery of a large crowd flowing into the cinema, their countless eyes looking around (Fig. 1). Over time, the Airone cinema lost its original function and

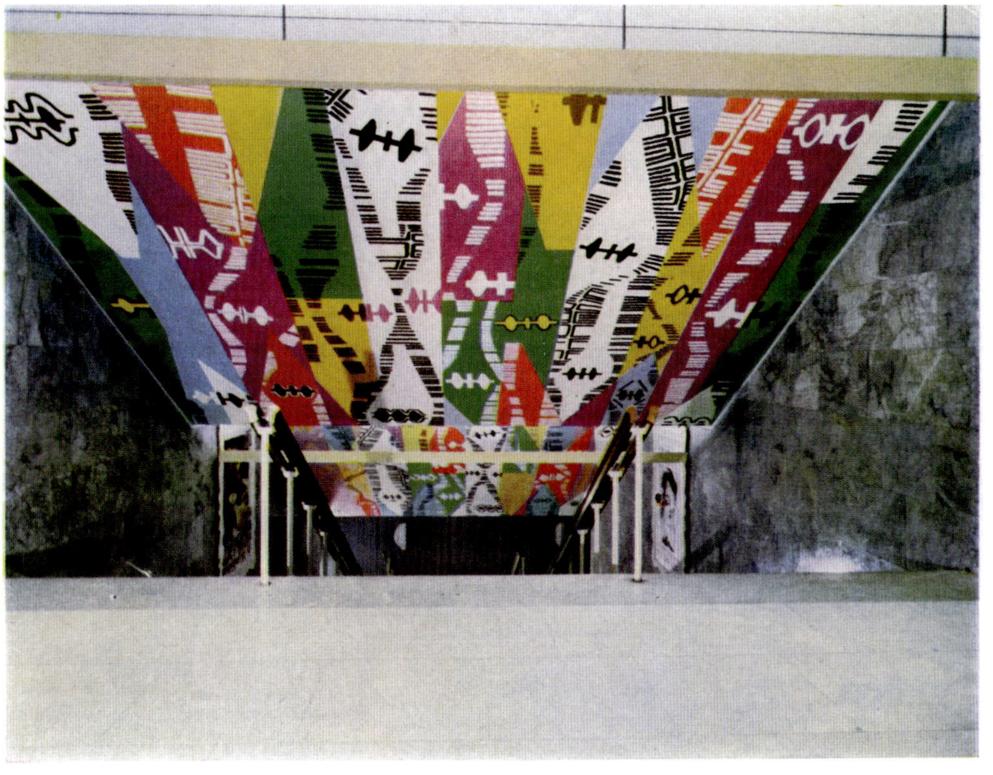

FIG. 1 Giuseppe Capogrossi, wall painting for the Airone cinema, 1954 (image first published by Aloi 1958, Capogrossi Foundation, all rights reserved).

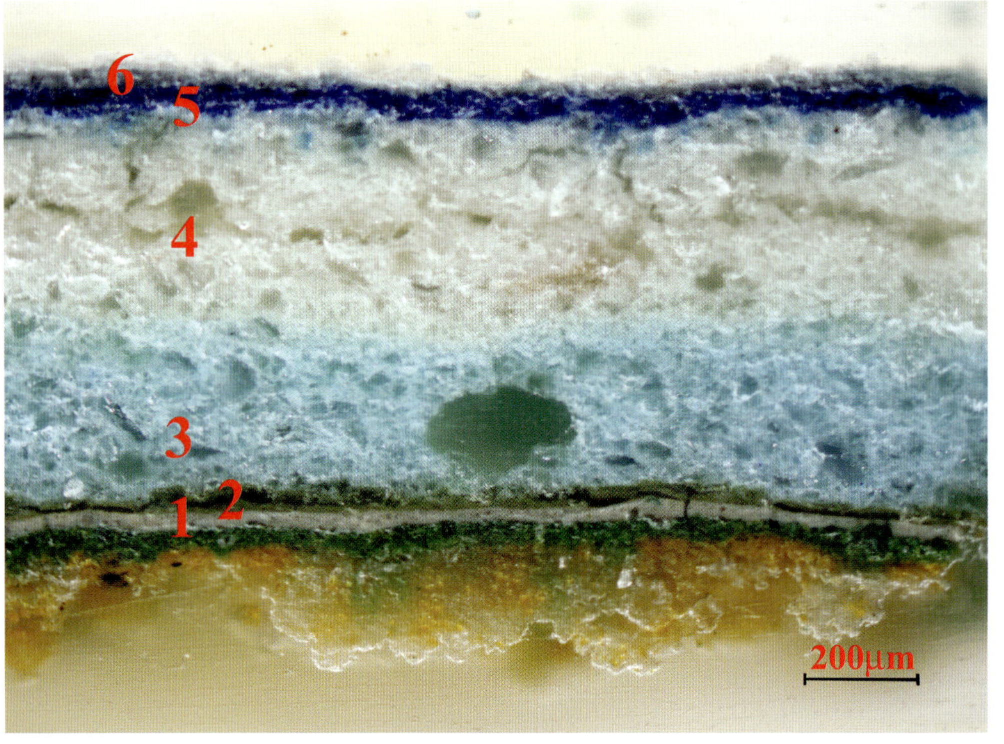

FIG. 2 Cross-section with six overpainted layers from the lower ceiling (image: Giancarlo Sidoti, ISCR, all rights reserved).

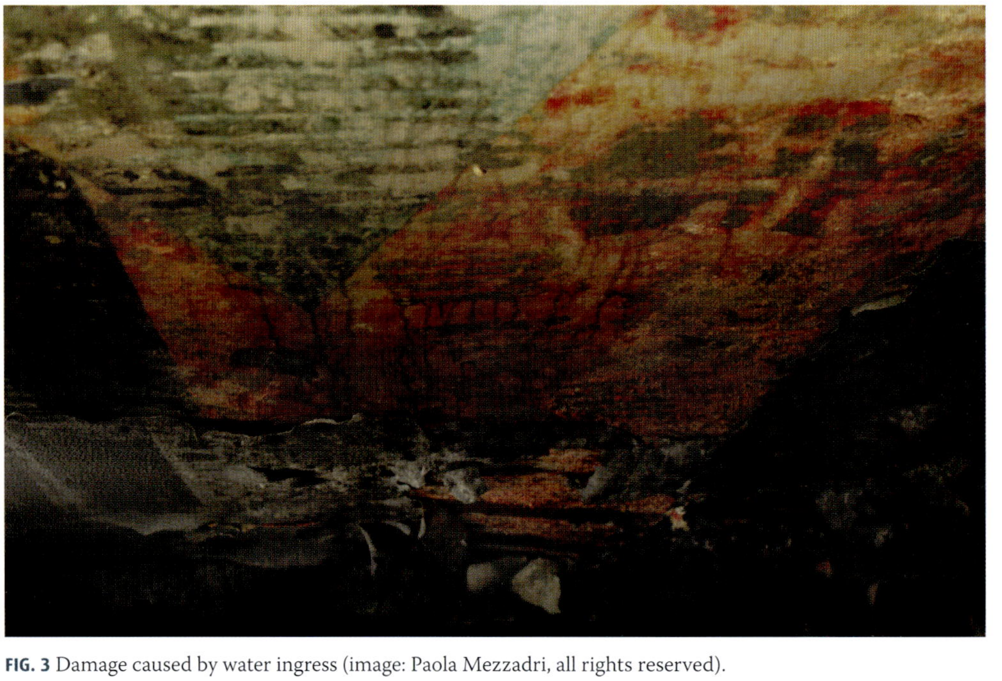

FIG. 3 Damage caused by water ingress (image: Paola Mezzadri, all rights reserved).

became successively a ballroom, a nightclub, a discotheque and eventually a strip club. The main architectural project was twisted, transformed and damaged. The painting was completely covered by synthetic coatings: nine layers were applied over the upper ceiling of the double staircase (De Cesare *et al.* 2015) and six (Fig. 2) over the lower ceiling (Mezzadri and Gaetani 2016). The building has been abandoned since 1999 and this restoration project, carried out on a sample area of the second ceiling, took place in 2015–16 (Mezzadri 2015).

Conservation issues

The painting has a complicated conservation history that includes water ingress and the changing use of the building. As confirmed by Fourier transform infrared (FTIR) microscopy, the overpainted layers were styrenated vinyl acetate, polyvinyl acetate (PVAc) and VeoVa-based PVAc. The closest layer to the original surface was an alkyd resin-based film. The combination of historical and technical information with materials analysis gave a deeper understanding of Capogrossi's

TABLE 1 Cleaning tests and gels application to the VeoVa-PVAc-based layer.

Gels	Cleaning system	Contact period (minutes)	Results
Semi-rigid agar 2% in deionized water (pH 7)	Adding a solution of TAC at 3%	15	FAIR Partial swelling Irregular cleaning Salts incorporated with the overpainted layers Inappropriate gel
Semi-rigid gellan 2% in deionized water (pH 7)	Adding a solution of TAC at 5%	15	FAIR Partial swelling Irregular cleaning Salts incorporated with the overpainted layers Inappropriate gel
Semi-rigid gellan 2% in deionized water (pH 7)	Adding an ethanol/*n*-octane/acetone (24:24:52)	15	GOOD Fair swelling Irregular cleaning Salts incorporated with the overpainted layers Inappropriate gel
Klucel G 3% in ethanol	Ethanol	15	NONE
Klucel G 5% in 1:1 ethanol/acetone	Ethanol/acetone (1:1)	15	FAIR Partial swelling Irregular cleaning Salts incorporated with the overpainted layers Inappropriate gel
Carbopol Ultrez 21 1% in deionized water, neutralized with Ethomeen C25 (pH 7)	Deionized water	10	GOOD Partial swelling of the VeoVa and alkyd layers Mechanical reduction of the concretion salts Appropriate gel
Carbopol Ultrez 21 1% in 1:1 ethanol/MEK, neutralized with Ethomeen C25 (pH 7)	Ethanol/MEK (1:1)	30	EXCELLENT Total swelling Mechanical reduction of the salts possible Combined cleaning action: complexing and swelling Appropriate gel

techniques and materials, mainly based on PVAc medium. The water ingress severely affected the lower ceiling, causing contraction in nearly all of the overpainted layers as a result of the initial softening of the layers, followed by evaporation of water. The loss of materials such as additives and surfactants (due to extraction by the water) also increased this contraction with the substantial rip and a consequential delamination of the original film beneath (Fig. 3). Other damages such as lacunae, swelling, mixing and loss of cohesion of the original colours were also a consequence of the water ingress, the latter connected to large quantities of soluble salts (potassium, sodium, magnesium and calcium sulphates as detected by X-ray diffraction), present both as subflorescence (within the original and later painted layers) and efflorescence on the surface.

Criteria in cleaning tests and gels application

Cleaning tests were carried out to identify the best and most appropriate method to swell the two synthetic overpainted layers closest to the original – a VeoVa plasticized PVAc and a (presumably oil-modified) alkyd resin (respectively layers 2 and 1 in the cross-section) – and to remove residues of the synthetic layers also combined with many accumulated salts, probably carbonates. Therefore, a small group of organic solvents (with low toxicity and retention time within the paint, selected by the Trisolv[1] system) and/or chelating agents were tested on this highly inhomogeneous surface first by swab and then within different gels, all fairly transparent, allowing for monitoring of the swelling action. The gels included Klucel G (hydroxypropylcellulose) (Borgioli and Cremonesi 2005: 63–105), agar (non-ionic polysaccharide agarose) (Campani *et al.* 2007; Anzani *et al.* 2008), Phytagel (gellan gum, anionic polysaccharide) (Talarico *et al.* 2001; Iannuccelli and Sotgiu 2012) and Carbopol Ultrez 21 (polyacrylic acid) (Borgioli and Cremonesi 2005: 63–105; Travers *et al.* 2010). The application of a gel system proved necessary for carrying the mixture of organic solvents and/or chelating agents selected for the final cleaning, for several reasons: the surface of the paint was irregular and rough, at the staircase the surface was inclined, and adhesion to a ceiling was required, all for a sufficiently

TABLE 2 Cleaning tests and gels application on the alkyd-based layer.

Gels	Cleaning system	Contact period (minutes)	Results
Klucel G 5% in 1:1 ethanol/acetone	Perfected by swab with an *n*-octane/acetone solution (1:1)	120	FAIR Aliphatic hydrocarbon required for swelling the alkyd paint Slow Inappropriate gel
Klucel G 5% in 1:1 ethanol/MEK	Perfected by swab with *n*-octane/acetone solution (1:1)	240	GOOD Aliphatic hydrocarbon required for swelling the alkyd paint Slow Inappropriate gel
Semi-rigid gellan 2% in deionized water (pH 7)	Adding an *n*-octane/MEK (1:1)	20	SCARCE Slow Inappropriate gel
Semi-rigid gellan 2% in deionized water (pH 7)	Adding an ethanol/*n*-octane /MEK (20:40:40)	20	FAIR/SCARCE Slow Inappropriate gel
Carbopol Ultrez 21 1% neutralized with Ethomeens C25 and C12 (pH 7)	In half part of deionized water and half part of an ethanol/*n*-octane/MEK solution (11:17:72)	Immediate	EXCELLENT BUT HAZARDOUS Swelling too rapid Combined action too aggressive Appropriate gel
Carbopol Ultrez 21 1% neutralized with Ethomeens C25 and C12 (pH 7)	In half part of deionized water and half part of an ethanol/*n*-Octane/MEK solution (20:40:40)	Immediate	EXCELLENT BUT HAZARDOUS Swelling too rapid Combined action too aggressive Appropriate gel
Carbopol Ultrez 21 1% in deionized water, neutralized with Ethomeens C25 (pH 7)	Perfected by swab with an ethanol/*n*-Octane/MEK solution (20:40:40)	5	EXCELLENT Swelling was rapid Separate actions: first complexing and then swelling Controllable cleaning action

long contact period to soften the thick and resistant alkyd resin that was being removed. Furthermore, the cleaning action had to be localized because adjacent original layers to those being removed had similar solubility.

Cleaning tests and gels application on the VeoVa-PVAc-based layer

After defining a possible solubility area of the VeoVa-PVAc-based layer in the Trisolv system, some mixtures of organic solvents were selected (Cremonesi 2004: 24, 83 ff., 102 ff.), as well as some solutions with chelating agents. The cleaning tests and gel application on the VeoVa-based layer are described in Table 1.

A 3% triammonium citrate solution (TAC) gelled in semi-rigid agar gel in deionized water (pH 7) was tested, but its chelating and swelling action was scarcely as much as that of a 5% TAC solution gelled in a semi-rigid gellan gel in deionized water (pH 7). The TAC cleaning action was poor compared to the solvent mixtures tested. Moreover, considering the irregular surface with the original layer often exposed or incorporated within hardened salts, this chelation method was excluded to avoid a possible interaction with the metallic cations of pigments and extenders present in the original film. Ethanol at 3% gelled in Klucel G was unsuccessful, while a 1:1 ethanol/acetone solution gelled in Klucel G at 5% was slightly better, even though the low viscosity of the gel applied made it useless in practice. The most successful solution for the removal of the VeoVa-PVAc-based layer was an ethanol/*n*-octane/acetone blend in a 24:24:52 ratio (calculated using TriSolv). The latter mixture, gelled in a semi-rigid gellan gel, showed a good swelling action but was ineffective where hardened salts had been incorporated into the overpainted layers. Moreover the application of semi-rigid gels, just as for Klucel G, was inappropriate due to the inclined ceiling of the staircase and the inhomogeneous rough surface, both of which prevented good adhesion. For this reason Carbopol Ultrez 21 was also tested, first gelled in deionized water, and then in a solvent mixture, to increase both the polarity of the gel system and the contact period. The use of polyacrylic acid neutralized with ethoxylated amines (Ethomeen C-25) showed the best results: first in terms of gel adhesion and then in terms of swelling action on the salts and the VeoVa-PVAc layer. Thanks to its high viscosity, Carbopol Ultrez 21 remained on the rough surface without any problems. Where the

VeoVa-PVAc layer was incorporated in some insoluble salts, a proper solvent gel was applied promoting both a complexing and swelling action: a 1:1 ethanol/methyl ethyl ketone (MEK) polar solution was gelled in Carbopol Ultrez 21 at 1% w/v. This was brushed onto the surface and then removed after 30 minutes, including agitating the surface with a stiff bristle brush and wiping all the residues (the later paint and the salts) from the surface with both cotton pads and a scalpel. Finally, the surface was rinsed twice with ligroin.

Cleaning tests and gel application to the oil-modified alkyd-based layer

This alkyd-based layer, adjacent to the original, was 40–50 µm thick, opaque, and difficult to remove. It had broad wrinkles included inorganic compounds in the paint formulation: calcium carbonate ($CaCO_3$), barium sulphate ($BaSO_4$), calcium sulphate ($CaSO_4{\cdot}2H_2O$) as extenders, and titanium white (TiO_2) and zinc white (ZnO) as white pigments. After defining a possible solubility area for alkyd paints in the TriSolv system, as the overlap of the polyester resin and polymerized linseed oil areas, several mixtures of organic solvents were selected. The cleaning tests on this layer are described in Table 2.

A 1:1 ethanol/acetone solution gelled at 5% in Klucel G showed fair results while a 1:1 ethanol/MEK gelled in Klucel G at 5% was good, albeit inappropriate due to the excessive contact period of 240 minutes that was required. Moreover this cleaning method had to be completed by swab with a 1:1 mixture of *n*-octane/acetone (the latter used because it is more volatile than MEK). In fact, the swelling of the alkyd-based layer was only possible with the introduction of a non-polar aliphatic hydrocarbon into the mixture. As Klucel G is unable to gel non-polar solvents it was rejected once again. This time, only one of the two semi-rigid gels was applied: the most transparent though less flexible gellan gel. The results were unsatisfactory due to the high viscosity of gel and the inclined wall surface, which made application very difficult. Carbopol was the most appropriate gel because its swelling action was more effective and rapid than the solvent mixtures tested by swab or with other types of gels. The best method for removal of the alkyd resin was gelling a 1:1 mixture of water and a 20:40:40 ethanol/*n*-octane/MEK in 1% w/v Carbopol Ultrez 21, neutralized by Ethomeens C-25 and C-12. This cleaning system was too hazardous with respect to the original PVAc layer beneath. Therefore, the solvent-gel action was carried out as a two-step process: first using only Carbopol – gelled in deionized water at 1% w/v and neutralized by Ethomeen C-25 – as a complexing agent to remove the bulk of the inorganic compounds and then using the cleaning system applied with a swab of the best working solvent mixture, namely 20:40:40 ethanol/*n*-octane/MEK.

Carbopol gel was brushed onto the surface and removed after 5 minutes with dry cotton pads. Then the surface was rinsed twice with ligroin. A semi-transparent veil of the alkyd layer, almost imperceptible, was left intentionally in order to avoid a possible interaction with the original paint, which was somewhat sensitive to the selected cleaning system.

Conclusion

This study highlights the necessity of cleaning tests with different gels to identify the most appropriate cleaning methods for a particularly damaged surface. Some materials were found to be more appropriate and effective than others. Carbopol Ultrez 21 was the most effective gel applied, both in terms of adhesion and swelling action on the layers to be removed; in fact it was used first as a solvent gel for the removal of the VeoVa-PVAc-based overpainted layer along with the hardened salts, and then utilized again for the final removal of the oil-modified alkyd resin layer. Full technical documentation of the site combined with materials analysis characterized both the original and non-original synthetic paints present, which proved necessary to clarify the complex layering sequence and to introduce a new approach for the removal of the non-original alkyd resin using Carbopol gel both as a complexing agent for salts and as a solvent gel. This allowed a specific cleaning system to be designed for the Capogrossi wall painting in the Airone cinema.

Acknowledgements

This conservation project was carried out by conservator Paola Mezzadri as her final dissertation for the Istituto Superiore per la Conservazione e il Restauro (ISCR), Rome, Italy, which was discussed in April 2016. The intellectual contribution and support of all the multidisciplinary ISCR team was crucial for both elucidating the issues and aiding treatment decision making. The ISCR team included: Maria Carolina Gaetani, conservator; Laura D'Agostino, art historian; Giancarlo Sidoti, Lucia Conti and Luigi Arceri, chemists; Marco Bartolini, biologist; Carlo Cacace and Fabio Aramini, physicists; Jacopo Russo, architect, for graphic and photographic documentation; and Annamaria Pandolfi, architect. For permission to examine and treat the wall painting, we are extremely grateful to Morena Costantini, art historian and ministerial authority for the Soprintendenza acheologia belle arti e paesaggio for the municipality of Rome; and Silvia Corsi, architect for the U.O. Città Storica di Roma and designer of the project 'Requalification of the Airone Cinema-Theatre'. We would also like to extend our gratitude to Gugliemo Capogrossi, Francesco Montuori and the late Giunio Tabacchi for their invaluable help.

Materials and suppliers

- Ethomeens C-12 and C-25, Carbopol Ultrez 21, Klucel G, Phytagel, TAC, ethanol, *n*-octane, acetone, MEK: IMAR, Rome, Italy, http://www.imaronline.com/
- Agar: D'Alena Remigio, Rome, Italy, https://aziende.virgilio.it/prodotti-chimici/roma-rm/alena-remigio

Note

1. See http://www.icr.beniculturali.it/flash/progetti/TriSolv/TriSolv.html. Trisolv (Interactive Solvent and Solubility Triangle): an online interactive software that reproduces the Teas Chart and its parameters. The system is graphically updated in real time. One of its most significant features is that it allows the user to calculate a precise volume percentage for solvent mixtures, having different values for volatility and with low toxicity, created by three solvents (alcohols, aliphatic hydrocarbons and ketones). For further information see http://www.icr.beniculturali.it/flash/progetti/TriSolv/dati/EN/TriSolv_EN.pdf.

References

Aloi, R. 1958. *Architetture per lo spettacolo*. Milan: Hoepli.

Anzani, M., Berzioli, M, Cagna, M., Campani, E., Casoli, A., Cremonesi, P., Fratelli, M., Rabbolini, D. and Riggiardi A. 2008. *Gel rigidi di Agar per il trattamento di pulitura di manufatti in gesso/Use of Rigid Agar Gels for Cleaning Plaster Objects, Quaderni del Cesmar7*, no. 6. Padua: Il Prato.

Bolpagni, P. 2013.'Capogrossi e una "idea" di Italian style', in M.L. Barbero (ed.), *Capogrossi. Una retrospettiva*, Venice: Marsilio, 344.

Bonavita, A. 2003. 'Cinema Airone', scheda C.23, in G. Remiddi and A. Greco (eds), *Adalberto Libera. Guida a 45 architetture*. Rome: Palombi.

Borgioli, P. and Cremonesi, P. 2005. *Le resine sintetiche utilizzate nel trattamento di opere policrome*. Padua: Il Prato.

Calandra, M. 1956. 'Cinema Airone a Roma', *L'architettura: cronache e storia* 1(5): 662–67.

Campani, E., Casoli, A., Cremonesi, P., Saccani, I. and Signorini, E. 2007. *L'uso di agarosio e agar per la preparazione di 'gel rigidi'/Use of Agarose and Agar for Preparing 'Rigid Gels'*, D. Kunzelman (tr.), *Quaderni del Cesmar7*, no. 4. Padua: Il Prato.

Capogrossi, G. 1957. 'Un pittore giudica l'architettura', *L'Architettura. Cronache e storia* 3(20): 132.

Cremonesi, P. 2004. *L'uso dei solventi organici nella pulitura di opere policrome*. Padua: Il Prato.

De Cesare, G., Gurgone, N., Putaturo, G., Miliani, C. and Rosi, F. 2015. 'Murals and architecture: the case of Capogrossi in Rome', in M. Sánchez Pons, W. Shank and L. Fuster López (eds), *Conservation Issues in Modern and Contemporary Murals*, 133–42. Available at: http://www.cambridgescholars.com/conservation-issues-in-modern-and-contemporary-murals (accessed 25 May 2017).

Garofalo, F. and Veresane, L. 1989. *Adalberto Libera*. Bologna: Zanichelli.

Iannuccelli, S. and Sotgiu S. (eds) 2012. *La pulitura a umido di opere d'arte su carta con gel rigidi di Gellano: presupposti teorici, metodologia applicativa e verifica analitica, Quaderni del Cesmar7*, no. 11. Padua: Il Prato.

Mezzadri, P. 2015. *Il dipinto murale di Giuseppe Capogrossi nell'ex cinema Airone: problemi conservativi e intervento di restauro*, unpublished final dissertation, Conservazione e Restauro dei Beni Culturali, ISCR, Rome.

Mezzadri, P. and Gaetani, M.C. 2016. 'Il dipinto murale di Giuseppe Capogrossi nell'ex cinema Airone: pronto intervento', in *Lo Stato dell'Arte 14*. Florence: Nardini and IGIIC, 265–77.

Montuori, E. 1981. *Eugenio Montuori architetto*. Milan: Over.

Pellicano, E. 2007. 'Cinema Airone, Adalberto Libera. Rilievo, recupero funzionale e riuso', in T. Carunchio, R. Sparvieri and S. Volterra (eds), *Progetti di restauro architettonico e dei monumenti: ricerca, sperimentazione, didattica*, 1: 299, 349–55.

Standeven, H.A.L. 2013. 'Synthetic resin housepaints from the 1920s to 1960s: properties, use and their reception by architects and painters', in G. Jean (ed.), *La conservazione delle policromie nell' architettura del XX secolo*. Florence: SUPSI-Nardini Editore, 214–27.

Talarico, F., Caldi, C., Valenzuela, M., Zaccheo, C., Zampa, A. and Nugari, M.P. 2001. 'Applicazioni dei gel come supportanti nel restauro', *Bollettino ICR* 3: 101–18.

Travers, K., Wolbers, R. and Tomkiewicz, C. 2010. 'Pemulen® case study: *Holy Innocents* mural project', *WAAC Newsletter* 32(3): 13–14.

Zevi, B. 1956. 'Critica d'arte e umanità architettonica', *L'Architettura. Cronache e storia* 1(5): 657.

Authors' addresses

- *Corresponding author*: Paola Mezzadri, Rome, Italy (p.mezzadri99@gmail.com)
- Giancarlo Sidoti, ISCR, Rome, Italy (giancarlo.sidoti@beniculturali.it)
- Maria Carolina Gaetani, ISCR, Rome, Italy (mariacarolina.gaetani@beniculturali.it)

Removing iron stains from wood and textile objects: assessing gelled siderophores as novel green chelators

Stavroula Rapti, Stamatis Boyatzis, Shayne Rivers, Athanasios Velios and Anastasia Pournou

Introduction

The ease of application, low toxicity and most importantly the potential of gels to clean in a more controlled and selective way has ensured their wide use in conservation (Wolbers *et al.* 1988; Wolbers 2000). Gels can be prepared using many different thickeners such as: cellulose ethers Klucel, methyl cellulose (MC), hydroxypropylmethycellulose (HPMC); polysaccharide xanthan gum; the high-viscosity polyacrylic acids Carbopol and Pemulen (Cremonesi 2013); and silicone-based bi-phasic gels, such as Velvesil Plus (Wolbers 2011).[1] These thickeners are often combined with other materials such as solvents, surfactants and chelators in order to facilitate the cleaning process.

Gel efficiency has been assessed initially on paintings, paper and textile objects; however their cleaning potential has not yet been extensively tested on many substrates encountered in museum objects. Such non-evaluated applications include the removal of iron corrosion products from composite museum objects, often consisting of several materials, such as wood, metal and textile. These artefacts start to show corrosion of their iron elements when situated in humid environments, causing staining of their neighbouring materials (Music *et al.* 1993; Gill 2007). Stains deriving from iron corrosion are problematic for artefacts, not only aesthetically, but also due to their potential to damage substrates. Corrosion products can promote further deterioration on cellulosic or proteinaceous substrates through the catalysis of various oxidation processes, leading to their depolymerization (Xu and Goodell 2001; Almkvist 2008). Iron corrosion, in combination with the deterioration of the surrounding materials, can lead to mechanical damage of objects and threaten their structural integrity (Baker 1974). Therefore, the removal of iron corrosion products is considered an essential process during the treatment of composite objects. However, even though there are many published studies on their removal with chelators from substrates, such as paper, textile, stone, metal, waterlogged wood and paintings (Burgess 1991; Phenix and Burnstock 1992; Margariti 2003; Rivers and Umney 2003; Almkvist *et al.* 2005; Fors 2008), studies on water-sensitive objects made of wood and textile are scarce. This is probably due to the fact that the chelating agents are usually water soluble and they are often applied using aqueous baths (Adler and Eaton 1995; Timár-Balázsy and Eastop 1998; Margariti 2003). Cleaning treatments using immersion procedures however, cannot be applied on dry porous hygroscopic and anisotropic materials such as wood. Wooden objects can undergo uneven dimensional changes resulting in distortion or warping of wood from its original shape, dislocation of their parts, splitting and loosening of their joints, etc. It therefore appears that finding a cleaning method for iron-stained composite museum wooden objects is a necessity. The ideal method should effectively remove the iron corrosion products, minimize the amount of water that will penetrate into wood and maintain the cleaning potency of the solution applied.

Therefore, this study examined different gel formulations combined with chelators, aiming to investigate their efficacy in successfully removing iron corrosion products from composite objects. A natural chelating agent, deferoxamine, was also included in the study as an innovative material in conservation, and was assessed in relation to conventional chelators. Deferoxamine (DFO) belongs to the group known as 'siderophores'.[2] These 'green' and natural chelators are low molecular weight compounds, synthesized and excreted by micro-organisms under conditions of iron deficiency (Raymond *et al.* 1984; Neilands 1995; Mossialos and Amoutzias 2009). DFO has been used in clinical medicine for patients suffering from transfusion-induced siderosis (Neilands 1995; Crisponi and Remelli 2008). DFO is an hexadentate ligand that demonstrates high affinity for Fe(III) and forms very stable Fe(III) chelate complexes, at alkaline pH, due to its pK_a values (8.39; 9.03; 9.70;

TABLE 1 Complexion parameters, chelators, formulations and application methods.

Chelators	Concentration	pH	Application method				
EDTA	3×10^{-2}M	6.2	cotton swab	Klucel G 10% w/v	agarose A0701 4% w/v	Carbopol 940 2.5% w/v +	
DTPA		6.8				Ethomeen C25 15% v/v	
Desferal		6.8					xanthan gum 2.2% w/v
Desferal		8.45					xanthan gum 2.2% w/v + TEAS 1.5% w/v

11.44) (Raymond *et al.* 1984; Ihnat and Robinson 1993). The efficiency of this siderophore, however, was also investigated at a lower pH value, since this would be more appropriate for wooden substrates. The formation constant of DFO for ferric iron, which reflects the stability of the complex and the efficiency of the chelator, is reported to be 30.6 (Miethke and Marahiel 2007), for ethylenediaminetetraacetic acid (EDTA) 25.1 and for diethylenetriaminepentaacetic acid (DTPA) 28.60. The efficacy of all chelators was assessed by colorimetry, scanning electron microscopy (SEM), and energy dispersive X-ray (EDX) spectroscopy.

Materials and methods

Preparation of mock-ups

Wooden mock-ups were made from maple wood (*Acer platanoides* L.) cut into 2 × 3 cm pieces in the tangential plane. Textile mock-ups comprised unbleached plain-weave cotton textile (149 g/m^2) of the same dimensions. All samples were artificially stained with ferric hydroxides, hydroxyl oxides or oxides.

Chelating agents

Aqueous solutions of three chelators – disodium EDTA, DTPA and DFO[3] – were prepared (Table 1). All chelators were applied to the surface of the mock-ups using gel formulations and cotton swabs. EDTA and DFO are hexadentate chelators, whereas DTPA is octadentate: all bind the ferric iron at a 1:1 molar ratio. The pH of the gels was adjusted to 6.2, 6.8, and 8.45, according to the acid dissociation constants of each chelator and substrate compatibility (Table 1). After each gel application, samples were cleaned with a cotton swab wetted in a 50% v/v ethanol aqueous solution, to remove gel residues and loosely connected complexes.

Although the temperature of the solution is an important factor for the performance of the chelating agents, experiments were carried out at room temperature (approximately 19–21 °C), as this temperature often prevails in conservation studios. In addition, the time of application was not set, since it varied depending on the time that each chelator required for the complexation. The gel was removed when coloured complexes had been formed;[4] fresh gels were applied until complexation was no longer observed.

Colorimetry

For measuring colour, an X–Rite portable sphere spectrophotometer (SP60 series) was used according to EN 15886:2010. Each colour measurement was replicated three times before and after cleaning, and the average was recorded. The total colour difference ΔE* was evaluated before and after each application indicating the cleaning efficacy.

SEM

Wooden and textile mock-up samples were mounted on aluminium disks, without a conductive coating, and examined before and after the cleaning procedure using a JEOL JSM–6510LV SEM and an accelerating voltage of 20 kV, in a low-vacuum mode with an air pressure of 30 Pa, in order to observe gel residues and the presence of iron corrosion products.

EDX

EDX was performed on wooden and textile mock-ups before and after the cleaning procedure with a JEOL JSM 6510LV SEM, using the low-vacuum mode with an air pressure of 30 Pa, and an X-act Pentafet Precision spectrometer with INCA software (Oxford Instruments) at an accelerating voltage of 20 kV. Results were used to evaluate the efficacy of the chelators and the applied gel formulations.

Results and discussion

Colorimetry

DFO at pH 6.8 showed the greatest colour change (ΔE*) when gelled with Klucel G, followed by xanthan gum (Fig. 1) when applied on textile. In contrast, DFO at pH 8.4 exhibited the greater ΔE*with xanthan gum instead of Klucel G and the second highest ΔE*of all the cleaning methods employed. Moreover, DTPA application provided similar ΔE*to DFO at 8.4. With regard to gel formulations, the greatest ΔE* value was obtained with Klucel G followed by Carbopol and agarose; EDTA was the least effective, with the exception of the agarose gel. Finally, swab application was the least efficient in comparison to gel applications.

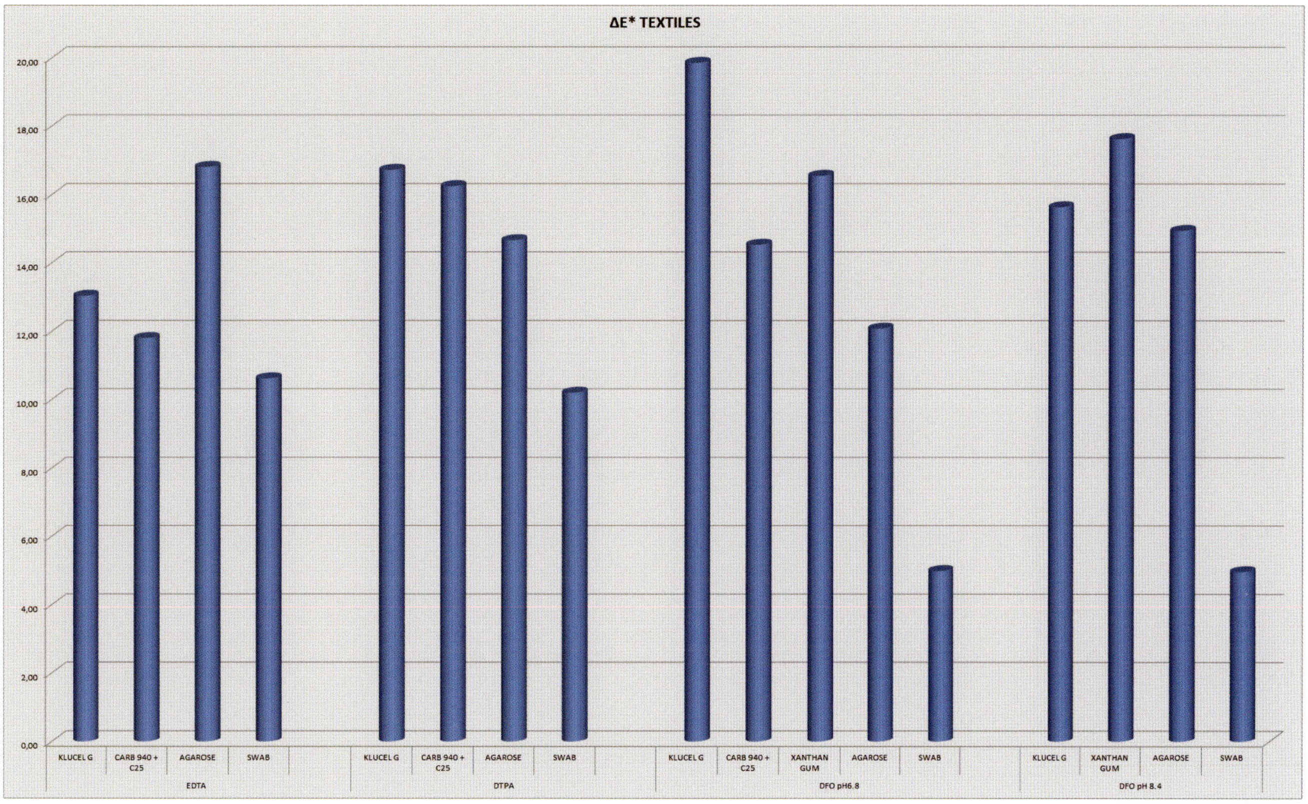

FIG. 1 ΔE^* values obtained after cleaning with different chelators and gel formulations.

It has to be noted that ΔE^* provides an overall estimation of the colour difference but in many cases this was not representative of the qualitative differences (L,a*,b*) of the colour changes. By interpreting the results obtained through the three colour coordinates, it appears that lightness (L) was increased in all samples after cleaning, as well as the b* coordinate, indicating that samples became yellower. In contrast, a* was reduced due to the partial removal of the red-brown iron corrosion products.

Analogous results on the chelators' efficacy were obtained on wooden mock-ups based on colorimetry measurements. However, it appears that all methods were more effective on wood than textile. Furthermore, ΔE^* values indicated that there were differences in the cleaning efficacy depending on the gelling agent in both substrates. It is not possible to draw conclusions as to how each agent affected cleaning as the results were not consistent for all chelators.

SEM

Based on the SEM results, it appears that more gel residues remained on the cleaned textile mock-ups than on the wooden ones (Fig. 2a–h). This might be due either to the heterogeneous surface of the textiles that enables trapping of the gel formulation or to the gel removal methodology, which does not permit easy or complete clearance (Stulik*et al.* 2002; Stavropoulou and Rapti 2011; Hernando and Domínguez 2013). It is evident from the SEM images that Klucel G and Carbopol gel left more residues compared to the other formulations on both substrates (Fig. 2a, b, e and f). Agarose appeared to be superior in limiting residue deposition (Fig. 2d and g).

EDX

Results indicate the weight percent of the detected iron before and after cleaning of wooden and textile mock-ups (Figs 3 and 4). Comparison of these results demonstrates that the wood was cleaned more effectively than the textile, since the amount of detected iron was lower after cleaning, even though its initial concentration was higher than on the textile samples. In addition, results indicate that all chelators perform equivalently for both substrates. More specifically, DFO at both pHs removed iron corrosion products most efficiently, especially considering the initial quantity of iron on the substrates. DTPA followed with good results. These results are in line with other research on the removal of ferric ions from waterlogged wood (Almkvist *et al.* 2005; Phillips and Godfrey 2012). On the other hand, EDTA appears to have the lowest cleaning efficacy for iron corrosion products. This result is in accordance with the literature where DFO was found to have a higher formation constant for iron(III) compared to DTPA and EDTA. According to the results obtained from EDX, the efficacy of the chelators was influenced by the gelling agents, in particular for textile cleaning (Fig. 4). In the case of EDTA, DTPA and DFO (pH 6.8), gels prepared with Klucel G seem to be more effective. For DFO (pH 8.4) xanthan gum was the most effective gelling agent. These results appear to be in agreement with the colorimetry measurements, where ΔE^* was high for both DFO and DTPA, followed by EDTA.

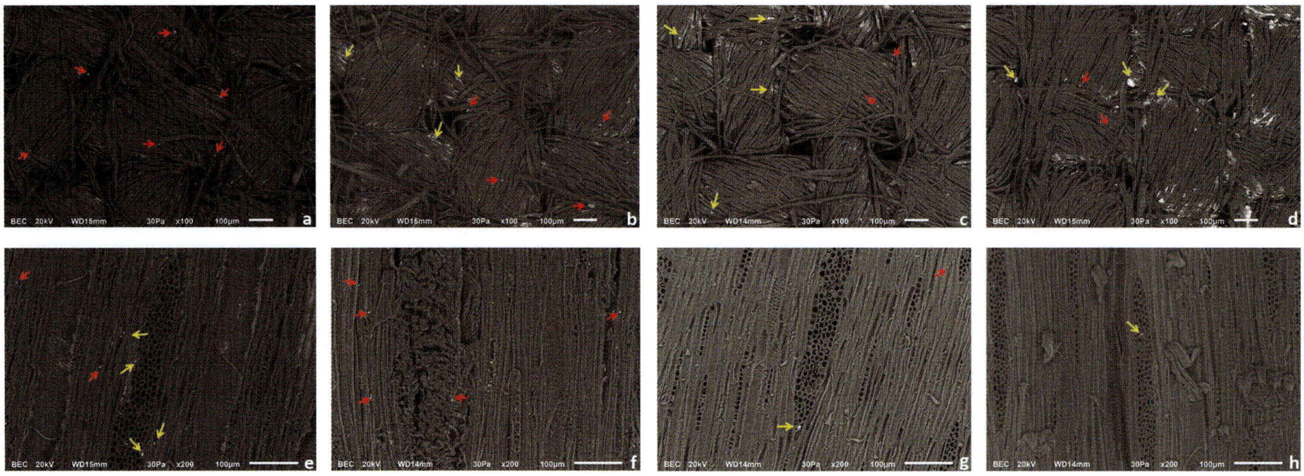

FIG. 2 SEM images showing the presence of iron (yellow arrow) and gel residues (red arrows), on textile and wooden substrates, after cleaning with chelators applied with different gel formulations: (a and e) Klucel G; (b and f) Carbopol; (c and g) xanthan gum; (d and h) agarose (© Stavroula Rapti).

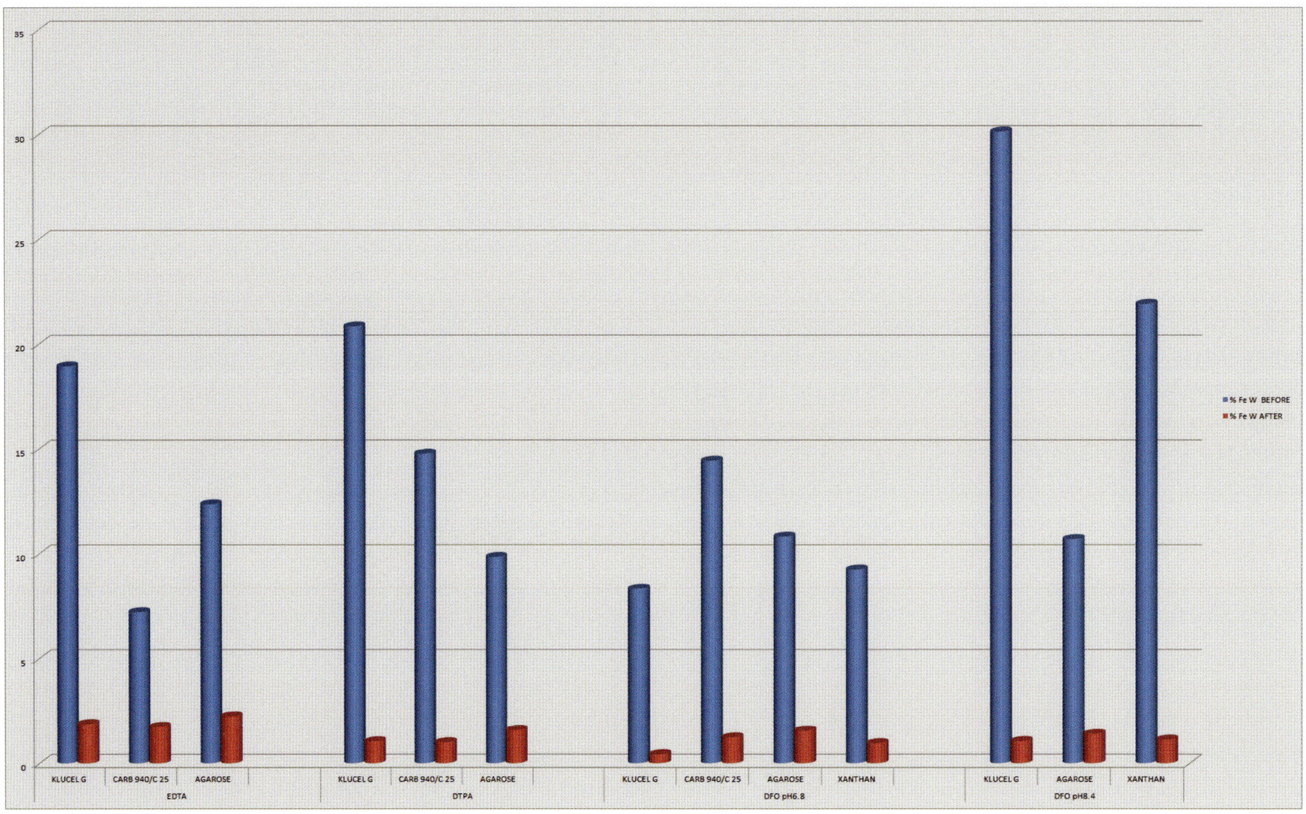

FIG. 3 EDX results for iron weight percentages before and after cleaning of wooden mock-ups with different chelators and gel formulations.

Conclusion

The efficacy of chelators appears to be the same for both wood and textile substrates. The novel 'green' chelator, deferoxamine, proved capable of removing iron corrosion products more effectively than the two conventional chelators EDTA and DTPA. Corrosion products were not removed from the textile substrates as successfully as from the wood, probably due to the fact that iron oxides were impregnated deeper within the fibres (Glenn *et al.* 2015). However, the overall appearance of both mock-ups was improved even though iron corrosion products were not removed completely. Moreover, the efficacy of the chelators was found to be influenced by their gelation. Formulations with Klucel G and Carbopol deposited more gel residues as their consistency did not allow for a complete removal. Moreover, the flexibility of textile samples treated with these gelling agents decreased, probably due to the penetration of the gel within the substrate. In contrast, agarose met the criteria and the objectives of the current study as a thickening agent for cleaning water-sensitive substrates due to its rigidity, which allowed for easy application and removal. The only drawback noted was that at high concentrations, the ability of the agarose to control and limit

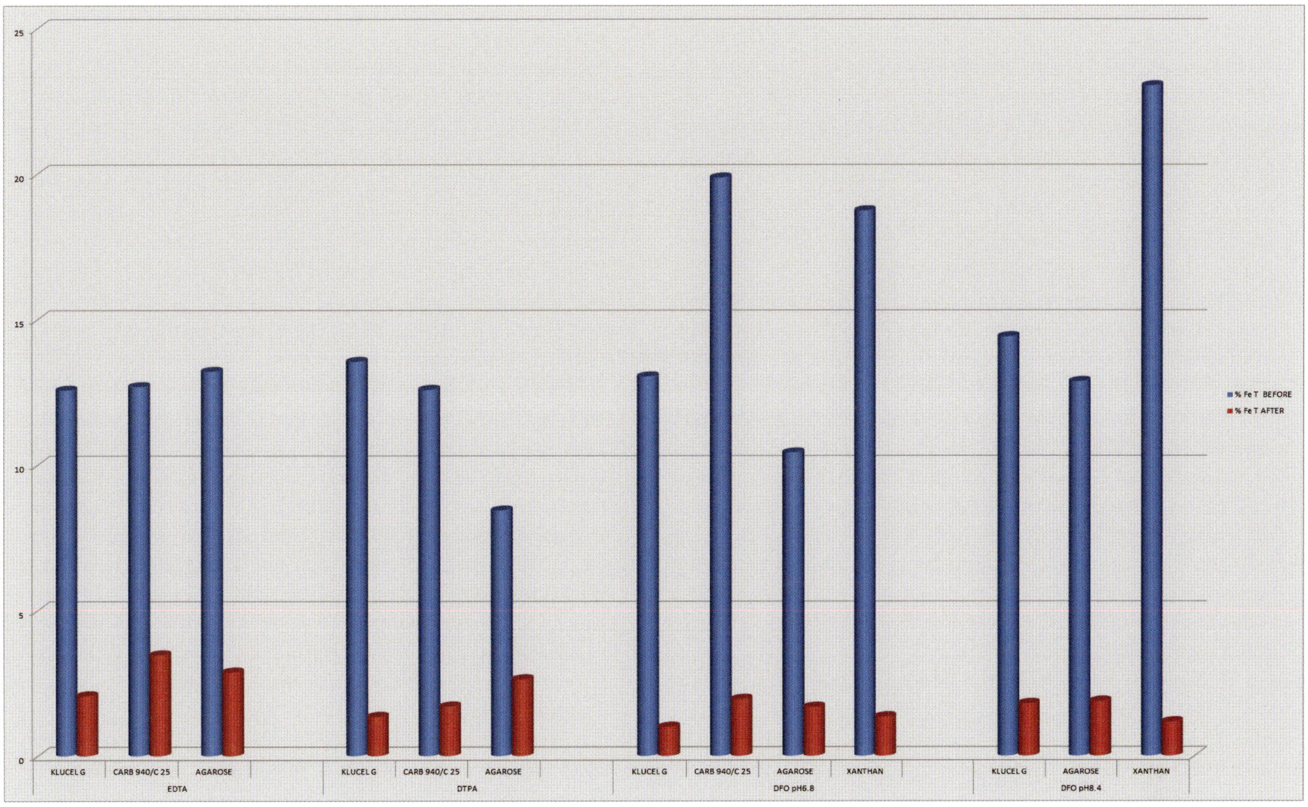

FIG. 4 EDX results for iron weight percentages before and after cleaning of textile mock-ups with different chelators and gel formulations.

moisture diffusion into the substrates resulted in the need for several reapplications of the gel. This is in accordance with other studies where an inverse relationship between concentrations and moisture penetration depth has been observed (Scott 2012).

Acknowledgements

The authors wish to thank A. Karabotsos, TEI of Athens, for technical assistance with the SEM.

Materials and suppliers

- Agarose A0701: Sigma Aldrich, K. Palama 36 14343 N. Chalkidona, Athens, Greece, info@lsc.gr,sales@lsc.gr
- Desferal: Novartis, Ethnikiodos Athinon Lamias, 120th Km Metamorfosi, Greece, orders.novartishellas@novartis.com
- DTPA, Carbopol 940, triethanolamine: Acro Organics, Tech Line S.A., Eleftherias 4, 14564 Kifisia, Greece, contact@techlinesa.com
- EDTA: Serva, M. Roumboulakis S.A., Pindou 43, Filadelfia 14342, Greece, roum@roum.gr
- Ethomeen C-25: Kremer, Exereton, Poulopoulou 18, 11851, Athens, Greece, info@exaireton.com
- Xanthan gum: Chemix A.E., L. Anthousas 44 Pallini, Greece, www.chemix.gr

Notes

1. R. Wolbers, 'Presentation Part I: wet cleaning'. Workshop held at Oaxaca, Mexico, 8–11 November 2011.
2. 'Siderophore' is derived from the Greek for 'iron carriers'.
3. Due to the small volume and the high costs of puredeferoxamine, Desferal, a drug manufactured by Novartis, was used instead.
4. EDTA ferric complex is pale yellow, while DTPA complex is yellow (Almkvist*et al.* 2005). DFO forms an orange-red complex with ferric iron (Domínguez-Vera 2004).

References

Adler, S. and Eaton, L. 1995. 'Chelating agents in wet cleaning systems', *AIC Textile Specialty Group Postprints* 5: 69.

Almkvist, G. 2008. *The Chemistry of the VASA: Iron, Acids and Degradation*, unpublished PhD dissertation, Department of Chemistry, Uppsala University, Uppsala, Sweden.

Almkvist, G., Dal, L. and Persson, I. 2005. 'Extraction of iron compounds from Vasa wood', in O. Hoffmann, K. Straetkvern, D. Gregory and J.A.Spriggs (eds), *9th WOAM ICOM-CC Conference Postprints, Copenhagen, 7–11 June 2004*. Paris: ICOM, 202–10.

Baker, A.J. 1974. 'Degradation of wood by products of metal corrosion', *Forest Service Research Paper*, FPL 229. Madison, WI: US Department of Agriculture Forest Service, Forest Products Laboratory.

Burgess, H. 1991. 'The use of chelating agents in conservation treatments', *The Paper Conservator* 15: 36–44.

Cremonesi, P. 2013. 'Rigid gels and enzyme cleaning', in M.E. Mecklenburg, E.A. Charola and R.J. Koestler (eds), *Cleaning 2010: New Insights into the Cleaning of Paintings*. Washington, DC: Smithsonian Institution Scholarly Press, 179–83.

Crisponi, G. and Remelli, M. 2008. 'Iron chelating agents for the treatment of iron overload', *Coordination Chemistry Reviews* 252(10–11): 1225–40.

Domínguez-Vera, J.M. 2004. 'Iron(III) complexation of Desferrioxamine B encapsulated in apoferritin', *Journal of Inorganic Biochemistry* 98(3): 469–72.

EN 15886:2010. 'Conservation of cultural property. Test methods. Colour measurement of surfaces'. Published European Standardfrom the European Committee for Standardization, CEN /TC 346-Conservation of Cultural Heritage.

Fors, Y. 2008. *Sulfur-Related Conservation Concerns for Marine Archaeological Wood: The Origin, Speciation and Distribution of Accumulated Sulfur with Some Remedies for the* VASA, PhD thesis, Department of Physical, Inorganic and Structural Chemistry, Stockholm University.

Gill, K. 2007. 'Evaluating X-radiography as a tool for examining upholstered furniture', inS. O'Connor and M. Brooks (eds), *X-Radiography of Textiles, Dress and Related Objects*. Oxford: Butterworth-Heinemann.

Glenn, S., Haldane, E.-A., Hackett, J. and Sung I. 2015. 'Borrowing from the neighbours: using the technology of other disciplines to treat difficult textile conservation problems', paper presented at ICON Textile Spring Forum, 13 April 2015, *Learning Curve: Education, Experience and Reflexion*.

Hernando, S. and Domínguez, M.M.B. 2013. 'Cleaning plaster surfaces with agar-agar gels: evaluation of the technique', *Grupo Espanol de conservation, Ge-Conservatión* 4: 111–26.

Ihnat, P.M. and Robinson, D.H. 1993. 'Potentiometric determination of the thermodynamic ionization constants of deferoxamine', *Journal of Pharmaceutical Science* 82(1): 110–12.

Margariti, C. 2003. 'The use of chelating agents in textile conservation: an investigation of the efficiency and effects on the use of three chelating agents for the removal of copper and iron staining from cotton textiles', in L. Dawson and M. Berkouwer (eds), *Dust, Sweat and Tears: Recent Advances in Cleaning Techniques*. London: UKIC, 28–38.

Miethke, M. and Marahiel M.A. 2007. 'Siderophore-based iron acquisition and pathogen control', *Microbiology and Molecular Biology Reviews* 71(3): 413–51.

Mossialos, D. and Amoutzias, G.D. 2009. 'Role of siderophores in cystic fibrosis pathogenesis: foes or friends?', *International Journal of Medical Microbiology* 299: 87–98.

Music, S., Gotic, M. and Popovic, S. 1993. 'X-ray diffraction and Fourier transform infrared analysis of the rust formed by corrosion of steel in aqueous solutions', *Journal of Materials Science*28: 5744–52.

Neilands, J.B. 1995. 'Siderophores: structure and function of microbial iron transport compounds', *Journal of Biological Chemistry* 270: 26723–6.

Phenix, A. and Burnstock, A. 1992. 'The removal of surface dirt on paintings with chelating agents', *The Conservator* 16: 28–38.

Phillips, E. and Godfrey, I.N. 2012. 'Removing iron compounds from a waterlogged woodengun-carriage using the chelating agent diethylenetriaminepentaacetic acid (DTPA)', in K. Straetkvern and E. Williams (eds), *11th WOAM ICOM-CC Conference Postprints, Greenville, North Carolina, USA, 24–28 May 2010*. Paris: ICOM, 505–10.

Raymond, K.N., Müller, G., Matzanke, B.F. 1984. 'Complexation of iron by siderophores a review of their solution and structural chemistry and biological function', in *Structural Chemistry*, Topics in Current Chemistry, vol. 123. Berlin and Heidelberg: Springer, 49–102.

Rivers, S. and Ummey, N. 2003. *Conservation of Furniture*. Oxford: Butterworth-Heinemann.

Scott, C.L. 2012. 'The use of agar as a solvent gel in objects conservation', *AIC Objects Speciality Group Preprints* 19: 71–83.

Stavropoulou, M. and Rapti, S. 2011. 'The use of SEM in investigating gel residues on cotton textiles, after spot cleaning treatments', in B. Szuhay (ed.), *8th NATCC Conference Preprints, Oaxaca, 8–11 November 2011*, 315–21.

Stulik, D., Khanjian, H., Dorge, V., Tagle, A., Maish, J., Considine, B., Miller, D. and Khandekar, N. 2002. 'Scientific investigation of surface cleaning processes: quantitative study of gel residues on porous and topographically complex surfaces', in R. Vontobel (ed.), *Preprints of the ICOM-CC 13th Triennial Meeting*. London: James & James, 245–51.

Timár-Balázsy, A. and Eastop D. 1998.*Chemical Principles in Textile Conservation*. Oxford: Butterworth-Heinemann.

Wolbers, R. 2000. *Cleaning Painting Surfaces: Aqueous Methods*. London: Archetype Publications.

Wolbers R., Sterman, N. and Stavroudis, C. 1988. *Notes for the Workshop on New Methods in the Cleaning of Paintings*. Los Angeles: Getty Conservation Institute.

Xu, G. and Goodell, B. 2001.'Mechanisms of wood degradation by brown-rot fungi: chelator-mediated cellulose degradation and binding of iron by cellulose', *Journal of Biotechnology* 87: 43–57.

Authors' addresses

- Stavroula Rapti, Department of Conservation of Antiquities & Works of Art, TEI of Athens, Athens, Greece (srapti@teiath.gr)
- Stamatis Boyatzis, Department of Conservation of Antiquities & Works of Art, TEI of Athens, Athens, Greece
- Shayne Rivers, West Dean College, Chichester, West Sussex, UK
- Athanasios Velios, Ligatus Research Centre, University of the Arts London, Chelsea College of Arts, London, UK
- Anastasia Pournou, Department of Conservation of Antiquities & Works of Art, TEI of Athens, Athens, Greece

Layer by layer: the removal of complex soiling on a collection of modern art bronzes using buffered pH-adjusted aqueous gels

Shelley M. Smith

Introduction

In 1929, The Field Museum in Chicago commissioned American artist Malvina Hoffman (1885–1966) to create 104 life-size bronze sculptures (full-length statues, busts, and heads) to illustrate 'The Living Races of Man'. The commission took five years to complete and involved a seven-month international expedition. The Exhibition Hall opened to great acclaim on June 6, 1933, to coincide with the opening of the International World's Fair in Chicago. More than two million people visited the Hall in its first year, and countless millions more until the exhibition was superseded by the Museum's 75th Anniversary Exhibition in 1967.

For the first 35 years, the sculptures were exhibited in the Hall as 'art' to entertain general audiences. During this time, they shared a similar maintenance history. But this changed when the Hall closed. During the next 47 years, two-thirds of the sculptures were treated like display furniture; scattered in offices and public thoroughfares, because of the lack of storage space, where they were over-exposed to the public and maintained by janitorial staff. The remaining third were crowded into closed-up light wells where they languished without dust protection or environmental controls for relative humidity and temperature.

Historical background

By the 1920s, Malvina Hoffman was a successful mid-career artist who owned studios in New York and Paris. She had studied at several prestigious academies and trained with renowned artists such as Gutzon Borglum and Auguste Rodin. Hoffman was also a writer and historian who published a treatise on sculpture techniques *Sculpture Inside and Out* (1939) and chronicled the Field Museum expedition in *Heads and Tales* (1936). Hoffman worked closely with French patineurs to create skin tones and hair color faithful to the live subjects who modeled for her. By 1984, however, when art critic Linda Nochlin visited the Museum she expressed her disappointment 'upon seeing the sculptures ... in many cases, the patina has been allowed to wear off the noses and other protrusions of the figures, with startling results' (Nochlin 1984: 106).

Maintenance history and its effect on physical condition

The Museum accessioned the sculptures in 1998, and until then they were maintained like exhibition furniture. Custodians wiped them with commercial cleaning or dusting products to re-saturate the dull, dusty surfaces and temporarily improve their appearance. The trapped dust initiated corrosion cells. The use of corrosive cleaning agents also caused the patinas on two of the sculptures to thicken, blister, and flake off. By the 1980s, conservators intervened and washed the statues still on display with an unspecified 'oleoresin' soap followed by a coat of Renaissance microcrystalline wax. Miscommunication about who was responsible for the sculpture's maintenance continued. In 1994, the Conservation Department sent a memo to housekeeping ordering a halt to the use of Wall-Glide Plus[1] a highly concentrated alkaline detergent.

Principal types of damage observed by the author in 2014 included:

- Non-malicious touching and rubbing by sweaty hands. Although mostly instinctive, such behavior is extremely damaging over time and causes permanent loss to the patinas. The surface details were sometimes eroded down to a mirror-like shine.
- Accidental and intentional damage was caused by individuals splashing soft drinks and sticking wads of

TABLE 1 Summary of cleaning tests for *Jakun Girl*.

Mixture no.	Deionized water	Buffer type & pH	Surfactant	Chelator	Observations	Results
1	2 mL	1 mL 8.5 pH bicine	1 mL deoxycholate	1 mL 8.5 pH EDTA	Removed slight amount of dirt on swab	No visible change on surface film
2	2 mL	1 mL 8.5 pH bicine	1 mL deoxycholate	1 mL 8.5 pH DTPA	Slightly better, small amount of dirt on swab than No. 1	Slight visible change in surface film
3	2 mL	1 mL 8.5 pH bicine	1 mL deoxycholate	1 mL 8.5 pH Citrate	Better result than No. 2	Visible cleaning action
4	2 mL	1 mL 8.5 pH bicine	1 mL Orvus	1 mL 8.5 pH EDTA	Good cleaning result. Most dirt removed so far on swab	Too aggressive. Good cleaning action but swab dried. Green tinge indicating dissolution of copper ions (surface)
5	2 mL	1 mL 8.5 pH bicine	1 mL Orvus	1 mL 8.5 pH DTPA	Good cleaning result. Similar amount of dirt pulled up as No. 4	
6	2 mL	1 mL 8.5 pH bicine	1 mL Orvus	1 mL 8.5 pH Citrate	Good cleaning result. Similar amount of dirt pulled up as No. 5	
7	3 mL	1 mL 8.5 pH bicine	1 mL deoxycholate	None added	Slight amount of dirt removed	
8	3 mL	1 mL 8.5 pH bicine	1 mL Ethofat	None added	Slight amount of dirt removed, more than No. 7	
9	3 mL	1 mL 8.5 pH bicine	1 mL Maypon 4C	None added	Slight amount of dirt removed, more than No. 8	
10	3 mL	1 mL 8.5 pH bicine	1 mL Triton 80XL-80N	None added	Good cleaning result.	
11*	3 mL	1 mL 8.5 pH bicine	1 mL Brij 700	None added	Good cleaning result. Faster action than No. 10	Best result without use of chelating agent. Removed film and left surface glossy. No patina removed
12	3 mL	1 mL 8.5 pH bicine	1 mL Orvus	None added	Good cleaning result. Same as No. 10. More foaming though	
13	3 mL	1 mL 9.5 pH glycine	1 mL Brij 700	None added	Good cleaning result but more aggressive. Dulled surface	
14	4 mL	None	1 mL Brij 700	None added	Little to no cleaning effect	
15	3 mL	7.5 Tris	1 mL Brij 700	None added	Little to no cleaning effect	

gum and decals on the sculptures. There was also a lot of paint spatter from exhibits painters.

- Intentional defacement and deliberate thefts are beyond the scope of this paper, but without protection from guards, every moveable accoutrement including jewelry, spears, arrows, and bows was either broken or stolen. Other vandalism included scratches from sharp objects, lipstick, pen and pencil marks, among others.

In general, the surface of each sculpture had acquired a unique stratigraphy of dirt, pitting or waxy green corrosion products and thick milky films. This tenacious milky film was mainly found on horizontal surfaces prone to dust accumulation and frequently wiped by housekeeping staff. A very thick, brackish, waxy green sludge of corrosion products formed in areas touched by generations of sweaty hands. The presence of an insoluble layer of crosslinked green wax was explained by conservation records dating from the 1980s, in which Museum restorers reported using Renaissance brand microcrystalline wax on some of the sculptures still on display. The Renaissance wax had become insoluble in everything but hot xylene. Moffett (1996) identified the polyethylene wax component as the culprit which can crosslink and become insoluble. This commercial product is still commonly found in museum conservation labs, but its use should be discontinued by conservators.

Gel cleaning to remove dirt, films, and corrosion products

Spot testing began with the two detergents most commonly used by objects conservators for cleaning bronze sculptures: Orvus and Triton XL-80N diluted to 1% concentration in distilled or deionized water. These detergents removed loose dirt but not the layers of soiling beneath. They did nothing to remove the copper corrosion products that diffused through the patina and deposited on top of the milky gray layer. Organic solvents were only partially effective in dissolving this layer and left a dull white haze, probably composed of partially solubilized deterioration products. Most of the original patinas were sensitive to organic solvents which limited the choice of retouching media within safe solubility parameters. Methyl ethyl ketone (MEK) was successful for

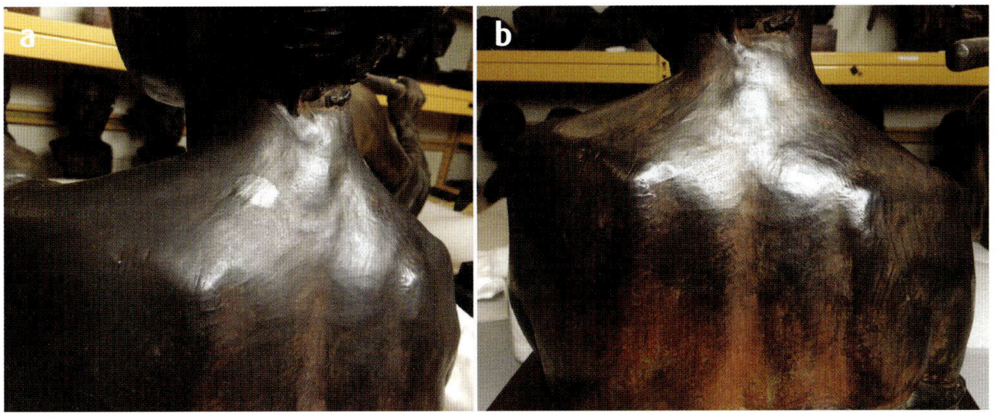

FIG. 1 *Jakun Girl* (Field Museum Chicago 337075). (a) A compact milky gray film was revealed after aqueous cleaning. The film was mainly present on horizontal surfaces, and most likely deposited when housekeepers wiped the sculptures with commercial dusting or cleaning products. (b) After the milky film was removed, white flecks of wall paint underneath it were revealed (photos: Michael Marozas).

localized removal of black spots on bronze, but overall the standard solvents and aqueous detergents used by objects conservators were unsatisfactory. No single solvent or solvent combination had any effect on either the milky gray film or the aged Renaissance wax, nor satisfactorily removed the thick brackish layers of copper corrosion products.

At this stage, one had to either accept these cleaning results, and proceed only to remove the upper layers of gross particulates by conventional aqueous washing after vacuum dusting, or try something new. The patinas were too thin to attempt mechanical removal of the films and corrosion products. The Modular Cleaning Program (MCP) developed by Chris Stavroudis (Stavroudis *et al.* 2005) is available as a stand-alone database in Filemaker Pro, offered at no charge to the conservation community. Based on user-defined cleaning parameters, a set of stock solutions is prepared including pH buffers, chelating agents, and detergents for methodical mixing and spot testing. When thickened, these gels are an excellent choice for cleaning vertical or highly detailed surfaces. A high viscosity carrier prevents the chemistry from migrating where it is not wanted and prevents it from drying on the surface. The dwell times can be controlled, and the use of a buffer can maintain the desired pH in an effective cleaning range that is safe for the substrate.

Spot tests were performed to establish the solubility parameters of the surface soiling in the range pH 7.0–10.0. A selection of detergents that target very different types of dirt was tested such as Deoxycholate, Ethofat, Maypon 4C, Triton XL80N, Orvus, and Brij700. Stavroudis (2009) explained the properties of these detergents in detail which helped to clarify the results. In general, the cleaning solutions below pH 8.0 had negligible effect while those between pH 8.0 and 9.0 removed the coatings and left a glossy, bright patina. But some spot tests above pH 9.5 dulled the surface. Only Orvus and Brij 700 yielded a visibly cleaner, glossy appearance. Chelating agents such as ethylenediaminetetraacetic acid (EDTA) and diethylenetriaminepentaacetic acid (DTPA) were tested in some of the solutions but yielded no detectable improvement and were excluded from further consideration.

Orvus and Brij 700 were carried forward for testing dwell times. A series of 1% detergent solutions adjusted in the range from pH 7.0 to 9.5 was thickened with hydroxypropyl methylcellulose (HPMC). Dwell times ranged from 30 seconds to 5 minutes.

Results of cleaning tests

Orvus (sodium lauryl sulfate) is a detergent commonly used by objects conservators. However, Brij700 (polyoxyethylene 100 stearyl ether) was novel. Several Orvus and Brij700 solutions were individually buffered to pH 8 and 8.5 and thickened with HPMC. Test spots for each solution were allowed to dwell for 30 seconds to 5 minutes. The tenacious gray milky films responded similarly to these mixtures as compared to liquid detergent solutions at a slightly higher pH. The gels were gently agitated with a soft round Taklon brush[2] before undergoing a rinse procedure. Rinsing was done with Saugwunder sponges dampened with deionized water. The sponges were cut into small wedges and rolled across the surface. Dirty sponges were washed and wrung out. Only clean sponges were dipped into the fresh rinse water. After the area was rinsed to satisfaction, it was blotted dry with a clean, soft cotton cloth and allowed to air dry. It is inevitable that trace amounts of the buffering agent and detergent residues were left on the surface, but for non-porous surfaces such as these, it was considered a negligible risk.

In the few areas that did not respond to gelled detergent solutions, Carbopol and Pemulen TR2 gels were adjusted in the pH 8–8.5 range and tested. While the Pemulen gel had negligible effect the Carbopol was a success. When adjusted to pH 8.5 Carbopol gels worked in instances when surface soiling did not respond to any cleaning solution in a safe pH range.[3] Carbopol, because it needs to be cleared with solvent, was used sparingly and only in areas where the original patina had been worn off. Even though Carbopol is technically an aqueous gel, the water is bound and clearing it with water alone can cause the Carbopol resin to precipitate out of the gel and leave a hazy

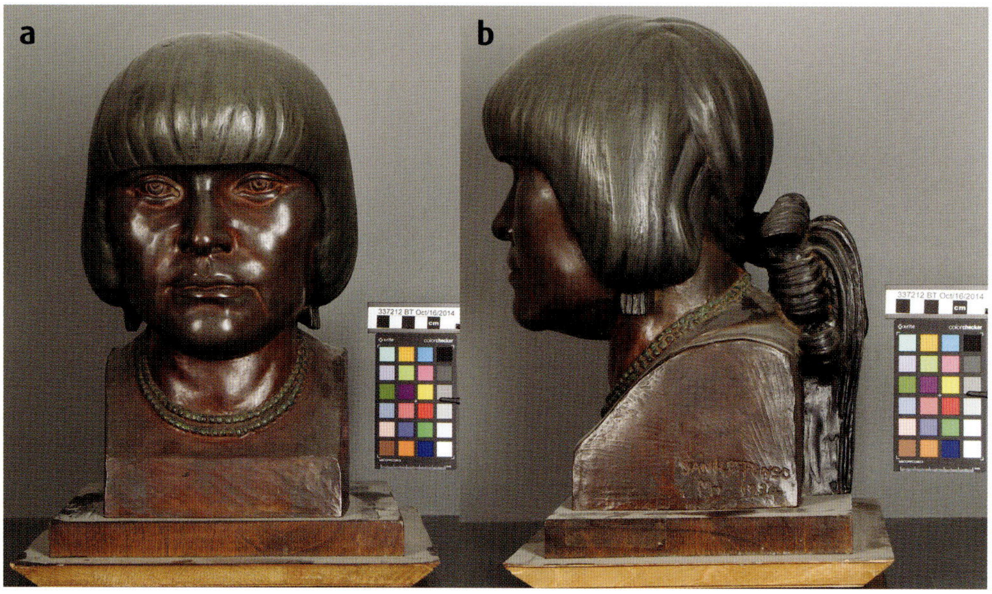

FIG. 2 *Desideria Montoya Sanchez* (Field Museum Chicago 337212). (a) Before treatment and aqueous cleaning, front view. (b) Side view showing the patination of the hair and bangs: a rich black color on the pony tail with the front bangs patinated a pastel green color to appear 'dusty' (photos: Michael Marozas).

residue. Deionized water and isopropanol were alternated to clear the Carbopol gel. A Carbopol benzyl alcohol[4] gel was sometimes successful in removing the brackish waxy/greasy corrosion build-ups in areas of severe damage.

Treatments for different types of damage

Jakun Girl[5] (Fig. 1)

This bust was placed in storage in 1968, and had acquired a dense, dark particulate layer. The sculpture was vacuum dusted, and after doing spot tests, it was cleaned with an unbuffered 1% solution of Brij 700 which removed a layer of dirt and revealed the thick, milky gray film. The film was removed locally with a 1% solution of Orvus gelled with HPMC buffered to pH 8.5.[6] The gel was applied in a small area, about the size of an inch, allowed to dwell from 1 to 3 minutes and gently agitated with a soft round Taklon brush. After the gray film was removed white paint spatters beneath it were revealed. These were removed with tiny dots of 50:50 acetone:Laponite gel applied with a toothpick (Neumann and Sansom 1970). Cleaning solutions were rinsed with the aqueous procedure. No retouching was needed since the worn areas in the patina had re-oxidized after many years in storage; undisturbed by sweaty hands.

Desideria Montoya Sanchez (Figs 2 and 3)

Hoffman depicted the American artist Desideria Sanchez of San Ildefonso, New Mexico with dusty bangs as a nod to her profession as a potter. Desideria's necklace had been painted bright green which made it sensitive to everything but plain deionized water. Her ponytail had a relatively thick black patina that responded best to aqueous cleaning with deionized water adjusted to pH 8.5,[7] whereas the top of her head, face, neck, and shoulder areas, responded best to a 1% Orvus detergent solution buffered to pH 8.5 and thickened with HPMC. It was necessary to use Carbopol on the degraded Renaissance microcrystalline wax coating on her skin, which as previously stated, is not ideal. According to the conservation records the wax had been applied in the 1990s. Careful testing, cleaning, and rinsing with deionized water revealed a rich, glossy reddish-brown patina. Black spots in her eyes were softened with a 50:50 MEK:Laponite gel and cleared with tightly rolled cotton swabs dampened with acetone. Only one area required retouching: a postage stamp size loss in the patina on top of her head that was retouched with Gamblin Conservation Colors.

Nobosodrou, a Mangbetu Woman from Belgian Congo[8] (Fig. 4)

After many years on continuous open display most of the patina was worn down to bare metal. From the 1980s to 1990s, Museum conservators extensively painted over the damage with oil and acrylic paints that had since been scratched through by vandals. The surface had developed brackish waxy green corrosion, a greasy layer from decades of touching, and residues from cleaning products.

The surface soiling responded well to the 1% Orvus detergent solution thickened with HPMC and adjusted to pH 8.5. After the dirt layer had been removed the actual extent of old retouching became apparent. Carbopol 934 gel adjusted to pH 8.5 was applied in small sections where the waxy green corrosion was thickest and allowed to dwell for up to 5 minutes. The Carbopol also swelled the overpaints which aided in their removal before rinsing with deionized water and isopropanol. Areas of loss in the patina were inpainted with Gamblin Conservation Colors.

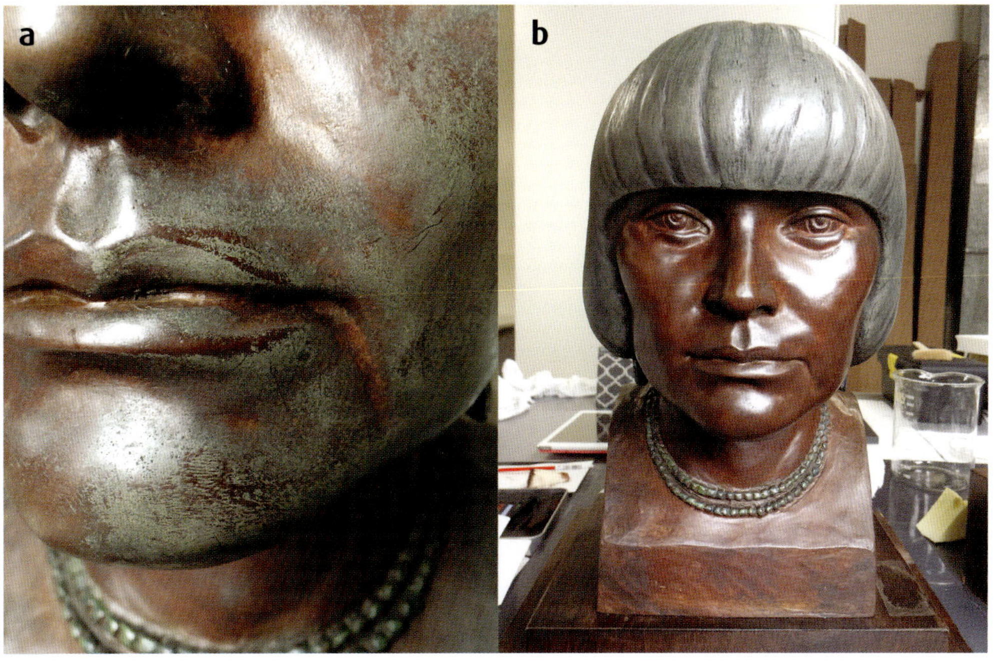

FIG. 3 *Desideria Montoya Sanchez* (Field Museum Chicago 337212). (a) Aqueous cleaning revealed an insoluble green layer of aged Renaissance microcrystalline wax. (b) After cleaning and removal of the Renaissance wax on the face. By contrast, the necklace was originally painted and lacquered green and the bangs were intentionally patinated a pastel green color to appear 'dusty'. (photos: Michael Marozas).

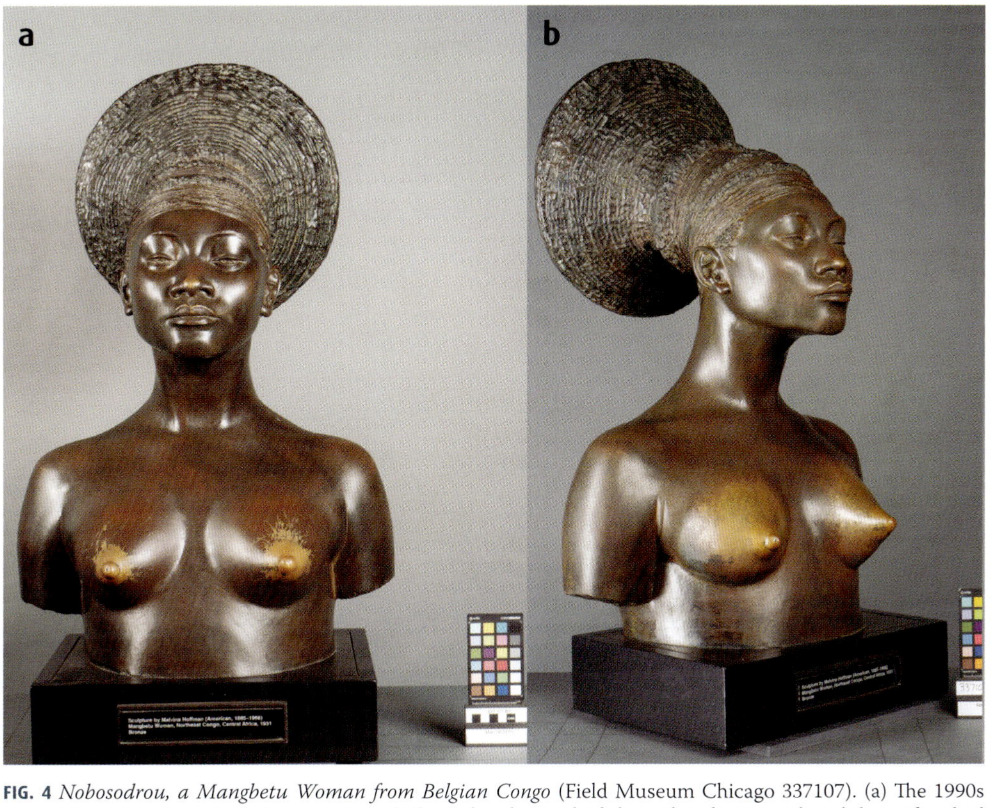

FIG. 4 *Nobosodrou, a Mangbetu Woman from Belgian Congo* (Field Museum Chicago 337107). (a) The 1990s retouching with oils and acrylic paints was disfigured and scratched through to bare metal, and the surface had greasy soiling overall. (b) After cleaning with aqueous and Carbopol gels (photos: Michael Marozas).

Eskimo Woman[9] (Fig. 5a)

For many years, this sculpture languished on continuous open display next to a garbage can. As a result, the sculpture's patina was extensively worn down and curiously dissolved leaving a patina transformed into a multicolored sludge several millimeters thick which bordered worn areas of brassy bronze. Unsightly layers of wax and dust intermingled with corrosion products and became further disturbed by extensive streaking from corrosive cleaning fluids.

The adhered gross particulate responded well to an unbuffered 1% solution of Brij 700 in deionized water. The unsightly

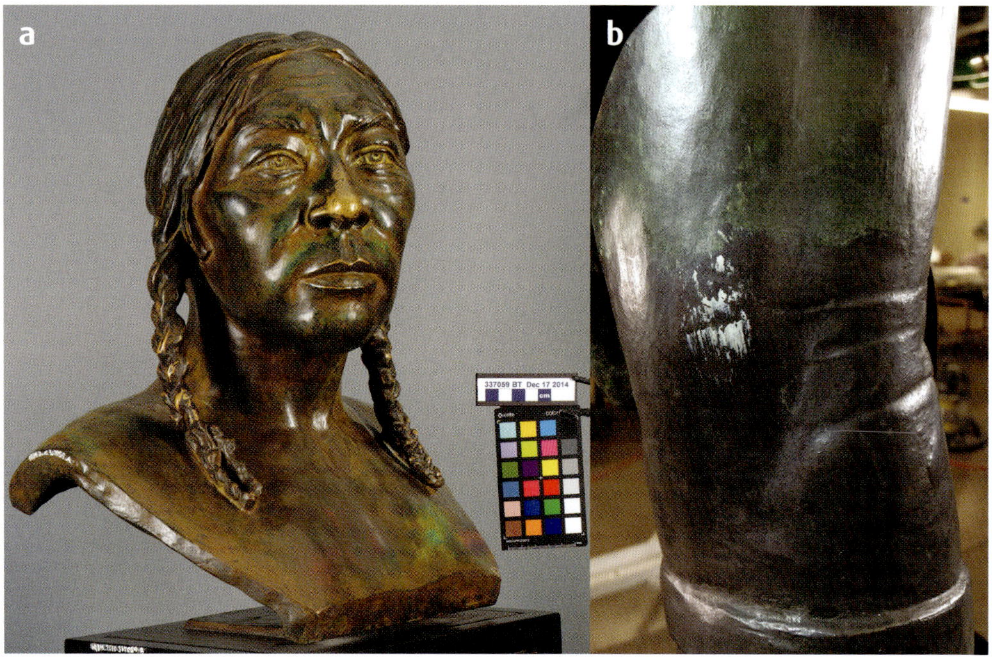

FIG. 5 (a) *Eskimo Woman* (Field Museum Chicago 337059). Green streaky patina with thick waxy green copper corrosion products. (b) *Nuer Warrior* (Field Museum Chicago 337117). A similarly thick, waxy green layer was removed from the leg with Carbopol gel, which revealed an old paint spatter that was removed with a 50:50 acetone:Laponite RD gel (photos: Michael Marozas).

green/black sludge of copper corrosion products responded best to alternating applications of Carbopol gel followed by a benzyl alcohol gel. The benzyl alcohol gel was the most effective on the thick green accretions while the Carbopol gel was applied after the benzyl alcohol gel to remove remaining thin, green residues.

These gels were applied in small sections, with a soft round Taklon brush and gently agitated in a circular motion for up to 2 minutes. This activity was repeated as necessary to achieve full removal. In especially tenacious spots, a cotton swab was rolled across the surface to provide a more targeted cleaning action before rinsing with deionized water and isopropanol. The sculpture was inpainted with Gamblin Conservation Colors.

Conclusion and future directions

Layer by layer, the heavy particulate, ingrained dirt, tough milky films, and corrosion products were removed from a selection of bronze sculptures by Malvina Hoffman in the collection of the Field Museum. Cleaning solutions and gels were adjusted according to the sensitivity or condition of the underlying soiling layer especially when the original patina was present. The varicolored patinas on the hair, skin, and clothing were unique to each sculpture and intended to demonstrate subtle variations in the skin and hair color of each sitter. The artist and artisans used hot and cold patinas in combination with waxes, dyes, and pigments such as Prussian blue, gold leaf, varnish, and paint. The different patinas on each sculpture were spot tested in an inconspicuous location before cleaning proceeded. Areas with loose pigments or flaking patina were not cleaned at all. Cleaning with heat spatulas, steam, or torch flames was inappropriate to these sculptures because in general, these techniques do not have an appropriate degree of control needed for these complex deterioration conditions combined with sensitive artisanal patinas as described. The aged Renaissance wax was especially challenging to remove without damaging the delicate patinas.

Retouching was limited to areas of damaged patina. Gamblin Conservation Colors are reversible in many solvents including isopropanol which had no effect on the original patinas, and as such, can be removed in future, if necessary, without damaging adjacent original patinas. Retouching was done at least 24 hours after cleaning was completed and only in areas of abrasion or disfiguring loss in the patina. While some of the worn areas have darkened through natural oxidation, this takes considerable time.

During a one-year project (2014–2015), 66 of the 104 sculptures were cleaned using a combination of the methods described in this paper. In January 2015, a selection from this group was placed on exhibition but once again, in a non-climate controlled public area. The Museum's decision to continue storing and displaying the collection in areas without environmental control are outside the scope of this paper. The biggest threat to these objects is continued visitor touching and the lack of climate control. Although sacrificial coatings such as waxes are suitable for outdoor bronze sculptures, these serve as sacrificial coatings to shed rainwater to protect underlying coatings and patinas. No coatings were applied overall throughout the course of this project. The Gamblin retouching media was only used on areas of damage or total loss of original patina.

Cold paste waxes were only applied over areas of retouching to adjust the gloss of the Gamblin Conservation Colors, which are matte. Cold waxes commonly applied overall to sculptures would only attract and hold dust during fluctuations in temperature and humidity. These sculptures will continue to undergo daily fluctuations while in storage and on display and would benefit the most from regular vacuum dusting with a soft brush.

If the remaining sculptures in this collection are to be cleaned in future, superior results might be achieved with microemulsions or silicon solvents introduced to conservators by Richard Wolbers, to reduce or eliminate water from cleaning systems for paintings. The chemistry and properties of microemulsions and solvents has been explained by Chris Stavroudis (Stavroudis 2012a). Some research in this area is needed and objects conservators working on metal alloys should explore ways to reduce or eliminate dependence on surfactants altogether. In 2017, an update to the Modular Cleaning Program incorporated rinse solutions that are specific and appropriate to each cleaning solution or gel formulation for optimal clearance. Residues left behind by some of the newer surfactants like Pluronic F127[10] are designed to break down and so dissipate. However, it is up to conservators to use their judgment regarding the suitability of the cleaning and rinse solutions for the substrate to be cleaned.

Materials and suppliers

- Benzyl alcohol gel: anhydrous, 99.8%, CAS Number 100-51-6; bicine: PharmaGrade Synonym: N, N-Bis(2-hydroxyethyl) glycine, CAS Number 150-25-4; Brij700 (polyoxyethylene (100) stearyl ether (C2H4O) nC18H38O); HPMC (hydroxypropyl methyl cellulose): Sigma-Aldrich, 3050 Spruce St., Saint Louis, MO 63103, USA
- Carbopol934 (crosslinked polyacrylate polymer): The Lubrizol Corporation, 9911 Brecksville Road Cleveland, OH 44141-3247, USA
- Ethomeen C-25 and C-12 resin neutralizer: Akzo Nobel Chemicals Inc., 617 W Johnson Ave # 2, Cheshire, CT 06410, USA
- Gamblin Conservation Colors (Laropal A-81, mineral spirits, lightfast pigments, and alumina hydrate to adjust tinting strength): 323 SE Division Pl., Portland, OR, 97202, USA
- Orvus WA Paste (sodium lauryl sulfate, anionic, neutral synthetic detergent): Procter and Gamble. 1 Procter & Gamble Plaza, Cincinnati, OH 45202, USA
- Pemulen TR2 The Lubrizol Corporation, 29400 Lakeland Boulevard, Wickliffe, OH 44092, USA
- Saugwunder sponges (poly(vinyl alcohol) sponge): https://www.amazon.de/LaserMax- Saugschwamm-Farbe-gelb/dp/B00AL01QT8/ref=pd_lpo_121_lp_t_3/255-0414207-1551414_encoding=UTF8&psc=1&refRID=PX2071XQNJS9YJA3GR5R
- Taklon Brushes (Economy Golden), Dick Blick 42 State Street, Chicago, IL 60603, USA, http://www.dickblick.com/products/blick-economy-golden-taklon-brushes/#photos
- Poly(vinyl alcohol) sponge (PVA), clean room industry grade, http://www.conservationsupportsystems.com/product/show/polyvinyl-alcohol-sponge- pva/sponges.

Notes

1. The MSDS for Wall-Glide Plus describes it as low foaming alkaline (pH 11.0) cleaner, registered in 1977 by Servicemaster Company, IL 60515.
2. Taklon is a type of synthetic filament that is dyed and baked to make it soft and absorbent. Taklon withstands the caustic nature of solvents and acrylic paints better than animal hair for greater durability overall.
3. For more information on the properties of Carbopol and Pemulen see Stavroudis 2012b.
4. To mix a benzyl alcohol gel (100 mL stock gel): add 12 mL Ethomeen C/25 and 5 mL Ethomeen C12, mix together. Add 100 ml benzyl alcohol and stir. Add 4 mL deionized water to form the gel.
5. Original title of the sculpture.
6. To make 100 mL Brij700 or Orvus adjusted to pH 8.5 with sodium hydroxide (10% solution) buffered with bicine and thickened with methyl cellulose: add 1 g bicine to the deionized water. Add 1 g Brij 700 or Orvus. Add 2 g methyl cellulose. Add approximately 1 mL sodium hydroxide (10% sol^{n}) dropwise to adjust the pH as necessary 8.5 by slowly adding sodium hydroxide (10% sol^{n}) while stirring and monitoring the pH. Bring the final volume to 100 mL.
7. To mix 500 mL buffered pH 8.5 adjusted water with sodium hydroxide: add 4 g bicine to 100 mL deionized water. Add 10% solution NaOH dropwise (about 6 mL) until pH 8.5. Bring the final volume to 500 mL.
8. Original title of the sculpture.
9. Original title of the sculpture.
10. Lecture notes from Chris Stavroudis Modular Cleaning Program Workshop in Saint Louis, Missouri March 13–16, 2017

References

Moffett, D.L. 1996. 'Wax coatings on ethnographic metal objects: justifications for allowing a tradition to wane', *Journal of the American Institute for Conservation* 35(1): 1–7.

Neumann, B.S. and Sansom, K.G. 1970. 'Laponite clay: a synthetic inorganic gelling agent for aqueous solutions', *Journal of the Society of Cosmetic Chemists* 21: 237–58.

Nochlin, L. 1984. 'Malvina Hoffman: a life in sculpture', *Arts Magazine* 59(3): 106–10.

Stavroudis, C., Doherty, T. and R. Wolbers. 2005. 'A new approach to cleaning: using mixtures of concentrated stock solutions and a database to arrive at an optimal aqueous cleaning system', *WAAC Newsletter* 27(2): 17–28.

Stavroudis, C. 2009. 'Sorting out surfactants', *WAAC Newsletter* 31(1): 18–21.

Stavroudis, C. 2012a. 'More from CAPS3: surfactants, silicone-based solvents, and microemulsions', *WAAC Newsletter* 34(3): 24–7.

Stavroudis, C. 2012b. 'Pemulen TR2 – the once and future king (of conservation)', *WAAC Newsletter* 36(2): 10–11.

Author's address

Shelley M. Smith, Brooklyn, NY, USA (orpiment@gmail.com)

The use of gel systems for cleaning water- and solvent-sensitive paintings

Carmen Ahedo Pino

Introduction

Since the mid-1980s, when solvent gels based on Carbopol and resin soaps were introduced by Richard Wolbers to the field of easel paintings conservation, new gel methods have been explored, including aqueous systems (Wolbers 2000) and more recently rigid gels of polysaccharides (Campani *et al.* 2007; Cremonesi 2013) and silicone-based materials[1] (Stavroudis 2012, 2016; Cremonesi 2016). Gel systems can act as tailored cleaning tools for surface cleaning, varnish and overpaint removal, and the removal of other film-forming materials, as well as a method for measuring surface pH and ionic strength. This paper presents a summary of the gel systems applied to several water- and solvent-sensitive Spanish paintings, and gives a brief description of the protocol, methodology and gel recipes used in the cleaning process.

Case studies

The first case study (CS-1) refers to a 16th-century large-format two-sided *tüchlein* painting – divided into predella, central body and attic – that covers the main doors of an altarpiece, on each side of the doors including the predella, which has doors that are articulated in two leaves. The paintings on all the inner doors formed the case study. This exceptional and colourful set of paintings was executed in glue size and oil. Pigments were bound to the hemp support with animal glue, without a ground layer; linseed oil binder was used to finish several parts of the composition and the paint surfaces were left mostly unvarnished.

The second artwork (CS-2.1) is a small 19th-century oil painting on paper with a secondary cardboard support (attached with animal glue-based adhesive) and a thin ground layer. The original varnish is composed of rosin and mastic resins. Its gold-coloured frame (CS-2.2) has a copper alloy metal leaf on top of a linseed oil and diterpenoid resin bole, with a natural resin coating. Both paintings appear to have been treated in the past. On the surface, analysis with Fourier transform infrared spectroscopy and gas chromatography-mass spectrometry indicates the presence of rosin and linseed oil (CS-1 for the earlier varnish, CS-2.1 for the grey uppermost overpaint layer) and large amounts of copper salts were also found in CS-2.2.

Cleaning protocol, methodology and formulations

Preliminary documentation was generated (condition report, historic research and analytical data); pH and conductivity were measured across all colours before treatment to inform the design of the cleaning system (range of 5–10× the surface conductivity).[2] For this purpose, different conductivity values of an aqueous cleaning system were compared with the painting surface measurements obtained in order to identify the values within the 5–10× range (for CS-2.1 see Fig. 1). An agarose gel was prepared using the microwave heating method and cut into small round pieces. Discs were placed on the surfaces and measurements were performed using handheld meters (Horiba Laqua Twin). Several preliminary tests were carried out (free organic solvents, buffered solutions, gels and emulsions) although this paper only refers to some of the gel systems used.

Surface cleaning

Some of the *tüchlein* paintings were essentially glue-bound (the inner doors located at the attic level, for CS-1), while the entire surface of CS-2.2 was covered with a dark green film of copper salts; both surfaces appeared to be quite water-sensitive when using a free aqueous solution. An agar[3] gel was selected to gradually release the aqueous phase and to avoid a clearance step which could affect the paint layer.

For CS-1, the agar was gelled as a thick film prior to use and then cut into pieces. According to the specific requirements

Interior of a Studio

Oil on paper + cardboard (secondary support)
19th century, Manuel Cabral y Aguado Bejarano
Museo del Romanticismo, Spain

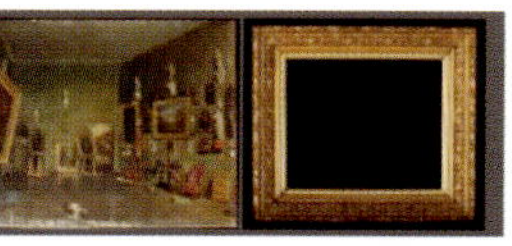

(a)

	COLOUR-LOCATION	pH	EC	EC values: aqueous cleaning systems							
				1500	2000	2500	3000	3500	4000	5000	6000
15	**red** (painting)	6.2	300	5x	6.7x	8.3x	10x	11.7x	13.3x	16.7x	20x
19	**white** (dog)	6.2	300	5x	6.7x	8.3x	10x	11.7x	13.3x	16.7x	20x
20	**blue** (painting)	6.4	310	4.8x	6.4x	8.1x	9.7x	11.3x	12.9x	16.1x	19.3x
31	overpaint	6.2	330	4.5	6.1x	7.6x	9.1x	10.6x	12.1x	15.1x	18.2x
6	**dark carmine** (painting)	6.2	340	4.4x	5.9x	7.3x	8.8x	10.3x	11.8x	14.7x	17.6x
11	overpaint (background)	6	370	4x	5.4x	6.8x	8.1x	9.5x	10.8x	13.5x	16.2x
7	**yellow** (frame)	6.1	400	3.7x	5x	6.2x	7.5x	8.7x	10x	12.5x	15x
12	wall	5.9	400	3.7x	5x	6.2x	7.5x	8.7x	10x	12.5x	15x
30	varnish (floor)	6.5	400	3.7x	5x	6.2x	7.5x	8.7x	10x	12.5x	15x
21	**white** (plate)	6.4	420	3.6x	4.8x	5.9x	7.1x	8.3x	9.5x	11.9x	14.3x
13	**blue** (painting)	6	480	3.1x	4.2x	5.2x	6.2x	7.3x	8.3x	10.4x	12.5x
16	**brown** (painting)	6.3	490	3.1x	4.1x	5.1x	6.1x	7.1x	8.2x	10.2x	12.2x
22	**black** (table)	6.3	500	3x	4x	5x	6x	7x	8x	10x	12x
24	**red** (table)	6.5	500	3x	4x	5x	6x	7x	8x	10x	12x
18	**brown** (dog)	6.3	510	2.9x	3.9x	4.9x	5.9x	6.9x	7.8x	9.8x	11.8x
23	**green** (table)	6.2	520	2.9x	3.8x	4.8x	5.8x	6.7x	7.7x	9.6x	11.5x
27	**ochre** (chest)	6.6	520	2.9x	3.8x	4.8x	5.8x	6.7x	7.7x	9.6x	11.5x
25	**green** (cushion, floor)	6.7	530	2.8x	3.8x	4.7x	5.7x	6.6x	7.5x	9.4x	11.3x
14	**light blue** (painting)	6.3	580	2.6x	3.4x	4.3x	5.2x	6x	6.9x	8.6x	10.3x
17	floor	6	590	2.5x	3.4x	4.2x	5.1x	5.9x	6.8x	8.5x	10.2x
9	**blue** (carpet)	6.1	600	2.5x	3.3x	4.2x	5x	5.8x	6.7x	8.3x	10x
4	**black** (frame overpaint)	6.3	630	2.4x	3.2x	4x	4.8x	5.6x	6.3x	7.9x	9.5x
10	floor	6.6	650	2.3x	3.1x	3.8x	4.6x	5.4x	6.1x	7.7x	9.2x
1	**white** (head statue)	6.3	690	2.2x	2.9x	3.6x	4.4x	5.1x	5.8x	7.2x	8.7x
8	**green** (cloth)	6.6	710	2.1x	2.8x	3.5x	4.2x	4.9x	5.6x	7x	8.4x
26	**carmine** (seat)	5.9	715	2.1x	2.8x	3.5x	4.2x	4.9x	5.6x	7x	8.4x
3	**white** (frame overpaint)	5.8	750	2x	2.7x	3.3x	4x	4.7x	5.3x	6.7x	8x
28	frame (covered part)	6	750	2x	2.7x	3.3x	4x	4.7x	5.3x	6.7x	8x
2	**gold** (frame overpaint)	6.6	860	1.7x	2.3x	2.9x	3.5x	4.1x	4.6x	5.8x	7x
5	**light carmine** (painting)	6.5	860	1.7x	2.3x	2.9x	3.5x	4x	4.6x	5.8x	7x
	AVERAGE	**6.3**	**533.5**								

(b) CLEANING TEST

VISIBLE LIGHT | UV LIGHT

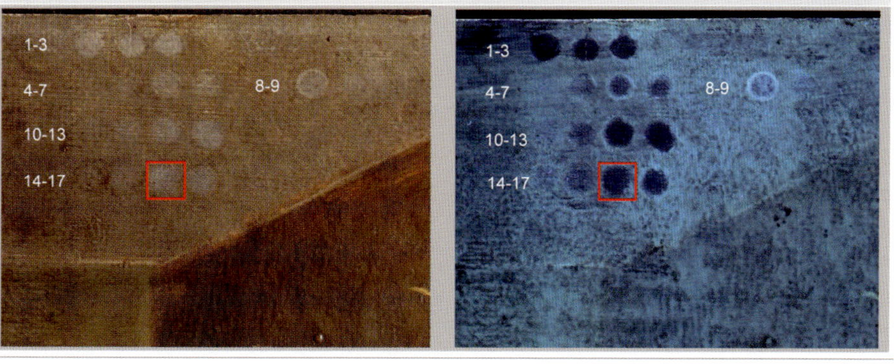

FIG. 1 CS-2.1, *Interior of a Studio*: (a) On the left, surface pH and electrical conductivity (EC) readings before cleaning, EC ascending order; on the right, EC aqueous systems values (5–10× surface EC, marked in green); (b) cleaning test (image: IPCE).

of the surface, the gel was placed either directly on the paint layer or on top of a thin piece of Japanese tissue. The gel was removed and the process was repeated, allowing each area to dry between treatments. For the frame (CS-2.2), a three-dimensional object, a semi-solid agar gel (with a low gelling temperature) was required for brush application (Cremonesi 2013: 181). After cooling to room temperature, the gel was removed mechanically with a minimum release of water into the metal surface (Fig. 2). For CS-2.2, the gel was redissolved after the initial solidification to achieve better water-retention properties (Cremonesi 2013: 181), and a buffered solution with a corrosion inhibitor (2-methylamino-ethanol, MAE) was used.[4] For CS-1, the agar pieces were immersed in a buffered solution before application. Observation of the gradual colouring of the gel was utilized to indicate ongoing action during cleaning.

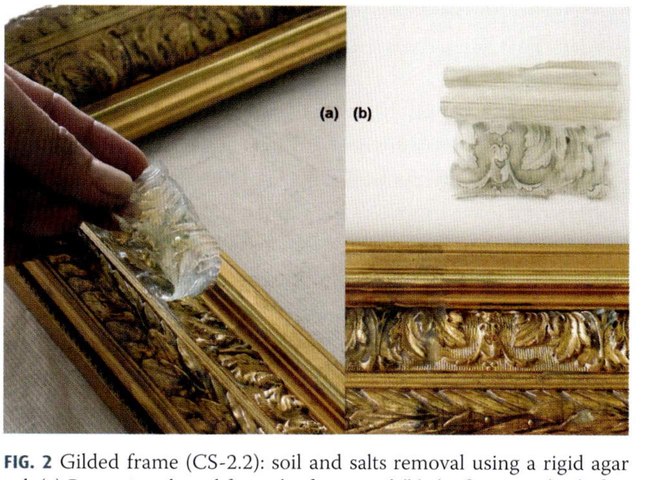

FIG. 2 Gilded frame (CS-2.2): soil and salts removal using a rigid agar gel. (a) Removing the gel from the frame and (b) the frame and gel after treatment (image: IPCE).

USE OF GELS IN CLEANING TREATMENTS: Surface Cleaning and Materials Removal

		16th century: Doors of an altarpiece (inner)	19th century: *Interior of a Studio*
CS-1	*tüchlein*	Pietro Morone Church of Saint Michael the Archangel, Ibdes, Spain	Manuel Cabral y Aguado Bejarano Museo del Romanticismo, Madrid, Spain
CS-2.1 CS-2.2	oil on paper gilded frame	• glue-size and oil on canvas • non ground layer • rosin resin, linseed oil	• oil on paper + cardboard (secondary support) • ground layer • rosin & mastic resins, linseed oil/ copper salts

(a) pH-EC[1] MEASUREMENT[2]

AGAROSE GEL (low EEO)	attic		c. body		predella		painting		frame	
	pH	EC	pH	EC	pH	EC	pH	EC	pH	EC
	7.6	355	6.9	274	7.2	349	6.3	533	6.2	6650

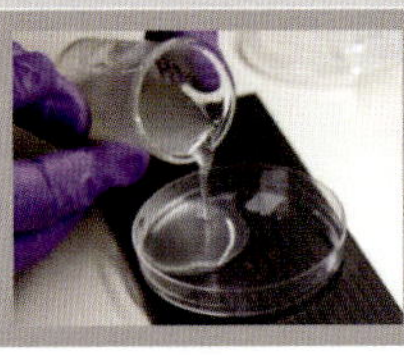

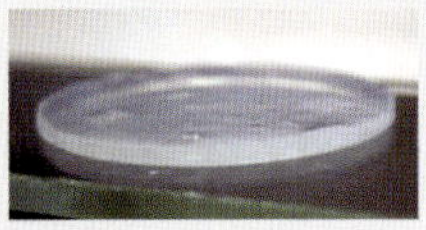

4-5 g in deonized water (w/v)
heated in microwave
round discs 4mm Ø, 2mm thick
5 minutes surface contact time
measurement (Horiba®), readings average

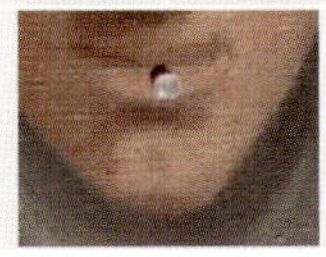

(b) SURFACE CLEANING

AGAR GEL

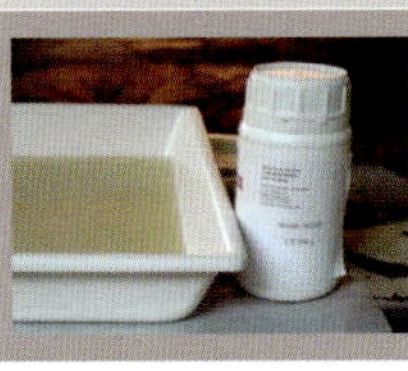

attic (inner paintings of the doors)		frame (copper alloy)	
soil · direct/tissue · pieces/rubber		soil, copper salts · direct · brushed	
4 g in deonized water (w/v) Buffered solution 100 ml deonized water 0.5 g citric acid 0.5 g EDTA[3] TEA	pieces 8mm thick immersed c.a.1 h c.a. 5 minutes 1 application pH 5.5 · EC 3000	3 g in buffered solution (w/v) Buffered solution 100 ml deonized water 0.5 g citric acid 0.5 g DTPA[4] 0.1 g MAE[5] TEA	redissolved 5-8 minutes 1-2 applications pH 6 · EC 6000

MATERIALS REMOVAL

VELVESIL PLUS emulsion

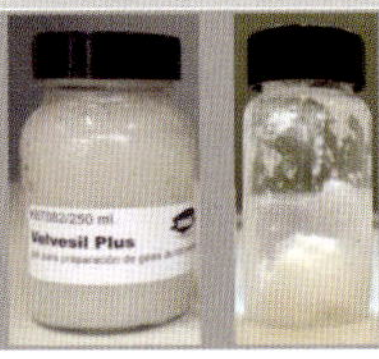

c. body & predella (inner paintings of the doors)		painting	
terpenoid resin & oil · D5 · brushed		terpenoid resin & oil · D4 · brushed	
5 g Velvesil® Plus 0.25 g buffered solution 5% 100 ml deonized water 0.5 g citric acid 0.5 g EDTA TEA 0.75-1 g BA[6] 10-15%	2-3 minutes 5 applications pH 6 · EC 2500	5 g Velvesil® Plus 0.50 g buffered solution 10% 100 ml deonized water 0.5 g citric acid 0.5 g DTPA TEA 0.25-0.5 g BA-IPP[7] (1:1) 5-10%	5 minutes 3 applications pH 7-7.5 · EC 3000

[1] EC: electrical conductivity (μS/cm)
[2] pH and EC average values
[3] EDTA: Ethylenediaminetetraacetic acid
[4] DTPA: Diethylenetriaminepentaacetic acid
[5] MAE: 2-(Methylamino)ethanol
[6] BA: Benzyl alcohol
[7] IPP: Isopropyl palmitate

(c)

CS-1

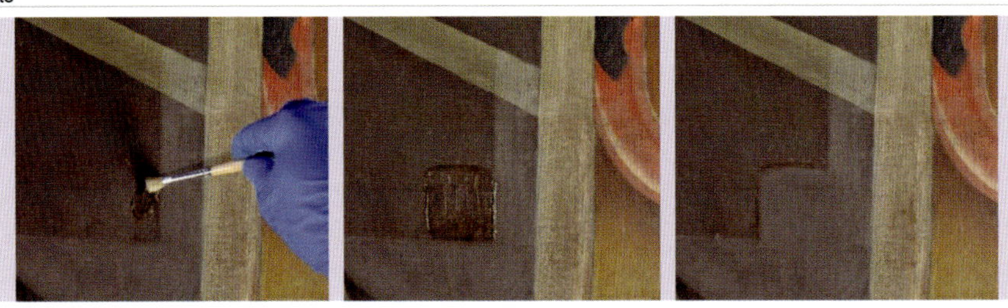

FIG. 3 Methodology and formulations of gels: (a) pH-EC measurements; (b) cleaning systems; (c) application methodology (image: IPCE).

Materials removal

A Velvesil Plus water-in-oil (w/o) emulsion gel was used for the removal of the natural resin and oil layers on both paintings. Emulsions were prepared by weight; the materials used, relative proportions between aqueous and solvent phase, pH and conductivity were modified to match each specific situation. For CS-2.1, several pigments proved to be sensitive to polar organic solvents. Because water is a poor solvent for a non-oxidized oil film, the pH was raised, while the aqueous phase was increased and the organic solvent phase was modified using a less polar solvent blend. Benzyl alcohol (BA) was mixed with isopropyl palmitate (IPP), a green solvent substitute for white spirit (having similar solubility parameters).[5]

Surfaces were first saturated with cyclomethicone solvents to create a hydrophobic barrier (Stavroudis 2016). Decamethylcylopentasiloxane (D5) or octamethylcylotetrasiloxane (D4) was selected depending on the temperature conditions *in situ* (Berthiaume 1999: 293), since the freezing point of pure D4 (17.6 °C) is a problem when the temperature in the workplace is low. Gels were applied with a soft brush and mechanically moved about the surface. Clearance was carried out using dry cotton swabs, followed by a final clearance with cotton swabs moistened in silicone solvent (Fig. 3).

Conclusion

Although both case studies were water- and solvent-sensitive, a customized approach to each painting produced very satisfactory cleaning results. Controlling the pH within a safe range of 5.5–8.5 through buffering, and the use of conductivity around 5–10× the isotonic condition on the surfaces was crucial to avoid the risk of saponification, swelling and softening these paint layers. The use of agarose gels to measure surface pH and conductivity before treatment was shown to be an essential method to inform design of an aqueous cleaning system. Since their rigid polysaccharide structure allows water to be in contact with surfaces while being retained within the gel structure, moisture impregnation of the paint layers was prevented. In CS-2.2, the methodology permitted the removal of copper salts in an efficient manner – surface conductivity measurements after cleaning showed a significant reduction of ions (6650 to 3000 μS/cm).

Silicone solvents proved to be a suitable hydrophobic barrier for water-sensitive surfaces, hindering the penetration of water and non-miscible solvents into paint layers. Velvesil Plus polymeric emulsion gel stabilizers are less likely to penetrate into surfaces (since they are macromolecules) and easily cleared with non-polar solvents; potential damage and residues can be avoided by using them with a silicone solvent barrier. By modifying the water-to-solvent ratio, the pH, conductivity, and polarity of the solvent phase of these emulsions, different cleaning issues may be addressed.

Materials and suppliers

- Agar CAS 9002-18-0, octamethylcyclotetrasiloxane CAS 556-67-2, decamethylcyclopentasiloxane CAS 541-02-6: Kremer Pigmente, http://www.kremer-pigmente.com/
- Agarose CAS 9012-36-6, benzyl alcohol CAS 100-51-6, boric acid CAS 10043-35-3, citric acid CAS 77-92-9, diethylenetriaminepentaacetic acid CAS 67-43-6, ethylenediaminetetraacetic acid CAS 60-00-4, isopropyl palmitate, CAS 142-91-6, 2-(methylamino)ethanol CAS 109-83-1, triethanolamine CAS 102-71-6: Sigma-Aldrich, https://www.sigmaaldrich.com
- Velvesil Plus Gel: Momentive/ Kremer, https://www.momentive.com/

Notes

1. Silicone-based gels were first introduced in 2007 by Richard Wolbers, in a workshop in Trento, Italy, sponsored by CESMAR7.
2. Richard Wolbers, workshop notes: UPV, Valencia, Spain, 1–2 July 2014; Museo del Prado, Madrid, Spain, 20–23 January, 2016; IPCE, Madrid, Spain, December 13–16, 2016.
3. For its slower rate of syneresis, a spontaneous extrusion of water through the surface of the gel, compared to agarose (Cremonesi 2013: 180).
4. Richard Wolbers, pers. comm., 26 August 2016.
5. Richard Wolbers, pers. comm., 3 August 2016.

References

Berthiaume, M.D. 1999. 'Silicones in cosmetics', in E.D. Goddard and J.V. Gruber (eds), *Principles of Polymer Science and Technology in Cosmetics and Personal Care*, Cosmetic Science and Technology Series, vol. 22. New York: Marcel Dekker, 275–324. Available at: https://rgmaisyah.files.wordpress.com/2009/02/principles-of-polymer-science-and-technology-in-cosmetics-and-personal-care.pdf.

Campani, E., Casoli, A., Cremonesi, P., Saccani, I. and Signorini, E. 2007. *L'uso di agarosio e agar per la preparazione di 'gel rigidi'/Use of Agarose and Agar for Preparing 'Rigid Gels'*, D. Kunzelman (tr.), *Quaderni del Cesmar7*, no. 4. Padua: Il Prato.

Cremonesi, P. 2013. 'Rigid gels and enzyme cleaning', in M.F. Mecklenburg, A.E. Charola and R.J. Koestler (eds), *Cleaning 2010: New Insights into the Cleaning of Paintings*. Washington, DC: Smithsonian Institution Scholarly Press, 179–84.

Cremonesi, P. 2016. *Proprietá ed esempi di utilizzo di materiali siliconici nel restauro di manufatti artistici* (Parte prima). Padua: Il Prato.

Stavroudis, C. 2012. 'More from CAPS3: surfactants, silicone-based solvents, and microemulsions', *WAAC Newsletter* 34(3): 24–7.

Stavroudis, C. 2016. 'Silicone-based solvents in conservation: as free solvents and components of gel systems and microemulsions', in V.E. Selva Bonino (ed.), *Dall'olio all'acrilico, dall'impressionismo all'arte contemporanea, Studi, ricerche, indagini ed interventi conservative [CESMAR7]*. Padua: Il Prato.

Wolbers, R. 2000. *Cleaning Painted Surfaces: Aqueous Methods*. London: Archetype Publications.

Author's address

Carmen Ahedo Pino, Spanish Cultural Heritage Institute (IPCE), Madrid, Spain (maria.ahedo@mecd.es)

Releasing St Peter

Claudia Alejandra Garza Villegas and Ramón Avendaño Esquivel

Conservation is one of the primary tasks of the National Museum of the Viceroyalty/Museo Nacional del Virreinato of the National Institute of Anthropology and History/Instituto Nacional de Antropología e Historia (INAH) in Tepotzotlán, Mexico. Its conservation department has existed since the foundation of the museum in 1964 by the INAH. The significant collection presents challenges that have resulted in work plans for adequate conservation, restoration, storage and loan conditions to other institutions and museums all over the world.

In May 2016, *The Release of St Peter*, an easel painting crafted during the 17th century in New Spain by a Mexican painter Pedro Ramírez the Younger, entered the conservation studio, having been on display since the foundation of the museum. First the work was housed in the Archbishop's Palace in Mexico City, but due to its aesthetic quality it was transferred to the old Museum of Religious Art, the Metropolitan Cathedral of Mexico City. Today the painting is part of the collection of the National Museum of the Viceroyalty. The painting underwent treatment that involved cleaning the varnish and eliminating previous restoration to the paint layers and the textile support.

Pedro Ramírez the Younger was descended from a line of altarpiece crafters: both his grandfather, Diego Ramírez, and his father, Pedro Ramírez the Elder, who arrived from Seville in the 16th century, followed the profession. Pedro Ramírez the Younger was considered to be Spanish because of the plastic qualities of his paintings, but there is very little information available on his life. However, his baptismal certificate was recently found, uncovering new information on his birth and baptism in Mexico City.

Pedro Ramírez' pictorial oeuvre, scanty but masterful, was so important during the Viceroyalty that his works travelled as far as Guatemala. Today, very few of his works have been

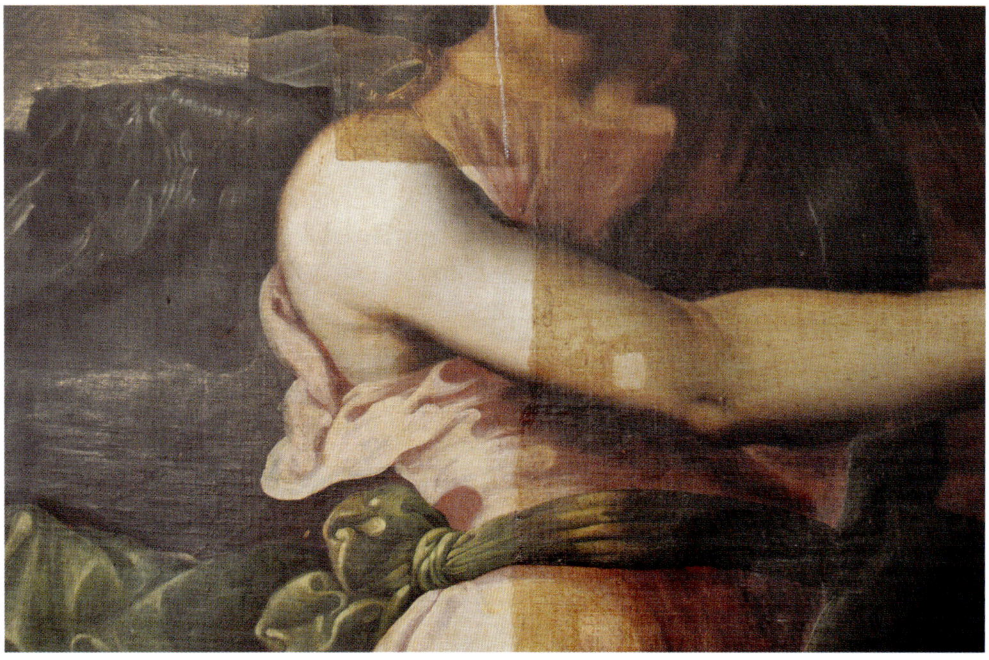

FIG. 1 Pedro Ramírez the Younger, *The Release of St Peter*: detail during cleaning, showing the extent to which the dark varnish has modified all the colours.

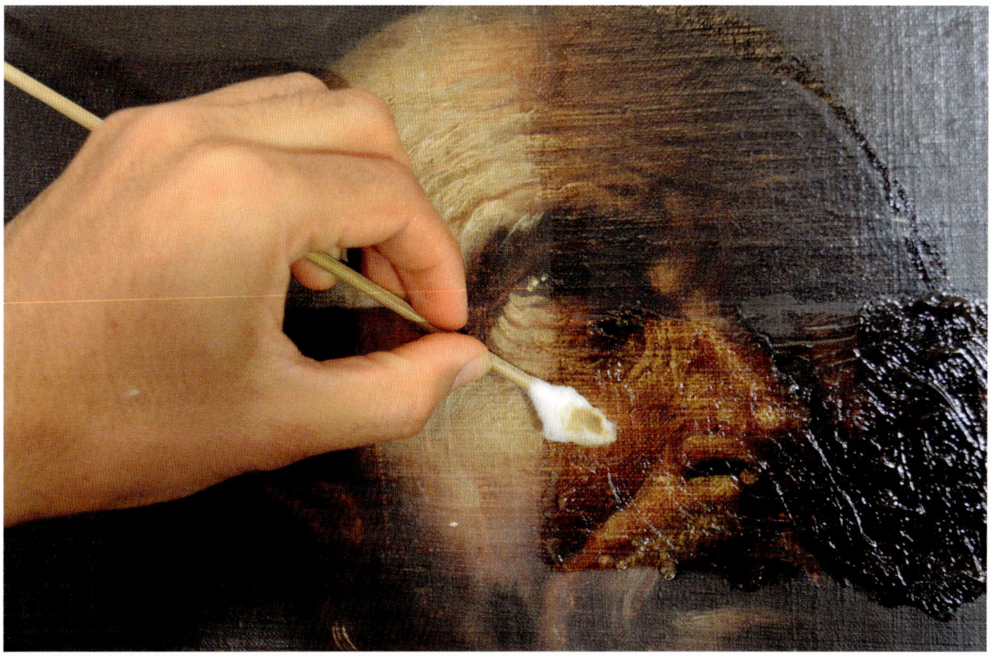

FIG. 2 Pedro Ramírez the Younger, *The Release of St Peter*: detail showing removal of gel applied to the painting, using a cotton swab.

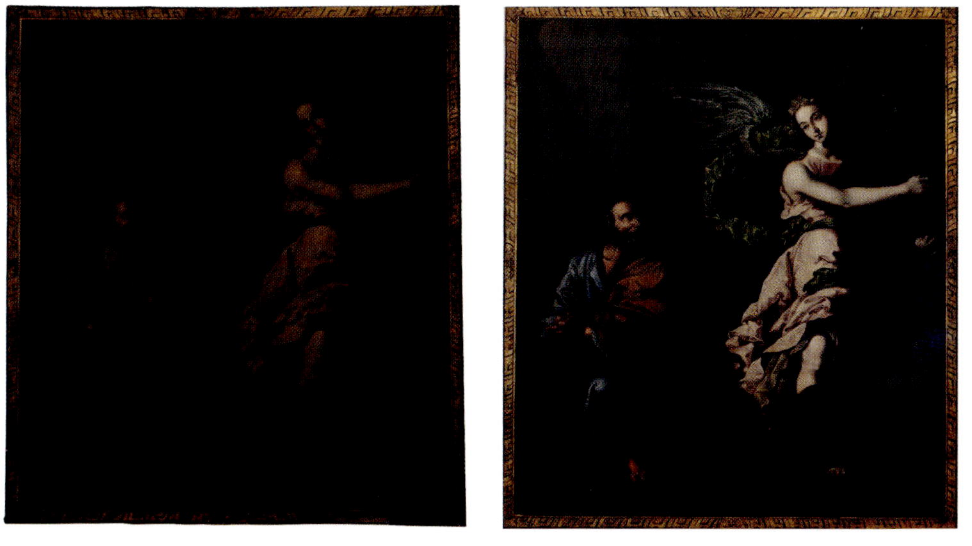

FIG. 3 Pedro Ramírez the Younger, *The Release of St Peter*, 17th century, oil on canvas, before and after treatment.

located therefore *The Release of St Peter* is noteworthy being one of his few autograph paintings. José Moreno Villa had discovered Ramírez' signature on cleaning the work in the 1930s.[1] Unfortunately, the signature had become illegible by the time the painting entered the conservation studio in 2016. From detailed observation of the painting we were able to locate it in the bottom right corner, although it was difficult to discern due to the darkening of the varnish and as similar pigments were used by the artist for both the background and the signature.

The painting shows the apostle Peter on the left side, sitting with his hands in prayer. He is wearing a blue tunic and an ochre mantle. He is barefoot, and below him some chains and shackles allude to his recent release. On the other side of the painting there is a blond angel looking at Peter, his hands pointing to an open gate. He is wearing a pink tunic and jewellery on his right arm, complemented by a green mantle and sandals. Beside these characters, some architectural details indicate an indoor setting. This scene clearly alludes to the biblical passage of the Acts (12: 3–8) when an angel releases St Peter after his incarceration by Herod Agrippa.

The painting presented both significant chromatic alterations and darkening caused by accumulated varnishes from a series of interventions over the years. Examination in ultraviolet (UV) light confirmed that it had an irregular layer of natural resin varnish, fluorescent and with a slightly whitish yellow hue. The varnish had become yellow and then brown over time and this dark layer concealed forms and modified volumes, dimensions and colours, while showing the entire composition with a darker tone (Fig. 1). UV examination also confirmed the presence of overpaint covering both earlier paint losses and part of the original paint layer. The overpaint comprises approximately 15% of the painting surface. Below the overpaint two kinds of fills were found, one white and one orange, indicating two different interventions. They both cover the original paint layer and modify its texture. In addition to the damage caused by these interventions, the painting was covered in dust and accumulated grime, which affected its appearance and made the work more vulnerable to further damage.

We started the cleaning tests using several solvents. In the end, a mixture of 1:1 isopropyl alcohol:ethyl alcohol was found to be the most efficient because of the nature of the varnish (natural resin) and the recent (20th-century) overpaintings. However, given that the paint layer was highly susceptible to the direct and persistent usage of these solvents, we chose to use a stable gel containing the two solvents so that the varnish could be removed without affecting the paint layers.

We prepared several gels using solvent-compatible Carbopol as a thickening agent. After trying different proportions while knowing that a mixture of isopropanol and ethanol would work well, we made a Carbopol gel with 3:1:1:1polymer:water: ethanol:isopropanol. The water assisted the application and was used for clearance. An efficient product for cleaning was obtained which yielded good results and successfully controlled the cleaning process. We placed the gel with a spatula in approximately 5–10 cm^2 areas and removed it almost as soon as we had finished its application, using a dry cotton swab followed by a cotton swab with isopropanol without pressure for further clearance (Fig. 2).

To avoid direct exposure to, and penetration by, the solvents, and to eliminate the previous restorations on the thin canvas, another rigid sustainable gel was prepared from agarose. This gel was used to remove the residues of the adhesive that was transferred to the textile during the application of strip lining and patches in a previous intervention. We obtained good results and avoided moistening the textile support. We chose a concentration of 3.5 wt% agarose, which made a rigid and concentrated gel, so that the amount of water diffusing into the support was reduced while being sufficient to soften the glue. The very small pores of the gel could absorb and remove most of the unwanted material from the textile.

The use of gels in this specific case was critical because both the paint layer and canvas were very thin, and using liquid solvents would have affected them by dissolving the glazes and causing deformation. The isopropanol-ethanol gel allowed the solvents to remain on the surface, so they acted only on the varnish and the overpaint. The agarose rigid gel yielded notable results because water did not migrate either to the support or the strata, so deformation and tidelines were both avoided.

Following completion of the conservation treatments, museum visitors were able to admire a released St Peter without all the foreign materials that had completely changed the reading of the painting (Fig. 3). The artwork can now be appreciated in the upper floor of the museum, in the Arts and Crafts display.

Acknowledgements

We are very grateful to the head of the Museum Conservation Department, Xochipilli Rossell Pedraza, for her interest and support for this publication. Thanks are also extended to Pablo Amador Marrero for his unconditional support and for sharing his knowledge with us.

Note

1. From the Aesthetics Research Institute, Universidad Nacional Autónoma de México. See R. Ruiz Gomar (2000), 'Nuevas noticias sobre los Ramírez, artistas novohispanos del siglo XVII', *Anales del Instituto de Investigaciones Estéticas* XXII(77).

Authors' addresses

- Claudia Alejandra Garza Villegas, Magisterial Vista Bella, Tlalnepantla, Estado de México, Mexico (c.a.garza.villegas@gmail.com)
- Ramón Avendaño Esquivel, Barrio Oxtopulco Universidad, Coyoacán, Ciudad de México, Mexico (avendanoesquivel@gmail.com)

The removal of Paraloid B-72 coatings with aqueous gelled systems: Roman frescoes from Empúries, Catalonia

Aleix Barberà Giné and Silvia Marín Ortega

Introduction

This is a practical example of the removal of Paraloid B-72 coatings from Roman wall paintings executed in true fresco. The resin was applied to adhere fabrics used to detach the paintings with the *stacco* technique. The applied coating was so thick that it was preventing the cleaning of the paintings as it covered concretion of salts, dirt and significant bioform colonization, as well as some other alterations.

Context

In 1999 the detachment of the wall paintings from two rooms of the archaeological Roman site at Empúries took place: *domus* 2B room 16 and *domus* 2A room 6. These paintings are the remains of the wall decorations of the rooms, painted in fresco, and represent decorations that imitate marble panels and architectonic framing in Pompeian style (Nieto 1980). It had been impossible to protect the rooms from adverse weather conditions following excavation, and the advanced state of deterioration of the paintings made it necessary to carry out detachment immediately, in order to guarantee their preservation (Rovira 2014). This process was performed with the *stacco* technique, using Paraloid B-72 as the adhesive for the facing. Once removed from their original support, the paintings were stored for 17 years before being transferred to a new support.

After preparing the reverse side of the paintings, the transfer of the paintings was carried out using traditional techniques (Soriano *et al.* 2008), to adhere new support fabrics to the reverse the paintings using a calcium caseinate preparation. After that, the facing material was removed using free solvents applied with cotton wool poultices (Fig. 1). This removal was possible because the Paraloid became tacky after a 25–30-minute application, permitting easy removal of the fabric. At this stage, it was then necessary to continue with a treatment for the complete removal of the acrylic resin.

Practical case

Utilizing traditional methods for resin removal, the first tests were based on the use of free solvents (Masschelein-Kleiner 2004; Baglioni and Chelazzi 2013). The Paraloid B-72 could still be dissolved using the solvents commonly used for applying it: tests were carried out using xylene, acetone and benzyl alcohol, all with very similar results. At the macroscopic level, the removal of the Paraloid using solvents was believed to be a fast and effective method. However, observation of samples with a digital microscope revealed that, despite having removed most of the resin from the surface, there was still quite a large amount of residue build-up in the pores and cracks.

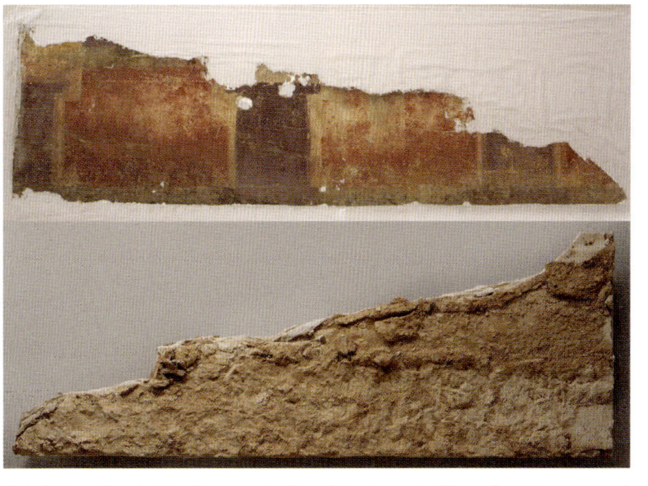

FIG. 1 A wall painting fragment after detachment. *Top*: after the removal of the facing and before further treatment. *Bottom*: reverse of the same fragment before transfer to a new fabric support (image: CRBMC).

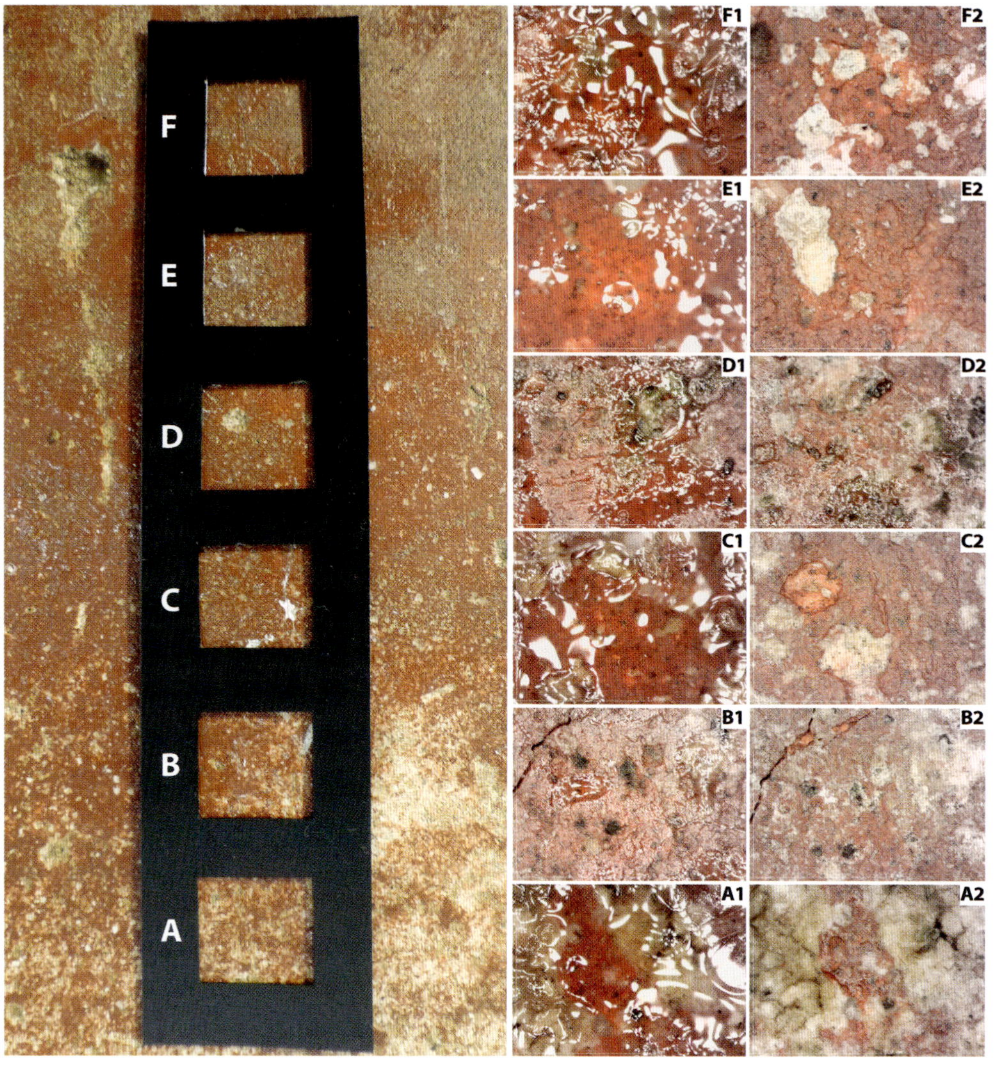

FIG. 2 Cleaning tests with different systems, showing the area of application on the left. On right side, details (×215), before treatment A1–F1, and after cleaning tests A2–F2, using free acetone [A], free benzyl alcohol [B], acetone gel (after two 15-minute applications) [C], XYL microemulsion in 2% Klucel G at first 30-minute application [D], EAPC microemulsion (after two 30-minute applications) [E], emulsion of 10% benzyl alcohol with buffer solution at pH 8.5 adjusted with Trizma in a 2% Vanzan gel (after two 10-minute applications) [F].

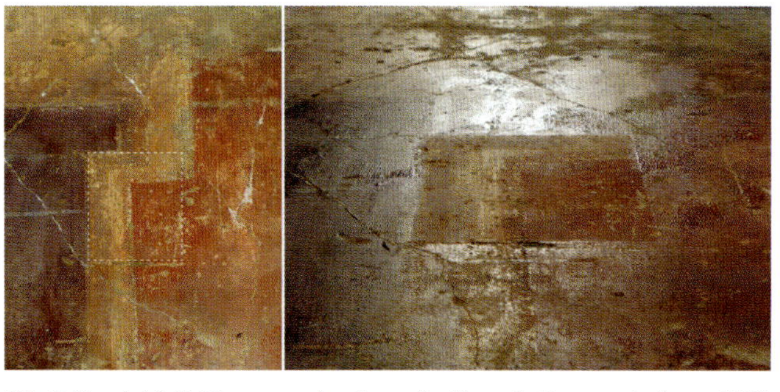

FIG. 3 Paraloid B-72 removal using a buffer solution made from 0.5% DTPA at pH 8.8 in a 2% Vanzan emulsion gel with 5% benzyl alcohol (right). The same area: raking light reveals satisfactory cleaning (left).

The high risk of not only finding traces of resin on the surface, but also resin penetrating into the wall painting, made it necessary to test gel systems. Tests were conducted using two solvent surfactant gels with Carbopol (Wolbers 2000), one with acetone and the other with xylene, with application times ranging from 5 to 15 minutes. It was observed that the removal was slightly slower and that the results could be improved by repeating the gel application process two or three times. Under the microscope, it was clear that the resin residues could be reduced in the hard-to-reach areas by increasing the application time and repeated treatments (Fig. 2).

Although the level of Paraloid removal from the surface could be improved by using such gels, this cleaning system had some drawbacks. It was necessary to use free solvents to clear the gel, therefore there was a risk of redissolved resin penetrating into the wall painting, and the procedure could be harmful to the conservator. Moreover, since this system does not comprise exclusively volatile materials, there was a real, albeit small, risk of leaving some cleaning residues (Dorge 2004).

To reduce both the toxicity and potential resin penetration, we considered the possibility of using microemulsions and oil-in-water (O/W) emulsions. Given that these are aqueous systems, we needed to know in advance the pH and conductivity values of the surface based on an absorption test with agar gel (3 wt% in deionized water). Results obtained with 15-minute applications showed a superficial pH ranging from 6.5 to 7.5, and a conductivity ranging from 80 to 100 μS, although some values went up to 400–500 μS in the areas with bioform colonization. Based on these results, we made a Trizma buffer solution at pH 8.5, with 0.5% concentration, resulting in a conductivity of 800 μS, a safe range for the wall painting, but not for the acrylic resin. This solution was to be the basis for preparing the subsequent microemulsions and emulsions.

At first, two microemulsion formulations to remove synthetic resins from porous substrates were tested (Baglioni *et al.* 2012; Carretti *et al.* 2003; Giorgi *et al.* 2009). The XYL (86.2 wt% Trizma buffer at pH 8.5; 2.6 wt% xylene; 4.1 wt% sodium dodecyl sulphate and 7.9 wt% pentanol) and EAPC (73.3 wt% Trizma buffer at pH 8.5; 8 wt% ethyl acetate; 8 wt% propylene carbonate; 3.7 wt% sodium dodecyl sulphate and 7 wt% pentanol) microemulsions were applied with cellulose poultices, according to the authors' description, although they were also applied in a 2 wt% Klucel-G gel. In both cases, the application times ranged from 30 to 60 minutes and were cleared with the same buffer solution. The removal of the resin was highly effective after the second application, although large amounts of foam were generated due to the surfactant residues which were very difficult, and sometimes impossible, to remove.

In view of these results, we considered the alternative of testing Vanzan (xanthan gum) emulsion gels to avoid the use of surfactants. This system has many benefits: removing the resin without the risk of it re-penetrating into the porous system, minimizing the presence of residues of the cleaning products used, and reducing as much as possible the use of solvents, resulting in systems which are safer both for conservators and the environment.

A benzyl alcohol and Trizma pH 8.5 buffer O/W emulsion with 2 wt% Vanzan was successful for Paraloid removal, with application times ranging from 5 to 10 minutes, with some slight mechanical action using a brush. The formula was optimized based on subsequent tests, which revealed that the use of a diethylenetriaminepentaacetic acid (DTPA) pH 8.8 buffer, with chelating properties, and a concentration of 0.5 wt%, was more effective than the Trizma solutions. Finally, the emulsion gel used to remove the Paraloid only had 5% solvent content and so aqueous clearance was possible (Fig. 3).

Conclusion

Although all the systems tested were successful to some degree for the effective removal of the Paraloid B-72 coating from the wall paintings, we came to the conclusion that the use of a Vanzan aqueous emulsion gel with just 5% of solvent offered the best results. The level of resin removal with this system was very effective, matching or even improving on traditional methods. Moreover, working without free solvents may prevent the redissolved resin from penetrating into the porous surface. The use of very small quantities of solvent, applied in an emulsion, reduced the toxicity of the process, both for the environment and for the conservator. Finally, the formulation of the cleaning system based on preliminary pH and conductivity measurements on the surface, the use of a buffered solution with pH and conductivity control to protect the fresco painting, made this cleaning system a very safe option.

Acknowledgements

Centre for the Restoration of Movable Cultural Heritage of Catalonia (CRBMC), Archeological Museum of Catalonia, Empúries, and Dr Richard Wolbers (University of Delaware).

References

Baglioni, P. and Chelazzi, D. 2013. *Nanoscience for the Conservation of Works of Art.* Cambridge: RSC Publishing.

Baglioni, M., Giorgi, R., Berti, D. and Baglioni, P. 2012. 'Smart cleaning of cultural heritage: a new challenge for soft nanoscience', *Nanoscale* 4(1): 42–53.

Carretti, E., Dei, L. and Baglioni, P. 2003. 'Solubilization of acrylic and vinyl polymers in nanocontainer solutions: application of microemulsions and micelles to cultural heritage conservation', *Langmuir* 19: 7867–72.

Dorge, V. (ed.) 2004. *Solvent Gels for Cleaning of Works of Art: The Residue Question.* Los Angeles: Getty Conservation Institute.

Giorgi, R., Baglioni, M., Berti, D. and Baglioni, P. 2009. 'New methodologies for the conservation of cultural heritage: micellar solutions, microemulsions, and hydroxide nanoparticles', *Accounts of Chemical Research* 43(6): 695–704.

Masschelein- Kleiner, L. 2004. *Los solventes.* Santiago de Chile: Publicaciones Centro Nacional de Conservación y Restauración DIBAM.

Nieto, J. 1980. 'Repertorio de la pintura mural romana de Ampurias', *Ampurias* 42: 279–312.

Rovira, P. 2014. *La conservación preventiva de las pinturas murales in situ y en su exposición.* Gijón: Ediciones Trea.

Soriano, M.P., Sánchez, M. and Roig, P. 2008. *Conservació i restauració de pintura mural: arrancaments, traspàs a nous suports i reintegració.* Valencia: Universitat Politècnica de València.

Wolbers, R. 2000. *Cleaning Painted Surfaces: Aqueous Methods.* London: Archetype Publications.

Authors' addresses

- Aleix Barberà Giné, Centre de Restauració de Béns Mobles de Catalunya (CRBMC) Valldoreix, Catalonia (abarbera@gencat.cat)
- Silvia Marín Ortega, Escola superior de conservació-restauració de béns culturals de Catalunya (ESCRBCC) Barcelona, Catalonia (marinortegasilvia@gmail.com)

Statistical evaluation of surface roughness measurements to evaluate rigid and mouldable hydrogels for cleaning unvarnished oil paint with heavy impasto

Agata Graczyk, Pauline Hélou-de La Grandière, Sigrid Mirabaud and Marine Page

Dated *c*.1910, *La Créole au perroquet* (*Creole with a Parrot*) is an unvarnished oil painting with heavy impasto which presented challenges for cleaning. French painter Alexis Mérodack-Jeaneau (1873–1919) was one of the first artists to squeeze paint directly from the tube onto the pre-primed canvas over a large part of the surface. The result is a boldly impasted but unvarnished painting, which now exhibits a significant layer of tenacious surface dirt adsorbed directly into the paint surface (Fig. 1). Different cleaning approaches were explored as part of a diploma thesis in paintings conservation at the Institut national du patrimoine, Paris, France, in order to find a safe and effective procedure for carrying out surface

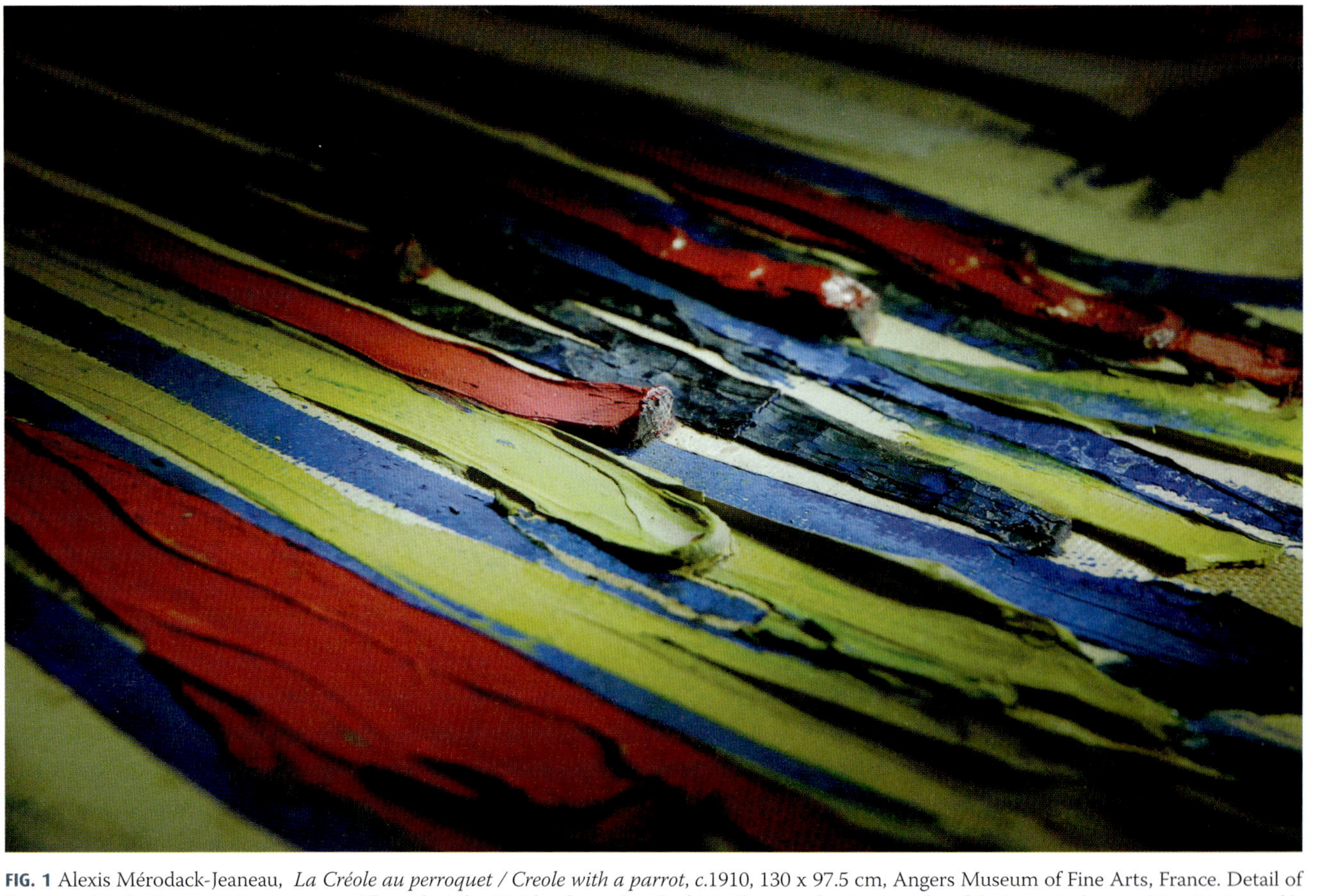

FIG. 1 Alexis Mérodack-Jeaneau, *La Créole au perroquet / Creole with a parrot*, *c*.1910, 130 x 97.5 cm, Angers Museum of Fine Arts, France. Detail of the parrot's tail, where paint was squeezed from the tube. Dirt has been absorbed into the complex topography of the impasto (image © Inp/A.Graczyk).

cleaning on this complex surface (Graczyk 2016). Typical conservation cleaning approaches involving solvents applied with cotton swabs or dry cleaning methods were unsuitable for the impasted passages, where accumulated dirt was often inaccessible within interstices of the paint. The water sensitivity of some of the colours used in *La Créole au perroquet* and the risk of residues were further considerations. Cleaning materials such as gels that have the potential to conform to the surface topography of impasto, restrict water diffusion into materials, and can be removed without leaving residues, were therefore explored.

Hydrogels restrict the amount of water delivered to a surface and adapt to surface topography, however removal of the gel, clearance and potential for residues are a concern. Fortunately, as so many parameters can be adjusted when working with gels, we were able to find what we were looking for: mouldable gels, i.e. gels that once set will not flow owing to their high viscosity (Mérel 2012). Although principally composed of liquid water, those behave as a solid because the polymer chains of the gel are closely intertwine. The main advantage of these gels is that they can be applied before gelation therefore they will match the topography of any given surface before they set. The high viscosity and other chemical properties of the gel slows diffusion of water into the substrate and maintains the cohesive properties of the gel, allowing it to be peeled off the surface without leaving significant residues (Anzani *et al.* 2008: 53; Carretti *et al.* 2010) and in all cases without harming the surface (Ianuccelli and Sotgiu 2010; Carretti *et al.* 2010).

Natural hydrocolloids such as gellan gum, agar and agarose (Phillips and Williams 2009: 16), in order of increasing viscosity, may be described as rigid and mouldable gels. Poly(vinyl alcohol) (PVAl)-borax gels may also reach the desired viscosity and can subsequently be peeled off (Angelova 2013: 15–36, 215). Therefore, in a first set of tests, gellan gum, agarose (which are polysaccharides), and PVAl-borax gels were evaluated in terms of their ability to conform to surface topography, efficacy of cleaning and ability to minimize residues. Empirical tests were carried out, firstly on mock-ups made of acrylic paint because it dries faster than oil paint. The polysaccharide gels were studied after being brushed onto the surface before gelation (at a temperature of ~ 35 °C) to allow the gel to adapt to the topography of the surface whilst still fluid. Indeed, it was noted in other studies that the cleaning action of the gel was more effective when the gel was brushed on before gelation rather than after (Anzani *et al.* 2008; Volk and van den Berg 2014). Solutions of agarose at 3% w/w and of PVAl at 4% w/w crosslinked with a 3% w/w borax solution in water were selected.

In a second phase, roughness measurements at the microscopic scale were used to evaluate the action of the two gels on oil paint mock-ups with impasto, compared to the action of a moistened cotton swab. Such measurements have been used previously to assess their cleaning potential on painted surfaces (Dorge 2004; Eipper and Frankowski 2004; Eipper 2010; Stein and Haugen 2010; Faust 2015). We used two different microscopy-based techniques.[1] The Stil CHR 150N micro-topographic station is a confocal microscope with a

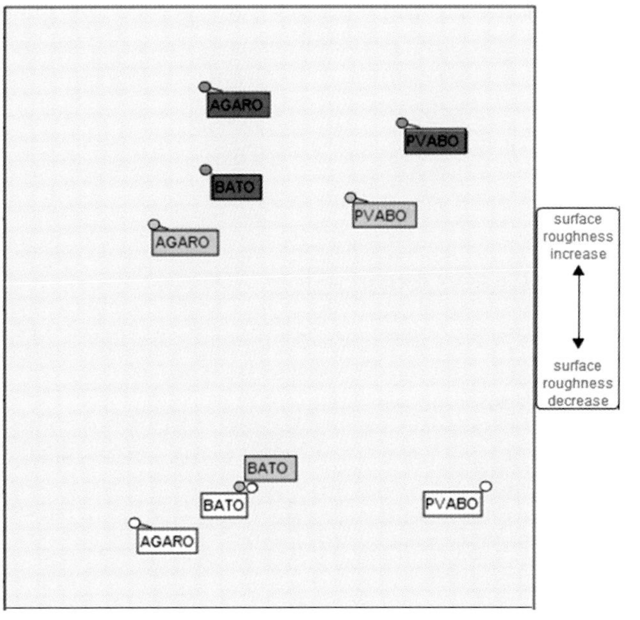

FIG. 2 Changes in roughness, before soiling (white), after soiling (dark grey) and after cleaning (light grey). AGARO = mock-ups cleaned with agarose, PVABO with PVAl-borax, and BATO with a moistened cotton swab.

dispersion-based system with chromatic coding which provides a colour topographical map of the surface. We used a micro-scale sensor working in the 0–300 µm range, which is the most suitable for determining the impact of the gels on surface roughness. This accurate technique was coupled with a Hirox KH-8700 high quality digital microscope. In its 3D mode, the microscope creates a digitally enhanced 3D image at a given magnification.

The surface roughness of oil paint impasto mock-ups was measured several times with both instruments and observed before and after soiling with artificial dirt (comprising pollution particles from Parisian windows mixed in water with soap, and brushed onto the surface), and subsequently after cleaning. The results were averaged and analysed in terms of roughness change. Principal component analysis (PCA) was used for data reduction, description and visualization. This well-established method indicates which groups of samples have the same behaviour (Fig. 2). The PCA results confirmed and refined the initial conclusions drawn from comparisons of roughness data and visual assessment.

Generally, soiling resulted in an increase in surface roughness, while cleaning resulted in a decrease in surface roughness. Based on changes in surface roughness measurements, the agarose gel appears to have a better cleaning action than the PVAl-borax gel, even though the changes were not perceptible to the naked eye. The mock-ups cleaned with the moistened cotton swab had the closest roughness value before soiling and after cleaning (Fig. 2). However, Hirox images showed a surface that was even dirtier after cleaning due to the dissolution and transportation of dirt by the cotton swab.

This study demonstrated the benefit of coupling roughness measurements and microscopic surface examination for evaluating cleaning of impasted paint using mouldable gels.

Unfortunately, it revealed that the hydrogels tested were not sufficiently effective at cleaning the impasted paint, which was confirmed by tests conducted on the artwork. The impasted areas were finally cleaned with a Velvesil Plus silicone gel carrying 5–10% water.

Materials and suppliers

- Agarose low EEO (Agarose Standard) A2114,0100: Applichem Panreac, https://www.applichem.com/home/
- Kelcogel CG-LA gellan gum: GMW, http://gmw-shop.de/
- Poly(vinyl alcohol) partially hydrolysed at 85%, MW ~ 160 000), no. CAS 25213-24-5: Merck, http://www.merck.fr/fr/index.html
- Sodium tetraborate decahydrate no. CAS 1303-96-4: Acros Organics, http://www.acros.com/

Note

1. Both available at the Centre de Recherche et de Restauration des Musées de France (C2RMF), Paris, France. Special thanks to Nicolas Mélard, archaeologist and curator, C2RMF.

References

Angelova, L.V. 2013. *Gels from Borate-crosslinked Partially Hydrolyzed Poly(vinyl acetate)s: Characterization of Physical and Chemical Properties and Applications in Art Conservation*, unpublished PhD dissertation, Georgetown University, Washington, DC.

Anzani, M., Berzioli, M, Cagna, M., Campani, E., Casoli, A., Cremonesi, P., Fratelli, M., Rabbolini, D. and Riggiardi A. 2008. *Gel rigidi di Agar per il trattamento di pulitura di manufatti in gesso/Use of Rigid Agar Gels for Cleaning Plaster Objects*, *Quaderni del Cesmar7*, no. 6. Padua: Il Prato.

Carretti, E., Natali, I., Matarrese, C., Bracco, P., Weiss, R.G., Baglioni, P., Salvini, A. and Dei, L. 2010. 'A new family of high viscosity polymeric dispersions for cleaning easel paintings', *Journal of Cultural Heritage* 11: 373–80.

Dorge, V. (ed.) 2004. *Solvent Gels for the Cleaning of Works of Art: The Residue Question*. Los Angeles: Getty Conservation Institute.

Eipper, P.-B. 2010. 'Untersuchung Unbehandelter Und Wässrig Behandelter Ölfarbenoberflächen', *Restauro* 2: 98–103.

Eipper, P.-B., and Frankowski G. 2004. 'Examination of cleaned oil paint surfaces by 3D-measurement technology', *The Picture Restorer* 2: 5–13.

Faust, E. 2015. *Entre Modernité et Tradition: étude et Conservation-Restauration de Secteur d'Argenteuil, Peint Par Jürg Kreienbühl En 1965 (Fonds Régional D'art Contemporain d'Ile-de-France). Étude Des Dégradations D'une Dispersion Vinylique à Plastification Externe Sur Panneau Dur et Comparaison de Différentes Méthodes de Nettoyage Aqueux*, unpublished thesis, Institut national du patrimoine, Paris.

Graczyk, A. 2016. *A pleins tubes! Le tube de peinture utilisé comme outil de mise en œuvre dans les années 1910 pour La Créole au perroquet par Alexis Mérodack-Jeaneau (musée des Beaux-Arts, Angers). Etude et Conservation-Restauration d'une œuvre témoin des matériaux commerciaux pour artistes du début du XXème siècle et de son châssis rudimentaire. Évaluation sous microtopographie de l'innocuité et de l'efficacité du décrassage par des gels rigides pelables d'une couche picturale à l'huile empâtée et non vernie*, unpublished thesis, Institut national du patrimoine, Paris.

Ianuccelli, S. and Sotgiu, S. 2010. 'A new methodology for wet conservation treatments of graphic art on paper with a rigid polysaccharide gel of gellam gum', *The Book and Paper Group Annual* 29: 25–39.

Mérel, A. 2012. *Le nettoyage des taches grasses sur support papier à l'aide d'enzymes en gel rigide*, unpublished thesis, Ecole Supérieure Nationale des Arts Visuels de La Cambre, Brussels.

Phillips, P.A., and Williams, O. (eds) 2009. *Handbook of Hydrocolloids*. Oxford: CRC Press and Woodhead Publishing.

Stein, M. and Haugen, A. 2010. 'Topographic registration of surfaces on canvas paintings: investigation with a mobile non-contact profilometer', *Restauro* 6: 396–401.

Volk, A. and van den Berg, K.J. 2014. 'Modern oil paints: formulations, organic additives and degradations: some case studies', in K.J. van den Berg, A. Burnstock, M. de Keijzer, J. Krueger, T. Learner, A. de Tagle and G. Heydenreich (eds), *Issues in Contemporary Oil Paint*. Cham: Springer, 311–32.

Authors' addresses

- Agata Graczyk, Paris, France (agata.graczyk.fr@gmail.com)
- Pauline Hélou-de La Grandière, Nîmes, France (lagrandiere@gmail.com)
- Sigrid Mirabaud, Institut national du patrimoine, Paris, France (sigrid.mirabaud@inp.fr)
- Marine Page, Centre de Recherche et de Restauration des Musées de France (C2RMF), Paris, France (marine.page@culture.gouv.fr)

Comparison of three hydrogels for cleaning tarnished silver threads using electrochemical treatment

Aline Létrange, Dominique Hourdet, Julie Guerrier and Emmanuelle Pons

Introduction

Tarnishing of silver or gilded silver threads is a recurring problem in textile conservation (Timár-Balázsy and Eastop 1998). Silver develops corrosion when exposed to atmospheres contaminated with sulphur or chlorinated compounds, or when handled with bare hands. The corroding metal surface can change from yellow to black depending on the thickness of the corrosion layer. In most cases silver sulphide (Ag_2S) is the principal compound. The dullness of the metal changes the perception of the object and causes an alteration in its aesthetic value. Depending on the object, its condition and its cultural value, cleaning of the metal threads may be considered. However mechanical and chemical cleaning cannot be performed on metal threads because of their delicacy and their proximity to textile or other organic material. Electrochemical cleaning, performed locally and with controlled humidification, might be a solution. We focused on localized electrochemical cleaning using hydrogels (Létrange 2016), a method that provides electrolytic reduction at the surface of the metal (Brett and Oliveira Brett 2002) and enables silver ions to be reduced to shiny metallic silver (Costa 2001).

Recently, various strategies have been proposed to localize electrolytic treatment: the Pleco[1] process developed by Degrigny (2014) and a technique specially developed by Guerrier (2014) for textile conservation (Fig. 1). In both cases the electrolytic solution is retained in a hydrophilic material in order to control moisture and to operate locally. The first technique uses a polyvinyl formaldehyde (PVFM) sponge and the second an agarose gel (2% w/w) (Guerrier and Pons 2015). As the agarose gel is rather rigid and not optimal for working on textured surfaces such as embroidery, the aim of this study was to extend the 'gelling' to other polymeric networks in order to increase surface contact between the swollen gel and textured surfaces.

Materials

The main goal was to develop softer gels which are compatible with the electrolytic solution and facilitate the electrolytic process, whilst limiting the release of water. Three hydrogels were selected:

- Iota-carrageenan at 2.5, 3 and 3.5% w/w, polysaccharide gel.
- Fully hydrolysed poly(vinyl alcohol) cryogels at 3.5 and 8% w/w: the polymer crystallizes in aqueous solution after 5 freeze/thaw cycles.
- Agarose at 2% w/w, with polysaccharide gel used as a reference.

The three polymers used in this study are known to form gels in aqueous solution at room temperature. These macromolecular networks, physically crosslinked with hydrogen

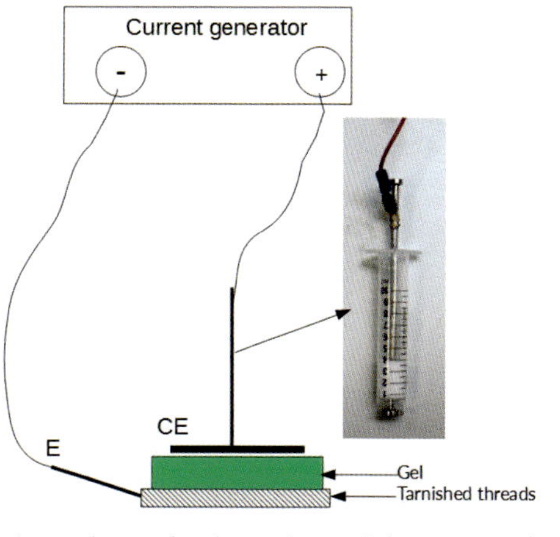

FIG. 1 Experimental setup for electrochemical cleaning. E is the working electrode, an inert needle in contact with the object, and CE is the inert counter-electrode. The inset image shows the tool made by Guerrier (2014) for the latter.

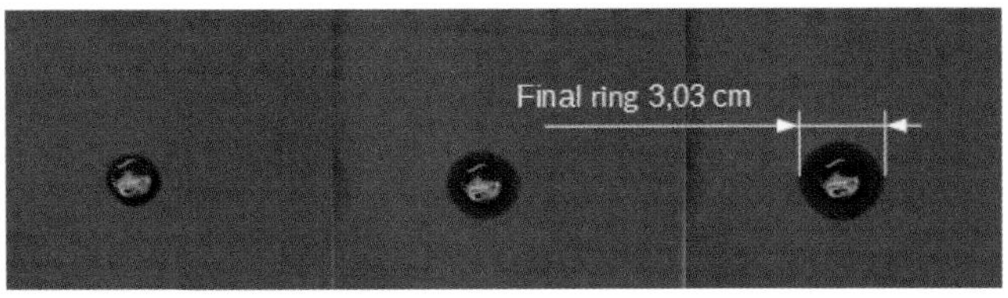

FIG. 2 Diffusion test.

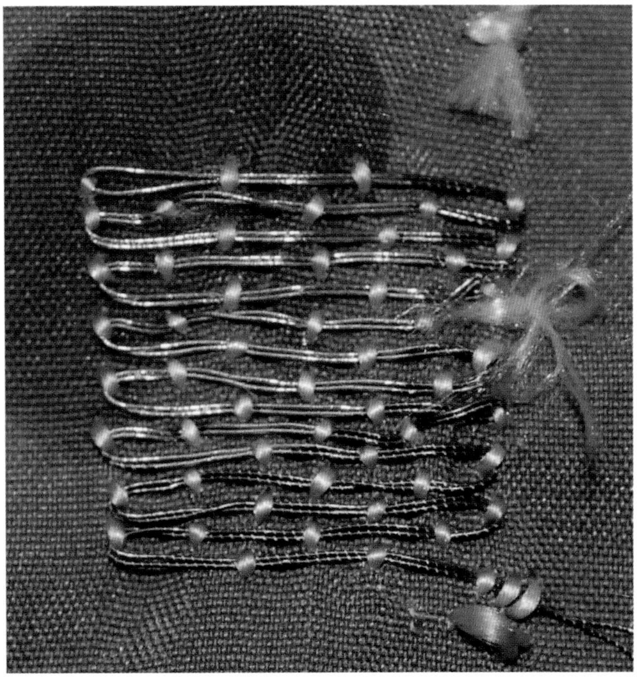

FIG. 3 Sample after partial cleaning.

bonds (Nijenhuis 1997), are reversible and can be melted at high temperature. Consequently, homogeneous solutions were initially prepared by dissolving the polymer in the electrolyte solution (sodium nitrate 0.1 mol/L in demineralized water) at a high temperature (75 °C).The hot solutions were then transferred into a syringe (for ease of delivery later) before cooling to room temperature, when gelation takes place. We carried out tests on artificially tarnished silver threads whose corrosion film was composed of silver sulphide.

Method

In order to compare and select the best hydrogel for local cleaning of textured tarnished silver embroideries, the stiffness of the gels was compared via their Young's modulus as determined by compression tests.[2] The ability of gels to retain water was evaluated by a'diffusion test' (Fig. 2): a disc of gel was placed on a piece of cloth and the ring of water that formed close to the contact area was measured at specific time intervals. The efficiency of electrochemical cleaning was evaluated by scanning electron microscopy (SEM) and characterized by energy dispersive X-ray (EDX) spectroscopy.[3]

Results

The stiffest gel is agarose prepared at 2% (Young's modulus E @ 400 kPa); the six other networks were two orders of magnitude less stiff. Poly(vinyl alcohol) cryogels and iota-carrageenan gels were therefore good candidates for this application. After 5 minutes of contact, the ring left by the agarose gel disk (2%) on the cloth was the largest, meaning that the other gels retain water more efficiently. The iota-carrageenan gel and poly(vinyl alcohol) cryogel at their lowest concentrations were discarded because they were too soft and were not very efficient in terms of water retention. Furthermore, the disc of poly(vinyl alcohol) cryogel left a stiff ring on the cloth, indicating that this gel deposited polymer residues. This would not be acceptable for a real object, so poly(vinyl alcohol) cryogel was eliminated. Consequently, we selected iota-carrageenan 3.5% and agarose 2% for further trials.

A sample of cloth embroidered with tarnished silver threads was studied before and after electrochemical cleaning. With the naked eye it could be seen that the metal became shiny (Fig. 3) regardless of the gel used. Nevertheless, from SEM images it appears that the cleaning was more efficient and homogeneous with agarose at 2%. This gel moistens more of the cloth, so the electrical contact is better and the cleaning more efficient. Similar results can be obtained with iota-carrageenan 3.5% if the treatment time is lengthened. EDX highlights the efficiency of the cleaning method: the level of sulphur was appreciably smaller after cleaning. EDX also indicated micro-residues of sodium on the cloth around the embroidery with silver threads. This is not surprising when sodium nitrate is the electrolyte, but it indicates that the treatment should be applied to a very limited area, close to the metal thread.

Conclusion

This study increases the possibilities for cleaning tarnished silver threads with non-immersive electrochemical treatment. Two gels have been found to be suitable for cleaning, depending on the object to be treated: agarose for smooth surfaces and iota-carrageenan for rougher surfaces.

Acknowledgements

We would like to thank Anne Genachte-Le Bail and Cécile Argenton of the Institut National de Patrimoine (INP), Paris, and Victor Hundsbuckler of Monnaie de Paris (the Paris Mint).

Notes

1. Potentiostatic cleaning method developed by Christian Degrigny and the Fablab, Neuchâtel, Switzerland.

2. Carried out at École Supérieure de Physique et de Chimie Industrielles (ESPCI), Paris, with the help of Dominique Hourdet.
3. SEM–EDX was carried out by Michel Mahé at Electricité de France, Paris (EDF) Research and Development laboratory, with the help of Emmanuelle Pons.

References

Brett, M.A. and Oliveira Brett, A.M. 2002. *Electrochemistry: Principles, Methods and Applications*. Oxford: Oxford University Press.

Costa, V. 2001. 'The deterioration of silver alloys and some aspects of their conservation', *Reviews in Conservation* 2: 18–34.

Degrigny, C. 2014. *Projet Saint-Maurice. Mise au point de méthodes de nettoyages électrolytiques innovantes pour la restauration de chefs-d'oeuvre d'orfèvrerie composites*, unpublished report, Haute École Arc.

Guerrier, J. 2014. *Étude et conservation-restauration de chaussures des 17è et 18è siècles (musée international de la chaussure, Romans). Le nettoyage des filés d'argent ternis brodés sur soie. Comparaison du gel d'agarose et de l'éponge PVA comme support poreux en vue d'une électrolyse locale sans immersion*, unpublished dissertation, Institut National du Patrimoine, Paris.

Guerrier, J. and Pons, E. 2015. 'Nettoyage des filés d'argent ternis dans le domaine textile: comparaison du gel d'agarose et de l'éponge dite PVA comme support poreux pour une électrolyse locale', *Technè* 42: 109–15.

Létrange, A. 2016. *Étude de trois bourses à jetons du XVIIIème siècle de la Monnaie de Paris. Conservation-restauration de la bourse verte. Comparaison de trois hydrogels pour le nettoyage des filés d'argent ternis par électrolyse locale*, unpublished dissertation, Institut National du Patrimoine, Paris.

Nijenhuis, K. te 1997. *Thermoreversible Networks: Viscoelastic Properties and Structure of Gels*, Advances in Polymer Science 130. Berlin and Heidelberg: Springer.

Timár-Balázsy, A. and Eastop, D. 1998. *Chemical Principles of Textile Conservation*. Oxford: Butterworth Heinemann.

Authors' addresses

- Aline Létrange, Brive-la-Gaillarde, France (aline.letrange@gmail.com)
- Dominique Hourdet, University Pierre et Marie Curie, Paris, France
- Julie Guerrier, Paris, France
- Emmanuelle Pons, EDF Research and Development Laboratory, Valectra Laboratory, Écuelles, France

The removal of dirt and coatings from heavily restored medieval painted ceramic *socarrats* from València, Spain, using gelled systems

Silvia Marín Ortega and Aleix Barberà Giné

Introduction

The treatment of seven *socarrats* (fired clay plates) in the collection of the Barcelona Design Museum (formerly the Barcelona Ceramics Museum), Barcelona, Spain, is described. Of unknown provenance, they may have been acquired by antique dealers in the early 20th century. Due to the diverse history of each *socarrat*, they had very heterogeneous layers, with dirt and damaged varnishes superimposed and concealing the original colouring. All the non-original layers were undocumented and difficult to characterize.

Socarrats are ceramic plates used to decorate the spaces between wooden beams on the medieval ceilings of expensive houses. Their use spread throughout the region of Spain

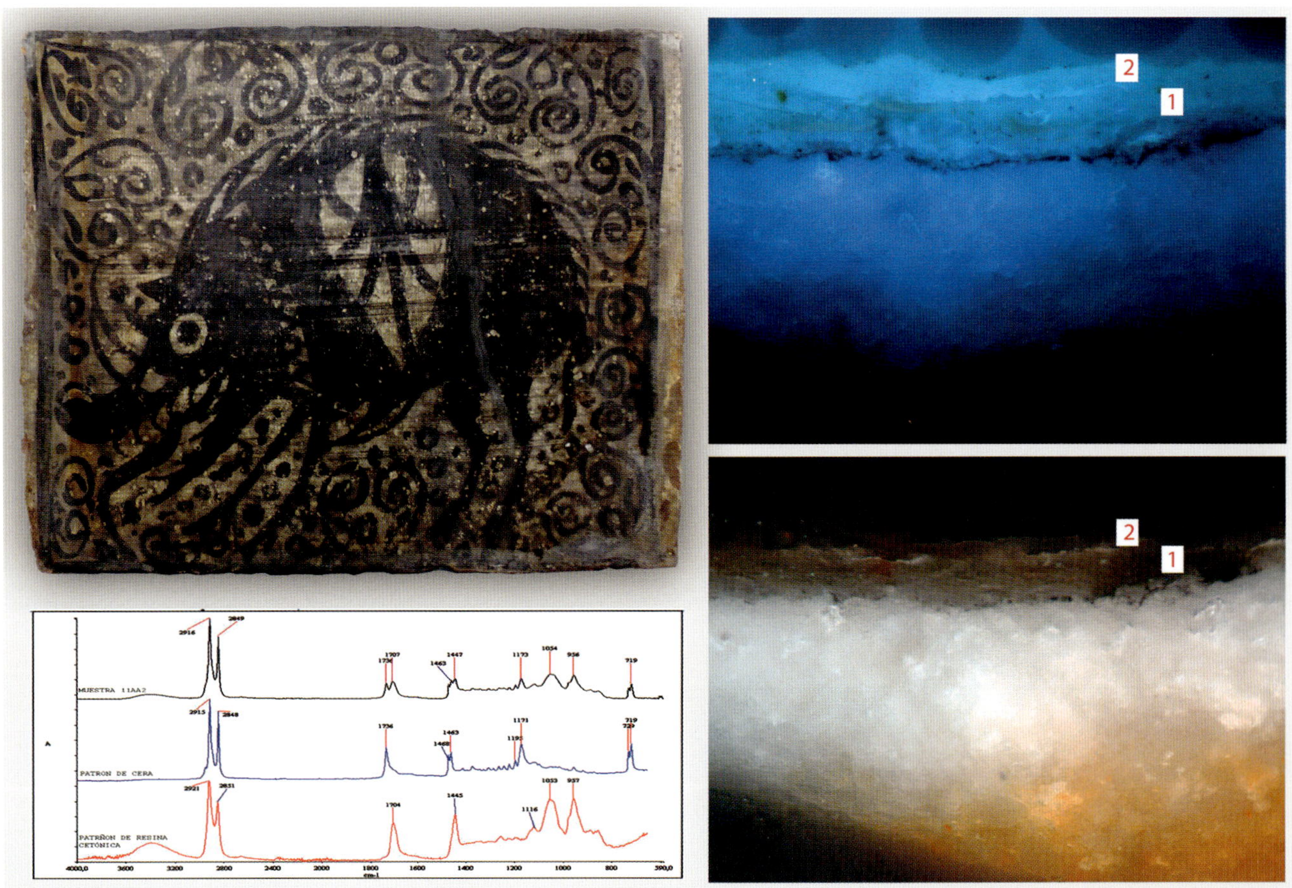

FIG. 1 *Socarrat* MCB 5535 before treatment (*upper left*). FTIR (*lower left*) indicated two layers: an irregular microcrystalline wax layer (1) and a thicker ketone resin layer (2). Cross-section imaged at ×20 in UV (*upper right*) and visible light (*lower right*) (images: Silvia Marín Ortega, Aleix Barberà Giné and Ledesma 2016).

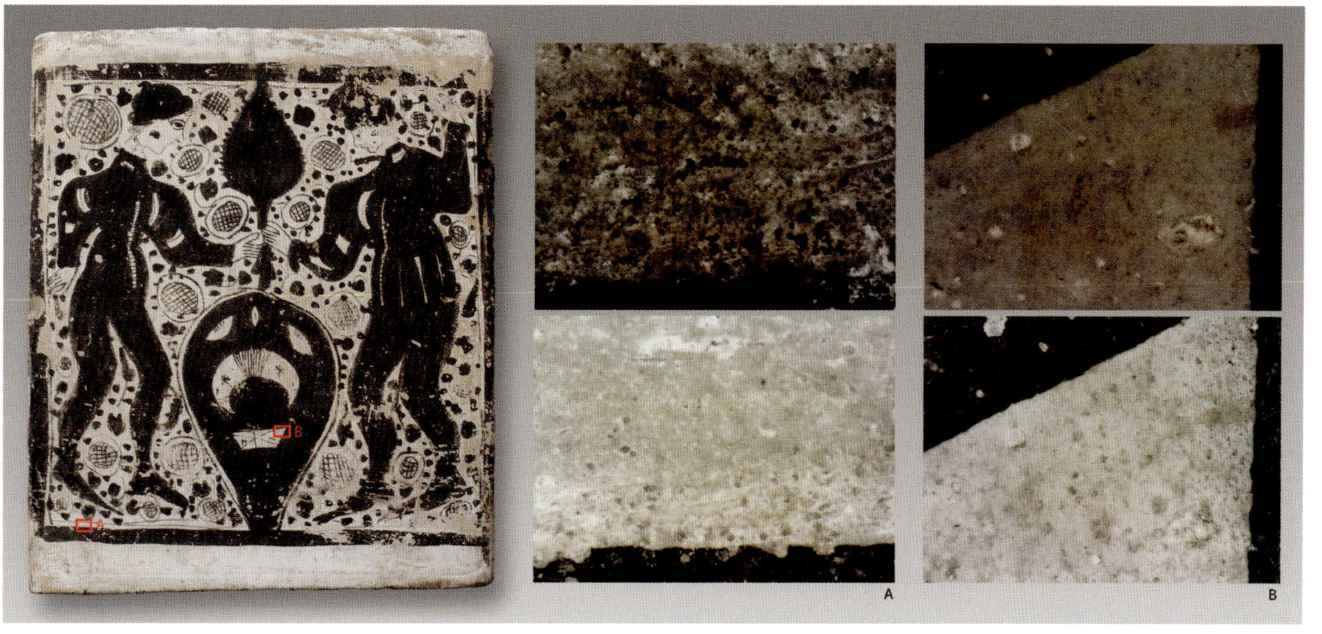

FIG. 2 *Socarrat* MCB 20120 after treatment (*left*), (*middle and right*) detail (×50) images with digital microscope and diffuse light comparing the same area before (*above*) and after the cleaning process (*below*) (images: Silvia Marín Ortega, Aleix Barberà Giné and Guillem Fernández Huerta, Museu del Disseny).

known as the Crown of Aragon (whose power was greatest between the 14th and 16th centuries), especially in the area of Valencia, where Paterna was the main production centre (Mesquida 2001). The ceramics were created individually using a mould. Once fired, they were covered with a thin layer of lime on which painted motifs were applied. The resulting plates have a white background with red, ochre and black decorations representing abstract, geometric and figurative motifs (Pinho 2007).

Case study

The seven examples under discussion, despite having a very similar structure, possess very different surface layers. *Socarrats* were very popular in the Spanish antique trade, due to their lightweight construction and historical and aesthetic interest, which has resulted in them undergoing both orthodox and unorthodox cleaning and repair processes aimed at embellishing and aging them. This is why each object required individual treatment, which began by characterization of the surface layers, sampling and then preparing cross-sections for observation with optical microscopy (using polarized, incident and transmitted light, visible and ultraviolet (UV) light) and Fourier transform infrared (FTIR) spectroscopy (Ledesma 2016). It was found that the surface layers did not homogeneously cover the surface of the objects, but had often been partially cleaned in the past or alternated with overpaint and dirt. This hampered the characterization of the layers, as well as their subsequent removal.

Despite this drawback, we were able to characterize several surface layers, two of which were present in most of the collection: a ketone varnish and a microcrystalline wax mixed with silicates and carbon black, possibly intended to create the appearance of an aged surface. We also detected the possible use of bitumen of Judea, and traces of alkyd resin. Finally, there was an extensive layer of well-absorbed and firmly attached soot and dirt on top of the paint layer (Fig. 1). These layers proved difficult to remove, not only due to the lack of reversibility usual for resins or waxes applied to porous substrates, but also because of the poor cohesion of the paint layer. Removing the multiple surface layers without affecting the original layers was a major challenge. It was important not to use free solvents to avoid any cleaning agents penetrating into the porous substrate.

Work began by determining the pH and surface conductivity of the *socarrats*, using 3 wt% agar gel applied for 15 minutes on several areas. Solvent tests were then performed to determine the polarity range of the wax and resin overlays (Masschelein-Kleiner 2004; Baglioni and Chelazzi 2013). Although the ketone resin still retained good solubility in a range of solvents, the layer of microcrystalline wax had become very insoluble in common solvents. The difficulty of solubilizing the wax was heightened by the incorporation of inorganic particles such as silicates and carbon black.

We evaluated the use of Carbopol solvent surfactant gels, agar gelled aqueous systems (Wolbers 2000) and microemulsions (Baglioni *et al.* 2012) as cleaning systems in order to: avoid penetration of the dissolved surface layers into the support base; avoid the presence of residues from the products used (Dorge 2004); and minimize the use of solvents. The goal was to work with safer systems for the conservators, the environment and the artwork itself. While the solvent surfactant gels were successful in removing the ketone resin, they were unable to remove the wax. It was decided to test systems which combine the action of water and solvents. A 0.5 wt% boric acid and 0.5 wt% ethylenediaminetetraacetic acid (EDTA) buffer at

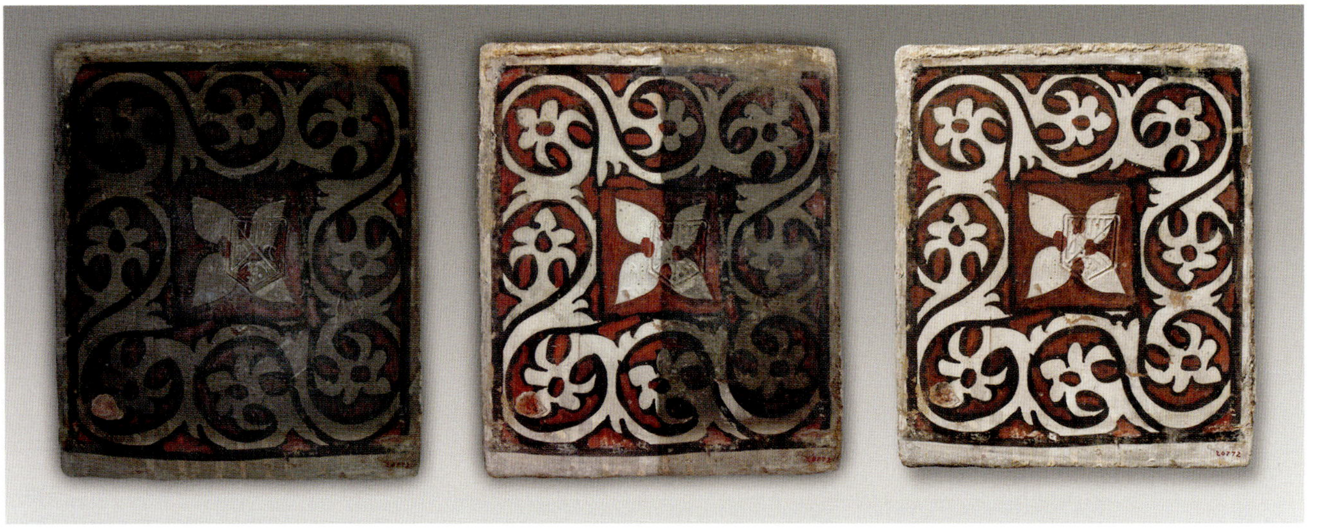

FIG. 3 *Socarrat* MCB 20772 before treatment (*left*), during cleaning (*middle*), and after treatment (*right*) (images: Silvia Marín Ortega, Aleix Barberà Giné and Guillem Fernández Huerta, Museu del Disseny).

pH 8.5 was used. The EDTA chelator aided the removal of the inorganic particles combined with the wax, while working at a safe pH for the surface of the *socarrats*, whose decoration is applied in acid-sensitive lime-based paint.

Various microemulsions (Baglioni *et al.* 2012, 2015; Giorgi *et al.* 2009; Carretti *et al.* 2003) were also tested, including XYL (86.2 wt% Trizma buffer at pH 8.5; 2.6 wt% xylene; 4.1 wt% sodium dodecyl sulphate and 7.9 wt% pentanol), which proved highly effective in removing wax. However, this was dismissed on the grounds that surfactant residues might be left on the surface. Whether they were applied with 2 wt% Klucel gel or absorbed into cellulose pulp poultices, there was no way of minimizing the soapy effect observed during the rinsing step. Further analysis detected residues of sodium dodecyl sulphate (SDS) in the tested areas, so the system was discarded (Ledesma 2017).

Finally, taking advantage of the relatively low glass transition temperature (T_g) of waxes, it was decided to incorporate a new factor to improve cleaning efficacy: temperature. The buffer and chelator solution was gelled with 3 wt% agar, and the mixture applied to the surface as a warm solution. It was necessary to control the application temperature with a thermometer and maintain a constant temperature of 45 °C in a bain-marie. The results obtained, with an application time of around 10 minutes, were entirely satisfactory for the removal of the wax with a completely aqueous system, without the need to add any type of solvent. The combination of the chelator and the temperature provided by the agar solution, in the specific case of this aged wax incorporating inorganic particles, proved optimal (Figs 2 and 3). Furthermore, no residue was detected (Ledesma 2017).

Conclusion

We tested two systems capable of removing extremely stubborn layers of wax, which had become insoluble due to their age and their incorporation of inorganic particles. It was found that a buffered aqueous system applied at an elevated temperature was the best option. Once the agar-buffer solution cooled on the surface and became a gel, we were able to work stratigraphically, and with maximum safety for both the objects and the conservators. By avoiding the use of free solvents we reduced and even prevented possible penetration of the redissolved layers into the porous substrate. Finally, a cleaning system based on prior analysis of the surface and the use of buffered solutions with controlled pH and conductivity in a range that was safe for lime-based painting was found to be a particularly appropriate cleaning system for *socarrats*. Treatment could not be accomplished using standard methods but required a search for a unique solution based on the condition and composition of the objects.

Acknowledgements

The authors would like to thank the Museu del Disseny de Barcelona and the Centre de Restauració de Béns Mobles de Catalunya, Barcelona, Spain.

References

Baglioni, P. and Chelazzi, D. 2013. *Nanoscience for the Conservation of Works of Art.* Cambridge: RSC Publishing.

Baglioni, M., Giorgi, R., Berti, D. and Baglioni, P. 2012. 'Smart cleaning of cultural heritage: a new challenge for soft nanoscience', *Nanoscale* 4(1): 42–53.

Baglioni, P., Chelazzi, D. and Giorgi, R. 2015. *Nanotechnologies in the Conservation of Cultural Heritage: A Compendium of Materials and Techniques.* New York and London: Springer.

Carretti, E., Dei, L. and Baglioni, P. 2003. 'Solubilization of acrylic and vinyl polymers in nanocontainer solutions: application of microemulsions and micelles to cultural heritage conservation', *Langmuir* 19: 7867–72.

Dorge, V. (ed.) 2004. *Solvent Gels for Cleaning Works of Art: The Residue Question*. Los Angeles: Getty Conservation Institute.
Giorgi, R., Baglioni, M., Berti, D. and Baglioni, P. 2009. 'New methodologies for the conservation of cultural heritage: micellar solutions, microemulsions, and hydroxide nanoparticles', *Accounts of Chemical Research* 43(6): 695–704.
Ledesma, A. 2016. *Estudio de los materiales presentes en dos micromuestras de recubrimiento tomadas de dos piezas cerámicas: MDU 20778 y MDU 5535*. Analytical report. Madrid: Arte-Lab SL.
Ledesma, A. 2017. *Estudio de los materiales presentes en dos micromuestras en las que puede haber residuos de los sistemas de limpieza. Caracterización de los diferentes productos empleados en la limpieza (potencialmente generadores de residuos)*. Analytical report. Madrid: Arte-Lab SL.
Masschelein- Kleiner, L. 2004. *Los solventes*. Santiago de Chile: Publicaciones Centro Nacional de Conservación y Restauración DIBAM.
Mesquida, M. 2001. *Socarrats y pavimentos medievales*. Paterna: Publicaciones Ayuntamiento de Paterna Concejalía de Cultura.
Pinho, L.M. 2007. *Los socarrats valencianos. Estudios de caracterización y técnicas de limpieza*. Lisbon: University of Lisbon.
Wolbers, R. 2000. *Cleaning Painted Surfaces: Aqueous Methods*. London: Archetype Publications.

Authors' addresses

- Silvia Marín Ortega, Escola Superior de Conservació i Restauració de Béns Culturals de Catalunya (ESCRBCC), Barcelona, Spain (marinortegasilvia@gmail.com)
- Aleix Barberà Giné, Centre de Restauració de Béns Mobles de Catalunya (CRBMC), Barcelona, Spain (aleixbg@gmail.com)

Paintings on canvas in the rooms of St Aloysius Gonzaga at the Roman College, Rome

Susanna Sarmati

Introduction

The Church of St Ignatius in Rome is located in a square with the same name and was built in 1626 in honour of St Ignatius of Loyola, the founder of the Society of Jesuits. The church was built near the Roman College in place of the 16th-century Church of St Maria Annunziata. The Roman College complex was inaugurated on 28 October 1584. It was conceived by its founder, Ignatius of Loyola, following the model of the Collège du Roi in Paris of 1530, and was intended for the theological and cultural formation of Jesuits and young members of the most important families of the time.

The 'retiring' rooms, the so-called 'rooms of St Aloysius Gonzaga', are on the top floor of the Roman College but are accessed through the Church of St Ignatius located directly behind it. These are the rooms referred to as having been 'inhabited by many young Jesuit students over the course of almost two centuries' (Croce 2011). There are seven rooms of varying sizes, of which only four are actually connected to St Aloysius Gonzaga himself, who lived there from 1587 until his death. At the top of the short staircase that leads to the rooms of St Aloysius Gonzaga is the Ricreazione room (Fig. 1). The large rectangular room measures 14 × 12 m and is completely decorated. The walls and ceiling were painted

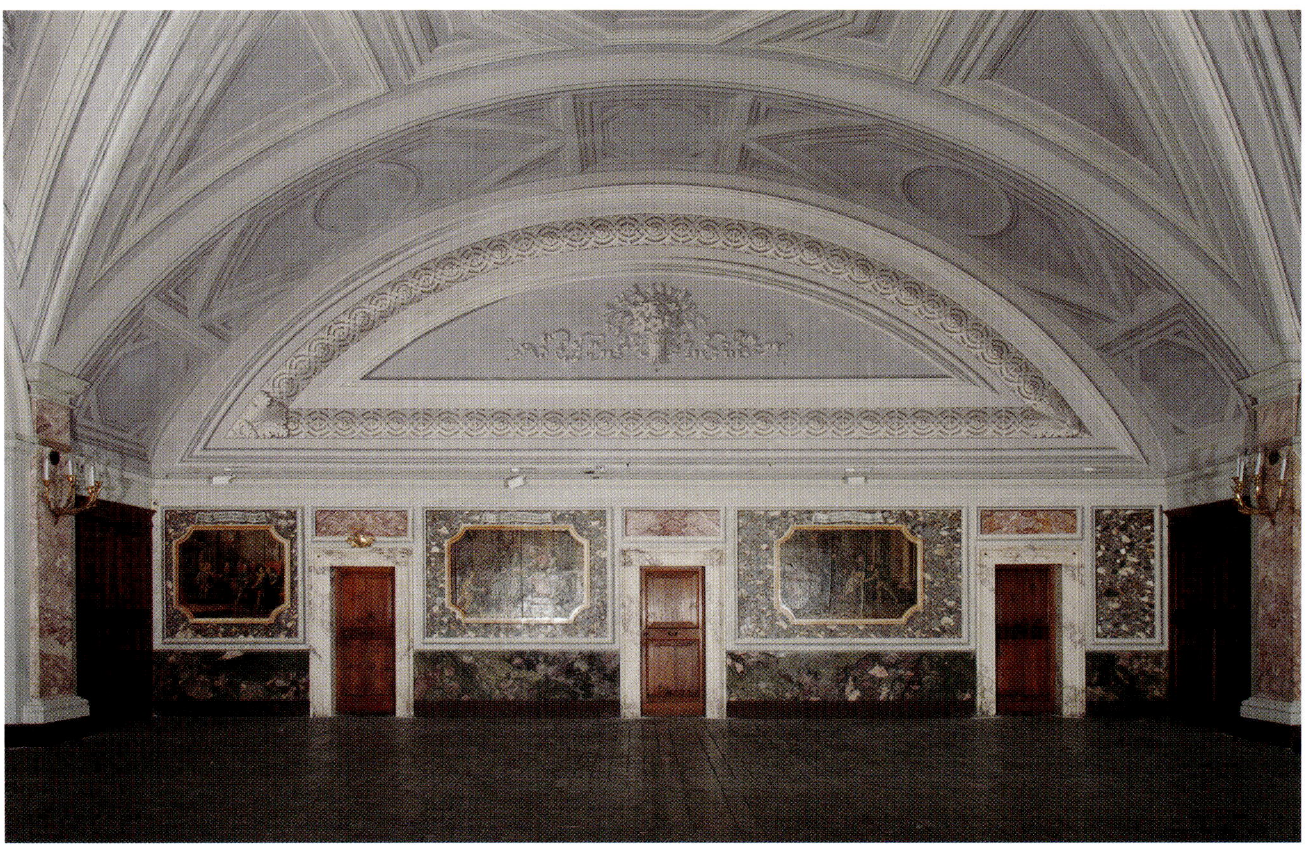

FIG. 1 The Ricreazione room (image: Pasquale Rizzi).

in 1719, as noted by an inscription at the entrance, at the expense of Carlo D'Aquino. There are lime-plaster decorations simulating frames and stucco reliefs on the ceiling, and a *trompe-l'oeil* decoration with faux-coloured marbles on the walls: Red African, Aleppo Breccia, Pavonazzetto, and antique yellow. During that same period (the early 1700s), 14 paintings depicting scenes from the life of St Aloysius Gonzaga were created on canvases glued to the walls.

The room has been extensively altered by deterioration caused by the passage of time and by a large fire in the Cappella dei Voti in 1849 that damaged all surrounding rooms. The restoration of the space – carried out at the end of the 19th century and documented by a painted plaque at the entrance to the Cappella dei Voti – completely distorted the room's original decoration. The original decoration on the vault is unknown as it was repainted in 1861 by Giovan Battista Caretti (Del Moro 2003). The paintings are in lime plaster with light colours, in full neoclassical style with faux plaster decorations.

The ceiling is supported by beams covered by a pitched roof. Ingress of rainwater from the roof and consequent cycles of wetting and drying have led to dimensional variations in the wooden structure of the vault, and caused orthogonal cracking in the wooden beams. In some places these cracks are many centimetres wide and have led to extensive detachment of plaster from the load-bearing structure. Rainwater has percolated along these cracks into the room, causing colour loss, stains and dark deposits of varying nature.

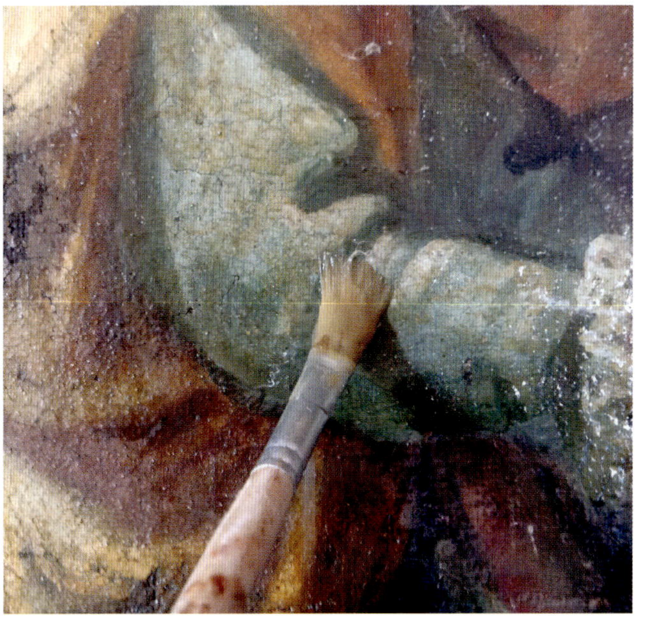

FIG. 2 The first stage: gel application.

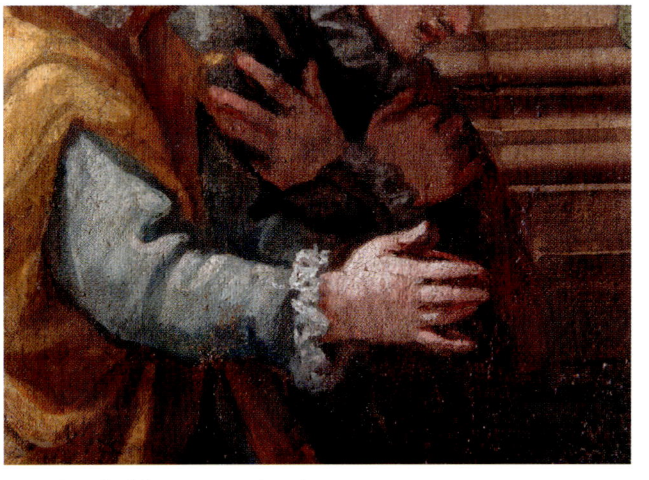

FIG. 3 Detail of the painting after cleaning.

The paintings on canvas

The paintings, oil on canvas, each depicting a scene from the life of St Aloysius Gonzaga, are placed on the four walls at a height of about 1 m from the floor, adorned on top with scrolls with painted writings in Latin that explain the subject of the scene depicted. There are three on the north wall, four on the south wall, four on the east wall and three more on the west wall. The different-sized paintings are rectangular with rounded corners with painted faux marble frames that partially cover the edges of the canvas.

Their support is an open-weave hemp canvas applied to the wall; only the outer edges of the canvases are glued to the plaster with animal glue. Once dry, these edges kept the canvases taut and slightly raised from the wall. The canvases all vary in size; the smaller ones are about 80 cm wide, which was the original width of a roll of hemp canvas. The larger ones, about 140–160 cm wide, comprise two or three pieces of canvas sewn together. The canvases all have the same white priming in gesso and glue, with a coarse grain size. The surface is not smooth, but slightly grainy. The application of this gesso primer is not the same on all canvases. In some cases, such as *Mary Magdalen in Ecstasy*, it was applied in a thick and compact layer to hide the canvas support, while in others, for example the 'kitchen scene', it is very thin and the weave of the canvas is readily seen.

Condition

In the mid-19th century, the paintings were in a poor condition – the ingress of rainwater from the ceiling and the room's state of neglect had led to the partial detachment of the canvases from the walls, and the formation of horizontal folds and tears as well as lifting of the paint. During the restoration work carried out around 1860, the paintings were hastily re-adhered to the walls and covered with animal glue. The gaps in the canvas were plastered over with a mixture of wax and plant resin. All the canvases were then extensively repainted to mask the gaps, the dark stains caused by water damage, and the minute but widespread craquelure. With time, the repainted layers on the surface had become very dull and dark due to oxidation of the binder, which completely obscured the original paint. The extremely light tone and rapid execution that is characteristic of these paintings had been completely concealed by this heavy-handed restoration, which had altered the textures. A more recent renovation took place in about 1960, but we have very little information on it.

Treatment

The cleaning took place in two stages: first, all the repainted layers on the pictorial surface had to be removed – only then could the animal glue residue on the surface be removed. The analyses of the surface showed the presence of acrylic-based binders, and after many tests, it was decided to use a gel made of acetone (100 ml), Ethomeen C-25 (25 ml), Carbopol (2 g) and deionized water (14 ml). The gel was applied with a brush to the vertical surface and left to act for 3 minutes (Fig. 2). It was then removed using a cotton swab together with the solubilized repainted layers. Residues in the treated area were always removed with a cotton swab and mineral spirits. The second stage of cleaning involved the removal of animal glue from the surfaces using Klucel G (4 g) and ethanol (100 ml). The gel residue was removed with ethanol (Fig. 3).

Conclusion

This two-stage cleaning permitted the removal of layered materials one at a time. The use of a viscous Carbopol gel allowed for the solubilization of the repainted layers without affecting the original pictorial surface, and the removal of the animal glue residue enabled the adhesion of the canvases to the walls to be verified. The cleaning process brought to light the original layers of paint with their figurative details made with quick, precise brushstrokes and pigments applied as layers of glazes. The preparatory sketch in red earth pigment, which is still clearly visible through the transparency colours, can now be seen. These data, together with archival research in progress, will make possible accurate comparisons to help us determine the provenance of the paintings.

References

Croce, A. 2011. *Le stanze di San Luigi*. Rome: Chiesa di Sant'Ignazio di Loyola, 14–25.

Del Moro, M. 2003. 'Vicende biografiche, nuovi documenti, inediti e notizie su Giovan Battista Caretti (1808–1878)', *Neoclassico* XXIII–XXIV: 120–38.

Author's address

Susanna Sarmati,Rome, Italy (mail@susannasarmati.com)

Appendix

Terminology and properties of selected gels

Richard Wolbers

Suppliers are given in brackets after commercial product names, as well as CAS numbers (the unique numerical identifiers assigned by the Chemical Abstracts Service to every chemical substance described in the open scientific literature) for specific compounds, where appropriate.

Agarose, agar (no CAS number, numerous suppliers)
In its native form, **agar** is one of the structural materials present in the cell wall of the seaweed species *Agarophytes algae*. Agar is comprised of two major carbohydrate fractions, now designated as **agaropectin** and agarose. These exist in the native cell wall material in roughly a 1:2 w/w ratio (although the exact proportion of these two polymers varies from species to species of seaweed).[1] Agaropectin in its 'native' form is a sulfated galactouronide (3%–10% sulfate by weight), but can also carry ester sulfate, D-glucuronic acid, and small amounts of pyruvic acid as well in the carbohydrate polymer. Acid groups can be found in salt form, with either Ca or Mg, or a combination of the two metal ions.

Agarose, on the other hand, is a 'neutral' carbohydrate comprised of β-1, 3-linked- D-galactose, α-1, 4-linked 3, 6-anhydro-L-galactose β -galactose and anhydrous β galactose.[2] Agarose is specifically the 'gelling' fraction of the crude mixture of the two carbohydrates. When it is purified from agar, agarose is theoretically 'pure' and a neutral material that has often been used as a medium for electrophoretic separations of bio-molecules like nucleic acids and proteins.

Because of the ionizable groups present in native agar (from agaropectin), in aqueous solutions it can carry a net negative charge above its lowest acid functional group (pK_a of 3.56 for the sulfonic acid groups).[3] Its solution stability depends upon its ionization: pHs below about 6.0 can result in precipitation (flocculation) of the agar. Prolonged exposure to high temperatures can degrade solutions of agar, resulting in a net lower gel strength after temperature decrease and gel formation. This effect is accelerated as pH decreases. To avoid decomposition of agar, high temperatures and pHs lower than 6.0 for prolonged periods of time should be avoided. As a natural biopolymer, agar is prone to biodeterioration and should be stored at cool temperatures (15–25 °C) and in dry conditions (<30%RH) to prevent contamination and biodeterioration.

Agarose has been isolated from agar by a number of methods.[4] These have included: gelation of agar, followed by freezing and aqueous extraction (agaropectin is soluble in water, at room temperature, agarose is not). Agaropectin is preferentially extracted from the crude gel once it has been frozen and ground or granulated. Agarose has also been prepared by a number of other specific solubilization methods, using selective precipitation of agaropectin with: quaternary ammonium salts (Hjerten 1962); ammonium sulfate (Azhitskii and Kobozev 1967); Rivanol (a tradename for ethacridine monolactate monohydrate) (Svridov *et al.* 1971); and Acrinol (again a trade name for ethacridine monolactate monohydrate) (Fuse and Goto 1971). Agaropectin has been directly extracted from agarose with solvents like polyethylene glycols (Russell *et al.*1964; Rolson 1965) and separated by various electrophoretic methods from agarose (Hjerten 1971). Agaropectin has been chemically modified (acetylated), and solvent extracted from agarose with chloroform (Araki 1937). These are just a few of the other methods that have been used to achieve separation of the agarose and agaropectin from agar.[5]

Agarose can exhibit a double helical structure. Double helices aggregate or 'interlace' to form a three-dimensional structure which holds the water molecules within it. Agarose can form what are often referred to as thermo-reversible gels. The gelling property of agarose is due to the three equatorial hydrogen atoms on the 3,6-anhydro-L-galactose residues, which constrain the molecule to form a helix. The interaction of the helices causes the formation of the gel.

Agar gels can be formed in very dilute solutions, containing as little as 0.5–1.0% of the biopolymer in solution.[6] These gels are rigid, brittle, have well-defined shapes, as well as sharp melting and gelling points. However, they are capable of **syneresis**, that is, they can spontaneously exude water through the surface of the gel under pressure. Water is held internally in the gel through a combination of capillary force as well as direct biopolymer solubility. They also exhibit a kind of 'temperature hysteresis' (meaning that there is a

Trade name	CTFA name	Manufacturer	Neutralization ratio base/ Carbopol polymer
NaOH (18%)	sodium hydroxide		2.3/1.0
ammonia (28%)	ammonium hydroxide		0.7/1.0
KOH (18%)	potassium hydroxide		2.7/1.0
L-arginine	arginine	Ajinomoto	4.5/1.0
AMP Ultra PC2000	aminomethyl propanol	Dow	0.9/1.0
Neutrol TE	tetrahydroxypropyl ethylenediamine	BASF	2.3/1.0
TEA (99%)	triethanolamine		1.5/1.0
Tris Amino (40%)*	tromethamine	Dow	3.3/1.0
Ethomeen C-25	PEG-15 cocamine	Akzo	6.2/1.0
diisopropanolamine	diisopropanolamine	Dow	1.2/1.0
triisopropanolamine	triisopropanolamine	Dow	1.5/1.0

* The 40% solution should be made from Tris Amino crystals from the manufacturer. Do not use the pre-dispersed solution from the manufacturer as it contains many impurities.

wide interval between melting and gelling temperatures). They must be melted at a high temperature (greater than about 85 °C), but will form a rigid gel above room temperature (usually between 32 and 45 °C). This temperature range between gel formation and melting is a novel property of agar that has found many uses in food chemistry. The gel strength of agar is influenced by concentration, time, pH, and sugar content when it is used as a food texturizing material. The pH noticeably affects the strength of the agar gel: as the pH decreases, the gel strength decreases. Sugar content also has a considerable effect on agar gel. Increasing levels of sugar make gels with a harder but a less cohesive texture.

The 'theoretical' neutral charge and lower degree of chemical complexity of agarose make it less likely to interact with biomolecules, such as nucleic acids or proteins. Hence it has found wide use as an electrophoretic medium in lab separations of various biological materials.[7] Gels made from purified agarose have a relatively uniform pore size, based on the agarose concentration (1–6% by weight), making them useful for size separation of large molecules, such as proteins or protein complexes with a molecular weight greater than 200 kilodaltons, or DNA fragments with more than 100 base pairs. Some pyruvate and sulfate groups will be present in agarose depending on the separation method from agaropectin. The 'neutrality' of agarose preparations has often been expressed by the electro-endo-osmosis (EEO) value assigned to it after purification from agaropectin. EEO values less than 0.1 indicate the highest purity, lowest inherently charged, material. EEO values greater than 0.25 are lower grade materials that carry an inherently higher internal charge and are less expensive. Gel strength is usually measured against a 1% w/v gel, with a penetration type test, and reported in g/cm^2. Again, higher quality gels might fall into a gel strength range of >2500g/cm^2, while lower grades would exhibit gel strengths less than 1200g/cm^2.

Agaropectin *see* **Agarose**

Armeen 2C CAS 61789-76-2 (Akzo Nobel)
Armeen 2C is a di-cocoalkylamine structure,[8] with a molecular weight of 390. As a cationic surfactant (it is an aliphatic amine) it has an **HLB** = 4.2 (pH >9, Davis Scale) or HLB = 14.6 (pH <8, Davis Scale). Armeen 2C is a solid at room temperature, and has a melting point of 46 °C. Under USA health and safety legislation (at the time of writing) it is non-toxic, non-carcinogenic and non-mutagenic.[9]

Canasol NF 1000 CAS 9016-45-9 (Canamex Quimicos)
This is a nonyl phenol polyoxyethylene (10) nonionic surfactant with HLB value 13.1.[10]

Carbopol (Lubrizol)
This can also be styled as a 'carbomer', a 'carboxyvinyl polymer', and as 'carboxy polymethylene'. Carbopol products are polymers of acrylic acid, crosslinked with poly alkenyl ethers or divinyl glycol.[11] They are hydrophilic, but are generally insoluble in water unless partially ionized. The pK_a of the acrylic acid groups is nominally in the range of about 6.7–7.0 which is slightly higher than the pK_a for either linear polyacrylic acid (5.5) or acrylic acid in the monomeric form (4.3) largely due to the crosslinking present in

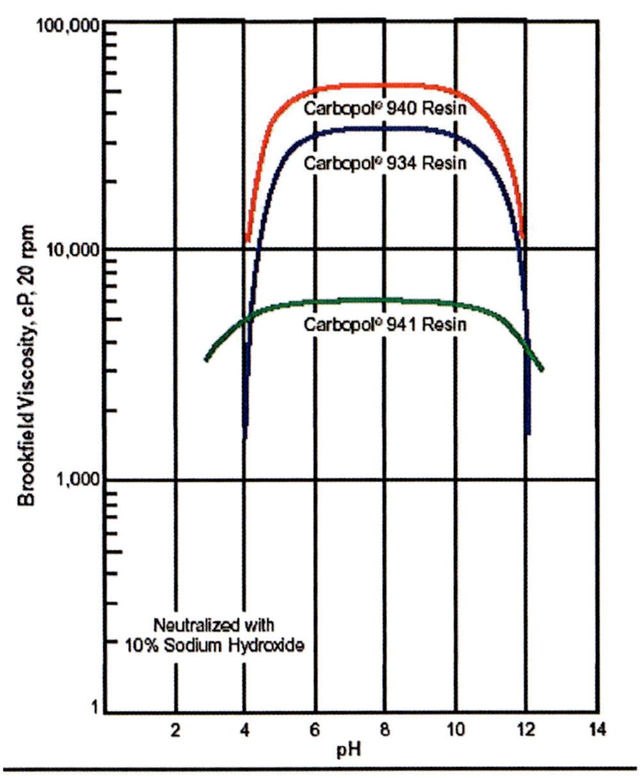

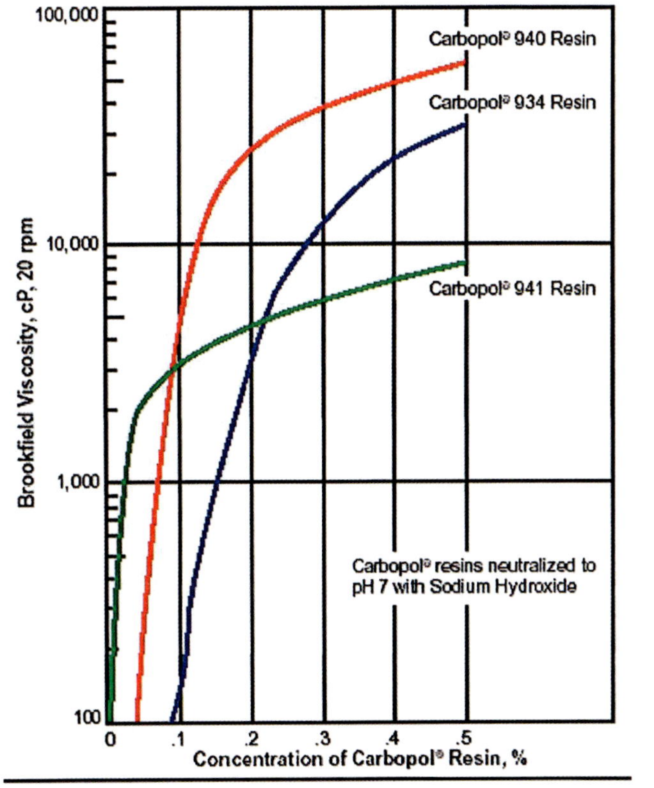

the commercial preparations,[12] As supplied the pH is slightly acidic (6.0), and they are poorly wetted by water until the pH is raised above 6. As supplied, the polymer is usually in a tightly coiled or 'particle' form (with the 'particles' being about 0.2 μm in diameter),[13] and upon neutralization it unfolds in aqueous solutions into an extended form which creates the thickening effect. Their neutralized aqueous dispersions can exhibit very high viscosities, which makes them fairly popular for controlling the flow properties of topically applied cosmetics and pharmaceuticals, as they are inexpensive, transparent, easy to prepare and relatively pure (Piau 2007). Any one of several bases can be employed to neutralize Carbopol polymers in aqueous preparations.[14] The table (opposite) lists just a few of these, with the weight ratio needed to form a stable gel.

When salts are added to Carbopol solutions, the crosslinked polymer pK_a is effectively lowered, and in reality, approaches the monomer value due to charge screening. As a polyacidic material Carbopol is a polyanionic material (polyelectrolyte) close to or above its pK_a. In this regard it has found commercial use as an effective pigment dispersal material in aqueous-based paints. But as a polyanionic material in solution, monovalent bases tend to form microgels (see below), but divalent ions like Ca and Mg tend to de-swell it and collapse the microgel structures.[15] As with other polyelectrolytes, osmotic and other ionic effects can also play a role in the swelling and de-swelling behavior of Carbopol gels in solution. And, as a polyelectrolyte, a Carbopol microgel can generate osmotic and ionic phenomena on materials it is applied to as well. Swelling is caused by an increase in solubility with increasing pH and electrostatic repulsion between chains. Upon titration with a basic solution, the carboxyl groups ionize and mutual ionic repulsion leads to the expansion of the microgel particles. Carbopol is very sensitive to changes in pH or ionic strength. The effects have been observed in both its bulk and microscopic properties using light-scattering techniques.

Upon ionization of these carboxylate groups on the polymer backbone, the result is an electrostatic repulsion among the negatively charged particles, and this repulsion effectively adds to the swelling and thickening capabilities of these polymers.[16] The un-neutralized dispersions have very low viscosities, especially with Carbopol ETD and Carbopol Ultrez polymers. Once neutralized, however, the thickening of aqueous solutions is dramatic (see Figs 1 and 2). A 0.1% w/v addition of Carbopol 940 can result in a 10k cP solution viscosity at the optimum swelling pH (see below). Optimum viscosity is typically achieved at a pH of 6.5–7.5, although high viscosities can be achieved across the pH range of 5.0–9.0.[17] The viscosity of Carbopol microgels (and Pemulen polymers, which share a similar polyacrylic acid chemistry) will begin to decrease at a pH of 9.0 and higher. This is simply an effect of the increased electrostatic repulsion between gel particles at higher pHs and ionization states. It is possible to achieve high viscosity systems at pH values below 5 and above 9, but the use level of the Carbopol (and Pemulen polymers) would have to be increased in order to compensate for the loss in viscosity due to pH extreme effects.

Carbopol polymers can be synthesized with and without crosslinking moieties or functional groups. Typically, the moiety alongside the acrylic acid monomer might be a low percentage of allylic epoxide monomer (which can be crosslinked later with a polyol like glycerol or carbohydrate such as sucrose). The amount of the latter would determine to some degree the crosslinking capacity in the polyacrylic resin. Each grade of Carbopol exhibits a different set of physical and chemical properties that makes it appropriate for different commercial applications.[18] In 'simple' non-crosslinked Carbopol polymers like Carbopol 907, the polymer is generally soluble in water and polar solvents, and thickening or increased viscosity results from simple polymer chain entanglement. In contrast, many types of Carbopol are crosslinked to some degree. Because of their effectively higher molecular weight due to crosslinking, and their increased resistance to solution solubility, they can ionize and absorb not only large volumes of water, but effectively form 'microgels' of swollen particles, which maintain highly rigid gel states at very low concentrations.[19] This neutralization step is considered the most common way to achieve the maximum thickening capability of Carbopol polymers. In aqueous preparations, as stated above, adding common bases like sodium hydroxide, or organic amines (such as triethanolamine) cause the carboxylic acid groups of these polymers to be converted into a 'salt' form. Upon ionization of these carboxylate groups on the polymer backbone, electrostatic repulsion among the negatively charged particles is observed, which adds to the swelling and thickening capabilities of these polymers.[20] Specifically, Carbopols 934, 940 and Ultrez 10 have been reported (Blanco-Fuente *et al.* 2002; Lu and Jun 1998; Fresno *et al.* 2002; Kim *et al.* 2007; Mou *et al.* 2008) in various pharmaceutical and cosmetic preparations as stable thickeners over a pH range of 5–9. Each, however, demonstrates different rheological properties and viscosity values, which are due to the difference between the particle sizes of each in solution, as well as the specific base used, and the degree and type of crosslinks in each, as well as the number of free terminal groups on the microgel particles.[21] Carbopol Ultrez 10 is a hybrid structure with properties intermediate between Carbopol 940 and 934. It shows increased dispersion and has a concomitantly wider range of commercial (and conservation) applications. Carbopol Ultrez 10 is easier to solubilize, using lower amounts of material and with less 'mixing' energy than either of the 934 or 940 resins (Welin-Berger *et al.* 2001; Fresno *et al.* 2002), to form stiffer gels.

Carbopol polymers can carry designations (954, 974, etc.) that also reflect the method or preparation (usually by emulsion polymerization techniques), and hence the resulting amount of residual solvent from the emulsion preparation. The higher the number, usually, the 'purer' or more solvent-free the ultimate product, as demanded by certain pharmacological, cosmetic, or even food industry applications. Carbopol resins are now often synthesized with ethyl acetate as the preparatory (continuous phase) solvent, and are commonly synthesized by what has been deemed a 'reverse emulsion' polymerization technique.[22] Specifically, this involves polymerization where monomer is dispersed in a small, limited, aqueous phase in which the resultant growing polymer is also synthesized and contained in the dispersed phase. In reverse emulsion polymerization the ratio of solvent to aqueous solution is reversed from 'normal' emulsion polymerization schemes (solvent or solvents comprise the continuous phase, while the dispersed phase, surrounded by surfactant, contains the monomer initially). Emulsion techniques are more rapid than bulk single phase polymerization methods, and allow for higher molecular weights to be obtained: this is true for both oil-in-water and water-in-oil emulsion techniques. In conventional emulsion polymerizations, monomer contained in micelles is volatilized by heating, and condensed back into the continuous phase to add to and grow the resultant polymer, which is also in a separate disperse phase, in the same reaction vessel. In reverse polymerizations, however, the monomer and polymer are contained in the same micelle. As in bulk polymerization, the monomer is added to the growing polymer. Initially the polymer is contained entirely in the micelle. As it grows in size, however, the polymer can extend well beyond the micelle and into the continuous phase medium. With crosslinked polymers like Carbopol, the crosslinking occurs in the micelle at a higher rate than in the continuous phase. The result is a particle of polyacrylic acid (PAA) that has a 'core' structure with a higher crosslink density than at its periphery. These 'cored' structures predominate as at this stage new smaller particles usually do not form, due to high surface tension costs. This gives Carbopol a characteristic 'microgel' structure in solution.[23]

Carbopol is also used in a variety of other commercial and industrial applications. For example, Carbopol microgels are used to stabilize detergents, to prevent the sedimentation of abrasives in scouring products, and to reduce splashing in bleach preparations. It has also been used in cosmetic tooth whitening gels that contain hydrogen peroxide in solution, in stabilized forms such as perborate, percarbonate, and carbamide peroxide (Reeve *et al.* 2006: 113–44).

Of the other novel uses for Carbopol resins, they can, when combined with organic bases, be soluble in, and form gels with, a range of organic solvents (Wolbers 1992, 2000). Linear polyacrylic acid polymers, as noted above, are soluble in polar solvents like acetone and ethanol, and therefore also in water, when they are neutralized, but are not soluble in low polarity solvents. The microgel structure of crosslinked Carbopols, like 934, additionally limits the solubility of these resins in polar solvents. However, when organic bases like Ethomeen C-25 (a tertiary amine ethoxylate, based on the primary cocoamine CAS 61791-14-8) are used as a base to neutralize or 'pair' up with some of the available acid groups on the polymer, the complex polymer/Ethomeen 'salt' can be made soluble in a wide range of organic solvents, commensurate with the miscibility of the aliphatic and polyethoxylate groups on the Ethomeen structure. The advantage is that 'salt' linked moieties generally tend to stay associated together in non-polar environments, so the 'salt' link is a convenient way of using the Carbopol resin as a kind of scaffolding to hold in place the Ethomeen surfactant. The Ethomeen aids in solubilizing the polymer, but it can also retain some of its surfactant-like character, and aid in the dispersion of insoluble materials that would remain insoluble in the bulk solvent phase.

The original formulation (Wolbers 1992) called for a ratio of Ethomeen:Carbopol of about 10:1 w/w to prepare stable solvent gels (Dorge 2004). This would represent about a 2:1 molar excess of Ethomeen to Carbopol. Each Carbopol polymer chain has the equivalent of 41666 potentially 'neutralizable' acrylic acid monomer units in a 3,000,000 dalton polymer like Carbopol 934, based on the acrylic acid monomer having had a molecular weight of 72 daltons. The original Wolbers' recipes that called for 20 g Ethomeen C-25 (MW Ethomeen-C25 = 860 Da = 0.023M), and 1 g Carbopol 934 (.012M) would have a molar ratio of about 2:1 of Ethomeen to Carbopol. This degree of neutralization represents an excess of Ethomeen to Carbopol. Full neutralization may not be achievable in any event in these kinds of preparations due to the microgel structure of Carbopol 934, as some of the 'internal' acrylic acid groups may in reality be inaccessible. Stavroudis and Blank (1995) for instance, revised the amount of Ethomeen needed to form stable gels in solvents downward to reflect this.

Carbopol polymers, as a class of materials, have received extensive review and toxicological evaluation.[24] Carbopol 934, 934P, 971P, and 974P are generally recognized as safe (acronym GRAS), a phrase with a specific health and safety implication in the USA when used in pharmaceuticals. As a result of the intensive testing and the properties offered by the polymers, they have gained wide acceptance in a variety of pharmaceutical, cosmetic, and detergent applications.

Colloid *see* **Laponite RD**

Cyclomethicone D4, D5

Cyclomethicone D4 (International Nomenclature of Cosmetic Ingredients (INCI) nomenclature) is a volatile poly-dimethylcyclosiloxane (CAS 556-67-2, Clearco) that is composed mainly of octamethylcyclotetrasiloxane. Cyclomethicone D5 (INCI nomenclature; CAS 541-02-6, Clearco) is composed of mainly decamethylcyclopentasiloxane. Cyclomethicone solvents like D4 and D5 are exempt from US Federal regulations for volatile organic compounds (VOCs)[25] and they do not contribute to ozone depletion or to global warming. They are being widely accepted as replacements for petroleum-based solvents that are not compliant with these regulations, as both 'carrying' agents (general solvents for the 'active' or principal components in cosmetics) and solvents in dry cleaning formulations, such as Cyclo-2244 (D4) and Cyclo-2245 (D5). Both D4 and D5 are especially effective for applications where an extremely low solubility parameter (Hanson parameter d = 5.0) is desirable.

As solvents, D4 and D5 have comparable evaporation rates to naphtha solvents, odorless mineral spirits (OMS in the USA, mineral spirits in the UK) as well as some of the iso-paraffinic solvents. They are 'compatible' with most substrates, including metals and plastics: this means they will not promote corrosion of metals or dissolve most plastics. Cyclomethicones are essentially clear, tasteless (which is significant for cosmetics), and odorless.[26] Due to their volatility (D4 is more volatile than D5), low surface tension (which confers high 'spreadability', the capacity to wet surfaces readily), and sensory feel, cyclomethicones are used as base fluids, carrying agents and wetting agents in a wide range of personal care products, including: room sprays, body sprays, antiperspirants, skin creams, skin lotions, suntan lotions, bath oils, hair care products, etc. The solvent miscibilities of D4 and D5 at room temperature are summarized in the table below.[27]

Solvent	Results
Toluene	
Xylene	
Solvent naphtha	
Industrial gasoline	
Mineral spirits	
Kerosene	
Cyclohexane	
n-hexane	
n-heptane	
Diisopropyl ether	
Hexyl ether	O
Ethyl acetate	
Butyl acetate	
Isopropyl laurate	
Isopropyl palmitate	
Isopropyl myristate	
Methyl ethyl ketone	
Methyl isobutyl ketone	
Lauryl alcohol	

Solvent	Results
Acetone	
Butanol	
2-ethylhexanol	
Amyl acetate	△
Glacial acetic acid	
Naphthene-based lubricating oils	
Methanol	
Ethanol	
Ethylene glycol	
Glycerin	
Diethylene glycol stearate	
Propylene glycol	
Liquid paraffin	
Paraffin wax	×
Petrolatums	
Lubricating oils	
Fatty acids (other than glacial acetic acid)	
Animal and vegetable oils	
Methyl phthalate	
Water	

O: Dissolves
△: Partially dissolves
×: Does not dissolve (results at room temperature)

In addition, D5 is being used as a federal VOC-exempted cleaning agent in dry cleaning applications. It acts as a carrying agent for the detergent and does not degrade the fabric or colors of the linens and clothing like other stronger solvents. Unlike D4 which has a freezing point of 17 °C that renders it unsuitable for outdoor use in cooler climates and seasons, D5 has a much higher 'pouring range', since the latter's freezing point is −40 °C.

Alternative, but comparable products to the Cyclo-2244 and Cyclo-2245 solvents include:[28] Cyclomethicone SF 1173 (Momentive, similar to cyclomethicone D4), Tetramére D4 (Bluestar Silicones), Silbione Tetramére D4 (Bluestar Silicones, D4), KF 994 (Shin Etsu), Baysilone COM 10.0000 (Momentive), Dow Corning 244 Fluid Cyclopentasiloxane (Dow, D5), AEC Cyclopentasiloxane (A and E Connock); Dow Corning 245 Fluid (Dow); KF995 (Shin Etsu); Mirasil CM 5 (Bluestar Silicones); and Wacker-Belsil CM 040 (Wacker), to list just a few.

The USA Cosmetics Ingredient Review Expert Panel has reviewed other siloxane polymers, such as polymethicone and other related polymethicone-containing ingredients, and concluded that those ingredients are safe as used in cosmetic products (Andersen 2003). The Panel also reviewed cyclic-siloxanes, the cyclomethicones, and concluded that they were safe as used in cosmetic products (Johnson *et al.* 2011).

Deoxycholic acid CAS 83-44-3 (numerous suppliers)

Deoxycholic acid (sodium deoxocholate $C_{24}H_{40}O_{5}$) has molecular weight 392.572, 0.024% solubility in water, pK_a = 6.58, aggregation number 3–12; critical micelle concentration (CMC) = 2–6 mM at 20 °C, and HLB = 16. It is a bile acid formed by bacterial action on sodium cholate.[29] It is usually conjugated with glycine or taurine in its native state. At physiological pHs (which are highly acidic for the digestion process), deoxycholic acid acts as a surfactant to solubilize fats for intestinal absorption. Bile acids, essential to all mammals for the absorption of dietary fats and vitamins, are generally methylsterol acids. These usually vary in structure from the parent methyl sterol molecule, with OH groups substituted at any or all of the 3, 7, or 12 carbons. In the presence of Ca, Mg bile acids are insoluble. The unique surfactant properties of bile acids are essential for the digestion and intestinal absorption of hydrophobic nutrients. At pHs below and around neutral (7.1–7.3) deoxycholates tend to aggregate and form gels. It has been postulated that above its pK_a, deoxycholic acid tends to form dimers that bind, in a ratio of 2:1, deoxycholate to free fatty acid (FFAs) complexes. Deoxycholates also tend to work synergistically with serum lipases to boost the activity of these triglyceride-degrading enzymes.

Ethomeen C-12 CAS 61791-31-9 (Akzo Nobel)
Ethomeen C-12 is a tertiary amine ethoxylate (polyoxyethylene (4)) based on a primary cocoamine.[30] It is a cationic/nonionic hybrid surfactant with HLB =12.1 on the Griffin scale, for pH > 9, or HLB = 21 on the Griffin scale, for pH < 8. Under USA health and safety legislation (at the time of writing) it is non-toxic, non-carcinogenic and non-mutagenic.[31]

Ethomeen C-25 CAS 61791-14-8 (Akzo Nobel)
Ethomeen C-25 is a tertiary primary cocoamine (polyoxyethylene (15)) based on a primary cocoamine.[32] It is a cationic/nonionic hybrid surfactant with HLB =16.4 on the Griffin scale, for pH > 9, or HLB = 27 on the Griffin scale, for pH < 8.

Gel
This is a generic term for a high viscosity colloidal dispersion. Gels have a high enough interaction between colloidal particles, polymers, etc. to exhibit physically more rigid forms, and very low flow or movement when no force is applied to them. The terms 'sol' and 'gel' describe dispersions at the extremes of colloid concentration: sols are very low in colloid concentration and have properties very close to those of a liquid, while gels, which are very high in colloid concentration, are rigid and do not flow.

Gellan gums
Gellan gum is a water-soluble, high molecular weight polysaccharide gum that is produced through fermentation by the bacterium *Pseudomonas* (now reclassified as *Sphingomonas elodea*).[33] Gellan gum is used primarily as a stabilizer, emulsifier, thickener and gelling agent in food preparation. In the commercial production process the bacteria are grown under carefully controlled conditions in cultures utilizing a carbon source (usually glucose), a nitrogen source (protein), and various minerals in the growth medium. During fermentation the bacteria converts an (algae-sourced) sugar to gellan gum as part of its natural metabolic process, the gum being exuded as a coating material around the bacterium. When fermentation is complete the growth medium is heated to kill the bacteria, the pH is raised to clarify the medium, and the gellan is then recovered by precipitation with isopropyl alcohol from the growth medium, dried, and milled to produce a powder.

The structure of gellan gum is given below. It is characterized by a straight chain polysaccharide with a repeating tetra-saccharide unit of b-D-glucose, b-D-glucuronic acid, b-D-glucose, and b-L-rhamnose.[34] In its native (or 'high acyl') form, two pendant groups can be additionally present on the initial b-D-glucose unit: an acetate and a glycerate group. On average, there is one glycerate per repeat tetra-saccharide unit and one acetate per every two repeats, for an average substitution of about 2:1 glycerate: acetate ratio. The acetate and glycerate groups can be 'stripped' off the polysaccharide by alkaline processing to form what is called a 'low acyl' type of gellan structure.[35]

Gellan Gum Structure

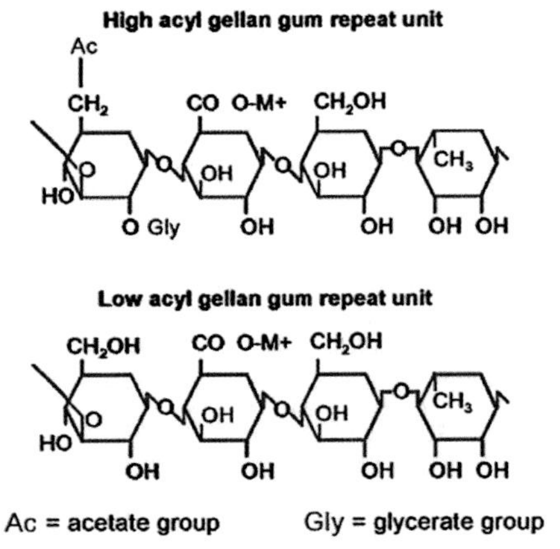

Like many other hydrocolloids, gellan gum is used as a gelling material with aqueous liquids. The powder is normally dispersed in the chosen liquid to dissolve it. Typical concentrations for gellan usage are in the 0.1–1.0% w/v range, depending on the type of gel that needs to be produced. In low acyl gellan gum, the acyl groups are removed completely. The acyl groups have a profound influence on gel characteristics. The high acyl form produces soft, elastic, non-brittle gels whereas the low acyl form produces firm, non-elastic, brittle gels.[36]

To form a gel with gellan, the material, which can only be swollen in cold water, is added as a dry powder to water, and heated to dissolve it. The melting point is usually between 85 and 95 °C. Once dissolved, the solution is then heated to facilitate liquid absorption and gelling by the hydrocolloid. As the gellan solution is cooled, gel formation begins between 80 °C and room temperature, depending whether it is the high or low acyl type. Gels thus formed are not thermo-reversible (unlike agarose gels, for instance) and can be heated and cooled without changing the gel form. The stability of the gels is due to two inherent properties:

- On cooling, the gellan polymer tends to take on a helical structure which intertwines to some degree, giving it its rigid structure.
- The glucuronic acid groups in the repeat unit are crosslinked with Ca, Mg or combinations of the two ions, creating additional polymer-to-polymer ionic bonds: the higher the amount of Ca in both high and low acyl types, the stiffer the resultant gel.

There are numerous retail suppliers of xanthan gum but the chief manufacturer is CP Kelco in the US. Gellan gum is also sold under the brand names Gelrite, Nanogel-TC, Gelrich, Grovgel, AppliedGel and Phytagel.

Syneresis is a common problem in polysaccharide gels (Glicksman 1978). Rigid gels tend to exhibit higher syneresis than softer, more flexible gels. Tang *et al.* (2001)[37] defined the 'critical Ca binding level' in gellans as the amount needed to ensure that half the available glucuronic acid groups were bound stoichiometrically with Ca. They also demonstrated that gellan gums (1% w/v) tended to exhibit an internal pore structure that falls into two ranges (0.1 μm and 1.0 μm) depending on the Ca concentration. Above the 'critical' binding level the pore sizes tend to the 1 μm size, while below the 'critical' binding level the pore size shifts towards the smaller value (0.1 μm).

Gellan gum first received approval for food use in Japan in 1988.[38] It is now approved for food, non-food, cosmetic and pharmaceutical use in the USA, Canada, Australia and many other countries in South America, Asia, and the EU, and can be regarded as non-toxic.

HLB

Surfactants are often characterized by their hydrophilic/lipophilic balance number (HLB). High HLB values indicate good solubility in water and polar solvents (the material is highly hydrophilic), while low HLB values are indicative of good solubility in non-polar systems, such as oil or hydrocarbon solvents. Structure plays a large role in determining HLB, specifically the proportion of polar to non-polar groups present in the surfactant molecule. Typical polar 'head' groups include amine, quaternary ammonium, ethoxylate, sulfate, phosphate and carboxylate, and zwitterionic functional groups. Typical non-polar groups might be linear, branched, or cyclic hydrocarbon structures, or aromatic groups. In the case of ethoxylate head groups, increasing the number of ethoxylate (EO) groups will generally tend to increase the HLB, while including or substituting propoxylate (PO) groups for ethoxylate will tend to lower the HLB. With anionic or cationic head groups, the pH will affect the HLB. Above the pK_a for an acid group, the HLB is raised; below the pK_a the HLB is lower. In addition, modifying the hydrophobic portion of the structure will also have an effect on the HLB. Increasing the mass of the hydrophobic 'tail' will tend to lower HLB.

HLB numbers have traditionally been empirically determined, or alternatively (and more recently), through the mass proportion of the hydrophilic proportion of the surfactant molecule to the whole molecule mass (Griffin 1954), and for more complex structures through an algorithm (Davies 1957). In the older empirical method, standardized emulsions were made with surfactant, water, and mineral oil. The time was then measured to phase separation, and the surfactant was assigned a number between 0 and 40 based on the length of time it remained stable. The Griffin method was simply to use the formula that 20 × [mass of the hydrophilic portion/ total mass of the surfactant molecule] = HLB number. This worked well for nonionic structures in particular, and gave rise to the now familiar Griffin scale for gauging the requisite HLB to create the desired stable emulsion type (but its applicability was limited essentially to structures with HLB = 0–20). Water-in-oil emulsions, for instance, require an HLB of at least 3–6; oil-in-water emulsions require a higher HLB (8–28). The Davies method is more recent, and essentially assigns a numerical value or 'weight' to typical polar and non-polar groups alike. The groups present in a surfactant structure are combined in terms of value, and the sum is the net HLB for the structure (giving a range of 0–40). The Davies method compares quite well to experimentally determined values.

Surfactants are often mixed for a variety of reasons. Nonionics for instance are blended with anionics, to lower the net surface tension at the interfacial boundary in emulsions. This is a technique often used to break 'bubbles' in commercial cleaning products like laundry detergents. Each surfactant system (pure or mixture) can be characterized by an HLB value. Each oil/water system can be characterized by a so-called 'required HLB value', which depends on the nature of the oil, and the product application. The required HLB value can be met either with a single surfactant or by mixing surfactants together. When surfactants are mixed the formula for calculating the net HLB is given by the equation below,[39] in which C is the weight % of each component surfactant, and C_{total} is the mass of all the surfactants summed together:

$$HLBmix = \frac{(C_1 \times HLB_1) + (C_2 \times HLB_2) + (C_3 \times HLB_3)\ldots}{C_{Total}}$$

The table below gives appropriate HLB values for a number of applications.[40]

Application	HLB
De-foaming of aqueous systems	1–3
Detergency and cleaning	12–15
W/O Emulsification	3–6
O/W Emulsification	8–28
Solubilization	11–18
Wetting	7–9

Laponite RD (BYK Additives)

This is a synthetized version of the magnesium lithium silicate clay hectorite.[41] The synthesis process involves combining salts of sodium, magnesium and lithium with sodium silicate at carefully controlled rates and temperatures to produce a sintered material. Post-sintering, the material is then crystallized at a very high temperature. The resulting product is filtered, washed, dried and milled to a fine white powder. Laponite can therefore be considered to be a ceramic material which exhibits a 2:1 type of clay structure (a lithium and/or magnesium octahedral oxide layer sandwiched between two silicate layers) with a layer of Na bound to (and situated between) each silicate face. When 'stacked' again and again (anywhere between 70 and 100 times) this 2:1 structure (with Na at the interface between each 2:1 layer) forms macroscopic disc-shaped particles (roughly 25 × 1 nm in dimension). The edge and face of each particle carry a net positive and net negative charge respectively. The 'face' negative charge is about 50–55 mmol/100 g, while the 'edge' charge is typically 4–5 mmol/100 g. Usually associated with these charged surfaces are counter-ions of Na (on the face) or hydroxyl ions (on the edge). The electrostatic repulsion of the Na ion layer on the surface of these particles would preclude any particle-to-particle associations in solution, but with the addition of other materials (e.g. simple salts, surfactants, coalescing solvents, soluble impurities and additives like pigments, fillers or binders etc.) the electrostatic repulsive forces can be overcome, allowing the weaker positive charges on the edge of the crystals to interact with the negative surfaces of adjacent particles. The process may continue to give a kind of 'house of cards' structure which, in a simple system of Laponite/water/salt, results in a highly thixotropic gel.

Particles in this size range are known as **colloids**. Colloids are very finely divided solid particles dispersed in a liquid medium. Other natural clay mineral thickeners such as bentonite and hectorite have similar particle structures, but are much larger in size. A high viscosity colloidal dispersion is termed a **gel**. A low viscosity liquid, colloidal dispersion is termed a **sol**.

Laponites are manufactured in different grades;[42] some are inherently gel-forming and some are sol-forming materials. Gel-forming grades of Laponite disperse readily in water, with agitation, to form clear and colorless dispersions. However the viscosity of such dispersions depends upon both the solids content and the electrolyte content of the aqueous solution. For instance: at 2% w/v the 'gel' grade of Laponite in (USA) tap water would appear to be highly thixotropic. The same grade of Laponite and at the same concentration in deionized water would appear as a sol or low viscosity material. Sol grades of Laponite can purposely contain dispersing agents which prevent the formation of a thixotropic gel structures, even at high concentrations of solids (30% or more solids w/v). Laponite RD is a gel-forming grade.

All Laponite grades (both gel-forming and sol-forming) must be added to water with mixing, and allowed to disperse and hydrate fully before any other components are added.[43] A typical concentration for this pre-gel might be 3% in water. In most formulations the pre-gel is made first, then diluted with other aqueous materials. Laponite gel types can be 'de-gelled' with a variety of materials (salts like tetrasodium phosphate ($Na_4P_2O_7$), or organic materials like low molecular weight glycols) to further adjust the viscosity before or after mixing has occurred. Polyphosphates in general can also impart a chelating effect: additionally, they allow the Laponite to function in high Ca or high Mg solutions, or 'hard' (high conductivity) surfaces.

Laponite is most useful in the pH range 6–13.[44] Laponite RD has been shown to form **Pickering emulsions** under specific ionic solution conditions (0.1M NaCl) (Teixeira *et al.* 2011).

Microemulsions

These are characterized by smaller, more thermodynamically stable arrangements of non-miscible liquids blended together than are found in 'macroemulsions' as typified by egg yolk/lecithin in tempera paints. Unlike macroemulsions, microemulsions are usually obtained only by mixing the non-miscible liquids together, with sufficient surfactant, and they do not require the high shear conditions (rapid stirring) necessary to generate and stabilize macroemulsions. Because they are considered thermodynamically stable, microemulsions represent a distinctly unique form of emulsion.

The salient difference between macroemulsions and microemulsions is reflected in the droplet size of the dispersed phase material. Microemulsions have droplets typically in the size range 10–100 nm in diameter, and macroemulsions generally exhibit droplet diameter sizes that are larger, in excess of 1 μm in diameter. In macroemulsions, visible light is easily scattered, so typical macroemulsions have a milky white appearance and obscure the surface on which they are placed. Microemulsions, because their droplet diameter falls below the wavelength range of visible light, do not scatter light and have a more transparent appearance (as do nanoemulsions, which have even smaller droplets).

The term 'microemulsion' was first introduced by Hoar and Schulman (1943) to describe a clear solution obtained when normal oil-in-water macroemulsions had medium chain length alcohols added to them. Undoubtedly the preparation of microemulsions dates back well into the 19th century in terms of cleaning preparations, and at least some historic preparations that included surfactant (soap), water, alcohol, and non-miscible solvents would fit the contemporary definition of a microemulsion. Danielson and Lindman (1981, 2000) define microemulsions as 'a system of water, oil, and amphiphile [defined below] which is [an] optically isotropic and thermodynamically stable liquid solution'.

Apart from droplet size, and transparency, their thermodynamic stability also distinguishes microemulsion from macroemulsion systems. In macroemulsions, surfactant properties such as HLB are critical to emulsion stability, as is the amount of surfactant (because each dispersed phase droplet must be 'covered' with surfactant at its interface with the continuous phase medium completely and efficiently), and also the density difference between the oil and water phases, the viscosity of the continuous phase, and gravitational forces impinging on the emulsion.

In microemulsions, thermodynamic stability is generated by lowering the surface tension between the dispersed and continuous phases. Typically in a microemulsion the interfacial surface tension is quite low, on the order of 10^{-2}–10^{-4} mN/m (the usual interfacial surface tension between oil in water, for instance is about 50 mN/m).[45] Such low surface tensions are possible because of the medium chain length surfactant addition to the emulsion system. Medium chain length alcohols (known as 'amphiphiles' or sometimes 'co-surfactants') tend to partition between continuous and dispersed phases, and reside at the interfacial boundary alongside the surfactant or soap used to create the emulsion. By co-existing with the surfactant at the interfacial boundary, the surface tension is lowered substantially in the surfactant layer. When the interfacial tension is thus lowered, the interaction energy between droplets is negligible, and a negative free energy of formation results, making the dispersion thermodynamically stable. The differences between microemulsions and macroemulsions are as follows:

Microemulsions	**Macroemulsions**
Thermodynamically stable	Thermodynamically unstable
Optically transparent	Cloudy colloidal systems
Interfacial tension 10^{-2}–10^{-4} mN m^{-1}	Interfacial tension 20–50 mN m^{-1}
May be single or multiple phase	Multiple phase only
No energy required in their formation	External energy required

Microemulsions can exist in the same arrangements physically as macroemulsions, i.e., as oil-in-water, water-in-oil and bi-continuous structures. To a large degree, the relative amounts of aqueous phase, oil phase, and surfactant required determine which of these forms the microemulsion will adopt. The exact aqueous conditions, solvent polarity, and surfactant and amphiphile can also determine whether or not microemulsions will form. Usually triangular or ternary phase diagrams are used to express these relationships. For example, in the figure below, the ternary plot for the common surfactant Triton X-100 is given.[46] Each corner of the triangle represents 100% of one of the three components in the formulation (here, Triton X-100, n-butanol (0.66:1.0 w/w) and water represent the surfactant and co-surfactant combination). Moving away from any corner reduces the volume fraction of that specific component: any point on one of the edges corresponds to a mixture of two of those components in a defined ratio. Any point inside the triangle represents a mixture of all three components in a defined ratio. In A the simplest system is plotted for Triton/butanol, iso-octane, and water. The blue shaded area represents all combinations of the three materials that will give rise to microemulsions. Outside of the blue area, other arrangements form (macroemulsions, phase separations, etc. predominate). As the specific components are changed as shown in B–F, when the aqueous phase is modified with salts, or other solutes, the extent of the blue shaded area changes, and the proportions of surfactant/co-surfactant, water, and solvent required to form microemulsions change as well. Förster *et al.* (1995) describe in detail the physical meaning of the phase behavior of ternary oil/surfactant/water systems, the concepts related to microemulsion formation, and the influence of additives on those microstructures.

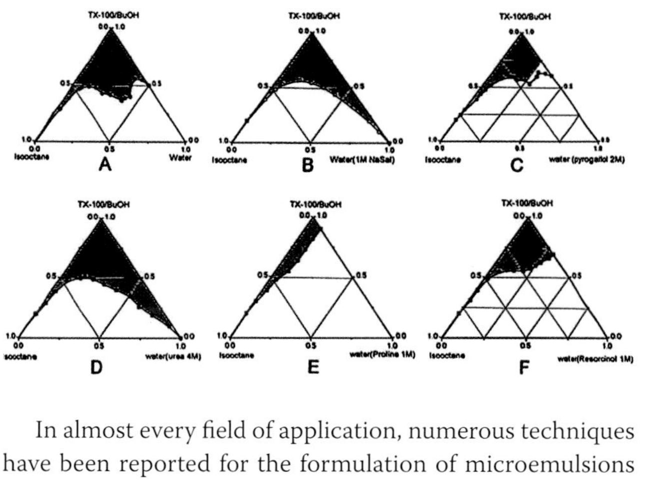

In almost every field of application, numerous techniques have been reported for the formulation of microemulsions for different applications. But most common among the techniques are what might be deemed 'pseudo-ternary diagram construction and titration methods' (Lawrence and Rees 2000). These are based on ternary plots like those cited for Triton X-100 above. Regardless of the type of microemulsion system, microemulsions can usually be formulated by first mixing the solvent or oil phase material with the surfactant/co-surfactant components. These should be miscible one in the other and form a clear solution. Then, as a second step, the water phase is added ('titrated') into this incrementally. Since microemulsions are thermodynamically stable systems, they will be formed spontaneously, as the water or aqueous phase is added, and will usually stay transparent. By 'titrating' or slowly building up the aqueous portion of the microemulsion in the solvent and surfactant/co-surfactant mixture, the resultant preparations will stay clear until an 'endpoint' is reached, when they suddenly form a turbid or cloudy preparation (which in structure is usually a macroemulsion, but may represent some other phase separation phenomenon). The major disadvantage of this method is that, although it can provide visual feedback on the phase boundary between the polyphasic and monophasic preparations ('clear' *vs.* 'non clear'), the different types – oil-in-water, water-in-oil or bi-continuous microemulsion – cannot be easily identified from the phase diagram without further characterization (Lawrence and Rees 2000). A wide variety of internal structures exists within microemulsion systems. They can be spherical, spheroid, or cylindrical rod-shaped micelles and may exist in cubic, hexagonal, or lamellar phases.

Pemulen (Lubrizol)
Pemulen is the trade name for a series of polymeric emulsifiers that are based on crosslinked polyacrylic acid co-polyacrylate ester block polymers (Simovic *et al.* 1999). The International Nomenclature of Cosmetic Ingredients (INCI) designation for these materials is generic: 'acrylates/C10-30 alkyl acrylate cross polymer'. These are microgels which share many of the same chemical and physical properties with those polymers licensed under the trademark **Carbopol.** These are water-soluble (with neutralization) microgel structures that are block copolymers of acrylic acid and C10–C30 alkyl acrylate esters crosslinked with allyl pentaerythritol.[47] As in the Carbopol series, small amounts of an allylic epoxide monomer are incorporated into the copolymer during polymerization.[48] These epoxide moieties are then reacted with polyols like pentaerythritol to form permanent covalent crosslinks. As in the Carbopol family of resins, they are made through a reverse emulsification technique that ultimately forms a microgel structure with a higher crosslinking density at the core of the microgel particle, and a lower density towards the outside of each microgel particle. But here the difference is the added ability (because of the block monomer structure) to form internally hydrophobic 'pockets' in between longer tracts of polyacrylic acid. Pemulens are generally sold as fluffy white powders, and are primarily used to form stable oil-in-water emulsions. Because they contain blocks of both hydrophilic (methyl acrylic acid) and hydrophobic (methyl acrylate aliphatic esters C10–30 in length) within the polymer, they can act to form dispersions of oil-in-water type spontaneously in the microgel structure as well as being capable of forming stable rigid gel structures. The hydrophobic portion of the polymer adsorbs at the oil or solvent surface and the hydrophilic portion swells in the water, forming a gel network around the oil droplets to provide emulsion stability *physically* through increased viscosity. This chemical structure ultimately allows these copolymers to function as primary emulsifiers in the oil-in-water sense; these are *physical* emulsions where oil or solvent is held in

droplet form *physically* in the gel structure. Polymeric emulsifiers avoid the complications and limitations of surfactant chemistry needed to stabilize emulsion systems in that they avoid the selection of an emulsifier based on an appropriate HLB, or critical micelle concentration (CMC), or solubility limitation (T_{cp}), as they are no longer of concern with these emulsifiers.

Pemulens are produced from primary polymer particles of about 0.2 μm diameter (dry). As with the Carbopol resins, in a simple linear polymeric arrangement, the thickening or gel-forming ability of Pemulens would be limited primarily to chain entanglement. But because these are crosslinked particles, they are only capable of swelling, rather than dissolving when neutralized in water or placed in polar solvents; they can form much more rigid structures in these environments. Pemulens can swell in water up to 1000 times their original volume to form a rigid 'microgel' structure, but they must be neutralized to bring them into an aqueous solution. The dry Pemulen polymers have an approximate pH range of 2.5–3.5. Optimum neutralization is achieved at a pH of 6.5–7.0. Viscosity begins to decrease after a pH of 9.0, and will continue to decrease if the pH is increased.[49] Emulsions can be created with very low concentrations of Pemulens; typically, stable gels can be made in the range 0.1–1.05 w/v of the polymer in solution.[50]

Essentially there are two 'types 'of Pemulens currently manufactured by Lubrizol: Pemulen TR-1 and TR-2. Pemulen TR-1 polymer can emulsify up to 30% of a solvent phase by weight, but the emulsifying ability of the polymer is pH dependent. In the (optimum) range of pH from 4 to 5.5,[51] the polymer can hold about 30% solvent by weight (based on a 1% w/v of the polymer in gel form). But over a broader pH range (3–11), the amount of solvent falls off from this value, and is more like 20% on average (again based on a 1% w/v gel), depending on the specific solvent. Pemulen TR-2 polymer can hold more solvent (40%) in the optimum pH range of 4–5.5, but this decreases above or below this optimum, and again would depend on the specific solvent being emulsified. In cosmetic preparations, Carbopol polymers have been used alongside the Pemulens to boost gel strength where higher viscosity emulsions have been needed. Pemulen gels will lose viscosity rapidly as the pH rises above about 9 or drops below about 5, similarly to Carbopol gels. Viscosity will also 'break' or drop quickly at high salt concentrations with either monovalent or divalent metal ions, and this may limit the application of Pemulen gels to lower conductivity surfaces, and may also indicate the need for chelators to be used alongside the polymer in cleaning preparations.

Generally, Pemulens are considered chemically stable in most cosmetic formulations.[52] However it should be noted that they are heat-sensitive materials. Heat may also be generated if the polymer comes in contact with concentrated basic materials such as ammonia, sodium hydroxide, potassium hydroxide or strongly basic amines, and care should be taken to formulate with these materials keeping this limitation in mind. In terms of regulatory status, Carbopol thickeners and Pemulens as polymeric emulsifiers are considered together as a class of polyacrylic acid materials used in cosmetics, and have received extensive review and toxicological evaluation accordingly. These are classified as Category 1 'generally recognized as safe' (GRAS) materials.[53] The toxicity of Carbopol polymers has been summarized by the Cosmetic Ingredient Review Expert Panel: this assessment and subsequent toxicology testing have demonstrated a low toxicity and irritation potential.

Pickering emulsions

These emulsions (Pickering 1907) are stabilized by the adsorption of small colloidal particles at the oil–water interface in an emulsion. These emulsifiers are distinct from traditional surfactant-based emulsifiers, where the chemistry, i.e. the combination of hydrophobic and hydrophilic structures in the surfactant moiety, determines their emulsifying behavior. The relative contribution of either portion – hydrophilic portion or hydrophobic portion – in the surfactant structure determines fundamental surfactant properties in solution such as the HLB, the critical micelle concentration (CMC), and cloud point temperature (T_{cp}) of a surfactant. Absorption onto various surfaces has been shown to be a function of these chemical or structural properties. In Pickering emulsions the oil phase is stabilized by a reduction of the bare oil–water interface by adsorption of small particles at the water–oil interface. The key property of a Pickering emulsifier is the contact angle between the oil and water phases (Ramsden 1903). It has been shown that for contact angles in the range 30–90°, oil-in-water emulsions are formed, whereas larger values of the contact angle lead to water-in-oil emulsions; 90° wetting in both oil and water phases allows for maximum stabilization to occur between the oil and water phases.

Additional stabilization can be achieved by modifying Pickering emulsifiers in various ways. For instance Velvesil Plus (Momentive) particle emulsifier is a crosslinked poly-dimethyl silicone particle, that forms a solid 'bead' or colloidal particle (in International Nomenclature of Cosmetic Ingredients (INCI) nomenclature this is a 'dimethicone cross-polymer) that carries on its surface hydrophilic pendant groups (short chains of combined polyethylene glycol, and polypropylene glycol (PEG/PPG) polymers, in a molar ratio of about 20:23 for the two glycol units) as well as internal aliphatic (cetearyl alcohol and stearyl alcohol-based) C30–45 crosslinks. The combination of internal crosslinking group and pendant group, alongside the dimethicone cross-polymer structure itself, give it its ultimate 90° wetting into oil and water phases. In contrast, Shin Etsu produces a particle emulsifier (KSG-350z) that is based on a silicone cross-polymer structure modified with an internal PEG crosslinking group (a PEG 15 glycol units in length) and external aliphatic pendant groups (lauryl or C12 chains). This latter arrangement is just the opposite of the **Velvesil Plus** structure, and allows for a higher volume of aqueous phase to be incorporated into the resultant emulsion. Stabilization of emulsions using microparticals or nanoparticles is not limited to using particle-only architectures. Recently, stabilizing foams using nanoparticles have also been reported. Stabilizing foams, however, is an 'active' process, i.e. one in which sufficient (mechanical) energy is used to prepare these particle-stabilized foams. Particle emulsifiers can also be found in food chemistry. The Swedish company Speximo has recently introduced a product called Eqmulse, a hydrophobically modified starch particle that can form Pickering-style emulsions of oil-in-water type. Pickering particle stabilized particles can also be used in concert with other materials, including surfactants. In commercial products, particle stabilization of emulsions can be controlled by using both surfactants and colloidal particles which lead to even newer hybrid classes of emulsions.

Sol
This is a low viscosity colloid dispersion. See the **Gel** definition above.

Syneresis
Syneresis is defined as the expression or loss of free liquid from a colloidal gel on standing or with an applied force.

Velvesil Plus (Momentive)
International Nomenclature of Cosmetic Ingredients (INCI) designation: C30-45 cetearyl dimethicone cross-polymer and PEG/PPG 20/23 dimethicone. Velvesil Plus[54] is a Pickering type of particle emulsifier based on a crosslinked and substituted poly-dimethylsilicone particle. The crosslinking groups are based on cetearyl and stearyl alcohols that are reacted with aliphatic-epoxide groups on the dimethicone polymer (A below). The silicone particles are further modified with pendant groups that are co-polymers of PEG and PPG in a ratio of about 20/23 molar amounts of each glycol (B below). As supplied, it is a gel-paste material that is a combination of the cross-polymer particles in **Cyclomethicone D5** solvent.

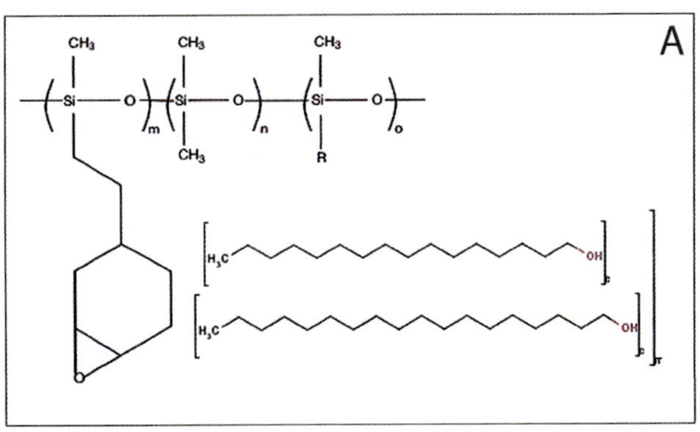

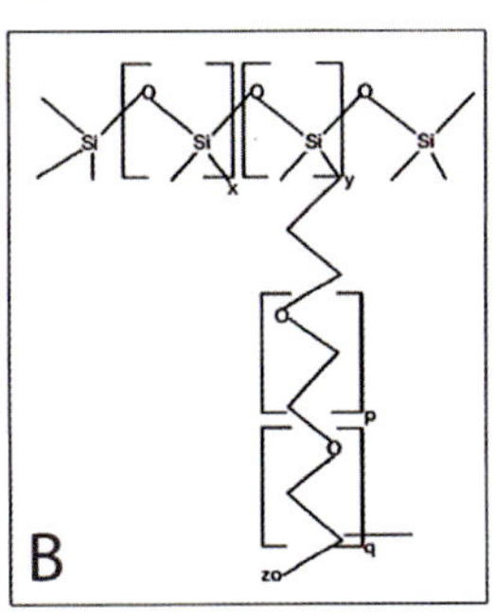

Xanthan gum
Xanthan gum is a natural exudate produced by the microorganism *Xanthomonas campestris* when it is grown as a culture under aerobic conditions, as a by-product of fermentation. Xanthan gums were first discovered, commercialized and patented in the 1950s. The bacterium strain is preserved in a freeze-dried state until it is cultured. It is activated by inoculation into a nutrient growth medium containing a carbon source (usually a carbohydrate), a nitrogen source (usually a protein), and various mineral salts. After an initial growth period, the culture is used to inoculate successive cultures to increase harvest amounts of the exuded gum. pH, O_2, and temperature are carefully controlled throughout the fermentation process to ensure a consistent product and yield. Once the carbohydrate is exhausted, the broth is sterilized. Xanthan gum is extracted by precipitation in an alcohol (isopropyl or ethanol) and the precipitate is separated, rinsed, pressed, dried and ground before being sold. The generic structure of the gum is given below:

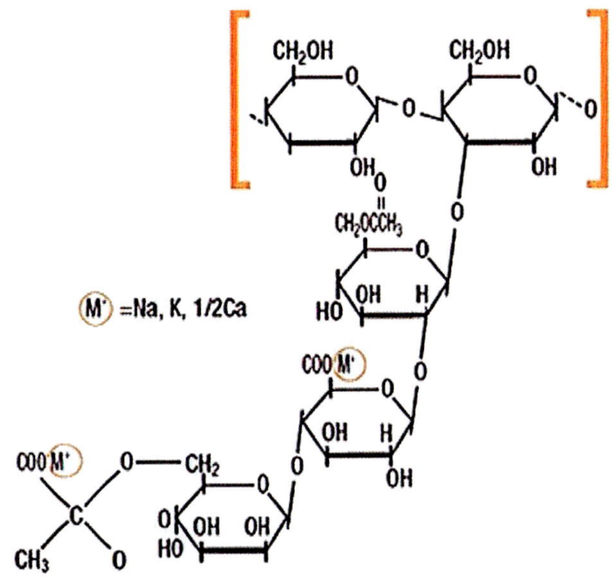

Xanthan gum is a hetero-polysaccharide with a very high molecular weight backbone polymer of repeating glucose units (1–4,000,000 daltons).[55] On every other glucose unit, pendant side chains of tri-saccharides 'branch' from the main polymer backbone. These tri-saccharide side chains are composed of α-D-mannose that contains an acetyl group, β-D-glucuronic acid, and a terminal β-D-mannose unit that is substituted with a pyruvate group. Depending on the sub-strain of the bacterium, each tri-saccharide unit will carry an acetate group, but may or may not carry a pyruvate group; the average ratio is about 2:2 of acetate to pyruvate. Both the glucuronic acid (pK_a=3.2) and the pyruvate group (pK_a 4.6) contribute to the overall configuration of the polymer in solution.[56] Above a pH of 6, both acidic moieties are fully ionized and the tri-saccharide branches extend well away from the polymer backbone (and increase inter-chain entanglement, and raise viscosity) through charge repulsion. Below a pH of about 4.6, the tri-saccharide chains are folded into the main backbone of the poly-glucosyl main chain, and are forced through hydrogen bonded into a more or less helical type of arrangement along with the main backbone polymer. This helical type of arrangement tends to also increase backbone polymer chain entanglement, and raise viscosity. As a function of pH, either one or the other or a mixture of both configurations can be present in the polymer, and a gel structure will form. One of the unique properties of xanthan gels is that they exhibit a wide tolerance to pH, forming gels of one type or the other from a pH of 0–14.[57]

Xanthan gels are thermally stable under a wide range of temperatures. They are soluble in cold and hot water; but at temperatures approaching 100 °C, the gel structure tends to be disrupted. Aqueous gels, however, are stable to low temperatures approaching 0 °C. The presence of anionic side chains on the xanthan gum molecules enhances hydration and makes xanthan gum soluble even in cold water. Low concentrations of divalent metal ions like Ca and Mg tend to stabilize the gel structure. Again, xanthan gels are tolerant to high divalent ion concentrations without precipitation.[58] Xanthan gels have been modified commercially into two forms: the native or intact material and an enzymatically 'clipped' (amorphous region removed) version referred to as a 'free flow' form.[59]

One of the other unique properties of xanthan gels is their ability to form hydrophobic 'pockets' within the gel structure.[60] The pockets develop naturally because there is always some helical and some branched or amorphous structure present in a xanthan gel regardless of the pH, ionic strength, or temperature. The amorphous areas tend to entangle, but this causes the helical branches in some cases to form hydrophobic 'pockets', which are more hydrophobic by virtue of the hydrogen bonding between tri-saccharide side chains and the backbone polymer, leaving the 'pocket' with fewer available hydroxyl groups to form water, thus making the pockets more hydrophobic in the sense that less polar material can more easily situate itself there. The 'size' of the pockets will generally be determined by the helical-to-amorphous ratio (which in turn is a function of solution parameters like pH conductivity, specific ionic materials, etc.). These 'pockets' are capable of suspending particulates, but also oil and/or solvents that are non-miscible with water. Xanthan gum is an example of a naturally occurring polymeric emulsifier, which means it is widely used in food, pharmaceutical, and cosmetic technologies. A 'typical' 2% w/v gel of xanthan in water at a pH of 6 can hold about 20% w/w of a non-miscible solvent in the gel structure as an emulsion. Miscible solvents can also be incorporated into xanthan gels as direct solutions: glycols, alcohols, acetone, and other water-miscible solvents can be dissolved into xanthan gels up to about 40–50% v/v directly, without loss of gel integrity.[61]

Xanthan gels tend to exhibit shear-thinning rheology, which means that stirring these gels imparts enough kinetic energy to break open the hydrophobic pockets and allow access to dispersed phase oils or solvents. Xanthan gels also tend to exhibit relatively 'flat' viscosity *vs.* concentration curves, making it easy to start with more concentrated gels, and then dilute them back or tailor them with other aqueous materials without huge losses in viscosity. Xanthan gums are generally regarded as safe (GRAS) materials.

Notes

1. See http://www.agargel.com.br/agar-tec-en.html.
2. See note 1.
3. http://www.fao.org/docrep/x5822e/x5822e03.htm
4. See note 1.
5. See note 3.
6. See https://www.quora.com/Whats-the-difference-between-jelly-agar-agar-and-lab-grade-agarose-gel.
7. See note 6.
8. See https://www.chemservice.com/media/product/msds/NG-S499.pdf.
9. See http://www.cir-safety.org/supplementaldoc/safety-assessment-pegs-cocamine-and-related-ingredients-used-cosmetics.
10. See http://en.quidelta.com.mx/Surfactants/Canasol-NF-1000.
11. See http://drug-dev.com/Main/Back-Issues/Carbopol174-Polymers-for-Pharmaceutical-Drug-Deliv-345.aspx#sthash.4qSgaD4R.dpuf and http://www.pjps.pk/wp-content/uploads/CD-PJPS-26-1-13/Paper-11.pdf.
12. See http://pemulentr2.pbworks.com/f/TDS-237_Neutralizing_Carbopol_Pemulen_in_Aqueous_Hydroalcoholic_Systems--PC.pdf.
13. Lubrizol Advanced Materials Inc. TDS-222, 2008. https://www.scribd.com/document/314116159/Bulletin-23-pdf.
14. See note 13; Lubrizol Advanced Materials Inc. TDS-237 2009; Lu and Jun 1998; Fresno *et al.* 2002.
15. Summit.sfu.ca/system/files/iritems1/11397/etd6172_IGutowski.pdf.
16. See Lubrizol Advanced Materials Inc. TDS-237 2009.
17. See https://www.researchgate.net/publication/305903042_PART_422_RHEOLOGY_MODIFIERS_AND_CONSUMER_PERCEPTION published by Lubrizol.
18. See A.H. Hosmani, 'Carbopol and its pharmaceutical significance: a review', available at http://www.pharmainfo.net/ reviews/carbopol-and-its-pharmaceutical significance-review.
19. See notes 13 and 16; Lu and Jun 1998; Fresno *et al.* 2002.
20. See note 14.
21. See note 13.
22. Ibid.
23. See note 16.
24. The potential toxicity of Carbopol polymers have been summarized by the USA Cosmetic Ingredient Review Expert Panel in their assessment of their safety for an ingredient in cosmetics. This assessment and subsequent toxicology testing have demonstrated low toxicity and a low irritation potential.
25. See https://ec.europa.eu/health/scientific_committees/consumer_safety/docs/sccs_o_029.pdf.
26. See http://www.clearcoproducts.com/cyclomethicones-cyclo-2244-d4.html.
27. Summarized from http://www.clearcoproducts.com/pdf/library/Solubility1.pdf.
28. See http://www.cir-safety.org/sites/default/files/PLYCTS122015TR.pdf.
29. See https://pubchem.ncbi.nlm.nih.gov/compound/deoxycholic_acid#section=Top.
30. See http://doc.ccc-group.com/spec/241035.pdf.
31. See note 9.
32. See http://doc.ccc-group.com/spec/241035.pdf.
33. See http://www.amazingfoodmadeeasy.com/define/molecular-gastronomy-glossary/what-is/gellan-gum.
34. See note 33.
35. Ibid.
36. Ibid.
37. http://sites.bsyse.wsu.edu/tang/main/publications/pdfdocs/gel/tang55.pdf.
38. See http://www.hawkinswatts.com/ingredient/gellan-gum.
39. See https://sc.akzonobel.com/en/fa/Documents/AkzoNobel_tb_HLBEmulsions.pdf.

40. Adapted from http://www.al-nasir.com/www/PharmCalc/mob_exec_calc.php?ID=hlb.
41. https://www.byk.com/fileadmin/byk/additives/product_groups/rheology/former_rockwood_additives/technical_brochures/BYK_B-RI21_LAPONITE_EN.pdf.
42. See note 41.
43. Ibid.
44. Ibid.
45. See http://pharmatips.doyouknow.in/Articles/Pharmaceutics/Emulsions/Introduction-Of-Microemulsions.aspx.
46. See http://cemunsel.unse.edu.ar/Pdf/r980.pdf.
47. Pemulen TR-1 Polymeric Emulsifier, Product Specifications, Lubrizol, 1997.
48. See note 47.
49. Noveon Pharmaceuticals, Noveon Polymers Technical Bulletins (2002).
50. See Summit.sfu.ca/system/files/iritems1/11397/etd6172_IGutowski.pdf.
51. See note 50.
52. See http://www.ordonearresearchlibrary.org/data/pdfs/ijdfr60.pdf.
53. See http://drug-dev.com/Main/Back-Issues/Carbopol174-Polymers-for-Pharmaceutical-Drug-Deliv-345.aspx.
54. See https://www.momentive.com/products/show-technical-datasheet.aspx?id=14438.
55. See https://www.cargill.com/food-bev/xanthan-gum.
56. See note 55.
57. See http://www.vanderbiltminerals.com/product_categories/product_listing/category/pharma-products-vanzan.
58. See note 57.
59. Ibid.
60. Ibid.
61. Ibid.

References

Andersen, F.A. 2003. 'Final report on the safety assessment of stearoxy dimenticone, dimethicone, methicone, amino bispropyl dimethicone, aminopropyl dimethicone, amodimethicone, amodimethicone hydroxystearate, behenoxy dimethicone, C24-28 alkyl methicone, C30- 45 alkyl methicone, C30-45 alkyl dimethicone, cetearyl methicone, cetyl dimethicone, dimethoxysilyl ethylenediaminopropyl dimethicone, heyxyl methicone, hydroxypropyldimethicone, stearamidopropyl dimethicone, stearyl dimethicone, stearyl methicone, vinyldimethicone', *International Journal of Toxicology* 22(Suppl. 2): 11–35.

Araki, C. 1937. 'Acetylation of agar-like Substance from *Gelidium amansii*', *Journal of the Chemistry Society of Japan* 58: 1338–50.

Azhitskii, G.Y. and Kobozev, B.V. 1967. 'Laboratornoe Delo', *Chem Abstracts* 66: 106120s.

Blanco-Fuente,H. , Esteban-Fernández, B., Blanco-Méndez, J. and Otero-Espinar, F.J. 2002. 'Use of β-cyclodextrins to prevent modifications of the properties of Carbopol hydrogels due to Carbopol–drug interactions', *Chemical and Pharmaceutical Bulletin* 50(1): 40–46.

Danielson. I. and Lindman, B. 1981. 'The definition of micro-emulsion', *Colloids and Surfaces* 3: 391–2.

Danielson. I. and Lindman, B. 2000. 'The definition of micro-emulsion', *Drug Delivery Review* 45: 89–121.

Davies, J.T. 1957. 'A quantitative kinetic theory of emulsion type I: physical chemistry of the emulsifying agent', in J.H. Schulman (ed.), *Gas/Liquid and Liquid/Liquid Interfaces*. London: Butterworth, 426–38.

Dorge, V. (ed.) 2004. *Solvent Gels for the Cleaning of Works of Art: The Residue Question*. Los Angeles: Getty Conservation Institute.

Förster, T., von Rybinski, W. and Wadle, A. 1995. 'Influence of micro-emulsion phases on the preparation of fine-disperse emulsions', *Advanced Colloid and Interface Science* 58(2–3): 119–49.

Fresno, M.J.C, Ramírez, A.D and Jiménez, M.M. 2002. 'Systematic study of the flow behaviour and mechanical properties of Carbopol® Ultrez™ 10 hydroalcoholic gels', *European Journal of Pharmaceutics and Biopharmaceutics* 54(3): 329–35.

Fuse, T. and Goto, F. 1971. 'Some properties of agarose and agaropectin isolated from various mucilaginous substances of red seaweeds', *Agricultural and Biological Chemistry* 35: 799–804.

Glicksman, M.1978. 'Gelling hydrocolloids in food applications', in J.M.V. Blanchard and J.R. Mitchell (eds), *Polysaccharides in Foods*. London: Butterworth, 185–204.

Griffin, W.C. 1954. 'Calculation of HLB values of nonionic surfactants', *Journal of the Society of Cosmetics Chemistry* 5: 249–56.

Hjerten, S. 1962. 'A new method for preparation of agarose for gel electrophoresis', *Biochimica et Biophysica Acta* 62: 445–9.

Hjerten, S. 1971. 'Some new methods for the preparation of agarose', *Journal of Chromatography* 61: 73–80.

Hoar, T.P. and Schulman, J.H. 1943. 'Transparent water-in-oil dispersions: the oleopathic hydro-Mmcelle', *Nature* 152(3847): 102–3.

Johnson Jnr,. W., Bergfeld, W., Belsito, D., Hill, R., Klaassen, C., Liebler, D., Marks Jnr., J., Shank, R., Slaga, T., Snyder, P. and Andersen, F. 2011. 'Safety assessment of cyclomethicone, cyclotetrasiloxane, cyclopentasiloxane, cyclohexasiloxane, and cycloheptasiloxane', *International Journal of Toxicology* 30 (Suppl. 3): 149S–227S.

Kim, T.H., Ahn, J.S., Choi, H.K., Choi, Y.J. and Cho, C.S. 2007. 'A novel mucoadhesive polymer film composed of Carbopol, poloxamer and hydroxypropylmethylcellulose', *Archives of Pharmaceutical Research* 30(3): 381–6.

Lawrence, J.M. and Rees, G.D. 2000. 'Micro-emulsion based media as novel drug delivery systems', *Advanced Drug Delivery Reviews* 45: 89–121.

Lu, Guangwei and Won Jun, H. 1998. 'Diffusion studies of methotrexate in Carbopol and Poloxamer gels', *International Journal of Pharmaceutics* 160(1): 1–9.

Mou, Dongsheng, Chen, Huabing, Du, Danrong, Mao, Chengwen, Wan, Jiangling, Xu, Huibi and Yang, Xiangliang 2008. 'Hydrogel-thickened nano-emulsion system for topical delivery of lipophilic drugs', *International Journal of Pharmaceutics* 353(1–2): 270–76.

Piau, J.M. 2007. 'Carbopol® gels: elastoviscoplastic and slippery glasses made of individual swollen sponges: meso- and macroscopic properties, constitutive equations and scaling laws', *Journal of Non-Newtonian Fluid Mechanics* 144(1): 1–29.

Pickering, S.U. 1907. 'Emulsions', *Journal of the Chemical Society: Transactions* 91: 2001–21.

Ramsden, W. 1903. 'Separation of solids in the surface layers of solutions and suspensions (observations on surface membranes, bubbles, emulsions and mechanical coagulations): preliminary account', *Philosophical Transactions of the Royal Society of London* 72: 156–64.

Rolson, A. 1965. *Fractionation of Mixtures of Agarose and Agaropectin*. UK Patent No. 1,006,259.

Russell, B., Mead, T.H. and Polson, A. 1964. 'A method for preparing agarose', *Biochimica et Biophysica Acta – General Subjects* 86(1): 169–74.

Simovic, S., Tamburic, S., Jela, M. and Dusan, R. 1999. 'An investigation into interactions between polyacrylic polymers and a non-ionic surfactant: an emulsion pre-formulation study', *International Journal of Pharmaceutics* 184(2): 207–17.

Stavroudis, C., Doherty, T. and Wolbers, R. 2005. 'A new approach to cleaning I: using mixtures of concentrated stock solutions and a database to arrive at an optimal aqueous cleaning system', *WAAC Newsletter* 27(2): 17–28.

Sviridov, S.M., Berdnikov, V.A. and Ivanov, V.N. 1971. 'Agarose isolation from agar', *Laboratornoe Delo* 1: 55–57.

Teixeira, R.F.A., McKenzie, H., Boyd, S. and Bon, S.A.F. 2011. 'Pickering emulsion polymerization using Laponite clay as stabilizer to prepare armored "soft" polymer latexes', *Macromolecules* 44(18): 7415–22.

Welin-Berger, K., Neelissen, J.A. and Bergenstahl, B. 2001. 'The effect of rheological behaviour of a topical anaesthetic formulation on the release and permeation rates of the active compound', *European Journal of Pharmaceutical Sciences* 13(3): 309–18.

Wolbers, R. 1992. 'Recent development in the use of gel formulations for the cleaning of paintings', in *Restoration 1992 Conference Preprints*. London: UKIC, 74–5.

Wolbers, R. 2000. *Cleaning Painted Surfaces: Aqueous Methods*. London: Archetype Publications.

Index